INDEX OF TABLES

Acceleration due to gravity, 135
Astronomical data, A-9
Atomic masses, 1065
Breaking stress, 301
Compressibility, 296
Conductivity, thermal, 376
Convection coefficient, 380
Critical point, 398
Density, 306
  of water, 350
Dielectric constant, 612
Elastic constants, 293
Electromagnetic spectrum, 827
Electron configurations, 1040
Electron shells, 1038
Expansion, thermal, 349
Friction, coefficient of, 93
Fundamental constants, A-9
Fundamental particles, 1088
Heat capacities, gases, 416
  metals, 361
Heat of fusion and vaporization, 365
Index of refraction, 843

International system of units, A-1–A-3
Magnetic susceptibility, 729
Mathematical formulas, A-4–A-5
Moment of inertia, 217
Nuclear compositions, 1064
Nuclear masses, 1065
Periodic table of elements, A-7
Poisson's ratio, 293
Prefixes, unit, 6
Quarks, 1090
Resistivity, 628
  temperature coefficient, 629
Sound intensity level, 516
Spectrum colors, 827
Stable nuclei, 1068
Surface tension, 315
Susceptibility, magnetic, 729
Thermal conductivity, 376
Thermal expansion coefficients, 349
Triple point, 397
Unit conversion factors, A-8
Viscosity, 326

VOLUME TWO

UNIVERSITY PHYSICS

SEVENTH EDITION

The end-of-chapter Problem Sets were revised for the *Seventh Edition* by A. Lewis Ford, Texas A&M University.

Contributions to the Problem Sets were made by Lawrence B. Coleman, University of California, Davis; James L. Monroe, Pennsylvania State University, The Beaver Campus; Terry F. O'Dwyer, Nassau Community College.

Craig Watkins, Massachusetts Institute of Technology, provided assistance to both Professor Young and to Professor Ford in the development of the manuscript.

VOLUME TWO

UNIVERSITY PHYSICS

SEVENTH EDITION

Francis W. Sears

Late Professor Emeritus
Dartmouth College

Mark W. Zemansky

Late Professor Emeritus
City College of the City University of New York

Hugh D. Young

Professor of Physics
Carnegie-Mellon University

ADDISON-WESLEY PUBLISHING COMPANY
Reading, Massachusetts ▪ Menlo Park, California
Don Mills, Ontario ▪ Wokingham, England ▪ Amsterdam
Sydney ▪ Singapore ▪ Tokyo ▪ Madrid ▪ Bogotá
Santiago ▪ San Juan

This book is in the
Addison-Wesley Series in Physics

Sponsoring Editor: Bruce Spatz, Debra Hunter, Steve Mautner
Developmental Editor: David M. Chelton
Production Supervisor: Marion E. Howe
Copy Editor: Jacqueline M. Dormitzer
Text Designer: Catherine L. Dorin
Layout Artist: Lorraine Hodsdon
Illustrators: Oxford Illustrators, Ltd.
Art Consultant: Loretta Bailey
Manufacturing Supervisor: Ann DeLacey
Cover: Marshall Henrichs

Photo credits: page 527, AT&T Bell Laboratories. 529, Lockyer Collection. 546, Education Development Center. 575, Wellcome Institute for the History of Medicine. 604, Chip Clark. 625, American Institute of Physics. 653 and 681, AT&T Bell Laboratories. 683, Fundamental Photographs. 714, Pacific Gas & Electric. 742, Chip Clark. 767, Varian Associates, Inc. 788, Pacific Gas & Electric. 813, NASA. 835, Bausch & Lomb Optical Co. 837, Corning Glass Works. 868, Chip Clark. 887, Palomar Observatory. 953, Lawrence Livermore National Laboratory. 955, U.S. Dept. of Energy. 980, IBM Corp. 1007, Manfred Kage/Peter Arnold, Inc. 1036, Chip Clark. 1060, Stanford Linear Accelerator.

Library of Congress Cataloging-in-Publication Data

Sears, Francis Weston, 1898–
    University physics.

    Includes index.
    1. Physics. I. Zemansky, Mark Waldo, 1900–
II. Young, Hugh D. III. Title.
QC21.2.S36 1986   530   85–28801
ISBN 0-201-06683-1

ABCDEFGHIJ-MU-8987

PREFACE

In this new edition of *University Physics*, we have tried to preserve those qualities and features that users of previous editions have found useful. Yet this is the most comprehensive revision in the long, successful history of the book. Physics courses and physics students have changed substantially in recent years, and new editions must keep pace with these changes.

Our basic goals have not changed. Our objective is to provide a broad, rigorous introduction to physics at the beginning college level. This book is appropriate for students of science and engineering who are taking an introductory calculus course concurrently. We place primary emphasis on physical principles and the development of problem-solving ability, rather than on historical background or specialized applications. The complete text may be taught in an intensive two- or three-semester course, and the book is also adaptable to a wide variety of shorter courses. It is available as a single volume or as two volumes. Volume I includes mechanics, heat, and mechanical waves, and Volume II includes electricity and magnetism, optics, and atomic and nuclear physics.

Here are some of the most important new features in this edition:

Table of Contents. After careful consideration and consultation with many users of our book, we have reorganized the chapters on mechanics. We now conform to the usual order in introductory courses, beginning with kinematics and dynamics and treating statics later as a special case of dynamics. Getting students into the study of motion immediately helps to build motivation for the study of physics, and it also helps to tie the physics course in with the often-concurrent calculus course.

Problems. The end-of-chapter problem collections have been extensively revised and augmented. We now group each collection into three categories: *Exercises,* single-concept problems that are keyed to specific sections of the text; *Problems,* usually requiring two or more nontrivial steps for their solution; and *Challenge Problems,* intended to challenge the strongest students. The number of problems has grown by 12%; the total number is now approximately 1800, of which over 25% are new. We have also added to the lists of thought-provoking questions at the ends of the chapters, about 700 questions in all. The revision of the problem collections was carried out by Professor A.

Lewis Ford (Texas A. & M. University), with the assistance of Mr. Craig Watkins (Massachusetts Institute of Technology). Additional problems were contributed by Professors Lawrence B. Coleman (University of California, Davis), James L. Monroe (Pennsylvania State University, The Beaver Campus), and Terry F. O'Dwyer (Nassau Community College).

Problem-Solving Strategies. The remark heard most often in the freshman Physics classroom is: "I understood the material, but I couldn't do the problems!" To respond to this universal cry for help, we have included in each chapter one or more sections called *Problem-Solving Strategy,* where we list suggestions for developing a methodical and systematic approach to solving problems. Our most important objective in this book is to help students learn to apply physical principles to a wide variety of problems, and these new strategy sections should be a substantial help.

Chapter Summaries. Each chapter concludes with a list of *Key terms,* which have been highlighted in boldface type in the text, and a *summary,* in prose and equations, of the most important principles presented in the chapter. These will be a useful aid for the student, especially in identifying and emphasizing the concepts and relationships that are of central importance.

Chapter Introductions and Perspectives. Each chapter now begins with an introductory paragraph summarizing briefly the content of the chapter and relating it to what has come before. In addition, a *Prospectus* and seven *Perspectives* are included at intervals throughout the book. Their object is to enhance continuity by looking both backward and forward to show how various areas of physics are interrelated and to exhibit as clearly as possible the beauty and fundamental unity of all branches of physics.

Study Notes. Brief notes inserted in the margins act as references for the reader and identify key ideas and concepts in the text. They help the student locate quickly a particular discussion or example, and they provide capsule summaries of paragraphs and sections of text, useful for re-study and review of troublesome areas.

Mathematical Level. The level of mathematical sophistication has not changed substantially, but we have increased somewhat the use of unit vectors and calculus in worked-out examples. There are also more problems providing opportunities to use unit vectors and calculus.

Changes in subject matter. Many topics have been added or treated in greater depth than in previous editions. A partial list includes:

> estimates and orders of magnitude
> gravitational field
> automotive power
> damped and forced oscillations
> Maxwell–Boltzmann distribution
> sign conventions for Kirchhoff's rules
> Maxwell's equations
> circular apertures and resolving power
> relativistic Doppler effect
> Planck radiation law
> superconductivity
> band theory of solids

The elasticity chapter has been rewritten so that each type of stress is introduced along with its corresponding strain. The treatment of fluid mechanics has been condensed and reduced from two chapters to one. The material on acoustic phenomena has been reduced, as has the treatment of magnetic materials. The discussion of polarization of light has been shortened and incorporated into the chapter on Nature and Propagation of Light. The material on refracting surfaces has been reorganized so that the thin-lens equation can be presented earlier.

Units and Notation. We have moved closer to 100% SI units. English units are retained in a few examples and problems in the first half of the text, but SI units are used exclusively in the second half. We use the joule as the standard unit of energy of all forms, including heat. In examples, units are always carried through all stages of numerical calculations. As usual, boldface symbols are used for vector quantities, and in addition boldface +, −, and = signs are used to remind the student at every opportunity of the crucial distinctions between operations with vectors and those with numbers.

Supplements. A textbook should stand on its own feet. Yet some students benefit from supplementary materials designed to be used with the text. With this thought in mind, we offer a *Study Guide* and a *Solutions Manual*. The *Study Guide,* prepared by Professors James R. Gaines and William F. Palmer, includes for each chapter & statement of objectives, a review of central concepts, problem-solving hints, additional worked-out examples, and a short quiz. The *Solutions Guide,* prepared by Professor A. Lewis Ford, includes completely worked-out solutions for about one third of the problems in the book, drawn from odd-numbered problems only. Answers to all odd-numbered problems are listed at the end of the text, and a booklet containing answers to all even-numbered problems can be obtained by instructors from the publisher.

Reviewers. Many of the changes in this edition are a direct result of recommendations from colleagues who have used earlier editions with their students. The views and suggestions collected through reviews, written questionnaires, telephone surveys, and discussion groups have been invaluable. In addition to those named in connection with the problem revisions, I gratefully acknowledge the very helpful and valuable contributions of the following reviewers and discussion-group participants:

Alex Azima, Lansing Community College
Dilip Balamore, Nassau Community College
Arun Bansil, Northeastern University
Albert Bartlett, University of Colorado
Lev I. Berger, San Diego State University
James Brooks, Boston University
Nicholas E. Brown, California Polytechnic University–San Luis Obispo
Hans Courant, University of Minnesota
Gayl Cook, University of Colorado
Bruce A. Craver, University of Dayton
Steve Detweiler, University of Florida
Lewis Ford, Texas A & M University
Walter S. Gray, University of Michigan
Graham D. Gutsche, U. S. Naval Academy–Annapolis
Michael J. Harrison, Michigan State University
Howard Hayden, University of Connecticut
Lorella Jones, University of Illinois
Jean P. Krisch, University of Michigan
Alfred Leitner, Rensselaer Polytechnic University
David Markowitz, University of Connecticut
Joseph L. McCauley, University of Houston
T. K. McCubbin, Jr., Pennsylvania State University
Thomas Meyer, Texas A & M University
Herbert Muether, S.U.N.Y.–Stony Brook
Jack Munsee, California State University–Long Beach

Lorenzo Narducci, Drexel University
Van E. Neie, Purdue University
David A. Nordling, U. S. Naval Academy–Annapolis
W. F. Parks, University of Missouri
Arnold Perlmutter, University of Miami
John S. Risley, North Carolina State University
Richard Roth, Eastern Michigan University
Rajarshi Roy, Georgia Institute of Technology
Russell A. Roy, Santa Fe Community College
Stan Shepherd, Pennsylvania State University
Malcolm Smith, University of Lowell
James Smith, U. S. Military Academy–West Point
Conley Stutz, Bradley University
G. David Toot, Alfred University
George Williams, University of Utah
John Williams, Auburn University
D. H. Ziebell, Manatee Community College
George O. Zimmerman, Boston University

With the departure of Professors Sears and Zemansky from this life, I have assumed sole responsibility for the book. I feel a little like a violinmaker who is asked to take a Stradivarius apart and repair it. It is an honor to be asked to do it, but it is also an awesome responsibility. I have taken great care to remain true to the original spirit of this text, while making it as useful for today's students as the first edition was for its users a few decades ago.

Acknowledgments. A special debt of gratitude is owed to the author's colleagues at Carnegie-Mellon, especially Professors Robert Kraemer, Bruce Sherwood, and Helmut Vogel, for many stimulating discussions about physics pedagogy, and to Professor Kraemer for major contributions to the high-energy physics material. An equally important debt of a different kind is owed to Dr. Michael Schur for his support and encouragement when they were most needed. Finally and most important, I offer my gratitude to my wife Alice and our children Gretchen and Rebecca for their love, support, and emotional sustenance during a difficult period in my life. May all men be blessed with love such as theirs.

As always, I welcome communications from students and professors, especially when they concern errors or deficiencies that may be found in this edition. I have written the best book I know how to write; I hope it will help you to teach and learn physics. In turn, you can help me by letting me know what still needs to be improved!

Pittsburgh, Pennsylvania   H. D. Y.
November 1986

CONTENTS

PART FIVE

ELECTROSTATICS AND ELECTRIC CURRENTS   527

24

COULOMB'S LAW   529

24–1	Electric charge	529
24–2	Atomic structure	530
24–3	Electrical conductors and insulators	532
24–4	Charging by induction	533
24–5	Coulomb's law	534
24–6	Applications of Coulomb's law	537
	Summary	541

25

THE ELECTRIC FIELD   546

25–1	Electric field and electrical forces	546
25–2	Electric-field calculations	551
25–3	Field lines	556
25–4	Gauss's law	557
25–5	Applications of Gauss's law	561
25–6	Charges on conductors	566
	Summary	568

26

ELECTRICAL POTENTIAL   575

26–1   Electrical potential energy   575
26–2   Potential   580
26–3   Calculation of potentials   583
26–4   Equipotential surfaces   587
26–5   Potential gradient   589
26–6   The Millikan oil-drop experiment   591
26–7   The electronvolt   593
26–8   The cathode-ray tube   594
         Summary   597

27

CAPACITANCE AND DIELECTRICS   604

27–1   Capacitors   604
27–2   The parallel-plate capacitor   605
27–3   Capacitors in series and parallel   607
27–4   Energy of a charged capacitor   609
27–5   Effect of a dielectric   611
27–6   Molecular model of induced charge   615
27–7   Gauss's law in dielectrics   617
         Summary   618

28

CURRENT, RESISTANCE, AND ELECTROMOTIVE FORCE   625

28–1   Current   625
28–2   Resistivity   628
28–3   Resistance   630
28–4   Electromotive force and circuits   632
28–5   Current–voltage relations   639
28–6   Energy and power in electric circuits   640
28–7   Theory of metallic conduction   644
         Summary   646

29

DIRECT-CURRENT CIRCUITS   653

29–1   Resistors in series and parallel   653
29–2   Kirchhoff's rules   656
29–3   Electrical instruments   659

29–4	The $R–C$ series circuit	662
29–5	Displacement current	666
29–6	Power distribution systems	667
	Summary	670

PART SIX

ELECTRODYNAMICS 681

30

MAGNETIC FIELDS AND MAGNETIC FORCES 683

30–1	Magnetism	683
30–2	Magnetic field	684
30–3	Magnetic field lines and magnetic flux	688
30–4	Motion of charged particles in a magnetic field	690
30–5	Thomson's measurement of e/m	692
30–6	Isotopes and mass spectroscopy	693
30–7	Magnetic force on a conductor	695
30–8	Force and torque on a current loop	697
30–9	The direct-current motor	702
30–10	The Hall effect	703
	Summary	704

31

SOURCES OF MAGNETIC FIELD 714

31–1	Magnetic field of a moving charge	714
31–2	Magnetic field of a current element	717
31–3	Magnetic field of a long straight conductor	719
31–4	Force between parallel conductors	721
31–5	Magnetic field of a circular loop	722
31–6	Ampere's law	724
31–7	Applications of Ampere's law	726
31–8	Magnetic materials	728
31–9	Magnetic field and displacement current	730
	Summary	732

32

ELECTROMAGNETIC INDUCTION 742

32–1	Induction phenomena	742
32–2	Motional electromotive force	743

32–3   Faraday's law   748

32–4   Induced electric fields   752

32–5   Lenz's law   754

32–6   Eddy currents   756

32–7   Maxwell's equations   757

        Summary   759

33

INDUCTANCE   767

33–1   Mutual inductance   767

33–2   Self-inductance   769

33–3   Energy in an inductor   771

33–4   The R–L circuit   773

33–5   The L–C circuit   775

33–6   The L–R–C circuit   778

        Summary   780

34

ALTERNATING CURRENTS   788

34–1   ac sources and phasors   788

34–2   Resistance, inductance, and capacitance   789

34–3   The L–R–C series circuit   793

34–4   Average and root-mean-square values   797

34–5   Power in ac circuits   798

34–6   Series resonance   800

34–7   Parallel resonance   803

34–8   Transformers   804

        Summary   806

35

ELECTROMAGNETIC WAVES   813

35–1   Introduction   813

35–2   Speed of an electromagnetic wave   815

35–3   Energy in electromagnetic waves   817

35–4   Electromagnetic waves in matter   820

35–5   Sinusoidal waves   821

35–6   Standing waves   824

35–7   The electromagnetic spectrum   826

35–8   Radiation from an antenna   828

        Summary   829

PART SEVEN

OPTICS 835

36

THE NATURE AND PROPAGATION OF LIGHT · 837

36–1   Nature of light   837

36–2   Sources of light   838

36–3   The speed of light   839

36–4   Waves, wave fronts, and rays   840

36–5   Reflection and refraction   841

36–6   Total internal reflection   845

36–7   Dispersion   847

36–8   Polarization   848

36–9   Polarizing filters   850

36–10   Scattering of light   854

36–11   Circular and elliptical polarization   855

36–12   Huygens' principle   857

        Summary   860

37

IMAGES FORMED BY A SINGLE SURFACE · 868

37–1   Reflection at a plane surface   868

37–2   Reflection at a spherical surface   870

37–3   Focus and focal length   874

37–4   Graphical methods   875

37–5   Refraction at a plane surface   877

37–6   Refraction at a spherical surface   879

        Summary   882

38

LENSES AND OPTICAL INSTRUMENTS · 887

38–1   The thin lens   887

38–2   Diverging lenses   889

38–3   Graphical methods   890

38–4   Images as objects   893

38–5   Lens abberations   895

38–6   The eye   897

38–7   Defects of vision   898

38–8   The magnifier   899

38–9   The camera   900

38–10 The projector   901
38–11 The compound microscope   902
38–12 Telescopes   903
       Summary   905

39

INTERFERENCE AND DIFFRACTION   **913**

39–1   Interference and coherent sources   913
39–2   Two-source interference   917
39–3   Intensity distribution in interference patterns   919
39–4   Interference in thin films   922
39–5   The Michelson interferometer   926
39–6   Fresnel diffraction   928
39–7   Fraunhofer diffraction from a single slit   930
39–8   The diffraction grating   934
39–9   X-ray diffraction   938
39–10 Circular apertures and resolving power   939
39–11 Holography   943
       Summary   945

PART EIGHT

MODERN PHYSICS 953

40

RELATIVISTIC MECHANICS   **955**

40–1   Invariance of physical laws   955
40–2   Relative nature of simultaneity   958
40–3   Relativity of time   958
40–4   Relativity of length   961
40–5   The Lorentz transformation   963
40–6   Momentum   966
40–7   Work and energy   967
40–8   Invariance   970
40–9   The Doppler effect   971
40–10 Relativity and newtonian mechanics   973
       Summary   974

41

PHOTONS, ELECTRONS, AND ATOMS   **980**

41–1   Emission and absorption of light   980
41–2   The photoelectric effect   982
41–3   Line spectra   985

41–4   Energy levels   988

41–5   Atomic spectra   989

41–6   The laser   992

41–7   Continuous spectra   994

41–8   X-ray production and scattering   996

       Summary   1002

42

QUANTUM MECHANICS   1007

42–1   The Bohr atom   1007

42–2   Wave nature of particles   1012

42–3   The electron microscope   1016

42–4   Probability and uncertainty   1019

42–5   Wave functions   1024

42–6   The Zeeman effect   1026

42–7   Electron spin   1028

       Summary   1030

43

ATOMS, MOLECULES, AND SOLIDS   1036

43–1   The exclusion principle   1036

43–2   Atomic structure   1039

43–3   Diatomic molecules   1041

43–4   Molecular spectra   1043

43–5   Structure of solids   1045

43–6   Properties of solids   1047

43–7   Band theory of solids   1049

43–8   Semiconductors   1051

43–9   Semiconductor devices   1052

43–10 Superconductivity   1055

       Summary   1056

44

NUCLEAR AND HIGH-ENERGY PHYSICS   1060

44–1   The nuclear atom   1060

44–2   Properties of nuclei   1062

44–3   Nuclear stability   1067

44–4   Radioactive transformations   1069

44–5   Nuclear reactions   1074

44–6   Nuclear fission   1076

44–7   Nuclear fusion   1079

43–8   Semiconductors   1051

43–9   Semiconductor devices   1052

43–10   Superconductivity   1055

         Summary   1056

APPENDIXES

A   The international system of units   A–1

B   Useful mathematical relations   A–4

C   The Greek alphabet   A–6

D   Periodic table of the elements   A–7

E   Unit conversion factors   A–8

F   Numerical constants   A–9

ANSWERS TO ODD-NUMBERED PROBLEMS   A–10

INDEX   I–1

ENDPAPERS

(Front) Appendix E   Unit conversion factors
        Index of Tables
(Rear) Appendix F   Numerical constants

ABRIDGED CONTENTS

PART ONE

MECHANICS AND FUNDAMENTALS

1	UNITS, PHYSICAL QUANTITIES, AND VECTORS	3
2	MOTION ALONG A STRAIGHT LINE	25
3	MOTION IN A PLANE	49
4	NEWTON'S LAWS OF MOTION	71
5	APPLICATIONS OF NEWTON'S LAWS—I	90
6	APPLICATIONS OF NEWTON'S LAWS—II	119

PART TWO

MECHANICS—FURTHER DEVELOPMENTS

7	WORK AND ENERGY	145
8	IMPULSE AND MOMENTUM	182
9	ROTATIONAL MOTION	210
10	EQUILIBRIUM OF A RIGID BODY	249
11	PERIODIC MOTION	263

PART THREE

MECHANICAL AND THERMAL PROPERTIES OF MATTER

12	ELASTICITY	291
13	FLUID MECHANICS	306
14	TEMPERATURE AND EXPANSION	340
15	QUANTITY OF HEAT	357
16	MECHANISMS OF HEAT TRANSFER	374
17	THERMAL PROPERTIES OF MATTER	389
18	THE FIRST LAW OF THERMODYNAMICS	403
19	THE SECOND LAW OF THERMODYNAMICS	424
20	MOLECULAR PROPERTIES OF MATTER	451

PART FOUR

WAVES

21	MECHANICAL WAVES	475
22	VIBRATING BODIES	496
23	ACOUSTIC PHENOMENA	512

UNIVERSITY
PHYSICS

SEVENTH EDITION

PART FIVE

ELECTROSTATICS AND ELECTRIC CURRENTS

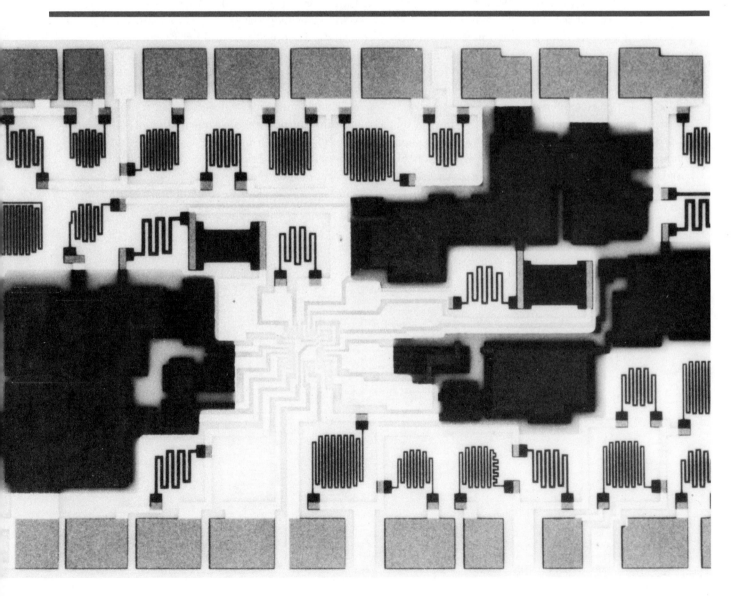

PERSPECTIVE

We are now at the halfway point in our study of physics. In the past three chapters we have been concerned with wave phenomena. We developed the most important wave concepts in the familiar context of mechanical waves, but we also tried to indicate their relevance in many other areas of physics and their role as a unifying concept throughout all areas of physics. In particular, we will explore in the last half of the book the indispensible role that wave phenomena play in optics; many optical phenomena simply cannot be understood without wave concepts. Equally important, as we will see later, wave concepts have proved to be an important gateway into the twentieth-century concepts of quantum mechanics, so vital to the understanding of atomic, molecular, and nuclear structure.

The next twelve chapters take us into the area of electricity and magnetism, a broad area of physics that includes a rich variety of fascinating physical phenomena and the fundamental principles on which they are based. We first introduce the concept of *electric charge;* we describe interactions among charges with the help of the concept of *electric field.* A study of the potential energies associated with these interactions leads to the concept of *electric potential.* Together, these concepts enable us to understand a wide range of phenomena in electrostatics, or "static electricity," as it is often called, and also to explore the subject of *capacitance,* so vital to contemporary electronics. Next we examine the *flow* of electric charge through materials we call *conductors.* A flow of charge is called an *electric current,* and this discussion leads us into a detailed study of electric *circuits.* We study the circuit concepts of resistance, voltage, current, and electromotive force and develop principles that enable us to analyze and understand a variety of practical devices and systems.

Electric circuits are at the heart of electric-power distribution systems, household wiring, the electrical systems of automobiles, and all of contemporary electronics, including radio, television, and computer circuitry and much of the instrumentation of present-day research in physics. Thus our study of electrostatics and of electric circuits has not only a great deal of inherent interest but also a broad range of practical applications to contemporary technology.

24

COULOMB'S LAW

INTERACTIONS BETWEEN ELECTRICALLY CHARGED BODIES ARE ONE OF THE four fundamental classes of interactions found in nature, as discussed at the beginning of Chapter 5. In this chapter we study the basic principles underlying the interactions of electric charges at rest, that is, *electrostatic* phenomena. The basic force law for interaction of electric charges at rest is called *Coulomb's law;* a variety of applications of this law are worked out in this chapter.

24–1 ELECTRIC CHARGE

The ancient Greeks discovered as early as 600 B.C. that when amber is rubbed with wool, it becomes able to attract other objects. Today we say that the amber has acquired an **electric charge,** or is *electrified*. Indeed, these terms are derived from the Greek word *elektron*, meaning "amber." A person may become electrified by scuffing shoes across a nylon carpet; a comb is electrified by passing it through dry hair; and so on.

For demonstrations of electric-charge interactions we often use plastic rods and fur. We can electrify a plastic rod by rubbing it with fur, and then touch the rod to two small, light balls of cork or pith, suspended by thin silk or nylon threads. We find that the rod then *repels* the balls and that they also repel each other.

We get the same results by rubbing a glass rod with silk. However, when a pith ball that has been in contact with electrified plastic is placed near one that has been in contact with electrified glass, the pith balls *attract* each other. We conclude that there are two kinds of electric charge, the kind on the plastic rod rubbed with fur and the kind on the glass rod rubbed with silk. Benjamin Franklin suggested calling these *negative* and *positive*, respectively, and these names are still used. These experiments lead to the fundamental conclusion that *like charges repel each other, and unlike charges attract each other*.

Two bodies may also interact by *magnetic* interaction; the most familiar example is the attraction of iron or steel objects to a permanent magnet. Magnetic forces were once believed to be a fundamentally different type of interaction, but it is now known that magnetic interactions are actually the

Objects can be electrified by rubbing them together.

There are two kinds of electric charge: Like charges attract, and unlike charges repel.

interactions of electrically charged particles in motion. An electromagnet shows *magnetic* interactions when an *electric* current passes through its coils. We will study magnetic interactions in detail in later chapters, but for now we concentrate on *charges at rest*, that is, on **electrostatics.**

Here is another fundamental experiment. We rub a plastic rod with fur and then touch it to a suspended pith ball. Both the rod and the ball then have negative charge. If we now bring the *fur* near the pith ball, the ball is *attracted*, showing that the fur is *positively* charged. Thus when plastic is rubbed with fur, *opposite* charges appear on the two materials. The same thing happens with glass and silk. The implication is that in these experiments, electric charge is not *created* but is *transferred* from one body to another. It is now known that the plastic rod acquires extra electrons, which have negative charge. These elecrons are taken from the fur, which thus acquires a net positive charge.

Electrifying a body involves transfer of charge from one body to another, not creation of charge.

24–2 ATOMIC STRUCTURE

The interactions responsible for the structure of atoms and molecules (and hence of all ordinary matter) are primarily *electrical* interactions between electrically charged particles. The fundamental building blocks of ordinary matter are three particles: the negatively charged **electron,** the positively charged **proton,** and the electrically neutral **neutron.** The negative charge of the electron has the same magnitude as the positive charge of the proton. Thus in one sense the charge of a proton or an electron is the fundamental natural unit of charge. In the presently accepted theory of fundamental particles, however, the proton and neutron are not truly fundamental but are combinations of other entities called *quarks,* which have charges of $\pm\frac{1}{3}$ and $\pm\frac{2}{3}$ times the electron charge. In this theory the electron *is* a truly fundamental particle. We return to the quark model of fundamental particles in Chapter 44.

Fundamental particles have electric charges.

These three particles are arranged in the same general way in all atoms. The protons and neutrons always form a closely packed, dense cluster called the **nucleus,** which thus has a positive charge. The particles in the nucleus are held together by strong interactions, mentioned in Section 5–1. The diameter of the nucleus, if we think of it as roughly spherical, is of the order of 10^{-14} m. Outside the nucleus, at distances of the order of 10^{-10} m from it, are the electrons. In a neutral atom the number of electrons equals the number of protons in the nucleus. Thus the net electric charge of such an atom (the algebraic sum of all the charges) is zero. If one or more electrons are removed, the remaining positively charged structure is called a **positive ion; a negative ion** is an atom that has *gained* one or more electrons. This gaining or losing of electrons is called *ionization.*

The nucleus is very much smaller than the overall size of the atom.

In a neutral atom, the negative charge of the electrons exactly balances the positive charge of the nucleus.

In an atomic model proposed by the Danish physicist Niels Bohr in 1913, the electrons were pictured as whirling about the nucleus in circular or elliptical orbits. We now know that the electrons are more accurately represented as spread-out distributions of electric charge, governed by the principles of quantum mechanics, which we will discuss in Chapter 42. Nevertheless, the Bohr model is still useful for visualizing the structure of an atom. The diameters of the electron charge distributions, which the Bohr model pictures as circular orbits, determine the overall size of the atom as a whole. The diame-

A crude but useful picture of an atom: electrons in orbits around the nucleus

ters of these orbits are of the order of 10^{-10} m, or about ten thousand times as great as the diameter of the nucleus. A Bohr atom is analogous to a miniature solar system, with electrical forces taking the place of gravitational forces. The massive, positively charged central nucleus corresponds to the sun, while the electrons, moving around the nucleus under the electrical force of its attraction, correspond to the planets moving around the sun under the influence of its gravitational attraction.

The masses of the proton and neutron are nearly equal, and the mass of the proton is about 1836 times that of the electron. Nearly all the mass of an atom, therefore, is concentrated in its nucleus. Since one mole of monatomic hydrogen consists of 6.022×10^{23} particles (Avogadro's number) and its mass is 1.008 g, the mass of a single hydrogen atom is

Most of the mass of an atom is concentrated in its nucleus.

$$\frac{1.008 \text{ g}}{6.022 \times 10^{23}} = 1.674 \times 10^{-24} \text{ g} = 1.674 \times 10^{-27} \text{ kg}.$$

The nucleus of a hydrogen atom is a single proton, and around it moves a single electron. Hence, of the total mass of the hydrogen atom, 1/1837 part is the mass of the electron, and the remainder is the mass of the proton. To four significant figures,

$$\text{Mass of electron} = 9.110 \times 10^{-31} \text{ kg},$$

$$\text{Mass of proton} = 1.673 \times 10^{-27} \text{ kg},$$

$$\text{Mass of neutron} = 1.675 \times 10^{-27} \text{ kg}.$$

After hydrogen, the atom with the next simplest structure is helium. Its nucleus consists of two protons and two neutrons, and it has two electrons outside the nucleus. When these two electrons are removed, the doubly charged helium ion (which is the helium nucleus itself) is often called an *alpha particle,* or α-particle. The next element, lithium, has three protons in its nucleus and thus has a nuclear charge of three units. In the un-ionized state the lithium atom has three extranuclear electrons. Each element has a different number of protons in its nucleus and therefore a different positive nuclear charge. In the *periodic table of elements* at the end of this book, each element occupies a box with an associated number called the **atomic number.**

> The atomic number is the number of nuclear protons; in the un-ionized state, this is equal to the number of extranuclear electrons.

Every material body contains a tremendous number of charged particles: positively charged protons in the nuclei of its atoms and negatively charged electrons outside the nuclei. When the total number of protons equals the total number of electrons, the body as a whole is electrically neutral. To give a body an excess negative charge, we may either *add negative* charges to a neutral body or *remove positive* charges from the body. Similarly, we can create an excess positive charge by either *adding positive* charge or *removing negative* charge. In most cases it is negative charge (electrons) that is added or removed, and a "positively charged body" is one that has lost some of its normal complement of electrons. When we speak of the charge of a body, we always mean its *net* charge. The net charge is always a very small fraction of the total positive or negative charge in the body.

Electrifying or charging a body usually involves adding or subtracting electrons.

The total electric charge in the universe is constant: Charge cannot be created or destroyed.

Implicit in the preceding statements is the principle of **conservation of charge.** This principle states that the algebraic sum of all the electric charges in any closed system is constant. Charge can be transferred from one body to another, but it cannot be created or destroyed. Conservation of charge is believed to be a *universal* conservation law; there is no experimental evidence for any violation of this principle.

Electrical interactions are responsible for the structure of atoms, molecules, and condensed matter.

Electrical interactions play a central and dominant role in most aspects of the structure of matter. The forces that hold atoms together in a molecule or in a solid crystal lattice, the adhesive force of glue, the forces associated with surface tension—all these are electrical in nature, arising from the electrical forces between the charged particles in the interacting atoms.

Nuclei are held together by strong interactions, despite the repulsions of their protons.

Electrical interactions alone are *not* sufficient to understand the structure of atomic *nuclei,* however. A nucleus consists of protons, which repel each other, and neutrons, which have no electric charge. For nuclei to be stable, there must be additional forces, attractive in nature, that hold the nucleus together despite the electrical repulsion. This additional interaction is called the *nuclear force;* it is an example of the *strong interaction* mentioned in Section 5–1. The nuclear force has a short range, of the order of nuclear dimensions, and its effects do not extend far beyond the nucleus. In analyzing the structure of atoms we can ignore the nuclear force, considering the nucleus as a rigid structure, and concentrate on the electrical interactions.

When the structure of the nucleus itself becomes the subject of our study, we do need to consider the strong interactions. Many phenomena associated with the stability or instability of nuclei arise from the competition between the repulsive electrical forces and the attractive nuclear forces. We will study these matters in greater detail in Chapter 44.

24–3 ELECTRICAL CONDUCTORS AND INSULATORS

Some materials permit electric charge to move from one region of the material to another, while others do not. For example, suppose we touch one end of a copper wire to an electrified plastic rod and the other end to a metal ball that is initially uncharged, as in Fig. 24–1. By studying the subsequent interaction of the ball with other charged objects, we discover that it has become charged. The copper wire is called a **conductor** of electricity. If we now repeat the experiment but use a rubber band or nylon thread in place of the wire, we find that *no* charge is transferred to the ball. These materials are called **insulators.** Conductors permit the passage of charge through them; insulators do not.

Electric charge can move through conductors but not through insulators.

In general, *metals* are good conductors, and most *nonmetals* are insulators. The positive valence of metallic elements and the fact that they form positive ions in solutions show that the atoms of a metal can easily lose one or more of their outer electrons. Within a metal, such as a copper wire, a few outer electrons become detached from each atom. These electrons can then move freely throughout the material, in much the same way that the molecules of a gas move freely through the spaces between grains of sand in a sand-filled container. In fact, these free electrons are sometimes described as an "electron gas." The positive nuclei and the other electrons remain fixed in their posi-

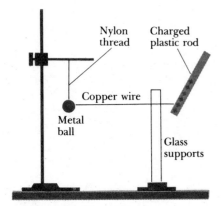

24–1 Copper is a conductor of electricity. Charge can move from the plastic rod through the wire to the metal ball.

tions within the material. In an insulator, however, there are no, or at most very few, free electrons.

24–4 CHARGING BY INDUCTION

When we charge a pith ball by touching it with a plastic rod that has been rubbed with fur, some of the extra electrons on the plastic are transferred to the ball, leaving the plastic with a smaller negative charge. However, we can use a different technique in which the plastic rod can give another body a charge of *opposite* sign without losing any of its own charge. This process, called charging by **induction,** is shown in Fig. 24–2.

In Fig. 24–2a two neutral metal spheres are in contact, each supported on an insulating stand. When we bring a negatively charged rod near one of the spheres, without actually touching it, as in (b), the free electrons in the metal spheres are repelled by the rod and shift slightly toward the right, away from the rod. The electrons cannot escape from the spheres because the supporting stands and the surrounding air are insulators. Thus excess negative charge must accumulate at the right surface of the right sphere, and a deficiency of negative charge (i.e., a net positive charge) arises at the left surface of the left sphere. These excess charges are called **induced charges.**

Of course, not *all* the free electrons are forced to the surface of the right sphere. As soon as any induced charge develops, it also exerts forces on the other free electrons. This force is toward the left; it consists of a repulsion by the induced negative charge on the right and an attraction toward the induced positive charge on the left. Thus the system reaches an equilibrium state. At each point, the force toward the right on an electron, due to the charged rod, is just balanced by the force toward the left on the electron, due to the induced charge. The induced charge remains on the surfaces of the spheres as long as the rod is held nearby. When the rod is removed, the free electrons shift back to the left and the original neutral condition is restored.

Now suppose we separate the spheres slightly, as shown in (c), while the plastic rod is near. If we now remove the rod, as in (d), we are left with two oppositely charged metal spheres whose charges attract each other. When the two spheres are separated by a great distance, as in (e), each of the two charges becomes uniformly distributed over its sphere. We note that the charge on the negatively charged rod has not changed during this process.

Figure 24–3 shows a variation of the technique of charging by induction. In this figure a single metal sphere (on an insulating stand) is charged by induction. The symbol labeled "ground" in part (c) simply means that the sphere is connected to the earth (a conductor). The earth takes the place of the second sphere in Fig. 24–2. In step (c), electrons are repelled to ground either through a conducting wire or along the moist skin of a person who touches the sphere. The earth thus acquires a negative charge equal to the induced positive charge remaining on the sphere.

The process taking place in Figs. 24–2 and 24–3 could be explained equally well if the mobile charges in the spheres were *positive* or, in fact, if *both* positive and negative charges were mobile. Although we now know that in a metallic conductor it is actually the *negative* charges that move, it is

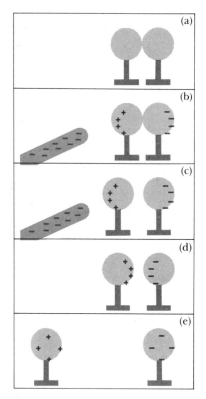

24–2 Two metal spheres are oppositely charged by induction.

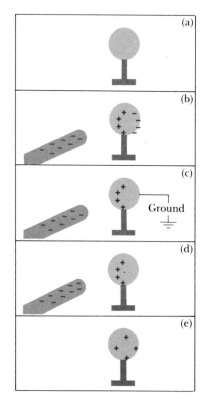

24–3 Charging a single metal sphere by induction.

often convenient to describe a process *as if* the positive charges moved. In ionic solutions, both positive ions and negative ions participate in the conduction process.

24–5 COULOMB'S LAW

Coulomb's law: a quantitative description of interactions between electric charges

We describe the electrical interaction between two charged particles in terms of the *forces* they exert on each other. A quantitative investigation of this interaction was carried out by Augustin de Coulomb (1736–1806) in 1784. For his force measurements he used a torsion balance similar to that used 13 years later by Cavendish to study the (much weaker) gravitational interaction, as we discussed in Section 6–3.

Coulomb studied the force of attraction or repulsion between two "point charges"; that is, charged bodies whose sizes are small compared with the distance between them. He found that the force grows weaker with increasing separation between the bodies. When the distance doubles, the force decreases to $\frac{1}{4}$ its initial value. That is, it varies inversely with the square of the distance. If r is the distance between the particles, then the force is proportional to $1/r^2$.

How to divide an electric charge into two equal parts

The force also depends on the quantity of charge on each body, which we will denote by q or Q. To explore this dependence, Coulomb devised a method for dividing a charge into two equal parts. He reasoned that if he brought a charged spherical conductor into contact with an identical conductor that was initially *uncharged*, then by symmetry the charge would be shared equally between the two conductors. Thus he could obtain one-half, one-quarter, and so on, of any given charge. He found that the force that each of two point charges q_1 and q_2 exerts on the other is proportional to each charge and is thus proportional to the *product* q_1q_2 of the two charges.

The magnitude F of the force between two point charges can be expressed as

$$F = k\frac{|q_1q_2|}{r^2},\qquad (24\text{–}1)$$

where k is a proportionality constant. The numerical value of k depends on the units in which F, q_1, q_2, and r are expressed. Equation (24–1) is the mathematical statement of what is known today as **Coulomb's law:**

> The force of attraction or repulsion between two point charges is directly proportional to the product of the charges and inversely proportional to the square of the distance between them.

The *direction* of the force on each particle is always along the line joining the two particles, pulling each particle toward the other in the case of attractive forces on unlike charges, and pushing them apart in the case of repulsive forces on like charges.

The exponent 2 in Coulomb's law appears to be exactly 2.

The proportionality of the electrical force to $1/r^2$ has been verified with great precision. There is no reason to suspect, for example, that the electrical force might vary as $1/r^{2.0001}$.

The charges q_1 and q_2 can be either positive or negative quantities, corresponding to the existence of two kinds of charge, but Eq. (24–1) gives the magnitude of the interaction force in all cases. When the charges are of like sign, the forces are repulsive; when they are unlike, the forces are attractive. In either case the forces obey Newton's third law; the force that q_1 exerts on q_2

is the negative of the force that q_2 exerts on q_1. The "absolute value" bars in Eq. (24–1) are needed because F, the magnitude of a vector quantity, is by definition always positive, while the product q_1q_2 is negative whenever the two charges have opposite signs.

24–4 The unit vector $\hat{r}$, pointing from q_1 toward q_2, is used to describe the directions of the forces F_1 and F_2 on the charges.

The form of Eq. (24–1) is the same as that of the law of gravitation, discussed in Section 6–3, but electrical and gravitational interactions are two distinct classes of phenomena. Electrical interactions depend on electric charges and can be either attractive or repulsive, while gravitational interactions depend on mass and are always attractive.

We can incorporate both the magnitude and direction of the interaction force between two charged particles into a single vector equation. To do this we use a unit vector $\hat{r}$ lying along the line joining q_1 and q_2, in the direction from q_1 to q_2, as shown in Fig. 24–4. The force F_2 on q_2 is then given by

$$F_2 = k\frac{q_1q_2}{r^2}\hat{r}, \qquad (24\text{–}2)$$

and the force F_1 on q_1 is given by

$$F_1 = -k\frac{q_1q_2}{r^2}\hat{r}. \qquad (24\text{–}3)$$

Unit vectors can be used to describe the magnitude and direction of an electrical force in a single equation.

When two charges exert forces simultaneously on a third charge, the total force experienced by that charge is found to be the *vector sum* of the forces that the two charges would exert individually. This important property, called the **principle of superposition,** also holds for any number of charges. It permits the application of Coulomb's law to arrays of charge of any degree of complexity, although the computational problems can be very great. Examples 24–1 and 24–4 (Section 24–6) show applications of the superposition principle.

Superposition: how to calculate the total force exerted by several charges

If matter is present in the space between the charges, the *net force* acting on each charge is altered because charges are induced in the molecules of the intervening material. We will describe this effect later. As a practical matter, Coulomb's law can be used as stated for point charges in air, since even at atmospheric pressure the effect of the air changes the electrical force from its value in vacuum by only about one part in two thousand.

In the chapters of this book dealing with electrical phenomena, we will use SI units exclusively. The SI electrical units include most of the common electrical units such as the volt, the ampere, the ohm, and the watt. The cgs system is also used, more so in scientific work than in commerce and industry. However, there is *no* British system of electrical units. This is one of many reasons for abandoning the British system and adopting metric units universally.

The coulomb: the SI unit of electric charge

To the three basic SI units (the meter, the kilogram, and the second) we now add a fourth: the unit of electric charge. This unit is called one **coulomb** (1 C). The electrical constant k in Eq. (24–1) is, in this system,

$$k = 8.98755 \times 10^9 \text{ N·m}^2\text{·C}^{-2}.$$

This value of k can be regarded as an operational definition of the coulomb, since in principle we can measure the interaction force between any two charges and use Coulomb's law to determine the charge. As a practical matter, it is better to define the coulomb instead in terms of a unit of electric current (charge per unit time), the *ampere*. We will return to this definition in Chapter 31.

The value of the constant k in Coulomb's law is not as arbitrary as it seems.

This value of k may seem arbitrary and strange, but it really is not. Later, when we study electromagnetic radiation, we will show that k is closely related to the *speed of light,*

$$c = 2.998 \times 10^8 \text{ m·s}^{-1}.$$

Specifically, the numerical value of k is given by

$$k = 10^{-7} c^2.$$

A different (and more usual) form for the constant in Coulomb's law

In the cgs system of electrical units (not used in this book) the constant k is defined to be unity, without units. This defines a unit of electric charge called the *statcoulomb* or the *esu* (electrostatic unit). The conversion factor is

$$1 \text{ C} = 2.998 \times 10^9 \text{ esu}.$$

In SI units the constant k in Eq. (24–1) is usually written not as k but as $1/4\pi\epsilon_0$, where ϵ_0 is another constant. This appears to complicate matters, but it actually simplifies many formulas to be encountered later. Thus Coulomb's law is usually written as

$$F = \frac{1}{4\pi\epsilon_0} \frac{|q_1 q_2|}{r^2}. \tag{24–4}$$

In the vector form of Eqs. (24–2) and (24–3) this becomes

$$\boldsymbol{F} = \frac{1}{4\pi\epsilon_0} \frac{q_1 q_2}{r^2} \hat{r}. \tag{24–5}$$

In Eqs. (24–4) and (24–5),

$$\frac{1}{4\pi\epsilon_0} = 8.98755 \times 10^9 \text{ N·m}^2\text{·C}^{-2}$$

and

$$\epsilon_0 = 8.854188 \times 10^{-12} \text{ C}^2\text{·N}^{-1}\text{·m}^{-2}.$$

In examples and problems we often use the approximate value

$$\frac{1}{4\pi\epsilon_0} = 9.0 \times 10^9 \text{ N·m}^2\text{·C}^{-2},$$

which is within about 0.1% of the correct value.

The charge of the electron: a fundamental unit of electric charge

As we mentioned in Section 24–2, the most fundamental unit of charge is the magnitude of the charge of an electron or a proton. This quantity is denoted by e; the most precise measurements to date yield the value

$$e = 1.602192 \times 10^{-19} \text{ C} \cong 1.60 \times 10^{-19} \text{ C}.$$

One coulomb therefore represents the negative of the total charge carried by about 6×10^{18} electrons. For comparison, the population of the earth is estimated to be about 5×10^9 persons, while a cube of copper 1 cm on a side contains about 2.4×10^{24} electrons.

In most contexts, the coulomb is a very large unit of charge.

In electrostatics problems, charges as large as one coulomb are unusual. Two charges of magnitude 1 C, a distance 1 m apart, would exert forces of magnitude 9×10^9 N on each other! A more typical range of magnitude is 10^{-9} to 10^{-6} C. The microcoulomb ($1 \ \mu\text{C} = 10^{-6}$ C) is often used as a practical unit of charge.

24–6 APPLICATIONS OF COULOMB'S LAW

In this section we present several examples of problems using Coulomb's law. There are no new principles here, but the problem-solving methods are helpful preparation for the next several chapters.

PROBLEM-SOLVING STRATEGY: Coulomb's law

1. As always, consistent units are essential. With the value of $k = 1/4\pi\epsilon_0$ given above, distances *must* be in meters and charge in coulombs; the force is then in newtons. If you are given distances in centimeters, inches, or furlongs, don't forget to convert!

2. When a charge has forces acting on it due to two or more other charges, the total force is the *vector sum* of the individual forces. Don't forget what you learned about vector addition at the beginning of the book. Components in an xy-coordinate system, with or without unit vectors, are often helpful. If you are in doubt, review the material on vector addition in Chapter 1. And be sure to use correct vector notation; if a symbol represents a vector quantity, underline it or

put an arrow over it. If you get sloppy with your notation, you will also get sloppy with your thinking. It is absolutely essential to distinguish between vector quantities and scalar quantities, and to treat vectors as vectors.

3. Some of the examples and problems in this and later chapters involve a continuous distribution of charge along a line or over a surface. In these cases the vector sum described in (2) becomes a vector integral, usually carried out by use of components. We divide the total charge distribution into infinitesimal pieces, use Coulomb's law for each piece, and then integrate to carry out the vector sum. Study Example 24–5 in this section carefully.

EXAMPLE 24–1 Two charges are located on the positive x-axis of a coordinate system, as shown in Fig. 24–5. Charge $q_1 = 2 \times 10^{-9}$ C is 2 cm from the origin, and charge $q_2 = -3 \times 10^{-9}$ C is 4 cm from the origin. What is the total force exerted by these two charges on a charge $q_3 = 5 \times 10^{-9}$ C located at the origin?

Using the superposition principle to find the total force on a charge due to two others

SOLUTION The total force on q_3 is the vector sum of the forces due to q_1 and q_2 individually. Converting distance to meters, we use Eq. (24–4) to find the magnitude F_1 of the force on q_3 due to q_1:

$$F_1 = \frac{(9.0 \times 10^9 \text{ N·m}^2\text{·C}^{-2})(2 \times 10^{-9} \text{ C})(5 \times 10^{-9} \text{ C})}{(0.02 \text{ m})^2} = 2.25 \times 10^{-4} \text{ N}.$$

This force has a negative x-component because q_3 is repelled (i.e., pushed in the negative x-direction) by q_1, which has the same sign. Similarly, the force due to q_2 is found to have magnitude

$$F_2 = \frac{(9.0 \times 10^9 \text{ N·m}^2\text{·C}^{-2})(3 \times 10^{-9} \text{ C})(5 \times 10^{-9} \text{ C})}{(0.04 \text{ m})^2} = 0.84 \times 10^{-4} \text{ N}.$$

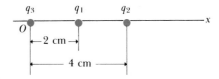

FIGURE 24–5

This force has a positive x-component because q_3 is attracted (i.e., pulled in the positive x-direction) by the opposite charge q_2. The sum of the x-components is

$$\Sigma F_x = -2.25 \times 10^{-4} \text{ N} + 0.84 \times 10^{-4} \text{ N} = -1.41 \times 10^{-4} \text{ N}.$$

There are no y- or z-components. Thus the total force on q_3 is directed to the left, with magnitude 1.41×10^{-4} N.

EXAMPLE 24–2 An α-particle is a nucleus of a helium atom. It has a mass m of 6.64×10^{-27} kg and a charge q of $+2e$ or 3.2×10^{-19} C. Compare the force of

Comparing the strengths of electrical and gravitational forces

the electrostatic repulsion between two α-particles with the force of gravitational attraction between them.

SOLUTION   The electrostatic force F_e is

$$F_e = \frac{1}{4\pi\epsilon_0} \frac{q^2}{r^2},$$

and the gravitational force F_g is

$$F_g = G\frac{m^2}{r^2}.$$

The ratio of the electrostatic to the gravitational force is

$$\frac{F_e}{F_g} = \frac{1}{4\pi\epsilon_0 G} \frac{q^2}{m^2} = \frac{(9.0 \times 10^9 \text{ N·m}^2\text{·C}^{-2})}{(6.67 \times 10^{-11} \text{ N·m}^2\text{·kg}^{-2})} \frac{(3.2 \times 10^{-19} \text{ C})^2}{(6.64 \times 10^{-27} \text{ kg})^2}$$
$$= 3.1 \times 10^{35}.$$

Thus the gravitational force is negligible compared to the electrostatic force. This is always the case for interactions of atomic and subatomic particles, while for objects the size of the earth the positive and negative charges are nearly equal, and the net *electrical* interactions are usually much smaller than the gravitational.

The Bohr model: a simple dynamic model of the structure of the hydrogen atom

EXAMPLE 24–3   The Bohr model of the hydrogen atom (mentioned in Section 24–2 and discussed in detail in Chapter 42) consists of a single electron having electric charge $-e$, revolving in a circular orbit about a single proton of charge $+e$. The electrostatic force of attraction between electron and proton provides the centripetal force that retains the electron in its orbit. Hence if v is the orbital speed, Newton's second law gives

$$\frac{1}{4\pi\epsilon_0} \frac{e^2}{r^2} = m\left(\frac{v^2}{r}\right). \tag{24–6}$$

This motion is completely analogous to the motion of a satellite around the earth, discussed in Section 6–5.

In the Bohr model, the angular momentum must be an integer multiple of some minimum value.

In Bohr's theory, the electron may revolve only in some one of a number of specified orbits. The orbit of smallest radius is that for which the angular momentum L of the electron is $h/2\pi$, where h is a universal constant called *Planck's constant*, equal to 6.625×10^{-34} J·s. Then,

$$L = mvr = \frac{h}{2\pi}. \tag{24–7}$$

When v is eliminated between the preceding equations, we find

$$r = \frac{\epsilon_0 h^2}{\pi m e^2},$$

and when numerical values are inserted, we find, for the radius of the *first Bohr orbit*,

$$r = 5.29 \times 10^{-11} \text{ m} = 0.529 \times 10^{-8} \text{ cm}.$$

This result corresponds roughly with other estimates of the "size" of a hydrogen atom obtained from deviations from ideal-gas behavior, the density of hydrogen in the liquid and solid states, and other observations.

EXAMPLE 24–4 In Fig. 24–6, two equal positive charges $q = 2.0 \times 10^{-6}$ C interact with a third charge $Q = 4.0 \times 10^{-6}$ C. Find the magnitude and direction of the total (resultant) force on Q.

SOLUTION The key word is *total;* we must compute the force each charge exerts on Q, and then obtain the *vector sum* of the forces. This is most easily accomplished by using components. The figure shows the force on Q due to the upper charge q. From Coulomb's law,

$$F = (9.0 \times 10^9 \text{ N·C}^{-2}\text{·m}^{-2})\frac{(4.0 \times 10^{-6} \text{ C})(2.0 \times 10^{-6} \text{ C})}{(0.5 \text{ m})^2}$$

$$= 0.29 \text{ N}.$$

The components of this force are given by

$$F_x = F \cos \theta = (0.29 \text{ N})\left(\frac{0.4 \text{ m}}{0.5 \text{ m}}\right) = 0.23 \text{ N},$$

$$F_y = -F \sin \theta = -(0.29 \text{ N})\left(\frac{0.3 \text{ m}}{0.5 \text{ m}}\right) = -0.17 \text{ N}.$$

The lower charge q exerts a force of the same magnitude, but in a different direction. From symmetry we see that its x-component is the same as that due to the upper charge, but its y-component is opposite. Hence

$$\sum F_x = 2(0.23 \text{ N}) = 0.46 \text{ N},$$

$$\sum F_y = 0,$$

or

$$\sum \mathbf{F} = (0.46 \text{ N}) \, \mathbf{i}.$$

The total force on Q is horizontal, with magnitude 0.46 N. How would this solution differ if the lower charge were *negative?*

EXAMPLE 24–5 An electric charge Q is distributed uniformly along a line of length $2a$, lying along the y-axis, as shown in Fig. 24–7. A point charge q lies on the x-axis, at a distance x from the origin. Find the total force Q exerts on q.

Using the superposition principle with vector addition to find the total force on a charge

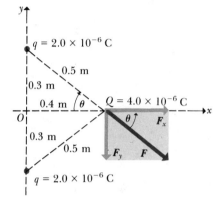
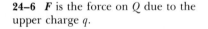

24–6 $\mathbf{F}$ is the force on Q due to the upper charge q.

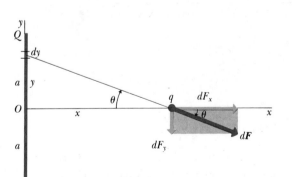
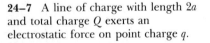

24–7 A line of charge with length $2a$ and total charge Q exerts an electrostatic force on point charge q.

Using the superposition principle for a continuous distribution of charge along a line

SOLUTION We divide the line charge into infinitesimal segments; let the length of a typical segment at height y be dy, as shown. To find the charge dQ in this segment, note that if the charge is distributed uniformly, the ratio of dQ to the total charge Q is equal to the ratio of dy to the total length $2a$. Thus

$$dQ/Q = dy/2a$$

and

$$dQ = Q\,dy/2a.$$

The distance r from this segment to q is $(x^2 + y^2)^{1/2}$, so the magnitude of force dF on q due to this segment is

$$dF = \frac{qQ}{4\pi\epsilon_0}\frac{dy}{2a(x^2 + y^2)}.$$

We now represent this force in terms of its x- and y-components, using the same procedure as in Example 24–4:

$$dF_x = dF\cos\theta,$$
$$dF_y = -dF\sin\theta.$$

We note that

$$\sin\theta = \frac{y}{\sqrt{x^2 + y^2}}, \qquad \cos\theta = \frac{x}{\sqrt{x^2 + y^2}}.$$

To find the components of the total force, we integrate the forces due to the separate charge elements.

Combining this with the above expression for dF, we find

$$dF_x = \frac{qQ}{4\pi\epsilon_0}\frac{x\,dy}{2a(x^2 + y^2)^{3/2}},$$
$$dF_y = -\frac{qQ}{4\pi\epsilon_0}\frac{y\,dy}{2a(x^2 + y^2)^{3/2}}.$$

To find the total force components F_x and F_y, we integrate these expressions, noting that to include all of Q we must integrate from $y = -a$ to $y = +a$. We invite you to work out the details of the integration; an integral table is helpful. The final results are

$$F_x = \int_{-a}^{a}\frac{qQ}{4\pi\epsilon_0}\frac{x\,dy}{2a(x^2 + y^2)^{3/2}} = \frac{qQ}{4\pi\epsilon_0}\frac{1}{x\sqrt{x^2 + a^2}},$$
$$F_y = -\int_{-a}^{a}\frac{qQ}{4\pi\epsilon_0}\frac{y\,dy}{2a(x^2 + y^2)^{3/2}} = 0,$$

or, in vector form,

$$\mathbf{F} = \frac{qQ}{4\pi\epsilon_0}\frac{1}{x\sqrt{x^2 + a^2}}\,\mathbf{i}.$$

We could have guessed from the symmetry of the situation that F_y would be zero; the upper half of Q pushes downward on q, but this is balanced by an equal upward push due to the lower half of Q (assuming both Q and q are positive). We also note that when x is much larger than a, we can neglect a in the denominator, and our result becomes approximately

$$\mathbf{F} \cong \frac{qQ}{4\pi\epsilon_0}\frac{1}{x^2}\,\mathbf{i},$$

which simply means that if the charge q is far away from the line charge, compared to its size, it looks like a point. When q and Q are both positive or both negative, the force is in the *positive* x-direction, corresponding to a repulsive interaction. If they have opposite signs, F_x is negative, corresponding to an attraction.

SUMMARY

The fundamental entity in electrostatics is electric charge. Charge can be transferred from one body to another by rubbing or other means, but it cannot be created or destroyed. There are two kinds of charge, positive and negative. Like charges repel each other; unlike charges attract.

The basic constituents of atoms are protons, neutrons, and electrons. The protons and neutrons are bound together by the nuclear force in a small, dense nucleus, with the electrons surrounding it at distances much greater than the nuclear size. The electrical interactions between the electrons and the positively charged nucleus are responsible for the structure of atoms, molecules, and solids.

Materials that permit electric charge to move within them are called conductors; those that do not are called insulators. Most metals are good conductors; most nonmetals are insulators. The presence of an electrically charged body near a conductor causes a redistribution of charge on the conductor, known as induced charge.

The basic law of interaction for point electric charges q_1 and q_2 separated by a distance r is Coulomb's law. The magnitude of the force is given by

$$F = k\frac{|q_1q_2|}{r^2}. \tag{24–1}$$

The force on each charge is directed along the line joining the two charges; if q_1 and q_2 have the same sign, the forces are repulsive; if opposite signs, attractive. If $\hat{r}$ is a unit vector in the direction from q_1 to q_2, then the force $\boldsymbol{F}_2$ on q_2 is given by

$$\boldsymbol{F}_2 = k\frac{q_1q_2}{r^2}\hat{r}, \tag{24–2}$$

and the force $\boldsymbol{F}_1$ on q_1 is given by

$$\boldsymbol{F}_1 = -k\frac{q_1q_2}{r^2}\hat{r}. \tag{24–3}$$

This pair of forces is an action–reaction pair and obeys Newton's third law.

The principle of superposition states that when two or more charges each exert a force on a certain charge, the total force on that charge is the vector sum of the forces exerted by the individual charges.

In SI units the unit of electric charge is the coulomb, abbreviated C. The numerical value of the constant k in Coulomb's law is

$$k = 8.98755 \times 10^9 \text{ N·m}^2\text{·C}^{-2}.$$

This constant is usually expressed in terms of another constant ϵ_0:

$$k = \frac{1}{4\pi\epsilon_0} = 8.98755 \times 10^9 \text{ N·m}^2\text{·C}^{-2},$$

KEY TERMS

electric charge

electrostatics

electron

proton

neutron

nucleus

positive ion

negative ion

atomic number

conservation of charge

conductor

insulators

induction

induced charges

Coulomb's law

principle of superposition

coulomb

and

$$\epsilon_0 = 8.854188 \times 10^{-12} \, \text{C}^2 \cdot \text{N}^{-1} \cdot \text{m}^{-2}.$$

The usual methods of vector addition, including use of components, are used in calculating vector sums for electrical forces due to two or more charges. When charge is distributed over a line or surface or through a volume, the vector sums become integrals.

QUESTIONS

24–1 Plastic food wrap can be used to cover a container by simply stretching the material across the top and pressing the overhanging material against the sides. What makes it stick? Does it stick to itself with equal tenacity? Why? Does it matter whether the container is metallic?

24–2 Bits of paper are attracted to an electrified comb or rod even though they have no net charge. How is this possible?

24–3 How do we know that the magnitudes of electron and proton charge are *exactly* equal? With what precision is this really known?

24–4 When you walk across a nylon rug and then touch a large metal object, you may get a spark and a shock. Why does this tend to happen more in the winter than the summer? Why do you not get a spark when you touch a *small* metal object?

24–5 The free electrons in a metal have mass and therefore weight and are gravitationally attracted toward the earth. Why, then, do they not all settle to the bottom of the conductor, as sediment settles to the bottom of a river?

24–6 Simple electrostatics experiments, such as picking up bits of paper with an electrified comb, never work as well on rainy days as on dry days. Why?

24–7 High-speed printing presses sometimes use gas flames to reduce electric charge buildup on the paper passing through the press. Why does this help? (An added benefit is rapid drying of the ink.)

24–8 What similarities do electrical forces have to gravitational forces? What are the most significant differences?

24–9 Given two identical metal objects mounted on insulating stands, describe a procedure for placing charges of equal magnitude and opposite sign on the two objects.

24–10 How do we know that protons have positive charge and electrons negative charge, rather than the reverse?

24–11 Gasoline transport trucks sometimes have chains that hang down and drag on the ground at the rear end. What are these for?

24–12 When a nylon sleeping bag is dragged across a rubberized-cloth air mattress in a dark tent, small sparks are sometimes seen. What causes them?

24–13 Atomic nuclei are made of protons and neutrons. This fact by itself shows that there must be another kind of interaction in addition to the electrical forces. Explain.

24–14 When transparent plastic tape is pulled off a roll and one tries to position it precisely on a piece of paper, it often jumps over and sticks where it is not wanted. Why does it do this?

24–15 When a thunderstorm is approaching, sailors at sea sometimes observe a phenomenon called "St. Elmo's fire," a bluish flickering light at the tips of masts and along wet rigging. What causes this?

EXERCISES

Section 24–2 Atomic Structure

24–1 What is the total positive charge, in coulombs, of all the protons in 1 mole of hydrogen atoms?

24–2 What is the total negative charge, in coulombs, of all the electrons in 20 g of aluminum?

Section 24–5 Coulomb's Law

Section 24–6 Applications of Coulomb's Law

24–3 A negative charge -0.50×10^{-6} C exerts a repulsive force of magnitude 0.20 N on an unknown charge 0.20 m away. What is the unknown charge (magnitude and sign)?

24–4 At what distance would the repulsive force between two electrons have a magnitude of 1 N? Between two protons?

24–5 Two small plastic balls are given positive electric charges. When they are 5 cm apart, the repulsive forces between them have magnitude 0.10 N. What is the charge on each ball

a) if the two charges are equal?

b) if one ball has twice the charge of the other?

24–6 How many excess electrons must be placed on each of two small spheres spaced 3 cm apart if the spheres are to have equal charge and if the force of repulsion between them is to be 10^{-19} N?

24–7 If 6.02×10^{23} atoms of monatomic hydrogen have a mass of 1 g, how far would the electron of a hydrogen atom have to be removed from the nucleus for the force of attraction to equal the weight of the atom?

24–8 If all the positive charges in 1 mole of hydrogen atoms were lumped into a single charge, and all the negative charges into a single charge, what force would the two lumped charges exert on each other at a distance of

a) 1 m

b) 10^7 m (comparable to the diameter of the earth)?

24–9 Two copper spheres, each having mass 1 kg, are separated by 1 m.

a) How many electrons does each sphere contain?

b) How many electrons would have to be removed from one sphere and added to the other to cause an attractive force of 10^4 N (roughly 1 ton)?

c) What fraction of all the electrons on a sphere does this represent?

24–10 The dimensions of atomic nuclei are of the order of 10^{-14} m. Suppose that two α-particles are separated by this distance.

a) What is the force exerted on each α-particle by the other?

b) What is the acceleration of each?

(See Example 24–2 for numerical data.)

24–11 Use the Bohr model to calculate the speed of the electron in a hydrogen atom when the electron is in the orbit of smallest radius.

24–12 Two point charges are located in the xy-plane as follows: A charge 2.0×10^{-9} C is at the point ($x = 0$, $y = 4$ cm), and a charge -3.0×10^{-9} C is at the point ($x = 3$ cm, $y = 4$ cm).

a) If a third charge of 4.0×10^{-9} C is placed at the origin, find the x- and y-components of the total force on this third charge.

b) Find the magnitude and direction of the total force on the charge at the origin in (a).

24–13 Two positive point charges, each of magnitude q, are located on the y-axis at points $y = +a$ and $y = -a$. A third positive charge of the same magnitude is located at some point on the x-axis.

a) What is the force exerted on the third charge when it is at the origin?

b) What are the magnitude and direction of the force on the third charge when its coordinate is x?

c) Sketch a graph of the force on the third charge as a function of x, for values of x between $+4a$ and $-4a$. Plot forces to the right upward, forces to the left downward.

d) How does the force on the third charge vary with x when x is very large?

24–14 A negative point charge of magnitude q is located on the y-axis at the point $y = +a$, and a positive charge of the same magnitude is located at $y = -a$. A third charge that is positive and of the same magnitude q is located at some point on the x-axis.

a) What are the magnitude and direction of the force exerted on the third charge when it is at the origin?

b) What is the force on the third charge when its coordinate is x?

c) Sketch a graph of the force on the third charge as a function of x, for values of x between $+4a$ and $-4a$.

24–15 The pair of equal and opposite charges in Exercise 24–14 is called an *electric dipole*.

a) Show that when the x-coordinate of the third charge in Exercise 24–14 is large compared with the distance a, the force on it is inversely proportional to the *cube* of its distance from the midpoint of the dipole.

b) Show that if the third charge is located on the y-axis, at a y-coordinate that is large compared with the distance a, the forces on it are also inversely proportional to the cube of its distance from the midpoint of the dipole.

24–16 A positive electric charge Q is distributed uniformly along the positive x-axis from $x = 0$ to $x = a$. A positive point charge q is located on the x-axis at $x = a + r$, so it is a distance r to the right of the end of Q. Calculate the magnitude and direction of the force that the charge distribution Q exerts on q.

24–17 Positive charge Q is distributed uniformly along the positive y-axis between $y = 0$ and $y = a$. A negative point charge $-q$ lies on the positive x-axis, a distance x from the origin. Calculate the x- and y-components of the force that the charge distribution Q exerts on q.

PROBLEMS

24–18 Each of two small spheres is positively charged, the combined charge totaling 4×10^{-8} C. What is the charge on each sphere if they are repelled with a force of 27×10^{-5} N when placed 0.1 m apart?

24–19 Two small balls, each of mass 10 g, are attached to silk threads 1 m long and hung from a common point. When the balls are given equal quantities of negative charge, each thread makes an angle of 20° with the vertical.

a) Draw a diagram showing all the forces on each ball.

b) Find the magnitude of the charge on each ball.

c) The two threads are now shortened to length $l = 0.5$ m, while the charge on the balls is held fixed. What will be the new angle θ that the threads each make with the vertical? (*Hint:* This part of the problem can be solved numerically by using trial values for θ and adjusting the values of θ until a self-consistent answer is obtained.)

24–20 One gram of monatomic hydrogen contains 6.02×10^{23} atoms, each consisting of an electron with charge -1.60×10^{-19} C and a proton with charge $+1.60 \times 10^{-19}$ C.

a) Suppose all these electrons could be located at the north pole of the earth and all the protons at the south pole. What would be the total force of attraction exerted on each group of charges by the other? The diameter of the earth is 12,800 km.

b) What would be the magnitude and direction of the force exerted by the charges in part (a) on a third positive charge, equal in magnitude to the total charge at one of the poles and located at a point on the equator? Draw a diagram.

24–21 According to the Bohr theory of the hydrogen atom (see Example 24–3), the electron orbits the hydrogen nucleus (a proton) in a circular path of radius 0.529×10^{-10} m when in the orbit of smallest radius. Consider an atom whose constituents are an electron and a positron (positronium). The positron is the antiparticle of the electron and has the same mass as the electron and a charge that is the same in magnitude but opposite in sign. If such a system conforms to the Bohr theory, the orbital angular momentum of the electron-positron system, for the orbit of smallest radius, must equal $h/2\pi$. Assuming that the electron and positron each follow circular orbits about the center of mass of the system, determine

a) the distance between the particles;

b) the magnitude of the velocity of each particle with respect to the center of mass.

24–22 A charge -3×10^{-9} C is placed at the origin of an xy-coordinate system, and a charge 2×10^{-9} C is placed on the positive y-axis, at $y = 4$ cm.

a) If a third charge 4×10^{-9} C is now placed at the point $(x = 3$ cm, $y = 4$ cm), find the components of the total force exerted on this charge by the other two.

b) Find the magnitude and direction of this force.

24–23 Point charges of 2×10^{-9} C are situated at each of three corners of a square whose side is 0.20 m. What would be the magnitude and direction of the resultant force on a point charge of -1×10^{-9} C if it were placed

a) at the center of the square?

b) at the vacant corner of the square?

24–24 A small ball having a positive charge q_1 hangs by an insulating thread. A second ball with a negative charge $q_2 = -q_1$ is kept at a horizontal distance a to the right of the first. (The distance a is large compared with the diameter of the ball.)

a) Show in a diagram all the forces on the hanging ball in its final equilibrium position.

b) You are given a third ball having a positive charge $q_3 = 2q_1$. Find at least two points at which this ball can be placed so that the first ball will hang vertically.

24–25 Positive charge $+Q$ is distributed uniformly along the $+x$-axis from $x = 0$ to $x = a$. Negative charge $-Q$ is distributed uniformly along the $-x$-axis from $x = 0$ to $x = -a$.

a) A positive point charge q lies on the positive y-axis, a distance y from the origin. Find the magnitude and direction of the force that the charge distribution exerts on q. Show that this force is proportional to y^{-3} as y becomes large.

b) Suppose instead that the positive point charge q lies on the positive x-axis, a distance $x > a$ from the origin. Find the magnitude and direction of the force that the charge distribution exerts on q. Show that this force is proportional to x^{-3} as x becomes large.

CHALLENGE PROBLEMS

24–26 Two identical 5-g pith balls are each attached to insulating threads of length $l = 0.5$ m and hung from a common point. One ball is given charge q_1 and the other a different charge q_2, which causes the balls to separate such that each thread makes an angle of 30° with the vertical. A small wire is then connected between the pith balls, allowing charge to be transferred from one ball to the other until the two balls have equal charges. The wire is removed, and the threads from which the balls are hanging now each make an angle of 40° with the vertical.

a) Determine the electrostatic force F between the balls before equalization of the charges.

b) Determine the product $q_1 q_2$ of the charges before equalization of the charges.

c) Determine the electrostatic force F' between the balls after equalization of the charges.

d) Determine the original charges q_1 and q_2. (*Hint:* The total charge on the pair of balls is conserved.)

24–27 Three charges are placed as shown in Fig. 24–8. It is known that the magnitude of q_1 is 4×10^{-6} C, but its sign and the value of the charge q_2 are not known. The charge q_3 equals $+1 \times 10^{-6}$ C, and the resultant force $\mathbf{F}$ on q_3 is measured to be entirely in the negative x-direction.

a) Considering the different possible signs of q_1 and q_2, there are four possible force diagrams representing the

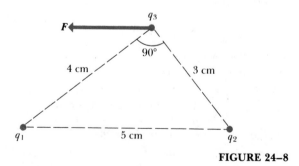

FIGURE 24–8

forces F_1 and F_2 that q_1 and q_2 exert on q_3. Sketch these four possible force configurations.

b) Using the sketches from (a) and the fact that the net force on q_3 has no y-component and a negative x-component, deduce the signs of the charges q_1 and q_2.

c) Calculate the magnitude of q_2.

d) Determine F.

24–28 Positive charge Q is distributed uniformly over the surface of a thin spherical shell of radius R. Calculate the force (magnitude and direction) that Q exerts on a positive point charge q located

a) a distance $r > R$ from the center of the shell (outside the shell);

b) a distance $r < R$ from the center of the shell (inside the shell).

(*Hint:* Divide the shell into infinitesimally thin coaxial rings, calculate the force due to each ring, and integrate to find the total force.)

25

THE ELECTRIC FIELD

THE ELECTRICAL INTERACTION BETWEEN CHARGED PARTICLES CAN BE reformulated by using the concept of *electric field*. We think of an electric charge as creating an electric field in the region of space surrounding it; that field in turn exerts a force on any other charge in that region. In this chapter we study methods for calculating the electric fields caused by various arrangements of charge. When a charge distribution has a high degree of symmetry, determining its electric field can be simplified by means of a principle called *Gauss's law*. This law is also useful in exploring several general properties of electric fields.

25–1 ELECTRIC FIELD AND ELECTRICAL FORCES

To introduce the concept of electric field, let us consider the mutual repulsion of two positively charged bodies A and B, as shown in Fig. 25–1a. In particular, consider the force on B, labeled F in the figure. This is an "action-at-a-distance" force; it can act across empty space and does not need any matter in the intervening space to transmit the force.

Now let us think of body A as having the effect of modifying some properties of the space in its vicinity. We remove body B and label its former position as point P (Fig. 25–1b). We say that the charged body A produces or causes an **electric field** at point P (and at all other points in the vicinity). Then when body B is placed at point P and experiences the force F, we take the point of view that the force is exerted on B *by the field at P*, rather than directly by A. Since B would experience a force at any point in space around A, the electric field exists at all points in the region around A. (We could equally well consider that body B sets up an electric field, and that the force on body A is exerted by the field due to B. The electric field due to B is of course not equal to that due to A.)

The concept of electric field is directly analogous to the concept of gravitational field, introduced in Section 6–4. We suggest you review that section as an aid to understanding what follows.

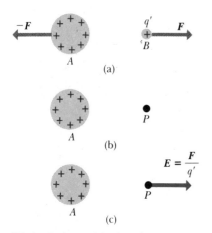

25–1 A charged body creates an electric field in the space around it.

The experimental test for the existence of an electric field at any point is simply to place a charged body, called a **test charge,** at the point. If the test charge experiences a force of electrical origin, then an electric field exists at that point.

Since force is a vector quantity, electric field is a *vector quantity.* We define the *electric field E* at a point as the quotient obtained when the force *F,* acting on a positive test charge, is divided by the magnitude q' of the test charge (Fig. 25–1c). Thus

$$E = \frac{F}{q'}$$

or

$$F = q'E. \qquad (25-1)$$

The direction of *E* is the direction of *F.* The force on a *negative* charge, such as an electron, has a direction *opposite* to that of the electric field.

The electric field is sometimes called *electric intensity* or *electric field intensity.* In SI units, where the unit of force is 1 N and the unit of charge is 1 C, the unit of electric field magnitude is one newton per coulomb (1 N·C^{-1}). Electric field may also be expressed in other equivalent units that will be introduced later.

The force experienced by the test charge q' varies from point to point, and so the electric field is also different at different points. Thus *E* is not a single vector quantity but an infinite set of vector quantities, one associated with each point in space. The electric field is an example of a **vector field.** Another example of a vector field is the description of motion of a flowing fluid. Different points in the fluid have different velocities, so the velocity is a vector field. If we use a rectangular coordinate system, then each component of *E* is a function of the coordinates (x, y, z) of a point in space. Vector fields are an important part of the mathematical language used in many areas of physics, particularly in electricity and magnetism.

One difficulty with our definition of electric field is that in Fig. 25–1 the force exerted by the test charge q' may change the charge distribution A, especially if the body is a conductor on which charge is free to move. The electric field around A when q' is present is not the same as when q' is absent. When q' is very small, however, the redistribution of charge on body A is also very small; thus the difficulty can be avoided by refining the definition of electric field to be *the limiting value of the force per unit charge on a test charge q' at the point, as the charge q' approaches zero:*

$$E = \lim_{q' \to 0} \frac{F}{q'}.$$

If an electric field exists within a *conductor,* a force is exerted on every charge in the conductor. The motion of the free charges brought about by this force is called a *current.* Conversely, if there is *no* current in a conductor, and hence no motion of its free charges, *the electric field in the conductor must be zero.*

In most instances, the magnitude and direction of an electric field vary from point to point. If the magnitude and direction are constant throughout a certain region, we say that the field is *uniform* in this region.

A test charge can be used to detect and explore an electric field.

Electric field is force per unit charge.
Electric field is a vector quantity.

Does a test charge disturb the electric field by its presence?

When all charges are at rest, the electric field inside a conductor is always zero.

PROBLEM-SOLVING STRATEGY: Electric field forces

1. To determine the electric field at a point, we use Coulomb's law to find the total force **F** on a test charge q' placed at the point. Then we divide **F** by q' to obtain **E**. If we assume q' is positive, **F** and **E** have the same direction, and the magnitude of **E** is the magnitude of **F** divided by q'.

2. To analyze the motion of a charged particle in an electric field, we need to use Newton's second law,

F = **ma**, with **F** given in this case by **F** = q**E**. If the field is uniform, the acceleration is constant; we find its components and then use the tools we developed back in Chapter 3. It wouldn't hurt to review that chapter now.

EXAMPLE 25–1   What is the electric field 30 cm from a charge $q = 4 \times 10^{-9}$ C?

SOLUTION   From Coulomb's law, the force on a test charge q' 30 cm from q has magnitude

$$F = \frac{(9 \times 10^9 \text{ N·m}^2\text{·C}^{-2})(4 \times 10^{-9} \text{ C})(q')}{(0.3 \text{ m})^2}$$
$$= (400 \text{ N·C}^{-1})(q').$$

Then from Eq. (25–1), the magnitude of **E** is

$$E = \frac{F}{q'} = 400 \text{ N·C}^{-1}.$$

The *direction* of **E** at this point is along the line joining q and q', away from q.

EXAMPLE 25–2   When the terminals of a 100-V battery are connected to two large parallel horizontal plates 1 cm apart, the electric field **E** in the region between the plates is very nearly uniform, with magnitude $E = 10^4$ N·C^{-1}. Suppose the direction of **E** is vertically upward. Compute the force on an electron in this field and compare it with the weight of the electron.

SOLUTION   We need the following data, found in Appendix F:

Electron charge $e = 1.60 \times 10^{-19}$ C,

Electron mass $m = 9.11 \times 10^{-31}$ kg.

From Eq. (25–1),

$$F_{\text{elec}} = eE = (1.60 \times 10^{-19} \text{ C})(10^4 \text{ N·C}^{-1}) = 1.60 \times 10^{-15} \text{ N};$$

$$F_{\text{grav}} = mg = (9.11 \times 10^{-31} \text{ kg})(9.8 \text{ m·s}^{-2}) = 8.93 \times 10^{-30} \text{ N}.$$

Electrical forces are often much larger than the weights of the particles.

The ratio of the electrical to the gravitational force is therefore

$$\frac{F_{\text{elec}}}{F_{\text{grav}}} = \frac{1.60 \times 10^{-15} \text{ N}}{8.93 \times 10^{-30} \text{ N}} = 1.8 \times 10^{14}.$$

The gravitational force is negligibly small compared to the electrical force.

EXAMPLE 25–3 If the electron of Example 25–2 is released from rest at the upper plate, what speed does it acquire while traveling 1 cm? What is then its kinetic energy? How much time is required for it to travel this distance?

SOLUTION The force is constant, so the electron moves with constant acceleration a given by

$$a = \frac{F}{m} = \frac{eE}{m} = \frac{1.60 \times 10^{-15} \text{ N}}{9.11 \times 10^{-31} \text{ kg}}$$
$$= 1.76 \times 10^{15} \text{ m·s}^{-2}.$$

From Eq. (2–13), its speed after traveling 1 cm, or 10^{-2} m, is

$$v = \sqrt{2ax} = 5.93 \times 10^6 \text{ m·s}^{-1}.$$

Its kinetic energy is

$$\frac{1}{2}mv^2 = 1.60 \times 10^{-17} \text{ J}.$$

The time required is

$$t = \frac{v}{a} = 3.37 \times 10^{-9} \text{ s}.$$

EXAMPLE 25–4 If the electron of Example 25–2 is projected into the electric field with an initial horizontal velocity v_0, as in Fig. 25–2, find the equation of its trajectory.

SOLUTION The direction of the field is upward in Fig. 25–2, so the force on the (negatively charged) electron is downward. We take the positive x-direction to be the direction of the initial velocity. The x-acceleration is zero; the y-acceleration is $-(eE/m)$. Hence after a time t,

$$x = v_0 t,$$
$$y = \frac{1}{2}a_y t^2 = -\frac{1}{2}\left(\frac{eE}{m}\right)t^2.$$

Elimination of t gives

$$y = -\frac{1}{2}\left(\frac{eE}{mv_0^2}\right)x^2,$$

which is the equation of a parabola. The motion is the same as that of a body projected horizontally in the earth's gravitational field, as we discussed in Section 3–4. The deflection of electrons by an electric field is used to control the direction of an electron beam in many electronic devices, such as the TV picture tube and cathode-ray oscilloscope, which will be discussed in Section 26–8.

EXAMPLE 25–5 Figure 25–3 shows a structure incorporating two point charges, $+q$ and $-q$, having equal magnitude but opposite sign, separated by a constant distance l. We call such a pair of charges an **electric dipole.** The electric-charge distributions of many molecules, such as the water molecule, can be represented approximately as electric dipoles. In Fig. 25–3 the dipole is in a uniform electric field E. The direction of E makes an angle θ with the line joining the two charges, called the *dipole axis*. A force F_1, of magnitude qE, in the direction of the field, is exerted on the positive charge; and a force F_2, of the same magnitude but

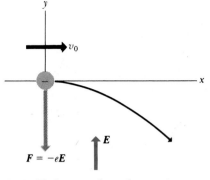

25–2 Trajectory of an electron in an electric field.

Electron trajectories in an electric field are very similar to ballistic trajectories.

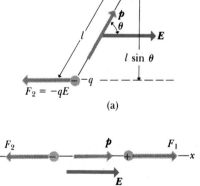

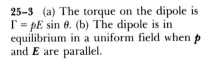

25–3 (a) The torque on the dipole is $\Gamma = pE \sin \theta$. (b) The dipole is in equilibrium in a uniform field when p and E are parallel.

in the opposite direction, is exerted on the negative charge. The resultant force on the dipole is zero, but since the two forces do not have the same line of action, they constitute a *couple* (see Section 10–4). The torque produced by a couple is

$$\Gamma = (qE)(l \sin \theta),$$

since $l \sin \theta$ is the perpendicular distance between the action lines of the forces.

Dipole in a uniform field: torque but no net force

The product ql of the charge q and the distance l is called the **electric dipole moment** and is denoted by p:

$$p = ql.$$

The torque exerted by the couple is therefore

$$\Gamma = pE \sin \theta. \tag{25–2}$$

We can also define a *vector dipole moment* $\boldsymbol{p}$, a vector of magnitude p lying along the dipole axis and pointing from negative toward positive charge. With this definition and the concept of vector torque introduced in Section 9–8, the vector torque $\boldsymbol{\Gamma}$ on the dipole is given by

$$\boldsymbol{\Gamma} = \boldsymbol{p} \times \boldsymbol{E}. \tag{25–3}$$

This torque tends to rotate the dipole to a position in which $\boldsymbol{p}$ is parallel to $\boldsymbol{E}$, as in Fig. 25-3b. In this position $\boldsymbol{\Gamma} = \boldsymbol{0}$, and if $\boldsymbol{E}$ is uniform, the dipole is in equilibrium.

If the electric field is *not* uniform, then the dipole experiences not only a torque but also a net force, because the two charges are located at points of slightly different electric field and the forces on them do not precisely cancel. In this case, the net force can be expressed in terms of derivatives of the components of $\boldsymbol{E}$ with respect to the coordinates.

25–2 ELECTRIC-FIELD CALCULATIONS

How to calculate electric fields from Coulomb's law

We can calculate the electric field at any point if we know the magnitudes and positions of all the charges that contribute to the field at that point. First we determine the electric field $\boldsymbol{E}$ at a point P caused by a single point charge q at a distance r from P. To do this we imagine a test charge q' at P. According to Coulomb's law, the force F on the test charge has magnitude

$$F = \frac{1}{4\pi\epsilon_0} \frac{qq'}{r^2},$$

so the electric field at P has magnitude

$$E = \frac{F}{q'} = \frac{1}{4\pi\epsilon_0} \frac{q}{r^2}.$$

The location of charge q is often called the *source point* and P the *field point*. We can define a unit vector $\hat{r}$ pointing in the direction from the source point to the field point; that is, from q to point P. In terms of $\hat{r}$, the field $\boldsymbol{E}$ at P due to charge q is given by

$$\boldsymbol{E} = \frac{1}{4\pi\epsilon_0} \frac{q}{r^2} \hat{r}. \tag{25–4}$$

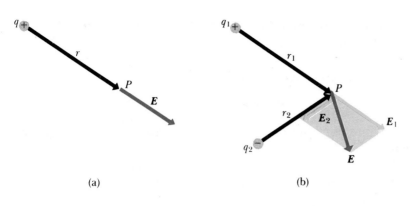

25–4 (a) The electric field E is in the same direction as the vector r when q is positive. (b) The resultant electric field at point P is the vector sum of E_1 and E_2.

When q is positive, the field is away from q, in the direction of $\hat{r}$; when it is negative, the field is toward q, opposite in direction to $\hat{r}$.

Now suppose the field at point P is caused by several point charges q_1, q_2, q_3, . . . , at distances r_1, r_2, r_3, . . . , from P, with associated unit vectors $\hat{r}_1$ pointing from q_1 to P, and so on. Each unit vector points from its associated source point to the field point P. Then from the principle of superposition introduced in Section 24–5, it follows that the *total* electric field at P is the *vector sum* of the fields E_1, E_2, and so on, of the individual charges. That is,

Superposition: combining fields caused by several point charges

$$E = E_1 + E_2 + E_3 + \cdots$$
$$= \frac{1}{4\pi\epsilon_0}\frac{q_1}{r_1^{\,2}}\hat{r}_1 + \frac{1}{4\pi\epsilon_0}\frac{q_2}{r_2^{\,2}}\hat{r}_2 + \cdots. \qquad (25\text{–}5)$$

An example with two charges, one positive and one negative, is shown in Fig. 25–4.

PROBLEM-SOLVING STRATEGY: *Electric-field calculations*

The strategy outlined in Section 24–6 is directly relevant here, too. We suggest you review it now. Briefly, the key points are:

1. Be sure to use a consistent set of units.

2. Use proper vector notation; distinguish clearly between vectors and components of vectors. Indicate your coordinate axes clearly on your diagram, and be certain the components are consistent with your choice of axes. Use the methods you learned in Chapter 1 for finding vector sums.

3. In some problems you will have a continuous distribution of charge along a line, over a surface, or through a volume. You must then define a small element of charge that can be considered as a point, find its electric field, and then integrate over the whole charge distribution to find the total electric field. Usually it is easiest to do the integration for each component of E separately.

EXAMPLE 25–6 Point charges q_1 and q_2 of $+12 \times 10^{-9}$ C and -12×10^{-9} C, respectively, are placed 0.1 m apart, as in Fig. 25–5. Compute the electric fields due to these charges at points a, b, and c.

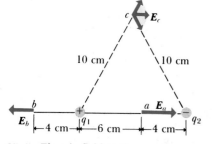

25–5 Electric field at three points, a, b, and c, in the field set up by charges q_1 and q_2.

A simple electric-field calculation: two point charges

SOLUTION At point a, the vector due to the positive charge q_1 is directed toward the right, and its magnitude is

$$E_1 = (9.0 \times 10^9 \text{ N·m}^2 \text{·C}^{-2}) \frac{(12 \times 10^{-9} \text{ C})}{(0.06 \text{ m})^2}$$

$$= 3.00 \times 10^4 \text{ N·C}^{-1}$$

The vector due to the negative charge q_2 is also directed toward the right; its magnitude is

$$E_2 = (9.0 \times 10^9 \text{ N·m}^2 \text{·C}^{-2}) \frac{(12 \times 10^{-9} \text{ C})}{(0.04 \text{ m})^2}$$

$$= 6.75 \times 10^4 \text{ N·C}^{-1}.$$

Hence, at point a,

$$E_a = (3.00 + 6.75) \times 10^4 \text{ N·C}^{-1}$$

$$= 9.75 \times 10^4 \text{ N·C}^{-1}, \qquad \text{toward the right.}$$

At point b, the vector due to q_1 is directed toward the left, with magnitude

$$E_1 = (9.0 \times 10^9 \text{ N·m}^2 \text{·C}^{-2}) \frac{(12 \times 10^{-9} \text{ C})}{(0.04 \text{ m})^2}$$

$$= 6.75 \times 10^4 \text{ N·C}^{-1}.$$

The vector due to q_2 is directed toward the right, with magnitude

$$E_2 = (9.0 \times 10^9 \text{ N·m}^2 \text{·C}^{-2}) \frac{(12 \times 10^{-9} \text{ C})}{(0.14 \text{ m})^2}$$

$$= 0.55 \times 10^4 \text{ N·C}^{-1}.$$

Hence, at point b

$$E_b = (6.75 - 0.55) \times 10^4 \text{ N·C}^{-1}$$

$$= 6.20 \times 10^4 \text{ N·C}^{-1}, \qquad \text{toward the left.}$$

At point c, the magnitude of each vector is

$$E = (9.0 \times 10^9 \text{ N·m}^2 \text{·C}^{-2}) \frac{(12 \times 10^{-9} \text{ C})}{(0.1 \text{ m})^2}$$

$$= 1.08 \times 10^4 \text{ N·C}^{-1}.$$

The directions of these vectors are shown in the figure; their resultant is easily seen to be

$$E_c = 1.08 \times 10^4 \text{ N·C}^{-1}, \qquad \text{toward the right.}$$

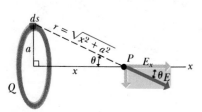

25–6 Electric field due to ring of charge.

EXAMPLE 25–7 A ring-shaped conductor of radius a carries a total charge Q. Find the electric field at a point P, a distance x from the center, along the line perpendicular to the plane of the ring, through its center.

SOLUTION The situation is shown in Fig. 25–6. Note the close resemblance of this problem to the gravitational-field calculation in Example 6–10 (Section 6–4). We represent the ring as made up of small segments ds, as shown. The charge of the segment ds is dQ. At the field point P the element of charge dQ causes an electric-field contribution dE having magnitude dE given by

$$dE = \frac{1}{4\pi\epsilon_0} \frac{dQ}{x^2 + a^2}.$$

The component dE_x of this field along the x-axis is

$$dE_x = dE \cos\theta = \frac{1}{4\pi\epsilon_0} \frac{dQ}{x^2 + a^2} \frac{x}{\sqrt{x^2 + a^2}}$$

$$= \frac{1}{4\pi\epsilon_0} \frac{x\, dQ}{(x^2 + a^2)^{3/2}}.$$

To find the total x-component of the field, E_x, we integrate this expression:

$$E_x = \int \frac{1}{4\pi\epsilon_0} \frac{x\, dQ}{(x^2 + a^2)^{3/2}}.$$

On the right side, x does not vary as we move from point to point on the ring. Thus everything except dQ may be taken outside the integral. The integral of dQ is simply the total charge Q, and we finally obtain

$$E_x = \frac{1}{4\pi\epsilon_0} \frac{Qx}{(x^2 + a^2)^{3/2}}. \tag{25–6}$$

In principle, this calculation should also be performed for the components perpendicular to the x-axis, but we can see from symmetry that these sum to zero.

Equation (25–6) shows that at the center of the ring ($x = 0$), the total field is zero, as might be expected; charges on opposite sides pull in opposite directions on a test charge at that point, and their fields cancel. When x is much larger than a, Eq. (25–6) becomes approximately

> At the center of the ring, the field is zero.

$$E_x = \frac{1}{4\pi\epsilon_0} \frac{Q}{x^2},$$

corresponding to the fact that at distances much greater than the dimensions of the ring it appears as a point charge.

EXAMPLE 25–8   An electric charge Q is distributed uniformly along a line with length $2a$, lying along the y-axis, as shown in Fig. 25–7. Find the electric field at point P on the x-axis, at a distance x from the origin.

SOLUTION   We encountered the same situation in Example 24–5 (Section 24–6). Looking back at that solution for the force $\boldsymbol{F}$, we see that the electric field $\boldsymbol{E}$ at P is just equal to $\boldsymbol{F}/q$:

$$\boldsymbol{E} = \frac{1}{4\pi\epsilon_0} \frac{Q}{x\sqrt{x^2 + a^2}}\, \boldsymbol{i}. \tag{25–7}$$

Again, when x is much larger than a, this becomes approximately

$$\boldsymbol{E} = \frac{1}{4\pi\epsilon_0} \frac{Q}{x^2}\, \boldsymbol{i},$$

that is, the field of a point charge Q at a distance x.

We get an added dividend from Eq. (25–7) if we express it in terms of the charge per unit length $Q/2a$. We call this quantity the *linear charge density* and

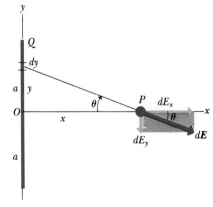

25–7 A line of charge with length $2a$ and total charge Q creates an electric field $\boldsymbol{E}$ at point P.

denote it by λ. In this case, $\lambda = Q/2a$. Substituting $Q = 2a\lambda$ into Eq. (25–7) and simplifying, we obtain

$$E = \frac{1}{2\pi\epsilon_0} \frac{\lambda}{x\sqrt{\dfrac{x^2}{a^2} + 1}} i. \qquad (25\text{–}8)$$

Now what happens if we make the line of charge longer and longer, adding charge in proportion to the total length so that λ, the charge per unit length, remains constant? What is E at a distance x from a *very* long line of charge?

To answer the question we take the *limit* of Eq. (25–8) as a becomes very large. In this limit, the term x^2/a^2 in the denominator becomes much smaller than unity and can be discarded. What remains is

$$E = \frac{\lambda}{2\pi\epsilon_0 x} i. \qquad (25\text{–}9)$$

From the symmetry of the situation, the field magnitude depends only on the distance of point P from the line of charge, so we can say that at any point P at a distance r from the line, the field has magnitude

$$E = \frac{\lambda}{2\pi\epsilon_0 r}. \qquad (25\text{–}10)$$

Its direction is radially outward from the line. Thus the electric field due to an infinitely long line of charge is proportional to $1/r$ rather than to $1/r^2$ as for a point charge.

Electric field caused by charge distributed over a plane

EXAMPLE 25–9   In Fig. 25–8, electric charge is distributed uniformly over the entire xy-plane. We call the charge per unit area the *surface charge density* and denote it by σ. What is the electric field at point P, at a distance a above the plane?

SOLUTION   Assume for the moment that σ is positive. We divide the charge into narrow strips of width dx, parallel to the y-axis. Each strip can be considered a *line* charge, and we can use the result of the preceding example.

25–8 Electric field created by a uniform infinite plane of charge.

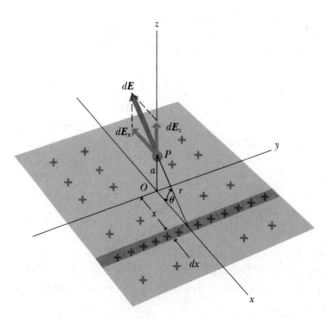

The area of a portion of a strip of length L is $L\,dx$, and the charge dq on the strip is

$$dq = \sigma L\,dx.$$

The charge per unit length, λ, is therefore

$$\lambda = \frac{dq}{L} = \sigma\,dx.$$

From Eq. (25–10), the strip sets up at point P a field $d\mathbf{E}$, lying in the xz-plane, of magnitude

$$dE = \frac{\sigma}{2\pi\epsilon_0}\frac{dx}{r}.$$

The field can be resolved into components dE_x and dE_z. By symmetry, the components dE_x will sum to zero when the entire sheet of charge is considered. (Be sure that you understand why.) The resultant field at P is therefore in the z-direction, perpendicular to the sheet of charge. From the diagram,

$$dE_z = dE\sin\theta.$$

The total z-component of $\mathbf{E}$ is given by

$$E = \int dE_z = \frac{\sigma}{2\pi\epsilon_0}\int_{-\infty}^{+\infty}\frac{\sin\theta\,dx}{r}.$$

But

$$\sin\theta = \frac{a}{r},$$

$$r^2 = a^2 + x^2,$$

and therefore

$$E = \frac{\sigma a}{2\pi\epsilon_0}\int_{-\infty}^{+\infty}\frac{dx}{a^2 + x^2} = \frac{\sigma a}{2\pi\epsilon_0}\left[\frac{1}{a}\arctan^{-1}\frac{x}{a}\right]_{-\infty}^{+\infty},$$

$$E = \frac{\sigma}{2\epsilon_0}. \tag{25–11}$$

Our final result does not contain the distance a from the plane. This correct but rather surprising result means that the electric field produced by an infinite-plane sheet of charge is *independent of the distance from the charge*. Thus the field is *uniform;* its direction is everywhere perpendicular to the sheet, away from it.

The field is independent of the distance from the plane.

If P is below the plane instead of above it, the result is the same except that the direction of $\mathbf{E}$ is vertically downward instead of upward. Also, if σ is a negative quantity, the directions of the fields both above and below the plane are toward the plane rather than away from it.

As we have seen, calculating the $\mathbf{E}$ fields of given charge distributions can be quite laborious. For problems having a lot of symmetry, such as the examples above, there is an alternative method using a principle called *Gauss's law;* we return to this important principle at the end of the chapter.

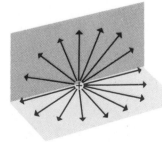

25–9 The direction of the electric field at any point is tangent to the field line through that point.

Field lines: a graphical representation of electric fields

Field lines begin and end at charges.

Field lines never cross.

25–3 FIELD LINES

Field lines are useful as an aid to visualizing electric (and also magnetic) fields. A **field line** is an imaginary line drawn through a region of space so that at every point it is tangent to the direction of the electric-field vector at that point. This idea is shown in Fig. 25–9. We will show in Section 25–4 that at every point the *number* of lines per unit cross-sectional area, perpendicular to *E*, is proportional to the magnitude of *E* at that point. The concept of field lines was introduced by Michael Faraday (1791–1867). He called them "lines of force," but the term *field lines* is preferable.

Thus field lines show the direction of *E* at each point, and their spacing gives a general indication of the magnitude of *E* at each point. Where the lines are bunched closely together, *E* is strong; where they are farther apart, *E* is weaker.

Figure 25–10 shows some of the field lines in two planes containing (a) a single positive charge; (b) two equal charges, one positive and one negative (an electric dipole); and (c) two equal positive charges. The direction of the resultant field at every point in each diagram is along the tangent to the field line passing through the point. Arrowheads on the field lines indicate the direction in which the tangent is to be drawn.

No field lines begin or end in the space surrounding a charge. Every field line in an *electrostatic* field is a continuous line terminated by a positive charge at one end and a negative charge at the other. Although we sometimes speak of an "isolated" charge and draw its field as in Fig. 25–10a, this simply means that the charges on which the lines end are at large distances from the charge under consideration. For example, if the charged body in Fig. 25–10a is a small sphere suspended by a thread from the laboratory ceiling, the negative charges on which its field lines end would be found on the walls, floor, or ceiling, as well as on other objects in the laboratory.

At any one point, the electric field can have but one direction. Hence only one field line can pass through each point of the field. In other words, *field lines never intersect*.

If a field line were drawn through every point of an electric field, all the space and the entire surface of a diagram would be filled with lines, and no individual line could be distinguished. By limiting the number of field lines, we can use them to indicate the *magnitude* of a field as well as its *direction*. In a region where the field magnitude is large, such as that between the positive

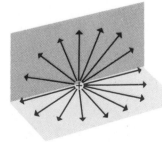

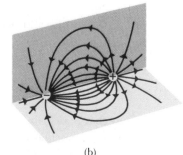

(a)   (b)   (c)

25–10 The mapping of an electric field with the aid of field lines.

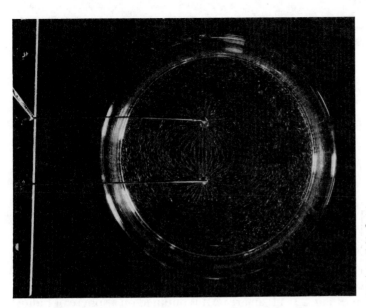

Electric field produced by two point charges. The pattern is formed by grass seeds floating in a liquid above the charges. (*PSSC Physics*, 2d. ed., 1965; D. C. Heath and Company, Inc., with Education Development Center, Inc., Newton, Mass.)

and negative charges of Fig. 25–10b, the field lines are closely spaced. In a region where the field magnitude is small, such as that between the two positive charges of Fig. 25–10c, the lines are widely separated. In a *uniform* field, the field lines are straight, parallel, and uniformly spaced.

25–4 GAUSS'S LAW

Gauss's law provides an alternative formulation of the relationship between electric charge and electric field. It is logically equivalent to Coulomb's law, but in some situations it is more convenient to use. It was formulated by Karl Friedrich Gauss (1777–1855), one of the greatest mathematical geniuses of all time. Many areas of mathematics, from number theory and geometry to the theory of differential equations, bear the mark of his influence, and he made equally significant contributions to theoretical physics. In this section we will derive Gauss's law, and in the following section we will use it to find the electric fields caused by several kinds of charge distributions of practical importance.

Gauss's law: an alternative formulation of the relation between charge and electric field

The content of Gauss's law is suggested by consideration of field lines, discussed in Section 25–3. The field of an isolated positive point charge q is represented by lines radiating out in all directions. Suppose we imagine this charge to be surrounded by a spherical surface of radius R, with the charge at its center. The area of this imaginary surface is $4\pi R^2$, so if the total number of field lines emanating from q is N, then the number of lines *per unit surface area* on the spherical surface is $N/4\pi R^2$. Now imagine a second sphere concentric with the first, but with radius $2R$. Its area is $4\pi(2R)^2 = 16\pi R^2$, and the number of lines per unit area on this sphere in $N/16\pi R^2$, one-fourth that of the first sphere. This corresponds to the fact that at distance $2R$ the field has only one-fourth the magnitude it has at distance R. It also verifies our statement in Section 25–3 that the number of lines per unit area is proportional to the magnitude of the field.

For a point charge, how many field lines pass through a given area?

The fact that the *total* number of lines at distance $2R$ is the same as at R can be expressed another way. The field is inversely proportional to R^2, but

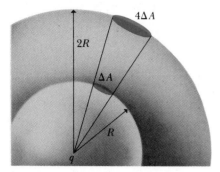

25–11 Projection of an element of area ΔA on a sphere of radius R, onto a sphere of radius $2R$. The projection multiplies each linear dimension by two, so the area element on the larger sphere is $4\,\Delta A$.

the *area* of the sphere is proportional to R^2. Thus the *product* of the two is independent of R. For a sphere of arbitrary radius r, the magnitude of E on the surface is

$$E = \frac{1}{4\pi\epsilon_0}\frac{q}{r^2},$$

the surface area is

$$A = 4\pi r^2,$$

and the product of the two is

$$EA = \frac{q}{\epsilon_0}. \tag{25–12}$$

This product is independent of r and depends *only* on the charge q. As we will see, this result is of crucial importance in the following development.

What is true of the entire sphere is also true of any portion of its surface. In the construction of Fig. 25–11, an area ΔA is outlined on a sphere of radius R and then projected onto the sphere of radius $2R$ by drawing lines from the center through points on the boundary of ΔA. The area projected on the larger sphere is clearly $4\,\Delta A$; thus again the product $E\,\Delta A$ is independent of the radius of the sphere.

This projection technique shows how this discussion may be extended to nonspherical surfaces. Instead of a second sphere, let us surround the sphere of radius R by a surface of irregular shape, as in Fig. 25–12a. Consider a small element of area ΔA; note that this area is *larger* than the corresponding element on a spherical surface at the same distance from q. If a line normal (perpendicular) to the irregular surface makes an angle θ with a radial line from q, two sides of the area projected onto the spherical surface are foreshortened by a factor $\cos\theta$, as shown in Fig. 25–12b. Thus the quantity corresponding to $E\,\Delta A$ for the spherical surface is $E\,\Delta A\cos\theta$ for the irregular surface.

Now we may divide the entire irregular surface into small elements ΔA, compute the quantity $E\,\Delta A\cos\theta$ for each, and sum the results. Each of these areas projects onto a corresponding element of area on the sphere, so summing the quantities $E\,\Delta A\cos\theta$ over the irregular surface must yield the same

25–12 (a) The outward normal makes an angle θ with the direction of $\mathbf{E}$. (b) The projection of the area element ΔA onto the spherical surface is $\Delta A\cos\theta$.

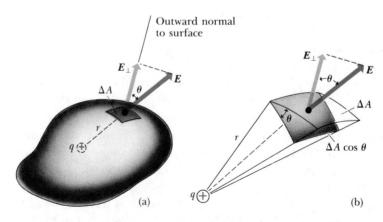

result as summing the quantities $E \, \Delta A$ over the sphere. But we have already performed that calculation; the result, given by Eq. (25–12), depends only on the charge q. Hence, for the irregular surface, the result is

$$\sum E \, \Delta A \cos \theta = \frac{q}{\epsilon_0}. \qquad (25\text{–}13)$$

This equation holds for *any* shape of surface, provided only that it is a *closed* surface enclosing the charge q. Correspondingly, for a closed surface enclosing *no* charge,

$$\sum E \, \Delta A \cos \theta = 0.$$

This is a mathematical statement of the fact that when a region contains no charge, any field lines (caused by charges outside the region) that enter on one side must leave again at some other point on the boundary surface of the region. *Field lines can begin or end inside a region of space only when there is charge in that region.*

Because the electric field varies from point to point on the irregular surface, Eq. (25–13) is strictly true only in the limit when the area elements become very small. In this limit, the sum becomes an integral called the *surface integral* of $E \cos \theta$, written

$$\oint E \cos \theta \, dA = \frac{q}{\epsilon_0}. \qquad (25\text{–}14)$$

The circle on the integral sign reminds us that the integral is always taken over a *closed* surface enclosing the charge q.

Since $E \cos \theta$ is the component of $\boldsymbol{E}$ perpendicular to the surface at each point, we may use the more compact notation $E_\perp = E \cos \theta$, and write

$$\oint E_\perp \, dA = \frac{q}{\epsilon_0}. \qquad (25\text{–}15)$$

If the point charge in Fig. 25–12 is negative, the $\boldsymbol{E}$ field is directed radially *inward;* the angle θ is then greater than 90°, its cosine is negative, and the integral in Eq. (25–15) is negative. But since q is also negative, Eq. (25–15) still holds.

Now suppose the point charge lies *outside* a closed surface; construct your own diagram. The electric field is then inward at some points of the surface and outward at others. It is not difficult to show that the positive and negative contributions to the integral over the surface exactly cancel, and the sum is zero. But the charge *inside* the closed surface is also zero, so again Eq. (25–15) is obeyed.

Although thus far we have considered only a single point charge, it is easy to generalize our results to *any* charge distribution. The total electric field $\boldsymbol{E}$ at a point on the closed surface is the vector sum of the fields produced by the individual charges, and the quantity $E \, dA \cos \theta$ is therefore the sum of the contributions from these charges. Since Eq. (25–15) holds for each point charge, a corresponding relation holds for the *total* $\boldsymbol{E}$ field and the *total* charge enclosed by the surface. That is,

$$\oint E_\perp \, dA = \frac{1}{\epsilon_0} \sum q, \qquad (25\text{–}16)$$

The total number of field lines coming out of a surface is proportional to the charge enclosed by the surface.

When no charge is enclosed, the numbers of field lines entering and leaving the enclosed volume are equal.

where $E_\perp$ is now the normal component of the total electric field and Σq represents the *algebraic* sum of all charges enclosed by the surface.

Equation (25–16) is the mathematical statement of Gauss's law. It states that when we multiply each element of area of a closed surface by the normal component of the total electric field $\boldsymbol{E}$ at the element, and sum over the entire surface, the result is a constant times the total charge inside the surface.

The notation can be simplified by use of the *vector area d$\boldsymbol{A}$,* defined as a vector whose magnitude equals dA and whose direction is that of the *outward* normal at dA. The product $E_\perp\, dA = E \cos \theta\, dA$ can then be written as the *scalar product* or *dot product* of the vectors $\boldsymbol{E}$ and $d\boldsymbol{A}$:

$$E_\perp\, dA = \boldsymbol{E} \cdot d\boldsymbol{A}.$$

Denoting the total charge enclosed by $Q = \Sigma q$, we may write Gauss's law more compactly as

$$\oint \boldsymbol{E} \cdot d\boldsymbol{A} = \frac{Q}{\epsilon_0}. \tag{25–17}$$

The quantity $E_\perp\, dA = E \cos \theta\, dA$ is also called the **electric flux** through the area dA. Electric flux may be denoted by Ψ, and an element of flux corresponding to a small area dA by $d\Psi$. Then

$$d\Psi = E_\perp\, dA = E \cos \theta\, dA = \boldsymbol{E} \cdot d\boldsymbol{A}. \tag{25–18}$$

The total flux Ψ through a finite surface area is

$$\Psi = \int E_\perp\, dA = \int E \cos \theta\, dA = \int \boldsymbol{E} \cdot d\boldsymbol{A}. \tag{25–19}$$

Thus Eq. (25–17) states that the total electric flux out of a closed surface is proportional to the total charge Q enclosed:

$$\Psi = \frac{Q}{\epsilon_0}. \tag{25–20}$$

As we have mentioned, the flux of $\boldsymbol{E}$ through a surface, as well as Gauss's law, can be interpreted graphically in terms of field lines. The number of field lines per unit area, perpendicular to the direction of the lines, is proportional to $\boldsymbol{E}$. Thus the surface integral of $E_\perp$ over a closed surface (equal to the flux Ψ through that surface) is proportional to the total number of lines crossing the surface in the outward direction, minus the number crossing in the inward direction. Furthermore, the net charge (algebraic sum of all charges) enclosed by the surface is proportional to this number.

Here is an example. Figure 25–13 shows the field produced by two equal and opposite point charges (an electric dipole). Surface A encloses the positive charge only, and 18 lines cross it in an outward direction. Surface B encloses the negative charge only; it is also crossed by 18 lines, but in an inward direction. Surface C encloses *both* charges. It is intersected by lines at 16 points; at 8 intersections the lines are outward, and at 8 they are inward. The *net* number of lines crossing in an outward direction is zero, and the net charge inside the surface is also zero. Surface D is intersected at 6 points; at 3 intersections the lines are outward, and at the other 3 they are inward. The net number of lines crossing in an outward direction and the enclosed charge are both zero.

Actually evaluating the integral in Eq. (25–17) may appear impossible. In quite a few cases, however, we can exploit the symmetry of a charge distribu-

A vector formulation of the electric-field lines passing through a surface

Electric flux: a quantity that depends on electric field and area

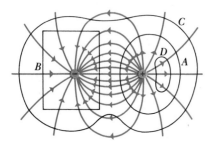

25–13 The net number of field lines leaving a closed surface is proportional to the total charge enclosed.

tion to accomplish the calculation quite simply. In these cases Gauss's law enables us to calculate the electric field of a charge distribution much more simply than with the methods used in Section 25–2. Several examples are worked out in the next section.

25–5 APPLICATIONS OF GAUSS'S LAW

In this section we present several examples of applications of Gauss's law. They are of two kinds: The first two are derivations of general properties of the electric field and the location of charges on conductors; the remainder are derivations of electric-field configurations caused by various specific charge distributions. In some of these examples we find that we can obtain much more simply some of the results derived in Section 25–2. The following problem-solving strategy is useful for both categories.

Gauss's law greatly simplifies some field calculations.

PROBLEM-SOLVING STRATEGY: *Gauss's law*

1. The first step is to select the surface (which we will often call a *Gaussian surface*) that you are going to use with Gauss's law. If you are trying to learn something about the field at a particular point, then obviously that point must lie on your Gaussian surface.

2. The Gaussian surface does not need to be a real physical surface, such as the surface of a solid body. In many applications of Gauss's law we use an imaginary geometric surface that may be in empty space, embedded in a solid body, or partly both.

3. The Gaussian surface must have enough *symmetry* to make it possible to evaluate the surface integral in Gauss's law. If the problem itself has cylindrical or spherical symmetry, the Gaussian surface will usually be cylindrical or spherical, respectively.

4. Often the surface can be thought of as several separate areas, such as the sides and ends of a cylinder. The integral of $E_\perp$ over the whole surface is always equal to the sum of the integrals over the separate areas. Some of these integrals may be zero, as in points 6 and 7 below.

5. If E is perpendicular to a surface A at every point, and if it also has the same *magnitude* at every point on the surface, then $E_\perp = E$ = constant, and the integral $\oint E \cdot dA$ over that surface is simply EA.

6. If E is *parallel* to a surface at every point, then $E_\perp = 0$, and the integral is zero.

7. If $E = 0$ at every point on a surface, the integral is zero.

8. Finally, in the integral $\oint E \cdot dA$, E is always the *total* electric field at each point on the surface. In general this field is caused partly by charges within the volume and partly by charges outside. Even when there is *no* charge within the volume, the field at points on the surface is not necessarily zero. In that case, however, the integral over the surface is always zero.

EXAMPLE 25–10 *Location of excess charge on a conductor.* We have remarked that the electric field E is zero at all points in the interior of a conducting material when the charges in the conductor are at rest. If E were *not* zero, the charges would move and we would not have an electrostatic situation. We may construct a Gaussian surface inside the conductor, such as surface A in Fig. 25–14a. Because $E = 0$ everywhere on this surface, Eq. (25–17) requires that the net charge inside the surface be zero. Now imagine shrinking the surface down like a collapsing balloon, until it encloses a region so small that we may consider it a point; then the charge at that point must be zero. We can do this anywhere inside the conductor, so *there can be no net charge at any point within the conductor.* Thus any excess charge on the conductor must be located only on its surface, as shown.

The charge on a conductor must be located entirely on its surface.

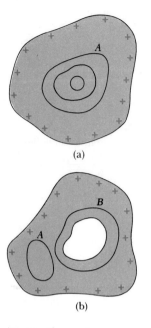

(a)

(b)

25–14 Excess charge on a conductor resides entirely on its outer surface.

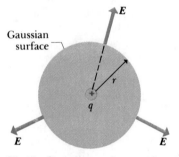

25–15 Gaussian surface used to derive Coulomb's law from Gauss's law.

Field of a charged conducting sphere: Outside the sphere, the field is the same as for a point charge.

If there is a cavity inside the conductor, as in Fig. 25–14b, the situation is somewhat different. We return to this case in Section 25–6.

EXAMPLE 25–11 *Coulomb's law.* We have considered Coulomb's law as the fundamental equation of electrostatics and have derived Gauss's law from it. An alternative procedure is to consider Gauss's law as a fundamental experiment relation. We can then derive Coulomb's law from Gauss's law by obtaining the expression for the electric field E due to a point charge.

Consider the electric field of a single positive point charge q, shown in Fig. 25–15. By symmetry, the field is everywhere radial; there is no reason why it should deviate to one side of a radial direction rather than to another. Also, its magnitude is the same at all points at the same distance r from the charge; any point at this distance is like any other. Hence, if we select as a Gaussian surface a spherical surface of radius r, $E_\perp = E = $ constant at all points of the surface. Then

$$\oint E_\perp dA = E_\perp \oint dA = EA = 4\pi r^2 E.$$

From Gauss's law

$$4\pi r^2 E = \frac{q}{\epsilon_0} \quad \text{and} \quad E = \frac{1}{4\pi\epsilon_0}\frac{q}{r^2}.$$

The force on a point charge q' at a distance r from the charge q is then

$$F = q'E = \frac{1}{4\pi\epsilon_0}\frac{qq'}{r^2},$$

which is Coulomb's law.

EXAMPLE 25–12 *Field of a charged conducting sphere.* Let us place a charge q on a solid conducting sphere. From Example 25–10 we know that all the charge is on the surface of the sphere, and from symmetry we know that it is distributed *uniformly* over the surface. In principle, we can find the field at any point outside the sphere by summing the contributions from elements of charge on the surface, but it is much easier to use Gauss's law.

As with the point charge, we conclude from spherical symmetry that the field is radial everywhere. Furthermore, the magnitude of the field is again uniform over a spherical surface of radius r, concentric with the conductor. Thus if we construct a Gaussian surface of radius r, where r is greater than the radius R of the sphere, and if q is the total charge on the sphere,

$$4\pi r^2 E = \frac{q}{\epsilon_0},$$

$$E = \frac{1}{4\pi\epsilon_0}\frac{q}{r^2}. \tag{25–21}$$

The field *outside* the sphere is, therefore, the same as though the entire charge was concentrated at a point at its center. Just outside the surface of the sphere, where $r = R$,

$$E = \frac{1}{4\pi\epsilon_0}\frac{q}{R^2}.$$

Inside the sphere, as in the interior of any conductor where no charge is in motion, the field is zero. Thus when r is less than R, $E = 0$.

Reasoning very similar to this can be used to prove the assertion in Chapter 6 that the *gravitational* field of any spherically symmetric mass distribution is the same, at any point outside the distribution, as though all the mass were concentrated at the center. This argument provides the justification for treating spherical bodies as points when calculating gravitational interactions.

The same argument may be applied to a conducting, hollow, spherical *shell* (a spherical conductor with a concentric spherical hole in the center), if there is no charge in the hole. This time we take a spherical Gaussian surface of radius r less than the radius of the hole. If there *is* a field inside the hole, it must be spherically symmetric (radial) as before, so again $E = q/4\pi\epsilon_0 r^2$. But this time $q = 0$, so E must also be zero.

Here is a challenge: Can you use this same technique to find the electric field in the interspace between a charged sphere and a concentric, hollow conducting sphere surrounding it?

EXAMPLE 25–13 *Field of a line charge and of a charged cylindrical conductor.* Consider next the electric field caused by a long, thin, uniformly charged wire. We solved this problem in Section 25–2 (Example 25–8) by a straightforward application of Coulomb's law, requiring a somewhat involved integration. Gauss's law makes it possible to find the field by an almost trivial calculation.

If the wire is very long and we are not too near either end, then, by symmetry, the field lines outside the wire are *radial* and lie in planes perpendicular to the wire. Also, the field has the same magnitude at all points at the same radial distance from the wire. This suggests that we use as a Gaussian surface a *cylinder* of arbitrary radius r and arbitrary length l, with its ends perpendicular to the wire, as in Fig. 25–16. If λ is the charge per unit length on the wire, the charge within the Gaussian surface is λl. Since $\mathbf{E}$ is at right angles to the wire, the component of $\mathbf{E}$ normal to the end faces is zero. Thus the end faces make no contribution to the integral in Gauss's law. At all points of the curved surface, $E_\perp = E = $ constant, and since the area of this surface is $2\pi rl$, we have

$$(E)(2\pi rl) = \frac{\lambda l}{\epsilon_0},$$

(25–22)

$$E = \frac{1}{2\pi\epsilon_0}\frac{\lambda}{r}.$$

Note that although the *entire* charge on the wire contributes to the field $\mathbf{E}$, only that portion of the total charge lying within the Gaussian surface is used when we apply Gauss's law. This may be puzzling at first; it appears as though we had somehow obtained the right answer by ignoring a part of the charge, and that the field of a *short* wire of length l would be the same as that of a very long wire. The existence of the entire charge on the wire *is*, however, taken into account when we consider the *symmetry* of this problem. If the wire had been short, we could not conclude by symmetry that the field at an end of the cylinder would equal that at the center, or that the field lines would everywhere be perpendicular to the wire. So the entire charge on the wire actually *is* taken into account, but in an indirect way.

Here is another challenge: (1) Show that the field outside a long, uniformly charged cylinder is the same, at points outside the cylinder, as though all the charge were concentrated on a line along its axis. (2) Calculate the electric field in the interspace between a charged cylinder and a coaxial hollow conducting cylinder surrounding it.

Gauss's law can be applied to gravitational fields.

Gauss's law applied to charge distributed along a line

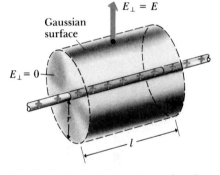

25–16 Cylindrical Gaussian surface for calculating the electric field due to a long charged wire.

A charged conducting cylinder: a field problem made much simpler by use of Gauss's law

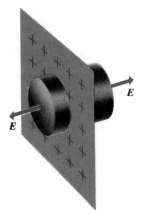

25–17 Gaussian surface in the form of a cylinder for finding the field of an infinite plane sheet of charge.

Field calculations with infinite planes of charge: Gauss's law makes it easy.

EXAMPLE 25–14 *Field of an infinite-plane sheet of charge.* To solve this problem, we construct the Gaussian surface shown by the shaded area in Fig. 25–17, consisting of a cylinder whose ends have an area A and whose walls are perpendicular to the sheet of charge. By symmetry, since the sheet is infinite, the electric field E has the same magnitude E on both sides of the surface, is uniform, and is directed normally away from the sheet of charge. No field lines cross the *side* walls of the cylinder; that is, the component of E normal to these walls is zero. At the ends of the cylinder the normal component of E is equal to E. The integral $\oint E_\perp \, dA$, calculated over the entire surface of the cylinder, therefore reduces to $2EA$. If σ is the charge per unit area in the plane sheet, the net charge within the Gaussian surface is σA. Hence

$$2EA = \frac{\sigma A}{\epsilon_0}, \qquad E = \frac{\sigma}{2\epsilon_0}. \qquad (25\text{–}23)$$

This is the same result obtained in Section 25–2 (Example 25–9) by a much more complicated method. The magnitude of the field is *independent* of the distance from the sheet and does *not* decrease inversely with the square of the distance. The field lines are straight, parallel, and uniformly spaced. This is because the sheet was assumed infinitely large.

Of course, nothing in nature can really be infinitely large; this infinitely large plane sheet is an idealization. Our result is still useful, however; the real meaning of Eq. (25–23) is that it is a very good approximation to the behavior of the electric field caused by a large but finite sheet of charge, at points that are not near its edge and are close to the plane compared to its dimensions. In such cases the field is very nearly uniform and perpendicular to the plane.

EXAMPLE 25–15 *Field of an infinite-plane charged conducting plate.* When a flat metal plate is given a net charge, this charge distributes itself over the entire outer surface of the plate. If the plate is of uniform thickness and is infinitely large (or if we are not too near the edges of a finite plate), the charge per unit area is uniform and is the same on both surfaces. Hence the field of such a charged plate arises from the superposition of the fields of *two* sheets of charge, one on each surface of the plate. By symmetry, the field is perpendicular to the plate; it is directed away from it (if the plate has a positive charge); and it is uniform. The magnitude of the electric field at any point can be found from Gauss's law or by using the results already derived for a sheet of charge.

Figure 25–18 shows a portion of a large charged conducting plate. Let σ represent the charge per unit area in the sheet of charge on *each* surface. (I.e., the total charge per unit area on both surfaces together is 2σ.) At point a, outside the plate at the left, the component of electric field E_1, due to the sheet of charge on the left face of the plate, is directed toward the left, and its magnitude is $\sigma/2\epsilon_0$. The component E_2 due to the sheet of charge on the right face of the plate is also toward the left, and its magnitude is also $\sigma/2\epsilon_0$. The magnitude of the *resultant* E is therefore

$$E = E_1 + E_2 = \frac{\sigma}{2\epsilon_0} + \frac{\sigma}{2\epsilon_0} = \frac{\sigma}{\epsilon_0}. \qquad (25\text{–}24)$$

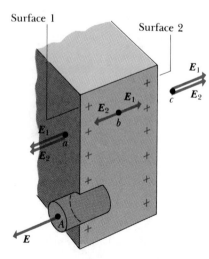

25–18 Electric field inside and outside a charged conducting plate.

At point b, inside the plate, the two components of electric field are in opposite directions and their resultant is zero, as it must be in any conductor where the charges are at rest. At point c, the components again add, and the magnitude of the resultant is σ/ϵ_0, directed toward the right.

To derive these results directly from Gauss's law, consider the cylinder shown in Fig. 25–18. Its end faces have area A; one end lies inside and one outside the plate. The field inside the conductor is zero. The field outside, by symmetry, is perpendicular to the plate, so the normal component of E is zero over the walls of the cylinder and is equal to E over the outside end face. Hence, from Gauss's law,

$$EA = \frac{\sigma A}{\epsilon_0}, \qquad E = \frac{\sigma}{\epsilon_0}. \qquad (25\text{--}25)$$

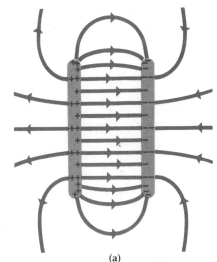

(a)

EXAMPLE 25–16 *Field between oppositely charged parallel conducting plates.* When two plane parallel conducting plates, having the size and spacing shown in Fig. 25–19, are given equal and opposite charges, the field between and around them is approximately as shown in Fig. 25–19a. Most of the charge accumulates at the opposing faces of the plates, and the field is nearly uniform in the space between them. A small quantity of charge resides on the outer surfaces of the plates, and some spreading or "fringing" of the field occurs at the edges of the plates.

As the plates are made larger and the distance between them smaller, the fringing becomes relatively less. Often the fringing is entirely negligible. We will usually assume that the field between two oppositely charged plates is uniform, as in Fig. 25–19b, and that the charges are distributed uniformly over the opposing surfaces.

The electric field at any point can be considered as the resultant of that due to two sheets of charge of opposite sign, or it may be found from Gauss's law. Thus at points a and c in Fig. 25–19b, the components E_1 and E_2 are each of magnitude $\sigma/2\epsilon_0$ but are oppositely directed, so their resultant is zero. At any point b between the plates, the components are in the same direction and their resultant is

$$E = \sigma/\epsilon_0.$$

We leave it to you to show that the same results follow from applying Gauss's law to the surface shown by broken lines.

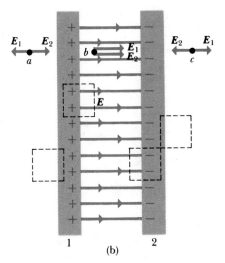

25–19 Electric field between oppositely charged parallel plates.

EXAMPLE 25–17 Electric charge is distributed uniformly throughout the volume of a sphere of radius R; the total charge is Q. Find the electric-field magnitude at a point P a distance r from the center of the sphere, where $r < R$.

SOLUTION We choose as our Gaussian surface a sphere of radius r, concentric with the charge distribution. The charge per unit volume, which we may call the volume charge density ρ, is given by

$$\rho = \frac{Q}{4\pi R^3/3}.$$

Charge distributed through a spherical volume: What is the electric field inside the charge distribution?

The volume V' enclosed by the Gaussian surface is $\frac{4}{3}\pi r^3$, so the total charge q enclosed by that surface is

$$q = \rho V' = \frac{Q}{4\pi R^3/3}\left(\frac{4}{3}\pi r^3\right) = Q\frac{r^3}{R^3}.$$

By symmetry, the electric-field magnitude has the same value E at every point on the Gaussian surface, and its direction at every point is radially outward. The total area of the surface is $4\pi r^2$, so the value of the integral in Gauss's law is simply $4\pi r^2 E$. We equate this to q/ϵ_0, with q given by the equation above. We find

$$4\pi r^2 E = \frac{Qr^3}{\epsilon_0 R^3},$$

or

$$E = \frac{1}{4\pi\epsilon_0} \frac{Qr}{R^3}.$$

The field magnitude is proportional to the distance of the field point from the center of the sphere. At the center ($r = 0$), $E = 0$, as we should expect from symmetry. At the surface of the sphere ($r = R$), the field magnitude is

$$E = \frac{1}{4\pi\epsilon_0} \frac{Q}{R^2}.$$

That is, at the surface the field has the same magnitude as though all the charge were concentrated at the center. As we have learned, this is also the case at any field point a distance greater than R from the center.

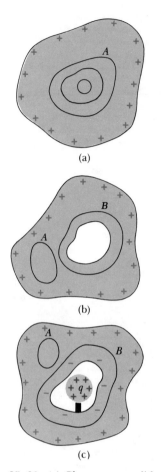

(a)

(b)

(c)

25–20 (a) Charge on a solid conductor resides entirely on its outer surface. (b) If there is no charge inside the cavity, the net charge on the surface of the cavity is zero. (c) If there is a charge q inside the cavity, the total charge on the cavity surface is $-q$.

A very sensitive test of the $1/r^2$ factor in Coulomb's law

25–6 CHARGES ON CONDUCTORS

In any electrostatic situation (i.e., when all charges are at rest) *the electric field at every point within a conductor is zero.* We have also shown in Section 25–5 (Example 25–10) that when excess charge is placed on a solid conductor, *the charge is located entirely on the surface of the conductor,* in Fig. 25–20a. We assume the conductor is mounted on an insulating stand after it is given an initial charge.

What if there is a cavity inside the conductor, as in Fig. 25–20b? If there is no charge in the cavity, we can use a Gaussian surface such as B to show that the net charge on the surface *of the cavity* must be zero, because $\boldsymbol{E} = \boldsymbol{0}$ everywhere on the Gaussian surface. In fact, we can prove that in this situation no charge is present *anywhere* on the cavity surface, but detailed proof of that statement will be postponed until Chapter 26.

Next, suppose a conductor is placed inside the cavity, carrying a charge q, as in Fig. 25–20c. Again $\boldsymbol{E} = \boldsymbol{0}$ everywhere on surface B, so according to Gauss's law the net charge inside this surface must be zero. Thus there must be a total charge $-q$ on the cavity surface. Since the *total* charge on the conductor cannot change, a charge $+q$ must appear on its outer surface. If the outer surface originally had a charge q', the total charge after the charge q is inserted into the cavity must be $q + q'$.

Now let us consider the experiment shown in Fig. 25–21. We place a conducting container, such as a tin can, on an insulating stand. The container is initially uncharged. Then we hang a charged metal ball from an insulating thread, lower it into the can, and put the lid on, as in Fig. 25–21b. As we have just discussed, charges are induced on the walls of the container, as shown. But now we let the ball touch the inner wall, as in Fig. 25–21c. The surface of the ball becomes, in effect, part of the cavity surface. The situation is now the same as Fig. 25–20b, and according to Gauss's law the net charge on this surface must be zero. Finally, we pull the ball out; we find that it has indeed lost all its charge.

This experimental result confirms the validity of Gauss's law, and therefore of Coulomb's law. So what is the point? Simply this: Coulomb's experimental methods, using a torsion balance and dividing charges, were not very precise. It is difficult to confirm with great precision the $1/r^2$ dependence of the electrostatic force by direct measurement. Here, by contrast, is an experi-

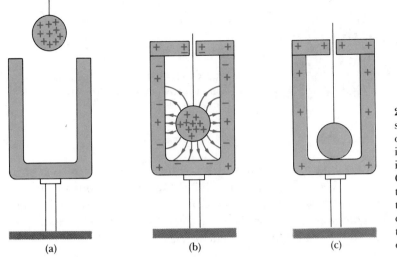

25–21 (a) A charged conducting ball suspended by an insulating thread outside a conducting container on an insulating stand. (b) The ball is lowered into the container and the lid put on. Charges are induced on the walls of the container. (c) When the ball is touched to the inner surface of the container, all its charge is transferred to the container and appears on its outer surface.

ment that tests the validity of Gauss's law and therefore of Coulomb's law with potentially much greater precision.

A contemporary version of this experiment is shown in Fig. 25–22. The details of the black box labeled "power supply" are not important; its function is to place charge on the outer sphere and remove it, on demand. The inner box with a dial is a sensitive electrometer, an instrument that can detect minute motion of charge between the outer and inner spheres. If Gauss's law is correct, there can never be any charge on the inner surface of the outer sphere, so there should be no flow of charge through the electrometer while the outer sphere is being charged and discharged. The fact that no flow is actually observed is a very sensitive confirmation of Gauss's law and therefore of Coulomb's law. The precision of the experiment is limited mainly by the sensitivity of the electrometer, which can be astonishing. The most recent (1971) experiments of this sort have shown that the exponent 2 in the $1/r^2$ expression in Coulomb's law does not differ from precisely 2 by more than 10^{-15}. So there is no reason to suspect that it is anything other than exactly 2.

This discussion also forms the basis for **electrostatic shielding,** shown in Fig. 25–23. Suppose we have a very sensitive electronic instrument that must be protected from stray electric fields that might give erroneous measurements. We surround the instrument with a conducting box, or we line the

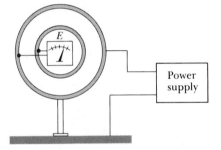

25–22 The outer surface can be alternately charged and discharged by the power supply to which it is connected. If there is any flow of charge between the inner and outer surfaces, it is detected by the electrometer inside the inner surface.

Electrostatic shielding: how to protect sensitive instruments from disruptive electric fields

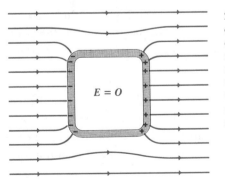

25–23 A conducting box in a uniform electric field. The field pushes electrons toward the left, leaving a net negative charge on the left side and a net positive charge on the right. The total electric field at every point inside the box is zero; the shapes of the exterior field lines near the box are somewhat changed.

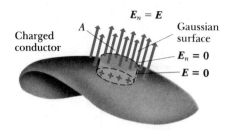

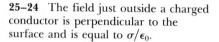

25-24 The field just outside a charged conductor is perpendicular to the surface and is equal to σ/ϵ_0.

The electric field just outside a conducting surface is proportional to the surface charge density.

walls, floor, and ceiling of the room with a conducting material such as sheet copper. The external electric field redistributes the free electrons in the conductor, leaving a net positive charge on the outer surface in some regions and a net negative charge in others, as shown in Fig. 25–23. This charge distribution causes an additional electric field such that the *total* field at every point inside the box is zero, as Gauss's law says it must be. The charge distribution on the box also alters the shapes of the field lines near the box, as the figure shows.

Finally, we note that there is a direct relation between the E field at any point just outside a conductor and the surface charge density σ on the conductor at that point. In general, σ varies from point to point on the surface. We will show in Chapter 26 that at any such point the direction of E is always perpendicular to the surface.

To find a relation between σ at any point on the surface and E at that point, we construct a Gaussian surface in the form of a small cylinder, as in Fig. 25–24. One end face, with area A, lies within the conductor, and the other lies just outside. The charge within the Gaussian surface is σA. The electric field is zero at all points within the conductor. Outside the conductor, the normal component of E is zero at the side walls of the cylinder (since E is normal to the conductor), while over the end face the normal component is equal to E. Hence from Gauss's law,

$$EA = \frac{\sigma A}{\epsilon_0}, \qquad E = \frac{\sigma}{\epsilon_0}. \qquad (25\text{–}26)$$

This equation agrees with the results already obtained for spherical, cylindrical, and plane surfaces. Just outside the surface of a sphere of radius R carrying a charge q, for example, the electric field is

$$E = \frac{1}{4\pi\epsilon_0} \frac{q}{R^2}.$$

But the surface density of charge on the sphere is $q/4\pi R^2$, so $E = \sigma/\epsilon_0$.

We also showed that the field outside an infinite charged conducting plate equals σ/ϵ_0. In this case, the field is the same at *all* distances from the plate, but in all other cases it decreases with increasing distance from the surface.

SUMMARY

KEY TERMS

electric field

test charge

Electric field is the force per unit charge exerted on a test charge at any point, provided the test charge is small enough so that it does not disturb the charges that cause the field. Electric field is a vector quantity; it is an example of a vector field, that is, a vector quantity associated with each point in a region of space (different vectors at different points).

The electric field of a point charge can be calculated from Coulomb's law; the electric field of any combination of charges is the vector sum of the fields caused by the individual charges. This principle is called the principle of superposition.

An electric dipole is a pair of electric charges of equal magnitude q but opposite sign, separated by a distance l. The electric dipole moment p is defined to be $p = ql$. The vector dipole moment is the vector $\boldsymbol{p}$ having this magnitude and a direction from negative toward positive charge. An electric dipole in an electric field experiences a torque Γ given by

$$\Gamma = pE \sin \theta, \tag{25–2}$$

where θ is the angle between the directions of $\boldsymbol{p}$ and $\boldsymbol{E}$. The vector torque Γ is given by

$$\boldsymbol{\Gamma} = \boldsymbol{p} \times \boldsymbol{E}. \tag{25–3}$$

Calculating the electric field caused by a continuous distribution of charge is carried out by dividing the distribution into small elements, calculating the field caused by each element, and then carrying out the vector sum, usually by integrating. It is often simplest to compute one component of $\boldsymbol{E}$ at a time. The concepts of linear charge density λ (charge per unit length), surface charge density σ (charge per unit area), and volume charge density ρ (charge per unit volume) are often helpful in characterizing charge distributions.

Field lines are a useful graphical representation of electric fields. A field line at any point in space is tangent to the direction of $\boldsymbol{E}$ at that point, and the number of lines per unit area (perpendicular to their direction) is proportional to the magnitude of $\boldsymbol{E}$ at the point. Electrostatic field lines always begin and end on charges, and they never cross.

Gauss's law is logically equivalent to Coulomb's law, but its use simplifies some problems considerably. It states that the surface integral of the component of $\boldsymbol{E}$ normal to the surface, over any closed surface, equals a constant times the total charge Q enclosed by the surface; symbolically,

$$\oint \boldsymbol{E} \cdot d\boldsymbol{A} = \frac{Q}{\epsilon_0}. \tag{25–17}$$

Electric flux Ψ is defined as

$$\Psi = \int \boldsymbol{E} \cdot d\boldsymbol{A}. \tag{25–19}$$

If the field is uniform and perpendicular to the area, then this equation becomes simply

$$\Psi = EA.$$

When excess charge is placed on a conductor, it resides entirely on the surface. In any electrostatic situation, $\boldsymbol{E} = \boldsymbol{0}$ everywhere in the interior of a conductor.

Table 25–1 lists electric fields caused by several simple charge distributions.

vector field
electric dipole
electric dipole moment
field line
Gauss's law
electric flux
electrostatic shielding

TABLE 25–1   Electric Fields around Simple Charge Distributions

Charge distribution responsible for the electric field	Arbitrary point in the electric field	Electric field at this point
Single point charge q	Distance r from q	$E = \dfrac{1}{4\pi\epsilon_0}\dfrac{q\hat{r}}{r^2}$
Several point charges, q_1, $q_2,\ldots$	Distance r_1 from q_1, r_2 from $q_2,\ldots$	$E = \dfrac{1}{4\pi\epsilon_0}\left(\dfrac{q_1\hat{r}_1}{r_1^{\,2}} + \dfrac{q_2\hat{r}_2}{r_2^{\,2}} + \cdots\right)$
Charge q uniformly distributed on the surface of a solid conducting sphere of radius R	(a) Outside, $r \geq R$ (b) Inside, $r < R$	(a) $E = \dfrac{1}{4\pi\epsilon_0}\dfrac{q\hat{r}}{r^2}$ (b) $E = 0$
Long cylinder of radius R, with charge per unit length λ	(a) Outside, $r \geq R$ (b) Inside, $r < R$	(a) $E = \dfrac{1}{2\pi\epsilon_0}\dfrac{\lambda}{r}$ (b) $E = 0$
Two oppositely charged conducting plates with charge per unit area σ	Any point between plates	$E = \dfrac{\sigma}{\epsilon_0}$
Any charged conductor	Just outside the surface	$E = \dfrac{\sigma}{\epsilon_0}$

QUESTIONS

25–1 It was shown in the text that the electric field inside a spherical hole in a conductor is zero. Is this also true for a cubical hole? Can the same argument be used?

25–2 Coulomb's law and Newton's law of gravitation have the same *form*. Can Gauss's law be applied to gravitational fields as well as electric fields? Is so, what modifications are needed?

25–3 By considering how the field lines must look near a conducting surface, can you see why the charge density and electric-field magnitude at the surface of an irregularly shaped solid conductor must be greatest in regions where the surface curves most sharply, and least in flat regions?

25–4 The electric field and the velocity field in a moving fluid are two examples of vector fields. Think of several other examples. There are also *scalar* fields, which associate a single number with each point in space. Temperature is an example; think of several others.

25–5 Consider the electric field caused by two point charges separated by some distance. Suppose there is a point where the field is zero; what does this tell you about the *signs* of the charges?

25–6 Does an electric charge experience a force due to the field that the charge itself produces?

25–7 A particle having electric charge and mass moves in an electric field. If it starts from rest, does it always move along the field line that passes through its starting point? Explain.

25–8 If the exponent 2 in the r^2 of Coulomb's law were 3 instead, would Gauss's law still be valid?

25–9 A student claimed that an appropriate unit for electric field magnitude is $1\ \text{J}\cdot\text{C}^{-1}\cdot\text{m}^{-1}$. Is this correct?

25–10 A certain region of space bounded by an imaginary closed surface contains no charge. Is the electric field always zero everywhere on the surface? If not, under what circumstances is it zero on the surface?

25–11 A student claimed that the electric field produced by a dipole is represented by field lines that cross each other. Is this correct? Is there any simple rule governing field lines for a superposition of fields due to point charges, if the field lines due to the separate charges are known? Do field lines *ever* cross?

25–12 Nineteenth-century physicists liked to give everything mechanical attributes. Faraday and his contemporaries thought of field lines as elastic strings that repelled each other and arranged themselves in equilibrium under the action of their elastic tension and mutual repulsion. Try this picture on several examples and decide whether it makes any sense. (These mechanical properties are of course now known to be completely fictitious.)

25–13 Are Coulomb's law and Gauss's law *completely* equivalent? Are there any situations in electrostatics where one is valid and the other is not?

25–14 The text states that, in an electrostatic field, every field line must start on a positive charge and terminate on a negative charge. But suppose the field is that of a single positive point charge. Then what?

25–15 Is the total (net) electric charge in the universe positive, negative, or zero?

25–16 A lightning rod is a pointed copper rod mounted on top of a building and welded to a heavy copper cable running down into the ground. Lightning rods are used in prairie country to protect houses and barns from lightning; the lightning current runs through the copper rather than through the barn. Why? Why should the end of the rod be pointed? (The answer to Question 25–3 may be helpful.)

25–17 When a high-voltage power line falls on your car, you are safe as long as you stay in the car, but when you step out you may be electrocuted. Why? What about a car as a safe place in a thunderstorm?

EXERCISES

Section 25–1 Electric Field and Electrical Forces

25–1 Find the magnitude and direction of the electric field at a point 0.5 m directly above a particle having an electric charge of $+2.0 \times 10^{-6}$ C.

25–2 The electric field caused by a certain point charge has a magnitude of 5.0×10^3 N·C^{-1} at a distance of 0.10 m from the charge. What is the magnitude of the charge?

25–3 At what distance from a particle of charge 5.0×10^{-9} C does the electric field of that charge have a magnitude of 2.00 N·C^{-1}?

25–4

a) What is the electric field of a gold nucleus, at a distance of 10^{-14} m from the nucleus?

b) What is the electric field of a proton, at a distance of 5.28×10^{-11} m from the proton?

25–5 A small object carrying a charge of -5×10^{-9} C experiences a downward force of 20×10^{-9} N when placed at a certain point in an electric field. What is the electric field at the point, in magnitude and direction?

25–6 What must be the charge (sign and magnitude) on a particle of mass 2 g for it to remain stationary in the laboratory when placed in a downward-directed electric field of 5000 N·C^{-1}?

25–7 What is the magnitude of an electric field in which the force on an electron is equal in magnitude to the weight of the electron?

25–8 A uniform electric field exists in the region between two oppositely charged plane parallel plates. An electron is released from rest at the surface of the negatively charged plate and strikes the surface of the opposite plate, 2 cm distant from the first, in a time interval of 1.5×10^{-8} s.

a) Find the electric field.

b) Find the velocity of the electron when it strikes the second plate.

25–9 An electron is projected with an initial velocity $v_0 = 1.0 \times 10^7$ m·s^{-1} into the uniform field between the parallel plates in Fig. 25–25. The direction of the field is vertically downward, and the field is zero except in the space between the two plates. The electron enters the field at a point

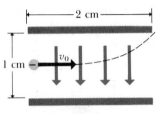

FIGURE 25–25

midway between the plates. If the electron just misses the upper plate as it emerges from the field, find the magnitude of the electric field.

Section 25–2 Electric-Field Calculations

25–10 Two particles having charges $q_1 = 1.00 \times 10^{-9}$ C and $q_2 = 2.00 \times 10^{-9}$ C are separated by a distance of 2.00 m. At what point is the total electric field due to the two charges equal to zero?

25–11 In a rectangular coordinate system a charge of 25×10^{-9} C is placed at the origin of coordinates, and a charge of -25×10^{-9} C is placed at the point $x = 6$ m, $y = 0$. What are the magnitude and direction of the electric field at

a) $x = 3$ m, $y = 0$?

b) $x = 3$ m, $y = 4$ m?

25–12 In a rectangular coordinate system, two positive point charges of 10^{-8} C each are fixed at the points $x = +0.1$ m, $y = 0$, and $x = -0.1$ m, $y = 0$. Find the magnitude and direction of the electric field at the following points:

a) the origin; b) $x = 0.2$ m, $y = 0$;

c) $x = 0.1$ m, $y = 0.15$ m; d) $x = 0$, $y = 0.1$ m.

25–13 Same as Exercise 25–12, except that the point charge at $x = +0.1$ m, $y = 0$ is positive and the other is negative.

25–14 A long, straight wire has charge per unit length 3.0×10^{-10} C·m^{-1}. At what distance from the wire is the electric field equal to 0.8 N·C^{-1}?

25–15 What must be the charge per unit area, in C·m^{-2}, of an infinite-plane sheet of charge if the electric field produced by the sheet of charge is to be 2.0 N·C^{-1}?

Section 25-4 Gauss's Law

25-16 The electric field E in Fig. 25–26 is everywhere parallel to the x-axis. The field has the same magnitude at all points in any given plane perpendicular to the x-axis (parallel to the yz-plane), but the magnitude is different for various planes. That is, E_x depends on x but not on y and z, and E_y and E_z are zero. At points *in* the yz-plane, $E_x = 400 \text{ N·C}^{-1}$. (The volume shown could be a section of a large insulating slab 1 m thick, with its faces parallel to the yz-plane and with a uniform volume charge distribution embedded in it.)

a) What is the value of $\int E_\perp dA$ over surface I in the diagram?

b) What is the value of the surface integral of E over surface II?

c) There is a positive charge of 26.6×10^{-9} C within the volume. What are the magnitude and direction of E at the face opposite I?

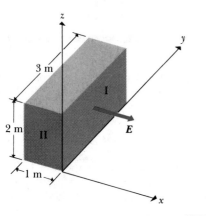

FIGURE 25-26

Section 25-5 Applications of Gauss's Law

Section 25-6 Charges on Conductors

25-17 How many excess electrons must be added to an isolated spherical conductor 0.10 m in diameter to produce a field of 1300 N·C^{-1} just outside the surface?

25-18 The earth has a net electric charge that causes a field at points near its surface of the order of 100 N·C^{-1}. If the earth is regarded as a conducting sphere of radius 6.38×10^6 m, what is the magnitude of its charge?

25-19 Prove that the electric field outside an infinitely long cylindrical conductor with a uniform surface charge is the same as if all the charge were on the axis.

25-20 Apply Gauss's law to the dotted Gaussian surfaces in Fig. 25–19b to calculate the electric field between and outside the plates.

25-21 The electric field in the region between a pair of oppositely charged plane parallel conducting plates, each 100 cm^2 in area, is 10^4 N·C^{-1}. What is the charge on each plate? Neglect edge effects.

25-22 A conducting sphere carrying charge q has radius a. It is inside a concentric hollow conducting sphere of inner radius b and outer radius c. The hollow sphere has no net charge. Calculate the electric field for

a) $r < a$,

b) $a < r < b$,

c) $b < r < c$,

d) $r > c$.

e) What is the charge on the inner surface of the hollow sphere?

f) On the outer surface?

25-23 A small conducting sphere of radius a, mounted on an insulating handle and having a positive charge q, is inserted through a hole in the walls of a hollow conducting sphere of inner radius b and outer radius c. The hollow sphere is supported on an insulating stand and is initially uncharged, and the small sphere is placed at the center of the hollow sphere. Neglect any effect of the hole.

a) Show that the electric field at a point in the region between the spheres, at a distance r from the center, is equal to

$$E = \frac{1}{4\pi\epsilon_0}\frac{q}{r^2}.$$

b) What is the electric field at a point outside the hollow sphere?

c) Sketch a graph of the magnitude of E as a function of r, from $r = 0$ to $r = 2c$.

d) Represent the charge of the small sphere by four + signs. Sketch the field lines of the system, within a spherical volume of radius $2c$.

PROBLEMS

25-24 A small sphere whose mass is 0.1 g carries a charge of 3×10^{-10} C and is attached to one end of a silk fiber 5 cm long. The other end of the fiber is attached to a large vertical conducting plate, which has a surface charge of 25×10^{-6} C·m^{-2} on each side. Find the angle the fiber makes with the vertical plate.

25-25 An electron is projected into a uniform electric field of 5000 N·C^{-1}. The direction of the field is vertically upward. The initial velocity of the electron is 1.0×10^7 m·s^{-1}, at an angle of 30° above the horizontal.

a) Find the maximum distance the electron rises vertically above its initial elevation.

b) After what horizontal distance does the electron return to its original elevation?

c) Sketch the trajectory of the electron.

25–26 A charge of 16×10^{-9} C is fixed at the origin of coordinates; a second charge of unknown magnitude is at $x = 3$ m, $y = 0$; and a third charge of 12×10^{-9} C is at $x = 6$ m, $y = 0$. What are the sign and magnitude of the unknown charge if the resultant field at $x = 8$ m, $y = 0$ is 20.25 N·C^{-1} directed to the right?

25–27 Two charges are placed as shown in Fig. 25–27. It is known that the magnitude of q_1 is 5×10^{-6} C, but its sign and the value of the other charge, q_2, are not known. The resultant electric field E at point P is measured to be entirely in the negative y-direction.

a) Considering the different possible signs of q_1 and q_2, there are four possible diagrams that could represent the electric fields E_1 and E_2 produced by q_1 and q_2. Sketch the four possible electric field configurations.

b) Using the sketches from (a) and the fact that the net electric field at P has no x-component and a negative y-component, deduce the signs of q_1 and q_2.

c) Determine the magnitude of the resultant field E.

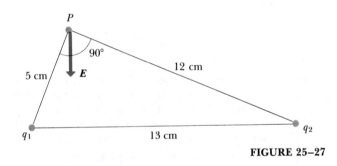

FIGURE 25–27

25–28 Electric charge is uniformly distributed around a semicircle of radius a, with total charge Q. What is the electric field at the center of curvature?

25–29 Negative electric charge is distributed uniformly around a quarter of a circle of radius a, with total charge $-Q$ (Fig. 25–28). What are the x- and y-components of the resultant electric field at the center of curvature (point P)?

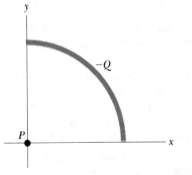

FIGURE 25–28

25–30 Electric charge is distributed uniformly along each of the sides of a square. Two adjacent sides have positive charge, with total charge $+Q$ on each.

a) If the other two sides have negative charge with total charge $-Q$ each (see Fig. 25–29), what are the x- and

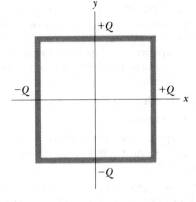

FIGURE 25–29

y-components of the resultant electric field at the center of the square? Each side of the square has length a.

b) Repeat the calculation of (a) if all four sides have positive charge $+Q$.

25–31 A uniform electric field, $E_1 = -9 \times 10^4$ N·C^{-1}i, is directed out of one face of a parallelopiped, and another uniform electric field, $E_2 = -11 \times 10^4$ N·C^{-1}i, is directed into the opposite face, as shown in Fig. 25–30. Assuming that there are no other electric-field lines crossing the surfaces of the parallelopiped, determine the net charge contained within.

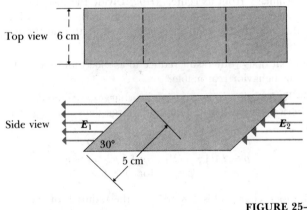

FIGURE 25–30

25–32 A conducting spherical shell of inner radius a and outer radius b has a positive point charge Q located at its center. The total charge on the shell is $-3Q$, and it is insulated from its surroundings.

a) Derive expressions for the electric-field magnitude in terms of the distance r from the center, for the regions $r < a$, $a < r < b$, and $r > b$.

b) What is the surface charge density on the inner surface of the conducting shell?

c) What is the surface charge density on the outer surface of the conducting shell?

d) Draw a sketch showing electric-field lines and the location of all charges.

e) Draw a graph of E as a function of r.

25–33 A long coaxial cable consists of an inner cylindrical conductor of radius a and an outer coaxial cylinder of inner radius b and outer radius c. The outer cylinder is mounted on insulating supports and has no net charge. The inner cylinder has a uniform positive charge λ per unit length. Calculate the electric field

a) at any point between the cylinders;

b) at any external point.

c) Sketch a graph of the magnitude of E as a function of the distance r from the axis of the cable, from $r = 0$ to $r = 2c$.

d) Find the charge per unit length on the inner surface of the outer cylinder, and on the outer surface.

25–34 Suppose that positive charge is uniformly distributed through a very long cylindrical volume of radius R, the charge per unit volume being ρ.

a) Derive the expression for the electric field inside the volume at a distance r from the axis of the cylinder, in terms of the charge density ρ.

b) What is the electric field at a point outside the volume, in terms of the charge per unit length λ in the cylinder?

c) Compare the answers to (a) and (b) when $r = R$.

d) Sketch a graph of the magnitude of E as a function of r, from $r = 0$ to $r = 3R$.

25–35 A nonuniform but spherically symmetric distribution of charge has a charge density ρ given as follows:

$$\rho = \rho_0(1 - r/R) \quad \text{for} \quad r \leq R,$$
$$\rho = 0 \quad \text{for} \quad r \geq R,$$

where $\rho_0 = 3Q/(\pi R^3)$ is a constant.

a) Show that the total charge contained in the charge distribution is Q.

b) Show that for the region defined by $r \geq R$ the electric field is identical to that produced by a point charge Q.

c) Obtain an expression for the electric field in the region $r \leq R$.

d) Compare your results in (b) and (c) for $r = R$.

CHALLENGE PROBLEMS

25–36 Electric charge is distributed uniformly over a disk of radius a, with total charge Q.

a) Find the electric field at a point on the axis of the disk, a distance x from its center. (*Hint:* Divide the disk into concentric rings; use the result of Example 25–7 to find the field due to each ring; and integrate to find the total field.)

b) What does your result reduce to as x becomes large? Is this behavior reasonable?

25–37 A region in space contains charge distributed spherically such that the volume charge density ρ is given by:

$$\rho = A \quad \text{for} \quad r \leq R/2,$$
$$\rho = 2A(1 - r/R) \quad \text{for} \quad R/2 \leq r \leq R,$$
$$\rho = 0 \quad \text{for} \quad r \geq R.$$

The total charge Q is 1×10^{-17} C, the radius R of the spherical charge distribution is 2×10^{-14} m, and A is a constant having units of C·m^{-3}.

a) Determine A in terms of Q and R, and also determine its numerical value.

b) Using Gauss's law, derive an expression for the electric field E as a function of the distance r from the center of the distribution. Do this separately for all three regions. Be sure to check that your results agree on the boundaries of the regions.

c) What fraction of the total charge is contained within the region where $r \leq R/2$?

d) If an electron of charge $q' = -e$ is oscillating back and forth about $r = 0$ (the center of the distribution) with an amplitude of $R/2$, show that the motion is simple harmonic. (*Hint:* Review the definition of simple

harmonic motion, as defined by Eq. (11–1). If it can be shown that the net force on the electron is of this form, then it follows that the motion is simple harmonic. Conversely, if the net force on the electron does not follow this form, the motion is not simple harmonic.)

e) For the motion in (d), what is the period? (Calculate a numerical value.)

f) If the amplitude of the motion described in (e) were greater than $R/2$, explain why the motion would no longer be simple harmonic.

25–38 A region in space contains charge distributed spherically such that the volume charge density ρ is given by:

$$\rho = 3Ar/(2R) \quad \text{for} \quad r \leq R/2,$$
$$\rho = A[1 - (r/R)^2] \quad \text{for} \quad R/2 \leq r \leq R,$$
$$\rho = 0 \quad \text{for} \quad r \geq R.$$

The total charge is Q, the radius R of the spherical charge distribution is 5×10^{-10} m, and $A = 6 \times 10^9$ C·m^{-3} and is constant.

a) Determine Q in terms of A and R, and also determine its numerical value.

b) Using Gauss's law, derive an expression for the electric field E as a function of the distance r from the center of the distribution. Do this separately for all three regions.

c) What fraction of the total charge is contained within the region where $R/2 \leq r \leq R$?

d) What is the magnitude of the electric field at $r = R/2$?

e) If an electron of charge $q' = -e$ is released from rest at any point in any of the three regions, the resulting motion will be oscillatory, although not simple harmonic. Why? (See Challenge Problem 25–37).

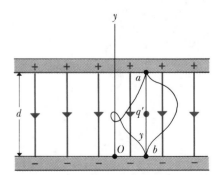

26

ELECTRICAL POTENTIAL

THE CONCEPT OF *POTENTIAL* IS DIRECTLY TIED TO POTENTIAL ENERGIES associated with electrical interactions. We studied the concepts of *work* and *energy* in Chapter 7; they provide an important method of analysis for many problems in mechanics. Energy plays an equally fundamental role in electricity and magnetism. In this chapter we apply work and energy considerations to the electric field. When a charged particle moves in an electric field, the electric-field force does *work* on the particle. This work can always be expressed in terms of a potential energy, which in turn is associated with a new concept called *electrical potential,* or simply *potential.* We will discuss several examples of calculations of potential and their applications to the motions of charged particles.

26–1 ELECTRICAL POTENTIAL ENERGY

When a charged particle moves in a region of space where an electric field is present, the field exerts a force on the particle and thus does *work* on it. Figure 26–1 shows a simple example; a pair of charged parallel metal plates sets up a uniform electric field of magnitude E in the region between them, and the resulting force on a test charge q' has magnitude $F = q'E$. When the charge moves from point a to point b, the work done by this force on the test charge is

$$W_{a \to b} = q'Ed. \tag{26–1}$$

This work can be represented by a **potential-energy** function, just as for gravitational potential energy, discussed in Section 7–5. If we take the potential energy to be zero at point b, then at point a it has the value $q'Ed$, and we may say

$$W_{a \to b} = -\Delta U = U_a - U_b. \tag{26–2}$$

More generally, the potential energy at any point a distance y above the bottom plate is given by

$$U(y) = q'Ey, \tag{26–3}$$

When an electric-field force acts on a moving charge, it does work.

26–1 A test charge q' moving from a to b experiences a force of magnitude $q'E$; the work done by this force is $W_{a \to b} = q'Ed$, and is independent of the particle's path.

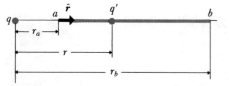

26–2 Charge q' moves along a straight line extending radially from charge q. As it moves from a to b, the distance varies from r_a to r_b.

The electric field is a conservative force field.

Work done on a point charge in the field of another point charge

and when the test charge moves from height y_1 to height y_2, the work done on the charge by the field is given by

$$W_{1 \to 2} = U(y_1) - U(y_2) = q'Ey_1 - q'Ey_2. \qquad (26\text{-}4)$$

When y_1 is greater than y_2, U decreases and the field does positive work; when y_1 is less than y_2, U increases and the field does negative work. This whole situation is directly analogous to the motion of a particle in a uniform gravitational field, with its associated work and potential energy.

Note that the work done by the electric-field force depends only on the change in the coordinate y and is independent of the *path* of the particle. Hence this force is an example of what we called in Section 7–7 a **conservative force field;** we suggest you review that discussion now. In fact, we will show very soon that *every* electric field produced by charges at rest is a conservative force field and thus can be associated with a potential-energy function.

Consider next the work done on a test charge q' moving in the electric field caused by a single stationary point charge q. Such a force field is clearly *not* uniform, and we have to integrate to calculate the work done on q'. We first calculate the work done on q' during a displacement along the *radial* line from point a to point b in Fig. 26–2. We introduce a unit vector $\hat{r}$ pointing from a toward b; the component of displacement in this direction is just dr, and the component of force on q' in this direction is

$$F = \frac{1}{4\pi\epsilon_0} \frac{qq'}{r^2}. \qquad (26\text{-}5)$$

The work done by this force on q' as it moves from r_a to r_b is given by

$$W_{a \to b} = \int_{r_a}^{r_b} F \, dr = \int_{r_a}^{r_b} \frac{1}{4\pi\epsilon_0} \frac{qq'}{r^2} \, dr$$

$$= \frac{qq'}{4\pi\epsilon_0} \left(\frac{1}{r_a} - \frac{1}{r_b} \right). \qquad (26\text{-}6)$$

Thus the work for this particular path depends only on the endpoints. We have not yet proved that the work is the same for *all possible* paths from a to b. To do this, we consider a more general displacement, in which a and b *do not* lie on the same radial line, as in Fig. 26–3. To find the work done by the field in the displacement from point a to point b we must now use the more general definition of work given by Eq. (7–8):

$$W = \int_a^b F \cos\theta \, dl = \int_a^b \boldsymbol{F} \cdot d\boldsymbol{l}. \qquad (26\text{-}7)$$

But from the figure, $\cos\theta \, dl = dr$. That is, the work done during a small displacement dl depends only on the change dr in the distance r between the charges, which is the *radial component* of the displacement. In any displacement in which r does not change, no work is done because $\boldsymbol{F}$ and $d\boldsymbol{l}$ are perpendicular, $\cos\theta = 0$, and $\boldsymbol{F} \cdot d\boldsymbol{l} = 0$.

Thus Eq. (26–6) gives the work done by the field even for this more general displacement. This result shows that the work done on q' by the $\boldsymbol{E}$ field depends only on the initial and final distances r_a and r_b and not on the path connecting these points. The work is the same for *any* path between these points or between any pair of points at distances r_a and r_b from the charge q.

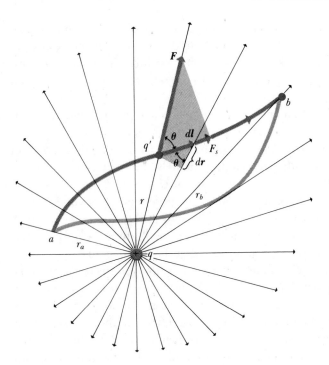

26–3 The work done by the electric-field force on charge q' depends only on the distances r_a and r_b.

Also, if the test charge returns from b to a along any path, the work done by the E field is the *negative* of that done in the displacement from a to b. So the total work done during a *loop* displacement returning to the starting point is always zero.

Comparing Eqs. (26–2) and (26–6), we see that the term $qq'/4\pi\epsilon_0 r_a$ is the potential energy U_a when q' is at point a, at distance r_a from q, and $qq'/4\pi\epsilon_0 r_b$ is the potential energy U_b when it is at point b, at distance r_b from q. Thus the potential energy U of the test charge q' at *any* distance r from charge q is given by

Potential energy of a point charge in the field of another point charge

$$U = \frac{1}{4\pi\epsilon_0}\frac{qq'}{r}. \tag{26–8}$$

If the field in which charge q' moves is not that of a single point charge but is due to a more general charge distribution, we can always divide the charge distribution into small elements and treat each element as a point charge. Then, since the total field is the *vector sum* of the fields due to the individual elements, and since the total work on q' is the sum of the contributions from the individual charge elements, we may conclude that *every electric field due to a static charge distribution is a conservative force field.* Furthermore, the potential energy of a test charge q' at point a in Fig. 26–4, due to a collection of charges $q_1, q_2, q_3,$ and so on, at distances $r_1, r_2, r_3,$ and so on, from the test charge q', is given by

Potential energy of a charge due to several point charges is the sum of the potential energies due to the individual charges.

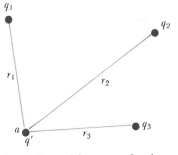

$$U = \frac{q'}{4\pi\epsilon_0}\left(\frac{q_1}{r_1} + \frac{q_2}{r_2} + \frac{q_3}{r_3} + \cdots\right)$$
$$= \frac{q'}{4\pi\epsilon_0}\sum\frac{q_i}{r_i}. \tag{26–9}$$

26–4 Potential energy of a charge q' at point a depends on charges $q_1, q_2,$ and q_3, and on their distances $r_1, r_2,$ and r_3 from point a.

At a second point b, the potential energy of q' is given by the same expression except that $r_1, r_2, \ldots$ now represent the distances from the respective charges to point b. The work done by the electric force in moving the test charge from a to b along any path is equal to the difference $U_a - U_b$ between its potential energies at a and at b.

The discussion of potential energy in Chapter 7 pointed out that it is always possible to add an arbitrary constant C to a potential-energy function U, since adding the same constant C to both U_a and U_b in Eq. (26–2) does not change the physically significant quantity, namely, the difference $U_a - U_b$. Thus we are always free to choose the constant C so that U is zero at some convenient reference position. In Chapter 7 the potential energy of a body in a uniform gravitational field was taken to be zero at some convenient reference level, often the surface of the earth. When the body is above this reference level, its potential energy is positive; when below, negative.

In Eqs. (26–8) and (26–9), the reference position for electrical potential energy (the position at which $U = 0$) has been chosen implicitly to be at a point where all the distances $r_1, r_2, \ldots$ are *infinite*. That is, the potential energy of the test charge is zero when it is very far removed from all the charges setting up the field. This reference level is the most convenient one for many electrostatic problems. When we consider electrical circuits, other reference levels are often more convenient.

From the preceding definitions, *the potential energy of a test charge at any point in an electric field is equal to the work done by the electric-field force when the test charge moves from the point in question to a zero reference level, often taken at infinity.*

Potential energy of two point charges can be defined to be zero when they are infinitely far apart.

EXAMPLE 26–1   A small plastic ball having mass 2.0 g and electric charge $+0.10 \mu C$ moves in the vicinity of a stationary metal ball with a charge of $+2.0 \mu C$. When the plastic ball is 0.1 m from the metal one, it is moving directly away from it with a speed of 5.0 m·s^{-1}.

(a)   What is its speed when the two balls are 0.2 m apart?

(b)   How would the situation change if the plastic ball had a charge of $-0.1 \mu C$?

SOLUTION

(a)   Because of the conservative nature of the force, total mechanical energy (kinetic plus potential) is conserved. Recalling the problem-solving strategy of Section 7–5, we list the initial and final kinetic and potential energies, K_1, K_2, U_1, and U_2, respectively:

A simple example of electrical potential energy

$$K_1 = \tfrac{1}{2}mv_1^2 = \tfrac{1}{2}(2.0 \times 10^{-3} \text{ kg})(5.0 \text{ m·s}^{-1})^2 = 0.025 \text{ J};$$

$$K_2 = \tfrac{1}{2}mv_2^2 = \tfrac{1}{2}(2.0 \times 10^{-3} \text{ kg})v_2^2;$$

$$U_1 = \frac{1}{4\pi\epsilon_0} \frac{q_1 q_2}{r_1}$$

$$= (8.99 \times 10^9 \text{ N·m}^2\text{·C}^{-2})(0.10 \times 10^{-6} \text{ C})(2.0 \times 10^{-6} \text{ C})/(0.1 \text{ m})$$

$$= 0.0180 \text{ J};$$

$$U_2 = (8.99 \times 10^9 \text{ N·m}^2\text{·C}^{-2})(0.10 \times 10^{-6} \text{ C})(2.0 \times 10^{-6} \text{ C})/(0.2 \text{ m})$$

$$= 0.0090 \text{ J}.$$

From conservation of energy, we have

$$K_1 + U_1 = K_2 + U_2,$$
$$0.025 \text{ J} + 0.0180 \text{ J} = (1.0 \times 10^{-3} \text{ kg})v_2{}^2 + 0.0090 \text{ J},$$

and finally

$$v_2 = 5.83 \text{ m·s}^{-1}.$$

The force is repulsive, and the plastic ball speeds up as it moves away from the stationary charge.

(b) If the moving charge is negative, the force on it is attractive rather than repulsive, and we expect it to slow down rather than speed up. The only difference in the calculations above is that both potential-energy quantities are negative; the conservation-of-energy equation becomes

$$0.025 \text{ J} - 0.0180 \text{ J} = (1.0 \times 10^{-3} \text{ kg})v_2{}^2 - 0.0090 \text{ J},$$

and the final result is

$$v_2 = 4.00 \text{ m·s}^{-1}.$$

We conclude this section with a comment about our treatment of electrical potential energy. We have spoken consistently about the work done *by the electric-field force* on the charged particle moving in the field. In a displacement of the particle from point a to point b, this work, which we denote as $W_{a \to b}$, is always given by

$$W_{a \to b} = U_a - U_b.$$

Relation between work and potential energy for electrical forces

When U_a is greater than U_b, this work is positive, and the potential energy decreases during the displacement. In this case, the field does positive work on the particle as the particle "falls" from a point of higher potential energy to a point of lower potential energy. This viewpoint is consistent with the approach of Chapter 7, where we always spoke of the work done on a particle by a gravitational field or by an elastic force.

An alternative viewpoint used in some books is that in order to "raise" a particle from a point b where the potential energy has the value U_b to a point a where it has a greater value U_a, it would be necessary to apply an additional force that *opposes* the electric-field force and does positive work. The potential-energy difference $U_a - U_b$ would then be defined as the work done *by that additional force* during a displacement from b to a. This viewpoint is not wrong, but in our view it may sometimes be confusing. We prefer always to deal with the work done *by the electric-field force*, without introducing any hypothetical additional forces that may or may not be present in a particular problem. We continue to use this viewpoint in the next section, which introduces the concept of *electrical potential*. If you use other books for reference, beware of the possible confusion about these two alternative viewpoints.

Alternative definitions of electrical potential energy: Don't get confused!

26–2 POTENTIAL

Potential is potential energy per unit charge.

In the preceding section we discussed the potential energy U of a charged particle. Now we introduce the more general concept of *potential energy per unit charge*. This quantity is called **potential.** The potential at any point of an electrostatic field is defined as *the potential energy per unit charge* for a test charge q' at that point. Potential is represented by the letter V:

$$V = \frac{U}{q'}, \quad \text{or} \quad U = q'V. \tag{26–10}$$

The volt: a unit of potential

Potential energy and charge are both scalars, so potential is a scalar quantity. Its basic SI unit is one *joule per coulomb* ($1 \text{ J} \cdot \text{C}^{-1}$). For brevity, a potential of $1 \text{ J} \cdot \text{C}^{-1}$ is called one **volt** (1 V), named in honor of the Italian scientist Alessandro Volta (1745–1827):

$$1 \text{ V} = 1 \text{ J} \cdot \text{C}^{-1}.$$

Commonly used multiples and submultiples of the volt are the kilovolt ($1 \text{ kV} = 10^3 \text{ V}$), the megavolt ($1 \text{ MV} = 10^6 \text{ V}$), the gigavolt ($1 \text{ GV} = 10^9 \text{ V}$), the millivolt ($1 \text{ mV} = 10^{-3} \text{ V}$), and the microvolt ($1 \text{ }\mu\text{V} = 10^{-6} \text{ V}$).

To put Eq. (26–2) on a "work-per-unit-charge" basis, we divide both sides by q', obtaining

$$\frac{W_{a \to b}}{q'} = \frac{U_a}{q'} - \frac{U_b}{q'} = V_a - V_b, \tag{26–11}$$

where $V_a = U_a/q'$ is the potential energy per unit charge at point a, and similarly for V_b. The terms V_a and V_b are called the *potential at point a* and *potential at point b*, respectively. From Eq. (26–9), the potential V at a point due to an arbitrary collection of point charges is given by

$$V = \frac{U}{q'} = \frac{1}{4\pi\epsilon_0} \sum \frac{q_i}{r_i}. \tag{26–12}$$

Potential difference: potential of one point with respect to another

The difference $V_a - V_b$ is called the *potential of a with respect to b*, and is often abbreviated V_{ab}. Note that potential, like electric field, is independent of the test charge q' used to define it.

The potential due to a collection of point charges is usually obtained most easily with Eq. (26–12), but in some problems where the E field is known, it is easier to work directly with the field. The force F on the test charge q' can be written as $F = q'E$. When this relation is applied in Eq. (26–7) and the result combined with Eq. (26–11), we obtain the useful result

$$V_{ab} = V_a - V_b = \int_a^b E \cdot dl = \int_a^b E \cos \theta \, dl. \tag{26–13}$$

Potential difference can be represented as a line integral of electric field.

Either integral on the right side is called the **line integral** of E. It represents the conceptual process of dividing a path into small elements dl, multiplying each magnitude dl by the component of E parallel to dl at that point, and summing the results for the entire path. We have already encountered the concept of line integral in our general definition of work in Section 7–3.

Equation (26–13) states that when a positive test charge moves from a region of high potential to one of lower potential (that is, $V_b < V_a$), the electric field does positive work on it. Thus a positive charge tends to "fall" from a high-potential region to a lower-potential one. The opposite is true for a negative charge.

A voltmeter measures potential difference (voltage).

An instrument that measures the difference of potential between two points is called a *voltmeter*. The principle of the common type of moving-coil voltmeter will be described later. There are also much more sensitive potential-measuring devices that utilize electronic amplification. Devices that can measure a potential difference of 1 μV are not unusual.

EXAMPLE 26–2 A particle having a charge $q = 3 \times 10^{-9}$ C moves from point a to point b along a straight line, a total distance $d = 0.5$ m. The electric field is uniform along this line, in the direction from a to b, with magnitude $E = 200$ N·C^{-1}. Determine (a) the force on q; (b) the work done on it by the field; and (c) the potential difference $V_a - V_b$.

SOLUTION

a) The force is in the same direction as the electric field, and its magnitude is given by

$$F = qE = (3 \times 10^{-9} \text{ C})(200 \text{ N·C}^{-1}) = 600 \times 10^{-9} \text{ N}.$$

b) The work done by this force is

$$W = Fd = (600 \times 10^{-9} \text{ N})(0.5 \text{ m}) = 300 \times 10^{-9} \text{ J}.$$

c) The potential difference is the work per unit charge, which is

$$V_a - V_b = \frac{W}{q} = \frac{300 \times 10^{-9} \text{ J}}{3 \times 10^{-9} \text{ C}} = 100 \text{ J·C}^{-1} = 100 \text{ V}.$$

Alternatively, since E is force per unit charge, the work per unit charge is obtained by multiplying E by the distance d:

$$V_a - V_b = Ed = (200 \text{ N·C}^{-1})(0.5 \text{ m}) = 100 \text{ J·C}^{-1} = 100 \text{ V}.$$

EXAMPLE 26–3 Point charges of $+12 \times 10^{-9}$ C and -12×10^{-9} C are placed 10 cm apart, as in Fig. 26–5. Compute the potentials at points a, b, and c.

A simple potential calculation with point charges

SOLUTION We must evaluate the *algebraic* sum

$$\frac{1}{4\pi\epsilon_0} \sum \frac{q_i}{r_i}$$

Potential is a scalar quantity; potentials combine algebraically, not vectorially.

at each point. At point a, the potential due to the positive charge is

$$(9.0 \times 10^9 \text{ N·m}^2\text{·C}^{-2}) \frac{12 \times 10^{-9} \text{ C}}{0.06 \text{ m}} = 1800 \text{ N·m·C}^{-1}$$

$$= 1800 \text{ J·C}^{-1} = 1800 \text{ V},$$

and the potential due to the negative charge is

$$(9.0 \times 10^9 \text{ N·m}^2\text{·C}^{-2}) \frac{-12 \times 10^{-9} \text{ C}}{0.04 \text{ m}} = -2700 \text{ J·C}^{-1} = -2700 \text{ V}.$$

Hence

$$V_a = 1800 \text{ V} - 2700 \text{ V} = -900 \text{ V}$$
$$= -900 \text{ J·C}^{-1}.$$

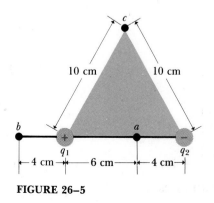

FIGURE 26–5

At point b, the potential due to the positive charge is $+2700$ V, and that due to the negative charge is -770 V. Hence

$$V_b = 2700 \text{ V} - 770 \text{ V} = 1930 \text{ V}$$
$$= 1930 \text{ J·C}^{-1}.$$

At point c the potential is

$$V_c = 1080 \text{ V} - 1080 \text{ V} = 0.$$

EXAMPLE 26–4 Refer to Fig. 26–5. Compute the potential energy of a point charge of $+4 \times 10^{-9}$ C if placed at points a, b, and c.

SOLUTION First, for any point charge q,

$$U = qV.$$

Hence at point a,

$$U = qV_a = (4 \times 10^{-9} \text{ C})(-900 \text{ J·C}^{-1})$$
$$= -36 \times 10^{-7} \text{ J}.$$

At point b,

$$U = qV_b$$
$$= (4 \times 10^{-9} \text{ C})(1930 \text{ J·C}^{-1})$$
$$= 77 \times 10^{-7} \text{ J}.$$

At point c,

$$U = qV_c = 0.$$

(All are relative to a point at infinity.)

EXAMPLE 26–5 In Fig. 26–6, a particle having mass $m = 5$ g and charge $q' = 2 \times 10^{-9}$ C starts from rest at point a and moves in a straight line to point b. What is its speed v at point b?

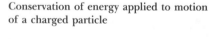
Conservation of energy applied to motion of a charged particle

SOLUTION Conservation of energy gives

$$K_a + U_a = K_b + U_b.$$

For this situation, $K_a = 0$ and $K_b = \frac{1}{2}mv^2$. The potential energies are given in terms of the potentials by Eq. (26–10): $U_a = q'V_a$ and $U_b = q'V_b$. Thus the energy equation becomes

$$0 + q'V_a = \frac{1}{2}mv^2 + q'V_b.$$

When solved for v, this yields

$$v = \sqrt{\frac{2q'(V_a - V_b)}{m}}.$$

We obtain the potentials just as we did in the preceding examples:

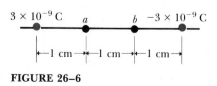

$$V_a = (9.0 \times 10^9 \text{ N·m}^2\text{·C}^{-2}) \left(\frac{3 \times 10^{-9} \text{ C}}{0.01 \text{ m}} + \frac{-3 \times 10^{-9} \text{ C}}{0.02 \text{ m}} \right) = 1350 \text{ V},$$

$$V_b = (9.0 \times 10^9 \text{ N·m}^2\text{·C}^{-2}) \left(\frac{3 \times 10^{-9} \text{ C}}{0.02 \text{ m}} + \frac{-3 \times 10^{-9} \text{ C}}{0.01 \text{ m}} \right) = -1350 \text{ V}.$$

FIGURE 26–6

Finally,

$$v = \sqrt{\frac{2(2 \times 10^{-9} \text{ C})(2700 \text{ V})}{5 \times 10^{-3} \text{ kg}}} = 4.65 \times 10^{-2} \text{ m·s}^{-1} = 4.65 \text{ cm·s}^{-1}.$$

We can check consistency of units by noting that $1 \text{ V} = 1 \text{ J·C}^{-1}$, and thus the numerator under the radical has units of J or kg·m^2·s^{-2}.

26-3 CALCULATION OF POTENTIALS

In principle we can always calculate the potential at a point by applying Eq. (26–12). We can then use the results of such calculations to find the potential *difference* between two points. An alternative approach, especially convenient when the electric field can be obtained easily, is to use Eq. (26–13) to calculate the needed potential difference from the known electric field. In some problems a combination of these two approaches is best. The following examples illustrate these comments.

Several different methods can be used to calculate potentials.

PROBLEM-SOLVING STRATEGY: Potential calculations

1. When you are dealing with a given continuous charge distribution, devise a way to divide it into infinitesimal elements identified by coordinates and coordinate differentials. For example, for a line charge along the y-axis, choose a segment of length dy. Next, determine how much charge dQ is contained in that element. If the total length is L and the charge is distributed uniformly, then $dy/L = dQ/Q$, where Q is the total charge. Then use Eq. (26–12), expressing the sum as an integral. Carry out the integration, using appropriate limits to include the entire charge distribution. Be careful about which geometric quantities vary in the integral and which are constant.

2. If the electric field is given or can be obtained with little effort, it is usually easier not to follow the above procedure but instead to use Eq. (26–13). When appropriate, make use of your freedom to define V to be zero at any convenient place. For point charges this will usually be at infinity, but for other distributions of charge (especially those that extend to infinity themselves) it is often necessary to set $V = 0$ at some finite distance from the charge distribution, say at point b. Then the potential at any other point, say a, can be found from Eq. (26–13), with $V_b = 0$

3. Remember that potential is a *scalar* quantity, not a *vector* quantity. Don't get carried away and try to use components. A scalar quantity does not have components, and to try to use components would be wrong.

EXAMPLE 26–6 *Charged spherical conductor.* Consider a solid conducting sphere of radius R, with total charge q. From Gauss's law, as discussed in Chapter 25, we conclude that at all points *outside* the sphere, the field is the same as that of a point charge q at the center of the sphere. *Inside* the sphere the field is zero everywhere; otherwise charge would move within the sphere.

As mentioned above, it is convenient to take the reference level of potential (the point where $V = 0$) at a very large distance from all charges. Since the field at any point r greater than R is the same as for a point charge, the work done by the E field on a test charge moving from a finite value of r to infinity is also the same as for a point charge. Thus the potential V at a radial distance r, relative to a point at

Potential of a charged conducting sphere: Outside the sphere, the potential is the same as for a point charge.

infinity, is

$$V = \frac{1}{4\pi\epsilon_0}\frac{q}{r}. \qquad (26-14)$$

The potential is positive if q is positive, negative if q is negative.

Equation (26–14) applies to the field of a charged spherical conductor only when r is greater than or equal to the radius R of the sphere. The potential at the surface is

$$V = \frac{1}{4\pi\epsilon_0}\frac{q}{R}. \qquad (26-15)$$

Inside the sphere, the potential is constant and the field is zero.

Inside the sphere the field is zero everywhere, and no work is done on a test charge displaced from any point to any other point in this region. Thus the potential is the same at all points inside the sphere and is equal to its value $q/4\pi\epsilon_0 R$ at the surface. The field and potential are shown as functions of r in Fig. 26–7. The electric field E at the surface has magnitude

$$E = \frac{1}{4\pi\epsilon_0}\frac{q}{R^2}. \qquad (26-16)$$

Air becomes conductive when very strong electric fields are present.

Here is a simple practical application of this result. The maximum potential to which a conductor in air can be raised is limited by the fact that air molecules become ionized, and hence the air becomes a conductor, at an electric-field magnitude of about 0.8×10^6 N·C^{-1}. Comparing Eqs. (26–15) and (26–16), we note that at the surface of a conducting sphere the field and potential are related by $V = RE$. Thus if E_m represents electric-field magnitude at which air becomes conductive (known as the *dielectric strength* of air), the maximum potential V_m to which a spherical conductor can be raised is

$$V_m = RE_m.$$

For a sphere 1 cm in radius, in air,

$$V_m = (10^{-2}\text{ m})(0.8 \times 10^6\text{ N·C}^{-1}) = 8000\text{ V},$$

and no amount of "charging" could raise the potential of a sphere of this size, in air, higher than about 8000 V. This fact necessitates the use of large spherical

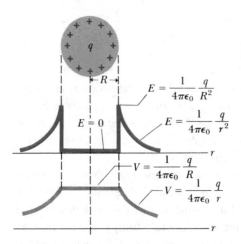

26–7 Electric-field magnitude E and potential V at points inside and outside a charged spherical conductor.

terminals on high-voltage machines. If we make $R = 2$ m, then

$$V_m = (2 \text{ m})(0.8 \times 10^6 \text{ N·C}^{-1}) = 1.6 \times 10^6 \text{ V} = 1.6 \text{ MV}.$$

At the other extreme is the effect produced by a surface of very *small* radius of curvature, such as a sharp point. Since the maximum potential is proportional to the radius, even relatively small potentials applied to sharp points in air will produce sufficiently high fields just outside the point to result in ionization of the surrounding air.

EXAMPLE 26–7 *Parallel plates.* We discussed the electric field between two parallel plates at the beginning of Section 26–1 and in Fig. 26–1. To obtain the potential difference between points y and b, also called the potential of y with respect to b, we note that the force on a test charge q' is $q'E$. The work done on the charge by the field during a displacement from y to b is

Finding the potential between two parallel plates.

$$W_{y \to b} = q'Ey. \tag{26–17}$$

Hence the potential difference, or work per unit charge, is

$$V_y - V_b = Ey. \tag{26–18}$$

The potential decreases linearly with y as we move from the upper to the lower plate. At point a, where $y = d$ and $V_y = V_a$,

$$V_a - V_b = Ed,$$

and

$$E = \frac{V_a - V_b}{d} = \frac{V_{ab}}{d}. \tag{26–19}$$

That is, *the electric field equals the potential difference between the plates divided by the distance between them.* This equation is a more useful expression for E than is Eq. (25–11), because the potential difference V_{ab} can be measured easily with a voltmeter. There are no instruments that read surface charge density directly.

Equation (26–19) also shows that the unit of electric field can be expressed as one *volt per meter* (1 V·m^{-1}), as well as 1 N·C^{-1}. In practice, the volt per meter is the most commonly used unit of E.

EXAMPLE 26–8 *Line charge and charged conducting cylinder.* We found in Example 25–13 (Section 25–5) that the field at a distance r from a long straight-line charge or a long charged conducting cylinder is given by

Finding the potential due to a line charge distribution.

$$E = \frac{1}{2\pi\epsilon_0} \frac{\lambda}{r},$$

where λ is the charge per unit length.

The potential of any point a with respect to any other point b, at radial distances r_a and r_b from the line of charge, is

$$V_a - V_b = \frac{\lambda}{2\pi\epsilon_0} \int_{r_a}^{r_b} \frac{dr}{r} = \frac{\lambda}{2\pi\epsilon_0} \ln \frac{r_b}{r_a}. \tag{26–20}$$

If we take point b at infinity and set $V_b = 0$, we find for the potential V_a,

$$V_a = \frac{\lambda}{2\pi\epsilon_0} \ln \frac{\infty}{r_a} = \infty.$$

The potential due to a charged cylinder: It is convenient to make the potential zero at the surface.

This does not make sense, and we are forced to conclude that we cannot use a reference point at infinity for this field! We can, however, set $V = 0$ at some arbitrary radius r_0. Then at any radius r,

$$V = \frac{\lambda}{2\pi\epsilon_0} \ln \frac{r_0}{r}. \tag{26-21}$$

Equations (26–20) and (26–21) give the potential in the field of a cylinder only for values of r equal to or greater than the radius R of the cylinder. If r_0 is taken as the cylinder radius R, so that the potential at the surface of the cylinder is zero, the potential at any external point, relative to that of the cylinder, is

$$V = \frac{\lambda}{2\pi\epsilon_0} \ln \frac{R}{r}, \tag{26-22}$$

where r is the distance from the axis of the cylinder.

EXAMPLE 26–9 Electric charge is distributed uniformly around a thin ring of radius a, with total charge Q, as shown in Fig. 26–8. Find the potential at a point along the line perpendicular to the plane of the ring, through its center, at a point P a distance x from the center of the ring.

Potential due to a ring of charge

SOLUTION This situation is the same as in Section 25–2 (Example 25–7). Referring back to that example, we note that the entire charge is at a distance $r = (x^2 + a^2)^{1/2}$ from point P. We conclude immediately that the potential at point P, which is a function of x, is given by

$$V(x) = \frac{1}{4\pi\epsilon_0} \frac{Q}{\sqrt{x^2 + a^2}}. \tag{26-23}$$

Potential is a *scalar* quantity; there is no need to consider components of vectors in this calculation, as we had to do in finding the electric field at P. Hence the potential calculation in this case is much simpler than the field calculation. Keep this comment in mind; we will return to it later in this chapter. Also note that when x is much larger than a, Eq. (26–23) becomes approximately equal to

$$V(x) = \frac{1}{4\pi\epsilon_0} \frac{Q}{x},$$

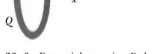

26–8 Potential at point P due to a ring of charge.

corresponding to the potential of a point charge Q at distance x; when we are far away from a ring, its charge looks like a point charge.

Potential calculations are often simpler than electric-field calculations for the same charge distributions.

EXAMPLE 26–10 Electric charge is distributed uniformly along a thin rod of length $2a$, with total charge Q. Find the potential at a point P along the perpendicular bisector of the rod, at a distance x from its center.

SOLUTION This situation is the same as in Example 24–5 (Section 24–6) and Example 25–8 (Section 25–2). In Fig 25–7 the element of charge dQ corresponding to an element dy on the rod is again given by $dQ = (Q/2a)dy$. The distance from dQ to P is $(x^2 + y^2)^{1/2}$, and the contribution it makes to the potential at P is

$$dV = \frac{1}{4\pi\epsilon_0} \frac{Q}{2a} \frac{dy}{\sqrt{x^2 + y^2}}.$$

The potential at P is obtained by integrating this over the length of the rod, from

$y = -a$ to $y = a$. The integral may be found in a table of integrals; the result is

$$V(x) = \frac{1}{4\pi\epsilon_0} \frac{Q}{2a} \int_{-a}^{a} \frac{dy}{\sqrt{x^2 + y^2}} = \frac{1}{4\pi\epsilon_0} \frac{Q}{2a} \ln \frac{\sqrt{a^2 + x^2} + a}{\sqrt{a^2 + x^2} - a}. \quad (26\text{–}24)$$

Again note that this problem is simpler than the calculation of E at point P because potential is a scalar quantity and no vector calculations are involved.

26–4 EQUIPOTENTIAL SURFACES

The potential at various points in an electric field may be represented graphically by **equipotential surfaces.** An equipotential surface is a surface for which the potential has the same value at all points on the surface. An equipotential surface may be constructed through every point of an electric field, but it is usual to show only a few equipotentials in a diagram.

Equipotential surface: a surface where the potential is constant

Since the potential energy of a charged body is the same at all points of a given equipotential surface, it follows that no work is done by the E field when a charged body moves over such a surface. Hence the equipotential surface through any point must be at right angles to the direction of the field at that point. If this were not so, the field would have a component tangent to the surface, and work would be done by the electric-field force when a charge moved in the direction of this component. Thus field lines and equipotential surfaces are mutually perpendicular. In general, the field lines of a field are curves, and the equipotentials are curved surfaces. For the special case of a uniform field, where the field lines are straight and parallel, the equipotentials are parallel planes perpendicular to the field lines.

Figure 26–9 shows several arrangements of charges. The field lines are represented by colored lines, and cross sections of the equipotential surfaces are shown as black lines. The actual field is, of course, three-dimensional. At each crossing of an equipotential and a field line, the two are perpendicular.

When charges at rest reside on the surface of a conductor, the electric field just outside the conductor must be perpendicular to the surface at every point. That is, at the surface there can never be a component of E *parallel* to the surface. If such a component did exist, it would do work on a charge moving close to and parallel to the surface. But then the charge could go through the surface and return to its starting point by a path close to the surface but inside the conductor. No work would be done in the return trip because E is zero everywhere inside the conductor. Thus when it returned to the starting point, the total work would not be zero; this statement contradicts the general principle that an electrostatic field is always conservative. We conclude that a parallel component of E at the surface is an impossibility.

The electric field just outside a conducting surface is always perpendicular to the surface.

It also follows that when a test charge is moved from point to point along the surface of a conductor, the electric field does no work on the charge. Hence when all charges are at rest, *a conducting surface is always an equipotential surface.* These principles are shown in Fig. 26–10.

When all charges are at rest, a conducting surface is always an equipotential surface.

Finally, we can now prove a theorem that we quoted without proof in Section 25–6. The theorem is as follows: In an electrostatic situation, if a conductor contains a cavity, and if no charge is present inside the cavity, then there can be no charge *anywhere* on the surface of the cavity. To prove this theorem, we first prove that *every point in the cavity is at the same potential.* In Fig.

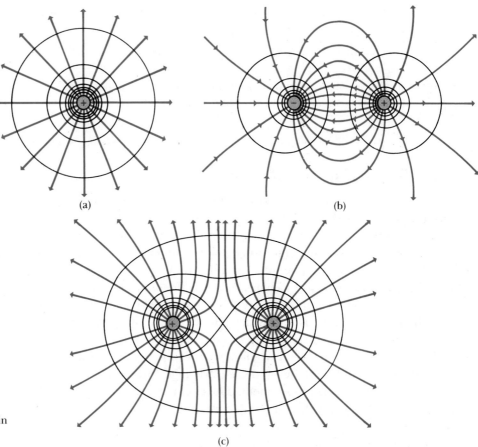

(a)   (b)

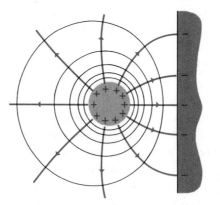

(c)

26–9 Equipotential surfaces (black lines) and field lines (colored lines) in the neighborhood of point charges.

26–10 Field lines always meet charged conducting surfaces at right angles.

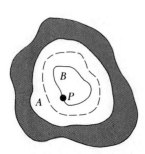

26–11 A cavity in a conductor. If the cavity contains no charge, every point in the cavity is at the same potential, the electric field is zero everywhere, and there is no charge on the surface of the cavity.

26–11, the surface A of the cavity is an equipotential surface, as we have just proved. Suppose point P in the cavity is at a different potential; then we can construct a different equipotential surface B including point P.

Now consider a Gaussian surface, shown as a broken line, between the two equipotential surfaces. Because of the relation between E and the equipotentials, we know that the field at every point between the equipotentials is from A toward B, or else at every point it is from B toward A, depending on which equipotential surface is at higher potential. In either case, the flux through this Gaussian surface is certainly not zero. But then Gauss's law says that the charge enclosed by the Gaussian surface cannot be zero. This conclusion contradicts our initial assumption that there is *no* charge in the cavity. So *P cannot* be at a different potential from the cavity wall.

We have proved that the entire region of the cavity is at the same potential. For this to be true, however, the electric field inside the cavity must be zero everywhere. Finally, because the electric field at any point on the surface of a conductor is proportional to the surface charge density σ at that point, according to Eq. (25–26), we conclude that *the surface charge density on the wall of the cavity is zero at every point.* This chain of reasoning may seem tortuous, but it is worth careful study.

26–5 POTENTIAL GRADIENT

Electric field and potential are closely related. Equation (26–13) expresses one aspect of that relationship; if we know E at various points, we can use Eq. (26–13) to calculate potential differences. Conversely, if we know the potential V at various points, we ought to be able to use it to determine E. This converse problem is the subject of this section. Regarding V as a function of the coordinates (x, y, z) of a point in space, we will show that the components of E are directly related to the *derivatives* of V with respect to x, y, and z.

In Eq. (26–13), $V_{ab} = V_a - V_b$ is the potential of a with respect to b, that is, the change of potential when a point moves from b to a. This may be written as

$$V_{ab} = \int_b^a dV = -\int_a^b dV,$$

where dV is the infinitesimal change of potential accompanying a small displacement dl. As usual, $E \cos \theta$ is the component of E parallel to dl. We may call this $E_\parallel$, then, using the above expression for V_{ab}, we may rewrite Eq. (26–13) as

$$-\int_a^b dV = \int_a^b E_\parallel \, dl.$$

Since these two integrals are equal for any pair of limits a and b, the *integrands* must be equal. Hence for any *infinitesimal* displacement dl,

$$-dV = E_\parallel \, dl,$$

or

$$E_\parallel = -\frac{dV}{dl}. \tag{26–25}$$

The derivative dV/dl is the rate of change of V for a displacement in the direction of dl. In particular, if dl is parallel to the x-axis, then the component of E parallel to dl is just the x-component of E; that is, E_x. Thus $E_x = -dV/dx$. Because V is, in general, also a function of y and z, we use partial derivative notation; a derivative in which only x varies is written $\partial V/\partial x$. The y- and z-components of E are related to the corresponding derivatives of V in the same way, so we have

$$E_x = -\frac{\partial V}{\partial x}, \qquad E_y = -\frac{\partial V}{\partial y}, \qquad E_z = -\frac{\partial V}{\partial z}. \tag{26–26}$$

Potential gradient: how to find the field when the potential is known

Derivatives of V in various directions

In terms of unit vectors, E may be written as

$$E = -\left(i\frac{\partial V}{\partial x} + j\frac{\partial V}{\partial y} + k\frac{\partial V}{\partial z}\right)$$

$$= -\left(i\frac{\partial}{\partial x} + j\frac{\partial}{\partial y} + k\frac{\partial}{\partial z}\right)V. \qquad (26\text{–}27)$$

In vector notation the operation

$$i\frac{\partial}{\partial x} + j\frac{\partial}{\partial y} + k\frac{\partial}{\partial z}$$

The gradient: a vector differential operation

is called the **gradient** and is denoted by the symbol ∇, pronounced "del." Thus in vector notation, Eqs. (26–26) are summarized compactly as

$$E = -\nabla V. \qquad (26\text{–}28)$$

This is read "E is the negative of the gradient of V," or "E equals negative del V."

Note again that the unit of electric field can be expressed in either of these equivalent forms:

$$1 \text{ V·m}^{-1} = 1 \text{ N·C}^{-1}.$$

EXAMPLE 26–11 We have shown that the potential at a radial distance r from a point charge q is

$$V = \frac{1}{4\pi\epsilon_0}\frac{q}{r}.$$

By symmetry, the electric field is in the radial direction, so

$$E = E_r = -\frac{dV}{dr} = -\frac{d}{dr}\left(\frac{1}{4\pi\epsilon_0}\frac{q}{r}\right) = \frac{1}{4\pi\epsilon_0}\frac{q}{r^2},$$

in agreement with Coulomb's law.

Some examples of the relation of potential to field

EXAMPLE 26–12 In Example 26–8 we found that the potential outside a charged conducting cylinder of radius R and charge per unit length λ is

$$V = \frac{\lambda}{2\pi\epsilon_0}\ln\frac{R}{r} = \frac{\lambda}{2\pi\epsilon_0}(\ln R - \ln r).$$

The electric field is radial, and its magnitude is given by

$$E = -\frac{dV}{dr} = \frac{\lambda}{2\pi\epsilon_0 r},$$

in agreement with our previous result.

EXAMPLE 26–13 In Example 26–9 we found that for a ring of charge with radius a and total charge Q, the potential at a point P, a distance x from the center of the ring on a line through the center perpendicular to the plane of the ring, is

$$V(x) = \frac{1}{4\pi\epsilon_0}\frac{Q}{\sqrt{x^2 + a^2}}.$$

From Eq. (26–26),

$$E_x = -\frac{\partial V}{\partial x} = \frac{1}{4\pi\epsilon_0}\frac{Qx}{(x^2 + a^2)^{3/2}}.$$

This expression agrees with the result we obtained in Section 25–2 (Example 25–7).

In this example, V is not a function of y, but it would *not* be correct to conclude that $\partial V/\partial y = 0$ and $E_y = 0$. The reason is that our expression for V is valid only for points on the x-axis. If we had the complete form of V, valid for all points in space, then we could use it to find $E_y = -\partial V/\partial y$ at any point, and so on.

EXAMPLE 26–14 A charge Q is uniformly distributed along a rod of length $2a$. In Example 26–10 (Section 26–3) we derived an expression for the potential at a point on the perpendicular bisector of the rod, at a distance x from its center:

$$V(x) = \frac{1}{4\pi\epsilon_0}\frac{Q}{2a}\ln\frac{\sqrt{a^2 + x^2} + a}{\sqrt{a^2 + x^2} - a}.$$

The negative of the derivative of this expression with respect to x is E_x. We challenge you to carry out the differentiation and show that the result is the same expression we found in Examples 24–5 (Section 24–6) and 25–8 (Section 25–2) by direct integration.

Several of these examples illustrate an important point. Often we can compute the electric field caused by a charge distribution in two ways; either directly, as in Section 25–2, or by first calculating the potential and then taking its gradient to find the field. The second method is often easier because potential is a *scalar* quantity, requiring at worst the integration of a scalar function, while the electric field is a *vector* quantity, requiring computation of components for each element of charge and a separate integration for each component. Thus, quite apart from its fundamental significance, potential offers a very useful computational technique in field calculations.

> Potential is a scalar, but electric field is a vector; that's why it's often easier to calculate potential.

Another important point is the relation of the *direction* of $\boldsymbol{E}$ to the behavior of V. Specifically, $\boldsymbol{E}$ always points in the direction in which V *decreases* most rapidly. In all these examples, the potential decreases as we move away from the charge distribution (assuming the charge is positive), and the electric field points away from it. If the charge is negative, the potential increases algebraically (becoming less negative) as we move away from the charge, and $\boldsymbol{E}$ points *toward* the charge. The situation is completely analogous to gravitational potential energy, which we studied in Section 7–5. Near the surface of the earth, for example, the gravitational field points down, the direction of most rapid decrease of gravitational potential energy.

26–6 THE MILLIKAN OIL-DROP EXPERIMENT

We can now describe one of the classic physics experiments of all time, the **Millikan oil-drop experiment.** In a brilliant series of investigations carried out at the University of Chicago in the period 1909–1913, Robert Andrews Millikan not only demonstrated conclusively the discrete nature of electric charge but actually measured the charge of an individual electron.

> How can you measure the charge of a single electron?

Millikan's apparatus is shown schematically in Fig. 26–12a. Two parallel horizontal metal plates, A and B, are insulated from each other and separated by a few millimeters. Oil is sprayed in fine droplets from an atomizer above the upper plate, and a few droplets are allowed to fall through a small hole in this plate. A beam of light is directed horizontally between the plates, and a telescope is set up with its axis at right angles to the light beam. The oil drops, illuminated by the light beam and viewed through the telescope, appear like tiny bright stars, falling slowly with a terminal velocity determined by their weight and by the viscous air-resistance force opposing their motion.

Studying the motion of electrically charged oil drops

Some of the oil droplets are electrically charged because of frictional effects. The drops can also be charged if the air in the apparatus is ionized by x-rays or a bit of radioactive material. Some of the electrons or ions then collide with the drops and stick to them. The drops are usually negatively charged, but occasionally one with a positive charge is found.

Free fall of oil drops in air: The viscous force just balances the weight.

In principle, the simplest method for measuring the charge on a drop is as follows. Suppose a drop has a negative charge and the plates are maintained at a potential difference such that a downward electric field of magnitude E $(= V_{AB}/l)$ is set up between them. The forces on the drop are then its weight mg and the upward force qE. By adjusting the field E, we can make qE equal to mg, so that the drop remains at rest, as indicated in Fig. 26–10b. Under these circumstances,

$$q = \frac{mg}{E}.$$

The mass of the drop equals the product of its density ρ and its volume, $4\pi r^3/3$, and $E = V_{AB}/l$, so

$$q = \frac{4\pi}{3}\frac{\rho r^3 g l}{V_{AB}}. \tag{26–29}$$

All the quantities on the right can be measured easily, with the exception of the drop radius r, which is of the order of 10^{-5} cm and is much too small to be measured directly. It can be calculated, however, by cutting off the electric field and measuring the terminal velocity v_{T} of the drop as it falls through a known distance d defined by reference lines in the eyepiece of the telescope.

At the terminal velocity, the weight mg is just balanced by the viscous force f. The viscous force on a sphere of radius r, moving with a velocity v through a fluid of viscosity η, is given by Stokes's law, discussed in Section 13–8, Eq. (13–29):

$$f = 6\pi\eta rv.$$

If Stokes's law applies, and the drop is falling with its terminal velocity v_{T}, then

$$mg = f,$$

$$\tfrac{4}{3}\pi r^3\rho g = 6\pi\eta rv_{\mathrm{T}},$$

and

$$r = 3\sqrt{\eta v_{\mathrm{T}}/2\rho g}.$$

When this expression for r is inserted into Eq. (26–29), we find

$$q = 18\pi\frac{l}{V_{AB}}\sqrt{\frac{\eta^3 v_{\mathrm{T}}{}^3}{2\rho g}}, \tag{26–30}$$

which expresses the charge q in terms of measurable quantities.

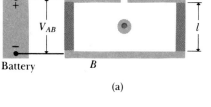

(a)

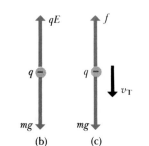

(b)   (c)

26–12 (a) Schematic diagram of Millikan apparatus. (b) Forces on a drop at rest. (c) Forces on a drop falling with its terminal velocity v_{T}.

In actual practice this procedure is modified somewhat. To correct for the buoyant force of the air through which the drop falls, the density ρ of the oil should be replaced by $(\rho - \rho_g)$, where ρ_g is the density of air. A correction to Stokes's law is also required, because air is not a continuous fluid but a collection of molecules separated by distances that are of the same order of magnitude as the dimensions of the drops.

Millikan and his coworkers measured the charges of thousands of drops and found that, within the limits of their experimental error, every drop had a charge equal to some small integer multiple of a basic charge e. That is, drops were observed with charges of e, $2e$, $3e$, and so on, but never with such values as $0.76e$ or $2.49e$. The evidence is conclusive that electric charge is not something that can be divided indefinitely, but that it exists in nature only in units of magnitude e. When a drop is observed with charge e, we conclude it has acquired one extra electron; if its charge is $2e$, it has two extra electrons, and so on. As we stated in Section 24–5, the best experimental value of the charge e is

$$e = 1.602192 \times 10^{-19} \text{ C} \cong 1.60 \times 10^{-19} \text{ C}.$$

The quark theory of fundamental particle structure, discussed in Section 44–11, incorporates particles called *quarks* having fractional charges $\pm e/3$ and $\pm 2e/3$. Physicists have conflicting views as to whether properties (such as charge) of an individual quark should be directly observable.

Correcting for buoyancy effects of the drop in air

The charge of the electron: a fundamental constant of nature

26–7 THE ELECTRONVOLT

The change in electrical potential energy of a particle having a charge q, when it moves from a point where the potential is V_a to a point where the potential is V_b, is

$$\Delta U = q(V_b - V_a) = qV_{ba}$$

In particular, if the charge q equals the electron charge $e = 1.60 \times 10^{-19}$ C, and the potential difference $V_{ba} = 1$ V, the change in energy is

$$\Delta U = (1.60 \times 10^{-19} \text{ C})(1 \text{ V}) = 1.60 \times 10^{-19} \text{ J}.$$

This quantity of energy is called one **electronvolt** (1 eV):

$$1 \text{ eV} = 1.60 \times 10^{-19} \text{ J}.$$

Other commonly used units are

$$1 \text{ keV} = 10^3 \text{ eV},$$
$$1 \text{ MeV} = 10^6 \text{ eV},$$
$$1 \text{ GeV} = 10^9 \text{ eV},$$
$$1 \text{ meV} = 10^{-3} \text{ eV}.$$

The electronvolt: an electron moving through a potential difference of one volt

The electronvolt is a convenient energy unit when we are dealing with the motions of electrons and ions in electric fields. For a particle having charge e, the change in potential *energy* between two points of the path of the particle, when expressed in electronvolts, is *numerically* equal to the potential difference between the points, in volts. If the charge is some multiple of e, say Ne, the change in potential energy in electronvolts is, numerically, N times the potential difference in volts. For example, if a particle having a charge $2e$ moves

The electronvolt is a convenient unit of energy in fundamental-particle calculations.

between two points for which the potential difference is 1000 V, the change in its potential energy is

$$\Delta U = qV_{ba} = (2)(1.60 \times 10^{-19} \text{ C})(10^3 \text{ V})$$
$$= 3.2 \times 10^{-16} \text{ J}$$
$$= 2000 \text{ eV}.$$

Although the electronvolt was defined above in terms of *potential* energy, energy of *any* form, such as the kinetic energy of a moving particle, can be expressed in terms of the electronvolt. Thus when we speak of a "one-million-volt electron," we mean an electron having a kinetic energy of one million electronvolts (1 MeV).

One of the principles of the special theory of relativity, to be developed in Chapter 40, is that the mass m of a particle is equivalent to a quantity of energy mc^2, where c is the speed of light. The rest mass of an electron is 9.11×10^{-31} kg, and the energy equivalent to this is

$$E_0 = mc^2 = (9.11 \times 10^{-31} \text{ kg})(3.00 \times 10^8 \text{ m·s}^{-1})^2$$
$$= 81.9 \times 10^{-15} \text{ J}.$$

Since $1 \text{ eV} = 1.60 \times 10^{-19}$ J, this is equivalent to

Rest mass in relativity theory, expressed in electronvolts

$$E_0 = 511{,}000 \text{ eV} = 0.511 \text{ MeV}.$$

When the kinetic energy of a particle becomes comparable in magnitude to its rest energy, Newton's laws of motion are not strictly valid and must be replaced by the more general relations of relativistic mechanics, developed in detail in Chapter 40. For example, an electron accelerated through a potential difference of 500 kV acquires a kinetic energy of 500 keV, approximately *equal* to its rest energy. A correct analysis of the motion of the particle requires the use of relativistic mechanics.

26–8 THE CATHODE-RAY TUBE

The cathode-ray tube: the heart and soul of oscilloscopes and television receivers

Figure 26–13 is a schematic diagram of the elements of a **cathode-ray tube.** Such tubes are found in oscilloscopes and computer terminal displays, and the principle of the TV picture tube is similar. A cathode-ray tube uses an electron beam that, before its basic nature was well understood, was called a *cathode-ray* beam. Cathode rays are now known to be electrons.

26–13 Basic elements of a cathode-ray tube.

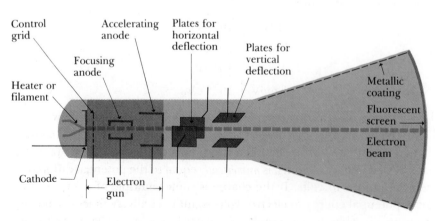

The interior of the tube is highly evacuated. Ordinarily the pressure is of the order of 0.01 Pa (10^{-7} atm); at any greater pressure, collisions of electrons with air molecules would scatter the electron beam excessively. The *cathode* at the left end is raised to a high temperature by the *heater*, and electrons evaporate from the surface of the cathode. The *accelerating anode*, which has a small hole at its center, is maintained at a high positive potential V_1 relative to the cathode, so that there is an electric field, directed from right to left, between the accelerating anode and cathode. This field is confined to the cathode–anode region, and electrons passing through the hole in the anode travel with constant horizontal velocity from the anode to the *fluorescent screen*. The area of impact of the electrons on the screen glows brightly.

The function of the *control grid* is to regulate the number of electrons that reach the anode (and hence the brightness of the spot on the screen). The *focusing anode* ensures that electrons leaving the cathode in slightly different directions are focused down to a narrow beam and all arrive at the same spot on the screen. These two electrodes need not be considered in the following analysis. The complete assembly of cathode, control grid, focusing anode, and accelerating electrode is referred to as an *electron gun*.

Creating an electron beam in a vacuum tube

The accelerated electrons pass between two pairs of *deflecting plates*. An electric field between the first pair of plates deflects the electrons horizontally, and an electric field between the second pair of plates deflects them vertically. If no such fields are present, the electrons travel in a straight line from the hole in the accelerating anode to the *fluorescent screen* and produce a bright spot on the screen where they strike it.

Deflecting the electron beam

To analyze the electron motion, we first calculate the speed v given to the electrons by the electron gun, just as in Example 26–5 (Section 26–2). We assume the electrons leave the cathode with zero initial speed. The electrons actually have some initial speed when they evaporate from the cathode, but it is very small compared to the final speed v and can be neglected. We find

$$v = \sqrt{\frac{2eV_1}{m}}. \qquad (26\text{–}31)$$

As a numerical example, if $V_1 = 2000$ V,

$$v = \sqrt{\frac{2(1.6 \times 10^{-19}\text{ C})(2 \times 10^3\text{ V})}{9.11 \times 10^{-31}\text{ kg}}}$$
$$= 2.65 \times 10^7 \text{ m·s}^{-1}.$$

Note that the kinetic energy of an electron at the anode depends only on the *potential difference* between anode and cathode, and not at all on the details of the fields within the electron gun or on the shape of the electron trajectory within the gun.

The speeds of electrons in the beam are determined by the accelerating voltage.

If there is no electric field between the plates for horizontal deflection, the electrons enter the region between the other plates with a speed equal to v and represented by v_x in Fig. 26–14. If there is a potential difference V_2 between the plates, and the upper plate is positive, a downward electric field of magnitude $E = V_2/l$ is set up between the plates. A constant upward force eE then acts on the electrons, and their upward acceleration is

$$a_y = \frac{eE}{m} = \frac{eV_2}{ml}. \qquad (26\text{–}32)$$

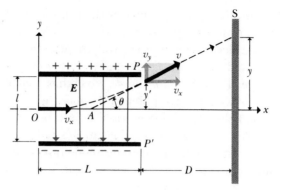

26–14 Electrostatic deflection of an electron beam in a cathode-ray tube.

The *horizontal* component of velocity v_x remains constant, so the time required for the electrons to travel the length L of the plates is

$$t = \frac{L}{v_x}. \tag{26-33}$$

In this time, they acquire an upward velocity component given by

$$v_y = a_y t. \tag{26-34}$$

Calculating the trajectories of deflected electrons

Combining this equation with Eqs. (26–32) and (26–33), we find

$$v_y = \left(\frac{eV_2}{ml}\right)\frac{L}{v_x}. \tag{26-35}$$

When the electrons emerge from the deflecting field, their velocity v makes an angle θ with the *x*-axis, where

$$\tan\theta = \frac{v_y}{v_x}. \tag{26-36}$$

From this point on, the electrons travel in a straight line to the screen. It is not difficult to show that this straight line, if projected backward, intersects the *x*-axis at a point A that is midway between the end of the plates. Then if y is the vertical component of the point of impact with screen S,

$$\tan\theta = \frac{y}{D + (L/2)}. \tag{26-37}$$

Combining this with Eq. (26–36), we find

$$\frac{y}{D + (L/2)} = \frac{v_y}{v_x},$$

or, using Eq. (26–35),

$$y = \left(D + \frac{L}{2}\right)\frac{eV_2 L}{mlv_x{}^2}. \tag{26-38}$$

Finally, v_x is given by Eq. (26–31); using this in Eq. (26–38), we obtain

$$y = \left[\frac{L}{2l}\left(D + \frac{L}{2}\right)\right]\frac{V_2}{V_1}. \tag{26-39}$$

The term in brackets is a purely geometrical factor. If the accelerating voltage V_1 is constant, *the deflection y is proportional to the deflecting voltage* V_2.

If a field is set up between the *horizontal* deflecting plates, the beam is also deflected in the horizontal direction, perpendicular to the plane of Fig. 26–14. The coordinates of the luminous spot on the screen are then proportional, respectively, to the horizontal and vertical deflecting voltages. This is the principle of the *cathode-ray oscilloscope*. If the horizontal deflection voltage sweeps the beam from left to right at a uniform rate, the beam traces out a graph of the vertical voltage as a function of time. Oscilloscopes are extremely useful laboratory instruments in many areas of pure and applied science.

The picture tube in a television set is similar, but the beam is deflected by magnetic fields (to be discussed in later chapters) rather than electric fields. The electron beam traces out the area of the picture 30 times per second in an array of 525 horizontal lines, as the intensity of the beam is varied to make bright and dark areas on the screen. The accelerating voltage in TV picture tubes (V_1 in the preceding discussion) is typically about 20 kV. Computer terminal displays operate on the same principle, using a magnetically deflected electron beam to trace out images on a fluorescent screen. In this context the device is called a CRT (cathode-ray tube) display or a VDT (video display terminal).

SUMMARY

The electric field caused by any collection of charges at rest is always a conservative force field. Thus the work W done by the electric-field force on a charged particle moving in a field can always be represented by a potential-energy function U:

$$W_{a \to b} = U_a - U_b. \qquad (26\text{–}2)$$

The potential energy of a charge q' in the electric field of a collection of charges q_i is given by

$$U = \frac{q'}{4\pi\epsilon_0} \sum \frac{q_i}{r_i}, \qquad (26\text{–}9)$$

where r_i is the distance from q_i to q'. This expression assumes that $U = 0$ at a point an infinite distance from all the charges.

Potential is potential energy per unit charge; it is denoted by V. The potential at any point due to a collection of charges q_i is given by

$$V = \frac{U}{q'} = \frac{1}{4\pi\epsilon_0} \sum \frac{q_i}{r_i}. \qquad (26\text{–}12)$$

The potential difference between two points a and b, also called the potential of a with respect to b, is given by the line integral of $\boldsymbol{E}$:

$$V_{ab} = V_a - V_b = \int_a^b \boldsymbol{E} \cdot d\boldsymbol{l} = \int_a^b E \cos\theta \, dl. \qquad (26\text{–}13)$$

Potentials can be calculated either by evaluating the sum in Eq. (26–12) or by first finding $\boldsymbol{E}$ and then using Eq. (26–13).

KEY TERMS

potential energy
conservative force field
potential
volt
line integral
equipotential surfaces
gradient
Millikan oil-drop experiment
electronvolt
cathode-ray tube

An equipotential surface is a surface such that the potential has the same value at every point on the surface. When a field line crosses an equipotential surface, the two are perpendicular at the point of intersection; that is, the tangent to the field line is perpendicular to the tangent plane to the equipotential surface at the point of intersection.

When all charges are at rest, the surface of a conductor is always an equipotential surface, and all points in the interior of a conductor are at the same potential. When a cavity within a conductor contains no charge, the entire cavity is an equipotential region, and there is no surface charge anywhere on the surface of the cavity.

If the potential is known as a function of the spatial coordinates x, y, and z, the electric field E at any point is given by

$$E = -\left(i\frac{\partial V}{\partial x} + j\frac{\partial V}{\partial y} + k\frac{\partial V}{\partial z}\right) = -\left(i\frac{\partial}{\partial x} + j\frac{\partial}{\partial y} + k\frac{\partial}{\partial z}\right)V, \quad (26\text{--}27)$$

which is the negative of the gradient of V:

$$E = -\nabla V. \quad (26\text{--}28)$$

Two equivalent sets of units for electric-field magnitude are the volt per meter ($V \cdot m^{-1}$) and the newton per coulomb ($N \cdot C^{-1}$). One volt is one joule per coulomb ($1\ V = 1\ J \cdot C^{-1}$).

The Millikan oil-drop experiment measured the electric charge of individual electrons by measuring the motion of electrically charged oil drops in an electric field. The size of the drop is determined by measuring its speed of free fall under gravity and the viscous force of air resistance given by Stokes's law.

The electronvolt, abbreviated eV, is the energy corresponding to a particle with a charge equal to that of the electron moving through a potential difference of one volt. The conversion factor is

$$1\ eV = 1.60 \times 10^{-19}\ J.$$

The cathode-ray tube uses an electron beam created by a set of electrodes, collectively called an electron gun, and deflected by two sets of deflecting plates. The beam strikes a fluorescent screen and forms an image on it. Such tubes are used in television sets, oscilloscopes, and video display terminals (VDT) for computers.

QUESTIONS

26–1 Are there cases in electrostatics where a conducting surface is *not* an equipotential surface? If so, give an example.

26–2 If the electrical potential at a single point is known, can the electric field at that point be determined?

26–3 If two points are at the same potential, is the electric field necessarily zero everywhere between them?

26–4 If the electric field is zero throughout a certain region of space, is the potential also zero in this region? If not, what *can* be said about the potential?

26–5 A student said: "Since electrical potential is always proportional to potential energy, why bother with the concept of potential at all?" How would you respond?

26–6 Is potential gradient a scalar quantity or a vector quantity?

26–7 A conducting sphere is to be charged by bringing in positive charge, a little at a time, until the total charge is Q. The total work required for this process is alleged to be proportional to Q^2. Is this correct? Why or why not?

26–8 The potential (relative to a point at infinity) midway between two charges of equal magnitude and opposite sign is zero. Can you think of a way to bring a test charge from infinity to this midpoint in such a way that no work is done in any part of the displacement?

26–9 A high-voltage dc power line falls on a car, so the entire metal body of the car is at a potential of 10,000 V with respect to the ground. What happens to the occupants

a) when they are sitting in the car?

b) when they step out of the car?

26–10 In electronics it is customary to define the potential of ground (thinking of the earth as a large conductor) as zero. Is this consistent with the fact that the earth has a net electric charge that is not zero? (Cf. Problem 25–18.)

26–11 A positive point charge is placed near a very large conducting plane. A professor of physics asserted that the field caused by this configuration is the same as would be obtained by removing the plane and placing a negative point charge of equal magnitude in the mirror-image position behind the initial position of the plane. Is this correct?

26–12 It is easy to produce a potential of several thousand volts on your body by scuffing your shoes across a nylon carpet; yet contact with a power line of comparable voltage would probably be fatal. What is the difference?

EXERCISES

Section 26–1 Electrical Potential Energy.

26–1 A point charge with charge $Q = +4.0 \ \mu C$ is held fixed at the origin.

a) A second point charge $q = -0.5 \ \mu C$ and mass 3×10^{-4} kg is placed on the x-axis, 0.8 m from the origin. What is the electrical potential energy of the pair of charges?

b) The second point charge is released at rest. What is its speed when it is 0.4 m from the origin?

26–2 Three equal point charges of 3×10^{-7} C are placed at the corners of an equilateral triangle whose side is 1 m. What is the potential energy of the system? Take as zero potential the energy of the three charges when they are infinitely far apart.

Section 26–2 Potential

26–3 There is a uniform electric field of magnitude E and directed in the positive x-direction. Consider point a at $x = 0.8$ m and point b at $x = 1.2$ m. The potential difference between these two points is 600 V.

a) Which point, a or b, is at the higher potential?

b) Calculate the magnitude E of the electric field.

c) A negative point charge of magnitude $q = -0.2 \ \mu C$ is moved from b to a. Calculate the work done on the point charge by the electric field.

26–4 The potential at a certain distance from a point charge is 600 V, and the electric field is 200 $N \cdot C^{-1}$.

a) What is the distance to the point charge?

b) What is the magnitude of the charge?

26–5 Two point charges $q_1 = +40 \times 10^{-9}$ C and $q_2 = -30 \times 10^{-9}$ C are 10 cm apart. Point A is midway between them; point B is 8 cm from q_1 and 6 cm from q_2. Find

a) the potential at point A;

b) the potential at point B;

c) the work that must be done to carry a charge of 25×10^{-9} C from point B to point A.

26–6 Two point charges whose magnitudes are $+2.0 \times 10^{-10}$ C and -1.2×10^{-10} C are separated by a distance of 5 cm. An electron is released from rest between the two charges, 1 cm from the negative charge, and moves along the line connecting the two charges. What is its velocity when it is 1 cm from the positive charge?

26–7 Two positive point charges, each of magnitude q, are fixed on the y-axis at the points $y = +a$ and $y = -a$.

a) Draw a diagram showing the positions of the charges.

b) What is the potential V_0 at the origin?

c) Show that the potential at any point on the x-axis is

$$V = \frac{1}{4\pi\epsilon_0} \frac{2q}{\sqrt{a^2 + x^2}}.$$

d) Sketch a graph of the potential on the x-axis as a function of x over the range from $x = +4a$ to $x = -4a$.

e) At what value of x is the potential one-half that at the origin?

26–8 A positive charge $+q$ is located at the point $(x = -a, y = -a)$, and an equal negative charge $-q$ is located at the point $(x = +a, y = -a)$.

a) Draw a diagram showing the positions of the charges.

b) What is the potential at the origin?

c) What is the expression for the potential at a point on the x-axis, as a function of x?

d) Sketch a graph of the potential as a function of x, in the range from $x = +4a$ to $x = -4a$. Plot positive potentials upward, negative potentials downward.

Section 26–3 Calculation of Potentials

26–9 Two large parallel metal plates carry opposite charges. They are separated by 0.1 m, and the potential difference between them is 500 V.

a) What is the magnitude of the electric field, if it is uniform, in the region between the plates?

b) Compute the work done by this field on a charge of 2.0×10^{-9} C as it moves from the higher-potential plate to the lower.

c) Compare the result of (b) to the change of potential energy of the same charge, computed from the electrical potential.

26–10 Two large parallel metal sheets carrying equal and opposite electric charges are separated by a distance of 0.05 m. The electric field between them is approximately uniform and has magnitude 600 $N \cdot C^{-1}$.

a) What is the potential difference between the plates?

b) Which plate is at higher potential?

26–11 A potential difference of 2000 V is established between parallel plates in air. If the air becomes electrically conducting when the electric field exceeds $0.8 \times 10^6 \, \text{N} \cdot \text{C}^{-1}$, what is the minimum separation of the plates?

26–12 Some cell walls in the human body have a double layer of surface charge, with a layer of negative charge inside and a layer of positive charge of equal magnitude on the outside. If the surface charge densities are $\pm 0.5 \times 10^{-3} \, \text{C} \cdot \text{m}^{-2}$ and the cell wall is 5×10^{-9} m thick, find

a) the electric-field magnitude in the wall, between the two charge layers;

b) the potential difference between inside and outside the cell. Which is at higher potential?

26–13 A total electric charge of 4×10^{-9} C is distributed uniformly over the surface of a sphere of radius 0.20 m. If the potential is zero at a point at infinity, what is the value of the potential

a) at a point on the surface of the sphere?

b) at a point inside the sphere, 0.1 m from the center?

26–14 A particle of charge $+3 \times 10^{-9}$ C is in a uniform electric field directed to the left. It is released from rest and moves a distance of 5 cm, after which its kinetic energy is found to be $+4.5 \times 10^{-5}$ J.

a) What work was done by the electrical force?

b) What is the magnitude of the electric field?

c) What is the potential of the starting point with respect to the endpoint?

26–15 A charge of 2.5×10^{-8} C is placed in an upwardly directed uniform electric field having magnitude $5 \times 10^4 \, \text{N} \cdot \text{C}^{-1}$. What work is done by the electrical force when the charge is moved

a) 0.45 m to the right?

b) 0.80 m downward?

c) 2.60 m at an angle of 45° upward from the horizontal?

26–16 A simple type of vacuum tube known as a *diode* consists essentially of two electrodes within a highly evacuated enclosure. One electrode, the *cathode,* is maintained at a high temperature and emits electrons from its surface. A potential difference of a few hundred volts is maintained between the cathode and the other electrode, known as the *anode,* with the anode at the higher potential. Suppose that a diode consists of a cylindrical cathode of radius 0.05 cm, mounted coaxially within a cylindrical anode 0.45 cm in radius. The potential of the anode is 300 V higher than that of the cathode. An electron leaves the surface of the cathode with zero initial speed. Find its speed when it strikes the anode.

26–17 Refer to Example 25–17.

a) From the expression for E obtained in Example 25–17, find the expression for the potential V as a function of r, both inside and outside the sphere, relative to a point at infinity.

b) Sketch graphs of V and E as functions of r from $r = 0$ to $r = 3R$, and compare with Fig. 26–7.

Section 26–5 Potential Gradient

26–18 A metal sphere of radius r_a is supported on an insulating stand at the center of a hollow metal sphere of inner radius r_b. There is a charge $+q$ on the inner sphere and a charge $-q$ on the outer.

a) Calculate the potential $V(r)$ for

i) $r < r_a$, ii) $r_a < r < r_b$, iii) $r > r_b$.

Use Eq. (26–14) and the fact that the net potential is the sum of the potentials due to the individual spheres.

b) Show that the potential of the inner sphere with respect to the outer is

$$V_{ab} = \frac{q}{4\pi\epsilon_0} \left(\frac{1}{r_a} - \frac{1}{r_b} \right).$$

c) Use Eq. (26–26) and the results from (a) to show that the electric field at any point between the spheres has the magnitude

$$E = \frac{V_{ab}}{(1/r_a - 1/r_b)} \cdot \frac{1}{r^2}.$$

d) Find the electric field at a point outside the larger sphere, at a distance r from the center, where $r > r_b$.

e) Suppose the charge on the outer sphere is not $-q$ but a negative charge of different magnitude, say $-Q$. Show that the answers for (b) and (c) are the same as before, but the answer for (d) is different.

26–19 Evaluate $-dV/dx$ for $V(x)$ given in Example 26–14. Show that this value gives an E_x that agrees with that calculated in Example 25–8.

Section 26–6 The Millikan Oil-Drop Experiment

26–20 In an apparatus for measuring the electronic charge e by Millikan's method, an electric field of $6.34 \times 10^4 \, \text{V} \cdot \text{m}^{-1}$ is required to maintain a certain charged oil drop at rest. If the plates are 1.5 cm apart, what potential difference between them is required?

26–21 An oil droplet of mass 3×10^{-14} kg and of radius 2×10^{-6} m carries ten excess electrons. What is its terminal velocity

a) when falling in a region in which there is no electric field?

b) when falling in an electric field of magnitude $3 \times 10^5 \, \text{N} \cdot \text{C}^{-1}$ directed downward?

(The viscosity of air is $180 \times 10^{-7} \, \text{N} \cdot \text{s} \cdot \text{m}^{-2}$. Neglect the buoyant force of the air.)

Section 26–7 The Electronvolt

26–22 Use the relation $E_0 = mc^2$ to find the energy equivalent of the rest mass of the proton; express your result in MeV.

26–23 Find the potential energy of the interaction of two protons at a distance of 1×10^{-15} m, typical of the dimensions of atomic nuclei. Express your result in MeV.

26–24

a) Prove that when a particle is accelerated from rest in an electric field, its final velocity is proportional to the square root of the potential difference through which it is accelerated.

b) What is the final velocity of an electron accelerated through a potential difference of 1136 V if it has an initial velocity of 1.0×10^7 m·s^{-1}?

26–25

a) What is the maximum potential difference through which an electron can be accelerated, from rest, if its kinetic energy is not to exceed 1% of the rest energy?

b) What is the speed of such an electron, expressed as a fraction of the speed of light, c?

c) Make the same calculations for a *proton*.

Section 26–8 The Cathode-Ray Tube

26–26 The electric field in the region between the deflecting plates of a certain cathode-ray oscilloscope is 30,000 N·C^{-1}.

a) What force is on an electron in this region?

b) What is the acceleration of an electron when acted on by this force?

26–27 In Fig. 26–15, an electron is projected along the axis midway between the plates of a cathode-ray tube with an initial velocity of 2×10^7 m·s^{-1}. The uniform electric field between the plates has a magnitude of 20,000 N·C^{-1} and is upward.

a) How far below the axis has the electron moved when it reaches the end of the plates?

b) At what angle with the axis is it moving as it leaves the plates?

c) How far below the axis will it strike the fluorescent screen S?

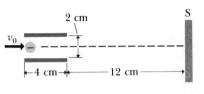

FIGURE 26–15

PROBLEMS

26–28 A small sphere of mass 0.2 g hangs by a thread between two parallel vertical plates 5 cm apart. The charge on the sphere is 6×10^{-9} C. What potential difference between the plates will cause the thread to assume an angle of 30° with the vertical?

26–29 In Exercise 26–14, suppose that another force in addition to the electrical force acts on the particle, so that when it is released from rest, it moves to the right. After it has moved 5 cm, the additional force has done 9×10^{-5} J of work and the particle has 4.5×10^{-5} J of kinetic energy.

a) What work was done by the electrical force?

b) What is the magnitude of the electric field?

c) What is the potential of the starting point with respect to the endpoint?

26–30 Consider the same distribution of charges as in Exercise 26–7.

a) Find the potential at a point on the y-axis, a distance y from the origin. Use your result to sketch a graph of the potential on the y-axis as a function of y, over the range from $y = +4a$ to $y = -4a$.

b) Discuss the physical meaning of the graph at the points $+a$ and $-a$.

c) At what point or points on the y-axis is the potential equal to its value at the origin?

d) At what points on the y-axis is the potential equal to half its value at the origin?

26–31 Consider the same distribution of charges as in Exercise 26–7.

a) Suppose a positively charged particle of charge q' and mass m is placed precisely at the origin and released from rest. What happens?

b) What will happen if the charge in part (a) is constrained not to move in the x-direction but is displaced slightly along the +y-axis and then released?

c) What will happen if it is displaced slightly in the direction of the +x-axis and then released?

26–32 Again consider the charge distribution of Exercise 26–7. Suppose a positively charged particle of charge q' and mass m is displaced slightly from the origin in the direction of the x-axis.

a) What is its speed at infinity?

b) Sketch a graph of the velocity of the particle as a function of x.

c) If the particle is projected toward the left along the x-axis from a point at a large distance to the right of the origin, with a velocity half that acquired in part (a), at what distance from the origin will it come to rest?

d) If a negatively charged particle, of charge $-q'$, were released from rest on the x-axis at a very large distance to the left of the origin, what would be its velocity as it passed the origin?

26–33 A potential difference of 1600 V is established between two parallel plates 4 cm apart. An electron is released from the negative plate at the same instant that a proton is released from the positive plate.

a) How far from the positive plate will they pass each other?

b) How do their velocities compare when they strike the opposite plates?

c) How do their kinetic energies compare when they strike the opposite plates?

26-34 In the Bohr model of the hydrogen atom, a single electron revolves around a single proton in a circle of radius R. Assume that the proton remains at rest.

a) By equating the electrical force to the electron mass times its acceleration, derive an expression for the electron's speed.

b) Obtain an expression for the electron's kinetic energy, and show that its magnitude is just half that of the electrical potential energy.

c) Obtain an expression for the total energy, and evaluate it by using $R = 5.29 \times 10^{-11}$ m.

Give your numerical result in joules and in electronvolts.

26-35 Refer to Problem 24–12.

a) Calcualte the potential at the origin, and at the point (3 cm, 0) due to the first two point charges.

b) Calculate the work the electric field would do on the third charge if it moved from the origin to the point (3 cm, 0).

26-36 Refer to Problem 24–22.

a) Calculate the potential at the point (3 cm, 0) and at the point (3 cm, 4 cm) due to the first two charges.

b) If the third charge moves from the point (3 cm, 0) to the point (3 cm, 4 cm), calculate the work done on it by the field of the first two charges. Comment on the *sign* of this work. Is your result reasonable?

26-37 A vacuum tube diode was described in Exercise 26–16. Because of the accumulation of charge near the cathode, the electrical potential between the electrodes is not a linear function of the position, even for planar geometry, but is given by

$$V = Cx^{4/3},$$

where, for given operating conditions, C is a constant, characteristic of a particular diode and operating conditions, and x is the distance from the cathode (negative electrode). If the distance between the cathode and anode (positive electrode) is 8 mm and the potential difference between electrodes is 160 V,

a) determine the value of C;

b) obtain a formula for the electric field between the electrodes as a function of x;

c) determine the force on an electron when the electron is halfway between the electrodes.

26-38 A long metal cylinder of radius r_a is supported on an insulating stand on the axis of a long, hollow metal cylinder of inner radius r_b. The positive charge per unit length on the inner cylinder is λ, and there is an equal negative charge per unit length on the outer cylinder.

a) Calculate the potential $V(r)$ for
   i) $r < r_a$,
   ii) $r_a < r < r_b$,
   iii) $r > r_b$.

Use the results of Example 26–8 and the fact that the net potential is the sum of the potentials due to the individual conductors. It is useful to take $V = 0$ at the same r_0 for both conductors.

b) Show that the potential of the inner cylinder with respect to the outer is

$$V_{ab} = \frac{\lambda}{2\pi\epsilon_0} \ln \frac{r_b}{r_a}.$$

c) Use Eq. (26–26) and the result from (a) to show that the electric field at any point between the cylinders has magnitude

$$E = \frac{V_{ab}}{\ln (r_b/r_a)} \cdot \frac{1}{r}.$$

d) What is the potential difference between the two cylinders if the outer cylinder has no net charge?

26-39 Refer to Problem 25–34.

a) From the expression for E obtained in Problem 25–34, find the expressions for the potential V as a function of r, both inside and outside the cylinder. Let $V = 0$ at the surface of the cylinder. In each case express your result in terms of the charge per unit length λ of the charge distribution.

b) Sketch graphs of V and E as functions of r, from $r = 0$ to $r = 3R$.

26-40 Four lines of charge are arranged to form a square with sides of length a. Calculate the potential at the center of the square for a zero of potential at infinity if

a) two opposite sides are positively charged with charge $+Q$ each and the other two sides are negatively charged with charge $-Q$ each;

b) if each side has positive charge $+Q$.

(*Hint:* Use the result of Example 26–10.)

26-41 Electric charge is distributed uniformly along a thin rod of length a, with total charge Q. Take the zero of potential to be at infinity. Find the potential at the following points (see Fig. 26–16):

a) point P, a distance x to the right of the rod;

b) point R, a distance y above the right-hand end of the rod.

c) In (a) and (b), what does your result reduce to as x or y becomes large?

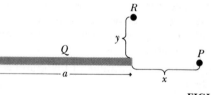

FIGURE 26–16

26-42 Electric charge is distributed uniformly along a semicircle of radius a. Calculate the potential at the center of curvature if the potential is assumed to be zero at infinity.

26-43 A charged oil drop, in a Millikan oil-drop apparatus, is observed to fall a distance of 1 mm in a time of 27.4 s, in the absence of any external field. The same drop can be held stationary in a field of 2.37×10^4 N·C^{-1}. How many excess electrons has the drop acquired? The viscosity of air is 180×10^{-7} N·s·m^{-2}. The density of the oil is 824 kg·m^{-3}, and the density of air is 1.29 kg·m^{-3}.

26–44 An alpha particle with kinetic energy 10 MeV makes a head-on collision with a gold nucleus at rest. What is the distance of closest approach of the two particles? (Assume that the gold nucleus remains stationary and that it may be treated as a point charge.)

26–45 Consider the charge distribution of Problem 25–35. Use the electric field calculated in that problem to do the following:

a) Show that for $r \geq R$ the potential is identical to that produced by a point charge Q. (Take the potential to be zero at infinity.)

b) Obtain an expression for the electrical potential valid in the region $r \leq R$.

CHALLENGE PROBLEMS

26–46 An electron with kinetic energy 100 MeV collides head-on with a gold nucleus at rest. Assume that the gold nucleus can be treated as a uniform distribution of charge through a sphere of radius 7×10^{-15} m and that the electron can penetrate into the nucleus, which remains at rest. What is the electron's kinetic energy when it reaches the center of the nucleus?

26–47 Two point charges are moving toward the right along the x-axis. Point charge 1 has charge $q_1 = 2$ μC, mass $m_1 = 6 \times 10^{-5}$ kg, and a velocity of magnitude v_1. Point charge 2 is to the right of q_1 and has charge $q_2 = -5$ μC, mass $m_2 = 3 \times 10^{-5}$ kg, and a velocity of magnitude v_2. At a particular instant the charges are separated by a distance of 9 mm and have velocities $v_1 = 700$ m·s^{-1} and $v_2 = 1200$ m·s^{-1}. The only forces on the particles are the forces they exert on each other.

a) Determine the velocity v_{cm} of the center of mass of the system.

b) The "relative energy" E_r of the system is defined as the total energy minus the kinetic energy contributed by the motion of the center of mass. That is,

$$E_r = E - \tfrac{1}{2}(m_1 + m_2)v_{cm}^2,$$

where $E = \tfrac{1}{2}m_1v_1^2 + \tfrac{1}{2}m_2v_2^2 + q_1q_2/4\pi\epsilon_0 r$ is the total energy of the system and r is the distance between the charges. Show that $E_r = \tfrac{1}{2}\mu v^2 + q_1q_2/4\pi\epsilon_0 r$, where μ is called the *reduced mass* of the system and is equal to $m_1m_2/(m_1 + m_2)$, and $v = v_2 - v_1$ is the relative velocity of the moving particles.

c) For the numerical values given above, calculate the numerical value of E_r.

d) The value of E_r can be used to determine whether or not the particles will "escape" from one another. For the conditions given above, will the particles escape from one another? Explain.

e) If the particles do escape, what will be their final relative velocity? That is, what will their relative velocity be when $r \rightarrow \infty$? If the particles do not escape, what will be their distance of maximum separation? That is, what will be the value of r when $v = 0$?

26–48 The electrical potential V in a region of space is given by

$$V = ax^2 + ay^2 - 2az^2,$$

where a is a constant.

a) Derive an expression for the electric field E, valid at all points in space.

b) The work done by the field when a 1 μC test charge moves from the point $(x, y, z) = (0, 0, 0.1$ m$)$ to the origin is measured to be -5.0×10^{-5} J. Determine a.

c) Determine the electric field at the point $(0, 0, 0.1$ m$)$.

d) Show that in every plane parallel to the xy-plane, the equipotential lines are circles.

e) What is the radius of the equipotential line corresponding to $V = 5000$ V and $z = \sqrt{2}$ m?

26–49 In a certain region a charge distribution exists that is spherical but nonuniform. That is, the volume charge density $\rho(r)$ depends on the distance r from the center of the distribution but not on the spherical polar angles θ and ϕ. The electric potential $V(r)$ due to this charge distribution is given by

$$V(r) = \left[\frac{\rho_0 a^2}{18\epsilon_0}\right][1 - 3(r/a)^2 + 2(r/a)^3] \quad \text{for} \quad r \leq a,$$

$$V = 0 \quad \text{for} \quad r \geq a,$$

where ρ_0 is a constant having units of C·m^{-3}, and a is a constant having units of meters.

a) Derive expressions for the electric field for the regions $r \leq a$ and $r \geq a$. (*Hint:* Use the gradient operator for spherical polar coordinates,

$$\nabla = i_r\frac{\partial}{\partial r} + i_\theta\frac{1}{r}\frac{\partial}{\partial\theta} + i_\phi\frac{1}{r\sin\theta}\frac{\partial}{\partial\phi},$$

where i_r, i_θ, and i_ϕ are unit vectors in the r-, θ-, and ϕ-directions.)

b) Derive an expression for $\rho(r)$ in each of the two regions $r \leq a$ and $r \geq a$. (*Hint:* Use Gauss's law for two spherical shells, one of radius r and the other of radius $r + dr$. Then use the fact that $dq = 4\pi r^2\rho\,dr$, where dq is the charge contained in the infinitesimal spherical shell of thickness dr.)

c) Show that the net charge contained in the volume of a sphere of radius greater than or equal to a is zero. (*Hint:* Integrate the expressions derived in (b) for ρ over a spherical volume of radius greater than or equal to a.)

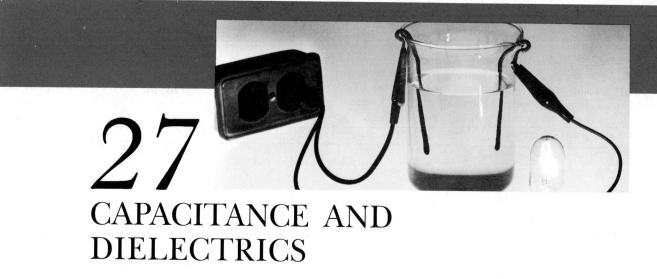

27

CAPACITANCE AND DIELECTRICS

Capacitors play a vital role in modern electronics.

A CAPACITOR IS A DEVICE CONSISTING OF TWO CONDUCTORS SEPARATED by vacuum or an insulating material. Capacitors are used in a wide variety of electric circuits and are a vital part of modern electronics. When charges of equal magnitude and opposite sign are placed on the conductors of a capacitor, an electric field is established in the region between them, with a corresponding potential difference between the conductors. The relations among charge, field, and potential can be analyzed by using the results of the two preceding chapters. For a given capacitor, the ratio of charge to potential difference is a constant, called the *capacitance*. Placing charges on the conductors requires an input of energy; this energy is stored in the capacitor and can be regarded as associated with the electric field in the space between conductors. When this space contains an insulating material (a dielectric) rather than vacuum, the capacitance is increased. This change can be understood on the basis of redistribution of charge within the material. We have previously studied mechanical and thermal properties of materials; electrical properties are another important class of properties of materials.

27–1 CAPACITORS

A capacitor: two charged conductors with a potential difference

Any two conductors separated by an insulator form a **capacitor.** In most cases of practical interest, the conductors usually have charges of equal magnitude and opposite sign, so that the *net* charge on the capacitor as a whole is zero. The electric field in the region between the conductors is proportional to the magnitude Q of charge on each conductor, and it follows that the *potential difference* V_{ab} between the conductors is also proportional to Q. If we double the magnitude of charge on each conductor, the charge density at each point doubles, the electric field at each point doubles, the potential difference between conductors doubles, and the ratio of charge to potential difference does not change.

When we speak of a capacitor as having charge Q, we mean that the conductor at higher potential has a charge Q and the conductor at lower potential has a charge $-Q$ (assuming Q is a positive quantity). This interpretation should be kept in mind in the following discussion and examples.

The **capacitance** C of a capacitor is defined as the ratio of the magnitude of the charge Q on *either* conductor to the magnitude of the potential difference V_{ab} between the conductors:

Capacitance: how much charge for a given voltage?

$$C = \frac{Q}{V_{ab}}. \qquad (27\text{–}1)$$

It follows from this definition that the unit of capacitance is one *coulomb per volt* (C·V^{-1}). A capacitance of one coulomb per volt is called one **farad** (1 F), in honor of Michael Faraday. A capacitor is represented by the symbol

The farad: one coulomb per volt, the SI unit of capacitance

Capacitors have numerous practical uses, and contemporary electronics could not exist without them. They are an essential element in tuning circuits in radio transmitters and receivers, in circuits that smooth and regulate the output of electronic power supplies, in engine ignition systems, and in other areas. The study of capacitors helps us develop insight into the behavior of electric fields and their interactions with matter. So there are many reasons to study capacitors!

Another term for capacitor is *condenser*. This term is seldom used except in reference to engine ignition systems, but you may find it occasionally in older literature.

27–2 THE PARALLEL-PLATE CAPACITOR

The most common type of capacitor consists of two conducting plates *parallel* to each other and separated by a distance that is small compared with the linear dimensions of the plates (see Fig. 27–1). Practically the entire field of such a capacitor is localized in the region between the plates, as shown. Some "fringing" of the field occurs at the edges of the plates, but if the plates are sufficiently close, the fringing may be neglected. The field between the plates is then uniform, and the charges on the plates are uniformly distributed over their opposing surfaces. This arrangement is known as a **parallel-plate capacitor.**

The electric-field magnitude between a pair of closely spaced parallel plates in vacuum, as discussed in Section 25–4, is

The capacitance of a parallel-plate capacitor is determined by its dimensions.

$$E = \frac{\sigma}{\epsilon_0} = \frac{Q}{\epsilon_0 A},$$

where σ is the magnitude of surface charge density on either plate, A is the area of each plate, and Q is the magnitude of total charge on each plate. Since the electric field (potential gradient) between the plates is uniform, the potential difference ("voltage") between the plates is

$$V_{ab} = Ed = \frac{1}{\epsilon_0}\frac{Qd}{A},$$

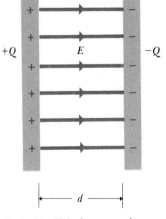

27–1 Parallel-plate capacitor.

where d is the separation of the plates. Hence the capacitance of a parallel-plate capacitor in vacuum is

$$C = \frac{Q}{V_{ab}} = \epsilon_0 \frac{A}{d}. \qquad (27\text{--}2)$$

Since ϵ_0, A, and d are constants for a given capacitor, the capacitance is a constant, independent of the charge on the capacitor, and is directly proportional to the area of the plates and inversely proportional to their separation.

The SI unit of capacitance is the *farad*, abbreviated F. In Eq. (27–2), if A is in square meters and d in meters, C is in farads. The units of ϵ_0 are $(C^2 \cdot N^{-1} \cdot m^{-2})$, so we see that

$$1 \text{ F} = 1 \text{ C}^2 \cdot N^{-1} \cdot m^{-1}.$$

A one-farad capacitor is huge!

As an example, let us compute the area of the plates of a 1-F parallel-plate capacitor if the separation of the plates is 1 mm and the plates are in vacuum:

$$C = \epsilon_0 \frac{A}{d},$$

$$A = \frac{Cd}{\epsilon_0} = \frac{(1 \text{ F})(10^{-3} \text{ m})}{8.85 \times 10^{-12} \text{ C}^2 \cdot N^{-1} \cdot m^{-2}}$$

$$= 1.13 \times 10^8 \text{ m}^2.$$

This area corresponds to a square whose side measures 10,600 m, or 34,800 ft, or about $6\frac{1}{2}$ miles!

Smaller and more practical units of capacitance

The farad is thus an extremely large unit of capacitance; units of more convenient size are the *microfarad* (1 μF = 10^{-6} F) and the *picofarad* (1 pF = 10^{-12} F). For example, a common radio contains in its power supply several capacitors whose capacitances are of the order of 10 or more microfarads, while the capacitances of the tuning capacitors are of the order of 10 to 100 picofarads.

Some numerical calculations for a parallel-plate capacitor

EXAMPLE 27–1 The plates of a parallel-plate capacitor are 5 mm apart and 2 m^2 in area. The plates are in vacuum. A potential difference of 10,000 V is applied across the capacitor. Compute (a) the capacitance, (b) the charge on each plate, and (c) the magnitude of electric field in the space between them.

SOLUTION

a) From Eq. (27–2),

$$C = \epsilon_0 \frac{A}{d} = \frac{(8.85 \times 10^{-12} \text{ C}^2 \cdot N^{-1} \cdot m^{-2})(2 \text{ m}^2)}{5 \times 10^{-3} \text{ m}}$$

$$= 3.54 \times 10^{-9} \text{ F} = 0.00354 \ \mu\text{F}.$$

b) The charge on the capacitor is

$$Q = CV_{ab} = (3.54 \times 10^{-9} \text{ C} \cdot V^{-1})(10^4 V)$$

$$= 3.54 \times 10^{-5} \text{ C} = 35.4 \ \mu\text{C}.$$

That is, the plate at higher potential has charge $+35.4 \ \mu$C, and the other plate has charge $-35.4 \ \mu$C.

c)   The electric field is

$$E = \frac{\sigma}{\epsilon_0} = \frac{Q}{\epsilon_0 A} = \frac{3.54 \times 10^{-5}\,\text{C}}{(8.85 \times 10^{-12}\,\text{C}^2 \cdot \text{N}^{-1} \cdot \text{m}^{-2})(2\,\text{m}^2)}$$
$$= 20 \times 10^5\,\text{N} \cdot \text{C}^{-1};$$

or, since the electric field equals the potential gradient,

$$E = \frac{V_{ab}}{d} = \frac{10^4\,\text{V}}{5 \times 10^{-3}\,\text{m}} = 20 \times 10^5\,\text{V} \cdot \text{m}^{-1}.$$

The newton per coulomb and the volt per meter are equivalent units.

27–3 CAPACITORS IN SERIES AND PARALLEL

The arrangement shown in Fig. 27–2a is called a **series** connection. Two capacitors are connected in series between points a and b, and a constant potential difference V_{ab} is maintained. The capacitors are both initially uncharged. In this connection, both capacitors always have the same charge Q. The lower plate of C_1 and the upper plate of C_2 cannot have charges different from those on the remaining two plates. If they did, the net charge on each capacitor would not be zero, and the resulting electric field in the conductor connecting the two capacitors would cause a current to flow until the total charge on each capacitor is zero. Hence, *in a series connection the magnitude of charge on all plates is the same.*

Referring again to Fig. 27–2a, we have

$$V_{ac} \equiv V_1 = \frac{Q}{C_1}, \qquad V_{cb} \equiv V_2 = \frac{Q}{C_2},$$

$$V_{ab} \equiv V = V_1 + V_2 = Q\left(\frac{1}{C_1} + \frac{1}{C_2}\right),$$

and

$$\frac{V}{Q} = \frac{1}{C_1} + \frac{1}{C_2}. \qquad (27\text{–}3)$$

The **equivalent capacitance** C of the series combination is defined as the capacitance of a *single* capacitor for which the charge Q is the same as for the combination, when the potential difference V is the same. For such a capacitor, shown in Fig. 27–2b,

$$Q = CV, \qquad \frac{V}{Q} = \frac{1}{C}. \qquad (27\text{–}4)$$

Hence, from Eqs. (27–3) and (27–4),

$$\frac{1}{C} = \frac{1}{C_1} + \frac{1}{C_2}.$$

Similarly, for any number of capacitors in series,

$$\frac{1}{C} = \frac{1}{C_1} + \frac{1}{C_2} + \frac{1}{C_3} + \cdots. \qquad (27\text{–}5)$$

Capacitors connected together: What is the equivalent capacitance?

Capacitors in series all have the same charge.

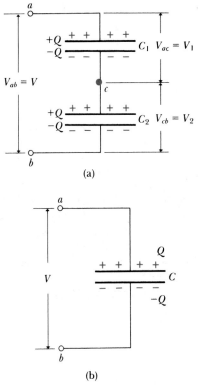

27–2 (a) Two capacitors in series, and (b) their equivalent.

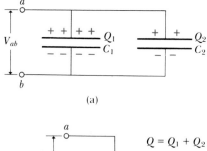

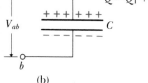

27–3 (a) Two capacitors in parallel, and (b) their equivalent.

Capacitors in parallel add directly.

The reciprocal of the equivalent capacitance of a series combination equals the sum of the reciprocals of the individual capacitances.

The arrangement shown in Fig. 27–3a is called a **parallel** connection. Two capacitors are connected in parallel between points a and b. In this case, the upper plates of the two capacitors are connected to form an equipotential surface, and the lower plates form another. The potential difference is the same for both capacitors and is equal to $V_{ab} = V$. The charges Q_1 and Q_2, not necessarily equal, are

$$Q_1 = C_1 V, \qquad Q_2 = C_2 V.$$

The *total* charge Q supplied by the source was

$$Q = Q_1 + Q_2 = V(C_1 + C_2),$$

and

$$\frac{Q}{V} = C_1 + C_2. \tag{27–6}$$

The *equivalent* capacitance C of the parallel combination is defined as that of a single capacitor, shown in Fig. 27–3b, for which the total charge is the same as in part (a). For this capacitor,

$$\frac{Q}{V} = C,$$

and hence

$$C = C_1 + C_2.$$

In the same way, for any number of capacitors in parallel

$$C = C_1 + C_2 + C_3 + \cdots. \tag{27–7}$$

The equivalent capacitance of a parallel combination equals the *sum* of the individual capacitances.

PROBLEM-SOLVING STRATEGY: *Equivalent capacitance*

1. Keep in mind that when we say a capacitor has a charge Q, we always mean that one plate has a charge Q and the other plate has a charge $-Q$.

2. When two capacitors are connected in series, as in Fig. 27–2a, they always have the same charge, provided they were initially uncharged before they were connected. The potential differences are *not* equal, unless the capacitances are equal. In this case the total potential difference across the combination is the sum of the individual potential differences.

3. When two capacitors have their plates connected as in Fig. 27–3a, so they are in parallel, the potential difference V is always the same for both. The charges on the two are *not* equal unless the capacitances are equal, but the total charge on the combination is the sum of the individual charges.

EXAMPLE 27–2 In Figs. 27–2 and 27–3, let $C_1 = 6\ \mu F$, $C_2 = 3\ \mu F$, and $V_{ab} = 18$ V. Find the equivalent capacitance, and find the charge and potential difference for each capacitor when the two capacitors are connected in (a) series, (b) parallel.

SOLUTION

a)   The equivalent capacitance of the series combination in Fig. 27–2a is given by

$$\frac{1}{C} = \frac{1}{6\ \mu F} + \frac{1}{3\ \mu F}, \qquad C = 2\ \mu F.$$

The charge Q is

$$Q = CV = (2\ \mu F)(18\ V) = 36\ \mu C.$$

The potential differences across the capacitors are

$$V_{ac} \equiv V_1 = \frac{Q}{C_1} = \frac{36\ \mu C}{6\ \mu F} = 6\ V, \qquad V_{cb} \equiv V_2 = \frac{Q}{C_2} = \frac{36\ \mu C}{3\ \mu F} = 12\ V.$$

The *larger* potential difference appears across the *smaller* capacitor.

b)   The equivalent capacitance of the parallel combination in Fig. 27–3a is

$$C = C_1 + C_2 = 9\ \mu F.$$

The charges Q_1 and Q_2 are

$$Q_1 = C_1 V = (6\ \mu F)(18\ V) = 108\ \mu C,$$

$$Q_2 = C_2 V = (3\ \mu F)(18\ V) = 54\ \mu C.$$

The potential difference across each capacitor is 18 V.

An example of capacitors in series and parallel

27–4 ENERGY OF A CHARGED CAPACITOR

To charge a capacitor we must transfer charge from the plate at lower potential to the plate at higher potential. This process requires work; energy is added to the capacitor and stored as *potential energy*. The positive and negative charges that are separated but attract each other are analogous to a stretched spring or a body lifted in the earth's gravitational field. The potential energy corresponds to the work done by the electrical forces when the capacitor becomes discharged, just as the spring or the earth's gravity does work when the system returns from its displaced position to the reference position.

Charging a capacitor requires the addition of energy.

To calculate the potential energy U of a charged capacitor, we calculate the work W required to charge it. The final charge Q and the final potential difference V are related by

$$Q = CV.$$

At a stage of the charging process at which the magnitude of the net charge on either plate is q, the potential difference v between the plates is $v = q/C$. The work dW required to transfer the next charge dq is

$$dW = v\ dq = \frac{q\ dq}{C}.$$

The total work W needed to increase the charge from zero to a final value Q is

$$W = \int dW = \frac{1}{C} \int_0^Q q\ dq = \frac{Q^2}{2C}.$$

The energy of a charged capacitor is proportional to the square of the charge.

If we define the potential energy of an *uncharged* capacitor to be zero, then W is equal to the potential energy U of the charged capacitor. The final potential difference V between the plates is $V = Q/C$. Thus we can write

$$U = W = \frac{Q^2}{2C} = \frac{1}{2}CV^2 = \frac{1}{2}QV. \qquad (27-8)$$

When Q is expressed in coulombs and V in volts (joules per coulomb), W is expressed in joules.

The last form of Eq. (27–8) also shows that the total work is equal to the *average* potential $V/2$ during the charging process, multiplied by the total charge Q transferred.

Potential energy in a charged capacitor: analogous to a stretched spring

A charged capacitor is the electrical analog of a stretched spring, whose elastic potential energy equals $\frac{1}{2}kx^2$. The charge Q is analogous to the elongation x, and the *reciprocal* of the capacitance, $1/C$, is analogous to the force constant k. The energy supplied to a capacitor in the charging process is stored by the capacitor and is released when the capacitor discharges.

It is often useful to consider the energy stored in a capacitor as localized in the *electric field* between the capacitor plates. The capacitance of a parallel-plate capacitor in vacuum, from Eq. (27–2), is

$$C = \epsilon_0 \frac{A}{d}.$$

Energy in a capacitor associated with the field between the conductors

The electric field fills the space between the plates, of volume Ad, and its magnitude E is

$$E = \frac{V}{d}.$$

The energy per unit volume, or the **energy density,** denoted by u, is

$$u = \text{Energy density} = \frac{\frac{1}{2}CV^2}{Ad}.$$

Making use of the preceding equations, we can express this relation as

$$u = \tfrac{1}{2}\epsilon_0 E^2. \qquad (27-9)$$

The energy per unit volume is proportional to the square of the field magnitude.

Although we have derived Eq. (27–9) only for one specific situation, it turns out to be valid in general. The energy per unit volume associated with *any* electric-field configuration is given by Eq. (27–9). We will need this result in Chapter 35 in connection with the energy transported by electromagnetic waves.

An example of energy relations in a capacitor

EXAMPLE 27–3 In Fig. 27–4 we charge a capacitor C_1 by connecting it to a source of potential difference V_0 (not shown in the figure). Let $C_1 = 8\ \mu\text{F}$ and $V_0 = 120\ \text{V}$. The charge Q_0 is

$$Q_0 = C_1 V_0 = 960\ \mu\text{C},$$

and the energy of the capacitor is

$$\tfrac{1}{2}Q_0 V_0 = \tfrac{1}{2}(960 \times 10^{-6}\ \text{C})(120\ \text{V}) = 0.0576\ \text{J}.$$

After we close the switch S, the positive charge Q_0 is distributed over the upper plates of both capacitors, and the negative charge $-Q_0$ is distributed over

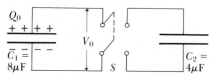

27–4 When the switch S is closed, the charged capacitor C_1 is connected to an uncharged capacitor C_2.

the lower plates of both. Let Q_1 and Q_2 represent the magnitudes of the final charges on the respective capacitors. Then

$$Q_1 + Q_2 = Q_0.$$

When the motion of charges has ceased, both upper plates are at the same potential; they are connected by a wire and so form a single equipotential surface. Both lower plates are at the same potential, different from that of the upper plates. The final potential difference between the plates, V, is therefore the same for both capacitors, and

$$Q_1 = C_1V, \qquad Q_2 = C_2V.$$

When we combine this with the preceding equation, we find

$$V = \frac{Q_0}{C_1 + C_2} = \frac{960 \ \mu C}{12 \ \mu F} = 80 \ V,$$
$$Q_1 = 640 \ \mu C, \qquad Q_2 = 320 \ \mu C.$$

The final energy of the system is

$$\tfrac{1}{2}Q_1V + \tfrac{1}{2}Q_2V = \tfrac{1}{2}Q_0V = \tfrac{1}{2}(960 \times 10^{-6} \ C)(80 \ V) = 0.0384 \ J.$$

This result is less than the original energy of 0.0576 J; the difference has been converted to energy of some other form. The conductors become a little warmer, and some energy is radiated as electromagnetic waves.

This process is exactly analogous to an inelastic collision of a moving car with a stationary car. In the electrical case, the charge $Q = CV$ is conserved. In the mechanical case, the momentum $p = mv$ is conserved. The electrical energy $\tfrac{1}{2}CV^2$ is *not* conserved, and the mechanical energy $\tfrac{1}{2}mv^2$ is *not* conserved.

27–5 EFFECT OF A DIELECTRIC

Most capacitors have a solid, nonconducting material or **dielectric** between their plates. A common type of capacitor incorporates strips of metal foil, forming the plates, separated by strips of wax-impregnated paper or plastic sheet such as Mylar, which serves as the dielectric. A sandwich of these materials is rolled up, forming a compact unit that can provide a capacitance of several microfarads in a relatively compact package.

Electrolytic capacitors have as their dielectric an extremely thin layer of nonconducting oxide between a metal plate and a conducting solution. Because of the thinness of the dielectric, electrolytic capacitors of relatively small dimensions may have a capacitance of the order of 100 or 1000 μF. (From Eq. [27–2], the capacitance is inversely proportional to the distance d between the plates.)

Placing a solid dielectric between the plates of a capacitor serves three functions. First, it solves the mechanical problem of maintaining two large metal sheets at an extremely small separation but without actual contact.

Second, any dielectric material, when subjected to a sufficiently large electric field, experiences *dielectric breakdown*, which is a partial ionization that permits conduction through a material that is supposed to insulate. Many insulating materials can tolerate stronger electric fields without breakdown than air can tolerate.

What happens when an insulating material is placed between the conductors?

The dielectric in a capacitor serves several purposes.

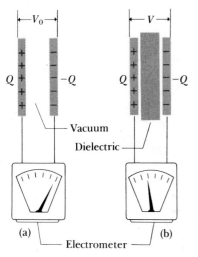

27–5 Effect of a dielectric between the plates of a parallel-plate capacitor. The electrometer measures potential difference. (a) With a given charge, the potential difference is V_0. (b) With the same charge, the potential difference V is smaller than V_0.

The dielectric constant: characterizing the effect of a dielectric on the capacitance and the field

Third, the capacitance of a capacitor of given dimensions is *larger* when a dielectric material is placed between the plates than when the plates are separated only by air or vacuum. This effect can be demonstrated with the aid of a sensitive electrometer, a device that can measure the potential difference between two conductors without permitting any charge to flow from one to the other. Figure 27–5a shows a charged capacitor, with magnitude of charge Q on each plate and potential difference V_0. When a sheet of dielectric, such as glass, paraffin, or polystyrene, is inserted between the plates, the potential difference is found to *decrease* to a smaller value V. When the dielectric is removed, the potential difference returns to its original value, showing that the original charges on the plates have not been affected by inserting the dielectric.

The original capacitance of the capacitor, C_0, was

$$C_0 = \frac{Q}{V_0}.$$

Since Q does not change and V is observed to be less than V_0, it follows that the capacitance C with the dielectric present is *greater* than C_0. The ratio of C to C_0 is called the **dielectric constant** of the material, K:

$$K = \frac{C}{C_0} \qquad (27\text{--}10)$$

Since C is always greater than C_0, the dielectric constants of all dielectrics are greater than unity. Some representative values of K are given in Table 27–1. For vacuum, of course, $K = 1$ by definition, and K for air is so nearly equal to 1 that for most purposes an air capacitor is equivalent to one in vacuum; the original measurement of V_0 in Fig. 27–5a could have been made with the plates in air instead of in vacuum.

With vacuum (or air) between the plates, the electric field E_0 in the region between the plates of a parallel-plate capacitor is

$$E_0 = \frac{V_0}{d} = \frac{\sigma}{\epsilon_0}.$$

The observed reduction in potential difference, when a dielectric is inserted between the plates, implies a reduction in the electric field, which in turn implies a reduction in the charge per unit area. Since no charge has leaked off the plates, such a reduction could be caused only by induced

TABLE 27–1   Dielectric Constant K at 20°C

Material	K	Material	K
Vacuum	1	Strontium titanate	310
Glass	5–10	Titanium dioxide (rutile)	173($\perp$),
Mica	3–6		86($\parallel$)
Mylar	3.1	Water	80.4
Neoprene	6.70	Glycerin	42.5
Plexiglas	3.40	Liquid ammonia (−78°C)	25
Polyethylene	2.25	Benzene	2.284
Polyvinyl chloride	3.18	Air (1 atm)	1.00059
Teflon	2.1	Air (100 atm)	1.0548
Germanium	16		

charges of opposite sign appearing on the two surfaces of the *dielectric*. That is, the dielectric surface adjacent to the positive plate must have an *induced negative charge*, and the surface adjacent to the negative plate must have an *induced positive charge of equal magnitude*, as shown in Fig. 27–6. These induced surface charges are a result of redistribution of charge within the dielectric material, a phenomenon called **polarization.**

If σ_i is the magnitude of the induced charge per unit area on the surfaces of the dielectric, then the *net* surface charge on each side of the capacitor that contributes to the electric field within the dielectric has magnitude $(\sigma - \sigma_i)$, and the electric field in the dielectric is

$$E = \frac{V}{d} = \frac{\sigma - \sigma_i}{\epsilon_0}. \qquad (27\text{–}11)$$

However,

$$K = \frac{C}{C_0} = \frac{Q/V}{Q/V_0} = \frac{V_0}{V} = \frac{E_0}{E} = \frac{\sigma}{\sigma - \sigma_i}; \qquad (27\text{–}12)$$

therefore

$$\sigma - \sigma_i = \frac{\sigma}{K}. \qquad (27\text{–}13)$$

Substituting Eq. (27–13) into Eq. (27–11), we obtain

$$E = \frac{\sigma}{K\epsilon_0}. \qquad (27\text{–}14)$$

The product $K\epsilon_0$ is called the **permittivity** of the dielectric and is represented by ϵ:

$$\epsilon = K\epsilon_0. \qquad (27\text{–}15)$$

The electric field within the dielectric may therefore be written

$$E = \frac{\sigma}{\epsilon}. \qquad (27\text{–}16)$$

Also,

$$C = KC_0 = K\epsilon_0\frac{A}{d},$$

hence the capacitance of a parallel-plate capacitor with a dielectric between its plates is

$$C = \epsilon\frac{A}{d}. \qquad (27\text{–}17)$$

In empty space, where $K = 1$, $\epsilon = \epsilon_0$, and therefore ϵ_0 may be described as the "permittivity of empty space" or the "permittivity of vacuum." Since K is a pure number, the units of ϵ and ϵ_0 are evidently the same, $C^2 \cdot N^{-1} \cdot m^{-2}$.

The derivation of Eq. (27–9) for the energy density in an electric field can be repeated for a dielectric, by using the relations presented above. The result is that the energy density u for the electric field in a dielectric is given by

$$u = \tfrac{1}{2}K\epsilon_0E^2 = \tfrac{1}{2}\epsilon E^2. \qquad (27\text{–}18)$$

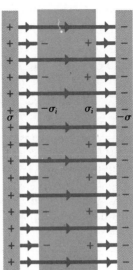

27–6 Induced charges on the faces of a dielectric in an external field.

Permittivity: an alternative description of the effect of a dielectric

Energy density in a dielectric

PROBLEM-SOLVING STRATEGY: *Dielectrics*

1. As usual, be careful with units. Distances must be in meters; remember that a microfarad is 10^{-6} farads, and so on. Don't confuse the numerical value of ϵ_0 with that of $1/4\pi\epsilon_0$. Several equivalent sets of units for electric-field magnitude exist including $N\cdot C^{-1}$ and $V\cdot m^{-1}$. Always check for consistency of units; it's a bit more of a nuisance with electrical quantities than it was in mechanics, but checking pays off!

2. In problems such as the following examples, it is easy to get lost in a blizzard of formulas. Ask yourself at each step what kind of quantity each symbol represents. For example, distinguish clearly between charges and charge densities, and between electric fields and potentials. In checking numerical values, remember that the capacitance with a dielectric present is always greater than without, and that the induced surface charge density σ_i on the dielectric is always less than the free charge density σ on the capacitor plates. With a given charge on a capacitor, the electric field and potential difference are less with a dielectric present than without it.

A detailed analysis of the effect of a dielectric on the field, potential, and surface charges

EXAMPLE 27–4 The parallel plates in Fig. 27–6 have an area of 2000 cm^2 (2×10^{-1} m^2) and are 1 cm (10^{-2} m) apart. The original potential difference between them, V_0, is 3000 V, and it decreases to 1000 V when a sheet of dielectric is inserted between the plates. Compute (a) the original capacitance C_0; (b) the charge Q on each plate; (c) the capacitance C after insertion of the dielectric; (d) the dielectric constant K of the dielectric; (e) the permittivity ϵ of the dielectric; (f) the induced charge Q_i on each face of the dielectric; (g) the original electric field E_0 between the plates; and (h) the electric field E after insertion of the dielectric.

SOLUTION

a) $C_0 = \epsilon_0 \dfrac{A}{d} = (8.85 \times 10^{-12} \text{ C}^2\cdot\text{N}^{-1}\cdot\text{m}^{-2}) \dfrac{2 \times 10^{-1} \text{ m}^2}{10^{-2} \text{ m}}$

$\quad = 17.7 \times 10^{-11} \text{ F} = 177 \text{ pF}.$

b) $Q = C_0 V_0 = (17.7 \times 10^{-11} \text{ F})(3 \times 10^3 \text{ V}) = 53.1 \times 10^{-8} \text{ C}.$

c) $C = \dfrac{Q}{V} = \dfrac{53.1 \times 10^{-8} \text{ C}}{10^3 \text{ V}} = 53.1 \times 10^{-11} \text{ F} = 531 \text{ pF}.$

d) $K = \dfrac{C}{C_0} = \dfrac{53.1 \times 10^{-11} \text{ F}}{17.7 \times 10^{-11} \text{ F}} = 3.$

The dielectric constant could also be found from Eq. (27–12),

$$K = \frac{V_0}{V} = \frac{3000 \text{ V}}{1000 \text{ V}} = 3.$$

e) $\epsilon = K\epsilon_0 = (3)(8.85 \times 10^{-12} \text{ C}^2\cdot\text{N}^{-1}\cdot\text{m}^{-2})$
$\quad = 26.6 \times 10^{-12} \text{ C}^2\cdot\text{N}^{-1}\cdot\text{m}^{-2}.$

f) $Q_i = A\sigma_i, \qquad Q = A\sigma,$

$\quad \sigma - \sigma_i = \dfrac{\sigma}{K}, \qquad \sigma_i = \sigma\left(1 - \dfrac{1}{K}\right),$

$\quad Q_i = Q\left(1 - \dfrac{1}{K}\right) = (53.1 \times 10^{-8} \text{ C})\left(1 - \dfrac{1}{3}\right)$

$\quad = 35.4 \times 10^{-8} \text{ C}.$

g) $E_0 = \dfrac{V_0}{d} = \dfrac{3000 \text{ V}}{10^{-2} \text{ m}} = 3 \times 10^5 \text{ V·m}^{-1}.$

h) $E = \dfrac{V}{d} = \dfrac{1000 \text{ V}}{10^{-2} \text{ m}} = 1 \times 10^5 \text{ V·m}^{-1};$

or

$E = \dfrac{\sigma}{\epsilon} = \dfrac{Q}{A\epsilon} = \dfrac{53.1 \times 10^{-8} \text{ C}}{(2 \times 10^{-1} \text{ m}^2)(26.6 \times 10^{-12} \text{ C}^2 \cdot \text{N}^{-1} \cdot \text{m}^{-2})}$

$= 1 \times 10^5 \text{ V·m}^{-1};$

or

$E = \dfrac{\sigma - \sigma_i}{\epsilon_0} = \dfrac{Q - Q_i}{A\epsilon_0}$

$= \dfrac{(53.1 - 35.4) \times 10^{-8} \text{ C}}{(2 \times 10^{-1} \text{ m}^2)(8.85 \times 10^{-12} \text{ C}^2 \cdot \text{N}^{-1} \cdot \text{m}^{-2})}$

$= 1 \times 10^5 \text{ V·m}^{-1};$

or, from Eq. (27–12),

$E = \dfrac{E_0}{K} = \dfrac{3 \times 10^5 \text{ V·m}^{-1}}{3} = 1 \times 10^5 \text{ V·m}^{-1}.$

As we mentioned earlier, when any dielectric material is subjected to a sufficiently strong electric field, it becomes a conductor, a phenomenon known as dielectric breakdown. The onset of conduction, associated with cumulative ionization of molecules of the material, is often quite sudden and may be characterized by spark or arc discharges. When a capacitor is subjected to excessive voltage, an arc may form through a layer of dielectric, burning or melting a hole in it. This hole permits the two metal foils to come in contact, creating a short circuit and rendering the device permanently useless as a capacitor.

Dielectric breakdown: when an insulator quits insulating

The maximum electric field a material can withstand without the occurrence of breakdown is called its **dielectric strength.** This quantity is affected significantly by impurities in the material, small irregularities in the metal electrodes, and other factors difficult to control. For this reason we can give only approximate figures for dielectric strengths. The dielectric strength of dry air is about $0.8 \times 10^6 \text{ V·m}^{-1}$. Typical values for plastic and ceramic materials commonly used to insulate capacitors and current-carrying wires are of the order of 10^7 V·m^{-1}. For example, a layer of such a material, 10^{-4} m in thickness, could withstand a maximum voltage of 1000 V, since $1000 \text{ V}/10^{-4} \text{ m} = 10^7 \text{ V·m}^{-1}$.

27–6 MOLECULAR MODEL OF INDUCED CHARGE

We now discuss briefly how surface charges on a dielectric, described in the preceding section, can come about. If the material were a *conductor*, the answer would be simple: Conductors contain charge that is free to move, and in the presence of an electric field some of the charge redistributes itself on the surface so that there is no electric field inside the conductor. But dielectrics have no charges that are free to move, so how can a surface charge occur?

The molecular basis of induced surface charge on dielectrics

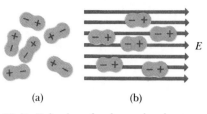

27–7 Behavior of polar molecules (a) in the absence and (b) in the presence of an electric field.

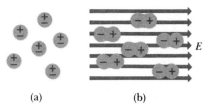

27–8 Behavior of nonpolar molecules (a) in the absence and (b) in the presence of an electric field.

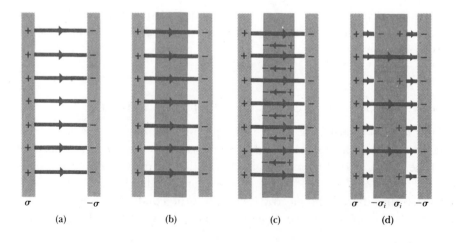

27–9 Polarization of a dielectric in an electric field gives rise to thin layers of bound charges on the surfaces.

To understand this phenomenon, we have to look at rearrangement of charge at the *molecular* level. First, some molecules, such as H_2O and N_2O, have equal amounts of positive and negative charge, but a lopsided distribution, with excess positive charge concentrated on one side of the molecule and negative on the other. Such a molecule is a little electric dipole, as defined in Example 25–5 (Section 25–1). It has an electric dipole moment and is called a *polar molecule*. When no electric field is present, the dipole moments of the molecules are randomly oriented. When polar molecules are placed in an electric field, however, they tend to orient themselves as in Fig. 27–7, as a result of the torques caused by the electric-field forces.

Even a molecule that is *not* polar acquires a dipole moment when placed in an electric field because the field forces cause some redistribution of charge within the molecule, as shown in Fig. 27–8. Such dipoles are called *induced* dipoles. With either polar or nonpolar molecules, an external field causes the formation of a layer of charge on each surface of the dielectric material, as shown in Fig. 27–9. These layers are the surface charges described in Section 27–5; their surface charge density is denoted by σ_i. The charges are not free to move indefinitely, as they would be in a conductor, because each charge is bound to a molecule. They are in fact called **bound charges** to distinguish them from the **free charges** that are added to and removed from the conducting capacitor plates. In the interior of the material the net charge per unit volume remains zero. We say that in this state the material is *polarized*.

The four parts of Fig. 27–10 show the behavior of a slab of dielectric when it is inserted in the field between a pair of oppositely charged capacitor plates. Part (a) shows the original field. Part (b) is the situation after the dielectric has

27–10 (a) Electric field between two charged plates. (b) Introduction of a dielectric. (c) Induced surface charges and their field. (d) Resultant field when a dielectric is between charged plates.

been inserted but before any rearrangement of charges has occurred. Part (c) shows by thinner lines the additional field set up in the dielectric by its induced surface charges. This field is *opposite* to the original field but not enough to cancel it completely, since the charges in the dielectric are not free to move indefinitely. The field in the dielectric is therefore decreased in magnitude. The resultant field is shown in Fig. 27–10d. Some of the field lines leaving the positive plate penetrate the dielectric; others terminate on the induced charges on the faces of the dielectric.

The charges induced on the surface of a dielectric in an external field afford an explanation of the attraction of an *uncharged* object such as a pith ball or bit of paper by a charged rod of rubber or glass. Figure 27–11 shows an uncharged dielectric sphere B in the radial field of a positive charge A. The induced positive charges on B experience a force toward the right, while the force on the negative charges is toward the left. Since the negative charges are closer to A and therefore in a stronger field than are the positive, the force toward the left exceeds that toward the right, and B, although its net charge is zero, experiences a resultant force toward A. The sign of A's charge does not affect the conclusion. Furthermore, the effect is not limited to dielectrics; a conducting sphere would be similarly attracted.

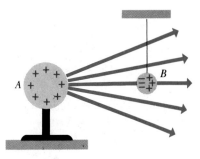

27–11 An uncharged dielectric sphere B in the radial field of a positive charge A.

How can an electric field exert a force on an uncharged object?

27–7 GAUSS'S LAW IN DIELECTRICS

With a slight extension of the analysis of Section 27–5, we can reformulate Gauss's law in a form that is particularly useful for dielectrics. Figure 27–12 shows the left capacitor plate and left surface of the dielectric in Fig. 27–6. Let us apply Gauss's law to the surface shown by the broken line; the surface area of each side is A. The left surface is embedded in the conductor forming the left capacitor plate, and so E everywhere on that surface is zero. The right surface is embedded in the dielectric, where the electric field has magnitude E. The total charge enclosed, including both the free charge on the capacitor plate and the bound charge on the dielectric surface, is $(\sigma - \sigma_i)A$, so Gauss's law gives us

$$EA = (\sigma - \sigma_i)A/\epsilon_0.$$

This is not very illuminating, as it stands, because it relates two unknown quantities, E in the dielectric and the induced surface charge density σ_i. But now we can use Eq. (27–13), developed for this same situation, to rewrite this equation as

$$EA = \sigma A/K\epsilon_0 \quad or \quad KEA = \sigma A/\epsilon_0. \qquad (27\text{–}19)$$

More generally, we can rewrite Gauss's law as

$$\int KE \cdot dA = Q/\epsilon_0. \qquad (27\text{–}20)$$

where Q is the total *free* charge (not bound charge) enclosed by the Gaussian surface. The significance of these results is that the right sides contain only the *free* charge, not the bound charge.

In fact, although we have not proved it, Eq. (27–20) remains valid when different parts of the Gaussian surface are embedded in dielectrics having different values of K.

Gauss's law: Both free and bound charge must be included.

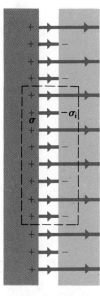

27–12 Gauss's law in a dielectric.

Another useful form of this relation, employing the notation $\epsilon = K\epsilon_0$ introduced in Section 27–5, is

$$\oint \epsilon \mathbf{E} \cdot d\mathbf{A} = Q. \tag{27–21}$$

Electric displacement: how to formulate Gauss's law using only free charges

This equation will be useful to our analysis of electromagnetic waves in matter in Chapter 35. For historical reasons of dubious validity, the quantity $\epsilon \mathbf{E}$ is given the name **electric displacement** and is denoted by $\mathbf{D}$. So still another form of Gauss's law in the presence of dielectric materials is

$$\oint \mathbf{D} \cdot d\mathbf{A} = Q, \tag{27–22}$$

where again Q represents only the *free* charge enclosed by the Gaussian surface, not the polarization charge.

SUMMARY

KEY TERMS

capacitor

capacitance

farad

parallel-plate capacitor

series

equivalent capacitance

parallel

energy density

dielectric

dielectric breakdown

dielectric constant

polarization

permittivity

dielectric strength

bound charges

free charges

electric displacement

A capacitor is any pair of conductors separated by an insulating material. When the capacitor is charged, with charges of equal magnitude Q and opposite sign on the two conductors, the potential difference V_{ab} between them is proportional to Q. The capacitance C is defined as

$$C = \frac{Q}{V_{ab}.} \tag{27–1}$$

A parallel-plate capacitor is made with two plates of area A, a distance d apart. If they are separated by vacuum, the capacitance is

$$C = \frac{Q}{V_{ab}} = \epsilon_0 \frac{A}{d}. \tag{27–2}$$

The SI unit of capacitance is the farad, abbreviated F. One farad is one coulomb per volt, and also $1 \ \text{C}^2 \cdot \text{N}^{-1} \cdot \text{m}^{-1}$. This is a very large unit; the microfarad ($1 \ \mu\text{F} = 10^{-6} \ \text{F}$) and the picofarad ($1 \ \text{pF} = 10^{-12} \ \text{F}$) are more commonly used.

When capacitors having capacitances $C_1, C_2, C_3, \ldots$ are connected in series, the equivalent capacitance C is given by

$$\frac{1}{C} = \frac{1}{C_1} + \frac{1}{C_2} + \frac{1}{C_3} + \cdots. \tag{27–5}$$

When they are connected in parallel, the equivalent capacitance is

$$C = C_1 + C_2 + C_3 + \cdots. \tag{27–7}$$

The energy required to charge a capacitor C to a potential difference V and a charge Q is equal to the energy stored in the capacitor and is given by

$$U = \frac{Q^2}{2C} = \frac{1}{2}CV^2 = \frac{1}{2}QV. \tag{27–8}$$

This energy can be thought of as residing in the electric field between the conductors, with an energy density (energy per unit volume) u given by

$$u = \tfrac{1}{2}\epsilon_0 E^2. \tag{27–9}$$

When the space between the conductors is filled with a dielectric material, the capacitance increases by a factor K, called the dielectric constant of the material. Surface charges are induced on the surface of the dielectric; these have the effect of decreasing the electric field and potential difference between conductors by a factor K. The surface charge results from a microscopic rearrangement of charge in the dielectric, called polarization.

Under sufficiently strong fields, dielectrics become conductors; this phenomenon is called dielectric breakdown. The maximum field that a material can withstand without breakdown is called its dielectric strength.

The microscopic basis of polarization of dielectrics is the reorientation of polar molecules in an applied E field, or the creation of induced dipole moments in nonpolar materials.

The energy density u in an electric field in a dielectric is given by

$$u = \tfrac{1}{2}K\epsilon_0 E^2 = \tfrac{1}{2}\epsilon E^2. \qquad (27\text{–}18)$$

Gauss's law can be reformulated for dielectrics, as follows:

$$\oint \epsilon E \cdot dA = \oint D \cdot dA = Q, \qquad (27\text{–}21)$$

where Q is only the free charge (not bound charge or polarization charge) enclosed by the Gaussian surface.

QUESTIONS

27–1 A student claimed that two capacitors connected in parallel always have an effective capacitance *greater* than that of either capacitor, but when they are in series the effective capacitance is always *less* than that of either capacitor. Confirm or refute these claims.

27–2 Could one define a capacitance for a single conductor? What would be a reasonable definition?

27–3 Suppose the two plates of a capacitor have different areas. When the capacitor is charged by connecting it to a battery, do the charges on the two plates have equal magnitude, or may they be different?

27–4 Can you think of a situation in which the two plates of a capacitor *do not* have equal magnitudes of charge?

27–5 A capacitor is charged by being connected to a battery, and is then disconnected from the charging agency. The plates are then pulled apart a little. How does the electric field change? The potential difference? The total energy?

27–6 According to the text, one can consider the energy in a charged capacitor to be located in the field between the plates. But suppose there is vacuum between the plates; can there be energy in vacuum?

27–7 The charged plates of a capacitor attract each other. To pull the plates farther apart therefore requires work by some external force. What becomes of the energy added by this work?

27–8 A solid slab of metal is placed between the plates of a capacitor without touching either plate. Does the capacitance increase, decrease, or remain the same?

27–9 The two plates of a capacitor are given charges $\pm Q$, and then they are immersed in a tank of oil. Does the electric field between them increase, decrease, or remain the same? How may this field be measured?

27–10 Is dielectric strength the same thing as dielectric constant?

27–11 Liquid dielectrics having polar molecules (such as water) always have dielectric constants that decrease with increasing temperature. Why?

27–12 A capacitor made of aluminum foil strips separated by Mylar film was subjected to excessive voltage, and the resulting dielectric breakdown melted holes in the Mylar. After this the capacitance was found to be about the same as before, but the breakdown voltage was much less. Why?

27–13 Two capacitors have equal capacitance, but one has a higher maximum voltage rating than the other. Which one is likely to be bulkier? Why?

27–14 A capacitor is made by rolling a sandwich of aluminum foil and Mylar, as described in Section 27–5. A student claimed that the capacitance when the sandwich is rolled up is twice the value when it is flat. Discuss this allegation.

EXERCISES

Section 27–2 The Parallel-Plate Capacitor

27–1 A parallel-plate air capacitor has a capacitance of 500 pF and a charge of magnitude 0.2 μC on each plate. The plates are 0.2 mm apart.

a) What is the potential difference between the plates?

b) What is the area of each plate?

c) What is the electric-field magnitude between plates?

d) What is the surface charge density on each plate?

27–2 The plates of a parallel-plate capacitor are 4 mm apart and each carries a charge of 8×10^{-8} C. The plates are in vacuum. The electric field between the plates has magnitude 40×10^5 V·m^{-1}.

a) What is the potential difference between the plates?

b) What is the area of each plate?

c) What is the capacitance?

27–3 A capacitor has a capacitance of 8.5 μF. How much charge must be removed to lower the potential difference of its plates by 50 V?

Section 27–3 Capacitors in Series and Parallel

27–4 In Fig. 27–2a let $C_1 = 4\ \mu$F, $C_2 = 6\mu$F, and $V_{ab} = 36$ V. Calculate

a) the charge on each capacitor;

b) the potential difference across each capacitor.

27–5 In Fig. 27–3a let $C_1 = 4\ \mu$F, $C_2 = 6\ \mu$F, and $V_{ab} = 36$ V. Calculate

a) the charge on each capacitor;

b) the potential difference across each capacitor.

27–6 In the circuit shown in Fig. 27–13, $C_1 = 2\ \mu$F, $C_2 = 4\ \mu$F, and $C_3 = 6\ \mu$F. The applied potential is $V_{ab} = 24$ V. Calculate

a) the charge on each capacitor;

b) the potential difference across each capacitor;

c) the potential difference between points a and d.

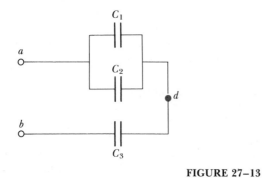

FIGURE 27–13

27–7 In Fig. 27–14 each capacitor has $C = 2\ \mu$F and $V_{ab} = 48$ V. Calculate

a) the charge on each capacitor;

b) the potential difference across each capacitor;

c) the potential difference between points a and d.

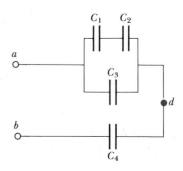

FIGURE 27–14

Section 27–4 Energy of a Charged Capacitor

27–8 An air capacitor is made from two flat parallel plates 0.5 mm apart. The magnitude of charge on each plate is 0.01 μC when the potential difference is 200 V.

a) What is the capacitance?

b) What is the area of each plate?

c) What maximum voltage can be applied without dielectric breakdown? (Dielectric breakdown for air occurs at an electric-field strength 8.0×10^5 V·m^{-1}.)

d) When the charge is 0.01 μC, what total energy is stored?

27–9 A parallel-plate air capacitor has a capacitance of 0.001 μF.

a) What potential difference is required for a charge of 0.5 μC on each plate?

b) In (a), what is the total stored energy?

c) If the plates are 1.0 mm apart, what is the area of each plate?

d) What potential difference is required for dielectric breakdown? (See Exercise 27–8c.)

27–10 An air capacitor consisting of two closely spaced parallel plates has a capacitance of 1000 pF. The charge on each plate is 1 μC.

a) What is the potential difference between the plates?

b) If the charge is kept constant, what will be the potential difference between the plates when the separation is doubled?

c) How much work is required to double the separation?

27–11 A 8-μF parallel-plate capacitor has a plate separation of 4 mm and is charged to a potential difference of 500 V. Calculate the energy density in the region between the plates, in units of J·m^{-3}.

27–12 A 20-μF capacitor is charged to a potential difference of 1000 V. The terminals of the charged capacitor are then connected to those of an uncharged 5-μF capacitor. Compute

a) the original charge of the system;

b) the final potential difference across each capacitor;

c) the final energy of the system;

d) the decrease in energy when the capacitors are connected.

27–13 A parallel-plate capacitor with plate area A and separation x is charged to a charge of magnitude q on each plate.

a) What is the total energy stored in the capacitor?

b) The plates are pulled apart an additional distance dx; now what is the total energy?

c) If F is the force with which the plates attract each other, then the difference in the two energies above must equal the work $dW = F\,dx$ done in pulling the plates apart. Show that $F = q^2/2\epsilon_0 A$.

d) Explain why F is not equal to qE, where E is the electric field between the plates.

Section 27–5 Effect of a Dielectric

27–14 Show that Eq. (27–18) holds for a parallel-plate capacitor with a dielectric material between the plates; use a derivation analogous to that used for Eq. (27–9). Explain why, even though a dielectric *decreases* E, the energy density *increases*.

27–15 A parallel-plate capacitor is to be constructed by using, as a dielectric, rubber with a dielectric constant of 3 and a dielectric strength of $2 \times 10^7\,\text{V}\cdot\text{m}^{-1}$. The capacitor is to have a capacitance of $0.15\ \mu\text{F}$ and must be able to withstand a maximum potential difference of 6000 V. What is the minimum area the plates of the capacitor can have?

27–16 The paper dielectric in a paper-and-foil capacitor is 0.005 cm thick. Its dielectric constant is 2.5, and its dielectric strength is $50 \times 10^6\,\text{V}\cdot\text{m}^{-1}$.

a) What area of paper and tinfoil is required for a 0.1-μF capacitor?

b) If the electric intensity in the paper is not to exceed one-half the dielectric strength, what is the maximum potential difference that can be applied across the capacitor?

27–17 Two parallel plates have equal and opposite charges. When the space between the plates is evacuated, the electric field is $2 \times 10^5\,\text{V}\cdot\text{m}^{-1}$. When the space is filled with dielectric, the electric field is $1.2 \times 10^5\,\text{V}\cdot\text{m}^{-1}$.

a) What is the charge density on the surface of the dielectric?

b) What is the dielectric constant?

27–18 Two oppositely charged conducting plates, with numerically equal quantities of charge per unit area, are separated by a dielectric 5 mm thick, of dielectric constant 3. The resultant electric field in the dielectric is $1 \times 10^6\,\text{V}\cdot\text{m}^{-1}$. Compute

a) the charge per unit area on the conducting plate;

b) the charge per unit area on the surfaces of the dielectric.

27–19 Two parallel plates of area $100\ \text{cm}^2$ are given equal and opposite charges of 1.0×10^{-7} C. The space between the plates is filled with a dielectric material, and the electric field within the dielectric is $3.3 \times 10^5\,\text{V}\cdot\text{m}^{-1}$.

a) What is the dielectric constant of the dielectric?

b) What is the total induced charge on either face of the dielectric?

PROBLEMS

27–20 A parallel-plate air capacitor is made from two plates 0.2 m square, spaced 1 cm apart. It is connected to a 50-V battery.

a) What is the capacitance?

b) What is the charge on each plate?

c) What is the electric field between the plates?

d) What is the energy stored in the capacitor?

e) If the battery is disconnected and the plates are pulled apart to a separation of 2 cm, what are the answers to parts (a), (b), (c), and (d)?

27–21 In Problem 27–20, suppose the battery remains connected while the plates are pulled apart. What are the answers to parts (a), (b), (c), and (d) after the plates have been pulled apart?

27–22 Several 0.5-μF capacitors are available. The voltage across each is not to exceed 400 V. A capacitor of capacitance $0.5\ \mu\text{F}$ must be connected across a potential difference of 600 V.

a) Show in a diagram how an equivalent capacitor having the desired properties can be obtained.

b) No dielectric is a perfect insulator, of infinite resistance. Suppose that the dielectric in one of the capacitors in your diagram is a moderately good conductor. What will happen?

27–23 In Fig. 27–15, each capacitance C_3 is $3\ \mu\text{F}$ and each capacitance C_2 is $2\ \mu\text{F}$.

a) Compute the equivalent capacitance of the network between points a and b.

b) Compute the charge on each of three capacitors nearest a and b when $V_{ab} = 900$ V.

c) With 900 V across a and b, compute V_{cd}.

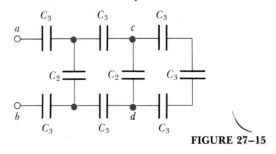

FIGURE 27–15

27–24 A 1-μF capacitor and a 2-μF capacitor are connected in series across a 1200-V supply line.

a) Find the charge on each capacitor and the voltage across each.

b) The charged capacitors are disconnected from the line and from each other and then reconnected, with terminals of like sign together. Find the final charge on each and the voltage across each.

27–25 A 1-μF capacitor and a 2-μF capacitor are connected in parallel across a 1200-V supply line.

a) Find the charge on each capacitor and the voltage across each.

b) The charged capacitors are disconnected from the line and from each other and then reconnected with terminals of unlike sign together. Find the final charge on each and the voltage across each.

27–26 In Fig. 27–3a, let $C_1 = 6$ μF, $C_2 = 3$ μF, and $V_{ab} = 18$ V. Suppose that the charged capacitors are disconnected from the source and from each other and then reconnected, with plates of *opposite* sign together. By how much does the energy of the system decrease?

27–27 Three capacitors having capacitances of 8, 8, and 4 μF are connected in series across a 12-V line.

a) What is the charge on the 4-μF capacitor?

b) What is the total energy of all three capacitors?

c) The capacitors are disconnected from the line and reconnected in parallel, with the positively charged plates connected together. What is the voltage across the parallel combination?

d) What is the total energy now stored in the capacitors?

27–28 The capacitors in Fig. 27–16 are initially uncharged and are connected as in the diagram with switch S open. The applied potential difference is $V_{ab} = +200$ V.

a) What is the potential difference V_{cd}?

b) What is the potential difference V_{ad} after switch S is closed?

c) How much charge flowed through the switch when it was closed?

27–29 A capacitor consists of two parallel plates of area 25 cm^2 separated by a distance of 0.2 cm. The material between the plates has a dielectric constant of 5. The plates of the capacitor are connected to a 300-V battery.

a) What is the capacitance of the capacitor?

b) What is the charge on either plate?

c) What is the energy in the charged capacitor?

d) What is the energy density in the dielectric?

27–30 An air capacitor is made from two flat plates of area A separated by a distance d. Then a metal slab with thickness a (less than d) and the same shape and size as the plates is inserted between them, parallel to the plates and not touching either plate.

a) What is the capacitance of this arrangement?

b) Express the capacitance as a multiple of the capacitance when the metal slab is not present.

27–31 A spherical capacitor consists of an inner metal sphere of radius r_a supported on an insulating stand at the center of a hollow metal sphere of inner radius r_b; there is a charge $+Q$ on the inner sphere and a charge $-Q$ on the outer. (See Exercise 26–18.)

a) What is the potential difference V_{ab} between the spheres?

b) Prove that the capacitance is

$$C = 4\pi\epsilon_0 \frac{r_a r_b}{r_b - r_a}.$$

c) If $r_b - r_a = d$, show that the equation obtained in (b) reduces to Eq. (27–2) when $d \ll r_a$, with A being the surface area of each sphere.

27–32 A coaxial cable consists of an inner solid cylindrical conductor of radius r_a supported by insulating disks on the axis of a conducting tube of inner radius r_b. The two cylinders are oppositely charged with a charge λ per unit length. (See Problem 26–38.)

a) What is the potential difference between the two cylinders?

b) Prove that the capacitance of a length L of the cable is

$$C = \frac{2\pi\epsilon_0 L}{\ln(r_b/r_a)}.$$

Neglect any effect of the supporting disks.

c) If $r_b - r_a = d$, show that the equation obtained in (b) reduces to Eq. (27–2) when $d \ll r_a$, with A being the surface area of each cylinder.

27–33 A parallel-plate capacitor has the space between the plates filled with two slabs of dielectric, one with constant K_1 and one with constant K_2. Each slab has thickness $d/2$, where d is the plate separation. Show that the capacitance is

$$C = \frac{2\epsilon_0 A}{d} \left(\frac{K_1 K_2}{K_1 + K_2} \right).$$

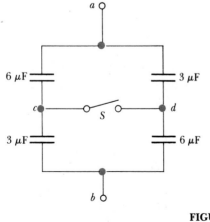

FIGURE 27–16

CHALLENGE PROBLEMS

27–34 A fuel gauge uses a capacitor to determine the height of the fuel in a tank. The effective dielectric constant K_{eff} changes from a value of 1 when the tank is empty to a value of K, the dielectric constant of the fuel, when the tank is full. The appropriate electronic circuitry can determine the effective dielectronic constant of the combined air and fuel between the capacitor plates. Each of the two rectangular plates has a width w (not shown) and a length l. (See Fig. 27–17.) The height of the fuel between the plates is h. Neglect any fringing effects.

a) Derive an expression for K_{eff} as a function of h.

b) What would be the effective dielectronic constant for a tank one-quarter full, one-half full, and three-quarters full if the fuel were gasoline ($K = 1.95$)?

c) Repeat part (b) for methanol ($K = 33$).

d) For which fuel would this fuel gauge be more practical?

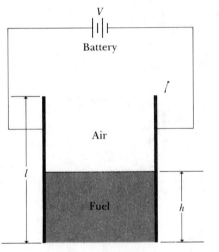

FIGURE 27–17

27–35 Three square metal plates A, B, and C, each 10 cm on a side and 3 mm thick, are arranged as in Fig. 27–18. The plates are separated by sheets of paper 0.5 mm thick and of dielectric constant 5. The outer plates are connected together and to point b. The inner plate is connected to point a.

a) Copy the diagram and show by + and − signs the charge distribution on the plates when point a is maintained at a positive potential relative to point b.

b) What is the capacitance between points a and b?

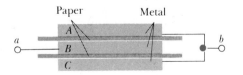

FIGURE 27–18

27–36 A parallel-plate capacitor consists of two horizontal conducting plates of equal area A. The bottom plate is resting on a fixed support, and the top plate is suspended by four springs of spring constant k, positioned at each of the four corners of the top plate. (See Fig. 27–19.) The plates, when uncharged, are separated by a distance z_0. A battery connected to the plates produces a potential difference V between them. This causes the plate separation to decrease to z. Neglect any fringing effects.

a) Show that the electrostatic force between the charged plates has a magnitude $\epsilon_0 A V^2/2z^2$. (*Note.* See Exercise 27–13.)

b) Obtain an expression that relates the plate separation z to the potential difference V between the plates. The resulting equation will be cubic in z.

c) Given the values $A = 0.25 \text{ m}^2$, $z_0 = 1 \text{ mm}$, $k = 25 \text{ N·m}^{-1}$, and $V = 100 \text{ V}$, find the two values of z for which the top plate will be in equilibrium. (*Hint:* You should solve this last part by using an iterative technique. Rearrange the equation to isolate z on the left side of the equation. Insert a trial value of z on the right side of the equation and calculate a new value of z. Insert the new value of z on the right side of the equation and calculate a new value again. Continue this process until the old and new values of z agree to within three significant figures. The third root of the cubic equation has a nonphysical negative value.)

d) For each of the two values of z found in (c), is the equilibrium stable or unstable? For stable equilibrium a small displacement of the object in question will give rise to a net force tending to return the object to the equilibrium position. For unstable equilibrium a small displacement gives rise to a net force that takes the object farther away from equilibrium.

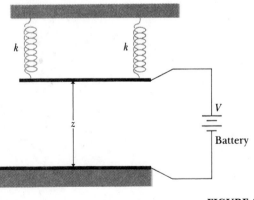

FIGURE 27–19

27–37 It is not always possible to combine capacitors in a simple series or parallel relationship. Consider the capacitors C_x, C_y, and C_z in the network depicted in Fig. 27–20a. Such a configuration of capacitors, referred to as a delta network, cannot be transformed into a single equivalent capacitor because three terminals a, b, and c exist in that network. It can be shown that as far as any effect on the external circuit is concerned, a delta network can be transformed into what is called a Y network. For example, the delta network of Fig. 27–20a can be replaced by the Y network of Fig. 27–20b.

a) Show that the transformation equations that give C_1, C_2, and C_3 in terms of C_x, C_y, and C_z are

$$C_1 = [C_xC_y + C_yC_z + C_zC_x]/C_x,$$
$$C_2 = [C_xC_y + C_yC_z + C_zC_x]/C_y,$$
$$C_3 = [C_xC_y + C_yC_z + C_zC_x]/C_z.$$

(*Hint:* The potential difference V_{ac} must be the same in both circuits, as must also be V_{bc}. Furthermore, the charge q_1 that flows from point a along the wire as indicated must be the same in both circuits, as must q_2. Obtain a relationship for V_{ac} as a function of q_1 and q_2 [and the capacitances] for each network and a separate relationship for V_{bc} as a function of the charges for each network. The coefficients of corresponding charges in corresponding equations must be the same for both networks.)

b) Determine the equivalent capacitance of the network of capacitors, between the terminals at the left-hand end of the network, for the network shown in Fig. 27–20c. (*Hint:* Use the delta-Y transformation derived in [a]. Use points a, b, and c to form the delta, and transform the delta into a Y. The capacitors can then be easily combined by using the relationships for series and parallel combinations of capacitors.)

c) Determine the charges of the 72-μF, 27-μF, 18-μF, 6-μF, 28-μF, and 21-μF capacitors and the potential differences across those capacitors.

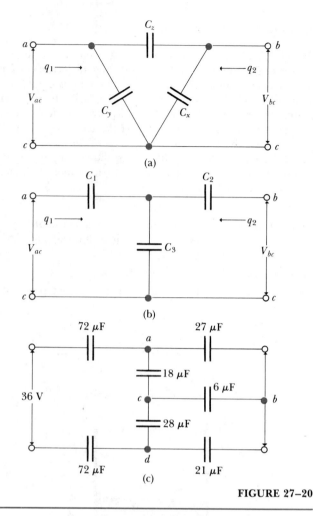

(a)

(b)

(c)

FIGURE 27–20

28

CURRENT, RESISTANCE, AND ELECTROMOTIVE FORCE

IN THIS CHAPTER WE STUDY THE BASIC PRINCIPLES OF ELECTRIC CURRENT (flow of charge within conducting materials) and electric circuits (continuous closed paths for electric currents). These principles form the basis of electric-circuit analysis, and as such underlie the whole fabric of our modern electronic age. We characterize the relation of current to potential difference or "voltage" by means of the concepts of resistivity and resistance and a relation called Ohm's law. Both resistivity and resistance are significantly temperature dependent. The basic principles of circuit analysis are contained in two principles called Kirchhoff's laws. Energy and power considerations are of primary importance in many electric circuits, and we study the energy relations in circuits. Finally, we discuss briefly a simple microscopic model of electrical conduction in metals.

28–1 CURRENT

Any motion of charge from one region of a conductor to another constitutes **current.** When an isolated conductor is placed in an electrostatic field, the mobile charges in the conductor rearrange themselves so as to make the entire interior of the conductor a field-free region and its surface an equipotential surface. The motion of charges during this rearrangement is a temporary, or *transient,* current. To maintain a *continuous* current, we must maintain a steady force on the mobile charge in the conductor, either with an electrostatic field or by other means to be described later in this chapter. For the present we assume that an electric field E is present within the conductor, so that a particle of charge q experiences a force $F = qE$.

The motion of a charged particle in a conductor is very different from the continuously accelerated motion that results when a constant field is applied to a charge in vacuum. In a conductor, a charge bumps along randomly, accelerating until it collides with a stationary particle. The charged particle thus gives up some of its kinetic energy, accelerates again until it bumps into something else, and so on. Thus there is a lot of back-and-forth motion, with a gradual *drift* in the direction of the electric-field force. The inelastic collisions

Current is charge in motion.

Motion of charge in a conductor: random motion plus a steady drift

28–1 All the particles, and only those particles, within the shaded cylinder will cross its base in time Δt.

with the stationary charges transfer energy to them; this increases their vibrational energy and hence the temperature of the conductor.

It is useful to consider the motion of charge in a conductor, across or through some imaginary area; the area might be, for example, a cross section of a wire used as a conductor. We define the current through an area as *the net charge flowing through the area per unit time.* Thus if a net charge ΔQ flows through an area in a time Δt, the current I through the area is

$$I = \frac{\Delta Q}{\Delta t}.$$

The rate of flow of charge may vary with time; in that case we generalize the definition of current in a natural way, using the derivative. We define the *instantaneous current* I as

$$I = \frac{dQ}{dt}. \qquad (28\text{--}1)$$

Current is a *scalar* quantity.

One ampere is one coulomb per second: the SI unit of current.

The SI unit of current, one *coulomb per second,* is called one **ampere** ($1\ \text{A} = 1\ \text{C}\cdot\text{s}^{-1}$), in honor of the French scientist André Marie Ampère (1775–1836). Small currents are more conveniently expressed in *milliamperes* ($1\ \text{mA} = 10^{-3}\ \text{A}$) or in *microamperes* ($1\ \mu\text{A} = 10^{-6}\ \text{A}$).

The current through an area can be expressed in terms of the drift velocity v of the moving charges. Consider a portion of a conductor of cross-sectional area A through which there is an electric field E directed from left to right. Suppose first that the conductor contains free *positively* charged particles; these move in the same direction as the field. A few positive particles are shown in Fig. 28–1. Suppose there are n such particles per unit volume, all moving with a drift velocity v. In a time Δt, each particle advances a distance $v\ \Delta t$. Hence all the particles within the shaded cylinder of length $v\ \Delta t$, and only those particles, flow through the end of the cylinder during time Δt. The volume of the cylinder is $Av\ \Delta t$, and the number of particles within it is $nAv\ \Delta t$. If each particle has a charge q, the charge ΔQ flowing through the end of the cylinder in time Δt is

$$\Delta Q = nqvA\ \Delta t.$$

How is current related to drift velocity and density of charge in the conductor?

The current carried by the positively charged particles is therefore

$$I = \frac{\Delta Q}{\Delta t} = nqvA. \qquad (28\text{--}2)$$

If the moving charges are negative rather than positive, the electric-field force is opposite to E, and the drift velocity is right to left, opposite to the direction shown in Fig. 28–1. But the current is still left to right; the reason is that negative charge moving right to left increases the positive charge at the right of the section, just as positive charge moving from left to right does. Thus the motion of *both* kinds of charge has the same effect, namely, to increase the positive charge at the right of the section. In both cases particles flowing out through an end of the cylindrical section are continuously replaced by particles flowing *in* through the opposite end.

A simple laboratory demonstration of electrolytic conductivity. The sodium chloride solution in the beaker is part of the circuit, and the bulb glows brightly. (Photo by Chip Clark.)

In general, a conductor may contain several different kinds of charged particles having charges $q_1, q_2, \ldots$; densities $n_1, n_2, \ldots$; and drift velocities v_1,

$v_2, \ldots$. The total current is then

$$I = A(n_1q_1v_1 + n_2q_2v_2 + \cdots). \qquad (28\text{–}3)$$

In metals, the moving charges are always (negative) electrons, while in an ionized gas (plasma) both electrons and positively charged ions are moving. In a semiconductor material such as germanium or silicon, conduction is partly by electrons and partly by motion of *vacancies*, also known as *holes;* these are sites of missing electrons and act like positive charges. Conduction by motion of holes in semiconductors is discussed in Section 43–8.

Current in conductors can be due to either positive or negative charge motion, or both.

The current *per unit cross-sectional area* is called the **current density** J. For each kind of charged particle, $J = I/A = nqv$, and in general,

Current density: current per unit area

$$J = \frac{I}{A} = n_1q_1v_1 + n_2q_2v_2 + \cdots. \qquad (28\text{–}4)$$

We can also define a vector current density $\boldsymbol{J}$ that includes the directions of the drift velocities:

$$\boldsymbol{J} = n_1q_1\boldsymbol{v}_1 + n_2q_2\boldsymbol{v}_2 + \cdots. \qquad (28\text{–}5)$$

The direction of the drift velocity $\boldsymbol{v}$ of a positive charge is the same as that of the electric field $\boldsymbol{E}$, and the direction of the velocity of a negative charge is opposite to $\boldsymbol{E}$. But because the charge q is negative, each of the vectors $nq\boldsymbol{v}$ is in the same direction as $\boldsymbol{E}$, and hence the *vector current density* $\boldsymbol{J}$ *always has the same direction as the field* $\boldsymbol{E}$. Even in a metallic conductor, where the moving charges are negative electrons only and move in the *opposite* direction to $\boldsymbol{E}$, the *vector* current density $\boldsymbol{J}$ is in the *same* direction as $\boldsymbol{E}$.

How is current density related to the electric field that causes the charge motion?

Thus the effect of a current is the same, whether it consists of positive charges moving in the direction of $\boldsymbol{E}$, of negative charges moving in the opposite direction, or a combination of the two. When the current consists of a flow of positive charges, the direction of the current is the same as that of the motion of the charges; when it is a flow of negative charges, the direction of the current is *opposite* to that of the motion of the charges. These statements are consistent with Eq. (28–5): When q is negative, $\boldsymbol{J}$ and $\boldsymbol{v}$ have opposite directions. In describing circuit behavior it is customary to describe currents as though they consisted entirely of positive charge flow, even in cases where the actual current is known to be due to electrons. We will follow this convention in the following sections. In Chapter 30, when we study the effect of a *magnetic* field on a moving charge, we will consider a phenomenon called the Hall effect, in which the sign of the moving charges *is* important.

When there is a steady current in a closed loop (a "complete circuit"), the total charge in every segment of the conductor is constant. The rate of flow of charge *out* at one end of a segment equals the rate of flow of charge *in* at the other end of the segment. In other words, *the current is the same at all cross sections.* Current is *not* something that squirts out of the positive terminal of a battery and is consumed or used up by the time it reaches the negative terminal. These statements are direct consequences of the principle of conservation of charge, introduced in Section 24–2.

Conservation of charge: The current is the same at all points in a conducting loop.

EXAMPLE 28–1 A copper conductor of square cross section 1 mm on a side carries a constant current of 20 A. The density of free electrons is 8×10^{28} electrons per cubic meter. Find the current density and the drift velocity.

An example of charge motion in a conductor

SOLUTION The current density in the wire is

$$J = \frac{I}{A} = 20 \times 10^6 \text{ A·m}^{-2}.$$

From Eq. (28–4),

$$v = \frac{J}{nq} = \frac{(20 \times 10^6 \text{ A·m}^{-2})}{(8 \times 10^{28} \text{ m}^{-3})(1.6 \times 10^{-19} \text{ C})}$$

$$= 1.6 \times 10^{-3} \text{ m·s}^{-1},$$

or about 1.6 mm·s^{-1}. At this speed an electron would require 625 s or about 10 min to travel the length of a wire 1 m long.

28–2 RESISTIVITY

Resistivity: the ratio of electric field to current density

The current density J in a conductor depends on the electric field E and on the nature of the conductor. In general, the dependence of J on E can be quite complex. For some materials, especially the metals, however, it can be represented quite well by a direct proportionality. For such materials the ratio of E to J is *constant*.

We define the **resistivity** ρ of a particular materials as the ratio of electric field to current density:

$$\rho = \frac{E}{J}. \qquad (28\text{–}6)$$

That is, the resistivity is the *electric field per unit current density*. The greater the resistivity, the greater the field needed to establish a given current density, or the smaller the current density caused by a given field. Representative values of resistivity are given in Table 28–1. The unit $\Omega \cdot \text{m}$ (ohm·meter) will be explained in the following section. A "perfect" conductor would have zero resistivity, and a "perfect" insulator would have infinite resistivity. Metals and alloys have the lowest resistivities and are the best conductors. The resistivities of insulators exceed those of the metals by a factor of the order of 10^{22}

Resistivities of conductors and insulators cover an enormous range of magnitudes.

Comparison with Table 16–1 shows that *thermal* insulators have thermal conductivities that differ from those of good thermal conductors by factors of only 10^3. By the use of electrical insulators, *electric* currents can be confined to well-defined paths in good electrical conductors, while it is impossible to con-

TABLE 28–1   Resistivities at Room Temperature

Substance		ρ, $\Omega \cdot$m	Substance		ρ, $\Omega \cdot$m
Conductors			Semiconductors		
Metals	Silver	1.47×10^{-8}	Pure	Carbon	3.5×10^{-5}
	Copper	1.72×10^{-8}		Germanium	0.60
	Gold	2.44×10^{-8}		Silicon	2300
	Aluminum	2.63×10^{-8}	Insulators		
	Tungsten	5.51×10^{-8}		Amber	5×10^{14}
	Steel	20×10^{-8}		Glass	10^{10}–10^{14}
	Lead	22×10^{-8}		Lucite	$>10^{13}$
	Mercury	95×10^{-8}		Mica	10^{11}–10^{15}
Alloys	Manganin	44×10^{-8}		Quartz (fused)	75×10^{16}
	Constantan	49×10^{-8}		Sulfur	10^{15}
	Nichrome	100×10^{-8}		Teflon	$>10^{13}$
				Wood	10^8–10^{11}

TABLE 28–2 Temperature Coefficients of Resistivity (Approximate Values Near Room Temperature)

Material	α, C$^{\circ -1}$	Material	α, C$^{\circ -1}$
Aluminum	0.0039	Lead	0.0043
Brass	0.0020	Manganin (Cu 84,	0.000000
Carbon	−0.0005	Mn 12, Ni 4)	
Constantan (Cu 60, Ni 40)	+0.000002	Mercury	0.00088
Copper (commercial	0.00393	Nichrome	0.0004
annealed)		Silver	0.0038
Iron	0.0050	Tungsten	0.0045

fine *heat* currents to a comparable extent. Just as good electrical conductors, such as the metals, are also good conductors of heat, poor electrical conductors, such as ceramic and plastic materials, are also poor thermal conductors. The free electrons in a metal that carry charge in electrical conductions also play an important role in the conduction of heat; hence we expect a correlation between electrical and thermal conductivity.

The *semiconductors* form a class of materials intermediate between the metals and the insulators. They are important not primarily because of their resistivities, but because of the way in which these are affected by temperature and by small amounts of impurities.

The discovery that J is proportional to E for a metallic conductor at constant temperature was made by G. S. Ohm (1789–1854) and is called **Ohm's law.** A material obeying Ohm 's law is called an *ohmic* conductor, or a *linear* conductor. If Ohm's law is *not* obeyed, the conductor is called *nonlinear*. Ohm's law, like the ideal-gas equation, Hooke's law, and many other relations describing the properties of materials, is an *idealized model* that describes the behavior of certain materials reasonably well but is by no means a general description of all matter.

The resistivity of all *metallic* conductors increases with increasing temperature, as shown in Fig. 28–2a. Over a moderate temperature range, the resistivity of a metal can be represented approximately by the equation

$$\rho_T = \rho_0[1 + \alpha(T - T_0)], \tag{28-7}$$

where ρ_0 is the resistivity at a reference temperature T_0 (often taken as 0°C or 20°C) and ρ_T is the resistivity at temperature T. The factor α is called the **temperature coefficient of resistivity.** Some representative values are given in Table 28–2. The resistivity of carbon (a nonmetal) *decreases* with increasing temperature, and its temperature coefficient of resistivity is negative. The resistivity of the alloy manganin is practically independent of temperature.

A number of materials have been found to exhibit the property of *superconductivity*. As the temperature is decreased, the resistivity at first decreases regularly, like that of any metal. At a certain transition temperature, usually in the range of 0.1 K to 20 K, a phase transition occurs, and the resistivity suddenly drops to zero, as shown in Fig. 28–2b. A current once established in a superconducting ring will continue indefinitely without the presence of any driving field. Superconductivity is discussed in greater detail in Section 43–10.

The resistivity of a *semiconductor* decreases rapidly with increasing temperature, as shown in Fig. 28–2c. A tiny bead of semiconducting material, called a *thermistor*, can be used as the temperature-sensing element in a sensitive electronic thermometer. That is, its resistivity is used as a thermometric property.

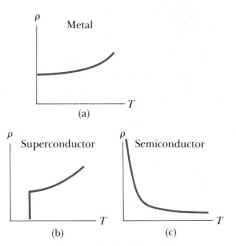

28–2 Variation of resistivity with temperature for three conductors: (a) an ordinary metal; (b) a superconducting metal, alloy, or compound; and (c) a semiconductor.

For most materials, resistivity increases with temperature.

Superconductivity: At very low temperatures, some materials have zero resistivity.

28–3 RESISTANCE

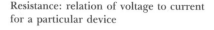

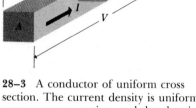

28–3 A conductor of uniform cross section. The current density is uniform over any cross section, and the electric field is constant along the length.

The current density J, at a point within a conductor where the electric field is E, is given by Eq. (28–6):

$$E = \rho J.$$

It is often difficult to measure E and J directly, so it is useful to state this relation in a form involving readily measured quantities, such as total current and potential difference. To do this, consider a conductor with uniform cross-sectional area A and length l, as shown in Fig. 28–3. Assuming a constant current density over a cross section, and a uniform electric field along the length of the conductor, the total current I is given by

$$I = JA,$$

and the potential difference V between the ends of the conductor is

$$V = El. \tag{28–8}$$

Solving these equations for J and E, respectively, and substituting the results in Eq. (28–6), we obtain

$$\frac{V}{I} = \frac{\rho l}{A}. \tag{28–9}$$

Thus the total current is proportional to the potential difference.

The quantity $\rho l / A$ for a particular specimen of material is called its **resistance** R:

$$R = \frac{\rho l}{A}. \tag{28–10}$$

Equation (28–9) then becomes

$$V = IR. \tag{28–11}$$

This relation is also called *Ohm's law;* in this form it refers to a specific piece of material, not to a general property of the material as with Eq. (28–6).

Equation (28–10) shows that the resistance of a wire or other conductor of uniform cross section is directly proportional to its length and inversely proportional to its cross-sectional area. It is of course also proportional to the resistivity of the material of which the conductor is made.

The SI unit of resistance is one *volt per ampere* (1 V·A^{-1}). This unit is given the name 1 **ohm** ($1 \ \Omega$). The unit of resistivity is therefore one *ohm·meter* ($1 \ \Omega\text{·m}$). Large resistances are conveniently expressed in *kilohms* ($1 \text{ k}\Omega = 10^3 \ \Omega$) or *megohms* ($1 \text{ M}\Omega = 10^6 \ \Omega$). Resistivities are also expressed in a variety of hybrid units, most common of which is the ohm·centimeter ($1 \ \Omega\text{·cm} = 10^{-2} \ \Omega\text{·m}$).

Because the resistance of any sample of material is proportional to its resistivity, which varies with temperature, resistance also varies with temperature. For temperature ranges that are not too great, this variation may be represented approximately as a linear relation analogous to Eq. (28–7):

$$R_T = R_0[1 + \alpha(T - T_0)]. \tag{28–12}$$

Here R_T is the resistance at temperature T, and R_0 is the resistance at temperature T_0, often taken to be 20°C or 0°C. Within the limits of validity of Eq. (28–12), the *change* in resistance resulting from a temperature change $T - T_0$ is given by $R_0\alpha(T - T_0)$, where α is the temperature coefficient of resistivity, given for several common materials in Table 28–2.

EXAMPLE 28–2 In Example 28–1 (Section 28–1), find (a) the electric field; (b) the potential difference between two points 100 m apart; and (c) the resistance of a 100-m length of this copper conductor.

SOLUTION

(a)   From Eq. (28–6), the electric field is given by

$$E = \rho J = (1.72 \times 10^{-8}\ \Omega\cdot\text{m})(20 \times 10^6\ \text{A}\cdot\text{m}^{-2})$$
$$= 0.344\ \text{V}\cdot\text{m}^{-1}.$$

An example of a resistance calculation

(b)   The potential difference is given by

$$V = El = (0.344\ \text{V}\cdot\text{m}^{-1})(100\ \text{m})$$
$$= 34.4\ \text{V}.$$

(c)   The resistance of a piece of this wire 100 m in length is

$$R = \frac{V}{I} = \frac{34.4\ \text{V}}{20\ \text{A}}$$
$$= 1.72\ \Omega.$$

This result can also be obtained directly from Eq. (28–10):

$$R = \frac{\rho l}{A} = \frac{(1.72 \times 10^{-8}\ \Omega\cdot\text{m})(100\ \text{m})}{(1 \times 10^{-3}\ \text{m})^2}$$
$$= 1.72\ \Omega.$$

EXAMPLE 28–3 In Example 28–2, suppose the resistance is 1.72 Ω at a temperature of 20°C. Find the reistance at 0°C and at 100°C.

SOLUTION We use Eq. (28–12). In this instance $T_0 = 20$°C and $R_0 = 1.72\ \Omega$. From Table 28–2, the temperature coefficient of resistivity of copper is $\alpha = 0.00393$ (C°)$^{-1}$. Thus at $T = 0$°C,

An example of the temperature variation of resistance

$$R = (1.72\ \Omega)[1 + (0.00393\text{C}^{°-1})(0°\text{C} - 20°\text{C})]$$
$$= 1.58\ \Omega,$$

and at $T = 100$°C,

$$R = (1.72\ \Omega)[1 + (0.00393\text{C}^{°-1})(100°\text{C} - 20°\text{C})]$$
$$= 2.26\ \Omega.$$

EXAMPLE 28–4 The space between two metallic coaxial cylinders of radii a and b is filled with a material of resistivity ρ. What is the resistance between the cylinders?

Radial current flow: an interesting variation

SOLUTION Equation (28–10) cannot be used directly because the cross section through which the charge travels varies from $2\pi a l$ at the inner cylinder to $2\pi b l$ at the outer cylinder. Instead, we consider a thin cylindrical shell of inner radius r and thickness dr. The area A is then $2\pi rl$, and the length of the current path through the shell is dr. Thus the resistance dR of the shell is

$$dR = \frac{\rho\,dr}{2\pi rl}$$

The current must pass successively through each one of these infinitesimal coaxial shells, so the total resistance between the cylinders is the *sum* of the resistances of the shells:

$$R = \frac{\rho}{2\pi l}\int_a^b \frac{dr}{r} = \frac{\rho}{2\pi l}\ln\frac{b}{a}.$$

In this last step we have used the fact that the total resistance of several resistors in series is the sum of their individual resistances. We will derive this relation in detail in Section 29–1.

28–4 ELECTROMOTIVE FORCE AND CIRCUITS

In order for a steady current to exist in a conducting path, that path must form a closed loop, or **complete circuit.** Otherwise charge would accumulate at the ends of the conductor, the resulting electric field would change with time, and the current could not be constant.

However, such a path cannot consist entirely of resistance. Current in a resistor requires an electric field and an associated potential difference. The field always does *positive* work on the charge, which always moves in the direction of *decreasing* potential. But after a complete trip around the loop, the charge returns to its starting point, and the potential there must be the same as when it left that point. This is impossible if the trip around the loop involves only *decreases* in potential.

We can compare this situation with that of a decorative water fountain. Water emerges from openings at the top, cascades down over terraces and spouts, and eventually reaches the basin in the bottom. It collects there and runs into a pump that lifts it back to the top for another trip. Without the pump, the water would not be able to circulate continuously.

Thus in the electric circuit there must be some part of the loop where a charge travels "uphill," from lower to higher potential, despite the fact that the electrostatic force is trying to push it from higher to lower potential. The influence that makes charge move from lower to higher potential is called **electromotive force.** Every complete circuit having a steady current must include some device that provides electromotive force, usually abbreviated emf and pronounced "ee-em-eff."

Batteries, electric generators, solar cells, thermocouples, and fuel cells are all examples of sources of emf. Any such device has the ability to convert energy of some form (mechanical, chemical, thermal, and so on) into electrical energy and transfer it into the circuit where it is connected. An ideal source of emf would maintain a constant potential difference between its terminals, independent of the current through it. As we will see, such an ideal source is a mythical beast, like the unicorn, the frictionless plane, and the ideal gas. But it is a useful idealized model, and we will discuss later how real-life sources of emf differ in their behavior from this model.

Electromotive force: Charge can't flow downhill forever.

Lightning striking the Eiffel Tower in Paris. The steel of the tower forms a good conducting path to ground. (Courtesy of the Lockyer Collection.)

Figure 28–4 is a schematic diagram of a source of emf that maintains a potential difference between conductors a and b, called the *terminals* of the device. Terminal a, marked $+$, is maintained at *higher* potential than terminal b, marked $-$. Associated with this potential difference is an electric field E in the region around the terminals, both inside and outside the source. The electric field inside the device is directed from a to b, as shown. A charge q within the source experiences an electrical force $F_e = qE$ caused by this field. But the source is itself a conductor, and if this were the *only* force on the free charges, then positive charge would move through the source from a toward b, and negative charge would move from b toward a. The excess charges on the conductors would decrease, and the potential difference would eventually decrease to zero.

But this is *not* the way batteries and generators actually behave; they maintain a potential difference even when there is a steady current through them from b to a. Thus in a source some additional influence must be present that tends to push positive charges from lower to higher potential, *opposite* to the direction of the electric-field force F_e. The origin of this additional influence and its associated energy depends on the nature of the source. In a generator it consists of an additional electrical force that results from redistribution of charge in a conductor; this redistribution is caused by magnetic-field forces on moving charges. In a battery or fuel cell it is associated with diffusion processes caused by varying electrolyte concentrations associated with chemical reactions and their energies. In an electrostatic machine such as a Van de Graaff or Wimshurst generator, mechanical force is applied by a moving belt or wheel.

In all these cases we can represent this additional influence in terms of an equivalent non-electrostatic force F_n that is present in addition to the electrostatic force F_e. In some cases, such as the Van de Graaff generator, F_n is an actual mechanical force. In other cases it is a simplified representation of a complex set of phenomena.

The potential V_{ab} of point a with respect to point b is defined, as always, as the work per unit charge performed by the electrostatic force F_e on a charge moving from a to b. Similarly, the emf of the source, denoted by $\mathcal{E}$, is the energy per unit charge supplied by the additional influence mentioned above during the "uphill" displacement from b to a. (It is the work per unit charge done by the non-electrostatic force F_n.) For a source on open circuit, the potential difference V_{ab} is equal to the electromotive force $\mathcal{E}$:

$$V_{ab} = \mathcal{E} \qquad \text{(source on open circuit).} \qquad (28–13)$$

The SI unit of emf is the same as that of potential or potential difference, namely, $1 \, \text{J} \cdot \text{C}^{-1}$ or $1 \, \text{V}$. There is a subtle but important distinction between emf and potential difference. Electromotive force refers to the energy supplied (per unit charge) by the non-electrostatic influences within a source (the work done by the non-electrostatic force). Potential difference is associated with electrostatic fields caused by distributions of electric charge. The emf may often be taken to be constant, independent of current, for a given source, while V_{ab} typically depends on current. In the following discussion we usually assume the emf of a source to be constant.

Now suppose that the terminals of a source are connected by a wire, as shown schematically in Fig. 28–5, forming a *complete circuit*. The driving force on the free charges *in the wire* is due solely to the electrostatic field E_e set up by

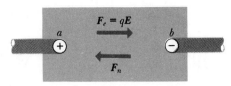

28–4 Schematic diagram of a source of emf in an "open-circuit" situation. The electric-field force $F_e = qE$ and the nonelectrostatic force F_n on a charge q are shown. The work done by F_n on a charge q moving from b to a is equal to $q\mathcal{E}$, where $\mathcal{E}$ is the electromotive force. In the open-circuit situation, F_e and F_n have equal magnitude.

Representing the effect of a source of emf in terms of an equivalent non-electrostatic field

When there is no current, the terminal potential difference of a source equals its emf.

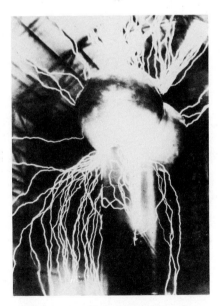

A Van de Graaff generator. Electric charge is carried to the dome at the top by a moving belt inside the vertical column. Potential differences as large as 10^6 can be developed. (American Institute of Physics.)

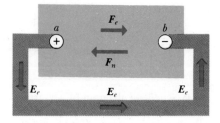

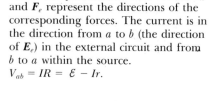

28–5 Schematic diagram of a source with a complete circuit. The vectors $\mathbf{F}_n$ and $\mathbf{F}_e$ represent the directions of the corresponding forces. The current is in the direction from a to b (the direction of $\mathbf{E}_e$) in the external circuit and from b to a within the source. $V_{ab} = IR = \mathcal{E} - Ir$.

Internal resistance of a source: why the terminal voltage isn't always equal to the emf

the charged terminals a and b of the source. This field sets up a current *in the wire* from a toward b. If the wire has resistance R, then from Eq. (28–11) the current I in the circuit is determined by

$$V_{ab} = IR. \qquad (28\text{–}14)$$

For an ideal source, where current can pass through the source without impediment, charge going out into the external circuit through terminal a is replaced immediately by charge flow through the source from terminal b. Thus the terminal potential difference V_{ab} is still equal to $\mathcal{E}$. But V_{ab} is also related to the current I and resistance R in the external circuit by Eq. (28–14), so we have

$$\mathcal{E} = V_{ab} = IR. \qquad (28\text{–}15)$$

Once $\mathcal{E}$ and R are known, this relation determines the current in the circuit. The current is the same at every point in the circuit, since charge cannot be created or destroyed.

Real sources do not behave exactly this way because charge cannot pass perfectly freely through any real source. Instead, every real source has some **internal resistance,** which we denote by r. The current through r has an associated drop in potential equal to Ir, and so the terminal potential difference under closed-circuit conditions is given by

$$V_{ab} = \mathcal{E} - Ir. \qquad (28\text{–}16)$$

The non-electrostatic force in the source must now be somewhat larger than the electrostatic force corresponding to V_{ab}, in order to push the charges through the internal resistance. The current in the external circuit is still determined by Eq. (28–14); combining it with Eq. (28–16), we find

$$\mathcal{E} - Ir = IR,$$

or

$$I = \frac{\mathcal{E}}{R + r}. \qquad (28\text{–}17)$$

What is a short circuit?

That is, the current equals the source emf divided by the *total* circuit resistance, external plus internal.

If the terminals of a source are connected by a conductor of zero (or negligible) resistance, the source is said to be *short circuited*. (This would be an *extremely* dangerous procedure to carry out with the storage battery of your car or with the terminals of the power line. Don't try it!) Then $R = 0$, and from the circuit equation the *short-circuit current* I_s is

$$I_s = \frac{\mathcal{E}}{r}. \qquad (28\text{–}18)$$

The terminal voltage is then zero:

$$V_{ab} = \mathcal{E} - \left(\frac{\mathcal{E}}{r}\right)r = 0. \qquad (28\text{–}19)$$

The *electrostatic* field within the source is zero, and the driving effect on the charges within it is only the *non-*electrostatic force.

Symbols used in circuit diagrams

Thus a source with its associated emf can be described in terms of two properties: an emf $\mathcal{E}$, which supplies a constant potential difference inde-

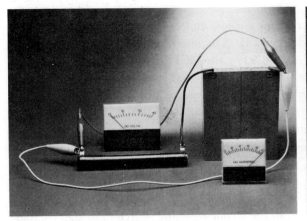 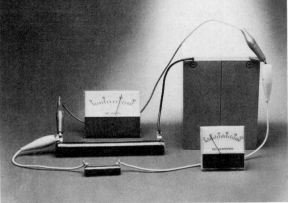

A battery on short circuit. The ammeter has very small resistance and provides a low-resistance current path. The current is very large (off the scale of the ammeter). The voltage drop across the internal resistance of the battery is nearly equal to its emf, and the terminal voltage of the battery is nearly zero.

A battery in a complete circuit. The resistor and the ammeter in series provide a complete circuit; the ammeter indicates the current through the resistor (about 2 A), and the voltmeter shows the terminal voltage of the battery (about 10 V). The battery's terminal voltage is somewhat less than its emf because of the voltage drop across its internal resistance. (Photos by Chip Clark.)

pendent of current, in series with an internal resistance r. These quantities can be determined (at least in principle) from measurements of the *open-circuit* terminal voltage, which equals $\mathcal{E}$, and the *short-circuit* current, which we can use to calculate r from Eq. (28–18).

In preceding sections we used pictorial diagrams to show the electric fields within sources and conductors. We now introduce the usual symbols used in electric-circuit diagrams. A resistor is represented by the symbol

Conductors having negligible resistance are shown by straight lines.

A variable resistor is called a *rheostat* or, particularly in electronics, a *potentiometer*. A common type consists of a resistor with a sliding contact that can be moved along its length. The symbol

is used for a variable resistor.

A source of emf is represented by the symbol

The longer vertical line always corresponds to the + terminal. We will modify this in the following examples to

Circuit symbol for a source with internal resistance

or

to show explicitly that a source has an internal resistance. The order of the two parts of this symbol is immaterial.

Examples of simple circuit calculations

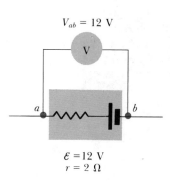

$V_{ab} = 12$ V

$\mathcal{E} = 12$ V
$r = 2\ \Omega$

28–6 A source on open circuit.

EXAMPLE 28–5 Consider a source whose emf $\mathcal{E}$ is constant and equal to 12 V, and whose internal resistance r is 2 Ω. (The internal resistance of a commercial 12-V lead storage battery is only a few thousandths of an ohm.) Figure 28–6 represents the source with a voltmeter V connected between its terminals a and b. A *voltmeter* reads the potential difference ("voltage") between its terminals. We assume that the resistance of the voltmeter is so large (essentially infinite) that it draws no appreciable current. The source is then on *open circuit,* corresponding to the source in Fig. 28–4, and the voltmeter reading V_{ab} equals the emf $\mathcal{E}$ of the source, or 12 V.

EXAMPLE 28–6 In Fig. 28–7, an ammeter A and a resistor of resistance $R = 4\ \Omega$ have been connected to the terminals of the source to form a complete circuit. The total resistance of the circuit is the sum of the resistance R, the internal resistance r, and the resistance of the ammeter. The ammeter resistance, however, can be made very small, and we will assume that it can be neglected. The ammeter (whatever its resistance) reads the current I through it. The circuit corresponds to that in Fig. 28–5.

The wires connecting the resistor to the source and the ammeter, shown by straight lines, have zero resistance; so no potential difference exists between their ends. Thus points a and a' are at the same potential, as are points b and b'. The potential differences V_{ab} and $V_{a'b'}$ are therefore equal.

The current I in the resistor (and hence at all points of the circuit) could be found from the relation $I = V_{ab}/R$, if the potential difference V_{ab} were known. However, V_{ab} is the terminal voltage of the source, equal to $\mathcal{E} - Ir$, and since this depends on I, it is unknown at the start. We can, however, calculate the current from the circuit equation:

$$I = \frac{\mathcal{E}}{R + r} = \frac{12\ \text{V}}{4\ \Omega + 2\ \Omega} = 2\ \text{A}.$$

The potential difference V_{ab} can now be found by considering a and b either as the terminals of the resistor or as the terminals of the source. If we consider them as the terminals of the resistor,

$$V_{a'b'} = IR = (2\ \text{A})(4\ \Omega) = 8\ \text{V}.$$

If we consider them as the terminals of the source,

$$V_{ab} = \mathcal{E} - Ir = 12\ \text{V} - (2\ \text{A})(2\ \Omega) = 8\ \text{V}.$$

The voltmeter therefore reads 8 V, and the ammeter reads 2 A.

$V_{ab} = V_{a'b'} = 8$ V

28–7 A source in a complete circuit.

a b

$\mathcal{E} = 12$ V
$r = 2\ \Omega$

I A $I = 2$ A

a' b'

$R = 4\ \Omega$

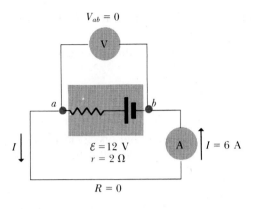

$V_{ab} = 0$

$\mathcal{E} = 12\ \text{V}$
$r = 2\ \Omega$

$I = 6\ \text{A}$

$R = 0$

28–8 A source on short circuit.

EXAMPLE 28–7 In Fig. 28–8, the source is short-circuited. The current is

$$I = \frac{\mathcal{E}}{r} = \frac{12\ \text{V}}{2\ \Omega} = 6\ \text{A}.$$

The terminal voltage is

$$V_{ab} = \mathcal{E} - Ir = 12\ \text{V} - (6\ \text{A})(2\ \Omega) = 0.$$

The ammeter reads 6 A, and the voltmeter reads zero.

Let us now restate the rule represented by Eq. (28–17), for finding the current in a circuit, in a form that is particularly useful in more complex circuits having several sources of emf or several branches. The electrostatic field is a *conservative* force field. Suppose we go around a loop, measuring potential differences across successive circuit elements. When we return to the starting point, we must find that the *algebraic sum* of these differences is zero; otherwise we could not say that the potential at this point has a definite value.

Thus, in Fig. 28–7, if we start at point *b* and travel counterclockwise (the same direction as *I*), we find a *rise* in potential (a *positive* change) due to the battery emf, a *drop* (a *negative* change) due to the battery's internal resistance, and an additional drop due to the 4-Ω resistor. The algebraic sum of these quantities must be zero. Thus

$$12\ \text{V} - I(2\ \Omega) - I(4\ \Omega) = 0, \qquad I = 2\ \text{A},$$

in agreement with the result obtained in Example 28–6.

To allow for the possibility of several sources of emf in the loop, we generalize this procedure as follows: *The algebraic sum of the potential differences around a complete circuit, including those corresponding to the emf's of the sources and those due to the IR products, must equal zero.* This principle is called **Kirchhoff's loop rule.** In applying it, we need some sign conventions. We first assume a direction for the current, and mark it on the diagram. Then, starting at any point in the circuit, we go around the circuit in the direction of the assumed current, adding emf's and *IR* products as we come to them. When we go through a source in the direction from − to +, the emf is considered *positive;* when going from + to −, it is negative. The *IR* products are all negative because the direction of current is always that of *decreasing* potential. We could also traverse the loop in the direction *opposite* to that of the assumed current. In this case, all the emf's have opposite sign, and all the *IR* products are *positive*

Rules for adding potential differences around a circuit

Kirchhoff's loop rule: an energy relation for charge moving around a complete circuit because in going in the opposite direction to the current we are going "uphill" from lower to higher potential.

For the circuit in Fig. 28–7, if we start at point b and go clockwise, the resulting equation is

$$I(4\ \Omega) + I(2\ \Omega) - 12\ \text{V} = 0,$$

which is the same as the previous equation except for an overall factor of -1, which does not change the value of I. But if we assume that I is clockwise, then starting at b and going counterclockwise around the circle yields the equation

$$12\ \text{V} + I(2\ \Omega) + I(4\ \Omega) = 0, \qquad I = -2\ \text{A}.$$

Here the negative sign on the result shows that our initial assumption about the current direction was wrong; the actual direction is counterclockwise.

PROBLEM-SOLVING STRATEGY: Electric circuits

1. Make an assumption about the direction of current in the circuit, and mark it clearly on your diagram. It doesn't matter whether the assumption is right or wrong. If it's wrong, your solution for I will give you a negative number, which means the actual current direction is opposite to your assumption. But you *must* pick a direction at the start and use it consistently when you apply Kirchhoff's loop rule.

2. Decide in which direction you will go around the loop when you add up the potential differences in applying Kirchhoff's loop rule. Again, it doesn't matter which way you go, but you *must* keep going the same way until you are back where you started.

3. Remember the sign rules: An emf is counted as positive if you go through the source from $-$ to $+$,

negative if from $+$ to $-$. The change in potential going through a resistor (IR) is negative if you go through in the same direction as the assumed current, and positive if in the opposite direction.

4. This same bookkeeping system may be used to find the potential difference between any two points a and b in a circuit. To find $V_{ab} = V_a - V_b$ (the potential of a with respect to b), start at b and add the potential changes you encounter in going from b to a. Use the same sign rules you used in step 3. An emf is positive when you go from $-$ to $+$, negative otherwise. An IR term is positive when you go "uphill," against the assumed current, and negative when "downhill," in the same direction as the current. Then the sum of these changes is V_{ab}.

An example of the use of Kirchhoff's loop rule

EXAMPLE 28–8   The circuit shown in Fig. 28–9 contains two batteries, each having an emf and an internal resistance, and two resistors. Find the current in the circuit and the potential difference V_{ab}.

SOLUTION   We assume a direction for the current, as shown. Then, starting at a and going counterclockwise, we add potential increases and decreases and equate the sum to zero. The resulting equation is

$$-I(4\ \Omega) - 4\ \text{V} - I(7\ \Omega) + 12\ \text{V} - I(2\ \Omega) - I(3\ \Omega) = 0.$$

Collecting terms containing I and solving for I, we find

$$8\ \text{V} = I(16\ \Omega) \quad \text{and} \quad I = 0.5\ \text{A}.$$

The result for I is positive, showing that our assumed current direction is correct. For exercise, try assuming the opposite direction for I; you should then get $I = -0.5\ \text{A}$, indicating that the actual current is opposite to this assumption.

To find V_{ab}, the potential at a with respect to b, we start at b and go toward a, adding potential changes. There are two possible paths from b to a; taking the lower one first, we find

$$V_{ab} = (0.5 \text{ A})(7 \text{ } \Omega) + 4 \text{ V} + (0.5 \text{ A})(4 \text{ } \Omega) = 9.5 \text{ V}.$$

Point a is at 9.5 V higher potential than b. All the terms in this sum are positive because each represents an *increase* in potential as we go from b toward a. If instead we use the upper path, the resulting equation is

$$V_{ab} = 12 \text{ V} - (0.5 \text{ A})(2 \text{ } \Omega) - (0.5 \text{ A})(3 \text{ } \Omega) = 9.5 \text{ V}.$$

Here the IR terms are negative because our path goes in the direction of the current, with potential decreases through the resistors. The result is the same as for the lower path, as it must be in order for the total potential change around the complete loop to be zero. In each case, potential rises are taken as positive, and drops as negative.

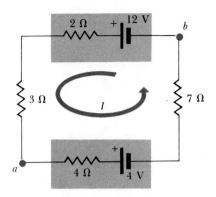

FIGURE 28-9

28-5 CURRENT-VOLTAGE RELATIONS

The amount of current in a device such as a resistor depends on the potential difference (voltage) between its terminals. For a device obeying Ohm's law, the current is *directly proportional* to voltage, as shown in Eq. (28-11). But in many devices current depends on voltage in a nonproportional way, and the current resulting from a given potential difference may depend on the *polarity* of the potential difference. This is the case with *diodes*, devices constructed deliberately to conduct much better in one direction than in the other.

It is often convenient to represent the current-voltage relation as a graph, and Fig. 28-10 shows several examples. Part (a) shows the behavior of a resistor that obeys Ohm's law, for which the graph is a straight line. Part (b) shows the relation for a vacuum diode, a vacuum tube used to convert high-voltage alternating current to direct current. For positive potentials of anode with respect to cathode, I is approximately proportional to $V^{3/2}$; for negative potentials, the current is several orders of magnitude smaller and for most purposes may be assumed to be zero. Semiconductor diode behavior (c) is somewhat different but still strongly asymmetric, acting as a one-way valve in a circuit. Diodes are used to convert alternating current to direct current and to perform a wide variety of logic functions in computer circuitry. We will study the microscopic basis of diode behavior in later chapters.

The current-voltage relation is temperature dependent for nearly all materials. At low temperatures the curve in Fig. 28-10c rises more steeply for

Current-voltage relations: Some devices don't obey Ohm's law.

A diode conducts better in one direction than in the other.

Current-voltage relations can be represented graphically.

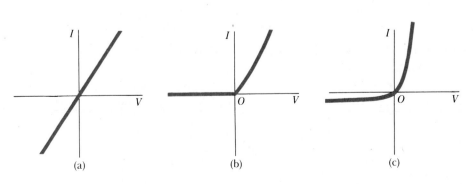

28-10 Current-voltage relations for (a) a resistor obeying Ohm's law; (b) a vacuum diode; (c) a semiconductor diode.

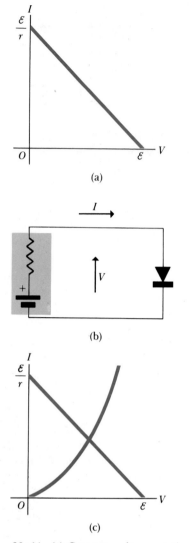

28–11 (a) Current–voltage relation for a source with emf $\mathcal{E}$ and internal resistance r; (b) a circuit containing a source and a nonlinear element; (c) simultaneous solution of I–V equations for this circuit.

A current transfers energy from one part of a circuit to another.

positive V than at higher temperatures, and at successively higher temperatures the asymmetry in the curve becomes less and less pronounced.

The current–voltage relation for a *source* may also be represented graphically. For a source represented by Eq. (28–16), that is,

$$V = \mathcal{E} - Ir,$$

the graph appears as in Fig. 28–11a. The intercept on the V-axis, corresponding to the open-circuit condition ($I = 0$), is at $V = \mathcal{E}$, and the intercept on the I-axis, corresponding to a short-circuit situation ($V = 0$), is at $I = \mathcal{E}/r$.

This graph may be used to find the current in a circuit containing a nonlinear device, as in Fig. 28–11b. The current–voltage relation is shown in Fig. 28–11c, and Eq. (28–16) is also plotted on this graph. Each curve represents a current–voltage relation that must be satisfied, so the intersection represents the only possible values of V and I. This amounts to a graphical solution of two simultaneous equations for V and I, one of which is nonlinear.

Finally, we remark that Eq. (28–16) is not always an adequate representation of the behavior of a source. What we have described as an internal resistance may actually be a more complex voltage–current relation. Nevertheless, the concept of internal resistance frequently provides an adequate description of batteries, generators, and other energy converters. The difference between a fresh flashlight battery and an old one is not in the emf, which decreases only slightly with use, but principally in the internal resistance, which may increase from a few ohms when fresh to as much as 1000 Ω or more after long use. Similarly, the current a car battery can deliver to the starter motor on a cold morning is less than when the battery is warm, not because the emf is appreciably less but because the internal resistance is temperature-dependent, increasing with decreasing temperature. Residents of northern Wisconsin have been known to soak their car batteries in warm water to provide greater starting power on very cold mornings!

28–6 ENERGY AND POWER IN ELECTRIC CIRCUITS

We are now ready for a detailed study of energy and power relations in electric circuits. The rectangle in Fig. 28–12 represents a portion of a circuit having current I and potential difference $V_a - V_b = V_{ab}$ between the two conductors leading to and from this point of the circuit. The detailed nature of this circuit element does not matter. As charge passes through the circuit element, the electric field does work on the charge. In a time interval Δt, an amount of charge $\Delta Q = I\,\Delta t$ passes through, and the work ΔW done by the electric field is given by the product of the potential difference (work per unit charge) and the quantity of charge:

$$\Delta W = V_{ab}\,\Delta Q = V_{ab}I\,\Delta t.$$

By means of this work the electric field transfers energy into this portion of the circuit.

The *rate* of the energy transfer is *power*, denoted by P. Dividing the above relation by Δt, we obtain the rate at which energy enters this part of the circuit:

$$P = \frac{\Delta W}{\Delta t} = V_{ab}I. \qquad (28-20)$$

28–12 The power input P to the portion of the circuit between a and b is $P = V_{ab}I$.

It may happen that the potential at b is higher than that at a; in this case, V_{ab} is negative. The charge then *gains* potential energy (at the expense of some other form of energy), and there is a corresponding transfer of electrical energy *out of* this portion of the circuit.

Equation (28–20) is the general expression for the magnitude of the electric-power input to (or the power output from) any portion of an electric circuit. The unit of V_{ab} is one volt, or one joule per coulomb, and the unit of I is one ampere, or one coulomb per second. The SI unit of power is therefore

$$(1 \text{ J·C}^{-1})(1 \text{ C·s}^{-1}) = 1 \text{ J·s}^{-1} = 1 \text{ W} = 1 \text{ watt}.$$

We now consider some special cases.

1. *Pure resistance.* If the portion of the circuit in Fig. 28–12 is a pure resistance, the potential difference is given by $V_{ab} = IR$ and

$$P = V_{ab}I = I^2R = \frac{V^2}{R}. \qquad (28\text{–}21)$$

The potential at a must be higher than at b, so there is a power *input* to the resistor. The circulating charges give up energy to the atoms of the resistor when they collide with them, and the temperature of the resistor increases unless there is a flow of heat out of it. We say that energy is *dissipated* in the resistor at a rate I^2R.

Because of this heat, every resistor has a maximum power rating, which is the maximum power that can be dissipated without overheating the device. When this rating is exceeded, the resistance may change unpredictably; in extreme cases, the resistor may melt or even explode. In practical applications, the power rating of a resistor is often as important a characteristic as its resistance value.

2. *Power output of a source.* The upper rectangle in Fig. 28–13 represents a source having emf $\mathcal{E}$ and internal resistance r, connected by ideal (resistanceless) conductors to an external circuit represented by the lower rectangle; the precise nature of the external circuit does not matter. We assume only

A resistor converts electrical energy to heat.

Power dissipation in a resistor is proportional to the square of the current.

Power relations for a source delivering energy to a circuit

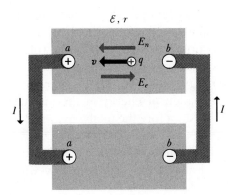

28–13 The rate of conversion of nonelectrical to electrical energy in the source equals $\mathcal{E}I$. The rate of energy dissipation in the source is I^2r. The difference $\mathcal{E}I - I^2r$ is the power output of the source.

that a current I is present in the circuit in the direction shown, from a to b in the external circuit and from b to a within the source. Point a is at higher potential than point b; $V_a > V_b$. The external circuit then corresponds to the rectangle in Fig. 28–12, and the power *input* to it is

$$P = V_{ab}I.$$

If a and b are considered to be the terminals of the source, then, as we have shown,

$$V_{ab} = \mathcal{E} - Ir,$$

and hence

$$P = V_{ab}I = \mathcal{E}I - I^2 r. \tag{28–22}$$

What do the terms $\mathcal{E}I$ and $I^2 r$ mean? The emf $\mathcal{E}$ has been defined as the work per unit charge performed on the charges by the non-electrostatic force as the charges are pushed "uphill" from b to a in the source. If a charge ΔQ flows in time Δt, this force does work (adds energy) $\Delta W = \mathcal{E}\,\Delta Q$, and its *rate* of doing work, or power, is

$$\frac{\Delta W}{\Delta t} = \mathcal{E}\,\frac{\Delta Q}{\Delta t} = \mathcal{E}I. \tag{28–23}$$

Hence the product $\mathcal{E}I$ is the rate at which work is done on the circulating charges by the agency that causes the non-electrostatic force in the source.

The term $I^2 r$ is the rate at which energy is *dissipated* in the internal resistance of the source, and the difference $\mathcal{E}I - I^2 r$ is the rate at which energy is delivered by the source to the remainder of the circuit. In other words, the power P in Eq. (28–22) represents the *net power output* of the source, or the power delivered to the remainder of the circuit.

Power input to a source, such as a battery being charged

3. *Power input to a source.* Suppose that the lower rectangle in Fig. 28–13 is itself a source, with an emf *larger* than that of the upper source and with its emf opposite to that of the upper source. The current I in the circuit is then *opposite* to that shown in Fig. 28–13. That is, the lower source pushes current backward through the upper source. The power output of the lower source is

$$P = V_{ab}I.$$

Considering a and b as the terminals of the upper source, we have

$$V_{ab} = \mathcal{E} + Ir,$$

and

$$P = V_{ab}I = \mathcal{E}I + I^2 r. \tag{28–24}$$

Storage batteries: You don't get back everything you put in.

We can interpret the terms $\mathcal{E}I$ and $I^2 r$ as follows. The charges in the upper source now move from left to right, in a direction opposite the field E_n, and the work done by the non-electrostatic force F_n is *negative*. That is, work is done *on* the agent maintaining the non-electrostatic field. The product $\mathcal{E}I$ equals the rate at which work is done on (energy transferred to) this agent, and the term $I^2 r$ again equals the rate of dissipation of energy in the internal resistance of the source. The sum $\mathcal{E}I + I^2 r$ is therefore the total *power input* to the upper source.

Thus the upper source has become a sink, *absorbing* electrical energy from the circuit and converting it to non-electrical energy. This is exactly what happens when a rechargeable battery (a storage battery) is connected to a charger. The charger supplies electrical energy to the battery; part of it is stored in the battery as chemical energy, to be reconverted later, and the remainder is dissipated as heat in the battery's internal resistance.

PROBLEM-SOLVING STRATEGY: *Power and energy in circuits*

1. A set of sign rules is useful in calculating the energy balance in a circuit. A source of emf puts positive power into a circuit when the current I runs through it from − to +; the energy is converted from chemical energy in a battery, from mechanical energy in a generator, or whatever. But when current passes through a source in the direction from + to −, the source is taking power out of the circuit (as in charging a storage battery, when electrical energy is converted back to chemical energy). In that case we count $\mathcal{E}I$ as negative. The internal resistance r of a source always removes energy from the circuit, converting it into heat at a rate I^2r, irrespective of the direction of the current. This represents a *negative* power input to the circuit. Similarly, a resistor R always removes energy from the circuit, at a rate given by $VI = I^2R = V^2/R$.

2. The energy balance can be expressed in one of two forms: either "net power input = net power output" or "the algebraic sum of the power inputs to the circuit is zero."

3. When every term in Kirchhoff's loop equation for a circuit is multiplied through by the current I in the circuit, the result is always an equation relating the power inputs and outputs in the circuit.

EXAMPLE 28–9 The rate of energy conversion in the source in Fig. 28–7 is

$$\mathcal{E}I = (12 \text{ V})(2 \text{ A}) = 24 \text{ W}.$$

Some examples of energy and power relations in simple circuits

The rate of dissipation of energy in the source is

$$I^2r = (2 \text{ A})^2(2 \text{ } \Omega) = 8 \text{ W}.$$

The power *output* of the source is the difference between these, or 16 W. The power output is also given by

$$IV_{ab} = (2 \text{ A})(8 \text{ V}) = 16 \text{ W}.$$

The power input to the resistor is

$$V_{a'b'}I = (2 \text{ A})(8 \text{ V}) = 16 \text{ W}.$$

This value equals the rate of dissipation of energy in the resistor:

$$I^2R = (2 \text{ A})^2(4 \text{ } \Omega) = 16 \text{ W}.$$

In terms of the sign convention suggested above, we may also write

$$\mathcal{E}I - I^2r - I^2R = 0,$$

or

$$(12 \text{ V})(2 \text{ A}) - (2 \text{ A})^2(2 \text{ } \Omega) - (2 \text{ A})^2(4 \text{ } \Omega) = 0.$$

Finally, Kirchhoff's loop equation for this circuit is

$$\mathcal{E} - Ir - IR = 0.$$

When we multiply this equation through by I, we again get the power-balance equation given above.

EXAMPLE 28–10   The rate of energy conversion in the source in Fig. 28–8 is

$$\mathcal{E}I = (12 \text{ V})(6 \text{ A}) = 72 \text{ W}.$$

The rate of dissipation of energy in the source is

$$I^2r = (6 \text{ A})^2(2 \text{ }\Omega) = 72 \text{ W}.$$

The power *output* of the source (also given by $V_{ab}I$) equals zero. *All* of the converted energy is dissipated within the source.

EXAMPLE 28–11   The rate of energy conversion in the upper source in Fig. 28–9 is

$$\mathcal{E}I = (12 \text{ V})(0.5 \text{ A}) = 6 \text{ W}.$$

The rate of dissipation of energy in this source is

$$I^2r = (0.5 \text{ A})^2(2 \text{ }\Omega) = 0.5 \text{ W}.$$

The power output of the upper source is 6 W − 0.5 W = 5.5 W.

The rates of dissipation of energy in the 3-Ω and 7-Ω resistors are, respectively,

$$(0.5 \text{ A})^2(3 \text{ }\Omega) = 0.75 \text{ W},$$

$$(0.5 \text{ A})^2(7 \text{ }\Omega) = 1.75 \text{ W}.$$

The rate of energy conversion in the lower source is

$$\mathcal{E}I = (4 \text{ V})(0.5 \text{ A}) = 2 \text{ W},$$

and the rate of energy dissipation in this source is

$$I^2r = (0.5 \text{ A})^2(4 \text{ }\Omega) = 1 \text{ W}.$$

The power *input* to the lower source is 2 W + 1 W = 3 W. Thus, of the 5.5-W output of the upper source, 2.5 W are dissipated in the two resistors and the remaining 3 W are partly converted, partly dissipated in the lower source. Conversion of energy in the sources is often partly *reversible,* as in a storage battery where electrical energy is converted to chemical energy for later retrieval as electrical energy. Energy dissipated in resistors is converted *irreversibly* to heat.

Finally, we write a power-balance equation, starting at point *b* and going counterclockwise:

$$(12 \text{ V})(0.5 \text{ A}) - (0.5 \text{ A})^2(2 \text{ }\Omega) - (0.5 \text{ A})^2(3 \text{ }\Omega)$$
$$- (0.5 \text{ A})^2(4 \text{ }\Omega) - (4 \text{ V})(0.5 \text{ A}) - (0.5 \text{ A})^2(7 \text{ }\Omega) = 0.$$

We invite you to add this up and verify that the power is accounted for.

28–7 THEORY OF METALLIC CONDUCTION

Microscopic picture of conduction in metals: electrons in random motion

We can gain additional insight into the phenomenon of conduction by examining the microscopic mechanisms of conductivity. Here we consider only a crude and primitive model that treats the electrons as classical particles and ignores their inherently quantum-mechanical behavior in solids. Thus this model is not entirely correct conceptually; yet it is useful in helping to develop an intuitive idea of the microscopic basis of conduction.

In the simplest microscopic model of metallic conduction, each atom in the crystal lattice is assumed to give up one or more of its outer electrons. These electrons are then free to move through the crystal lattice, colliding at intervals with the stationary positive ions. The motion of the electrons is like that of the molecules of gas in a container, and they are often referred to as an "electron gas." In the absence of an electric field, the electrons move in straight lines between collisions; but if an electric field is present, the paths are slightly curved, as in Fig. 28–14, which represents schematically a few free paths of an electron in an electric field directed from right to left. We assume that at each collision, the electron loses any energy it may have acquired from the field and makes a fresh start. The energy given up in these collisions increases the thermal energy of vibration of the positive ions.

A force $F = eE$ is exerted on each electron by the field and produces an acceleration a in the direction of the force given by

$$a = \frac{F}{m} = \frac{eE}{m},$$

where m is the electron mass. Let u represent the average *random* speed of an electron, and let λ be the *mean free path* (the average distance traveled between collisions). The average time t between collisions, called the *mean free time*, is

$$t = \frac{\lambda}{u}.$$

In this time, the electron acquires a final velocity component v_f in the direction of the force, given by

$$v_f = at = \frac{eE}{m} \frac{\lambda}{u}.$$

Its *average* velocity v in the direction of the force, which is superposed on its random velocity and which we interpret as the *drift* velocity, is one-half the final velocity, so

$$v = \frac{1}{2}v_f = \frac{1}{2}\frac{e\lambda}{mu}E.$$

The drift velocity is therefore proportional to the electric field E.

The current density is

$$J = nev = \frac{ne^2\lambda}{2mu}E,$$

and the resistivity is

$$\rho = \frac{E}{J} = \frac{2mu}{ne^2\lambda}. \qquad (28–25)$$

This is the theoretical expression for the resistivity, and it is in *qualitative* agreement with experiment. At a given temperature, the quantities m, u, n, e, and λ are constant. The resistivity is then constant, and Ohm's law is obeyed. When the temperature is increased, the random speed u increases, so the theory predicts that the resistivity of a metal increases with increasing temperature.

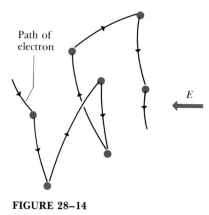

Path of electron

E

FIGURE 28–14

How far does an electron travel between collisions? How much time does it take?

The drift velocity is proportional to the electric field.

What is the difference between a
conductor and a semiconductor?

In a semiconductor, the number of charge carriers per unit volume, n, increases rapidly with increasing temperature. The increase in n far outweighs any increase in u, and the resistivity decreases. At low temperatures, n is very small and the resistivity becomes so large that the material can be considered an insulator.

The modern theory of superconductivity predicts that, in effect, at temperatures below the critical temperature, the electrons move freely throughout the lattice. The mean free path λ then becomes very large and the resistivity very small. We discuss superconductivity in greater detail in Section 43–10.

SUMMARY

KEY TERMS
current
ampere
current density
resistivity
Ohm's law
temperature coefficient of resistivity
resistance
ohm
complete circuit
electromotive force
internal resistance
Kirchhoff's loop rule

Current is the amount of charge flowing through a boundary surface, per unit time. The SI unit of current is the ampere, equal to one coulomb per second ($1 \text{ A} = 1 \text{ C·s}^{-1}$). In terms of the charges q_i and drift velocities v_i of the charge carriers in a material, the current I is given by

$$I = A(n_1 q_1 v_1 + n_2 q_2 v_2 + \cdots). \tag{28–3}$$

Current density J is current per unit cross-sectional area; $J = I/A$. In terms of the above quantities, current density is given by

$$J = n_1 q_1 v_1 + n_2 q_2 v_2 + \cdots. \tag{28–4}$$

The corresponding vector current density $\boldsymbol{J}$ is given by

$$\boldsymbol{J} = n_1 q_1 \boldsymbol{v}_1 + n_2 q_2 \boldsymbol{v}_2 + \cdots. \tag{28–5}$$

Current is described in terms of a flow of positive charge, even when the actual charge carriers are negative or of both signs.

Ohm's law, obeyed approximately by many materials, states that current density is proportional to electric-field magnitude. For such materials, resistivity is defined as electric-field magnitude per unit current density. The SI unit of resistivity is the ohm-meter ($1 \text{ }\Omega\text{·m}$). Good conductors have small resistivity; good insulators have large resistivity. Resistivity usually increases with temperature; for small temperature changes this variation can be represented approximately by

$$\rho_T = \rho_0[1 + \alpha(T - T_0)], \tag{28–7}$$

where α is the temperature coefficient of resistivity.

For materials obeying Ohm's law, the potential difference V across a particular sample of material is proportional to the current I through the material:

$$V = IR, \tag{28–11}$$

where R is the resistance of the sample. In terms of the resistivity ρ, length l, and cross-sectional area A of the sample, the resistance is given by

$$R = \frac{\rho l}{A}. \tag{28–10}$$

The SI unit of resistance is the ohm ($1 \text{ }\Omega = 1 \text{ V·A}^{-1}$).

A complete circuit is a conductor in the form of a loop, providing a continuous current-carrying path. A complete circuit carrying a steady current must contain a source of electromotive force (emf), which is an influence within a source that makes charge move from a region of lower potential (the negative [−] terminal) to a region of higher potential (the positive [+] terminal) within the device. The SI unit of electromotive force is the volt (1 V). An ideal source of emf maintains a constant potential difference, independent of current through the device, but every real source of emf has some internal resistance r. The terminal potential difference then depends on current:

$$V_{ab} = \mathcal{E} - Ir. \qquad (28\text{--}16)$$

Kirchhoff's loop rule states that the algebraic sum of potential differences around any closed circuit loop is zero. This rule follows from the conservative nature of the electrostatic field.

Many devices do not obey Ohm's law but show a more complicated voltage–current relationship; often this must be represented graphically.

A circuit element with a potential difference V and a current I puts energy into a circuit if the current direction is from lower to higher potential in the device, and takes energy out if the current is opposite. The power P (rate of energy transfer) is given by $P = VI$. A resistor R always takes energy out of a circuit, converting it to heat at a rate given by

$$P = V_{ab}I = I^2R = \frac{V^2}{R}. \qquad (28\text{--}21)$$

When current in a source is from − to +, the source puts energy into the circuit at a rate given by

$$P = V_{ab}I = \mathcal{E}I - I^2r. \qquad (28\text{--}22)$$

In this equation $\mathcal{E}I$ is the power corresponding to the emf, and I^2r is the power dissipated in the internal resistance.

The microscopic basis of conduction in metals is the motion of electrons that move freely through the crystal lattice sites, bumping into ion cores in the lattice. In a crude classical model of this motion, the resistivity of the material can be related to the electron mass, charge, speed of random motion, density, and mean free path between collisions.

QUESTIONS

28–1 A rule of thumb used to determine the internal resistance of a source is that it is the open-circuit voltage divided by the short-circuit current. Is this correct?

28–2 The energy that can be extracted from a storage battery is always less than the energy that goes into it while it is being charged. Why?

28–3 In circuit analysis one often assumes that a wire connecting two circuit elements has no potential difference between its ends; yet there must be an electric field within the wire to make the charges move, and so there must be a potential difference. How do you resolve this discrepancy?

28–4 Long-distance electric-power transmission lines always operate at very high voltage, sometimes as much as 750 kV. What are the advantages of such high voltages? The disadvantages?

28–5 Ordinary household electric lines usually operate at 120 V. Why is this a desirable voltage, rather than a value considerably larger or smaller? What about cars, which usually have 12-V electrical systems?

28–6 What is the difference between an emf and a potential difference?

28–7 Electric power for household and commercial use always uses *alternating current*, which reverses direction 120 times each second. A student claimed that the power conveyed by such a current would have to average out to zero, since it is going one way half the time and the other way the other half. What is your response?

28–8 As discussed in the text, the drift velocity of electrons in a good conductor is very slow. Why then does the light come on so quickly when the switch is turned on?

28–9 A fuse is a device designed to break a circuit, usually by melting, when the current exceeds a certain value. What characteristics should the material of the fuse have?

28–10 What considerations determine the maximum current-carrying capacity of household wiring?

28–11 The text states that good thermal conductors are also good electrical conductors. If so, why don't the cords used to connect toasters, irons, and similar heat-producing appliances get hot by conduction of heat from the heating element?

28–12 Eight flashlight batteries in series have an emf of about 12 V, similar to that of a car battery. Could they be used to start a car with a dead battery?

28–13 High-voltage power supplies are sometimes designed intentionally to have rather large internal resistance, as a safety precaution. Why is such a power supply with a large internal resistance safer than one with the same voltage but lower internal resistance?

28–14 How would you expect the resistivity of a good insulator such as glass or polystyrene to vary with temperature? Why?

28–15 A would-be inventor proposed to increase the power supplied by a battery to a light bulb by using thick wire near the battery and thinner wire near the bulb. In the thin wire, the electrons from the thick wire would become more densely packed, more electrons per second would reach the bulb, and the energy received by the bulb would be greater than that emitted by the battery. What do you think of this scheme?

EXERCISES

Section 28–1 Current

28–1 A silver wire 1 mm in diameter transfers a charge of 90 C in 1 hr and 15 min. Silver contains 5.8×10^{28} free electrons per m^3.

a) What is the current in the wire?

b) What is the drift velocity of the electrons in the wire?

28–2 When a sufficiently high potential difference is applied between two electrodes in a gas, the gas ionizes; electrons move toward the positive electrode, and positive ions move toward the negative electrode.

a) What is the current in a hydrogen discharge if, in each second, 4×10^{18} electrons and 1.5×10^{18} protons move in opposite directions past a cross section of the tube?

b) What is the direction of the current?

28–3 The current in a wire varies with time according to the relation

$$I = 4 \text{ A} + (2 \text{ A·s}^{-2})t^2.$$

a) How many coulombs pass a cross section of the wire in the time interval between $t = 5$ s and $t = 10$ s?

b) What constant current would transport the same charge in the same time interval?

Section 28–2 Resistivity

Section 28–3 Resistance

28–4 A copper wire has a square cross section 2.0 mm on a side. It is 4 m long and carries a current of 10 A. The density of free electrons is 8×10^{28} m^{-3}.

a) What is the current density in the wire?

b) What is the electric field?

c) How much time is required for an electron to travel the length of the wire?

28–5 In household wiring, a copper wire commonly known as 12-gauge is often used. Its diameter is 2.05 mm. Find the resistance of a 50-m length of this wire.

28–6 What length of copper wire 1.0 mm in diameter would have a resistance of 1.00 Ω?

28–7 An aluminum bar 2.5 m long has a rectangular cross section 1 cm by 5 cm.

a) What is its resistance?

b) What would be the length of a copper wire 1.5 mm in diameter having the same resistance?

28–8 What diameter must an aluminum wire have if its resistance is to be the same as that of an equal length of copper wire of diameter 2.0 mm?

28–9

a) What is the resistance of a Nichrome wire at 0°C, if its resistance is 100.00 Ω at 12°C?

b) What is the resistance of a carbon rod at 30°C, if its resistance is 0.0150 Ω at 0°C?

28–10 A certain resistor has a resistance of 150.4 Ω at 20°C and a resistance of 162.4 Ω at 40°C. What is its temperature coefficient of resistivity?

28–11 A carbon resistor is to be used as a thermometer. On a winter day when the temperature is 0°C, its resistance is 217.3 Ω. What is the temperature on a hot summer day when the resistance is 214.2 Ω?

28–12 Refer to Example 28–4. Let the resistivity of the material between the cylinders be 10 Ω ·m and let $r_a = 10$ cm, $r_b = 20$ cm, and $l = 5$ cm.

a) Find the resistance between the cylinders.

b) Find the current between the cylinders if $V_{ab} = 8$ V.

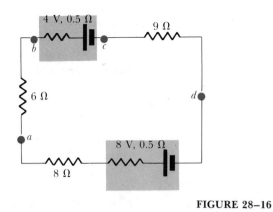

FIGURE 28–16

Section 28–4 Electromotive Force and Circuits

28–13 When switch S is open, the voltmeter V, connected across the terminals of the dry cell in Fig. 28–15, reads 1.52 V. When the switch is closed, the voltmeter reading drops to 1.37 V and the ammeter A reads 1.5 A. Find the emf and internal resistance of the cell. Neglect meter corrections.

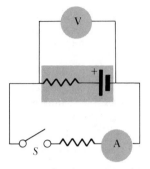

FIGURE 28–15

28–14 A closed circuit consists of a 12-V battery, a 3.7-Ω resistor, and a switch. The internal resistance of the battery is 0.3 Ω. The switch is opened. What would a high-resistance voltmeter read when placed

a) across the terminals of the battery?

b) across the resistor?

c) across the switch?

d) Repeat (a), (b), and (c) for the case when the switch is closed.

28–15 The internal resistance of a dry cell increases gradually with age, even though the cell is not used. The emf, however, remains fairly constant at about 1.5 V. Dry cells may be tested for age at the time of purchase by connecting an ammeter directly across the terminals of the cell and reading the current. The resistance of the ammeter is so small that the cell is practically short-circuited.

a) The short-circuit current of a fresh No. 6 dry cell (1.5-V emf) is about 30 A. Approximately what is the internal resistance?

b) What is the internal resistance if the short-circuit current is only 10 A?

c) The short-circuit current of a 6-V storage battery may be as great as 1000 A. What is its internal resistance?

28–16

a) What is the potential difference V_{ad} in the circuit of Fig. 28–16?

b) What is the terminal voltage of the 4-V battery?

c) A battery of emf 17 V and internal resistance 1 Ω is inserted in the circuit at d, with its positive terminal connected to the positive terminal of the 8-V battery. What is now the difference of potential V_{bc} between the terminals of the 4-V battery?

Section 28–5 Current–Voltage Relations

28–17 The following measurements of current and potential difference were made on a resistor constructed of Nichrome wire:

I, A	V_{ab}, V
0.5	2.18
1.0	4.36
2.0	8.72
4.0	17.44

a) Make a graph of V_{ab} as a function of I.

b) Does Nichrome obey Ohm's law?

c) What is the resistance of the resistor, in ohms?

28–18 The following measurements were made on a Thyrite resistor:

I, A	V_{ab}, V
0.5	4.76
1.0	5.81
2.0	7.05
4.0	8.56

a) Make a graph of V_{ab} as a function of I. Does Thyrite have a constant resistance?

b) Construct a graph of the resistance $R = V_{ab}/I$.

Section 28–6 Energy and Power in Electric Circuits

28–19 Consider a resistor of length L, uniform cross section A, and uniform resistivity ρ that is carrying a current of uniform current density J. Use Eq. (28–21) to find the electric power dissipated per unit volume, p. Express your result in terms of

a) E and J,

b) J and ρ,

c) E and ρ.

28–20 A resistor develops heat at the rate of 360 W when the potential difference across its ends is 180 V. What is its resistance?

28–21 A "660-W" electric heater is designed to operate from 120-V lines.

a) What is its resistance?

b) What current does it draw?

c) If the line voltage drops to 110 V, what power does the heater take, in watts? (Assume the resistance to be constant. Actually, it will change because of the change in temperature.)

28–22 A typical small flashlight contains two batteries, each having an emf of 1.5 V, connected in series with a bulb having resistance 15 Ω.

a) If the internal resistance of the batteries is negligible, what power is delivered to the bulb?

b) If the batteries last for five hours, what is the total energy delivered to the bulb?

c) The resistance of real batteries increases as they run down. If the initial internal resistance is negligible, what is the internal resistance of each battery when the power to the bulb has decreased to half its initial value?

28–23 The capacity of a storage battery, such as those used in automobile electrical systems, is rated in ampere-hours (A·hr). A 50-A·hr battery can supply a current of 50 A for one hour, or 25 A for two hours, and so on.

a) What total energy is stored in a 12-V, 50-A·hr battery if its internal resistance is negligible?

b) What volume (in liters) of gasoline has a total heat of combustion equal to the energy obtained in (a)? (See Section 15–5, near the end. Take the density of gasoline to be 900 kg·m^{-3}.)

c) If a windmill-powered generator has an average electric-power output of 300 W, how much time will be required for it to charge the battery fully?

28–24 In the circuit in Fig. 28–17, find

a) the rate of conversion of internal (chemical) energy to electrical energy within the battery;

b) the rate of dissipation of electrical energy in the battery;

c) the rate of dissipation of electrical energy in the external resistor.

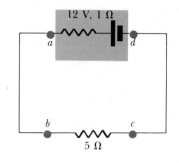

FIGURE 28–17

28–25 A person with body resistance between hands of 10 kΩ accidentally grasps the terminals of a 20-kV power supply.

a) If the internal resistance of the power supply is 1000 Ω, what is the current through the person's body?

b) What is the power dissipated in the body?

c) If the power supply is to be made safe by increasing its internal resistance, what should the internal resistance be for the maximum current in the above situation to be 0.001 A or less?

28–26 The average bulk resistivity of the human body (apart from surface resistance of the skin) is about 5 Ω·m. The conducting path between the hands can be represented approximately as a cylinder 1.6 m long and 0.1 m in diameter. The skin resistance can be made negligible by soaking the hands in salt water (or sea water).

a) What is the resistance between the hands if the skin resistance is negligible?

b) What potential difference between the hands is needed for a lethal shock current of 100 mA?

c) With the current in (b), what power is dissipated in the body?

d) Does the result of (b) increase your respect for electrical-shock hazards?

PROBLEMS

28–27 In the Bohr model of the hydrogen atom, the electron makes about 6×10^{15} rev·s^{-1} around the nucleus. What is the average current at a point on the orbit of the electron?

28–28 A vacuum diode (see Exercise 26–16) can be approximated by a plane cathode and a plane anode, parallel to each other and 5 mm apart. The area of both cathode and anode is 2 cm^2. In the region between cathode and anode the current is carried solely by electrons. If the electron current is 50 mA, and the electrons strike the anode surface with a speed of 1.2×10^7 m·s^{-1}, find the

number of electrons per cubic millimeter in the space just outside the surface of the anode.

28–29 A certain electrical conductor has a square cross section 2.0 mm on a side and is 12 m long. The resistance between its ends is 0.072 Ω.

a) What is the resistivity of the material?

b) If the electric-field magnitude in the conductor is 0.12 V·m^{-1}, what is the total current?

c) If the material has 8.0×10^{28} free electrons per cubic meter, find the average drift velocity under conditions of part (b).

28–30 The current in a wire varies with time according to the relation

$$I = (20\ \text{A}) \sin (377\ \text{s}^{-1}t).$$

a) How many coulombs pass a cross section of the wire in the time interval between $t = 0$ and $t = \frac{1}{120}$ s?

b) In the interval between $t = 0$ and $t = \frac{1}{60}$ s?

c) What constant current would transport the same charge in each of the intervals above?

28–31 Two parallel plates of a capacitor have equal and opposite charges Q. The dielectric has a dielectric constant K and a resistivity ρ. Show that the "leakage" current carried by the dielectric is given by the relationship $i = Q/K\epsilon_0\rho$.

28–32 The region between two concentric conducting spheres of radii r_a and r_b is filled with a conducting material of resistivity ρ.

a) Show that the resistance between the spheres is given by

$$R = \frac{\rho}{4\pi}\left(\frac{1}{r_a} - \frac{1}{r_b}\right).$$

b) Derive an expression for the current density as a function of radius, if the potential difference between the spheres is V_{ab}.

c) Show that the result in (a) reduces to Eq. (28–10) when the separation $l = r_b - r_a$ between the spheres is small.

28–33 A piece of wire has a resistance R. It is cut into three pieces of equal length, and the pieces are twisted together in parallel. What is the resistance of the resulting wire?

28–34 A toaster using a Nichrome heating element operates on 120 V. When it is switched on at 0°C, it carries an initial current of 1.5 A. A few seconds later the current reaches the steady value of 1.33 A. What is the final temperature of the element? The average value of the temperature coefficient of Nichrome over the temperature range is $0.00045(\text{C}°)^{-1}$.

28–35 The potential difference across the terminals of a battery is 8.5 V when there is a current of 3 A in the battery from the negative to the positive terminal. When the current is 2 A in the reverse direction, the potential difference becomes 11 V.

a) What is the internal resistance of the battery?

b) What is the emf of the battery?

28–36 The open-circuit terminal voltage of a source is 10 V and its short-circuit current is 4.0 A.

a) What will be the current when the source is connected to a linear resistor of resistance 2 Ω?

b) What will be the current in the Thyrite resistor of Exercise 28–18 when connected across the terminals of this source?

c) What is the terminal voltage at this current?

28–37 A certain 12-V storage battery has a capacity of 60 A·hr. (See Exercise 28–23.) Its internal resistance is 0.2 Ω. The battery is charged by passing a 15-A current through it for 4 hr.

a) What is the terminal voltage during charging?

b) What total electrical energy is supplied to the battery during charging?

c) What electrical energy is dissipated in the internal resistance during charging?

The battery is now completely discharged through a resistor, again with a constant current of 15 A.

d) What is the external circuit resistance?

e) What total electrical energy is supplied to the external resistor?

f) What total electrical energy is dissipated in the internal resistance?

g) Why are the answers to (b) and (e) not equal?

28–38 Repeat Problem 28–37 with charge and discharge currents of 30 A. The charging and discharging times will now be 2 hr rather than 4 hr. What differences in performance do you see?

28–39 In the circuit of Fig. 28–18, find

a) the current through the 8-Ω resistor;

b) the total rate of dissipation of electrical energy (conversion to heat) in the 8-Ω resistor and in the internal resistance of the batteries.

c) In one of the batteries chemical energy is being converted into electrical energy. In which one is this happening, and at what rate?

d) In one of the batteries electrical energy is being converted into chemical energy. In which one is this happening, and at what rate?

e) Show that the overall rate of production of electrical energy equals the overall rate of consumption of electrical energy in the circuit.

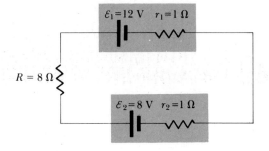

FIGURE 28–18

CHALLENGE PROBLEMS

28–40 A source whose emf is $\mathcal{E}$ and whose internal resistance is r is connected to an external circuit.

a) Show that the power output of the source is maximum when the current in the circuit is one-half the short-circuit current of the source.

b) If the external circuit consists of a resistance R, show that the power output is maximum when $R = r$, and that the maximum power is $\mathcal{E}^2/4r$.

28–41 The definition of α, the temperature coefficient of resistivity, is given by

$$\alpha = \left(\frac{1}{\rho}\right)\frac{d\rho}{dT},$$

where ρ is the resistivity at the temperature T. Eq. (28–7) then follows if α is assumed constant and much smaller than $(T - T_0)^{-1}$. (See Challenge Problem 14–36 for the analogous case of the coefficient of linear expansion.)

a) If α is not constant but is given by $\alpha = -n/T$, where T is the Kelvin temperature and n is a constant, show that the resistivity ρ is given by $\rho = a/T^n$ where a is a constant.

b) From Fig. 28–2c, it is seen that such a relation might be used as a rough approximation for a semiconductor. Using the values of ρ for carbon from Table 28–1 and of α from Table 28–2, determine a and n. (In Table 28–1 assume that "room temperature" means 293 K.)

c) Using your result from part (b), determine the resistivity of carbon at the freezing and normal boiling points of water. (Remember to express T in kelvins.)

28–42 A semiconductor diode is a nonlinear device whose current–voltage relation is described by

$$I = I_0[\exp(eV/kT) - 1],$$

where I and V are, respectively, the current through and the voltage across the diode; I_0 is a constant characteristic of the device; e is the electron charge; k is Boltzmann's constant; and T is the Kelvin temperature. Such a diode is connected in series with a 1-Ω resistor and a 2-V battery. The polarity of the battery is such that the current through the diode is in the forward direction. (See Fig. 28–19.)

a) Obtain an equation for V. Note that V cannot be solved for analytically.

b) Since V cannot be solved for analytically, the value of V must be obtained by using a numerical method, such as an iterative technique. Using $I_0 = 0.5$ A and $T = 293$ K, obtain a solution (accurate to three significant figures) for the voltage drop V across the diode and the current I through it. (For a brief discussion of how to do an iterative calculation, see Challenge Problem 27–36, part [c]. This procedure can also be very effectively done on a computer or programmable calculator.)

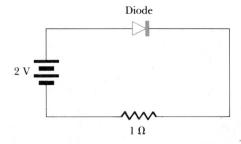

FIGURE 28–19

29

DIRECT-CURRENT CIRCUITS

IN THE LAST CHAPTER WE STUDIED SOME OF THE BASIC PRINCIPLES UNDER-
lying the behavior of electric currents in simple circuits. But many circuits
used in the real world contain several sources, resistors, and other circuit elements such as capacitors and motors, interconnected in a *network*. In this chapter we study general methods for analyzing networks, including finding
unknown voltages, currents, and properties of circuit elements.

When several resistors are connected in series or in parallel, they can
always be represented as a single equivalent resistor. To analyze more general
networks we need two rules called *Kirchhoff's rules*. One is basically the principle of conservation of charge applied to a junction; the other is based on the
sum of potential differences around a closed loop. We also discuss instruments
for measuring various electrical quantities. Finally, we examine the concept of
displacement current in the charging of capacitors.

29–1 RESISTORS IN SERIES AND PARALLEL

Figure 29–1 shows four different ways in which three resistors having resistances R_1, R_2, and R_3 might be connected between points a and b. In (a), the
resistors provide only a single path between these points. Any number of
circuit elements such as resistors, cells, and motors are in **series** with one
another between two points if they are connected as in (a) so as to provide only
a single current path between the points. The *current* is the same in each
element.

The resistors in Fig. 29–1b are said to be in **parallel** between points a and
b. Each resistor provides an alternative path between the points, and any num-
ber of circuit elements similarly connected are in parallel with one another.
The *potential difference* is the same across each element.

In Fig. 29–1c, resistors R_2 and R_3 are in parallel with each other, and this
combination is in series with the resistor R_1. In Fig. 29–1d, R_2 and R_3 are in
series, and this combination is in parallel with R_1.

It is always possible to find a single resistor that could replace a combination of resistors and leave unaltered both the potential difference between the

Combining resistors in a circuit: What is
the equivalent resistance of a
combination?

What do we mean by equivalent
resistance?

653

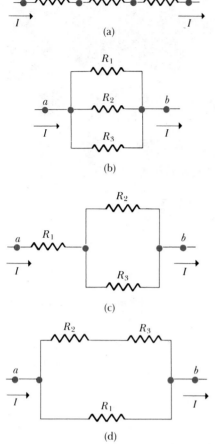

29–1 Four different ways of connecting three resistors.

terminals of the combination and the current in the rest of the circuit. The resistance of this single resistor is called the **equivalent resistance** of the combination. If any one of the networks in Fig. 29–1 were replaced by its equivalent resistance R, we could write

$$V_{ab} = IR \quad \text{or} \quad R = \frac{V_{ab}}{I},$$

where V_{ab} is the potential difference between terminals a and b of the network and I is the current at point a or b. To compute an equivalent resistance, we assume a potential difference V_{ab} across the actual network, compute the corresponding current I, and take the ratio of one to the other.

If the resistors are in *series*, as in Fig. 29–1a, the current I must be the same in all of them. Hence

$$V_{ax} = IR_1, \qquad V_{xy} = IR_2, \qquad V_{yb} = IR_3,$$

and

$$V_{ab} = V_{ax} + V_{xy} + V_{yb} = I(R_1 + R_2 + R_3), \qquad \frac{V_{ab}}{I} = R_1 + R_2 + R_3.$$

But V_{ab}/I is, by definition, the equivalent resistance R. Therefore

$$R = R_1 + R_2 + R_3. \tag{29–1}$$

The equivalent resistance of *any number* of resistors in series equals the sum of their individual resistances.

If the resistors are in *parallel*, as in Fig. 29–1b, the potential difference between the terminals of each must be the same and equal to V_{ab}. If the currents in each are denoted by I_1, I_2, and I_3, respectively,

$$I_1 = \frac{V_{ab}}{R_1}, \qquad I_2 = \frac{V_{ab}}{R_2}, \qquad I_3 = \frac{V_{ab}}{R_3}.$$

Charge is delivered to point a by the current I and removed from a by the currents I_1, I_2, and I_3. Since charge is not accumulating at a, it follows that

$$I = I_1 + I_2 + I_3 = V_{ab}\left(\frac{1}{R_1} + \frac{1}{R_2} + \frac{1}{R_3}\right), \quad \text{or} \quad \frac{I}{V_{ab}} = \frac{1}{R_1} + \frac{1}{R_2} + \frac{1}{R_3}.$$

But

$$\frac{I}{V_{ab}} = \frac{1}{R},$$

so

$$\frac{1}{R} = \frac{1}{R_1} + \frac{1}{R_2} + \frac{1}{R_3}. \tag{29–2}$$

For *any number* of resistors in parallel, the *reciprocal* of the equivalent resistance equals the *sum of the reciprocals* of their individual resistances.

For the special case of *two* resistors in parallel,

$$\frac{1}{R} = \frac{1}{R_1} + \frac{1}{R_2} = \frac{R_2 + R_1}{R_1 R_2}$$

and

$$R = \frac{R_1 R_2}{R_1 + R_2}.$$

For resistors in series, the currents are all the same.

Resistances in series add directly.

For resistors in parallel, the voltages are all the same.

Resistances in parallel add reciprocally.

Also, since $V_{ab} = I_1 R_1 = I_2 R_2$,

For two resistors in parallel, the current in each is inversely proportional to its resistance.

$$\frac{I_1}{I_2} = \frac{R_2}{R_1}, \tag{29–3}$$

and the currents carried by two resistors in parallel are *inversely proportional* to their resistances.

The equivalent resistances of the networks in Figs. 29–1c and 29–1d could be found by the same general method, but it is simpler to consider them as combinations of series and parallel arrangements. Thus in (c) the combination of R_2 and R_3 in parallel is first replaced by its equivalent resistance, which then forms a simple series combination with R_1. In (d) the combination of R_2 and R_3 in series forms a simple parallel combination with R_1.

PROBLEM-SOLVING STRATEGY: *Resistors in series and parallel*

1. It helps to remember that when two or more resistors are connected in series, the total potential difference across the combination is the sum of the individual potential differences. When they are connected in parallel, the potential difference is the same for every resistor, and the total potential difference across the combination is equal to that for each individual resistor.

2. Also keep in mind the analogous statements for current. When two or more resistors are connected in series, the current is the same through every resistor, and this is also equal to the current through the combination. When resistors are connected in parallel, the total current through the combination is equal to the sum of currents through the individual resistors.

EXAMPLE 29–1 Compute the equivalent resistance of the network in Fig. 29–2, and find the current in each resistor.

Examples of combining resistors in series and parallel

SOLUTION Successive stages in the reduction to a single equivalent resistance are shown in parts (b) and (c). From Eq. (29–2), the 6-Ω and the 3-Ω resistors in parallel in part (a) are equivalent to the single 2-Ω resistor in part (b), and the series combination of this with the 4-Ω resistor results in the single equivalent 6-Ω resistor in part (c).

In the simple series circuit of (c), the current is 3 A, and hence the current in the 4-Ω and 2-Ω resistors in part (b) is also 3 A. The potential difference V_{cb} is therefore 6 V, and since it must be 6 V in part (a) as well, the currents in the 6-Ω and 3-Ω resistors in part (a) are 1 A and 2 A, respectively.

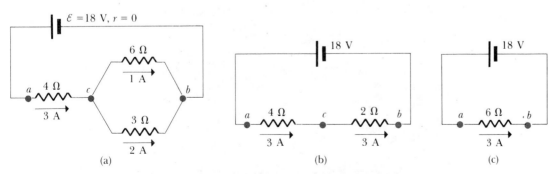

29–2 Steps in reducing a combination of resistors to a single equivalent resistor.

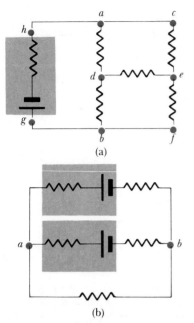

(a)

(b)

29–3 Two networks that cannot be reduced to simple series–parallel combinations of resistors.

29–2 KIRCHHOFF'S RULES

Not all networks can be reduced to simple series–parallel combinations. An example is a resistance network with a cross connection, as in Fig. 29–3a. A circuit like that in Fig. 29–3b, which contains sources in parallel paths, is another example. No new *principles* are required to compute the currents in these networks, but a number of techniques are available that help us handle such problems systematically. We will describe one of these, first developed by Gustav Robert Kirchhoff (1824–1887).

We first define two terms. A **branch point** in a network is a point where three or more conductors are joined. A **loop** is any closed conducting path. In Fig. 29–3a, for example, points *a*, *b*, *d*, and *e* are branch points but *c* and *f* are not. The circuit in Fig. 29–3b has only two branch points, *a* and *b*. Some possible loops in Fig. 29–3a are the closed paths *aceda*, *defbd*, *hadbgh*, and *hadefbgh*.

Kirchhoff's rules consist of the following two statements. **Point rule:** *The algebraic sum of the currents toward any branch point is zero:*

$$\sum I = 0. \qquad (29\text{–}4)$$

Loop rule: *The algebraic sum of the potential differences in any loop*, including those associated with emf's and those of resistive elements, *must equal zero.*

The point rule is an application of the principle of *conservation of electric charge*. Since no charge can accumulate at a branch point, the total current entering the point must equal the total current leaving, or (considering those entering as positive and those leaving as negative) the algebraic sum of currents into a point must be zero.

We have already seen the loop rule in Section 28–4. It is an expression of an *energy* relationship; as a charge goes around a loop and returns to its starting point, the algebraic sum of the changes in potential must be zero. Rises in potential are associated with sources of emf, and drops are associated with resistors and other circuit elements. No matter what the detailed nature of the circuit elements is, or whether Ohm's law is obeyed, the algebraic sum of potential differences around every closed loop must be zero.

These basic rules are all we need to solve a wide variety of network problems. Usually some of the emf's, currents, and resistances are known, and others are unknown. We must always obtain from Kirchhoff's rules a number of equations equal to the number of unknowns, so we can solve the equations simultaneously. Often the hardest part of the solution is not in understanding the basic principles but in keeping track of algebraic signs!

A printed circuit using deposited thin films for conducting paths. This circuit cannot be reduced to simple series and parallel combinations, but it can be analyzed using Kirchhoff's rules. (Courtesy of AT&T Bell Laboratories.)

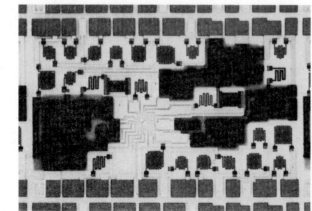

PROBLEM-SOLVING STRATEGY: *Kirchhoff's Rules*

1. Draw a *large* circuit diagram so you have plenty of room for labels. Label all quantities, known and unknown, including an assumed sense of direction for each unknown current and emf. Often you will not know in advance the actual direction of an unknown current or emf, but this doesn't matter. Carry out your solution, using the assumed direction. If the actual direction of a particular quantity is opposite to your assumption, the result will come out with a negative sign. Hence if you use Kirchhoff's rules correctly, they give you the directions as well as magnitudes of unknown currents and emf's. We have made this point before, in Section 28–4; you may want to review that discussion now. We will illustrate it further in the following examples.

2. Usually, when you label currents it is best to use the point rule immediately to express the currents in terms of as few quantities as possible. For example, Fig. 29–4a shows a circuit correctly labeled; Fig. 29–4b shows the same circuit, relabeled by applying the point rule to point a to eliminate I_3.

3. Choose any closed loop in the network, and designate a direction (clockwise or counterclockwise) to traverse the loop in applying the loop rule.

4. Go around the loop in the designated direction, adding potential differences as you cross them. An emf is counted as positive when it is traversed from $-$ to $+$, and negative when from $+$ to $-$. An IR product is negative if your path passes through the resistor in the *same* direction as the assumed current, positive if in the opposite direction. "Uphill" potential changes are always positive, "downhill" changes negative.

5. Equate the sum in step 4 to zero.

6. If necessary, choose another loop to get a different relation among the unknowns, and continue until you

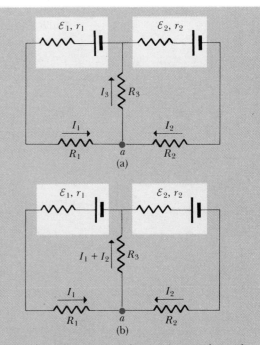

29–4 Application of the point rule to point a reduces the number of unknown currents from three to two.

have as many equations as unknowns, or until every circuit element has been included in at least one of the chosen loops.

7. Finally, solve the equations simultaneously to determine the unknowns. This step is algebra, not physics, but it can be fairly complex. Be careful with algebraic manipulations; one sign error is fatal to the entire solution.

EXAMPLE 29–2 In the circuit shown in Fig. 29–5, find the unknown current I, resistance R, and emf $\mathcal{E}$.

Some examples of circuit problems using Kirchhoff's rules

SOLUTION Application of the point rule to point a yields the relation

$$I + 1\,\text{A} - 6\,\text{A} = 0,$$
$$I = 5\,\text{A}.$$

To determine R we apply the loop rule to the loop labeled (1), obtaining

$$18\,\text{V} - (5\,\text{A})R + (1\,\text{A})(2\,\Omega) = 0,$$
$$R = 4\,\Omega.$$

The term for resistance R is negative because our loop traverses that element in the same direction as the current and hence finds a potential *drop*, while the term

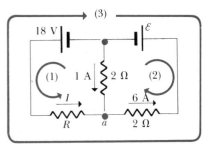

29–5 Circuit for Example 29–2.

for the 2-Ω resistor is positive because in traversing it in the direction opposite to the current we find a potential *rise*. If we had chosen to traverse loop (1) in the opposite direction, every term would have had the opposite sign, and the result for R would have been the same.

To determine $\mathcal{E}$, we apply the loop rule to loop (2):

$$\mathcal{E} + (6 \text{ A})(2 \text{ }\Omega) + (1 \text{ A})(2 \text{ }\Omega) = 0,$$

$$\mathcal{E} = -14 \text{ V}.$$

This result shows that the actual polarity of this emf is opposite to that assumed, and that the positive terminal of this source is really on the left side. Alternatively, we could use loop (3), obtaining the equation

$$\mathcal{E} + (6 \text{ A})(2 \text{ }\Omega) + (5 \text{ A})(4 \text{ }\Omega) - 18 \text{ V} = 0,$$

from which again $\mathcal{E} = -14 \text{ V}$.

EXAMPLE 29–3 In Fig. 29–6, find the current in each resistor and the equivalent resistance of the network.

SOLUTION As pointed out at the beginning of this section, it is not possible to represent this network in terms of series and parallel combinations. There are five different currents to determine, but by applying the point rule to junctions a and b we represent them in terms of three unknown currents, as indicated in the figure. The current in the battery is $(I_1 + I_2)$.

We now apply the loop rule to the three loops shown, obtaining the following three equations:

$$13 \text{ V} - I_1(1 \text{ }\Omega) - (I_1 - I_3)(1 \text{ }\Omega) = 0; \qquad (1)$$

$$-I_2(1 \text{ }\Omega) - (I_2 + I_3)(2 \text{ }\Omega) + 13 \text{ V} = 0; \qquad (2)$$

$$-I_1(1 \text{ }\Omega) - I_3(1 \text{ }\Omega) + I_2(1 \text{ }\Omega) = 0. \qquad (3)$$

This is a set of three simultaneous equations for the three unknown currents. They may be solved by various methods; one straightforward procedure is to solve the third for I_2, obtaining $I_2 = I_1 + I_3$, and then substitute this expression into the first two equations to eliminate I_2. When this is done we obtain the two equations

$$13 \text{ V} = I_1(2 \text{ }\Omega) - I_3(1 \text{ }\Omega), \qquad (1')$$

$$13 \text{ V} = I_1(3 \text{ }\Omega) - I_3(5 \text{ }\Omega), \qquad (2')$$

Now I_3 may be eliminated by multiplying the first of these by 5 and adding the two equations, obtaining

$$78 \text{ V} = I_1(13 \text{ }\Omega),$$

from which $I_1 = 6$ A. This result may be substituted back into (1′) to obtain $I_3 = -1$ A, and finally from (3) we find $I_2 = 5$ A. Note that the direction of I_3 is opposite that of the initial assumption.

The total current through the network is $I_1 + I_2 = 11$ A, and the potential drop across it is equal to the battery emf, namely, 13 V. Thus the equivalent resistance of the network is

$$R = \frac{13 \text{ V}}{11 \text{ A}} = 1.18 \text{ }\Omega.$$

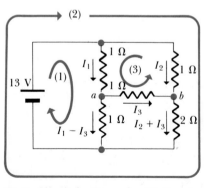

29–6 Circuit for Examples 29–3 and 29–4.

Once the currents in a circuit are determined, the potential difference between any two points a and b can be found by the procedure discussed in Section 28–4, just preceding Example 28–8. To find $V_{ab} = V_a - V_b$ (the potential at a with respect to b), we start at b and add the potential differences as we go from b to a. An emf is considered positive when we go from $-$ to $+$, negative otherwise. An IR term is positive when we go "uphill," against the current direction, negative when in the same direction as the current.

How to find the potential difference between two points when the currents are known

---

EXAMPLE 29–4   In the circuit of Example 29–3 (Fig. 29–6) find the potential difference V_{ab}.

SOLUTION   Starting at point b, we follow a path to point a, adding potential rises and drops as we go. The simplest path is through the center 1-Ω resistor. We have found $I_3 = -1$ A, showing that the actual current direction in this branch is from right to left. Thus as we go from b to a there is a drop of potential of magnitude $IR = (1\ \text{A})(1\ \Omega) = 1$ V, and $V_{ab} = -1$ V. Alternatively, we may go around the lower loop. We then have

$$I_2 + I_3 = 5\ \text{A} + (-1\ \text{A}) = 4\ \text{A},$$

$$I_1 - I_3 = 6\ \text{A} - (-1\ \text{A}) = 7\ \text{A},$$

and

$$V_{ab} = -(4\ \text{A})(2\ \Omega) + (7\ \text{A})(1\ \Omega) = -1\ \text{V}.$$

We suggest you try some other paths from b to a to verify that they also give this result.

29–3 ELECTRICAL INSTRUMENTS

Many familiar instruments for measuring potential difference (voltage), current, or resistance use a device called a **d'Arsonval galvanometer.** A pivoted coil of fine wire is placed in the magnetic field of a permanent magnet, as shown in Fig. 29–7. When there is a current in the coil, the magnetic field exerts on the coil a *torque* that is proportional to the current. (This magnetic interaction is discussed in detail in Chapter 30.) The torque is opposed by a spring, similar to the hairspring on the balance wheel of a watch, which exerts a restoring torque proportional to the angular displacement.

The d'Arsonval galvanometer: a common current-measuring instrument

Thus the angular deflection of the pointer attached to the pivoted coil is directly proportional to the coil current, and the device can be calibrated to measure current. The maximum deflection for which the meter is designed, typically 90° to 120°, is called *full-scale deflection*. The current required to produce full-scale deflection (typically of the order of 10 μA to 10 mA) and the resistance of the coil (typically of the order of 10 to 1000 Ω) are the essential electrical characteristics of the meter.

The meter deflection is proportional to the *current* in the coil, but if the coil obeys Ohm's law, the current is proportional to the *potential difference* between the terminals of the coil. Thus the deflection is also proportional to this potential difference. For example, consider a meter whose coil has a resistance of 20 Ω, and which deflects full scale with a current of 1 mA in its coil. The

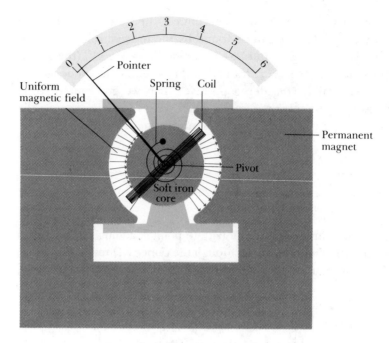

29–7 A d'Arsonval meter movement, showing pivoted coil with attached pointer, permanent magnet supplying uniform magnetic field, and spring to provide restoring torque, which opposes magnetic-field torque.

corresponding potential difference is

$$V_{ab} = IR = (10^{-3} \text{ A})(20 \ \Omega) = 0.020 \text{ V} = 20 \text{ mV}.$$

A current-measuring instrument is usually called an **ammeter** (or milliammeter, microammeter, etc., depending on the range). Such a device always measures the current passing through it. A meter such as the one just described can be adapted to measure currents larger than its full-scale reading by connecting a resistor in parallel with it, as in Fig. 29–8a, so that some of the current bypasses the meter. The parallel resistor is called a **shunt resistor** or simply a *shunt*, symbol R_{sh}.

For example, suppose we need an ammeter with a range of 0 A to 10 A, based on the 1-mA meter described above. We must choose a shunt such that the total current I through both meter and shunt is 10 A when the current through the meter itself is 1 mA = 0.001 A. Thus the current in the shunt is 9.999 A at full-scale deflection. The potential difference across the shunt is the same as that across the meter, namely 0.020 V, so the shunt resistance must be (from Ohm's law)

Using a d'Arsonval meter to measure larger currents: using a shunt to bypass some of the current

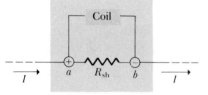

$$R_{\text{sh}} = \frac{0.020 \text{ V}}{9.999 \text{ A}}$$

$$= 0.00200 \ \Omega.$$

The equivalent resistance R of the instrument is

$$\frac{1}{R} = \frac{1}{R_c} + \frac{1}{R_{\text{sh}}},$$

and

$$R = 0.00200 \ \Omega.$$

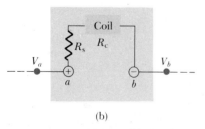

(b)

29–8 (a) Internal connections of an ammeter. (b) Internal connections of a voltmeter.

Thus we have a low-resistance instrument with the desired range of 0 to 10 A. Of course, if the current I is *less* than 10 A, the coil current and the deflection are correspondingly less.

An *ideal* ammeter would have zero resistance, so that including it in a branch of a circuit would not affect the current in that branch. Real ammeters always have some finite resistance, but it is always desirable for an ammeter to have the smallest practical resistance.

This same 1-mA meter may also be used to measure potential difference or *voltage;* any voltage-measuring device is usually called a **voltmeter** (or millivoltmeter, etc., depending on range). A voltmeter always measures the potential difference between two points, and its terminals must be connected to these points. Our 1-mA meter may be used as a voltmeter, but the maximum voltage it can measure is 20 mV. The range may be extended by connecting a resistor R_s in series with the meter, as in Fig. 29–8b, so that only some fraction of the total potential difference appears across the meter itself, and the remainder across R_s.

For example, suppose we need a voltmeter with a maximum range of 10 V. Then when the voltage across the meter is 20 mV = 0.020 V, the voltage across the series resistor R_s must be 10 V − 0.020 V, or 9.98 V. The current through the meter at full-scale deflection is still 1 mA or 0.001 A, so from Ohm's law the value of R_s must be

$$R_s = \frac{9.98 \text{ V}}{0.001 \text{ A}} = 9980 \ \Omega.$$

The equivalent resistance of the device is then

$$R = R_c + R_s = 10,000 \ \Omega.$$

An ideal voltmeter would have infinite resistance, so that connecting it between two points in a circuit would not alter any of the currents. Real voltmeters always have finite resistance; to be useful, however, a voltmeter must have large enough resistance so that connecting it in a circuit does not change the other currents appreciably.

A voltmeter and an ammeter can be used together to measure *resistance* and *power*. The resistance of a resistor equals the potential difference V_{ab} between its terminals, divided by the current I:

$$R = \frac{V_{ab}}{I},$$

and the power input to any portion of a circuit equals the product of the potential difference across this portion and the current:

$$P = V_{ab}I.$$

The most straightforward method of measuring R or P is therefore to measure V_{ab} and I simultaneously.

In Fig. 29–9a, ammeter A reads correctly the current I in the resistor R. Voltmeter V, however, reads the *sum* of the potential difference V_{ab} across the resistor and the potential difference V_{bc} across the ammeter.

If we transfer the voltmeter terminal from c to b, as in Fig. 29–9b, the voltmeter reads correctly the potential difference V_{ab}, but the ammeter now reads the *sum* of the current I in the resistor and the current I_V in the voltmeter. Thus, whichever connection is used, we must correct the reading of one instrument or the other to obtain the true values of V_{ab} or I (unless, of course, the corrections are small enough to be neglected).

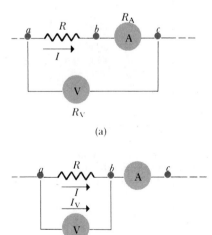

(a)

(b)

29–9 Ammeter–voltmeter method for measuring resistance or power.

Examples of resistance measurements using a voltmeter and an ammeter

EXAMPLE 29–5 Suppose we want to measure an unknown resistance R by using the circuit of Fig. 29–9a. The meter resistances are $R_V = 10,000\ \Omega$ and $R_A = 2.0\ \Omega$. If the voltmeter reads 12.0 V and the ammeter reads 0.10 A, what is the true resistance?

SOLUTION If the meters were ideal (i.e., $R_V = \infty$ and $R_A = 0$), the resistance would be simply $R = V/I = (12.0\ \text{V})/(0.10\ \text{A}) = 120\ \Omega$. But the voltmeter reading includes the potential V_{bc} across the ammeter as well as that (V_{ab}) across the resistor. We have $V_{bc} = IR_A = (0.10\ \text{A})(2.0\ \Omega) = 0.2\ \text{V}$, so the actual potential drop V_{ab} across the resistor is $12.0\ \text{V} - 0.2\ \text{V} = 11.8\ \text{V}$, and the resistance is

$$R = \frac{V_{ab}}{I} = \frac{11.8\ \text{V}}{0.10\ \text{A}} = 118\ \Omega.$$

EXAMPLE 29–6 Suppose the meters of Example 29–5 are connected to a different resistor, in the circuit shown in Fig. 29–9b, and the readings noted above are obtained. What is the true resistance?

SOLUTION In this case the voltmeter measures the potential across the resistor correctly; the difficulty is that the ammeter measures the voltmeter current I_V as well as the current I in the resistor. We have $I_V = V/R_V = (12.0\ \text{V})/(10,000\ \Omega) = 1.2\ \text{mA}$; so the actual current I in the resistor is $I = 0.10\ \text{A} - 0.0012\ \text{A} = 0.0988\ \text{A}$. Thus the resistance is

$$R = \frac{V_{ab}}{I} = \frac{12.0\ \text{V}}{0.0988\ \text{A}} = 121.5\ \Omega.$$

The ohmmeter: a simple but not very precise resistance-measuring instrument

An alternative method for measuring resistance is to use a d'Arsonval meter in an arrangement called an **ohmmeter.** It consists of a meter, a resistor, and a source (often a flashlight cell) connected in series, as in Fig. 29–10. The resistance R to be measured is connected between terminals x and y.

The series resistance R_s is chosen so that when terminals x and y are short-circuited (that is, when $R = 0$), the meter deflects full scale. When the circuit between x and y is open (that is, when $R = \infty$), the galvanometer shows no deflection. For a value of R between zero and infinity, the meter deflects to some intermediate point depending on the value of R, and hence the meter scale can be calibrated to read the resistance R. Larger currents correspond to smaller resistances, so this scale reads backward compared to the current scale.

In situations where high precision is required, instruments containing d'Arsonval meters have been supplanted by electronic instruments with direct digital readouts. These are more precise, stable, and mechanically rugged than d'Arsonval meters, but they are also considerably more expensive. Digital voltmeters can be made with extremely high internal resistance, of the order of 100 MΩ.

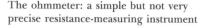
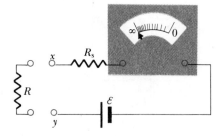

29–10 Ohmmeter circuit. The backward scale on the meter is calibrated to read resistance directly.

29–4 THE R–C SERIES CIRCUIT

What happens when a battery charges a capacitor through a resistor?

Thus far in our discussion of circuits, we have assumed that the emf's and resistances are constant, so that all potentials and currents are constant, independent of time. Figure 29–11 shows a simple example of a circuit in which the current and voltages are *not* constant. The capacitor is initially uncharged; at some initial time $t = 0$ we close the switch, completing the circuit and permitting current around the loop to begin charging the capacitor. The current begins at the same instant in every part of the circuit, and at each instant the current is the same in every part.

Because the capacitor is initially uncharged, the initial potential difference across it is zero. The entire battery voltage appears across the resistor, causing an initial current $I = V/R$. As the capacitor charges, its voltage increases, and the potential difference across the resistor decreases, corresponding to a decrease in current. After a long time the capacitor becomes fully charged and the entire battery voltage V appears across the capacitor. There is then no potential difference across the resistor, and the current is zero.

We let q represent the charge on the capacitor and i the current in the circuit at some time t after the switch has been closed. It is customary in circuit analysis to use small letters for quantities that vary with time, and capital letters for constant quantities; we follow that convention here. We note that i and q are related by $i = dq/dt$; the instantaneous potential differences v_{ac} and v_{cb} are

$$v_{ac} = iR, \qquad v_{cb} = \frac{q}{C}. \tag{29–5}$$

Therefore

$$V_{ab} = V = v_{ac} + v_{cb} = iR + \frac{q}{C}, \tag{29–6}$$

where V is the terminal voltage of the battery, assumed to be constant. (That is, we neglect the internal resistance of the battery; V is then equal to its emf.) Solving this equation for i, we find

$$i = \frac{V}{R} - \frac{q}{RC}. \tag{29–7}$$

At time $t = 0$, when the switch is first closed, the capacitor is uncharged, so $q = 0$. The *initial* current, which we call I_0, is, from Eq. (29–7), given by $I_0 = V/R$. This would be the (constant) current if the capacitor were not in the circuit.

As the charge q increases, the term q/RC becomes larger, and the current decreases and eventually becomes zero. When $i = 0$,

$$\frac{V}{R} = \frac{q}{RC}, \qquad q = CV = Q_{\mathrm{f}},$$

where Q_{f} is the final charge on the capacitor. The behavior of the current and capacitor charge as functions of time are shown in Fig. 29–12. At the instant the switch is closed ($t = 0$), the current jumps from zero to its initial value $I_0 = V/R$, and then gradually decreases to zero. The capacitor charge starts at zero and gradually approaches the final value $Q_{\mathrm{f}} = CV$.

We can derive expressions for the charge q and current i as functions of time. To do this, we replace i in Eq. (29–7) with dq/dt, obtaining

$$\frac{dq}{dt} = \frac{V}{R} - \frac{q}{RC}, \tag{29–8}$$

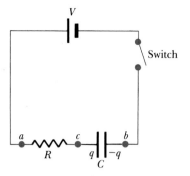

29–11 A capacitor C charged by a circuit containing a battery V and a resistor R.

The initial voltage across the capacitor is zero.

The final current in the circuit is zero.

Applying Kirchhoff's loop law to the capacitor-charging circuit

A differential equation for capacitor charge as a function of time

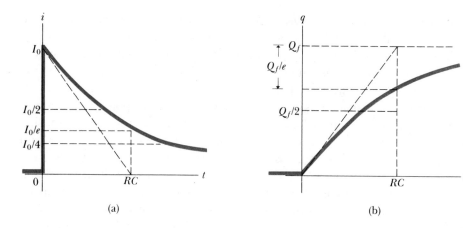

29–12 Current and capacitor charge as functions of time for the circuit of Fig. 29–11.

(a)   (b)

which may be rearranged as

$$\frac{dq}{VC - q} = \frac{dt}{RC}.$$

Integrating both sides, we obtain

$$-\ln(VC - q) = t/RC + \text{constant}.$$

To evaluate the constant, note that at $t = 0$, $q = 0$, so

$$-\ln(VC - 0) = 0 + \text{constant}.$$

Rearranging again, we obtain

The current is an exponentially decreasing function of time.

$$\ln(VC - q) - \ln VC = \ln\frac{VC - q}{VC} = -\frac{t}{RC},$$

$$1 - \frac{q}{VC} = e^{-t/RC}, \tag{29–9}$$

$$q = VC(1 - e^{-t/RC}) = Q_f(1 - e^{-t/RC}).$$

The time derivative of this expression is the current:

$$i = \frac{dq}{dt} = \frac{V}{R}e^{-t/RC} = I_0 e^{-t/RC}. \tag{29–10}$$

The time constant is the time needed for the current to drop to $1/e$ of its initial value.

The charge and current are therefore both *exponential* functions of time. Figure 29–12a is a graph of Eq. (29–10), and Fig. 29–12b is a graph of Eq. (29–9). At a time $t = RC$, the current has decreased to $1/e$ (about 0.368) of its initial value and the charge has increased to *within* $1/e$ of its final value. The product RC is called the **time constant,** or the **relaxation time,** of the circuit, denoted by τ:

$$\tau = RC. \tag{29–11}$$

EXAMPLE 29–7 A resistor of resistance $R = 10$ MΩ is connected in series with a capacitor of capacitance 1 μF. The time constant is

$$\tau = RC = (10 \times 10^6 \ \Omega)(10^{-6} \ \text{F}) = 10 \ \text{s}.$$

But if $R = 10 \ \Omega$, the time constant is only 10×10^{-6} s, or 10 μs.

Suppose now that after the capacitor in Fig. 29–11 has acquired some charge Q_0, we remove the battery from the circuit and connect points a and b directly together. We reset our stopwatch so the connection is made at time $t = 0$. The capacitor then *discharges* through the resistor, and its charge eventually decreases to zero.

A capacitor discharges through a resistor: The behavior is similar to the charging process.

Again let i and q represent the time-varying current and charge at some instant after the switch is thrown. Since V_{ab} is now zero, we have, from Eq. (29–6),

$$0 = v_{ac} + v_{cb}.$$

The direction of the current in the resistor is now from c to a, so $v_{ca} = -v_{ac} = iR$, and

$$i = \frac{q}{RC}. \tag{29–12}$$

When $t = 0$, $q = Q_0$ and the initial current I_0 is

$$I_0 = \frac{Q_0}{RC} = \frac{V_0}{R},$$

where V_0 is the initial potential difference across the capacitor. As the capacitor discharges, both q and i decrease.

The same procedures as above can be followed to obtain $i(t)$ and $q(t)$. If we replace i in Eq. (29–12) by $-dq/dt$ (the charge q is now *decreasing*), we get

$$\frac{dq}{dt} = -\frac{q}{RC}. \tag{29–13}$$

Integration of this equation gives $q(t)$, and by differentiation we find $i(t)$.

The charge and current are exponentially decreasing functions of time.

Alternatively, differentiation of Eq. (29–12) gives

$$\frac{di}{dt} = -\frac{i}{RC}, \tag{29–14}$$

from which we can get $i(t)$ and, by a second integration, get $q(t)$. We leave it to you to show that

$$i = I_0 e^{-t/RC}, \tag{29–15}$$

$$q = Q_0 e^{-t/RC}. \tag{29–16}$$

Both the current and the charge decrease exponentially with time. Comparing these results with Eqs. (29–9) and (29–10), we note that the expressions for the current are identical. The capacitor charge approaches zero asymptotically in Eq. (29–16), while the *difference* between q and Q_f approaches zero asymptotically in Eq. (29–9).

The time constant is the same for charging and discharging.

Energy considerations offer us additional insight into the behavior of an RC circuit. The instantaneous rate at which the battery delivers energy to the circuit is $P = Vi$. The instantaneous rate at which energy is dissipated in the resistor is i^2R, and the rate at which energy is stored in the capacitor is $iv_{cb} = qi/C$. Multiplying Eq. (29–6) by i, we find

$$Vi = i^2R + iq/C. \tag{29–17}$$

This expression means that of the power Vi supplied by the battery, part (i^2R) is dissipated in the resistor and part (iq/C) is stored in the capacitor.

Energy relations for a charging or
discharging capacitor

The *total* energy supplied by the battery during charging of the capacitor
equals the battery potential difference V multiplied by the total charge Q_f, or
$Q_f V$. The total energy stored in the capacitor, from Eq. (27–8), is $Q_f V/2$. Thus
of the energy supplied by the battery, *exactly half* is stored in the capacitor, and
the other half is dissipated in the resistor. It is a little surprising that this
half-and-half division of energy does not depend on C, R, or V. This result can
also be verified in detail by taking the integral over time of each of the power
quantities mentioned above; we leave this calculation for your amusement.

29–5 DISPLACEMENT CURRENT

For a *conducting* circuit the total current *into* any given portion must equal the
current *out of* that portion. This statement forms the basis of Kirchhoff's point
rule, discussed in Section 29–2. However, this rule is *not* obeyed for a capacitor that is being charged. In Fig. 29–13, a conduction current goes *into* the left
plate, but no conduction current comes *out of* this plate; similarly, there is
conduction current out of the right plate, but none into it.

James Clerk Maxwell (1831–1879) showed that it is possible to generalize
the definition of current so that we can still say that the current out of each
plate is equal to the current into it. As the capacitor charges, the conduction
current increases the charge on each plate, and this, in turn, increases the
electric field between the plates. The *rate* of increase of field is proportional to
the conduction current. Maxwell's idea was to associate an equivalent current
density with this rate of increase of field.

Consider a parallel-plate capacitor with uniform charge density σ on the
plates. The field magnitude E between the plates is then given by

$$E = \frac{\sigma}{\epsilon_0} = \frac{Q}{\epsilon_0 A}. \tag{29–18}$$

In a time interval dt, the charge Q increases by $dQ = I_C \, dt$; the corresponding
change in E is

$$dE = \frac{dQ}{\epsilon_0 A} = \frac{I_C \, dt}{\epsilon_0 A},$$

and the *rate* of change of E is

$$\frac{dE}{dt} = \frac{I_C}{\epsilon_0 A}. \tag{29–19}$$

We now define an *equivalent current density* J_D between the plates as

$$J_D = \epsilon_0 \frac{dE}{dt}. \tag{29–20}$$

Then the total equivalent current between the plates, which we may call I_D, is

$$I_D = J_D A = \epsilon_0 \left(\frac{dE}{dt} \right) A = I_C. \tag{29–21}$$

Thus, if we include this equivalent current as well as the conduction current,
the current *into* the region bounded by the broken line in Fig. 29–13 equals
the current *out of* this region. The subscript D is chosen because Maxwell gave
this equivalent current the name **displacement current.**

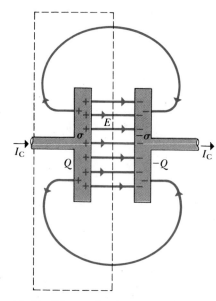

29–13 The conduction current into
the left plate of the capacitor equals
the displacement current between the
plates.

Maxwell's breakthrough: associating a
changing electric field with an effective
current density

When both conduction and displacement
currents are included, Kirchhoff's point
rule is obeyed for one plate of a
capacitor.

Displacement current may also be expressed in terms of *electric flux*, defined in Section 25–4, Eqs. (25–18) and (25–19). For a uniform field E perpendicular to an area A, $\Psi = EA$, and Eq. (29–21) may be rewritten as

$$I_D = \epsilon_0 \frac{d\Psi}{dt}. \tag{29–22}$$

If the space between the capacitor plates contains a dielectric, Eq. (29–18) must be replaced by the more general relation derived in Section 27–5:

$$E = \frac{\sigma}{\epsilon} = \frac{Q}{\epsilon A} = \frac{Q}{K\epsilon_0 A}. \tag{29–23}$$

The corresponding modification of the definition of displacement current density is

$$J_D = \epsilon\left(\frac{dE}{dt}\right) = K\epsilon_0\left(\frac{dE}{dt}\right). \tag{29–24}$$

Maxwell's generalized view of current and current density may appear to be merely an artifice introduced to preserve Kirchhoff's current rule even in cases where charge is accumulating in a certain region of space, such as a capacitor plate. However, it is much more than this. In Chapter 31 we will study the role of current as a source of *magnetic* field, and we will see that a *displacement* current sets up a magnetic field in exactly the same way as an ordinary *conduction* current. Thus displacement current, far from being an artifice, is a fundamental fact of nature, and Maxwell's discovery of it was the bold step of an extraordinary genius. As we will see, the concept of displacement current provided the necessary basis for the understanding of electromagnetic waves in the last third of the nineteenth century.

Displacement current isn't just a gimmick; it is a basic fact of nature.

29–6 POWER DISTRIBUTION SYSTEMS

We conclude this chapter with a brief discussion of practical household and automotive electric-power distribution systems. Automobiles use direct-current (dc) systems, while nearly all household, commercial, and industrial systems use alternating current (ac). Most of the same basic wiring concepts apply to both. Alternating-current circuits are discussed in greater detail in Chapter 34.

Parallel circuits are the usual thing in household wiring.

The various lamps, motors, and other appliances to be operated are always connected in *parallel* to the power source—the wires from the power company in the case of houses, the battery and alternator for a car. The basic idea of house wiring is shown in Fig. 29–14. One side of the "line," as the pair of conductors is called, is always connected to "ground." For houses this is an actual electrode driven into the earth (usually a good conductor) and also connected to the household water pipes. Electricians speak of the "hot" side and the "ground" side of the line.

The voltage–current–power relations are the same as those we saw in Section 28–5. Household voltage is nominally 120 V in the United States and Canada, often 240 V in Europe. (These are actually the root–mean–square voltages, as we will discuss in Section 34–4). The current in a 100-W light

Voltage, current, and resistance for an ordinary light bulb

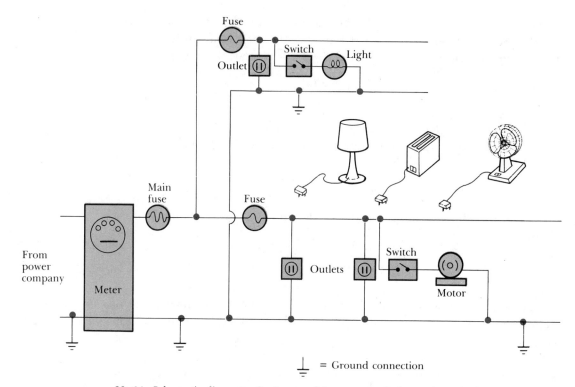

29–14 Schematic diagram of a house wiring system. Only two branch circuits are shown; an actual system might have 4 to 30 branch circuits. The conventional symbol for ground is shown. Lamps and appliances may be plugged into the outlets. The grounding conductors, which normally carry no current, are not shown.

bulb, for example, can be determined from Eq. (28–21):

$$I = \frac{P}{V} = \frac{100 \text{ W}}{120 \text{ V}} = 0.833 \text{ A}.$$

The resistance of this bulb at operating temperature is given by

$$R = \frac{V}{I} = \frac{120 \text{ V}}{0.833 \text{ A}} = 144 \text{ }\Omega,$$

or

$$R = \frac{V^2}{P} = \frac{(120 \text{ V})^2}{100 \text{ W}} = 144 \text{ }\Omega.$$

Similarly, a 1500-W waffle iron draws a current of $(1500 \text{ W})/(120 \text{ V}) = 12.5 \text{ A}$ and has a resistance, at operating temperature, of 9.6 Ω.

The maximum current available from an individual circuit is determined by the resistance of the wires; the power loss in the wires from I^2R heat causes them to become hot, and in extreme cases this heat can cause a fire or melt the wires. Ordinary lighting and outlet wiring in houses is usually 12-gauge wire, which has a diameter of 2.05 mm and can carry a maximum current of 20 A without overheating. Larger sizes such as 8-gauge (3.26 mm) or 6-gauge (4.11 mm) are used for high-current appliances such as ranges and clothes dryers, and 2-gauge (6.54 mm) or larger is used for the main entrance lines.

Protection against overloading and overheating of circuits is provided by fuses or circuit breakers. A *fuse* contains a link of lead–tin alloy with a very low

The safe current-carrying capacity of a wire is limited by resistive heating; bigger wires can carry more current.

Fuses and circuit breakers: protecting wiring against dangerous overloads

melting temperature; the link melts and breaks the circuit when its rated current is exceeded. A *circuit breaker* is an electromechanical device that performs the same function, using an electromagnet or a bimetallic strip to "trip" the breaker and interrupt the circuit when the current exceeds a specified value. Circuit breakers have the advantage that they can be reset after they are tripped; a blown fuse must be replaced, but fuses are somewhat more reliable in operation than circuit breakers.

You may have had the experience of blowing a fuse by plugging too many high-current appliances into the same outlet or into two outlets fed by the same circuit. *Do not* replace the fuse with one of larger rating; to do so risks overheating the wires and starting a fire. The only safe solution is to distribute the appliances among several circuits. Modern kitchens often have three or four separate 20-A circuits.

Accidental contact between the hot and ground sides of the line causes a **short circuit.** Such a situation, which can be caused by faulty insulation or by any of a variety of mechanical malfunctions, provides a very low-resistance current path, permitting a very large current that would quickly melt the wires and ignite their insulation if the current were not interrupted by a fuse or circuit breaker. The opposite situation, a broken wire that interrupts the current path, creates an **open circuit.** An open circuit can be hazardous because of the sparking that can occur at the point of intermittent contact.

Dangers of short circuits and open circuits

In approved wiring practice, a fuse or breaker is placed *only* in the hot side of the line, never in the ground side. Otherwise, if a short circuit should develop because of faulty insulation or other malfunction, the ground-side fuse could blow, leaving the hot side alive and posing a shock hazard to a person who touched the live conductor and a grounded object such as a water pipe. For similar reasons, the wall switch for a light fixture is always in the hot side of the line, never the ground side.

Further protection against shock hazard is provided by a third conductor included in all present-day wiring. This conductor (corresponding to the long, round prong of the three-prong connector plug on an appliance or power tool) normally carries no current, but it connects the metal case or frame of the device to ground. If a conductor in the hot side of the line accidentally contacts the frame or case, the grounding conductor provides a current path and the fuse blows. Without the ground wire, the frame could become "live," that is, at a potential 120 V above ground; then touching it and a water pipe (or even a damp basement floor) at the same time would offer a shock hazard. In some special situations, especially outlets located outdoors or near a sink or other water pipes, a special kind of circuit breaker called a *ground-fault interrupter* is used. This device senses the difference in current between the hot and neutral conductors (which is normally zero) and trips when this difference exceeds some very small value, typically 10 mA.

Grounding conductors: What is the third wire for?

All the discussion above can be applied directly to automobile wiring. The voltage is 12 V; the power is supplied by the battery and by the alternator, which charges the battery when the engine is running. The ground side of the circuit is connected to the body and frame of the vehicle. There is no separate grounding conductor, and indeed the frame itself sometimes provides the ground side of the circuit. The fuse or circuit breaker arrangement is the same in principle as in household wiring. Because of the lower voltage, more current is required for the same power; a 100-W headlight bulb requires a current of (100 W)/(12 V) = 8.3 A.

Automotive wiring: same principles, lower voltage

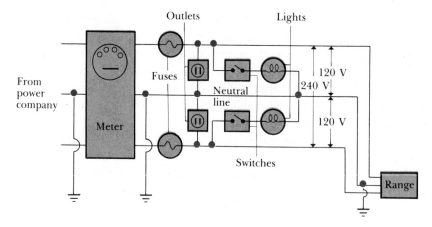

29–15 Simplified schematic diagram of a 120–240-V house wiring system. Only one circuit on each side is shown; in actual systems there would be several 120-V circuits on each side of the neutral line. Grounding wires are not shown.

Color codes for wiring: not always consistent

Power-company employees read a "light meter." This meter records the total input of electrical energy into the facility being supplied; the meter indicates the energy input directly in kilowatt-hours. (Courtesy of Pacific Gas and Electric.)

The cost of electrical energy

KEY TERMS

series

parallel

equivalent resistance

branch point

loop

Kirchhoff's point rule

Kirchhoff's loop rule

To help prevent wiring errors, household wiring uses a standardized color code in which the hot side of a line has black or red insulation, the ground side has white insulation, and the grounding conductor is bare or has green insulation. In electronic devices and equipment, however, the ground side of the line is usually black. Beware!

Most household wiring systems actually use a slight elaboration of the system described above. The power company provides *three* conductors. One is grounded or "neutral"; the other two are both at 120 V with respect to the neutral, but with opposite polarity, giving a voltage between them of 240 V. Thus 120-V lamps and appliances can be connected between neutral and either hot conductor, and high-power devices requiring 240 V are connected between the two hot lines, as shown in Fig. 29–15. Ranges and dryers usually are designed for 240-V power input.

Although we have spoken of *power* in this discussion, what we buy from the power company is *energy*. Power is energy transferred per unit time, so energy is power multiplied by time. The usual unit of energy sold by the power company is the kilowatt-hour (1 kWh):

$$1 \text{ kWh} = (10^3 \text{ W})(3600 \text{ s}) = 3.6 \times 10^6 \text{ W·s} = 3.6 \times 10^6 \text{ J}.$$

One kilowatt-hour typically costs 5 to 10 cents, depending on location and quantity of energy purchased. To operate a 1500-W (1.5-kW) waffle iron for one hour requires 1.5 kWh of energy and costs 7.5 to 15 cents. The cost of operating any lamp or appliance for a specified time can be calculated in the same way if the power rating is known.

SUMMARY

When several resistors $R_1, R_2, R_3, \ldots$, are connected in series, the equivalent resistance R is the sum of the individual resistances;

$$R = R_1 + R_2 + R_3 + \cdots. \tag{29–1}$$

When several resistors are connected in parallel, the equivalent resistance R is given by

$$\frac{1}{R} = \frac{1}{R_1} + \frac{1}{R_2} + \frac{1}{R_3}. \tag{29–2}$$

Kirchhoff's point rule states that the algebraic sum of the currents into any branch point must be zero. This rule follows from conservation of electric charge. Kirchhoff's loop rule states that the algebraic sum of potential differences around any loop must be zero. This rule follows from conservation of energy and the conservative nature of electrostatic fields.

Many simple electrical measuring instruments incorporate a d'Arsonval galvanometer. The deflection is proportional to the current in the coil. For a larger current range, a shunt resistor is added, so some of the current bypasses the meter. Such an instrument is called an ammeter. If the coil and any additional series resistance included obey Ohm's law, the meter can also be calibrated to read potential difference, or voltage. The instrument is then called a voltmeter.

When a capacitor is charged by a battery in series with a resistor, the current and capacitor charge are not constant. The charge approaches its final value asymptotically, and the current approaches zero asymptotically. The charge and current in the circuit are given by exponential functions. The time τ at which the charge has approached within $1/e$ of its final value is called the time constant or the relaxation time, given by $\tau = RC$. Discharge of a capacitor through a resistor is also characterized by exponential variation of the charge and current; the time constant is the same for charging and discharging.

A single plate of a charging or discharging capacitor does not obey Kirchhoff's point rule because there is current in the conductor attached to the plate but none in the gap between plates. Maxwell generalized the definition of current by postulating an equivalent current density J_D between the plates, given by

$$J_D = \epsilon_0 \frac{dE}{dt}. \qquad (29\text{--}20)$$

With this definition, the conduction current in the conductor is equal to the displacement current in the gap. When a dielectric is present between the plates, displacement current density is defined as

$$J_D = \epsilon\left(\frac{dE}{dt}\right) = K\epsilon_0\left(\frac{dE}{dt}\right). \qquad (29\text{--}34)$$

Displacement current acts as a source of magnetic field, just as conduction current does. Thus it is not an artifice but a real fact of nature.

In household wiring systems, the various electrical devices are connected in parallel across the power line, which consists of a pair of conductors, one "hot" and the other "ground." The current capacity of a circuit is determined by the size of the wires and the maximum temperature they can tolerate. Protection against excessive current and the resulting fire hazard is provided by fuses or circuit breakers.

d'Arsonval galvanometer
ammeter
shunt resistor
voltmeter
ohmmeter
time constant (relaxation time)
displacement current
short circuit
open circuit

QUESTIONS

29–1 Can the potential difference between terminals of a battery ever be opposite in direction to the emf?

29–2 Why do the lights on a car become dimmer when the starter is operated?

29–3 What determines the maximum current that can be carried safely by household wiring? (Typical limits are 15 A for 14-gauge wire, 20 A for 12-gauge, and so on.)

29–4 Lights in a house often dim momentarily when a

motor, such as a washing machine or a power saw, is turned on. Why does this happen?

29–5 Compare the formulas for resistors in series and parallel with those for capacitors in series and parallel. What similarities and differences do you see? Sometimes in circuit analysis one uses the quantity *conductance*, denoted as g and defined as the reciprocal of resistance: $g = 1/R$. What is the corresponding comparison for conductance and capacitance?

29–6 Is it possible to connect resistors together in a way that cannot be reduced to some combination of series and parallel combinations? If so, give examples; if not, state why not.

29–7 In a two-cell flashlight, the batteries are usually connected in series. Why not connect them in parallel?

29–8 Some Christmas-tree lights have the property that, when one bulb burns out, all the lights go out, while with others only the burned-out bulb goes out. Discuss this difference in terms of series and parallel circuits.

29–9 What possible advantage could there be in connecting several identical batteries in parallel?

29–10 Two 120-V light bulbs, one 25-W and one 200-W, were connected in series across a 240-V line. It seemed like a good idea at the time, but one bulb burned out almost instantaneously. Which one burned out, and why?

29–11 When the direction of current in a battery reverses, does the direction of its emf also reverse?

29–12 Under what conditions would the terminal voltage of a battery be zero?

29–13 What sort of meter should be used to test the condition of a dry cell (such as a flashlight battery) having constant emf but internal resistance that increases with age and use?

29–14 For very large resistances, it is easy to construct *RC* circuits having time constants of several seconds or minutes. How might this fact be used to measure very large resistances, too large to measure by more conventional means?

EXERCISES

Section 29–1 Resistors in Series and Parallel

29–1 A 60-Ω resistor and a 90-Ω resistor are connected in parallel, and the combination is connected across a 120-V dc line.

a) What is the resistance of the parallel combination?

b) What is the total current through the parallel combination?

c) What is the current through each resistor?

29–2 Three resistors having resistances of 1 Ω, 2 Ω, and 3 Ω are connected in series to a 12-V battery having negligible internal resistance.

a) Find the equivalent resistance of the combination.

b) Find the current in each resistor.

c) Find the total current through the battery.

d) Find the voltage across each resistor.

e) Find the power dissipated in each resistor.

29–3 In Exercise 29–2, the same three resistors are connected in parallel to the same battery. Answer the same questions for this situation.

29–4 A 25-W, 120-V light bulb and a 100-W, 120-V light bulb are connected in series across a 240-V line. Assume that the resistance of each bulb does not vary with current.

a) Find the current through the bulbs.

b) Find the power dissipated in each bulb.

c) One bulb burns out very quickly. Which one? Why?

29–5

a) The power rating of a 10,000-Ω resistor is 2 W. (The power rating is the maximum power the resistor can safely dissipate without too great a rise in temperature.) What is the maximum allowable potential difference across the terminals of the resistor?

b) A 20,000-Ω resistor is to be connected across a potential difference of 300 V. What power rating is required?

c) It is desired to connect an equivalent resistance of 1000 Ω across a potential difference of 200 V. A number of 10-W, 1000-Ω resistors are available. How should they be connected?

Section 29–2 Kirchhoff's Rules

29–6 Find the emf's $\mathcal{E}_1$ and $\mathcal{E}_2$ in the circuit of Fig. 29–16, and the potential difference between points a and b.

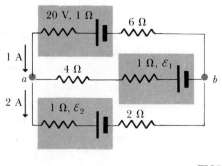

FIGURE 29–16

29–7 In the circuit shown in Fig. 29–17, find

a) the current in resistor R;

b) the resistance R;

c) the unknown emf $\mathcal{E}$.

d) If the circuit is broken at point x, what is the current in the 28-V battery?

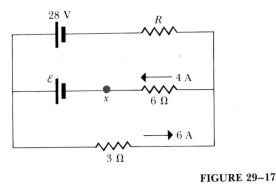

FIGURE 29–17

29–8 In the circuit shown in Fig. 29–18, find

a) the current in each branch;

b) the potential difference V_{ab}.

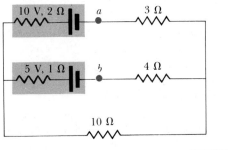

FIGURE 29–18

29–9

a) Find the potential difference between points a and b in Fig. 29–19.

b) If a and b are connected, find the current in the 12-V cell.

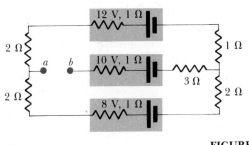

FIGURE 29–19

Section 29–3 Electrical Instruments

29–10 The resistance of a galvanometer coil is 50 Ω, and the current required for full-scale deflection is 500 μA.

a) Show in a diagram how to convert the galvanometer to an ammeter reading 5 A full scale, and compute the shunt resistance.

b) Show how to convert the galvanometer to a voltmeter reading 150 V full scale, and compute the series resistance.

29–11 The resistance of the coil of a pivoted-coil galvanometer is 10 Ω, and a current of 0.02 A causes it to deflect full scale. We want to convert this galvanometer to an ammeter reading 10 A full scale. The only shunt available has a resistance of 0.04 Ω. What resistance R must be connected in series with the coil? (See Fig. 29–20.)

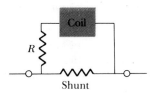

FIGURE 29–20

29–12 The resistance of the moving coil of the galvanometer G in Fig. 29–21 is 25 Ω; the meter deflects full scale with a current of 0.010 A. Find the magnitudes of the resistances R_1, R_2, and R_3 required to convert the galvanometer to a multirange ammeter deflecting full scale with currents of 10 A, 1 A, and 0.1 A.

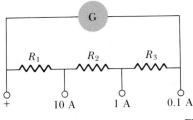

FIGURE 29–21

29–13 Figure 29–22 shows the internal wiring of a "three-scale" voltmeter whose binding posts are marked +, 3 V, 15 V, 150 V. The resistance of the moving coil, R_G, is 15 Ω, and a current of 1 mA in the coil causes it to deflect full scale. Find the resistances R_1, R_2, R_3, and the overall resistance of the meter on each of its ranges.

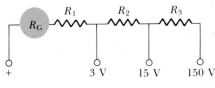

FIGURE 29–22

29–14 A 100-V battery has an internal resistance of $r = 5\ \Omega$.

a) What is the reading of a voltmeter having a resistance of $R_V = 500\ \Omega$ when placed across the terminals of the battery?

b) What maximum value may the ratio r/R_V have if the error in the reading of the emf of a battery is not to exceed 5%?

29–15 Two 150-V voltmeters, one with resistance 15,000 Ω and the other with resistance 150,000 Ω, are connected in series across a 120-V dc line. Find the reading of each voltmeter.

29–16 In the ohmmeter in Fig. 29–23, M is a 1-mA meter having a resistance of 100 Ω. The battery B has an emf of 3 V and negligible internal resistance. R is so chosen that, when terminals a and b are shorted ($R_x = 0$), the meter reads full scale. When a and b are open ($R_x = \infty$), the meter reads zero.

a) What should be the value of the resistor R?

b) What current would indicate a resistance R_x of 600 Ω?

c) What resistances correspond to meter deflections of 1/4, 1/2, and 3/4 full scale, if the deflection is proportional to the current through the galvanometer?

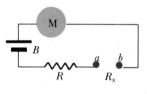

FIGURE 29–23

Section 29–4 The R–C Series Circuit

29–17 A 10-μF capacitor is connected through a 1-MΩ resistor to a constant potential difference of 100 V.

a) Compute the charge on the capacitor at the following times after the connections are made: 0, 5 s, 10 s, 20 s, 100 s.

b) Compute the charging current at the same instants.

c) How much time would be required for the capacitor to acquire its final charge if the charging current remained constant at its initial value?

d) Find the time required for the charge to increase from zero to 5×10^{-4}C.

e) Construct graphs of the results of parts (a) and (b) for a time interval of 20 s.

29–18 A capacitor is charged to a potential of 10 V and is then connected to a voltmeter having an internal resistance of 1.0 MΩ. After a time of 5 s, the voltmeter reads 5 V. What is the capacitance?

29–19 A 0.05-μF capacitor is charged to a potential of 200 V and is then permitted to discharge through a 10-MΩ resistor. How much time is required for the charge to decrease to

a) $1/e$

b) $1/e^2$

of its initial value?

29–20

a) The differential equation for the instantaneous charge q of a capacitor a moment after its terminals have been disconnected from a source and connected to a resistance R is given by Eq. (29–13), namely,

$$\frac{dq}{dt} = -\frac{q}{RC}.$$

Show that

$$q = Q_0 e^{-t/RC}.$$

b) The current in the circuit of part (a) is given by Eq. (29–14), namely,

$$\frac{di}{dt} = -\frac{i}{RC}.$$

Show that

$$i = I_0 e^{-t/RC}.$$

Section 29–5 Displacement Current

29–21 In Fig. 29–13 the capacitor plates have area 4 cm^2 and separation 3 mm. The plates are in vacuum. The charging current I_C has a *constant* value of 2 mA. At $t = 0$ the charge on the plates is zero.

a) Calculate the charge on the plates, the electric field between the plates, and the potential difference between the plates when $t = 5.0 \times 10^{-6}$ s.

b) Calculate dE/dt, the time rate of change of the electric field between the plates. Does dE/dt vary with time?

c) Calculate the displacement current density J_D between the plates, and from this the total displacement current I_D. How do I_C and I_D compare?

29–22 Suppose that the parallel plates in Fig. 29–13 have an area of 2 cm^2 and are separated by a sheet of dielectric 1 mm thick, of dielectric constant 3. (Neglect edge effects.) At a certain instant, the potential difference between the plates is 100 V and the current I_C equals 2 mA.

a) What is the charge Q on each plate at this instant?

b) What is the rate of change of charge on the plates?

c) What is the displacement current in the dielectric?

Section 29–6 Power Distribution Systems

29–23 A 100-W driveway light is left on night and day for a year.

a) What total energy is required? Express your results in joules and kilowatt-hours.

b) What does this energy cost if the power company's rate is 10 cents per kWh?

29–24 An electric dryer is rated at 5.0 kW when connected to a 240-V line.

a) What is the current in the dryer? Is 12-gauge wire large enough to supply this current?

b) What is the resistance of the dryer's heating element?

c) At 10 cents per kWh, how much does it cost to operate the dryer for 1 hr?

29–25 A 1500-W toaster, a 1200-W electric frypan, and a 100-W lamp are plugged into the same outlet in a 20-A, 120-V circuit.

a) What current is drawn by each device?

b) Will this combination blow a fuse?

29–26 How many 100-W light bulbs can be connected to a 20-A, 120-V circuit without tripping the circuit breaker?

PROBLEMS

29-27 Prove that when two resistors are connected in parallel, the equivalent resistance of the combination is always smaller than that of either resistor.

29-28

a) A resistance R_2 is connected in parallel with a resistance R_1. What resistance R_3 must be connected in series with the combination of R_1 and R_2 so that the equivalent resistance is equal to the resistance R_1? Draw a diagram.

b) A resistance R_2 is connected in series with a resistance R_1. What resistance R_3 must be connected in parallel with the combination of R_1 and R_2 so that the equivalent resistance is equal to R_1? Draw a diagram.

29-29 A 1000-Ω 2-W resistor is needed, but only several 1000-Ω 1-W resistors are available.

a) How can the required resistance and power rating be obtained by a combination of the available units?

b) What power is then dissipated in each resistor?

29-30

a) Calculate the equivalent resistance of the circuit of Fig. 29-24 between x and y.

b) What is the potential difference between x and a if the current in the 8-Ω resistor is 0.5 A?

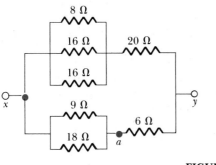

FIGURE 29-24

29-31 Each of three resistors in Fig. 29-25 has a resistance of 2 Ω and can dissipate a maximum of 18 W without becoming excessively heated. What is the maximum power the circuit can dissipate?

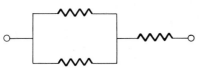

FIGURE 29-25

29-32 Three equal resistors are connected in series. When a certain potential difference is applied across the combination, the total power dissipated is 10 W. What power would be dissipated if the three resistors were connected in parallel across the same potential difference?

29-33 Calculate the three currents indicated in the circuit diagram of Fig. 29-26.

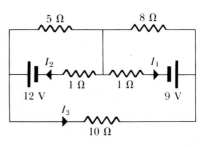

FIGURE 29-26

29-34 Find the current in each branch of the circuit shown in Fig. 29-27.

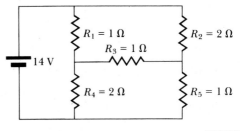

FIGURE 29-27

29-35

Note. Figure 29-28a employs a convention often used in circuit diagrams. The battery (or other power supply) is not shown explicitly. It is understood that the point at the top, labeled "36 V," is connected to the positive terminal of a 36-V battery having negligible internal resistance, and the "ground" symbol at the bottom is connected to its negative terminal. The circuit is completed through the battery, even though it is not shown on the diagram.

a) In Fig. 29-28a, what is the potential difference V_{ab} when the switch S is open?

b) What is the current through switch S when it is closed?

c) In Fig. 29-28b, what is the potential difference V_{ab} when the switch is open?

d) What is the current through switch S when it is closed?

e) What is the equivalent resistance in Fig. 29-28b when switch S is open?

f) When switch S is closed?

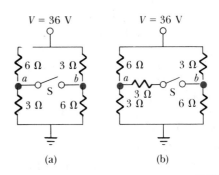

FIGURE 29-28

29–36 (See note with Problem 29–35.)

a) What is the potential of point a with respect to point b in Fig. 29–29 when switch S is open?

b) Which point, a or b, is at the higher potential?

c) What is the final potential of point b when switch S is closed?

d) How much does the charge on each capacitor change when S is closed?

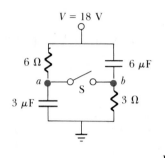

FIGURE 29–29

29–37 (See note with Problem 29–35.)

a) What is the potential of point a with respect to point b in Fig. 29–30 when switch S is open?

b) Which point, a or b, is at the higher potential?

c) What is the final potential of point b when switch S is closed?

d) How much charge flows through switch S when it is closed?

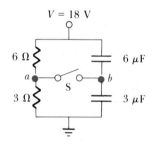

FIGURE 29–30

29–38 A certain galvanometer has a resistance of 200 Ω and deflects full scale with a current of 1 mA in its coil. We want to replace this with a second galvanometer whose resistance is 50 Ω and which deflects full scale with a current of 50 μA in its coil. Devise a circuit incorporating the second galvanometer such that

a) the equivalent resistance of the circuit equals the resistance of the first galvanometer;

b) the second galvanometer will deflect full scale when the current into and out of the circuit equals the full-scale current of the first galvanometer.

29–39 A 600-Ω resistor and a 400-Ω resistor are connected in series across a 90-V line. A voltmeter across the 600-Ω resistor reads 45 V.

a) Find the voltmeter resistance.

b) Find the reading of the same voltmeter if connected across the 400-Ω resistor.

29–40 Point a in Fig. 29–31 is maintained at a constant potential of 300 V above ground. (See note with Problem 29–35.)

a) What is the reading of a voltmeter of the proper range and of resistance 3×10^4 Ω when connected between point b and ground?

b) What would be the reading of a voltmeter of resistance 3×10^6 Ω?

c) What would be the reading of a voltmeter of infinite resistance?

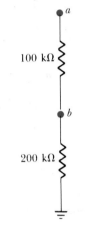

FIGURE 29–31

29–41 A 150-V voltmeter has resistance of 20,000 Ω. When connected in series with a large resistance R across a 110-V line, the meter reads 5 V. Find the resistance R. (This problem illustrates one method of measuring large resistances.)

29–42 Let V and I represent the readings of the voltmeter and ammeter, respectively, shown in Fig. 29–9, and R_V and R_A their equivalent resistances.

a) When the circuit is connected as in Fig. 29–9a, show that

$$R = \frac{V}{I} - R_A.$$

b) When the connections are as in Fig. 29–9b, show that

$$R = \frac{V}{I - (V/R_V)}.$$

c) Show that the power delivered to the resistor in part (a) is $IV - I^2 R_A$, and in part (b) $IV - (V^2/R_V)$.

29–43 In Fig. 29–32, a resistor of resistance 75 Ω is connected between points a and b. The resistance of the galvanometer G is 90 Ω. What should be the resistance between b and the sliding contact c if the galvanometer current I_G is to be ⅓ of the current I?

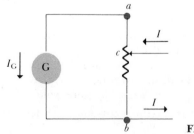

FIGURE 29–32

29–44 The current shown in Fig. 29–33, called a *Wheatstone bridge,* is used to determine the value of an unknown resistor X by comparison with three resistors M, N, and P whose resistance can be varied. For each setting, the resistance of each resistor is precisely known. With switches K_1 and K_2 closed, these resistors are varied until the current in the galvanometer G is zero; the bridge is then said to be *balanced.*

a) Show that under this condition the unknown resistance is given by $X = MP/N$. (This method permits very high precision in comparing resistors.)

b) If the galvanometer G shows zero deflection when $M = 1000\ \Omega$, $N = 10.00\ \Omega$, and $P = 27.49\ \Omega$, what is the unknown resistance X?

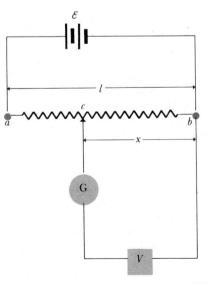

FIGURE 29–33

29–45 The circuit shown in Fig. 29–34 is called a *potentiometer.* It permits measurements of potential difference without drawing current from the circuit being measured, and hence acts as an infinite-resistance voltmeter. The resistor between a and b is a uniform wire of length l, with a sliding contact c at a distance x from b. An unknown potential difference V is measured by sliding the contact until the galvanometer G reads zero.

a) Show that under this condition the unknown potential difference is given by $V = (x/l)\ \mathcal{E}$.

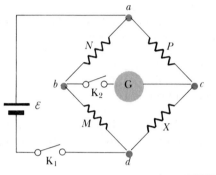

FIGURE 29–34

b) Why is the internal resistance of the galvanometer not important?

Now suppose $\mathcal{E} = 12.00$ V and $l = 1.000$ m. The galvanometer G reads zero when $x = 0.793$ m.

c) What is the potential difference V?

d) Suppose V is the emf of a battery. Can its internal resistance be determined by this method?

29–46 The current in a discharging capacitor is given by Eq. (29–15).

a) Using Eq. (29–15), derive an expression for the instantaneous power $P = i^2R$ dissipated in the resistor.

b) Integrate the expression for P to find the total energy dissipated in the resistor, and show that this value is equal to the total energy initially stored in the capacitor.

29–47 The current in a charging capacitor is given by Eq. (29–10).

a) The instantaneous power supplied by the battery is Vi. Integrate this expression to find the total energy supplied by the battery.

b) The instantaneous power dissipated in the resistor is i^2R. Integrate this expression to find the total energy dissipated in the resistor.

c) Find the final energy stored in the capacitor, and show that this value equals the total energy supplied by the battery, less the energy dissipated in the resistor, as obtained in parts (a) and (b).

d) What fraction of the energy supplied by the battery is stored in the capacitor? How does this fraction depend on R?

29–48 Two capacitors are charged in series by a 12-V battery that has an internal resistance of 1 Ω. There is a 5-Ω resistor in series between the capacitors (Fig. 29–35).

a) What is the time constant of the charging circuit?

b) After the switch has been closed for the time determined in (a), what is the voltage across the 6-μF capacitor?

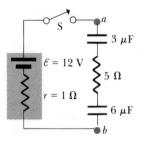

FIGURE 29–35

29–49 In a certain copper conductor ($\rho = 1.72 \times 10^{-8}\ \Omega\cdot$m) carrying a current, the electric field varies sinusoidally with time according to $E = E_0 \sin \omega t$, where $E_0 = 0.1$ V$\cdot$m^{-1}, $\omega = 2\pi f$, and the frequency is $f = 60$ Hz.

a) Find the magnitude of the maximum conduction current density in the wire.

b) Assuming $\epsilon = \epsilon_0$, find the maximum displacement current density in the wire, and compare it with the result of part (a).

29–50

a) Repeat the calculations of Problem 29–49 for a rod of pure silicon having $\rho = 2300 \ \Omega \cdot m$.

b) At what frequency f would the maximum conduction and displacement current densities become equal, if $\epsilon = \epsilon_0$ (which is not actually the case)?

c) At the frequency determined in part (b), what is the relative *phase* of the conduction and displacement currents?

CHALLENGE PROBLEMS

29–51 Prove that the resistance of the infinite network shown in Fig. 29–36 is equal to $(1 + \sqrt{3})r$.

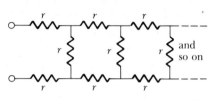

FIGURE 29–36

29–52 Suppose a resistor R lies along each edge of a cube (12 resistors in all) with connections at the corners. Find the equivalent resistance between two diagonally opposite corners of the cube.

29–53 A parallel-plate capacitor is made from two plates, each having area A, spaced a distance d apart. The space between plates is filled with a material having dielectric constant K. The material is not a perfect insulator but has resistivity ρ. The capacitor is initially charged with charge of magnitude Q_0 on each plate, which gradually discharges by conduction through the dielectric. Show that at any instant the displacement current density in the dielectric is equal in magnitude to the conduction current density but is opposite in direction, so that the *total* current density is zero at every instant.

29–54 The capacitance of a capacitor can be affected by dielectric material that, although not inside the capacitor, is near enough to the capacitor to be polarized by the fringing electric field that exists near a charged capacitor. This effect is usually of the order of picofarads (pF), but it can be used with appropriate electronic circuitry to detect a change in the dielectric material surrounding the capacitor. Such a dielectric material might be the human body, and the effect described above might be used in the design of a burglar alarm.

Consider the simplified circuit shown in Fig. 29–37. The voltage source has emf $\mathcal{E}_0 = 1000$ V, and the capacitor has capacitance $C = 10$ pF. The electronic circuitry for detecting the current, represented as an ammeter in the diagram, has negligible resistance and is capable of detecting a current that persists at a level of at least 1 μA for at least 200 μs after the capacitance has changed abruptly from C to C'. The burglar alarm is to be designed to be activated if the capacitance changes by 10%.

a) Determine the charge on the 10-pF capacitor when it is fully charged.

b) If the capacitor is fully charged before the intruder is detected, and assuming that the time taken for the capacitance to change by 10% is small enough to be neglected, derive an equation that expresses the current through the resistor R as a function of the time t since the capacitance has changed.

c) Determine the range of values of the resistance R that will meet the design specifications of the burglar alarm. What happens if R is too small? Too large? (*Hint:* You will not be able to solve this part analytically but must use numerical methods. R can be expressed as a logarithmic function of R plus known quantities. Use a trial value of R and calculate from the expression a new value. Continue to do this until the input and output values of R agree to within three significant figures.)

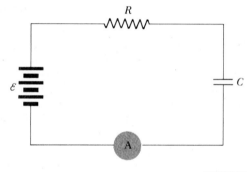

FIGURE 29–37

29–55 Thevenin's theorem, as it applies to a dc circuit, can be stated as follows: *Any linear, two-terminal network branch can be replaced by an equivalent source of emf in series with an equivalent resistance.* This theorem can be used to simplify the analysis of a circuit if we want to modify one part of the circuit and then see what effect that modification will have on that part of the circuit without having to use Kirchhoff's rules to reanalyze the entire circuit each time. Consider the circuit shown in Fig. 29–38a.

a) Using Kirchhoff's rules, determine the branch currents in each of the three branches.

b) According to Thevenin's theorem, the part of the circuit to the left of points a and b can be replaced by an equivalent emf and, in series with that emf, by an equivalent resistance, as depicted in Fig. 29–38b. Find the equivalent emf by considering that circuit as an open circuit between a and b. That is, remove the branch containing the 60-Ω resistor and the 150-V source and do not replace the branch with anything. The voltage drop between a and b is the emf of the equivalent voltage source.

c) Find the equivalent resistance that must be in series with the source found in part (b). Do this once again by removing the 60-Ω resistor and the 150-V source, but this time provide a short circuit between a and b. The current through that short circuit must be the equivalent emf divided by the equivalent resistance.

d) Suppose we want to replace the 150-V source with another source such that the current through the 60-Ω resistor will be 1 A. Using the Thevenin-equivalent circuit elements found in parts (b) and (c), determine the emf of the source that should replace the 150-V source.

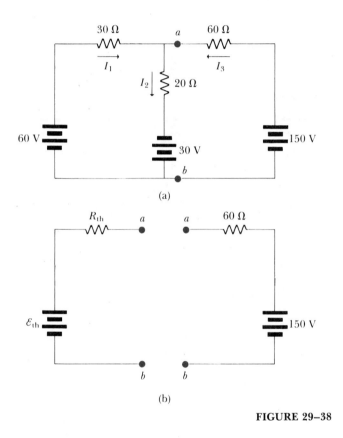

(a)

(b)

FIGURE 29–38

29–56 Norton's theorem, as it applies to a dc circuit, can be stated as follows: *Any linear, two-terminal network branch can be replaced by an equivalent current source in parallel with an equivalent shunt resistance.* As with Thevenin's theorem, Norton's theorem can be used to simplify the analysis of a circuit if we want to modify one part of the circuit and then see what effect that modification will have on that part of the circuit without having to use Kirchhoff's rules to reanalyze the entire circuit each time. Consider the circuit shown in Fig. 29–39a.

a) Using Kirchhoff's rules, determine the branch currents in each of the three branches.

b) According to Norton's theorem, the part of the circuit to the left of points a and b can be replaced by an equivalent current source and, in parallel with that current source, by an equivalent resistance, as depicted in

Fig. 29–39b. Find the equivalent current source by providing a short circuit between a and b. That is, remove the branch containing the 60-Ω resistor and the 30-V source and replace it with a short circuit. The current that passes through the short circuit between a and b is the current produced by the equivalent current source.

c) Find the equivalent resistance that must be in parallel with the source found in part (b). Do this by once again removing the 60-Ω resistor and the 30-V source, but this time consider that circuit as an open circuit between a and b. That is, remove the branch containing the 60-Ω resistor and the 150-V source and do not replace it with anything. The voltage drop across the open circuit must be the product of the current of the equivalent current source and the resistance of the equivalent resistance.

d) If a 20-Ω resistor is connected between a and b in the circuit of Fig. 29–39a, what will be the current through the 60-Ω resistor? Use the Norton-equivalent-circuit elements found in parts (b) and (c) to answer this part.

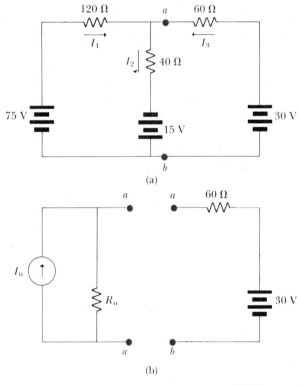

(a)

(b)

FIGURE 29–39

29–57 According to the theorem of superposition, the response (current) in a circuit is proportional to the stimulus (voltage) that causes it. This is true even if there are multiple sources in a circuit. The theorem can be used to analyze a circuit without resorting to Kirchhoff's rules by considering the currents in the circuit to be the superposition of currents caused by each source independently. In this way, the circuit can be analyzed by computing equivalent resistances rather than by using the (sometimes) more cumbersome method of Kirchhoff's rules.

Furthermore, by using the superposition theorem, it is possible to examine how the modification of a source in one part of the circuit will affect the currents in all parts of the circuit without having to use Kirchhoff's rules to recalculate all the currents.

Consider the circuit shown in Fig. 29–40. If the circuit were redrawn with the 55-V and 57-V sources replaced by short circuits, the circuit could be analyzed by the method of equivalent resistances without resorting to Kirchhoff's rules, and the current in each branch could be found in a simple manner. Similarly, by redrawing the circuit with the 92-V and the 55-V sources replaced by short circuits, the circuit could again be analyzed simply. Finally, by replacing the 92-V and the 57-V sources with a short circuit, the circuit could also be analyzed simply. By superposing the respective currents found in each of the branches using the three simplified circuits, the actual current in each branch can be found.

a) Using Kirchhoff's rules, find the branch currents in the 140-Ω, 210-Ω, and 35-Ω resistors.

b) Using a circuit similar to that in Fig. 29–40, but with the 55-V and 57-V sources replaced by a short circuit, determine the current in each resistance.

c) Repeat part (b) by replacing the 92-V and 55-V sources by short circuits, leaving the 57-V source intact.

d) Repeat part (b) by replacing the 92-V and 57-V sources by short circuits, leaving the 55-V source intact.

e) Verify the superposition theorem by taking the currents calculated in parts (b), (c), and (d) and comparing them with the currents calculated in part (a).

f) If the 57-V source is replaced by a 100-V source, what will be the new currents in all branches of the circuit? (*Hint:* Applying the superposition theorem, recalculate the partial currents calculated in part [c] using the fact that those currents are proportional to the source that is being replaced. Then superpose the new partial currents with those found in parts [b] and [d].)

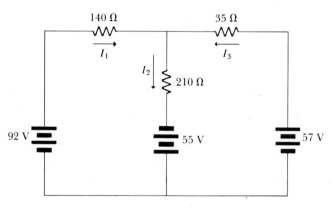

FIGURE 29–40

PART SIX

ELECTRODYNAMICS

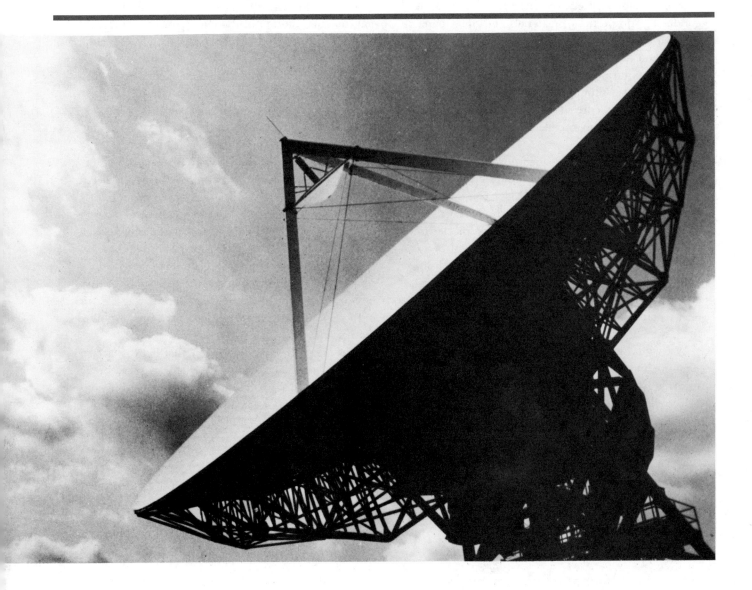

PERSPECTIVE

Electromagnetism is the study of the interaction of electric charges. We began our study with *electrostatics*, the study of charges *at rest*. In this case the forces are *conservative* forces; we described them and their associated potential energies using the concepts of *electric field* and *electric potential*. We analyzed several fundamental experiments and practical devices that use electrostatic principles. We also analyzed the behavior of *capacitors* and some aspects of the electrical properties of materials.

Next we studied electric *circuits;* we described the conducting properties of materials and their temperature variation, and we introduced the concepts of *resistance* and *electromotive force*. Circuits are the heart of all contemporary electronics and electric-power distribution systems. Although we confined our study mostly to *constant* currents, many of the same principles are applicable in circuits with time-varying currents.

Now we are ready to broaden our study to include electric charges in motion. The forces now depend on the *velocities* of the charged particles. These interactions are best described in terms of *magnetic field*. Our analysis takes the same two-part path as for electric field. First, moving charge (such as a current in a conductor) acts as a *source* of magnetic field, establishing a magnetic field in the surrounding region. Second, a charged particle moving in a magnetic field experiences a *force* that depends on the magnitude of the field, the particle's speed, and its charge.

We begin by taking a magnetic field as a given quantity and describing the force it exerts on a moving charge. This force provides the basis for a broad variety of phenomena, including electric motors, the deflections of electron beams in TV picture tubes, and experiments with fundamental particles. Next we discuss in detail how magnetic fields are *produced* by currents and moving charges. This discussion enables us to understand electromagnets, motors, and many other practical devices. We also look at some magnetic properties of materials, including *magnetization*, which alters the magnetic field inside a material.

An essential aspect of electromagnetic interactions is *electromagnetic induction*, in which a time-varying magnetic field acts as a source of electric field. An electric current can be induced in a circuit by a changing magnetic field, whether caused by a moving permanent magnet, motion of the circuit, or changing current in a nearby electromagnet. To analyze induction phenomena we introduce the concept of *magnetic flux*, closely related to magnetic field. Induction is the basis for inductors and transformers, two indispensible elements in modern electronics and power-distribution systems. We study the role of these and other circuit elements in alternating-current (ac) circuits.

Finally, a changing *electric* field acts as a source of *magnetic* field. All the fundamental relations of electrodynamics can be summarized in four compact but general equations called *Maxwell's equations*. These equations, comparable in power and elegance to Newton's laws in mechanics, show once again the beauty and generality of physics. The culminating achievement of Maxwell's equations is the prediction of the existence of *electromagnetic waves:* time-varying electric and magnetic fields that *propagate* from one region to another.

Thus Maxwell's equations show that light is electromagnetic radiation. Electromagnetic waves also include radio and TV transmission, infrared, ultraviolet, x-rays, and gamma rays, and they play a significant role in nearly every area of contemporary science and technology.

30

MAGNETIC FIELD AND MAGNETIC FORCES

THE MOST FAMILIAR ASPECTS OF MAGNETISM ARE THOSE ASSOCIATED WITH permanent magnets, which attract unmagnetized iron objects and can also attract or repel other magnets. The fundamental nature of magnetism, however, is to be found in interactions involving electric charges in motion. A magnetic field is established by a permanent magnet, by an electric current in a conductor, or by moving charges. This magnetic field, in turn, exerts forces on moving charges and current-carrying conductors. In this chapter we study these magnetic forces, and in Chapter 31 we examine the ways in which magnetic fields are produced by moving charges. Magnetic forces are an essential aspect of the interactions of electrically charged particles. Electric motors, TV picture tubes, high-energy particle accelerators, magnetrons in microwave ovens, and many other devices depend in part on magnetic forces for their operation.

30–1 MAGNETISM

Magnetic phenomena were first observed at least 2000 years ago, in fragments of magnetized iron ore found near the ancient city of Magnesia. It was found that when an iron rod was brought in contact with a natural magnet, the rod also became a magnet. Such a rod, when suspended by a string from its center, tended to line itself up in a north–south direction, like a compass needle; magnets have been used for navigation at least since the eleventh century.

Before the relation of magnetic interactions to moving charges was understood, the interactions of bar magnets and compass needles were described in terms of **magnetic poles.** The end of a compass needle that points north is called a *north pole* or *N-pole,* and the other end is a *south* or *S-pole.* Two opposite poles attract each other, and two like poles repel each other. The concept of magnetic poles is of limited usefulness and is somewhat misleading because a single isolated magnetic pole has never been found; poles always appear in pairs. If a bar magnet is broken in two, each broken end becomes a pole. The existence of an isolated magnetic pole, or *magnetic monopole,* would have

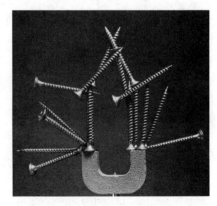

Screws made of steel (a ferromagnetic material) are attracted to a permanent magnet. Screws made of brass or aluminum (nonferromagnetic materials) would not be attracted. (Photo by Chip Clark.)

(a) A compass needle (a permanent magnet) deflects in the magnetic field of another permanent magnet. Magnetization is the result of motion of electrons in the magnetized material. (b) A compass needle deflects in the magnetic field produced by the current in the conductor. The field exerts torques on the circulating electric charges in the needle associated with its magnetization. (Photos by Chip Clark.)

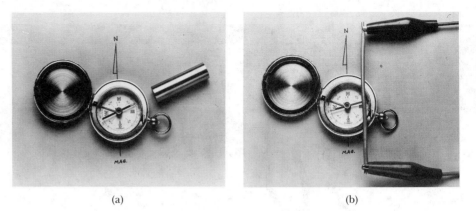

(a)   (b)

A compass responds to the earth's magnetic field.

sweeping implications for theoretical physics; extensive searches for magnetic monopoles have been carried out, so far without success.

A compass needle points north because the earth is a magnet; its north geographic pole is a magnetic *south* pole. The earth's magnetic axis is not quite parallel to its geographic axis (the axis of rotation), so a compass reading deviates somewhat from geographic north; this deviation, which varies with location, is called *magnetic declination*. A sketch of the earth's magnetic field is shown in Fig. 30–1. The lines show the direction a compass would point at each location. These lines are actually magnetic-field lines, to be discussed in Section 30–3. The direction of the field at any point can be defined as the direction of the force the field would exert on a magnetic north pole. In Section 30–2 we will introduce a more fundamental way to define magnetic field.

Magnetic effects of an electric current

In 1819 the Danish scientist Hans Christian Oersted discovered that a compass needle was deflected by a current-carrying wire. A few years later Michael Faraday in England and Joseph Henry in the United States discovered that moving a magnet near a conducting loop can cause a current in the loop, and that a changing current in one conducting loop can cause a current in another separate loop. These observations were the first evidence of the relationship of magnetism to moving charges.

30–2 MAGNETIC FIELD

In describing the interaction between two charges *at rest*, we introduced the concept of *electric field*, and we described the interaction in two stages:

1. One charge sets up or creates an electric field E in the space surrounding it.
2. The electric field E exerts a force $F = qE$ on a charge q placed in the field.

We will follow the same pattern in describing the interactions of charges in motion:

1. A *moving* charge or a current sets up or creates a **magnetic field** in the space surrounding it.
2. The magnetic field exerts a *force* on a *moving* charge or a current in the field.

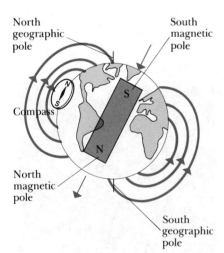

30–1 A sketch of the earth's magnetic field. A compass placed at any point in this field would point in the direction of the field line at that point. Representing the earth's field as that of a tilted bar magnet is only a crude approximation of the actual fairly complex field configuration.

Like electric field, magnetic field is a *vector field,* that is, a vector quantity associated with each point in space. We use the symbol **B** for magnetic field.

In this chapter we consider the *second* aspect of the interaction: Given the presence of a magnetic field, what force does it exert on a moving charge or a current? In Chapter 31 we return to the problem of how magnetic fields are *created* by moving charges and currents.

Some properties of the magnetic force on a moving charge are analogous to corresponding properties of the electric-field force. The magnitude of the magnetic force is proportional to the charge. If a 1-μC charge and a 2-μC charge move through a given magnetic field with the same velocity, the force on the 2-μC charge is twice as great as that on the 1-μC charge. The force is also proportional to the magnitude or "strength" of the field; if a given charge moves with the same velocity in two magnetic fields, one having twice the magnitude (and the same direction) as the other, the charge experiences twice as great a force in the larger field as in the smaller.

The magnetic force also depends on the particle's motion; this is quite different from the electric-field force, which is the same whether the charge is moving or not. The *magnetic* force is found to have a magnitude that is directly proportional to the particle's speed. A particle at rest experiences no magnetic force at all. The *direction* of the force is determined by the directions of the magnetic field **B** and the velocity **v** in an interesting way. The force **F** *does not* have the same direction as **B,** but instead is always *perpendicular* to both **B** and **v.** The magnitude F of the force is found to be proportional to the component of **v** perpendicular to the field; when that component is zero (that is, when **v** and **B** are parallel or antiparallel) there is *no* force!

These characteristics of the magnetic force can be summarized, with reference to Fig. 30–2, as follows: The direction of **F** is perpendicular to the plane containing **v** and **B;** its magnitude is given by

$$F = qv_{\perp}B = qvB \sin \phi, \qquad (30\text{--}1)$$

where q is the charge and ϕ is the angle between the vectors **v** and **B,** as shown in the figure.

This description does not specify the direction of **F** completely; there are always two directions perpendicular to the plane of **v** and **B** but opposite to each other. To complete the description we use the same right-hand-thread rule used to define the vector product in Section 1–9 and for angular mechanical quantities in Section 9–14. Imagine turning **v** into **B,** through the smaller of the two possible angles; the direction of **F** is the direction in which a right-hand-thread screw would advance if turned the same way. Alternatively, wrap the fingers of your right hand around the line perpendicular to the plane of **v** and **B** so that they curl around with this sense of rotation from **v** to **B;** your thumb then points in the direction of **F.**

Thus the force on a charge q moving with velocity **v** in a magnetic field **B** is given, both in magnitude and in direction, by

$$F = qv \times B. \qquad (30\text{--}2)$$

This is the first of several applications of the vector product we will encounter in our study of magnetic-field relationships.

Spatial relations for magnetic forces can be troublesome; the following review of the definition of the vector product may help. First, always draw the

The magnetic force on a charged particle depends on its velocity.

The magnetic force on a charged particle is always perpendicular to its velocity.

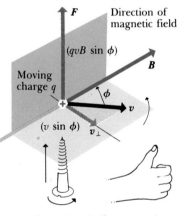

30–2 The magnetic force **F** acting on a charge q moving with velocity **v** is perpendicular to both the magnetic field **B** and to **v.**

A right-hand rule for magnetic force on a moving charge

The magnetic force on a moving charge can be expressed neatly as a vector product.

vectors v and B from a common point, that is, with their tails together. Second, visualize and if possible draw the *plane* in which these two vectors lie. The force vector always lies along a line perpendicular to this plane. To determine its direction along this line, imagine rotating v into B, as in Fig. 30–2. The direction of F is the direction of advance of a right-hand-thread screw rotated in this direction, or the direction of advance of the extended thumb of your right hand if your fingers are curled around the perpendicular line with this direction of rotation.

An alternative way to describe the magnitude of the magnetic force on a moving charge

Finally, we note that Eq. (30–1) can be interpreted in a different but equivalent way. Recalling that ϕ is the angle between the direction of vectors v and B, we may interpret ($B \sin \phi$) as the component of B perpendicular to v, that is, $B_\perp$. With this notation, the force expression becomes

$$F = qvB_\perp. \qquad (30\text{–}3)$$

Although equivalent to Eq. (30–1), this form is sometimes more convenient, especially in problems involving currents rather than individual particles. We will discuss forces on currents in conductors later in this chapter.

The tesla: the SI unit of magnetic field

The *units* of $\mathcal{B}$ can be deduced from Eq. (30–1); they must be the same as the units of F/qv. Therefore the SI unit of B is 1 N·s·C^{-1}·m^{-1}, or, since one ampere is one coulomb per second (1 A $= 1$ C·s^{-1}), 1 N·A^{-1}·m^{-1}. This unit is called one **tesla** (1 T), in honor of Nikolai Tesla, the prominent nineteenth-century Russian scientist and inventor. The cgs unit of B, the **gauss** (1 G $= 10^{-4}$ T) is also in common use. Instruments for measuring magnetic field are often called gaussmeters. To summarize,

$$1 \text{ T} = 1 \text{ N·A}^{-1}\text{·m}^{-1} = 10^4 \text{ G.}$$

In this discussion we assumed that q is a *positive* charge. If q is negative, the direction of F is opposite to that shown in Fig. 30–2 and given by the right-hand rule. Thus if two charges of equal magnitude and opposite sign move in the same B field with the same velocity, the forces on the two charges have equal magnitude and opposite direction.

To explore an unknown magnetic field, we can measure the magnitude and direction of the force on a *moving* test charge and use the relationship $F = qv \times B$. The cathode-ray tube, discussed in Section 26–8, is a convenient device for making such measurements. The electron gun shoots out a narrow beam of electrons at a known speed. If there is no deflecting force on the beam, it strikes the center of the screen.

Using a cathode-ray tube to explore a magnetic field

In the presence of a magnetic field, the electron beam is in general deflected. If the beam is parallel or antiparallel to the field, however, then $v \times B = 0$; then there is no force and no deflection. If we find that the electron beam is undeflected when its direction is parallel to the z-axis, as in Fig. 30–3, the B-vector must point either up or down.

When we turn the tube 90°, so that its axis is along the x-axis in Fig. 30–3, the beam is deflected in a direction showing a force perpendicular to the plane of B and v. We can perform additional experiments in which the angle between B and v is between zero and 90° to confirm Eq. (30–1) and the accompanying discussion. We note that the electron has a negative charge, and the force in Fig. 30–3 is opposite in direction to the force on a positive charge.

When a charged particle moves through a region of space where *both* electric and magnetic fields are present, both fields exert forces on the parti-

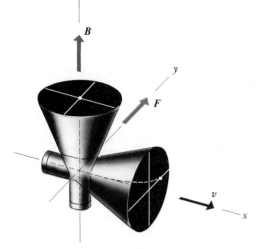

30–3 The electron beam of the cathode-ray tube is undeflected when the beam is parallel to the z-axis. The **B**-vector then points either up or down. When the tube axis is parallel to the x-axis, the beam is deflected in the positive y-direction. Then the **B**-vector points upward, and the force **F** on the electrons points along the positive y-axis, opposite to the rule of Fig. 30–2 because q is negative.

cle, and the total force is the vector sum of the electric-field and magnetic-field forces:

$$F = q(E + v \times B). \qquad (30\text{–}4)$$

PROBLEM-SOLVING STRATEGY: *Magnetic-field forces*

1. The biggest difficulty is in relating the directions of the vector quantities. In evaluating $v \times B$, draw the two vectors with their tails together so you can visualize and draw the plane in which they lie. This also helps to identify the angle ϕ between the two vectors and to avoid getting its complement, or some other erroneous angle. Then remember that **F** is always perpendicular to this plane. The direction is determined by the right-hand rule; keep referring to Fig. 30–2 and back to Fig. 1–13 until you're sure you have this down cold.

2. Whenever possible, do the problem two ways. Do it directly from the geometric definition of the vector product. Then find the components of the vectors in some convenient axis system and calculate the vector product from the components. Check that both methods give the same result.

EXAMPLE 30–1 A proton beam moves through a region of space where there is a uniform magnetic field, of magnitude 2.0 T, directed along the positive z-axis, as in Fig. 30–4. The protons have velocity of magnitude 3×10^5 m·s^{-1}, in the xz-plane at an angle of 30° to the positive z-axis. Find the force on a proton ($q = 1.6 \times 10^{-19}$ C).

SOLUTION The right-hand rule shows that the direction of the force is along the negative y-axis. The magnitude of the force, from Eq. (30–1), is

$$
\begin{aligned}
F &= qvB \sin \phi \\
  &= (1.6 \times 10^{-19}\text{ C})(3 \times 10^5 \text{ m·s}^{-1})(2.0 \text{ T})(\sin 30°) \\
  &= 4.8 \times 10^{-14} \text{ N}.
\end{aligned}
$$

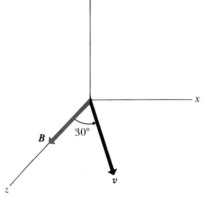

30–4 Directions of v and **B** for Example 30–1.

Alternatively, in vector language, with Eq. (30–2),

$$\boldsymbol{v} = (3 \times 10^5 \text{ m·s}^{-1})(\sin 30°)\boldsymbol{i} + (3 \times 10^5 \text{ m·s}^{-1})(\cos 30°)\boldsymbol{k},$$
$$\boldsymbol{B} = (2.0 \text{ T})\boldsymbol{k},$$
$$\boldsymbol{F} = q\boldsymbol{v} \times \boldsymbol{B} = (1.6 \times 10^{-19} \text{ C})(3 \times 10^5 \text{ m·s}^{-1})(2.0 \text{ T})$$
$$\cdot (\sin 30°\boldsymbol{i} + \cos 30° \boldsymbol{k}) \times \boldsymbol{k}$$
$$= (-4.8 \times 10^{-14} \text{ N})\boldsymbol{j},$$

since $\boldsymbol{i} \times \boldsymbol{k} = -\boldsymbol{j}$ and $\boldsymbol{k} \times \boldsymbol{k} = \boldsymbol{0}$.

If the beam consists of *electrons* rather than protons, the charge is negative ($q = -1.6 \times 10^{-19}$ C), and the direction of the force is reversed. The solution proceeds just as before; the force is now directed along the *positive* y-axis: $\boldsymbol{F} = +(4.8 \times 10^{-14} \text{ N})\boldsymbol{j}$.

30–3 MAGNETIC-FIELD LINES AND MAGNETIC FLUX

Magnetic-field lines: a useful graphical representation of magnetic fields

A magnetic field can be represented by lines, just as we represented an electric field by lines in Section 25–3. We draw the lines so that the line through any point is tangent to the magnetic-field vector $\boldsymbol{B}$ at that point, and so that the number of lines per unit area (perpendicular to the lines at a given point) is proportional to the magnitude of the field at that point. We call these lines **magnetic-field lines.** They are sometimes called magnetic lines of force, but this term is unfortunate because, unlike electric-field lines, they *do not* point in the direction of the force on a charge.

The field lines for a uniform field are equally spaced parallel straight lines.

In a uniform magnetic field, where the $\boldsymbol{B}$ vector has the same magnitude and direction at every point in a region, the field lines are straight and parallel. If the poles of an electromagnet are large, flat, and close together, the magnetic field in the region between them is very nearly uniform. Magnetic-field lines produced by a few examples of magnetic-field sources are shown in Figs. 30–5 and 30–6. The magnetic field of the earth is shown in Fig. 30–1; the field lines resemble those of a bar magnet with its axis tilted somewhat with respect to the earth's axis of rotation. The earth's field changes with time, and there is geologic evidence that it has actually changed direction many times in the last 100 million years.

Magnetic flux: the magnetic analog of electric flux

We define the **magnetic flux** Φ through a surface in direct analogy with the electric-field flux used with Gauss's law in Section 25–4. We can divide any surface into elements of area dA, as shown in Fig. 30–7. For each element, we determine the components of $\boldsymbol{B}$ normal and tangent to the surface at the position of that element, as shown. In general, these components will vary from point to point on the surface. From the figure, $B_\perp = B \cos \theta$. We define the magnetic flux $d\Phi$ through this area as

$$d\Phi = B_\perp \, dA = B \cos \theta \, dA = \boldsymbol{B} \cdot d\boldsymbol{A}. \tag{30–5}$$

The *total* magnetic flux through the surface is the sum of the contributions from the individual area elements, given by

$$\Phi = \int B_\perp \, dA = \int \boldsymbol{B} \cdot d\boldsymbol{A}. \tag{30–6}$$

In the special case where $\boldsymbol{B}$ is uniform over a plane surface with total area A,

$$\Phi = B_\perp A = BA \cos \theta. \tag{30–7}$$

If **B** happens to be perpendicular to the surface, cos $\theta = 1$, and this expression reduces to $\Phi = BA$. The chief utility of the concept of magnetic flux is in our study of electromagnetic induction in Chapter 32.

The definition of magnetic flux given above has an ambiguity of sign associated with the direction of the vector area element $d\mathbf{A}$. In Gauss's law we

A sign convention for magnetic flux

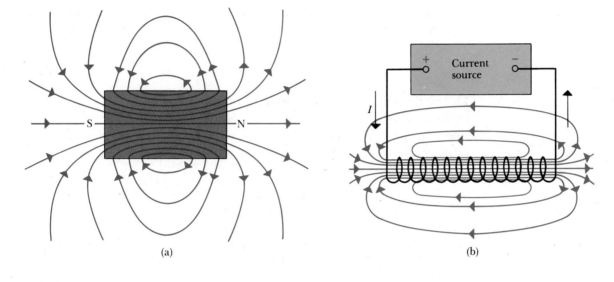

(a)   (b)

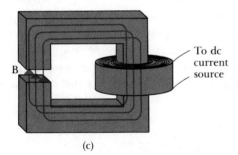

(c)

30–5 Magnetic-field lines produced by (a) a bar-shaped permanent magnet, (b) a coil of wire wound on a cylindrical form, that is, a solenoid, and (c) a laboratory electromagnet with an iron core. In (c) the magnetic field is nearly uniform in the gap in the core. In all three cases the field lines are continuous curves closing on themselves; there are no endpoints.

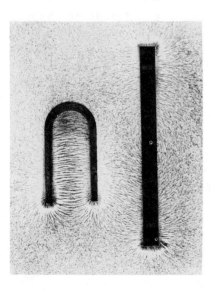

30–6 Magnetic fields can be visualized by use of iron filings, which orient themselves parallel to the field direction. This photograph shows the concentration of the fields near the poles. (Photo by Fundamental Photographs, New York.)

30–7 The magnetic flux through an area element dA is defined to be $\Phi = B_\perp\, dA$.

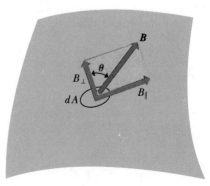

always took *d**A*** as pointing *out of* a closed surface. Some applications of *magnetic* flux involve an *open* surface with an edge. In these cases we define the direction of *d**A*** as follows: Walk around the edge, keeping the surface on your left. The direction of *d**A*** is from your feet toward your head.

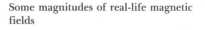

The SI unit of magnetic field ***B*** is one tesla = one newton per ampere meter; hence the unit of magnetic flux is one *newton meter per ampere* (1 N·m·A^{-1}). In honor of Wilhelm Weber (1804–1890), 1 N·m·A^{-1} is called one **weber** (1 Wb). In the cgs system, the unit of magnetic flux is one **maxwell.**

If the element of area *d**A*** in Eq. (30–5) is at right angles to the field lines, $B_\perp = B$; in this case,

$$B = \frac{d\Phi}{dA}. \tag{30–8}$$

That is, the magnetic field equals the *flux per unit area* across an area at right angles to the magnetic field. Magnetic field ***B*** is sometimes called **flux density.** Since the unit of flux is one weber, the unit of field, one tesla, is equal to one *weber per square meter* (1 Wb·m^{-2}). Similarly, in cgs units, one gauss equals one maxwell per square centimeter:

$$1\ T\ =\ 1\ Wb \cdot m^{-2},$$
$$1\ G\ =\ 1\ maxwell \cdot cm^{-2}.$$

We can picture the total flux through a surface as proportional to the number of field lines crossing the surface, and the field (the flux density) as the number of lines *per unit area.*

The magnetic field of the earth is of the order of 10^{-4} T, or 1 G. Magnetic fields of the order of 10 T(10^5 G) occur in the interior of atoms and are important in the analysis of atomic spectra. The largest values of steady magnetic field that have been achieved in the laboratory are of the order of 30 T = 300,000 G. Some pulsed-current electromagnets can produce fields of the order of 120 T = 1.2×10^6 G for short time intervals of the order of a millisecond.

30–4 MOTION OF CHARGED PARTICLES IN A MAGNETIC FIELD

Here is a simple example of the motion of a charged particle in a magnetic field. In Fig. 30–8 a particle with positive charge *q* is at point *O*, moving with velocity *v* in a uniform magnetic field ***B*** directed into the plane of the figure. An upward force ***F*** = *q**v*** × ***B***, of magnitude *qvB*, acts on the particle at this point. The force is always perpendicular to *v*, so it cannot change the *magnitude* of the velocity, but only its direction. Thus the magnitudes of both ***F*** and *v* are constant. At points such as *P* and *Q* the directions of force and velocity have changed as shown; the magnitude of the force is constant, since the magnitudes of *q*, *v*, and ***B*** are constant. The particle therefore moves under the influence of a force whose *magnitude* is constant but whose *direction* is always at right angles to the velocity of the particle. The orbit of the particle is therefore a *circle* described with constant tangential speed *v*. Since the centripetal acceleration is v^2/R, as shown in Section 3–5, we have, from Newton's second law,

$$F = qvB = m\left(\frac{v^2}{R}\right),$$

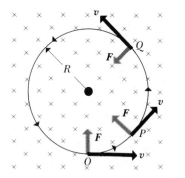

30–8 The orbit of a charged particle in a uniform magnetic field is a circle when the initial velocity is perpendicular to the field. The crosses represent a uniform magnetic field directed *away from* the reader.

The weber: the SI unit of magnetic flux

Magnetic-field magnitude is magnetic flux per unit area.

Some magnitudes of real-life magnetic fields

Circular motion of a charged particle in a uniform magnetic field

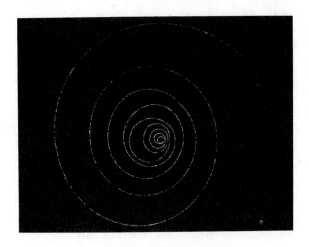

30–9 A track in a liquid-hydrogen bubble chamber made by an electron with initial kinetic energy of about 27 MeV. The magnetic field strength is 1.8 T. As the particle loses energy, the radius of curvature decreases; the maximum radius is about 5 cm. (Courtesy of Lawrence Berkeley Laboratory, University of California.)

where m is the mass of the particle. The radius of the circular orbit is

$$R = \frac{mv}{Bq}. \qquad (30\text{–}9)$$

If the direction of the initial velocity is *not* perpendicular to the field, the velocity component parallel to the field remains constant and the particle moves in a helix. In this case, v in Eq. (30–9) is the component of velocity perpendicular to the field.

Note that the radius is proportional to the *momentum* of the particle, mv. Note also that the magnetic force acting on a charged particle can never do *work*, because the force is always at *right angles* to the motion. The only effect of a magnetic force is to change the *direction* of motion, never to increase or decrease the *magnitude* of the velocity. Thus, *motion of a charged particle under the action of a magnetic field alone is always motion with constant speed.*

Figure 30–9 is a photograph of the track made in a liquid-hydrogen bubble chamber by a high-energy electron moving in a magnetic field perpendicular to the plane of the paper. As the particle loses energy (and speed), the radius of curvature decreases according to Eq. (30–9). Figure 30–10 shows electron–positron pair production, also in a bubble chamber. Similar experiments using a cloud chamber provided the first experimental evidence in 1932 for the existence of the *positron,* or positive electron.

30–10 Electron–positron pair production in a liquid-hydrogen bubble chamber. A high-energy gamma ray coming in from above scatters on an atomic electron. The paths of the recoil electron and of the electron–positron pair can be seen; the directions of curvature in the magnetic field show the signs of the charges. (Courtesy of Lawrence Berkeley Laboratory, University of California.)

PROBLEM-SOLVING STRATEGY: *Motion in magnetic fields*

1. In analyzing the motion of a charged particle in electric and magnetic fields, you are combining the use of Newton's laws of motion with what you have learned about electric and magnetic forces. So you are still dealing with $F = ma$, with F given by $q(E + v \times B)$. Many of the problems are similar to the trajectory problems we encountered in Sections 3–4, 3–5, and 6–1, and it wouldn't do any harm to review those sections.

2. Often the use of components is the most efficient approach. Set up a coordinate system, and then express all the vector quantities (including E, B, v, F, and a) in terms of their components in this system. Then use $F = ma$ in component form: $F_x = ma_x$, and so forth. This approach is particularly useful when you have both electric and magnetic fields present.

3. The next two sections, although not explicitly labeled as examples, are in fact applications of the principles introduced in this chapter and this strategy. Study them carefully!

30–5 THOMSON'S MEASUREMENT OF e/m

Measuring the ratio of charge to mass for the electron: a crucially important experiment at the turn of the century

One of the landmark experiments in modern physics at the turn of the century was the measurement of the charge-to-mass ratio of an electron, e/m. This quantity was first measured by Sir J. J. Thomson in 1897 at the Cavendish Laboratory in Cambridge, England. The discovery that this ratio is *constant* provided the best experimental evidence available at that time of the *existence* of electrons, particles with definite mass and charge. Thomson's term for these particles was "cathode corpuscles." His experiment offers an important and interesting example of magnetic-field forces.

The electron speed can be computed from the accelerating voltage.

Thomson's apparatus (Fig. 30–11) is very similar in principle to the cathode-ray tube discussed in Section 26–8. It consists of a highly evacuated glass tube into which several metal electrodes are sealed. Electrons from the hot cathode C are formed into a beam by the anodes A and A′, and the beam passes into the region between the two plates P and P′. After passing between the plates, the electrons strike the end of the tube, where they cause fluorescent material at S to glow. The speed of the electrons depends on the accelerating potential V; the kinetic energy $\frac{1}{2}mv^2$ equals the loss of potential energy eV (where e is the magnitude of the electron charge):

$$\tfrac{1}{2}mv^2 = eV, \quad \text{or} \quad v = \sqrt{\frac{2eV}{m}}. \tag{30–10}$$

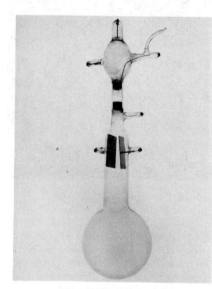

If a potential difference is established between the two deflecting plates P and P′, as shown, the resulting downward electric field deflects the negatively charged electrons *upward*. Alternatively, we may impose a *magnetic* field directed into the plane of the figure, as shown by the × × ×'s; this field results in a *downward* deflection of the beam. (Can you verify this direction?) Finally, if *both* **E** and **B** fields are applied simultaneously, we can adjust their relative magnitudes so that the two forces cancel and the beam is undeflected. The condition that must be satisfied, obtained by equating the two force magnitudes, is

$$eE = evB, \quad \text{or} \quad v = \frac{E}{B}. \tag{30–11}$$

Electron-beam apparatus used by J. J. Thomson (1856–1940) for his crucial e/m experiment in 1897. A sketch of the apparatus is shown in Fig. 30–11. (Copyright Cavendish Laboratory, University of Cambridge.)

Finally, we may combine this with Eq. (30–10) to eliminate v and obtain an expression for the charge-to-mass ratio e/m in terms of the other quantities:

$$\frac{E}{B} = \sqrt{\frac{2eV}{m}}, \qquad \frac{e}{m} = \frac{E^2}{2B^2V}. \tag{30–12}$$

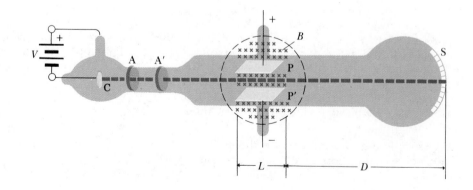

30–11 Thomson's apparatus for measuring the ratio e/m for cathode rays.

All the quantities on the right side can be measured, so e/m can be determined. Note that it is *not* possible to measure e or m separately by this method, but only their ratio.

Thomson measured e/m for his "cathode corpuscles" and found a unique value for this quantity that was independent of the cathode material and the residual gas in the tube. This independence indicated that cathode corpuscles are a common constituent of all matter. The most precise value of e/m available at present is $(1.758803 \pm 0.000003) \times 10^{11}$ C·kg^{-1}. Thus Thomson is credited with discovery of the first subatomic particle, the electron. He also found that the speed of the electrons in the beam was about one-tenth the speed of light, much larger than any previously measured material-particle speed.

The charge and mass of the electron: two fundamental constants of nature

Fifteen years after Thomson's experiments, Millikan succeeded in measuring the charge of the electron with his famous oil-drop experiment, described in Section 26–6. The magnitude of the electron charge e is 1.6022×10^{-19} C. Thus the *mass* of the electron can be obtained:

$$m = \frac{1.6022 \times 10^{-19} \text{ C}}{1.7588 \times 10^{11} \text{ C·kg}^{-1}} = 9.110 \times 10^{-31} \text{ kg}.$$

30–6 ISOTOPES AND MASS SPECTROSCOPY

Thomson devised a method similar to the above e/m measurement, for measuring the charge-to-mass ratio for positive ions. In Thomson's day it was difficult to produce a beam of positive ions all having the same speed. Because the e/m electron experiment depends on the particles having a common speed (in order for the electric- and magnetic-field forces to balance), this method is not directly applicable for a beam of particles having various velocities. Thomson's idea was to make the electric and magnetic fields *parallel*, so that the deflections due to these fields are in perpendicular directions. The net deflection can then never be zero, but it turns out that the relation between the x- and y-deflections for a beam permits determination of the charge-to-mass ratio of the particles.

Using magnetic-field forces to measure molecular masses

Thomson assumed that each positive ion had a charge equal in magnitude to that of the electron because each ion was an atom that had lost one electron. He could then identify particular values of q/m with particular ions. Positive ions move more slowly than electrons and have lower values of q/m because they are much more massive. The *largest* q/m for positive particles is that for the *lightest* element, hydrogen. From the value of q/m it was found that the mass of the *hydrogen ion* or *proton* is 1836.13 ± 0.01 times the mass of an electron. This showed for the first time that electrons contribute only a small fraction of the mass of material objects.

The most striking result of these experiments was that certain chemically pure gases had *more than one* value of q/m. Most notable was the case of neon, which has atomic mass 20.2 g·mol^{-1}. Thomson obtained *two* values of q/m, corresponding to 20 and 22 g·mol^{-1}, and after trying and discarding various explanations he concluded that there must be two kinds of neon atoms with different masses.

Some elements have more than one molecular mass: the discovery of isotopes.

Soon afterward, Francis Aston, a student of Thomson, succeeded in actually separating these two atomic species. Aston permitted the gas to diffuse repeatedly through a porous plug between two containers. He made use of

Varieties of mass spectrometers, and their uses to identify isotopes of many elements

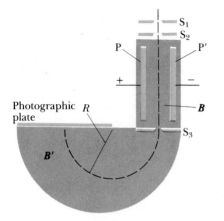

30–12 Bainbridge's mass spectrometer, utilizing a velocity selector.

Velocity selectors: obtaining a beam of electrons that all have the same speed

the fact that at a given temperature T, each atom has an average kinetic energy equal to $\frac{3}{2}kT$, independent of the atom's mass, as discussed in Section 20–4. If the average of $\frac{1}{2}mv^2$ is the same for all atoms, then the more massive atoms must have, on the average, somewhat smaller speeds. Because of this speed difference, the gas emerging from the plug had a slightly greater concentration of the less massive atoms than the gas entering the plug. Thus Aston demonstrated directly the existence of two species of neon atoms with different masses.

Since these experiments in the early twentieth century, many other elements have been shown to have several kinds of atoms, identical in their chemical behavior but differing in mass. Such forms of an element are called **isotopes.** We will see in Chapter 44 that the mass differences are due to differing numbers of neutrons in the *nuclei* of the atoms.

A detailed search for the isotopes of all the elements required precise experimental technique. Aston built the first of many instruments called **mass spectrometers** in 1919; his instrument had a precision of one part in 10,000. A variation built by Bainbridge incorporates a velocity selector to produce a beam of ions all with the same speed. In Fig. 30–12, a source of ions (not shown) is situated above S_1. The ions under study pass through slits S_1 and S_2 and move down into the electric field between the two plates P and P'. In this region there is also a magnetic field $\textbf{B}$, perpendicular to the page. Thus the ions enter a region of crossed electric and magnetic fields like those used in the Thomson e/m experiment. Only those ions whose speed is equal to E/B pass straight through this region; ions with other speeds are deflected and blocked by slit S_3. Thus all ions emerging from S_3 have the same velocity. The region of crossed fields is called a *velocity selector.*

Below S_3 the ions enter a region where there is another magnetic field $\textbf{B}'$, perpendicular to the page, but no electric field. Here the ions move in circular paths of radius R. From Eq. (30–9), we find that

$$ m = \frac{qB'R}{v}. \qquad (30\text{–}13) $$

Usually the charges on all the ions are the same, so the mass of each ion is proportional to the radius of its path. Ions of different isotopes converge at different points on the photographic plate. The relative abundance of the isotopes is measured from the densities of the photographic images they produce.

The atomic mass unit: a convenient unit for masses of atoms and molecules

In present-day mass spectrometers the photographic plate is replaced by a more sophisticated particle detector. Figure 30–13 shows a modern mass spectrometer and a typical isotope analysis. For each isotope, the number given represents the **mass number,** equal to the total number of protons and neutrons in the nucleus of the atom. Different isotopes of an element have the same number of protons but differing numbers of neutrons in the nucleus.

Masses of atoms are often expressed in **atomic mass units.** By definition, one atomic mass unit (1 u) is $\frac{1}{12}$ the mass of one atom of the most abundant isotope of carbon, ^{12}C. Since the mass of an atom in grams is equal to its atomic mass divided by Avogadro's number, it follows that

$$ 1\text{ u} = \frac{(1/12)(12\text{ g·mol}^{-1})}{6.022 \times 10^{23}\text{ mol}^{-1}} $$
$$ = 1.661 \times 10^{-24}\text{ g} = 1.661 \times 10^{-27}\text{ kg.} $$

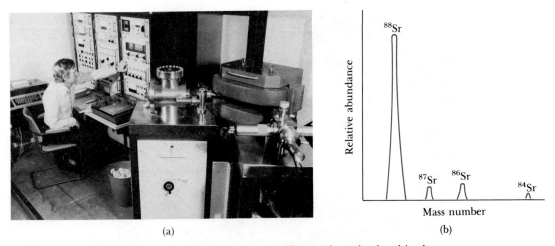

(a)   (b)

30–13 (a) A modern solid-source mass spectrometer. The specimen is placed in the
cylindrical chamber in the center; the ion beam is deflected by the large magnet toward
the collector and detector (right front). The instrument is controlled and the data
processed by the equipment in front of the operator. (Courtesy of Institute of
Geological Studies.) (b) A typical isotope analysis, showing several isotopes of strontium.

30–7 MAGNETIC FORCE ON A CONDUCTOR

When a current-carrying conductor lies in a magnetic field, the field exerts
magnetic forces on the moving charges within the conductor. These forces are
transmitted to the material of the conductor, and the conductor as a whole
experiences a force distributed along its length. The electric motor and the
moving-coil galvanometer both depend for their operation on the magnetic
forces on conductors carrying currents.

Figure 30–14 represents a segment of a conducting wire, with length l and
cross-sectional area A, in which the current density J is from left to right. The
wire is in a magnetic field B, perpendicular to the plane of the diagram, and
directed *into* the plane. For generality, we assume the wire contains moving
charges of both signs. In metals the moving charges are always negatively
charged electrons, but in semiconductors charges of both signs may be present. A positive charge q_1 within the wire, moving with its drift velocity v_1, is
acted on by an upward force F_1 given by Eq. (30–2), $F_1 = q_1(v_1 \times B)$. As the
figure shows, the direction of this force is upward, and since in this case v_1 and
B are perpendicular, the magnitude of the force is $F_1 = q_1 v_1 B$. Similarly, a
negative charge q_2, with drift velocity v_2 in a direction opposite to that of J,
experiences a force $F_2 = q_2 v_2 \times B$. Because q_1 and q_2 have opposite signs and
v_1 and v_2 have opposite directions, F_2 has the *same* direction as F_1, as shown.

Magnetic force on a moving charge in a
conductor

30–14 Forces on the moving charges
in a current-carrying conductor. The
forces on both positive and negative
charges are in the same direction.

An example of magnetic levitation in railroads. This German system using attractive magnetic forces with conventional electromagnetic technology can achieve speeds of 400 km/hr (250 mi/hr). (Courtesy of Transit America, Inc.)

The *total* force on all the moving charges in a length l of conductor can be expressed in terms of the current, using the considerations of Section 28–1, Eqs. (28–3), (28–4), and (28–5). Let n_1 and n_2 represent the numbers of positive and negative charges, respectively, per unit volume. The numbers of charges in the portion having length l are then n_1Al and n_2Al. The total force $\boldsymbol{F}$ on all charges (and hence the total force on the wire) has magnitude

$$
\begin{aligned}
F &= (n_1Al)(q_1v_1B) + (n_2Al)(q_2v_2B) \\
  &= (n_1q_1v_1 + n_2q_2v_2)AlB.
\end{aligned}
$$

The magnetic force on the moving charges in a conductor can be expressed in terms of the current.

But $n_1q_1v_1 + n_2q_2v_2$ (or more generally, Σnqv) equals the current density J, and the product JA equals the current I, so finally

$$F = IlB. \tag{30–14}$$

If the $\boldsymbol{B}$ field is not perpendicular to the wire but makes an angle ϕ with it, the situation is just like that discussed in Section 30–2 for a single charge. The component of $\boldsymbol{B}$ parallel to the wire (and thus to the drift velocities of the charges) exerts no force; the component perpendicular to the wire is given by $B_\perp = B \sin \phi$. Thus for this more general case

$$F = IlB_\perp = IlB \sin \phi. \tag{30–15}$$

Expressing the magnetic force on a conductor using the vector product

To find the direction of the force on a current-carrying conductor placed in a magnetic field, we may use the same right-hand rule that we used for a moving positive charge. Rotate a right-hand-thread screw from the direction of I toward $\boldsymbol{B}$; the direction of advance is the direction of $\boldsymbol{F}$. The situation is the same as that shown in Fig. 30–2, but with the direction of v replaced by the direction of I.

Thus this magnetic force, like the force on a single moving charge, may be expressed as a vector product. We represent the section of wire by a vector $\boldsymbol{l}$ along the wire and in the direction of the current. Then the force on this section is given by

$$\boldsymbol{F} = I\boldsymbol{l} \times \boldsymbol{B}. \tag{30–16}$$

EXAMPLE 30–2   A straight horizontal wire carries a current of 50 A from west to east in a region between the poles of a large electromagnet, which provides a horizontal magnetic field toward the northeast (i.e., 45° north of east) with magnitude 1.2 T. Find the magnitude and direction of the force on a 1-m section of wire.

SOLUTION   The angle ϕ between the directions of current and field is 45°. From Eq. (31–2) we obtain

$$F = (50 \text{ A})(1 \text{ m})(1.2 \text{ T})(\sin 45°) = 42.4 \text{ N}.$$

An example of the magnetic-field force on a conductor

Consistency of units may be checked by observing, as noted in Section 30–2, that $1 \text{ T} = 1 \text{ N} \cdot \text{A}^{-1} \cdot \text{m}^{-1}$. The direction of the force is perpendicular to the plane of the current and the field, both of which lie in the horizontal plane. Thus the force must be along the vertical; the right-hand rule shows that it is vertically upward.

Alternatively, we can use a coordinate system with the *x*-axis pointing east, the *y*-axis north, and the *z*-axis up. Then we have

$$\boldsymbol{l} = (1 \text{ m})\boldsymbol{i},$$

$$\boldsymbol{B} = (1.2 \text{ T})(\cos 45° \, \boldsymbol{i} + \sin 45° \, \boldsymbol{j}),$$

$$\boldsymbol{F} = I\boldsymbol{l} \times \boldsymbol{B}$$

$$= (50 \text{ A})(1 \text{ m})(1.2 \text{ T})\boldsymbol{i} \times (\cos 45° \, \boldsymbol{i} + \sin 45° \, \boldsymbol{j})$$

$$= (42.4 \text{ N})\boldsymbol{k}.$$

30–8 FORCE AND TORQUE ON A CURRENT LOOP

We can represent any current-carrying conductor in terms of straight-line segments; curved portions can be represented by a large number of very small segments. Thus we can use Eq. (30–15) or (30–16) to calculate the magnetic-field force on *any* conductor. In general, this requires integration; but when the magnetic field is uniform, the calculation is quite simple. As an example, we will consider a rectangular current loop in a uniform field. We will find that the total *force* on the loop is zero but that there is a *torque* that tends to turn the loop to a particular orientation with respect to the field.

Figure 30–15 shows a rectangular loop of wire having sides of lengths *a* and *b*. The normal to the plane of the loop makes an angle α with the direction of a uniform magnetic field, and the loop carries a current *I*. (Provision must be made for leading the current into and out of the loop, or for inserting a source of emf. This is omitted from the diagram, for simplicity.)

The total magnetic-field force on a closed current loop in a uniform field is zero.

The force $\boldsymbol{F}$ on the right side of the loop (length *a*), is in the direction of the *x*-axis, toward the right, as shown. On this side $\boldsymbol{B}$ is perpendicular to the current direction, and the total force on this side (the sum of the forces distributed along the side) has magnitude

$$F = IaB.$$

A force of the same magnitude but in the opposite direction acts on the opposite side, as shown in the figure.

The forces on the sides of length *b*, represented by the vectors $\boldsymbol{F}'$ and $-\boldsymbol{F}'$, have magnitude

Torque on a current loop in a magnetic field

$$IbB \sin (90° - \alpha), \quad \text{or} \quad IbB \cos \alpha.$$

The lines of action of both lie along the *y*-axis.

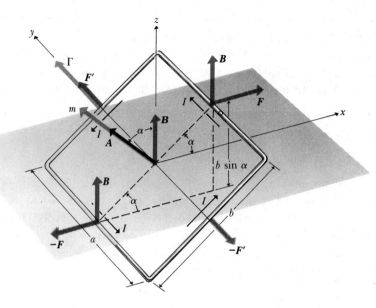

30–15 Forces on the sides of a current-carrying loop in a magnetic field. The resultant of the set of forces is a couple of moment $\Gamma = IAB \sin \alpha$.

The total force on the loop is zero, since the forces on opposite sides cancel out in pairs. The two forces $\boldsymbol{F}'$ and $-\boldsymbol{F}'$ lie along the same lines and so have no torque with respect to any axis, but the two forces $\boldsymbol{F}$ and $-\boldsymbol{F}$ constitute a *couple*, as defined in Section 10–4. The torque of a couple has the same value with respect to any point, and is given by the magnitude of either force multiplied by the distance between the lines of action of the two forces. From Fig. 30–15, this distance is $b \sin \alpha$, so the torque is

$$\Gamma = (IBa)(b \sin \alpha). \tag{30–17}$$

The torque is greatest when $\alpha = 90°$ (i.e., when the plane of the coil is parallel to the field), and it is zero when α is zero or $180°$ and the plane of the coil is perpendicular to the field. (One of the latter positions is a position of *stable* equilibrium, the other of *unstable* equilibrium; which is which?)

Since ab is the area A of the coil, Eq. (30–17) may also be written

$$\Gamma = IBA \sin \alpha. \tag{30–18}$$

Magnetic moment: a convenient way to represent the interaction of a current loop with a magnetic field

The product IA is called the **magnetic moment** m of the loop:

$$m = IA.$$

This is analogous to the electric dipole moment introduced in Section 25–1. The torque on a magnetic moment associated with a current loop can be expressed simply as

$$\Gamma = mB \sin \alpha. \tag{30–19}$$

Because of the directional relations indicated, the torque Γ tends to rotate the loop in the direction of *decreasing* α; that is, toward its equilibrium position, in which it lies in the xy-plane, perpendicular to the direction of the field $\boldsymbol{B}$.

Using the vector product to describe the torque on a current loop in a magnetic field

We can express the interaction between a conducting loop and a magnetic field more compactly in terms of vector torque, introduced in Section 9–8. We first define a vector area $\boldsymbol{A}$ (as in Section 25–4) having magnitude A and direction perpendicular to its plane; the sense is determined by the right-hand rule applied to the direction of circulation of current around the loop, as shown in Fig. 30–15. Then, as Eq. (30–18) and the figure show, the vector torque Γ is given by

$$\Gamma = I\mathbf{A} \times \mathbf{B}. \qquad (30\text{--}20)$$

We may also define a vector magnetic moment

$$\mathbf{m} = I\mathbf{A}; \qquad (30\text{--}21)$$

in terms of $\mathbf{m}$ we can express the torque simply as

$$\Gamma = \mathbf{m} \times \mathbf{B}. \qquad (30\text{--}22)$$

The magnetic moment of a current loop depends on its area and the current.

In the equilibrium position of the loop, its vector magnetic moment $\mathbf{m}$ is parallel to the field $\mathbf{B}$. The torque Γ tends to rotate the loop toward this position. The torque is greatest when $\mathbf{m}$ and $\mathbf{B}$ are perpendicular, and zero when they are parallel or antiparallel. Equation (30–22) is the analog of Eq. (25–3) (Section 25–1) for the torque on an *electric* dipole in an electric field.

When a magnetic dipole changes its orientation in a magnetic field, the field does work on it, given by $\int \Gamma\, d\theta$, and there is a corresponding potential energy. As this discussion suggests, the potential energy is least when $\mathbf{m}$ and $\mathbf{B}$ are parallel and greatest when they are antiparallel. If the potential energy U is taken as zero when $\alpha = \pi/2$, then the potential energy $U(\alpha)$ at any other position is given by

Potential energy of a current loop in a magnetic field depends on its orientation.

$$U(\alpha) = \int_{\alpha}^{\pi/2} \Gamma\, d\theta.$$

We can express the torque Γ in terms of the variable angle θ by use of Eq. (30–19):

$$\Gamma = -mB \sin\theta.$$

The extra minus sign is included here because the torque is in the direction of *decreasing* θ. Combining the relations given above and evaluating the integral, we find

$$\begin{aligned} U(\alpha) &= -mB \int_{\alpha}^{\pi/2} \sin\theta\, d\theta \\ &= -mB \cos\theta, \end{aligned}$$

or

$$U = -\mathbf{m} \cdot \mathbf{B}. \qquad (30\text{--}23)$$

Although we have derived Eqs. (30–18), (30–22), and (30–23) for a rectangular current loop, it is not hard to show that all these relations are valid for a plane loop of any shape at all. A *circular* loop, for example, may be approximated as closely as we wish by a very large number of rectangular loops. If these loops all carry equal currents in the same sense, then the forces on the sides of two loops adjacent to each other cancel out, and the only forces that do not cancel are around the boundary. Pursuing this line of reasoning, we can prove that all the relations above are valid not only for a circular loop but for *a plane loop of any shape*. In particular, for a circular loop of radius R,

Current loops of arbitrary shape.

$$\Gamma = \pi I B R^2 \sin\alpha \qquad (30\text{--}24)$$

and

$$\Gamma = \pi I R^2 \mathbf{n} \times \mathbf{B}, \qquad (30\text{--}25)$$

where n is a unit vector perpendicular to the plane of the loop as determined by the sense of circulation of current.

An arrangement of particular interest is the **solenoid,** which is a helical winding of wire, such as a coil wound on a circular cylinder. If the windings are closely spaced, the solenoid can be approximated by a number of circular loops lying in planes at right angles to its long axis. The total torque on a solenoid in a magnetic field is simply the sum of the torques on the individual turns. Hence, for a solenoid of N turns in a uniform field B,

$$\Gamma = NIAB \sin \alpha, \qquad (30\text{--}26)$$

where α is the angle between the axis of the solenoid and the direction of the field. The torque is maximum when the magnetic field is parallel to the planes of the individual turns or perpendicular to the long axis of the solenoid. The effect of this torque, if the solenoid is free to turn, is to rotate it into a position in which each turn is perpendicular to the field and the axis of the solenoid is parallel to the field.

The behavior of a solenoid in a magnetic field resembles that of a bar magnet or compass needle; both the solenoid and the magnet, if free to turn, orient themselves with their axes parallel to a magnetic field. The torque on a solenoid could be incorporated into an alternative *definition* of magnetic field. The behavior of a bar magnet or compass needle is sometimes described in terms of magnetic forces on "poles" at its ends; for the solenoid, no such concept is needed. In fact, the moving electrons in a bar of magnetized iron play the same role as the current in the windings of a solenoid, and the observed torque arises from the same cause in both instances. We will return to this subject later.

EXAMPLE 30–3   A circular coil of wire 0.05 m in radius, having 30 turns, lies in a horizontal plane, as shown in Fig. 30–16. It carries a current of 5 A, in a counterclockwise sense when viewed from above. The coil is in a magnetic field directed toward the right, with magnitude 1.2 T. Find the magnetic moment and the torque on the coil.

SOLUTION   The area of the coil is

$$A = \pi r^2 = \pi (0.05 \text{ m})^2 = 7.85 \times 10^{-3} \text{ m}^2.$$

The magnetic moment of each turn of the coil is

$$m = IA = (5 \text{ A})(7.85 \times 10^{-3} \text{ m}^2) = 3.93 \times 10^{-2} \text{ A·m}^2,$$

and the total magnetic moment of all 30 turns is

$$m = (30)(3.93 \times 10^{-2} \text{ A·m}^2) = 1.18 \text{ A·m}^2.$$

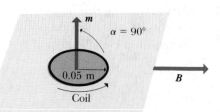

30–16 A circular coil of wire and its associated magnetic moment, in a magnetic field (Example 30–3).

The angle α between the direction of B and the normal to the plane of the coil is 90°, and from Eq. (30–18) the torque on each turn of the coil is

$$\Gamma = IBA \sin \alpha = (5 \text{ A})(1.2 \text{ T})(7.85 \times 10^{-3} \text{ m}^2)(\sin 90°)$$
$$= 0.047 \text{ N·m},$$

and the total torque on the coil is

$$\Gamma = (30)(0.047 \text{ N·m}) = 1.41 \text{ N·m}.$$

Alternatively, from Eq. (30–19),

$$\Gamma = mB \sin \alpha = (1.18 \text{ A·m}^2)(1.2 \text{ T})(\sin 90°)$$
$$= 1.41 \text{ N·m}.$$

The direction of the torque is such as to tend to push the right side of the coil down and the left side up and to rotate it into a position where the normal to its plane is parallel to B.

EXAMPLE 30–4 If the coil in Example 30–3 rotates from its initial position to a position where its magnetic moment is parallel to B, what is the change in potential energy?

An example of energy relations for a current loop in a magnetic field

SOLUTION From Eq. (30–23), the initial potential energy U_1 is

$$U_1 = -(1.18 \text{ A·m}^2)(1.2 \text{ T})(\cos 90°) = 0,$$

and the final potential energy U_2 is

$$U_2 = -(1.18 \text{ A·m}^2)(1.2 \text{ T})(\cos 0°) = -1.41 \text{ J}.$$

Thus the change in potential energy is -1.41 J.

EXAMPLE 30–5 What vertical forces applied to the left and right edges of the coil of Fig. 30–16 would be required to hold it in equilibrium in its initial position?

SOLUTION An upward force of magnitude F at the right side and a downward force of equal magnitude on the left side would have a total torque of

$$\Gamma = (2)(0.05 \text{ m})F.$$

This must be equal to the magnitude of the magnetic-field torque of 1.41 N·m, and we find that the required forces have magnitude 14.1 N.

The d'Arsonval galvanometer, which we described in Section 29–3, makes use of a magnetic-field torque on a coil carrying a current. As Fig. 29–7 shows, the magnetic field is not uniform but is *radial*, so the side thrusts on the coil are always perpendicular to its plane. Thus the magnetic-field torque is directly proportional to the current, no matter what the orientation of the coil. A restoring torque proportional to the angular displacement of the coil is provided by two hairsprings, which also serve as current leads to the coil. When current is supplied to the coil, it rotates, along with its attached pointer, until the restoring spring torque just balances the magnetic-field torque. Thus the pointer deflection is proportional to the current. Such instruments can measure currents as small as 10^{-7} A, and modified versions currents as small as 10^{-10} A. For extreme sensitivity and reliability, however, d'Arsonval instruments have been mostly replaced by digital electronic instruments.

The d'Arsonval galvanometer: an application of magnetic torque on a current loop

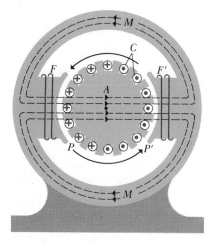

30–17 Schematic diagram of a dc motor. The armature or rotor A rotates on a shaft through its center, perpendicular to the plane of the figure. The conductors on the rotor are shown in cross section; those with dots at their centers carry current out of the plane; those with crosses, into the plane.

30–9 THE DIRECT-CURRENT MOTOR

No one needs to be reminded of the importance of electric motors in contemporary society. Their operation depends on magnetic-field forces on current-carrying conductors. As an example, let us consider the operation of one type of direct-current motor, shown schematically in Fig. 30–17. The center part A is the *armature*, or *rotor*; it is a cylinder of soft steel mounted on a shaft so that it can rotate about its axis (perpendicular to the plane of the figure).

Embedded in slots in the rotor surface (parallel to its axis) are insulated copper conductors C. Current is led into and out of these conductors through graphite brushes making contact with a cylinder of the shaft called the *commutator* (not shown in Fig. 30–17). The commutator is an automatic switching arrangement that maintains the currents in the conductors in the directions shown in the figure, whatever the position of the rotor. The current in the field coils F and F' sets up a magnetic field in the motor frame M and in the gap between the pole pieces P and P' and the rotor. Some of the magnetic field lines are shown as broken lines. With the directions of field and rotor currents shown, the side thrust on each conductor in the rotor is such as to produce a *counterclockwise* torque on the rotor.

> The rotor and field windings can be connected in various ways.

If the rotor and the field windings are connected in series, we have a *series motor*; if they are connected in parallel, we have a *shunt motor*. In some motors the field windings are in two parts, one in series with the rotor and the other in parallel with it; such a motor is called *compound*.

A motor converts electrical energy to mechanical energy or work, and thus requires electrical energy input. If the potential difference between its terminals is V_{ab} and the current is I, then the power input is $P = V_{ab}I$. Even if the motor coils have negligible resistance, there must be a potential difference between the terminals if P is to be different from zero. This potential difference results principally from magnetic forces exerted on the charges in the conductors of the rotor as they rotate through the magnetic field. These forces are an example of nonelectrostatic forces; the resulting electromotive force $\mathcal{E}$ is called an *induced* emf or sometimes a *back* emf, referring to the fact that its sense is opposite to that of the current. Induced emf's resulting from motion of conductors in magnetic fields will be considered in greater detail in Chapter 32.

> Voltage–current relations for a direct-current motor

In a series motor having internal resistance r, V_{ab} is greater than $\mathcal{E}$, and the difference is the drop Ir across the internal resistance. Thus, for either a motor or a battery being charged,

$$V_{ab} = \mathcal{E} + Ir. \qquad (30\text{–}27)$$

Because of the nature of the magnetic-field force, $\mathcal{E}$ is *not* constant but depends on the speed of rotation of the rotor.

The behavior of a dc motor is analogous to that of a battery being charged, as discussed in Section 28–4. In that case, electrical energy is converted to chemical rather than mechanical energy.

> An example of a series dc motor

EXAMPLE 30–6 A dc motor with its rotor and field coils connected in series has an internal resistance of 2.0 Ω. When running at full load on a 120-V line, it draws a current of 4.0 A.

a) What is the emf in the rotor?

$$V_{ab} = \mathcal{E} + Ir,$$

$$120 \text{ V} = \mathcal{E} + (4.0 \text{ A})(2.0 \text{ }\Omega),$$

$$\mathcal{E} = 112 \text{ V}.$$

b) What is the power delivered to the motor?

$$P = V_{ab}I = (120 \text{ V})(4.0 \text{ A}) = 480 \text{ W}.$$

c) What is the rate of dissipation of energy in the resistance of the motor?

$$P = I^2 r = (4.0 \text{ A})^2 (2.0 \text{ }\Omega) = 32 \text{ W}.$$

d) What is the mechanical power developed?

The mechanical power output is the electrical power input, minus the rate of dissipation of energy in the motor's resistance:

$$P = 480 \text{ W} - 32 \text{ W} = 448 \text{ W}.$$

30–10 THE HALL EFFECT

The reality of the forces acting on the moving charges in a conductor in a magnetic field is strikingly demonstrated by the **Hall effect.** The conductor in Fig. 30–18 is in the form of a flat strip. The charges within it are driven toward the upper edge of the strip by the magnetic force qvB exerted on them. Here v is the drift velocity of the moving charges in the material.

If the charge carriers are electrons, as in Fig. 30–18a, an excess negative charge accumulates at the upper edge of the strip, leaving an excess positive charge at its lower edge. This accumulation continues to a point where the resulting transverse electrostatic field E_e causes a force of magnitude qE_e that is equal and opposite to the magnetic field force of magnitude qvB. Then there is no longer any net force to deflect the moving charges sideways. Associated with this electric field is a potential difference between opposite edges

The Hall effect: sideways potential difference in a conductor, caused by a magnetic field

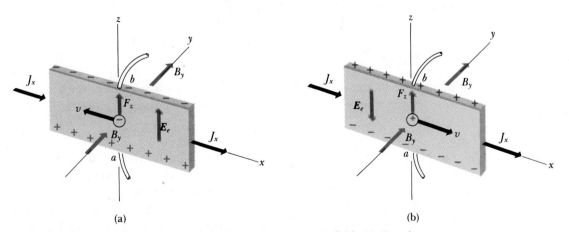

(a)   (b)

30–18 Forces on charge carriers in a conductor in a magnetic field. (a) Negative current carriers (electrons) are pushed toward top of slab, leading to charge distribution as shown. Point a is at higher potential than point b. (b) Positive current carriers; polarity of potential difference is opposite to that of (a).

of the strip. This potential difference, which can be measured with a potentiometer, is called the *Hall voltage,* or the *Hall emf.* Experiments show that for metals, the upper edge of the strip in Fig. 30–18a *does* become negatively charged, showing that the charge carriers in a metal are indeed negative electrons.

> Hall-effect measurements can be used to determine whether the moving charges in a conductor are positive or negative.

However, if the charge carriers are *positive,* as in Fig. 30–18b, then *positive* charge accumulates at the upper edge, and the potential difference is *opposite* to that resulting from the deflection of negative charges. Soon after the discovery of the Hall effect, in 1879, it was observed that some materials, notably the *semiconductors,* exhibit a Hall emf opposite to that of the metals, *as if* their charge carriers were *positively* charged. We now know that these materials conduct by a process known as *hole conduction.* Within such a material there are locations, called *holes,* that would normally be occupied by an electron but are actually empty, and a *missing negative* charge is equivalent to a *positive* charge. When an electron moves in one direction to fill a hole, it leaves another hole behind it, and the result is that the hole (equivalent to a positive charge) migrates in the direction *opposite* to that of the electron.

In terms of the coordinate axes in Fig. 30–18a, the electrostatic field E_e is in the z-direction, and we write it as E_z. The magnetic field is in the y-direction, and we write it as B_y. The magnetic-field force (in the $-z$-direction) is qvB_y. The current density J_x is in the x-direction. In the final steady state, when the forces qE_z and qvB_y are equal,

$$E_z = vB_y.$$

The current density J_x is

$$J_x = nqv.$$

When v is eliminated, we have

> Hall-effect measurements can be used to determine charge density in a conductor.

$$nq = \frac{J_x B_y}{E_z}. \qquad (30\text{--}28)$$

Thus, from measurements of J_x, B_y, and E_z, we can compute the product nq. In both metals and semiconductors, q is equal in magnitude to the electron charge, so the Hall effect permits a direct measurement of n, the density of current-carrying charges in the material.

SUMMARY

KEY TERMS

magnetic poles

magnetic field

tesla

gauss

magnetic-field lines

magnetic flux

weber

maxwell

flux density

isotopes

mass spectrometer

mass number

Magnetic field, denoted by **B,** is a vector field. A particle with charge q moving with velocity v in a magnetic field **B** experiences a force **F** given by

$$\boldsymbol{F} = q\boldsymbol{v} \times \boldsymbol{B}. \qquad (30\text{--}2)$$

The SI unit of magnetic field is the tesla (1 T). The direction of **B** is defined by Eq. (30–2); it can be defined alternatively in terms of the equilibrium orientation of a compass needle or a solenoid permitted to rotate freely. A magnetic field can be represented graphically by magnetic-field lines; at each point a line is tangent to the direction of **B** at that point, and the number of lines per unit area perpendicular to **B** is proportional to the magnitude of **B**.

Magnetic flux through an area is defined as

$$\Phi = \int B_\perp \, dA = \int \boldsymbol{B} \cdot d\boldsymbol{A}. \qquad (30\text{--}6)$$

The SI unit of magnetic flux is the weber (1 Wb). Magnetic field is also called magnetic flux density, and $1\text{ T} = 1\text{ Wb·m}^{-2}$.

Because the magnetic-field force is always perpendicular to v, it does no work on the particle and cannot change the magnitude of v. Thus a particle moving under the action of a magnetic field alone moves with constant speed.

J. J. Thomson used crossed electric and magnetic fields to measure the charge-to-mass ratio (e/m) for electrons. The electric- and magnetic-field forces exactly cancel when $v = E/B$.

Different atoms of an element having different atomic masses are called isotopes; molecular masses are measured with a mass spectrometer. For each isotope, the mass number is the total number of protons and neutrons in the nucleus. Masses of atoms are often expressed in atomic mass units. One atomic mass unit (1 u) is defined to be $1/12$ the mass of one atom of ^{12}C.

The force $\boldsymbol{F}$ on a segment $\boldsymbol{l}$ of a conductor carrying current I in a magnetic field $\boldsymbol{B}$ is given by

$$\boldsymbol{F} = I\boldsymbol{l} \times \boldsymbol{B}. \tag{30–16}$$

A current loop of vector area $\boldsymbol{A}$ carrying current I in a uniform magnetic field $\boldsymbol{B}$ experiences no net force but a torque $\boldsymbol{\Gamma}$ given by

$$\boldsymbol{\Gamma} = I\boldsymbol{A} \times \boldsymbol{B}. \tag{30–20}$$

The magnetic moment $\boldsymbol{m}$ of the loop is defined as

$$\boldsymbol{m} = I\boldsymbol{A}. \tag{30–21}$$

The torque can also be expressed as

$$\boldsymbol{\Gamma} = \boldsymbol{m} \times \boldsymbol{B}. \tag{30–22}$$

The potential energy U of a magnetic moment $\boldsymbol{m}$ in a magnetic field $\boldsymbol{B}$ is given by

$$U = -\boldsymbol{m} \cdot \boldsymbol{B}. \tag{30–23}$$

The magnetic moment of a loop depends only on the current and the area, and is independent of the shape of the loop.

In a d'Arsonval galvanometer, the magnetic forces on the conducting loop are perpendicular to its plane. The magnetic-field torque is proportional to the current in the coil; the spring restoring torque is proportional to angular displacement, so the pointer deflection is proportional to current.

In a dc motor the magnetic field set up by the field windings exerts tangential forces on the currents in the rotor. In a series motor the rotor and field coils are in series; in a parallel motor they are in parallel. Motion of the rotor through the magnetic field causes an induced emf. For a series motor the terminal voltage is the sum of the induced emf and the drop Ir across the internal resistance.

The Hall effect is a potential difference perpendicular to the direction of current in a conductor, when the conductor is placed in a magnetic field. The Hall potential is determined by the requirement that the associated electric field must just balance the magnetic-field force on a moving charge. Hall-effect measurements can be used to determine the density n of charge carriers and their sign.

atomic mass unit
magnetic moment
solenoid
Hall effect

QUESTIONS

30–1 Does the earth's magnetic field have a significant effect on the electron beam in a TV picture tube?

30–2 If an electron beam in a cathode-ray tube travels in a straight line, can you be sure there is no magnetic field present?

30–3 If the magnetic-field force does no work on a charged particle, how can it have any effect on the particle's motion? Are there other examples of forces that do no work but have a significant effect on a particle's motion?

30–4 A permanent magnet can be used to pick up a string of nails, tacks, or paper clips, even though these are not magnets by themselves. How can this be?

30–5 How could a compass be used for a *quantitative* determination of magnitude and direction of magnetic field at a point?

30–6 The direction in which a compass points (magnetic north) is not in general exactly the same as the direction toward the north pole (true north). The difference is called *magnetic declination;* it varies from point to point on the earth and also varies with time. What are some possible explanations for magnetic declination?

30–7 Can a charged particle move through a magnetic field without experiencing any force? How?

30–8 Could the electron beam in an oscilloscope tube (cathode-ray tube) be used as a compass? How? What advantages and disadvantages would it have, compared with a conventional compass?

30–9 Does a magnetic field exert forces on the electrons within atoms? What observable effect might such interaction have on the behavior of the atom?

30–10 How might a loop of wire carrying a current be used as a compass? Could such a compass distinguish between north and south?

30–11 How could the direction of a magnetic field be determined by making only *qualitative* observations of the magnetic force on a straight wire carrying a current?

30–12 Do the currents in the electrical system of a car have a significant effect on a compass placed in the car?

30–13 A student claimed that if lightning strikes a metal flagpole, the force exerted by the earth's magnetic field on the current in the pole can be large enough to bend it. Typical lightning currents are of the order of 10^4 to 10^5 A; is the student's opinion justified?

30–14 A student tried to make an electromagnetic compass by suspending a coil of wire from a thread (with the plane of the coil vertical) and passing a current through it. He expected the coil to align itself perpendicular to the horizontal component of the earth's magnetic field; but instead, the coil went into what appeared to be angular simple harmonic motion (cf. Section 11–4), turning back and forth past the expected direction. What was happening? Was the motion truly simple harmonic?

30–15 When the polarity of the voltage applied to a dc motor is reversed, the direction of rotation *does not* reverse. Why not? How *could* the direction of rotation be reversed?

30–16 If an emf is produced in a dc motor, would it be possible to use the motor somehow as a *generator* or *source*, taking power out of it instead of putting power into it? How might this be done?

30–17 Hall-effect voltages are much *larger* for relatively poor conductors such as germanium than for good conductors such as copper, for comparable currents, fields, and dimensions. Why?

30–18 Is it possible that, in a Hall-effect experiment, *no* transverse potential difference will be observed? Under what circumstances might this happen?

EXERCISES

Section 30–2 Magnetic Field

30–1 In a magnetic field directed vertically upward, a particle initially moving north is deflected toward the west. What is the sign of the charge of the particle?

30–2 A particle with mass 0.001 kg and charge 1.2×10^{-8} C has at a given instant a velocity $v = (2 \times 10^5 \text{ m·s}^{-1})j$. What are the magnitude and direction of the force exerted on the particle by a uniform magnetic field $B = -(0.8 \text{ T})i$?

30–3 A particle with charge -2.5×10^{-8} C is moving with instantaneous velocity

$$v = (-3 \times 10^4 \text{ m·s}^{-1})i + (5 \times 10^4 \text{ m·s}^{-1})j.$$

What is the force exerted on this particle by a magnetic field

a) $B = (2 \text{ T})i;$ b) $B = (2 \text{ T})k?$

30–4 A particle having a mass of 0.5 g carries a charge of 2.5×10^{-8} C. The particle is given an initial horizontal velocity of 6×10^4 m·s^{-1}. What are the magnitude and direction of the magnetic field needed to keep the particle moving in a horizontal direction? ~~P9 (30-3)~~

30–5 Each of the lettered circles at the corners of the cube in Fig. 30–19 represents a positive charge q moving with a velocity of magnitude v in the direction indicated. The region in the figure is a uniform magnetic field B, parallel to the x-axis and directed toward the right. Copy the figure, find the magnitude and direction of the force on each charge, and show the force in your diagram.

~~P5 30-1~~

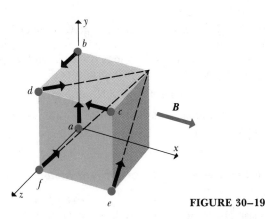

FIGURE 30–19

Section 30–3 Magnetic-Field Lines and Magnetic Flux

30–6 The magnetic field **B** in a certain region has magnitude 2 T, and its direction is that of the positive *x*-axis in Fig. 30–20.

a) What is the magnetic flux across the surface *abcd* in the figure?

b) What is the magnetic flux across the surface *befc*?

c) What is the magnetic flux across the surface *aefd*?

d) What is the net flux through all five surfaces that enclose the shaded volume?

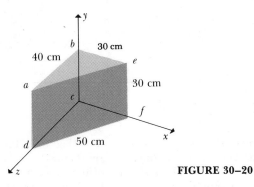

FIGURE 30–20

Section 30–4 Motion of Charged Particles in a Magnetic Field

30–7 An electron at point *A* in Fig. 30–21 has a speed v_0 of 1.0×10^7 m·s^{-1}. Find

a) the magnitude and direction of the magnetic field that will cause the electron to follow the semicircular path from *A* to *B*;

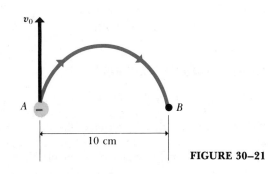

FIGURE 30–21

b) the time required for the electron to move from *A* to *B*.

30–8 In Exercise 30–7, suppose the particle is a proton rather than an electron. Answer the same questions as in that problem.

30–9 A deuteron, an isotope of hydrogen whose mass is very nearly 2 u, travels in a circular path of radius 0.40 m in a magnetic field of magnitude 1.5 T.

a) Find the speed of the deuteron.

b) Find the time required for it to make one-half a revolution.

c) Through what potential difference would the deuteron have to be accelerated to acquire this speed?

30–10 A singly charged ^{7}Li ion has a mass of 1.16×10^{-26} kg. It is accelerated through a potential difference of 500 V and then enters a magnetic field of 0.4 T, moving perpendicular to the field. What is the radius of its path in the magnetic field?

30–11 In a TV picture tube, an electron in the beam is accelerated by a potential difference of 20,000 V. Then it passes through a region of transverse magnetic field, where it moves in a circular arc with radius 0.12 m. What is the magnitude of the field?

30–12 An electron moves in a circular path of radius 1.2 cm perpendicular to a uniform magnetic field. The speed of the electron is 1×10^6 m·s^{-1}. What is the total magnetic flux encircled by the orbit?

Section 30–5 Thomson's Measurement of e/m

Section 30–6 Isotopes and Mass Spectroscopy

30–13

a) What is the velocity of a beam of electrons when the simultaneous influence of an electric field of 34×10^4 V·m^{-1} and a magnetic field of 2×10^{-2} T, both fields being normal to the beam and to each other, produces no deflection of the electrons?

b) Show in a diagram the relative orientation of the vectors **v**, **E**, and **B**.

c) What is the radius of the electron orbit when the electric field is removed?

30–14 In the Bainbridge mass spectrometer (Fig. 30–12), suppose the magnetic field *B* in the velocity selector is 1.0 T, and ions having a speed of 4.0×10^6 m·s^{-1} pass through undeflected.

a) What should be the electric field between the plates P and P′?

b) If the separation of the plates is 0.5 cm, what is the potential difference between plates?

30–15 The electric field between the plates of the velocity selector in a Bainbridge mass spectrometer is 1.20×10^5 V·m^{-1}, and the magnetic field in both regions is 0.6 T. A stream of singly charged neon moves in a circular path of 0.728-m radius in the magnetic field. Determine the mass number of the neon isotope.

Section 30–7 Magnetic Force on a Conductor

30–16 An electromagnet produces a magnetic field of 1.2 T in a cylindrical region of radius 5 cm between its poles. A wire carrying a current of 20 A passes through this region, intersecting the axis of the cylinder and perpendicular to it. What force is exerted on the wire?

30–17 A horizontal rod 0.2 m long is mounted on a balance and carries a current. In the vicinity of the rod is a uniform horizontal magnetic field of magnitude 0.05 T, perpendicular to the rod. The magnetic force on the rod is measured by the balance and is found to be 0.24 N. What is the current?

30–18 In Exercise 30–17, suppose the magnetic field is horizontal but makes an angle of 30° with the rod. What is the current in the rod?

30–19 A wire along the x-axis carries a current of 5 A in the positive direction. Calculate the force (expressed in terms of unit vectors) on a 1 cm section of the wire exerted by the following magnetic fields:

a) $B = -(0.6 \text{ T})j$　　b) $B = +(0.5 \text{ T})k$　　c) $B = -(0.3 \text{ T})i$

d) $B = +(0.2 \text{ T})i - (0.3 \text{ T})k$　　e) $B = +(0.9 \text{ T})j - (0.4 \text{ T})k$

Section 30–8 Force and Torque on a Current Loop

30–20 What is the maximum torque on a rectangular coil 5 cm × 12 cm and of 600 turns when carrying a current of 1×10^{-5} A in a uniform field of magnitude 0.1 T?

30–21 A circular coil of wire 8 cm in diameter has 12 turns and carries a current of 5 A. The coil is in a region where the magnetic field is 0.6 T.

a) What is the maximum torque on the coil?

b) In what position would the torque be one-half as great as in (a)?

30–22 The plane of a rectangular loop of wire 5 cm × 8 cm is parallel to a magnetic field of magnitude 0.15 T.

a) The loop carries a current of 10 A. What torque acts on it?

b) What is the magnetic moment of the loop?

c) What is the maximum torque that can be obtained with the same total length of wire carrying the same current in this magnetic field?

30–23 A circular coil of area A and N turns is free to rotate about a diameter that coincides with the x-axis. Current I is circulating in the coil. There is a uniform magnetic field B in the positive y-direction. Calculate the magnitude and direction of the torque $\boldsymbol{\Gamma}$ and the value of the potential energy U, as given in Eq. (30–23), when the coil is oriented as shown in parts (a)–(d) of Fig. 30–22.

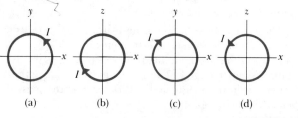

(a)　　　(b)　　　(c)　　　(d)

FIGURE 30–22

30–24 The coil of a pivoted-coil galvanometer has 50 turns and encloses an area of 6 cm². The magnetic field in the region in which the coil swings is 0.01 T and is radial. The torsional constant of the hairsprings is

$$k' = 1.0 \times 10^{-8} \text{ N·m·(degree)}^{-1}.$$

Find the angular deflection of the coil for a current of 1 mA.

Section 30–9 The Direct-Current Motor

30–25 In a shunt-wound dc motor (Fig. 30–23), the resistance R_f of the field coils is 150 Ω, and the resistance R_r of the rotor is 2 Ω. When a difference of potential of 120 V is applied to the brushes, and the motor is running at full speed delivering mechanical power, the current supplied to it is 4.5 A.

a) What is the current in the field coils?

b) What is the current in the rotor?

c) What is the induced emf developed by the motor?

d) How much mechanical power is developed by this motor?

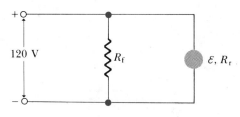

FIGURE 30–23

30–26 A shunt-wound dc motor (Fig. 30–23) operates from a 120-V dc power line. The resistance of the field windings, R_f, is 240 Ω. The resistance of the rotor, R_r, is 3 Ω. When the motor is running, the rotor develops an emf $\mathcal{E}$. The motor draws a current of 4.5 A from the line. Compute

a) the field current,

b) the rotor current,

c) the emf $\mathcal{E}$,

d) the rate of development of heat in the field windings,

e) the rate of development of heat in the rotor,

f) the power input to the motor,

g) the efficiency of the motor,

if friction losses amount to 50 W.

Section 30–10 The Hall Effect

30–27 Figure 30–24 shows a portion of a silver ribbon with $z_1 = 2$ cm and $y_1 = 1$ mm, carrying a current of 200 A in the positive x-direction. The ribbon lies in a uniform magnetic field, in the y-direction, of magnitude 1.5 T. If there are 7.4×10^{28} free electrons per m³, find

a) the drift velocity of the electrons in the x-direction;

b) the magnitude and direction of the electric field in the z-direction due to the Hall effect;

c) the Hall emf.

30–28 Let Fig. 30–24 represent a strip of copper of the same dimensions as those of the silver ribbon of the preceding exercise. When the magnetic field is 5 T and the current is 100 A, the Hall emf is found to be 45.4 μV. What is the density of free electrons in the copper?

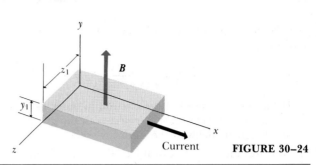

FIGURE 30–24

PROBLEMS

30–29 A particle with initial velocity $v_0 = (4 \times 10^3 \text{m·s}^{-1})i$ enters a region of uniform electric and magnetic fields. The magnetic field in the region is $B = -(0.3 \text{ T})j$. Calculate the magnitude and direction of the electric field in the region if the particle is to pass through undeflected, for a particle of charge

a) $+0.4 \times 10^{-8}$ C   b) -0.4×10^{-8} C

Neglect the weight of the particle.

30–30 Estimate the effect of the earth's magnetic field on the electron beam in a TV picture tube. Suppose the accelerating voltage is 20,000 V; calculate the approximate deflection of the beam over a distance of 0.4 m from the electron gun to the screen, under the action of a transverse field of magnitude 5.0×10^{-5} T (comparable to the magnitude of the earth's field), assuming there are no other deflecting fields. Is this deflection significant?

30–31 A particle carries a charge of 4×10^{-9} C. When it moves with a velocity v_1 of 3×10^4 m·s^{-1} at 45° above the x-axis in the xy-plane, a uniform magnetic field exerts a force F_1 along the negative z-axis. When the particle moves with a velocity of v_2 of 2×10^4 m·s^{-1} along the z-axis, there is a force F_2 of 4×10^{-5} N exerted on it along the x-axis. What are the magnitude and direction of the magnetic field? (See Fig. 30–25.)

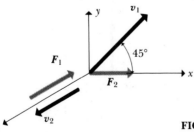

FIGURE 30–25

30–32 A particle having a charge $q = 2 \times 10^{-6}$ C is traveling with a velocity $v = (1 \times 10^3 \text{ m·s}^{-1})j$. The particle experiences a force $F = 2 \times 10^{-4}(3i - 4k)$N due to a magnetic field B.

a) Determine F, the magnitude of F.

b) Determine B_x, B_y, and B_z, or at least as many of the three components as possible, from the information given.

c) If it is given in addition that the magnitude of the magnetic field is 0.5 T, determine the remaining components of B.

30–33 An electron and an alpha particle (a doubly ionized helium atom) both move in circular paths in a magnetic field with the same tangential speed. Compare the number of revolutions they make per second. The mass of the alpha particle is 6.64×10^{-27} kg.

30–34 A particle of mass m and charge q moves with velocity v in a magnetic field B. The velocity of the particle is perpendicular to the field, and the particle moves in a circle whose radius is given by Eq. (30–9).

a) Calculate the period of the motion, the time it takes for the particle to complete one revolution.

b) From your answer in (a) obtain the frequency $f = 1/\tau$ of the circular motion. This frequency is called the cyclotron frequency.

30–35 Suppose the electric field between the plates P and P' in Fig. 30–12 is 1.5×10^4 V·m^{-1}, and the magnetic field is 0.5 T. If the source contains the three isotopes of magnesium, ^{24}Mg, ^{25}Mg, and ^{26}Mg, and the ions are singly charged, find the distance between the lines formed by the three isotopes on the photographic plate. Assume the atomic masses of the isotopes are equal to their mass numbers.

30–36 An electron is moving in a circular path of radius $r = 4$ cm in the space between two concentric cylinders. The inner cylinder is a positively charged wire of radius $a = 1$ mm and the outer cylinder is a negatively charged cylinder of radius $b = 5$ cm. The potential difference between the inner and outer cylinders is $V_{ab} = 100$ V, with the wire being at the higher potential. See Fig. 30–26. The electric field E in the region between the cylinders is

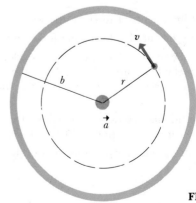

FIGURE 30–26

radially outward and was shown in Problem 26–38 to have the magnitude

$$E = \frac{V_{ab}}{r \ln(b/a)}.$$

a) Determine the speed of the electron in order for it to maintain its circular orbit. Neglect both the gravitational and magnetic fields of the earth.

b) Now include the effect of the earth's magnetic field. If the axis of symmetry of the cylinders is positioned parallel to the magnetic field of the earth, at what speed must the electron move in order to maintain the same circular orbit? Assume that the magnetic field of the earth has magnitude 1.0×10^{-4} T, and that its direction is out of the plane of the paper in Fig. 30–26.

c) Redo the calculation of part (b) for the case in which the magnetic field is in the opposite direction to that of part (b).

30–37 A particle of positive charge q and mass $m = 1 \times 10^{-15}$ kg is traveling through a region containing a uniform magnetic field $\boldsymbol{B} = -(0.1\text{ T})\boldsymbol{k}$. At a particular instant of time, the velocity of the particle is given by

$$\boldsymbol{v} = (4\boldsymbol{i} - 3\boldsymbol{j} + 12\boldsymbol{k}) \times 10^6\text{ m·s}^{-1}$$

and the force $\boldsymbol{F}$ on the particle has a magnitude of 2 N.

a) Determine the charge q.

b) Determine the acceleration $\boldsymbol{a}$ of the particle.

c) Explain why the path of the particle is a helix and determine the radius of curvature R of the circular component of the helical path.

d) Determine the cyclotron frequency of the particle. (See Problem 30–34.)

e) Although helical motion is not periodic in the full sense of the word, it is periodic with respect to the motion projected onto the xy-plane. If the coordinates of the particle at $t = 0$ are $(x, y, z) = (R, 0, 0)$, determine the coordinates of the particle at a time $t = 2\tau$, where τ is the "period" of the motion in the xy-plane.

30–38 A wire 0.5 m long lies along the y-axis and carries a current of 10 A in the $+y$-direction. The magnetic field is uniform and has components

$$B_x = 0.3\text{ T}, \qquad B_y = -1.2\text{ T}, \qquad \text{and } B_z = 0.5\text{ T}.$$

a) Find the components of force on the wire.

b) What is the magnitude of the total force on the wire?

30–39 The cube in Fig. 30–27, 0.5 m on a side, is in a uniform magnetic field of 0.6 T, parallel to the x-axis. The wire $abcdef$ carries a current of 4 A in the direction indicated.

a) Determine the magnitude and direction of the force acting on each of the segments ab, bc, cd, de, and ef.

b) What are the magnitude and direction of the total force on the wire?

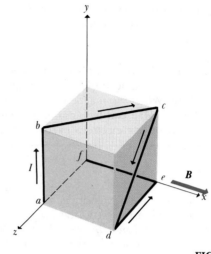

FIGURE 30–27

30–40 The rectangular loop in Fig. 30–28 is pivoted about the y-axis and carries a current of 10 A in the direction indicated.

a) If the loop is in a uniform magnetic field of magnitude 0.2 T, parallel to the x-axis, find the torque required to hold the loop in the position shown.

b) Same as (a), except that the field is parallel to the z-axis.

c) For each of the above magnetic fields, what torque would be required if the loop were pivoted about an axis through its center, parallel to the y-axis?

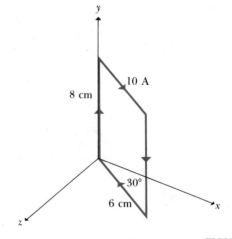

FIGURE 30–28

30–41 The rectangular loop of wire in Fig. 30–29 has a mass of 0.1 g per centimeter of length and is pivoted about side ab as a frictionless axis. The current in the wire is 10 A in the direction shown. Find the magnitude and sense of the magnetic field, parallel to the y-axis, that will cause the loop to swing up until its plane makes an angle of 30° with the yz-plane.

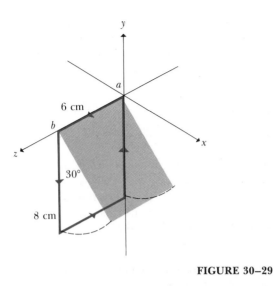

FIGURE 30–29

30–42 The neutron is a particle with zero charge but with a nonzero magnetic moment of magnitude $m = 9.66 \times 10^{-27}$ J·T^{-1}. If the neutron is considered to be a fundamental entity with no internal structure, the two properties listed above seem to be contradictory. According to present theory in particle physics, a neutron is composed of three more fundamental particles called "quarks." (See the discussion of quarks in Section 44–10.) In this model the neutron consists of an "up quark" having a charge of $+2e/3$ and two "down quarks" each having a charge of $-e/3$. The combination of the three quarks produces a net charge of $2e/3 - e/3 - e/3 = 0$, as required, and if the quarks are in motion they could also produce a nonzero magnetic moment. As a very simple model, suppose the up quark is moving in a counterclockwise circular path and the down quarks are moving in a clockwise path, all of radius r and all with the same speed v. (See Fig. 30–30).

a) Obtain an expression for the current due to the circulation of the up quark.

b) Obtain an expression for the magnitude m_u of the magnetic moment due to the circulating up quark.

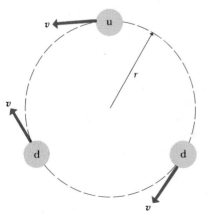

FIGURE 30–30

c) Obtain an expression for the magnitude of the magnetic moment of the three-quark system. (Be careful to use the correct magnetic moment directions.)

d) With what speed v must the quarks move if this model is to reproduce the magnetic moment of the neutron? For the radius of the orbits use $r = 1.2 \times 10^{-15}$ m, the radius of the neutron.

30–43 A wire of length 35 cm and mass $m = 9.79 \times 10^{-5}$ kg is bent into the shape of an inverted U such that the horizontal part is of length $l = 25$ cm. The bent ends of the wire, which are each 5 cm long, are completely immersed in two pools of mercury, and the entire structure is in a region containing a magnetic field of magnitude 0.01 T and a direction into the page. (See Fig. 30–31.) An electrical connection from the mercury pools is made through wires that are connected to a 1.5-V battery and a switch S. Switch S is closed, and the wire jumps 1.00 m into the air, measured from its initial position.

a) Determine the speed v of the wire as it leaves the mercury.

b) Assuming that the current I through the wire was constant from the time the switch was closed until the wire left the mercury, determine I.

c) Neglecting the resistance of the mercury and the circuit wires, determine the resistance of the moving wire.

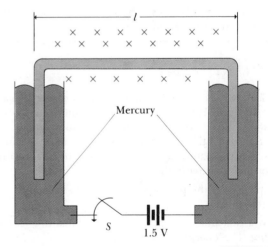

FIGURE 30–31

30–44 We can derive Eq. (30–25) explicitly without great difficulty. Consider a wire ring in the xy-plane, with its center at the origin. The ring carries a counterclockwise current I (Fig. 30–32). Let the magnetic field $\boldsymbol{B}$ be in the x-direction, $\boldsymbol{B} = B_x\boldsymbol{i}$. (The result is easily extended to $\boldsymbol{B}$ in an arbitrary direction.)

a) In Fig. 30–32 show that the element

$$d\boldsymbol{l} = R \, d\theta(-\boldsymbol{i} \sin \theta + \boldsymbol{j} \cos \theta),$$

and find $d\boldsymbol{F} = I \, d\boldsymbol{l} \times \boldsymbol{B}.$

b) Integrate $d\boldsymbol{F}$ around the loop to show that the net force is zero.

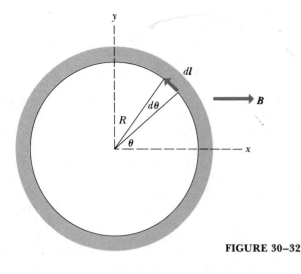

FIGURE 30–32

c) From part (a), find $d\mathbf{\Gamma} = \mathbf{r} \times d\mathbf{F}$, where

$$\mathbf{r} = R(\mathbf{i} \cos \theta + \mathbf{j} \sin \theta).$$

(Note that $d\mathbf{l}$ is perpendicular to $\mathbf{r}$.)

d) Integrate $d\mathbf{\Gamma}$ over the loop to find the total torque $\mathbf{\Gamma}$ on the loop. Show that the result can be written as

$$\mathbf{\Gamma} = I A \mathbf{n} \times \mathbf{B}.$$

(*Note.* $\int \cos^2 x \, dx = \frac{1}{2}x + \frac{1}{4} \sin 2x$, $\int \sin^2 x \, dx = \frac{1}{2}x - \frac{1}{4} \sin 2x$, and $\int \sin x \cos x \, dx = \frac{1}{2} \sin^2 x$.)

30–45 A circular loop of wire of area 10 cm² carries a current of 10 A. The loop lies in the xy-plane. As viewed in toward the origin along the z-axis, the current is circulating counterclockwise. The torque produced by an external magnetic field $\mathbf{B}$ is given by $\mathbf{\Gamma} = (-6 \times 10^{-3}\mathbf{i} + 8 \times 10^{-3}\mathbf{j})\,\text{N·m}$, and for this orientation of the loop the magnetic potential energy $U = -\mathbf{m} \cdot \mathbf{B}$ is negative. The magnitude of the magnetic field is 2.6 T.

a) Determine the magnetic moment of the current loop.

b) Determine the components B_x, B_y, and B_z of $\mathbf{B}$.

30–46 A circular ring of area 10 cm² and negligible mass is carrying a current of 50 A. The ring is free to rotate about a diameter. The ring, initially at rest, is immersed in a region of magnetic field where $\mathbf{B}$ is given by

$$\mathbf{B} = (3\mathbf{i} - 4\mathbf{j} - 12\mathbf{k}) \times 10^{-2}\,\text{T}.$$

The ring is positioned initially such that its magnetic moment $\mathbf{m}_i$ is given by

$$\mathbf{m}_i = m(-0.6\mathbf{i} + 0.8\mathbf{j}),$$

where m is the (positive) magnitude of the magnetic moment. The ring is released and turns through an angle of 90°, at which point its magnetic moment is given by

$$\mathbf{m}_f = -m\mathbf{k}.$$

a) Determine the decrease in potential energy of the ring.

b) If the moment of inertia of the ring about a diameter is 1.70×10^{-6} kg·m², determine the angular velocity of the ring as it passes through the second position.

CHALLENGE PROBLEMS

30–47 A particle of mass m and charge $+q$ starts from rest at the origin in Fig. 30–33. There is a uniform electric field $\mathbf{E}$ in the positive y-direction and a uniform magnetic field $\mathbf{B}$ directed toward the reader. It is shown in more advanced books that the path is a *cycloid* whose radius of curvature at the top points is twice the y-coordinate at that level.

a) Explain why the path has this general shape and why it is repetitive.

b) Prove that the speed at any point is equal to $\sqrt{2qEy/m}$. (*Hint:* Use energy conservation.)

c) Applying Newton's second law at the top point and taking as given that the radius of curvature here equals $2y$, prove that the speed at this point is $2E/B$.

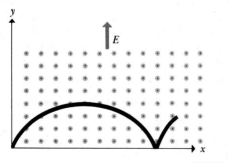

FIGURE 30–33

30–48 Two positive ions having the same charge q but different masses, m_1 and m_2, are accelerated horizontally from rest through a potential difference V. They then enter a region where there is a uniform magnetic field $\mathbf{B}$ normal to the plane of the trajectory.

a) Show that if the beam entered the magnetic field along the x-axis, the value of the y-coordinate for each ion at any time t is approximately

$$y = Bx^2 \left(\frac{q}{8mV} \right)^{1/2},$$

provided y remains much smaller than x.

b) Can this arrangement be used for isotope separation?

30–49 A particle of charge $q = 4 \times 10^{-6}$ C and mass $m = 1 \times 10^{-11}$ kg is initially traveling in the y-direction with a speed $v_0 = 2 \times 10^5$ m·s⁻¹. It then enters a region containing a uniform magnetic field, directed away from the reader and perpendicular to the page in Fig. 30–34. The magnitude of the field is 0.5 T. The region extends a distance of 25 cm along the initial direction of travel; 75 cm from the point of entry into the magnetic field region is a wall. The length of the field-free region is thus 50 cm. When the charged particle enters the magnetic field, it will follow a curved path whose radius of curvature is R. It then leaves the magnetic field after a time t_1, having been

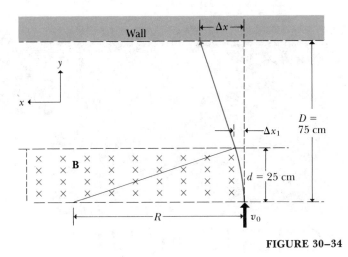

FIGURE 30–34

deflected a distance Δx_1. The particle then travels in the field-free region and strikes the wall after undergoing a total deflection Δx.

a) Determine the radius R of the curved part of the path.

b) Determine t_1, the time the particle spends in the magnetic field.

c) Determine Δx_1, the horizontal deflection at the point of exit from the field.

d) Determine Δx, the total horizontal deflection.

30–50 Magnetic forces acting on conducting fluids provide a convenient means of pumping these fluids. This problem deals with such an electromagnetic pump. A horizontal tube of rectangular cross section (height h, width w) is placed at right angles to a uniform magnetic field of magnitude B, so that a length l is in the field. (See Fig. 30–35.) The tube is filled with liquid sodium and an electric current of density J is maintained in the third mutually perpendicular direction.

a) Show that the difference of pressure between a point in the liquid on a vertical plane through ab (Fig. 30–35) and a point in the liquid on another vertical plane through cd, under conditions in which the liquid is prevented from flowing, is $\Delta p = JlB$.

b) What current density would be needed to provide a pressure difference of 1 atm between these two points if $B = 1$ T and $l = 0.1$ m?

30–51 A wire of length 60 cm, mass $m = 1.27 \times 10^{-5}$ kg, and resistance $R = 2.01\ \Omega$ is bent into the shape of an inverted U such that the horizontal part is of length $l = 40$ cm. The bent ends of the wire, which are each 10 cm long, are completely immersed in two pools of mercury, and the entire structure is in a region containing a magnetic field of magnitude $B = 4$ T and directed into the page in Fig. 30–36. (See Problem 30–43.) An electrical connection from the mercury pools is made through wires connected to a capacitor whose capacitance C is 20 μF and then to a switch S. With a potential difference of 1000 V across the capacitor, switch S is closed and the resulting upward magnetic force causes the wire to jump into the air.

a) Derive an expression which relates the speed with which the wire leaves the mercury pool to the charge Δq that passes through the wire. You may assume that the magnetic force on the wire is much greater than the weight.

b) Show that

$$md = lBq_0(\Delta t - RC\,[1 - e^{-\Delta t/RC}]),$$

where q_0 is the initial charge on the capacitor, d is the length of the bend in the wire at each end and is 10 cm, and Δt is the time it takes for the wire to leave the pool of mercury, that is, the time to travel the first 10 cm. It should be assumed that all resistances can be neglected except that of the moving wire.

c) Solve for a numerical value for Δt. (*Hint:* You will not be able to solve for Δt analytically but must use a numerical method such as iteration. Rearrange the equation with $\Delta t/(RC)$ by itself on the left side. Use a trial guess for $\Delta t/(RC)$ (make an astute guess) in the right-hand side of the equation. Compute a revised value for $\Delta t/(RC)$ and use this new value to recompute still another value. Continue to do this until a self-consistent value of $\Delta t/(RC)$, to within three significant figures, is found.)

d) Determine the total charge Δq that passed through the wire.

e) Determine the speed v of the wire as it leaves the mercury.

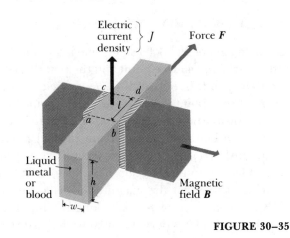

FIGURE 30–35

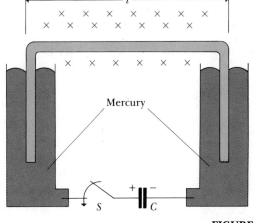

FIGURE 30–36

31

SOURCES OF MAGNETIC FIELD

IN CHAPTER 30 WE STUDIED ONE ASPECT OF THE MAGNETIC INTERACTION of moving charges: the *forces* exerted on moving charges and on currents in conductors when a magnetic field is present. We did not worry about how the magnetic field got there, but simply took its existence as a given fact. Now we are ready to return to the other aspect of this interaction, the principles that describe how magnetic fields are *produced* by moving charges and by currents.

The most basic relationship is the magnetic field produced by a single moving point charge. This relationship can be used to derive an equation for the field produced by a small segment of a current-carrying conductor. From this equation, called the law of Biot and Savart, we derive relationships for the magnetic fields produced by several specific shapes of conductor, including long, straight wires; circular loops; and solenoids. Ampere's law, the magnetic analog of Gauss's law in electrostatics, forms a useful alternative formulation of the relation of magnetic fields to their sources. Finally, we study the role of displacement current, which we encountered in Section 29–5, as a source of magnetic field. This relationship will be one of the key elements in our study of electromagnetic waves in Chapter 35.

31–1 MAGNETIC FIELD OF A MOVING CHARGE

A moving charge creates a magnetic field.

We begin with the basics, the magnetic field of a single moving point charge q. We call the location of the charge the **source point** and the point P where we want to find the field the **field point.** In our study of *electric* fields in Chapter 25, we found that the E field of a point charge q, at a field point located a distance r from the charge, is proportional to q and to $1/r^2$, and that its direction (for positive q) is along the line from source point to field point. The corresponding relationship for the *magnetic* field B of a point charge q moving with velocity v has some similarities to this relationship and some interesting differences. First, experiments show that the magnitude B is again proportional to q and to $1/r^2$. But the *direction* of B is *not* along the line from the source point to the field point; instead it is perpendicular to the plane containing this line and the particle's velocity vector v, as shown in Fig. 31–1. Furthermore, the field magnitude is proportional to the sine of the angle θ between these two directions. Thus the magnitude of the magnetic field at point P is given by

The magnetic field of a moving charge shows inverse-square behavior.

$$B = k' \frac{qv \sin \theta}{r^2}, \qquad (31\text{–}1)$$

where k' is a proportionality constant.

We can incorporate both the magnitude and direction of $\boldsymbol{B}$ into a single vector equation using the vector product. First we introduce a unit vector $\hat{r} = r/r$, pointing in the direction from charge q to point P, that is, from the source point to the field point. Then the $\boldsymbol{B}$ field of a moving point charge is

$$\boldsymbol{B} = k' \frac{qv \times \hat{r}}{r^2}. \qquad (31\text{–}2)$$

Figure 31–1 shows the magnetic field $\boldsymbol{B}$ at several points in the vicinity of the charge. At all points along a line through the charge parallel to the velocity v, the field is zero, because $\sin \theta = 0$ at all such points. At any distance r from q, $\boldsymbol{B}$ has its greatest magnitude at points lying in the plane through the charge perpendicular to v, because at all such points $\theta = 90°$ and $\sin \theta = 1$. The charge also produces an *electric* field in its vicinity; the electric-field vectors are not shown in the figure.

$\boldsymbol{B}$ is perpendicular to both v and r.

Recall that the field lines for the *electric* field of a point charge radiate outward from the charge. The *magnetic*-field lines are completely different in character; review of the above discussion shows that for a point charge moving with velocity v, the magnetic-field lines are *circles* with centers along the line of v, lying in planes perpendicular to this line. The directions of these lines are given by a right-hand rule: Grasp the velocity vector v with your right hand, so that your right thumb points in the direction of v; your fingers then curl around the line of v in the same sense as the magnetic-field lines.

As we discussed in Section 30–2, the units of B are

The tesla: the SI unit of magnetic field

$$1 \text{ T} = 1 \text{ N·s·C}^{-1}\text{·m}^{-1} = 1 \text{ N·A}^{-1}\text{·m}^{-1}.$$

Using this with Eq. (31–1) or (31–2) and solving for k', we find that the units

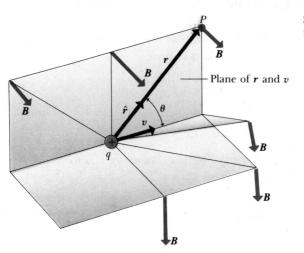

31–1 Magnetic field vectors due to a moving positive point charge q.

of the constant k' are

$$1 \text{ N·s}^2\text{·C}^{-2} = 1 \text{ N·A}^{-2} = 1 \text{ Wb·A}^{-1}\text{·m}^{-1}$$
$$= 1 \text{ T·A}^{-1}\text{·m}.$$

An arbitrary constant in magnetic-field relations

In SI units, the numerical value of k' is arbitrarily assigned to be exactly 10^{-7}. Thus

$$k' = 10^{-7} \text{ N·s}^2\text{·C}^{-2} = 10^{-7} \text{ N·A}^{-2}$$
$$= 10^{-7} \text{ Wb·A}^{-1}\text{·m}^{-1} = 10^{-7} \text{ T·A}^{-1}\text{·m} \qquad \text{(exactly)}.$$

In electrostatics we found it convenient to express electric-field relations not in terms of the constant k in Eq. (24–1) but in terms of ϵ_0, where $k = 1/4\pi\epsilon_0$. Similarly, in magnetic-field relations it is convenient to introduce a constant μ_0, defined by the relation

$$k' = \frac{\mu_0}{4\pi}.$$

Thus

$$\mu_0 = 4\pi \times 10^{-7} \text{ Wb·A}^{-1}\text{·m}^{-1} = 4\pi \times 10^{-7} \text{ T·A}^{-1}\text{·m}.$$

Magnetic field of a moving charge can be expressed as a vector product.

In terms of μ_0, Eqs. (31–1) and (31–2) become

$$B = \frac{\mu_0}{4\pi} \frac{qv \sin\theta}{r^2}, \tag{31–3}$$

$$\boldsymbol{B} = \frac{\mu_0}{4\pi} \frac{qv \times \hat{r}}{r^2}. \tag{31–4}$$

Interaction force between two moving charges

We can now write the expression for the magnetic *force* between two point charges, both of which are in motion relative to an observer. A charge q', moving with velocity v' in a magnetic field $\boldsymbol{B}$, experiences a force

$$\boldsymbol{F} = q'v' \times \boldsymbol{B},$$

and if the field $\boldsymbol{B}$ is set up by a charge q moving with velocity v, then

$$\boldsymbol{F} = \frac{\mu_0}{4\pi} \frac{qq'v' \times (v \times \hat{r})}{r^2}. \tag{31–5}$$

This equation is analogous to Coulomb's law for the *electrical* force between the charges. In Eq. (31–5), $\boldsymbol{F}$ is the force exerted on q' by q, and $\hat{r}$ is in the direction from q toward q'.

Recall from Section 24–5 that the electrical constant k has the value 8.98755×10^9 N·m^2·C^{-2}. The ratio k/k' is therefore

$$\frac{k}{k'} = \frac{8.98755 \times 10^9 \text{ N·m}^2\text{·C}^{-2}}{10^{-7} \text{ N·s}^2\text{·C}^{-2}}$$
$$= 8.98755 \times 10^{16} \text{ m}^2\text{·s}^{-2},$$

which is equal to the square of the speed of light, c. We invite you to verify that the corresponding relation between ϵ_0 and μ_0 is

$$\epsilon_0\mu_0 = \frac{1}{c^2}. \tag{31–6}$$

Thus there is a close relationship between these constants and the nature of light. We will return to this matter in Chapter 35.

The arbitrary constant isn't really so arbitrary.

What if we have several point charges? For electric fields, we could take the vector sum of the *E* fields of the individual charges, and experiments show that this also works for magnetic fields. That is, the magnetic field also obeys the **superposition principle:** *The total magnetic field caused by several moving charges is the vector sum of the fields caused by the individual charges.* We can use this fact to calculate the field caused by a current in a circuit, in terms of the fields caused by individual segments of conductor. We develop this relationship in the next section.

Superposition principle for magnetic fields: finding the total field caused by several moving charges

31–2 MAGNETIC FIELD OF A CURRENT ELEMENT

The magnetic field produced at any field point P by the current in a conductor is the vector sum of the fields due to all the moving charges in the conductor; this statement is the *principle of superposition.* We can think of the conductor as divided into short segments of length dl; a typical segment is shown in Fig. 31–2. We can use the relations of Section 31–1 to find the magnetic field caused by the current in this segment.

The volume of the segment is $A\,dl$, where A is the cross-sectional area. Suppose we have n charges q per unit volume; then the total moving charge dQ in the segment is

The moving charge in an element of a current-carrying conductor

$$dQ = nqA\,dl.$$

The moving charges are therefore equivalent to a single charge dQ, traveling with a velocity equal to the *drift* velocity v. (Fields due to the *random* velocities of the carriers will, on average, cancel out at every point.) From Eq. (31–3), the magnitude of the resulting field dB at any point is

$$dB = \frac{\mu_0}{4\pi}\frac{dQ\,v\sin\theta}{r^2} = \frac{\mu_0}{4\pi}\frac{nqvA\,dl\sin\theta}{r^2}.$$

But $(nqvA)$ equals the current I in the element, so

The law of Biot and Savart: magnetic field caused by an element of a current-carrying conductor

$$dB = \frac{\mu_0}{4\pi}\frac{I\,dl\sin\theta}{r^2}, \tag{31–7}$$

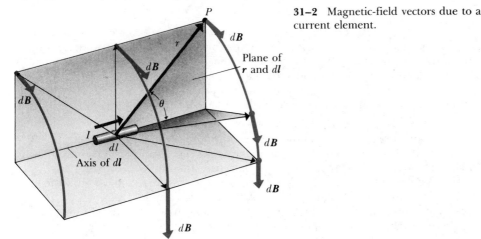

31–2 Magnetic-field vectors due to a current element.

or in vector form,

$$d\boldsymbol{B} = \frac{\mu_0}{4\pi} \frac{I\, d\boldsymbol{l} \times \hat{\boldsymbol{r}}}{r^2}, \tag{31-8}$$

Integrating to find the total magnetic field caused by a current-carrying conductor

where $d\boldsymbol{l}$ is a vector of length dl, in the same direction as the current in the conductor, and $\hat{\boldsymbol{r}}$ is a unit vector in the direction from the current element to point P. As shown in Fig. 31–2, the field vectors $d\boldsymbol{B}$ are exactly like those set up by a positive charge dQ moving in the direction of the drift velocity $\boldsymbol{v}$.

Equations (31–7) and (31–8) are called the **law of Biot and Savart.** To find the total magnetic field $\boldsymbol{B}$ any point in space due to the current in a complete circuit, we have to integrate one of these expressions; symbolically,

$$\boldsymbol{B} = \frac{\mu_0}{4\pi} \int \frac{I\, d\boldsymbol{l} \times \hat{\boldsymbol{r}}}{r^2}. \tag{31-9}$$

We will learn how to do this in specific cases in the next several sections.

The magnetic-field lines due to an element of a conductor are concentric circles.

The field lines of the magnetic field produced by a current element $I\, d\boldsymbol{l}$ look just like those of a single moving point charge; they are circles in planes perpendicular to $d\boldsymbol{l}$, centered on the line of $d\boldsymbol{l}$. Their directions are given by the same right-hand rule we introduced for point charges in Section 31–1. A few field lines are shown in Fig. 31–2.

An individual element of a conductor cannot be isolated.

To be completely honest, we must point out that it is impossible to verify Eq. (31–8) directly because we can never experiment with an isolated segment of a current-carrying circuit. The only thing we can measure experimentally is the *total* $\boldsymbol{B}$ for a complete circuit. But we can still verify Eq. (31–8) indirectly by calculating $\boldsymbol{B}$ for various current configurations and comparing the results with experimental measurements.

If matter is present in the space around a conductor, the field at a field point P in its vicinity will have an additional contribution resulting from the *magnetization* of the material. We return to this point in Section 31–8. Unless the material is iron or some other ferromagnetic material, however, the additional field is so small that it is usually negligible.

The next three sections include several applications of the law of Biot and Savart to particular conductor shapes. The following comments will help you understand these examples and also help you with additional problems.

PROBLEM-SOLVING STRATEGY: *Magnetic-field calculations*

1. Be careful about the directions of vector quantities. The current element $d\boldsymbol{l}$ always points in the direction of the current. The unit vector $\hat{\boldsymbol{r}}$ is always directed *from* the current element *toward* the point P at which the field is to be determined, that is, from the source point toward the field point.

2. In some problems the $d\boldsymbol{B}$'s at point P have the same direction for all the current elements. In that case, the magnitude of the total $\boldsymbol{B}$ field is the sum of the magnitudes of the $d\boldsymbol{B}$'s. But often the $d\boldsymbol{B}$'s have different directions for different current elements; then you have to set up a coordinate system and represent each $d\boldsymbol{B}$ in terms of its components. The integral for

the total $\boldsymbol{B}$ is then expressed in terms of an integral for each component. Sometimes you can see from the symmetry of the situation that one component integrates to zero. Always be on the lookout for ways to use symmetry to simplify the problem.

3. Look for ways to use the superposition principle. If you know the fields produced by certain simple conductor shapes, and if you encounter a complex shape that can be represented as a combination of simple shapes, then it is easy to use superposition to find the field of the complex shape. Examples: a rectangular loop, or a semicircle with straight-line segments on both sides.

31–3 MAGNETIC FIELD OF A STRAIGHT CONDUCTOR

Our first application of the law of Biot and Savart is finding the magnetic field of a conductor of length $2a$ carrying a current I, at a point on its perpendicular bisector, at a distance x from the conductor. The situation is shown in Fig. 31–3; the geometry is reminiscent of the electric-field problem of Example 25–8 (Section 25–2), where we found the electric field caused by a line of charge of length $2a$. Although the *shapes* of the sources are the same in these two problems, the behavior of the magnetic field is completely different from that of the electric field.

We begin by identifying an element of the conductor, having length $dl = dy$. To use Eq. (31–7), we note that $r = \sqrt{x^2 + y^2}$ and $\sin\theta = \sin(\pi - \theta) = x/\sqrt{x^2 + y^2}$. The *direction* of $d\boldsymbol{B}$ is perpendicular to the plane of the figure, into the plane, and in this case the directions of the $d\boldsymbol{B}$'s from all elements of the conductor are the same. Thus in integrating over the conductor, we can just add their magnitudes.

Putting the pieces together, we find that the magnitude B of the total $\boldsymbol{B}$ is

$$B = \frac{\mu_0 I}{4\pi} \int_{-a}^{a} \frac{x\, dy}{(x^2 + y^2)^{3/2}}. \tag{31–10}$$

We can integrate this by trigonometric substitution or by using an integral table. We challenge you to fill in the details and show that the final result is

$$B = \frac{\mu_0 I}{4\pi} \frac{2a}{x\sqrt{x^2 + a^2}}. \tag{31–11}$$

An alternative route to Eq. (31–11) uses Eqs. (31–8) and (31–9) with the appropriate vectors. We have $d\boldsymbol{l} = dy\,\boldsymbol{j}$, where $\boldsymbol{j}$ is the usual unit vector in the y-direction. Also,

$$\hat{\boldsymbol{r}} = \frac{\boldsymbol{r}}{r} = \frac{x\boldsymbol{i} - y\boldsymbol{j}}{\sqrt{x^2 + y^2}}. \tag{31–12}$$

Putting the pieces together, we find

$$\begin{aligned}
\boldsymbol{B} &= \frac{\mu_0 I}{4\pi} \int_{-a}^{a} \frac{(dy\boldsymbol{j}) \times (x\boldsymbol{i} - y\boldsymbol{j})}{(x^2 + y^2)^{3/2}} \\
&= \frac{\mu_0 I}{4\pi} \int_{-a}^{a} \frac{(-\boldsymbol{k})x\, dy}{(x^2 + y^2)^{3/2}} \\
&= \frac{\mu_0 I}{4\pi} \frac{2a(-\boldsymbol{k})}{x\sqrt{x^2 + a^2}}.
\end{aligned} \tag{31–13}$$

In a right-handed axis system, the z-axis points out of the plane of Fig. 31–3. Thus the negative sign in Eq. (31–13) shows that the direction of $\boldsymbol{B}$ at point P is *into* the plane of the figure. At points in the xy-plane to the left of the y-axis, of course, the direction of $\boldsymbol{B}$ is opposite, *out of* the plane.

When the length $(2a)$ of the conductor is very large compared to its distance x from point P, we can consider it to be infinitely long. When a is much larger than x, $\sqrt{x^2 + a^2}$ is approximately equal to a, and so in the limit as $a \to \infty$, Eq. (31–11) becomes

$$B = \frac{\mu_0 I}{2\pi x}. \tag{31–14}$$

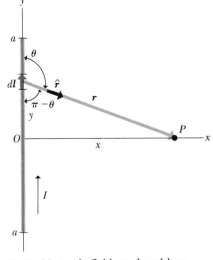

31–3 Magnetic field produced by a straight conductor of length $2a$. The $\boldsymbol{B}$ field at point P is directed into the plane of the figure.

Finding the magnetic field of a long, straight conductor by integrating over its length

Using unit vectors to evaluate vector products in magnetic-field calculations

For a long, straight wire, B is inversely proportional to the distance from the wire.

Because the physical situation has axial symmetry about the y-axis, B has the same *magnitude* at all points on a circle centered on the conductor and lying in a plane perpendicular to the conductor, and its *direction* is everywhere tangent to such a circle. Thus, at all points on a circle of radius r around the conductor, the magnitude B is given by

$$B = \frac{\mu_0 I}{2\pi r} \qquad \text{(long, straight wire)}. \qquad (31\text{--}15)$$

EXAMPLE 31–1 A long, straight conductor carries a current of 100 A. At what distance from the conductor is the magnetic field caused by the current equal in magnitude to the earth's magnetic field in Pittsburgh (about 0.5×10^{-4} T)?

SOLUTION We use Eq. (31–15). Everything except r is known, so we solve for r and insert the appropriate numbers

$$r = \frac{\mu_0 I}{2\pi B} = \frac{(4\pi \times 10^{-7}\ \text{T·m·A}^{-1})(100\ \text{A})}{(2\pi)(0.5 \times 10^{-4}\ \text{T})} = 0.4\ \text{m}.$$

At smaller distances, the field becomes stronger; for example, when $r = 0.2$ m, $B = 1.0 \times 10^{-4}$ T, and so on.

Magnetic-field lines for a long, straight wire are circles centered on the wire.

Part of the magnetic field around a long, straight conductor is shown in Fig. 31–4. The shape of the magnetic-field lines in this situation is completely different from that of the electric-field lines in the analogous electrical situation. Electric-field lines radiate outward from the charges that are their sources (or inward for negative charges). By contrast, magnetic-field lines *encircle* the current that acts as their source. Electric-field lines begin and end at charges, while experiments have shown that magnetic-field lines *never* have endpoints, irrespective of the shape of the conductor that sets up the field. If lines *did* begin or end at a point, this point would correspond to a "magnetic charge," or a single magnetic pole. There is no experimental evidence that such entities exist.

31–4 Magnetic field around a long, straight conductor. The field lines are circles, with directions determined by the right-hand rule.

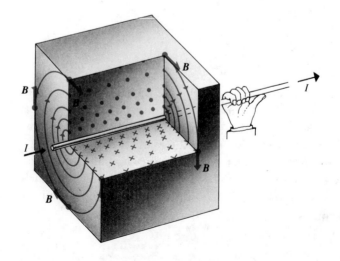

Thus if we construct an imaginary closed surface in a magnetic field, no field line can start or end inside this surface. The number of lines emerging from the surface must equal the number entering it. We have shown that the number of lines crossing a surface is proportional to the flux Φ across the surface. Hence in a magnetic field, the flux across any *closed* surface is zero, or

The total magnetic flux out of a closed surface is zero.

$$\oint B_\perp \, dA = \oint \boldsymbol{B} \cdot dA = 0. \qquad (31\text{--}16)$$

This equation should be compared with Gauss's law for electrostatic fields, in which the surface integral of $\boldsymbol{E}$ over a closed surface equals $1/\epsilon_0$ times the enclosed charge. Thus Eq. (31–16) is an alternative statement of the fact that there is no such thing as "magnetic charge" to act as a source of $\boldsymbol{B}$. The sources of $\boldsymbol{B}$ are moving electric charges, as outlined above.

31–4 FORCE BETWEEN PARALLEL CONDUCTORS

We are now ready to examine the interaction force between two long current-carrying conductors. This problem comes up in a variety of practical problems, and it also has fundamental significance in connection with the definition of the ampere. Figure 31–5 shows segments of two long, straight, parallel conductors separated by a distance r and carrying currents I and I', respectively, in the same direction. Since each conductor lies in the magnetic field set up by the other, each experiences a force. The diagram shows some of the field lines set up by the current in the lower conductor.

From Eq. (31–15), the magnitude of the $\boldsymbol{B}$-vector at the upper conductor is

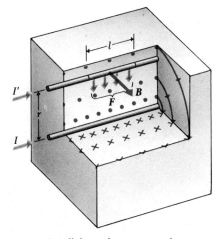

$$B = \frac{\mu_0 I}{2\pi r}.$$

From Eq. (30–14), the force on a length l of the upper conductor is

$$F = I'lB = \frac{\mu_0 II' l}{2\pi r},$$

and the force *per unit length* is therefore

31–5 Parallel conductors carrying currents in the same direction attract each other.

$$\frac{F}{l} = \frac{\mu_0 II'}{2\pi r}. \qquad (31\text{--}17)$$

The right-hand rule shows that the direction of the force on the upper conductor is downward. An equal and opposite force per unit length acts on the lower conductor, as may be seen by considering the field set up by the upper conductor. Hence the conductors *attract* each other.

Two parallel currents attract each other.

If the direction of either current is reversed, the forces reverse also. Parallel conductors carrying currents in *opposite* directions *repel* each other.

The fact that two straight, parallel conductors exert forces of attraction or repulsion on one another is the basis of the official SI definition of the ampere. The ampere is defined as follows:

Defining the ampere in terms of forces between parallel conductors

> One **ampere** is that unvarying current which, if present in each of two parallel conductors of infinite length and one meter apart in empty space, causes each conductor to experience a force of exactly 2×10^{-7} newton per meter of length.

It follows from this definition and Eq. (31–17) that, by definition, the constant μ_0 is exactly $4\pi \times 10^{-7}$ N·A^{-2}, as we stated in Section 31–1.

From the definition above, the ampere can be established, in principle, with the help of a meter stick and a spring balance. For high-precision standardization of the ampere, coils of wire are used instead of straight wires, and their separation is made only a few centimeters. The complete instrument, which is capable of measuring currents with a high degree of precision, is called a *current balance*.

Note that this definition of the ampere is an *operational definition;* it gives us an actual experimental procedure for measuring current and defining the unit of current. With this definition, we can now return to electrostatics and define the *coulomb* as *the quantity of charge that in one second crosses a section of a circuit in which there is a constant current of one ampere.* Finally, we have to regard the constant $k = 1/4\pi\epsilon_0$ in Coulomb's law as an empirical constant, determined by experiment. As we mentioned in Section 31–1, however, ϵ_0 and μ_0 are related to the speed of light. This fact provides the most precise determination of ϵ_0. We return to this relationship in Chapter 35.

Mutual forces of attraction exist not only between *wires* carrying currents in the same direction, but also between the longitudinal elements of a single current-carrying conductor. If the conductor is a liquid or an ionized gas (a plasma), these forces result in a constriction of the conductor, as if its surface were acted on by an external, inward pressure. The constriction of the conductor is called the *pinch effect*. The high temperature produced by the pinch effect in a plasma has been used in one technique to bring about nuclear fusion.

31–5 MAGNETIC FIELD OF A CIRCULAR LOOP

In many practical devices where a current is used to establish a magnetic field, such as an electromagnet or transformer, the wire carrying the current is wound into a *coil*, often consisting of many circular loops. Thus it is useful to derive an expression for the magnetic field produced by a single circular conducting loop carrying a current. Figure 31–6 shows a circular conductor of radius a, carrying a current I. The current is led into and out of the loop

Margin notes:

Defining the coulomb in terms of forces between parallel conductors

Forces between current elements in an ionized gas

Finding the magnetic field of a circular loop

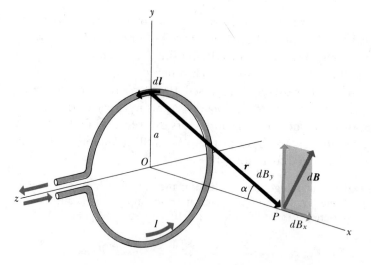

31–6 Magnetic field of a circular loop. The segment dl causes the field $d\mathbf{B}$, lying in the xy-plane. Other dl's have different components perpendicular to the x-axis; these add to zero, while the x-components combine to give the total $\mathbf{B}$ field at point P.

through two long, straight wires side by side; the currents in these straight wires are in opposite directions, and their magnetic fields cancel each other.

We can use the law of Biot and Savart, Eq. (31–7) or (31–8), to find the magnetic field at a point P along a line through the center of the loop perpendicular to its plane, at a distance x from the center. As the figure shows, dl and r are always perpendicular, and the direction of the field $d\boldsymbol{B}$ caused by element dl lies in the xy-plane. Also, $r = \sqrt{x^2 + y^2}$. Thus for the element dl we find

$$dB = \frac{\mu_0 I}{4\pi} \frac{dl}{x^2 + a^2}. \tag{31–18}$$

In terms of components,

$$dB_x = dB \sin \alpha = \frac{\mu_0 I}{4\pi} \frac{dl}{x^2 + a^2} \cdot \frac{a}{(x^2 + a^2)^{1/2}}, \tag{31–19}$$

$$dB_y = dB \cos \alpha = \frac{\mu_0 I}{4\pi} \frac{dl}{x^2 + a^2} \cdot \frac{x}{(x^2 + a^2)^{1/2}}. \tag{31–20}$$

Because of symmetry about the x-axis, the components of $\boldsymbol{B}$ perpendicular to this axis must sum to zero. To see why this must be, consider an element dl on the opposite side of the loop, having direction opposite to that of the dl shown. This element gives a contribution to the field having the same x-component as given by Eq. (31–19) but an *opposite* y-component. Similarly, the components of $\boldsymbol{B}$ perpendicular to the x-axis set up by pairs of diametrically opposite elements cancel, leaving only the x-components. To obtain the total x-component, we integrate Eq. (31–19). Everything is constant in this integration except dl, and the integral of dl for the entire loop is simply the circumference of the circle, $2\pi a$. Thus we obtain

Using symmetry to prove that the component perpendicular to the axis is zero

$$B_x = \int \frac{\mu_0 I}{4\pi} \cdot \frac{a\,dl}{(x^2 + a^2)^{3/2}} = \frac{\mu_0 I a}{4\pi(x^2 + a^2)^{3/2}} \int dl,$$

or, since $\int dl = 2\pi a$,

$$B_x = \frac{\mu_0 I a^2}{2(x^2 + a^2)^{3/2}} \qquad \text{(circular loop).} \tag{31–21}$$

We can use unit vectors to obtain Eqs. (31–19) and (31–20) somewhat more compactly. We write $dl = dl\,\boldsymbol{k}$, $r = x\boldsymbol{i} - a\boldsymbol{j}$, and

Using unit vectors in the circular-loop calculation

$$\frac{dl \times \hat{r}}{r^2} = \frac{dl \times r}{r^3} = \frac{(dl\,\boldsymbol{k}) \times (x\boldsymbol{i} - a\boldsymbol{j})}{(x^2 + a^2)^{3/2}}$$

$$= \frac{x\,dl\,\boldsymbol{j} + a\,dl\,\boldsymbol{i}}{(x^2 + a^2)^{3/2}}.$$

As before, the components perpendicular to the x-axis sum to zero, and only the term containing $\boldsymbol{i}$ survives. Using the above expression in Eq. (31–9) and again using the fact that $\int dl = 2\pi a$, we obtain

$$\boldsymbol{B} = \int \frac{\mu_0 I}{4\pi} \frac{a\,dl\,\boldsymbol{i}}{(x^2 + a^2)^{3/2}} = \frac{\mu_0 I a^2 \boldsymbol{i}}{2(x^2 + a^2)^{3/2}}, \tag{31–22}$$

which is equivalent to Eq. (31–21). At the center of the loop, $x = 0$, and Eq. (31–21) reduces to

$$B_x = \frac{\mu_0 I}{2a} \qquad \text{(center of circular loop).} \tag{31–23}$$

If the coil has N turns, the magnetic field is multiplied by N.

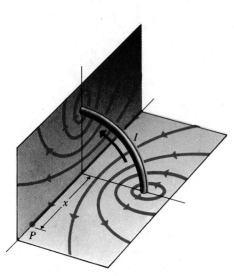

31–7 Field lines surrounding a circular loop.

Line integral of magnetic field around a closed path

The integration path need not be formed by a physical object.

If, instead of a single loop as in Fig. 31–6, we have a coil of N closely spaced loops, all having the same radius, each loop contributes equally to the field, and Eq. (31–23) becomes

$$B_x = \frac{\mu_0 NI}{2a} \quad (N \text{ circular loops}). \quad (31-24)$$

Similarly, we can adapt Eqs. (31–21) and (31–22) to the case of N loops by inserting a factor N.

Some of the magnetic-field lines surrounding a circular loop and lying in planes through the axis are shown in Fig. 31–7. Again we see that the field lines encircle the conductor, and that their directions are given by the right-hand rule, as for a long, straight conductor. The field lines for the circular loop are *not* circles, but they are closed curves that link the conductor.

31–6 AMPERE'S LAW

Ampere's law provides an alternative formulation of the relation between a magnetic field and its sources. In some problems it is more convenient than the law of Biot and Savart. The role of Ampere's law for magnetic fields is analogous to that of Gauss's law for electric fields, which we studied in Chapter 25. Recall that Gauss's law relates the integral of the normal component of electric field over a closed surface to the total electric charge enclosed by that surface. Ampere's law relates the *tangential* component of *magnetic* field at points on a closed *curve* to the current passing through the *area* bounded by that curve.

Ampere's law is formulated in terms of the line integral of $\boldsymbol{B}$ around a closed path, denoted by

$$\oint \boldsymbol{B} \cdot d\boldsymbol{l}.$$

This is the same sort of integral we used to define work in Chapter 7 and electric potential in Chapter 26. We divide the path into infinitesimal segments $d\boldsymbol{l}$ and calculate the scalar product of $\boldsymbol{B}$ and $d\boldsymbol{l}$ for each segment. In general $\boldsymbol{B}$ varies from point to point, and we must use the $\boldsymbol{B}$ at the location of each $d\boldsymbol{l}$. An alternative notation is $\oint B_\parallel \, dl$, where $B_\parallel$ is the component of $\boldsymbol{B}$ parallel to $d\boldsymbol{l}$ at each point. The circle on the integral sign indicates that this integral is always computed for a *closed* path whose beginning and end points are the same. As with Gauss's law, the path used for Ampere's law need not be the outline of any actual physical object; usually it is a purely geometric curve that we construct in order to apply Ampere's law to a specific situation.

To introduce the basic idea, we consider a long, straight conductor carrying a current I. Imagine a circle of radius r in a plane perpendicular to the conductor and centered on it. From Eq. (31–15), the magnetic field has magnitude $\mu_0 I/2\pi r$ at every point on the circle, and it is tangent to the circle at each point. Thus for every point on the circle $B_\parallel = B = \mu_0 I/2\pi r$. The line integral of $\boldsymbol{B}$ around the circle is

$$\oint \boldsymbol{B} \cdot d\boldsymbol{l} = \oint \frac{\mu_0 I}{2\pi r} dl = \frac{\mu_0 I}{2\pi r} \oint dl,$$

or, since $\oint dl$ is just the circumference $(2\pi r)$ of the circle,

$$\oint \boldsymbol{B} \cdot d\boldsymbol{l} = \mu_0 I. \quad (31-25)$$

That is, the line integral of B is equal to μ_0 multiplied by the current passing through the area bounded by the circle.

We can also derive this result for a more general integration path, such as the one in Fig. 31–8. At the position of the line element dl, the angle between dl and B is θ, and

$$B \cdot dl = B \, dl \cos \theta.$$

From the figure, $dl \cos \theta = r \, d\phi$, where $d\phi$ is the angle subtended by dl at the position of the conductor. Thus

$$\oint B \cdot dl = \oint \left(\frac{\mu_0 I}{2\pi r} \right)(r \, d\phi) = \frac{\mu_0 I}{2\pi} \oint d\phi.$$

But $\oint d\phi$ is just the total angle swept out by the radial line from the conductor to dl during a complete trip around the path, that is, 2π. Thus

$$\oint B \cdot dl = \mu_0 I. \qquad (31\text{–}26)$$

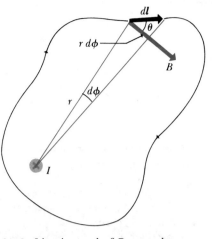

31–8 Line integral of B around a closed path linking a long, straight conductor, carrying current I *into* the plane of the figure. The conductor is seen end-on.

The line integral does not depend on the shape of the path or on the position of the wire within it. If the current in the wire is opposite to that shown, the integral has the opposite sign.

Now suppose several long, straight conductors pass through the surface bounded by the integration path. The total magnetic field at any point on the path is the vector sum of the fields produced by the individual conductors. Thus the line integral of the total B equals μ_0 times the *algebraic sum* of the currents, with currents going into the page in Fig. 31–8 considered as positive and those coming out of it considered negative. In general, the positive direction for current is determined by a right-hand rule. Considering a line perpendicular to the surface, we wrap the fingers of the right hand around this line so our fingers curl around in the same direction we plan to go around the integration path; then the thumb indicates the positive current direction.

If the integration path does not encircle the wire (or if a wire lies outside the path), the line integral of the B field of that wire is zero, because then the angle ϕ does not sweep through 2π as it did before; instead, its net change is zero. Hence if there are other conductors present that do not pass through a given path, they may contribute to the value of B at every point, but the *line integrals* of their fields are zero.

It follows that if we interpret I in Eq. (31–26) to mean the *algebraic sum* of the currents through an area bounded by a closed path, this equation implies all of the above statements. Thus the general form of Ampere's law is

> The line integral of the magnetic field B around any closed path is equal to μ_0 times the net current through any area bounded by the path.

Although we have derived Ampere's law only for the special case of the field of a number of long, straight, parallel conductors, it is true for conductors and paths of any shape. The general derivation is no different in principle from what we have presented, but it is more complicated geometrically.

Finally, we emphasize that the statement $\oint B \cdot dl = 0$ does *not* necessarily mean that $B = 0$ everywhere along the path, but only that the total current through an area bounded by the path is zero.

Ampere's law: The line integral of B around a closed path is proportional to the current through the area enclosed by the path.

Sign and direction conventions used with Ampere's law

31–7 APPLICATIONS OF AMPERE'S LAW

Ampere's law is particularly helpful when we can make use of the symmetry of a situation to evaluate the line integral of **B**. Following are several examples. Note that the problem-solving strategy we suggest is directly analogous to the strategy proposed for applications of Gauss's law in Section 25–5; we invite you to review that strategy now and compare the two methods.

PROBLEM-SOLVING STRATEGY: *Ampere's law*

1. The first step is to select the integration path you will use with Ampere's law. If you want to learn something about the magnetic field at a certain point, then the path must pass through that point.

2. The integration path doesn't have to be any actual physical boundary. Usually it is a purely geometric curve; it may be in empty space, embedded in a solid body, or some of each.

3. The integration path has to have enough *symmetry* to make evaluation of the integral possible. If the problem itself has cylindrical symmetry, the integration path will usually be a circle coaxial with the cylinder axis.

4. If **B** is tangent to the integration path and has the same magnitude B at every point, then its line integral equals B multiplied by the circumference of the path.

5. If **B** is everywhere perpendicular to the path, for all or some portion of the path, that portion of the path makes no contribution to the integral.

6. In the integral $\oint \boldsymbol{B} \cdot d\boldsymbol{l}$, **B** is always the *total* magnetic field at each point on the path. In general this field is caused partly by currents enclosed by the path and partly by currents outside. Even when *no* net current is enclosed by the path, the field at points on the path need not be zero. In that case, however, $\oint \boldsymbol{B} \cdot d\boldsymbol{l}$ is always zero.

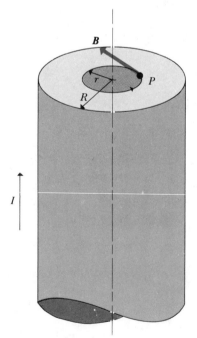

31–9 The current through the color area is $(r^2/R^2)I$. To find the magnetic field at point P we apply Ampere's law to the circle around the color area.

EXAMPLE 31–2 *Field of a long, straight conductor:* Because of the way we derived Ampere's law in Section 31–6, we could reverse the process and use it to derive Eq. (31–15) for a long, straight conductor. We take as our integration path a circle of radius r centered on the conductor and in a plane perpendicular to it. Then in Eq. (31–26), $\oint \boldsymbol{B} \cdot d\boldsymbol{l} = 2\pi r B$, and Eq. (31–15) follows immediately.

EXAMPLE 31–3 *Field inside a long, cylindrical conductor.* To find the magnetic field inside a cylindrical conductor of radius R, at a point a distance r from the axis, as in Fig. 31–9, we consider a circle of radius r. By symmetry, **B** has the same magnitude at every point on this circle and is tangent to it. Thus the line integral is simply $B(2\pi r)$. To find the current enclosed by the path, note that the current *density* J (current per unit area) is $J = I/\pi R^2$, so the current through the path is $I_r = J(\pi r^2) = Ir^2/R^2$. Finally, Ampere's law gives

$$2\pi r B = \mu_0 \, Ir^2/R^2,$$

or

$$B = \frac{\mu_0 I}{2\pi} \frac{r}{R^2}. \tag{31–27}$$

Note that at $r = R$, that is, at the surface of the conductor, Eq. (31–15) (derived for $r > R$) and Eq. (31–27) (derived for $r < R$) agree.

EXAMPLE 31–4 *Field of a solenoid:* A *solenoid* consists of a helical winding of wire around the surface of a cylindrical form, usually circular in cross section. Ordinarily the turns are so closely spaced that each one is very nearly a circular loop. There may be several layers of windings. For simplicity, we have shown a solenoid in Fig. 31–10 as having only a few turns. All turns carry the same current I, and the total B field at every point is the vector sum of the fields caused by the individual turns. The figure shows field lines in the xy- and yz-planes. Exact calculations show that for a long, closely wound solenoid, half the lines passing through a cross section at the center emerge from the ends, and half "leak out" through the windings between center and end.

If the length of the solenoid is large compared with its cross-sectional diameter, the *internal* field near its center is very nearly uniform and parallel to the axis, and the *external* field near the center is very small. The internal field at or near the center can then be found by use of Ampere's law.

We choose as our integration path the broken-line rectangle $abcd$ in Fig. 31–10. Side ab, of length l, is parallel to the axis of the solenoid. Sides bc and da are taken to be very long, so that side cd is far from the solenoid, and the field at this side is negligibly small.

By symmetry, the B field along side ab is parallel to this side and is constant, so for this side, $B_\parallel = B$ and

$$\int \boldsymbol{B} \cdot d\boldsymbol{l} = Bl.$$

Along sides bc and da, $B_\parallel = 0$ since B is perpendicular to these sides; and along side cd, $B_\parallel = 0$ also since $B = 0$. The sum around the entire closed path therefore reduces to Bl.

Let n be the number of turns *per unit length* in the windings. The number of turns in length l is then nl. Each of these turns passes once through the rectangle $abcd$ and carries a current I, where I is the current in the windings. The total current through the rectangle is then nlI and, from Ampere's law,

$$Bl = \mu_0 nlI,$$

$$B = \mu_0 nI \qquad \text{(solenoid)}. \qquad (31\text{–}28)$$

Since side ab need not lie on the axis of the solenoid, the field is uniform over the entire cross section.

A solenoid is a coil wound on a cylindrical form.

What is the magnetic field inside a solenoid?

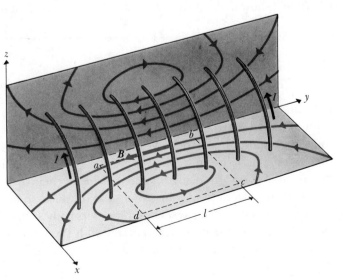

31–10 Magnetic-field lines surrounding a solenoid. The dashed rectangle $abcd$ is used to compute the flux density B in the solenoid from Ampere's law.

Toroidal solenoid: a solenoid that meets itself coming back

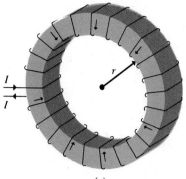

(a)

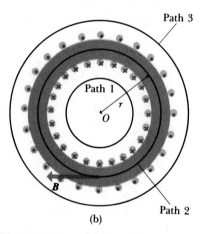

(b)

31–11 (a) A toroidal winding. (b) Closed paths (black circles) used to compute the magnetic field **B** set up by a current in a toroidal winding. The field is very nearly zero at all points except those within the space enclosed by the windings.

What if the space around a conductor contains a material?

Magnetized materials make additional contributions to magnetic field.

EXAMPLE 31–5 *Field of a toroidal solenoid:* Figure 31–11a shows a **toroidal solenoid** wound with wire carrying a current I. The black lines in Fig. 31–11b are paths to which we wish to apply Ampere's law. Consider first path 1. By symmetry, if there is any field at all in this region, it will be *tangent* to the path at all points, and $\oint \boldsymbol{B} \cdot d\boldsymbol{l}$ will equal the product of B and the circumference $l = 2\pi r$ of the path. The current through the path, however, is zero; hence, from Ampere's law, the field **B** must be zero.

Similarly, if there is any field at path 3, it will also be tangent to the path at all points. Each turn of the winding passes *twice* through the area bounded by this path, carrying equal currents in opposite directions. The *net* current through this area is therefore zero, and hence $B = 0$ at all points of the path. *The field of the toroidal solenoid is therefore confined wholly to the space enclosed by the windings.* The toroid may be thought of as a solenoid that has been bent into a circle.

Finally, consider path 2, a circle of radius r. Again by symmetry, the **B** field is tangent to the path, and $\oint \boldsymbol{B} \cdot d\boldsymbol{l}$ equals $2\pi r B$. Each turn of the winding passes *once* through the area bounded by path 2, and the total current through the area is NI, where N is the *total* number of turns in the winding. Then, from Ampere's law,

$$2\pi r B = \mu_0 NI,$$

and

$$B = \frac{\mu_0 NI}{2\pi r} \qquad \text{(toroidal solenoid).} \qquad (31\text{–}29)$$

The magnetic field is *not* uniform over a cross section of the core, because the path length l is larger at the outer side of the section than at the inner side. However, if the radial thickness of the core is small compared with the toroid radius r, the field varies only slightly across a section. In that case, considering that $2\pi r$ is the circumferential length of the toroid and that $N/2\pi r$ is the number of turns per unit length n, the field may be written as

$$B = \mu_0 nI,$$

just as at the center of the long, *straight* solenoid.

The questions derived above for the field in a closely wound straight or toroidal solenoid are strictly correct only for windings in *vacuum*. For most practical purposes, however, they can be used for windings in air or on a core of any non-ferromagnetic material. We will show in the next section how they are modified if the core is of iron.

31–8 MAGNETIC MATERIALS

In the preceding discussion of magnetic fields caused by currents, we assumed that the space surrounding the conductors contained only vacuum. If matter is present in the surrounding space, the magnetic field is changed. The atoms that make up all matter contain electrons in motion, and these electrons form microscopic current loops capable of producing magnetic fields of their own. In many materials these currents are randomly oriented so as to cause no net magnetic field. But in some materials the presence of an externally caused field can cause the loops to become oriented preferentially with the field, so that their magnetic fields *add* to the external field. The material is said to become *magnetized*.

A material showing this behavior is said to be **paramagnetic.** The result is that the magnetic field at any point in such a material is greater by a factor K_m then it would be if the material were not present. The value of K_m is different for different materials; it is called the *relative permeability* of the material. Values of K_m for common paramagnetic materials are typically 1.0001 to 1.0020.

All the equations in this chapter that relate magnetic fields to their sources can be adapted to the situation where the conductor is embedded in a paramagnetic material by simply replacing μ_0 everywhere by $K_m\mu_0$. This product is usually denoted as μ; it is called the **permeability** of the material:

$$\mu = K_m\mu_0. \tag{31–30}$$

The amount by which the relative permeability differs from unity is called the **magnetic susceptibility,** symbol χ_m:

$$\chi_m = K_m - 1. \tag{31–31}$$

Values of magnetic susceptibility for several materials are given in Table 31–1.

In some materials the total field due to the electrons in each atom sums to zero when there is no external field, and these materials have *no* net atomic current loops. But even in these materials magnetic effects are present because an external field alters the electron motions to cause additional current loops. In this case the additional field caused by these current loops is always *opposite* in direction to the external field. Such materials are said to be **diamagnetic;** they always have relative permeabilities very slightly less than unity, typically of the order of 0.99990 to 0.99999. The susceptibility of a diamagnetic material is negative, as Table 31–1 shows.

There is a third class of materials, called **ferromagnetic** materials, in which the atomic current loops tend to line up parallel to each other even when no external field is present. This cooperative phenomenon leads to a relative permeability that is *much larger* than unity, typically of the order of 1000 to 10,000. Iron, cobalt, nickel, and many alloys containing these elements are ferromagnetic.

Ferromagnetic materials differ in their behavior from paramagnetic and diamagnetic materials in two other important ways. First, the additional field caused by the microscopic current loops is not directly proportional to the external field except at relatively small field magnitudes. Thus for a ferromagnetic material, K_m is not constant. As the external field increases, a point is reached where nearly all the microscopic current loops have their axes parallel to the external field. This condition is called *saturation magnetization;* after it is reached, a further increase in the external field causes no increase in magnetization or in the additional field caused by the material.

Second, some ferromagnetic materials retain their magnetization even when there is no externally caused field at all. These materials can thus become *permanent magnets.* Many kinds of steel and many alloys, such as Alnico, are commonly used for permanent magnets. The magnetic field in such a material, when it is magnetized to near saturation, is typically of the order of 1 T.

More generally, for many ferromagnetic materials the relation of magnetization to magnetic field is different when the field is decreasing from when it is increasing. Thus the magnetization for a given field depends somewhat on the past history of the material. This phenomenon is called **hysteresis.** One consequence is the magnetization that remains when the field is reduced to

A paramagnetic material makes the field a little stronger.

A diamagnetic material makes the field a little weaker.

A ferromagnetic material makes the field much stronger.

Magnetization of ferromagnetic materials is nonlinear.

TABLE 31–1 Magnetic Susceptibilities of Paramagnetic and Diamagnetic Materials, at $T = 20°C$

Materials	$\chi_m = K_m - 1$
Paramagnetic	
Iron ammonium	
alum	66×10^{-5}
Uranium	40×10^{-5}
Platinum	26×10^{-5}
Aluminum	2.2×10^{-5}
Sodium	0.72×10^{-5}
Oxygen gas	0.19×10^{-5}
Diamagnetic	
Bismuth	-16.6×10^{-5}
Mercury	-2.9×10^{-5}
Silver	-2.6×10^{-5}
Carbon (diamond)	-2.1×10^{-5}
Lead	-1.8×10^{-5}
Rock salt	-1.4×10^{-5}
Copper	-1.0×10^{-5}

Hysteresis: The material may stay magnetized when the external magnetic field is removed.

zero, as discussed above. Magnetizing and demagnetizing a material that has hysteresis involves the dissipation of energy, and the temperature of such a material increases in this process.

Ferromagnetic materials are widely used in electromagnets, transformer cores, and motors and generators, where it is desirable to have as large a magnetic field as possible for a given current. In these applications it is usually desirable for the material *not* to have permanent magnetization; that is, for it to have as little hysteresis as possible. Soft iron is often used; it has high permeability without appreciable hysteresis or permanent magnetization.

Electrical machines often require high permeability and low hysteresis.

31–9 MAGNETIC FIELD AND DISPLACEMENT CURRENT

When we discussed Ampere's law in Section 31–6, we assumed that the current I was a *conduction* current. However, experimental evidence shows that in situations that include time-varying electric fields, *displacement current* I_D, as defined in Section 29–5, must also be included in Ampere's law. That is, in using Ampere's law in problems involving the relation of magnetic field to its sources (currents), we must treat conduction and displacement currents on an equal footing; both act as sources of magnetic field. Thus the general form of Ampere's law is

Displacement current: an additional source of magnetic field

$$\oint \boldsymbol{B} \cdot d\boldsymbol{l} = \mu_0 (I_C + I_D). \tag{31–32}$$

Here is an example of the role of displacement current in Ampere's law in a magnetic-field calculation. Figure 31–12 is a cross-sectional view of two circular conducting plates of radius R, separated by a narrow gap, in vacuum. They are, in other words, a parallel-plate capacitor. If the gap is small compared to R, the $\boldsymbol{E}$ field caused by the charge distributions on the plates is confined almost entirely to the region between the plates and is uniform, as shown. Suppose we want to calculate the $\boldsymbol{B}$ field at some point in the midplane, shown in the figure as a thin line. In principle this field can be obtained by applying the law of Biot and Savart to the conduction currents in the plates and the wires leading to them. Because of the radial currents in the circular plates, this is a fairly involved calculation. Instead, we may take advantage of the axial symmetry of the situation and use Ampere's law.

Generalizing Ampere's law to include displacement current

There is no *conduction* current in the region between plates, but there is *displacement current*. From the definition of displacement current in Section 29–5, the total displacement current out of the left plate is equal to the conduction current into the plate. Also, by symmetry, the $\boldsymbol{B}$ field lines are circles with centers on the axis. Thus, at any point on a circle perpendicular to the axis and passing through points such as a and b, the $\boldsymbol{B}$-vector has the same magnitude and is tangent to the circle. At point a, the $\boldsymbol{B}$-vector points out of the page, and at point b it points into the page, as indicated by the symbols $\odot$ and $\otimes$.

Using displacement current to find the magnetic field between the plates of a charging capacitor.

The displacement current density J_D is given by Eq. (29–20):

$$J_D = \epsilon_0 \frac{dE}{dt}. \tag{31–33}$$

At any instant E is uniform over the region between the plates, so J_D is also uniform. The *total* displacement current I_D between the plates is just J_D times the area πR^2 of the plates. As we remarked above, this must also equal the

Displacement-current density is proportional to rate of change of electric field.

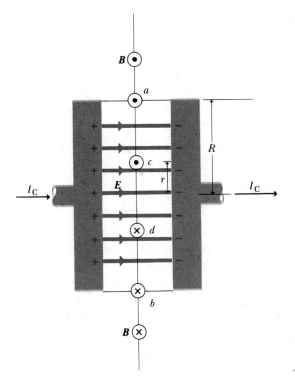

31–12 A capacitor being charged by a current I_C has a displacement current I_D between its plates equal to I_C, with displacement-current density $J_D = \epsilon_0 \, dE/dt$.

total *conduction* current I_C. Thus

$$I_D = I_C = \pi R^2 J_D, \tag{31–34}$$

$$J_D = \frac{I_C}{\pi R^2}. \tag{31–35}$$

Now we may find the magnetic field at a point in the region between plates, at a distance r from the axis. We apply Ampere's law to a circle of radius r passing through the point; such a circle passes through points c and d in Fig. 31–12. The total current enclosed by the circle is J_D times its area, or $(I_C/\pi R^2)(\pi r^2)$. The integral $\oint \boldsymbol{B} \cdot d\boldsymbol{l}$ in Ampere's law is just B times the circumference $2\pi r$ of the circle, and Ampere's law becomes

$$\oint \boldsymbol{B} \cdot d\boldsymbol{l} = 2\pi r B = \mu_0 \left(\frac{r^2}{R^2}\right) I_C,$$

or

$$B = \frac{\mu_0}{2\pi}\left(\frac{r}{R^2}\right) I_C. \tag{31–36}$$

This result shows that in the region between the plates $\boldsymbol{B}$ is zero at the axis, increasing linearly with distance from the axis. A similar calculation shows that *outside* the region between the plates, $\boldsymbol{B}$ is the same as though the wire were continuous and the plates not present at all.

This calculation bears a strong resemblance to the one in Example 31–3 (Section 31–7), where we used Ampere's law to find the magnetic field inside a cylindrical conductor. In a certain sense we can think of the space between the conducting plates as a cylindrical conductor. Of course, there is no actual charge in motion to constitute a conduction current, but there is a displacement current. Thus it should not be too surprising that Eq. (31–36) is in fact identical to Eq. (31–27).

Ampere's law with displacement current
expressed in terms of electric flux

The role of displacement current as a source of magnetic field can be expressed in a different way by using the concept of *electric flux* Ψ, which we introduced in Sections 25–4 and 29–5. As stated in Eq. (29–22), the displacement current may be expressed as

$$I_\mathrm{D} = \epsilon_0 \frac{d\Psi}{dt}.$$

Thus in empty space, where there is no conduction current ($I_\mathrm{C} = 0$), Ampere's law, Eq. (31–32), may be rewritten

$$\oint \boldsymbol{B} \cdot d\boldsymbol{l} = \mu_0 \epsilon_0 \frac{d\Psi}{dt}. \tag{31–37}$$

We will see in Chapter 32 that this equation bears a close resemblance to another relation, *Faraday's law*, which describes the *electric* field induced by a changing *magnetic* field.

The magnetic field caused by displacement current plays a vital role in electromagnetic-wave propagation.

The fact that *displacement* current acts as a source of magnetic field plays an essential role in the understanding of electromagnetic *waves*. A changing *electric* field in a region of space induces a *magnetic* field in neighboring regions, even when no conduction current and no matter are present. This relationship, first proposed by Maxwell in 1865, provides the key to a theoretical understanding of electromagnetic radiation and of light as a particular example of this radiation. We return to this topic in Chapter 35.

SUMMARY

The magnetic field $\boldsymbol{B}$ produced by a charge q moving with velocity v is described by

$$B = \frac{\mu_0}{4\pi} \frac{qv \sin \theta}{r^2}, \tag{31–3}$$

$$\boldsymbol{B} = \frac{\mu_0}{4\pi} \frac{qv \times \hat{r}}{r^2}, \tag{31–4}$$

where r is the vector from the source point (the location of q) to the field point P, and $\hat{r}$ is a unit vector in the direction of r. Magnetic fields produced by moving charges obey the superposition principle; the total $\boldsymbol{B}$ produced by several moving charges is the vector sum of the fields produced by the individual charges.

The magnetic interaction force $\boldsymbol{F}$ exerted on a charge q' moving with velocity v' by a charge q moving with velocity v is given by

$$\boldsymbol{F} = \frac{\mu_0}{4\pi} \frac{qq'v' \times (v \times \hat{r})}{r^2}, \tag{31–5}$$

where r points from q toward q'.

An element $d\boldsymbol{l}$ of a conductor carrying a current I produces a magnetic field $d\boldsymbol{B}$ described by

$$dB = \frac{\mu_0}{4\pi} \frac{I\,dl \sin \theta}{r^2}, \tag{31–7}$$

or in vector form,

$$d\boldsymbol{B} = \frac{\mu_0}{4\pi} \frac{I\,d\boldsymbol{l} \times \hat{r}}{r^2}. \tag{31–8}$$

KEY TERMS

source point
field point
superposition principle
law of Biot and Savart
ampere
Ampere's law
toroidal solenoid
paramagnetic
permeability
magnetic susceptibility
diamagnetic
ferromagnetic
hysteresis

These relations are called the law of Biot and Savart. The total field of a finite conductor is obtained by integrating these equations over the length of the conductor; this usually requires vector calculations, often best carried out by use of components.

Magnetic-field lines never have endpoints, and the surface integral of B over any closed surface is always zero. This is the magnetic analog of Gauss's law and shows that there are no isolated magnetic poles.

The interaction force, per unit length, between two long, parallel conductors carrying currents I and I' is

$$\frac{F}{l} = \frac{\mu_0 I\, I'}{2\pi r}. \qquad (31\text{--}17)$$

This relation is the basis for the operational definition of the ampere.

The magnetic field B at a distance r from a long, straight wire carrying a current I has magnitude

$$B = \frac{\mu_0 I}{2\pi r} \qquad \text{(long, straight wire).} \qquad (31\text{--}15)$$

The magnetic-field lines for a long, straight wire are circles coaxial with the wire, with directions given by the right-hand rule.

The magnetic field produced by a circular conducting loop of radius a, carrying current I, at a distance x from its center, along its axis, is given by

$$B_x = \frac{\mu_0 I a^2}{2(x^2 + a^2)^{3/2}} \qquad \text{(circular loop).} \qquad (31\text{--}21)$$

At the center of the loop, where $x = 0$, this equation reduces to

$$B_x = \frac{\mu_0 I}{2a} \qquad \text{(center of circular loop).} \qquad (31\text{--}23)$$

If there are N loops, this expression is multiplied by N.

Ampere's law states that the line integral of B around any closed path equals μ_0 times the net current through the area enclosed by the path. The positive sense of current is determined by a right-hand rule.

The magnetic field in the interior of a long, cylindrical conductor of radius R carrying current I is given by

$$B = \frac{\mu_0 I}{2\pi}\frac{r}{R^2}. \qquad (31\text{--}27)$$

The magnetic field at the center of a long solenoid with n turns per unit length, carrying current I, is

$$B = \mu_0 n I. \qquad (31\text{--}28)$$

The magnetic field inside a toroidal solenoid with N turns, carrying current I, at a distance r from its axis, is

$$B = \frac{\mu_0 N I}{2\pi r} \qquad \text{(toroidal solenoid).} \qquad (31\text{--}29)$$

When magnetic materials are present, the magnetization of the material causes an additional contribution to B. For paramagnetic and diamagnetic materials, all the relations between magnetic field and its sources are still correct if we replace μ_0 by $\mu = K_m\mu_0$, where μ is the permeability of the material and K_m is its relative permeability. The magnetic susceptibility χ_m is defined $\chi_m = K_m - 1$. Magnetic susceptibilities for paramagnetic materials are small positive quantities; those for diamagnetic materials are small negative quantities. For ferromagnetic materials K_m is much larger than unity and is not constant. Some ferromagnetic materials also can retain magnetization even after the external magnetic field is removed.

Displacement current acts as a source of magnetic field in exactly the same way as conduction current. Ampere's law is generalized to include displacement current, as follows:

$$\oint \boldsymbol{B} \cdot d\boldsymbol{l} = \mu_0(I_C + I_D). \tag{31–32}$$

The concept of displacement current as a source of magnetic field is essential to the understanding of electromagnetic waves.

QUESTIONS

31–1 Streams of charged particles emitted from the sun during unusual sunspot activity create a disturbance in the earth's magnetic field. How does this happen?

31–2 A topic of current interest in physics research is the search (thus far unsuccessful) for an isolated magnetic pole, or magnetic *monopole*. If such an entity were found, how could it be recognized? What would its properties be?

31–3 What are the relative advantages and disadvantages of Ampere's law and the law of Biot and Savart for practical calculations of magnetic fields?

31–4 A student proposed to obtain an isolated magnetic pole by taking a bar magnet (N pole at one end, S at the other) and breaking it in half. Would this work?

31–5 Pairs of conductors carrying current into and out of the power-supply components of electronic equipment are sometimes twisted together to reduce magnetic-field effects. Why does this help?

31–6 The text discusses the magnetic field of an infinitely long, straight conductor carrying a current. Of course, there is no such thing as an infinitely long *anything*. How do you decide whether a particular wire is long enough to be considered infinite?

31–7 Suppose one has three long, parallel wires, arranged so that in cross section they are at the corners of an equilateral triangle. Is there any way to arrange the currents so that all three wires attract each other? So that all three wires repel each other?

31–8 Two parallel conductors carrying current in the same direction attract each other. If they are permitted to move toward each other, the forces of attraction do work. Where does the energy come from? Does this contradict the assertion in Chapter 30 that magnetic forces on moving charges do no work?

31–9 Considering the magnetic field of a circular loop of wire, would you expect the field to be greatest at the center, or is it greater at some points in the plane of the loop but off-center?

31–10 Two concentric circular loops of wire, of different diameter, carry currents in the same direction. Describe the nature of the forces exerted on the inner loop.

31–11 A current was sent through a helical coil spring. The spring appeared to contract, as though it had been compressed. Why?

31–12 Using the fact that magnetic-field lines never have a beginning or an end, explain why it is reasonable for the field of a toroidal solenoid to be confined entirely to its interior, while a straight solenoid *must* have some field outside it.

31–13 Can one have a displacement current as well as a conduction current within a conductor?

31–14 A character in a popular comic strip has at various times proposed the possibility of "harnessing the earth's magnetic field" as a nearly inexhaustible source of energy. Comment on this concept.

31–15 Why should the permeability of a paramagnetic material be expected to decrease with increasing temperature?

31–16 In the discussion of magnetic forces on current loops in Section 30–8, it was found that no net force is exerted on a complete loop in a uniform magnetic field, only a torque. Yet magnetized materials, which contain atomic current loops, certainly *do* experience net forces in magnetic fields. How is this discrepancy resolved?

31–17 What features of atomic structure determine whether an element is diamagnetic or paramagnetic?

31–18 The magnetic susceptibility of paramagnetic materials is quite strongly temperature dependent, while that of diamagnetic materials is nearly independent of temperature. Why the difference?

EXERCISES

$$1\ T = 1\ Wb \cdot m^{-2} = 1\ N \cdot A^{-1} \cdot m^{-1},$$
$$\mu_0 = 4\pi \times 10^{-7}\ N \cdot A^{-2} = 4\pi \times 10^{-7}\ T \cdot A^{-1} \cdot m$$
$$= 4\pi \times 10^{-7}\ Wb \cdot A^{-1} \cdot m^{-1}.$$

Section 31–1 Magnetic Field of a Moving Charge

31–1 A positive point charge of magnitude $q = 5\ \mu C$ has velocity $v = (8 \times 10^6\ m \cdot s^{-1})i$. At the instant when the point charge is at the origin, what is the magnetic field vector B it produces at the following points:

a) $x = 0.5\ m$, $y = 0$, $z = 0$?

b) $x = 0$, $y = -0.5\ m$, $z = 0$?

c) $x = 0$, $y = 0$, $z = +0.5\ m$?

d) $x = 0$, $y = -0.5\ m$, $z = +0.5\ m$?

31–2 Two positive point charges q and q' are moving relative to an observer as shown in Fig. 31–13.

a) What is the direction of the force that q' exerts on q?

b) What is the direction of the force that q exerts on q'?

c) If $v = v' = 5 \times 10^6\ m \cdot s^{-1}$, what is the ratio of the magnitude of the magnetic force to that of the Coulomb force between the two charges?

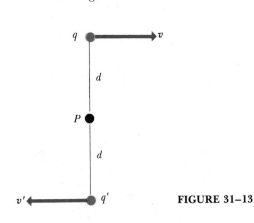

FIGURE 31–13

31–3 A pair of point charges, where $q = +5\mu C$ and $q' = -3\mu C$, are moving as shown in Fig. 31–14. At this instant, what are the magnitude and direction of the magnetic field produced at the origin (point P)? Take $v = v' = 4 \times 10^5\ m \cdot s^{-1}$.

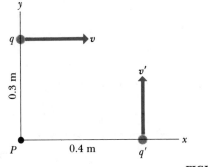

FIGURE 31–14

Section 31–2 Magnetic Field of a Current Element

31–4 A long, straight wire, carrying a current of 200 A, runs through a cubical wooden box, entering and leaving through holes in the centers of opposite faces, as in Fig. 31–15. The length of each side of the box is 20 cm. Consider an element of the wire 1 cm long at the center of the box. Compute the magnitude of the magnetic field ΔB produced by this element at points a, b, c, d, and e in Fig. 31–15. Points a, c, and d are at the centers of the faces of the cube; point b is at the midpoint of one edge; and point e is at a corner. Copy the figure and show by vectors the directions and relative magnitudes of the field vectors. (*Note.* Assume that Δl is small compared to the distances from the current element to the points where B is to be calculated.)

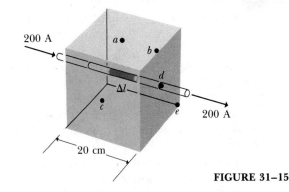

FIGURE 31–15

31–5 Calculate the magnitude and direction of the magnetic field at point P due to the current in the semicircular section of wire shown in Fig. 31–16. (Does the current in the long, straight section of the wire produce any field at P?)

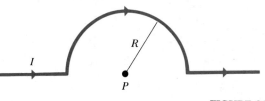

FIGURE 31–16

31–6 Calculate the magnetic field at point P of Fig. 31–17 in terms of R, I_1, and I_2. What does your expression give when $I_1 = I_2$?

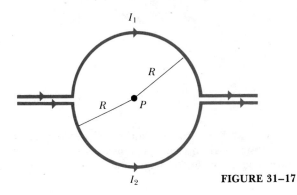

FIGURE 31–17

Section 31–3 Magnetic Field of a Straight Conductor

31–7 Two hikers are reading a compass under an overhead transmission line that is 5.0 m above the ground and carries a current of 400 A in a direction from south to north.

a) Find the magnitude and direction of the magnetic field at a point on the ground directly under the conductor.

b) One hiker suggests they walk on another 50 m to avoid inaccurate compass readings caused by the current. Considering that the magnitude of the earth's field is of the order of 0.5×10^{-4} T, is the current really a problem?

31–8 A magnetic field of magnitude 5.0×10^{-4} T is to be produced at a distance of 0.05 m from a long, straight wire.

a) What current is required to produce this field?

b) With the current found in (a), what is the magnitude of the field at a distance of 0.10 m from the wire? At 0.20 m?

31–9 A long, straight telephone cable contains six wires, each carrying a current of 0.5 A. The distances between wires can be neglected.

a) If the currents in all six wires are in the same direction, what is the magnitude of the magnetic field 0.10 m from the cable?

b) If four wires carry currents in one direction and the other two in the opposite direction, what is the field magnitude 0.10 m from the cable?

31–10 A long, straight wire carries a current of 10 A directed along the y-axis, as shown in Fig. 31–18. A uniform magnetic field $\mathbf{B}_0$ of magnitude 1×10^{-6} T is directed parallel to the x-axis. What is the resultant field at the following points:

a) $x = 0$, $z = 2$ m?

b) $x = 2$ m, $z = 0$?

c) $x = 0$, $z = -0.5$ m?

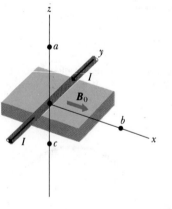

FIGURE 31–18

31–11 Two long, straight, horizontal parallel wires, one above the other, are separated by a distance $2a$. If the wires carry equal currents of magnitude I in opposite directions,

what is the field magnitude in the plane of the wires at a point

a) midway between them?

b) at a distance a above the upper wire?

If the wires carry equal currents in the same direction, what is the field magnitude in the plane of the wires at a point

c) midway between them?

d) at a distance a above the upper wire?

Section 31–4 Force between Parallel Conductors

31–12 Two long, parallel wires are separated by a distance of 0.4 m, as shown in Fig. 31–19. The currents I_1 and I_2 have the directions shown.

a) Calculate the magnitude and direction of the force exerted by each wire on a 1-m length of the other.

b) If each current is doubled, so that I_1 becomes 10 A and I_2 becomes 4 A, what is now the magnitude of the force each wire exerts on a 1-m length of the other?

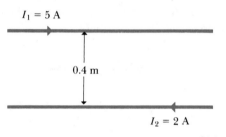

FIGURE 31–19

31–13 Two long, parallel wires are separated by a distance of 0.1 m. The force per unit length that each wire exerts on the other is 3.0×10^{-5} N·m^{-1}, and the wires repel each other. If the current in one wire is 2 A,

a) what is the current in the second wire?

b) are the two currents in the same or in opposite directions?

31–14 Three parallel wires each carry current I in the directions shown in Fig. 31–20. If the separation between adjacent wires is d, calculate the magnitude and direction of the resultant magnetic force per unit length on each wire.

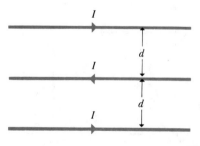

FIGURE 31–20

31–15 A long, horizontal wire AB rests on the surface of a table. (See Fig. 31–21.) Another wire CD vertically above the first is 1.0 m long and free to slide up and down on the two vertical metal guides C and D. The two wires are connected through the sliding contacts and carry a current of 50 A. The mass per unit length of the wire CD is 0.005 kg·m^{-1}. To what equilibrium height h will the wire CD rise, assuming the magnetic force on it to be due wholly to the current in the wire AB?

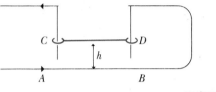

FIGURE 31–21

Section 31–5 Magnetic Field of a Circular Loop

31–16 Refer to Fig. 31–6. Sketch a graph of the magnitude of the B field on the axis of the loop, from $x = -3a$ to $x = +3a$.

31–17 A circular coil of radius 5 cm has 200 turns and carries a current 0.2 A. What is the magnetic field

a) at the center of the coil?

b) at a point on the axis of the coil 10 cm from its center?

Section 31–7 Applications of Ampere's Law

31–18 A solenoid of length 20 cm and radius 3 cm is closely wound with 200 turns of wire. The current in the winding is 5 A. Compute the magnetic field at a point near the center of the solenoid.

31–19 A wooden ring of mean diameter 0.10 m is wound with a closely spaced toroidal winding of 500 turns. Compute the field at a point on the mean circumference of the ring when the current in the windings is 0.3 A.

31–20 A solenoid is to be designed to produce a magnetic field of 0.1 T at its center. The radius is to be 5 cm and the length 50 cm, and the available wire can carry a maximum current of 10 A.

a) What is the minimum number of turns per unit length the solenoid must have?

b) What total length of wire is required?

31–21 A coaxial cable consists of a solid conductor of radius R_1, supported by insulating disks on the axis of a tube of inner radius R_2 and outer radius R_3. If the central conductor and tube carry equal currents I in opposite directions, find the magnetic field

a) at points outside the central solid conductor but inside the tube;

b) at points outside the tube.

31–22 Repeat Exercise 31–21 for the case where the current in the central solid conductor is I_1 and in the tube I_2. Also take these currents to be in the same direction rather than in opposite directions.

Section 31–8 Magnetic Materials

31–23 Experimental measurements of the magnetic susceptibility of iron ammonium alum are given in Table 31–2. Make a graph of $1/\chi$ against Kelvin temperature; what conclusion can you draw?

TABLE 31–2

T,°C	χ
−258.15	129 × 10^{-4}
−173	19.4 × 10^{-4}
−73	9.7 × 10^{-4}
27	6.5 × 10^{-4}

31–24 A toroidal solenoid having 500 turns of wire and a mean circumferential length of 0.50 m carries a current of 0.3 A. The relative permeability of the core is 600.

a) What is the magnetic field in the core?

b) What part of the magnetic field is due to atomic currents?

31–25 A toroidal solenoid with 400 turns is wound on a ring having a cross section of 2 cm^2 and a mean circumference of 30 cm. Find the current in the winding that is required to set up a magnetic field of 0.1 T in the ring

a) if the ring is of annealed iron ($K_m = 1400$);

b) if the ring is of silicon steel ($K_m = 5200$).

Section 31–9 Magnetic Field and Displacement Current

31–26 A copper wire with circular cross section of area 4 mm^2 carries a current of 50 A. The resistivity of the material is 2.0×10^{-8} Ω·m.

a) What is the electric field in the material?

b) If the current is changing at the rate of 5000 A·s^{-1}, at what rate is the electric field in the material changing?

c) What is the displacement-current density in the material in (b)?

d) If the current is changing as in (b), what is the magnetic field 5 cm from the wire? Note that both the conduction current and the displacement current must be included in the calculation of B. Is the contribution from the displacement current significant?

PROBLEMS

31-27 A long, straight wire carries a current of 1.5 A. An electron travels with a speed of $5 \times 10^4 \text{m·s}^{-1}$ parallel to the wire, 0.10 m from it, and in the same direction as the current. What are the magnitude and direction of the force that the magnetic field of the current exerts on the moving electron?

31-28 Figure 31-22 is an end view of two long, parallel wires perpendicular to the xy-plane, each carrying a current I but in opposite directions.

a) Copy the diagram, and show by vectors the $\mathbf{B}$ field of each wire and the resultant $\mathbf{B}$ field at point P.

b) Derive the expression for the magnitude of $\mathbf{B}$ at any point on the x-axis, in terms of the coordinate x of the point.

c) Construct a graph of the magnitude of $\mathbf{B}$ at points on the x-axis.

d) At what value of x is $\mathbf{B}$ a maximum?

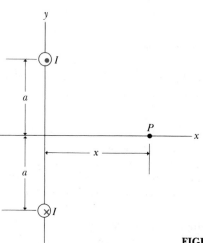

FIGURE 31-22

31-29 Same as Problem 31-28, except that the current in both wires is *away* from the reader.

31-30 In Fig. 31-22, suppose a third long, straight wire, parallel to the other two, passes through point P, and that each wire carries a current $I = 20$ A. Let $a = 0.30$ m and $x = 0.40$ m. Find the magnitude and direction of the force per unit length on the third wire

a) if the current in it is away from the reader;

b) if current is toward the reader

31-31 Two long, straight, parallel wires are 1.0 m apart, as in Fig. 31-23. The upper wire carries a current I_1 of 6 A into the plane of the paper.

a) What must be the magnitude and direction of the current I_2 for the resultant field at point P to be zero?

b) What is then the resultant field at Q?

c) At S?

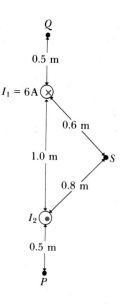

FIGURE 31-23

31-32 Two long, parallel wires are hung by cords 4 cm long from a common axis, as shown in Fig. 31-24. The wires have a mass per unit length of 0.050 kg·m^{-1} and carry the same current in opposite directions. What is the current if the cords hang at an angle of 6° with the vertical?

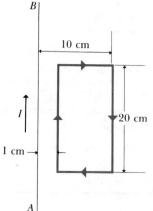

FIGURE 31-24

31-33 The long, straight wire AB in Fig. 31-25 carries a current of 20 A. The rectangular loop whose long edges are parallel to the wire carries a current of 10 A. Find the magnitude and direction of the resultant force exerted on the loop by the magnetic field of the wire.

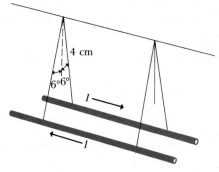

FIGURE 31-25

31–34 The long, straight wire AB in Fig. 31–26 carries a constant current I.

a) What is the magnetic field at the shaded area a perpendicular distance x from the wire?

b) What is the magnetic flux $d\Phi$ through the shaded area?

c) What is the total flux through the rectangle $CDEF$?

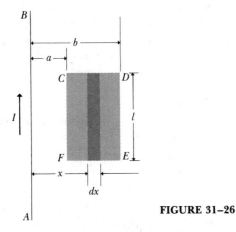

FIGURE 31–26

31–35 In Fig. 31–27, calculate the magnitude and direction of the magnetic field produced at point P by current I in the rectangular wire loop. (Point P is at the center of the rectangle.) (*Hint:* The gap on the left-hand side where the wires enter and leave the rectangle is so small that this side of the rectangle can be taken to be a continuous wire of length b.)

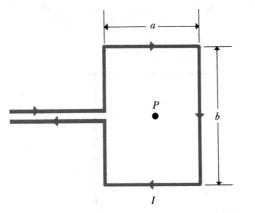

FIGURE 31–27

31–36 The wire semicircles in Fig. 31–28 have radii a and b. Calculate the resultant magnetic field (magnitude and direction) at point P.

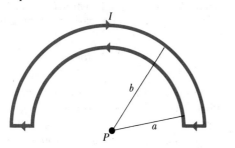

FIGURE 31–28

31–37 Figure 31–29 is a sectional view of two circular coils of radius a, each wound with N turns of wire carrying a current I, circulating in the same direction in both coils. The coils are separated by a distance a equal to their radii.

a) Derive the expression for the magnetic field at a point on the axis a distance x to the right of point P that is midway between the coils.

b) From (a) obtain an expression for the magnetic field at point P.

c) Calculate the magnitude of $\boldsymbol{B}$ at P if $N = 100$ turns, $I = 5$ A, and $a = 0.30$ m.

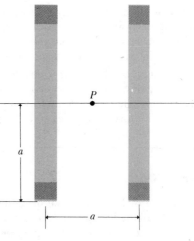

FIGURE 31–29

31–38 A conductor is made in the form of a hollow cylinder with inner and outer radii a and b, respectively. It carries a current I, uniformly distributed over the cross section. Derive expressions for the magnetic field in the regions

a) $r < a$,

b) $a < r < b$,

c) $r > b$.

31–39 The current in the windings on a toroid is 2.0 A. There are 400 turns, and the mean circumferential length is 0.40 m. The magnetic field is found to be 1.0 T. Calculate

a) the relative permeability;

b) the magnetic susceptibility

31–40 Consider the coaxial cable of Exercise 31–21. Assume the currents to be uniformly distributed over the cross sections of the solid conductor and the tube. Calculate the magnetic field at

a) $r < R_1$ (inside the central conductor; does your answer give the expected result for $r = R_1$?);

b) $R_2 < r < R_3$ (inside the outer tube; does your answer give the expected results for $r = R_3$ and for $r = R_2$?).

31–41 A long, straight wire with circular cross section of radius R carries current I. Assume that the current density is not constant across the cross section of the wire but rather varies as $J = \alpha r$, where α is a constant.

a) By the requirement that J integrated over the cross section of the wire gives the total current I, calculate the constant α in terms of I and R.

b) Use Ampere's law to calculate the magnetic field $B(r)$ for
   i) $r \leq R$;   ii) $r \geq R$.

31–42 A capacitor has two parallel plates of area A separated by a distance d. The capacitor is given an initial charge Q but, because the damp air between the plates is slightly conductive, the charge slowly leaks through the air. The charge initially changes at a rate dQ/dt.

a) In terms of dQ/dt, what is the initial rate of change of the electric field between the plates?

b) Show that the displacement-current density has the same magnitude as the conduction-current density, but the opposite direction. Hence show that the magnetic field in the air between the plates is exactly zero at all times.

31–43 A long, straight, solid cylinder, oriented with its axis in the z-direction, carries a current whose current density is $\boldsymbol{J}$. The current density, although symmetrical about the cylinder axis, is not uniform but varies according to the relation

$$J = \frac{2I_0}{\pi a^2}[1 - (r/a)^2]\boldsymbol{k} \quad \text{for} \quad r \leq a,$$
$$J = 0 \quad\quad\quad\quad\quad\quad\quad \text{for} \quad r \geq a,$$

where a is the radius of the cylinder, r is the radial distance from the cylinder axis, and I_0 is a constant having units of amperes.

a) Show that I_0 is the total current passing through the entire cross section of the wire.

b) Using Ampere's law, derive an expression for the magnetic field $\boldsymbol{B}$ in the region $r \geq a$.

c) Obtain an expression for the current I contained in a circular cross section of radius $r \leq a$ and centered at the cylinder axis.

d) Using Ampere's law, derive an expression for the magnetic field $\boldsymbol{B}$ in the region $r \leq a$.

31–44 Integrate B_x as given in Eq. (31–21) from $-\infty$ to $+\infty$. That is, calculate

$$\int_{-\infty}^{+\infty} B_x \, dx.$$

Explain the significance of your result.

CHALLENGE PROBLEMS

31–45 A thin disk of dielectric material of radius a has a total charge $+Q$ distributed uniformly over its surface. It rotates n times per second about an axis perpendicular to the surface of the disk and passing through its center. Find the magnetic field at the center of the disk. (*Hint:* Divide the disk into concentric rings of infinitesimal width.)

31–46 Wire in the shape of a semicircle of radius a is oriented in the zy- plane, with its center of curvature at the origin. (See Fig. 31–30.) If the current in the wire is I, calculate the magnetic-field components produced at point P, a distance x out along the x-axis. (*Note:* Do not forget the contribution from the straight wire at the bottom of the

semicircle that runs from $z = -a$ to $z = +a$. The fields of the two antiparallel currents at $z > a$ cancel.)

31–47 The wire in Fig. 31–31 is infinitely long and carries current I. Calculate the magnitude and direction of the magnetic field this current produces at point P.

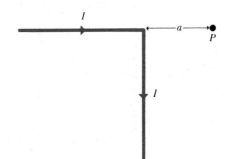

FIGURE 31–31

31–48 A long, straight, solid cylinder, oriented with its axis in the z-direction, carries a current whose current density is $\boldsymbol{J}$. The current density, although symmetrical about the cylinder axis, is not uniform but varies according to the relation

$$J = (b/r)e^{(r-a)/\delta}\boldsymbol{k} \quad \text{for} \quad r \leq a,$$
$$J = 0 \quad\quad\quad\quad\quad\quad \text{for} \quad r \geq a,$$

where a is the radius of the cylinder and is equal to 10 cm, r is the radial distance from the cylinder axis, b is a constant equal to 400 A·m^{-1}, and δ is a constant equal to 5 cm.

a) Let I_0 be the total current passing through the entire cross section of the wire. Obtain an expression for I_0 in

FIGURE 31–30

terms of b, δ, and a. Evaluate your expression to obtain a numerical value for I_0.

b) Using Ampere's law derive an expression for the magnetic field B in the region $r \geq a$. Express your answer in terms of I_0 rather than b.

c) Obtain an expression for the current I contained in a circular cross section of radius $r \leq a$ and centered at the cylinder axis. Express your answer in terms of I_0 rather than b.

d) Using Ampere's law, derive an expression for the magnetic field B in the region $r \leq a$.

e) Evaluate the magnitude of the magnetic field at $r = \delta$, a, and $2a$.

31–49 Two long, straight conducting wires of linear mass density $\lambda = 1.75 \times 10^{-3} \text{kg·m}^{-1}$ are suspended from cords so that they are each horizontal, parallel to each other, and a distance $d = 2$ cm apart. The back ends of the wires are connected to one another by a slack low-resistance connecting wire. The positive side of a 1.00-μF capacitor that has been charged by a 1000-V source is connected to the front end of one of the wires, and the negative side of the capacitor is connected to the front end of the other wire. See Fig. 31–32. When the connection is made, the wires are pushed aside by the repulsive force between the wires, and each wire has an initial horizontal velocity v_0. It can be assumed that the time for the capacitor to discharge is negligible compared to the time it takes for any appreciable displacement in the position of the wires to occur.

a) Show that the initial velocity v_0 of either wire is given by

$$v_0 = \frac{\mu_0 Q_0^2}{4\pi\lambda RCd},$$

where Q_0 is the initial charge on the capacitor of capacitance C, and R is the total resistance of the circuit.

b) Determine v_0 numerically if $R = 0.01 \, \Omega$.

c) To what height h will each wire rise as a result of the circuit connection?

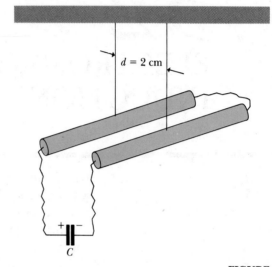

FIGURE 31–32

ELECTROMAGNETIC INDUCTION

WE RETURN IN THIS CHAPTER TO THE SUBJECT OF ELECTROMOTIVE FORCE, a topic we studied in Chapter 28 in connection with current, energy, and power in electric circuits. Here our concern is electromotive force of *magnetic* origin. The present large-scale production, distribution, and use of electrical energy all depend directly on the principles underlying magnetically induced emf's; none of these would be possible if we had to depend on chemical sources of emf, such as batteries. Magnetically induced emf's are essential to the understanding of energy-conversion devices such as electric motors, generators, and transformers.

We begin by studying the emf developed in a conductor moving in a magnetic field. Then we develop the relationship between emf and changing magnetic flux in a stationary conducting loop; this relationship is called Faraday's law. We also discuss Lenz's law, which helps us determine the directions of induced emf's and currents. Finally, we discuss the summing up of all the essential principles of electromagnetism in four equations called Maxwell's equations. These pave the way to the analysis of electromagnetic waves in Chapter 35.

32–1 INDUCTION PHENOMENA

When a conductor moves in a magnetic field, an emf is induced.

Our study of magnetically induced electromotive force begins with a look at several pioneering experiments carried out in the 1830s by Michael Faraday in England and Joseph Henry (first director of the Smithsonian Institution) in the United States. Figure 32–1a shows a coil of wire connected to a current-measuring device, such as a d'Arsonval galvanometer of the sort described in Section 29–3. Near the coil is a bar magnet. When the magnet is stationary, the galvanometer shows no current flow, as we would expect with no source of emf in the circuit. But if we move the magnet either toward or away from the coil, the meter shows current in the circuit *while the magnet is moving.* If we hold the magnet stationary and move the coil and galvanometer, the same thing happens. So something about the *changing* magnetic field through the coil is

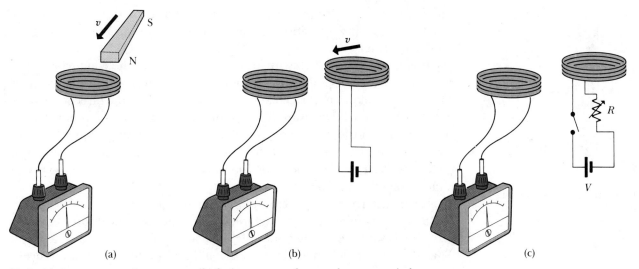

32–1 (a) A magnet moving near a coil of wire connected to a galvanometer induces a current in the coil. (b) A second coil carrying a current moves toward the coil connected to the galvanometer, inducing a current in it. (c) A changing current in the second coil induces a current in the first coil.

causing a current in the circuit. We call this an **induced current,** and the corresponding emf that has to be present to cause this current is called an **induced emf.**

The same effect occurs when the source of magnetic field is current in another circuit, as in Fig. 32–1b. We find that when the second coil is stationary, there is no current in the first coil. But when the second coil is moved toward or away from the first, or if the first coil is moved toward or away from the second, there is current in the first coil during the motion. With these experiments, as well as those with the bar magnet, the essential feature seems to be the *relative* motion of the two parts of the setup.

Finally, using the two-coil setup in Fig. 32–1c, we keep both coils stationary and vary the current in the second coil, either by opening and closing the switch or by changing the resistance R in its circuit. We find that as we open or close the switch, there is a momentary current pulse in the first circuit. When we vary the current in the second coil, there is an induced current in the first circuit *while the current in the second circuit is changing.*

What do all these experiments have in common? The answer, which we will explore in detail in this chapter, is that *changing magnetic flux* in the galvanometer circuit, from whatever cause, induces a current in that circuit. This statement forms the basis of Faraday's law of induction, the main subject of this chapter.

A changing magnetic flux through a closed circuit induces an emf and a current in the circuit.

32–2 MOTIONAL ELECTROMOTIVE FORCE

As the first step in our detailed study of magnetically induced electromotive force, we consider a conductor moving in a magnetic field. Figure 32–2 shows a conductor of length l in a uniform magnetic field perpendicular to the plane of the figure, directed *into* the page. We give the conductor a constant velocity

Magnetic forces on charges in a moving conductor

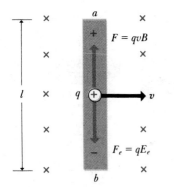

32–2 Conducting rod in uniform magnetic field.

Finding the emf due to motion of a straight conductor in a uniform magnetic field

Motional emf in a circuit with resistance is analogous to a battery with emf and internal resistance.

v to the right; a charged particle q in the conductor then experiences a force $\boldsymbol{F}$ equal to $q\boldsymbol{v} \times \boldsymbol{B}$. If q is positive, the direction of this force is upward along the conductor, from b toward a.

Under the action of this force the free charges in the conductor move, creating an excess of positive charge at end a and of negative charge at end b. This charge buildup in turn creates an electric field $\boldsymbol{E}$ in the direction from a to b, which exerts a downward force $q\boldsymbol{E}$ on the charges. The accumulation of charge at the ends of the conductor continues until E is large enough for the downward electric-field force qE to cancel exactly the upward magnetic-field force $q\boldsymbol{v} \times \boldsymbol{B}$. In this condition, the charges are in equilibrium, with point a at higher potential than point b.

Let us determine the magnitude of this potential difference. Because v and $\boldsymbol{B}$ are perpendicular, the magnitude of the magnetic-field force is simply qvB, and as we have pointed out, this is equal in the equilibrium state to qE. Thus the electric-field magnitude is $E = vB$, and the potential difference is

$$V_{ab} = El = vBl. \tag{32–1}$$

Now suppose the moving conductor slides along a stationary U-shaped conductor, as in Fig. 32–3. No magnetic force acts on the charges in the stationary conductor, but it lies in the electrostatic field caused by the charge accumulations at a and b. Under the action of this field, a current is established in the counterclockwise sense around this complete circuit. The moving conductor has become a source of electromotive force; charge moves within it from lower to higher potential, and in the remainder of the circuit charge moves from higher to lower potential. We call this a **motional electromotive force,** denoted by $\mathcal{E}$; we can write

$$\mathcal{E} = vBl. \tag{32–2}$$

If the resistance of the sliding bar is negligible, then $\mathcal{E}$ is also equal to the potential difference V_{ab}. For a real conductor with finite resistance r, there is a potential drop Ir, and the potential difference between the ends of the bar is *less* than $\mathcal{E}$ by this amount:

$$V_{ab} = \mathcal{E} - Ir. \tag{32–3}$$

Note that this equation is exactly the same form of relationship that we used for the emf, terminal potential difference, and current in a battery, in Section 28–4, Eq. (28–16), although the origins of the two emf's are completely different.

If the velocity v of the conductor is not perpendicular to the field $\boldsymbol{B}$ but makes an angle ϕ with it, then we replace v in Eq. (32–2) by $v \sin \phi$, and the induced emf is

$$\mathcal{E} = v \sin \phi\, Bl. \tag{32–4}$$

More generally, we can define the electromotive force $\mathcal{E}$ as the line integral of the associated non-electrostatic force, divided by the charge q to put it on a "force-per-unit-charge" basis. This corresponds directly to our definition of potential difference as the line integral of $\boldsymbol{E}$, the electrostatic force per unit charge. In the present situation the non-electrostatic force per unit charge is

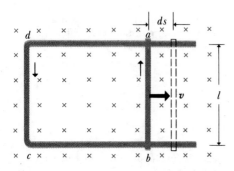

32–3 Current produced by the motion of a conductor in a magnetic field.

$v \times \boldsymbol{B}$, and we define the emf as

$$\mathcal{E} = \int_{b}^{a} v \times \boldsymbol{B} \cdot d\boldsymbol{l}. \tag{32–5}$$

In Fig. 33–2, v and $\boldsymbol{B}$ are perpendicular; both are constant over the length of the bar, and their vector product is parallel to $d\boldsymbol{l}$. Thus in this case Eq. (32–5) reduces to vBl, in agreement with Eq. (32–2).

As we have mentioned, the emf associated with the moving conductor in Fig. 32–3 is analogous to that of a battery with its positive terminal at a and negative terminal at b. In each case a non-electrostatic force acts in the direction from b to a. When the device is connected to an external circuit, the direction of current is from a to b in the external circuit and from b to a in the device. Electromotive force is not a vector quantity, but the quantity defined by Eq. (32–5) may be positive or negative, depending on the directions of v and $\boldsymbol{B}.$

If v is expressed in m·s^{-1}, B in T, and l in meters, $\mathcal{E}$ is in joules per coulomb or volts; we invite you to verify this statement.

Determining the sign of the induced emf in a moving conductor

---

EXAMPLE 32–1   Let the length l in Fig. 32–3 be 0.1 m; the velocity v, 0.1 m·s^{-1}; the resistance of the loop, 0.01 Ω; and B, 1 T. Find $\mathcal{E}$, the induced current, the force acting on the loop, and the mechanical power needed to keep the loop moving.

SOLUTION   The emf $\mathcal{E}$ is

$$\begin{aligned} \mathcal{E} &= vBl = (0.1 \text{ m·s}^{-1})(1 \text{ T})(0.1 \text{ m}) \\ &= 0.01 \text{ V}. \end{aligned}$$

The current in the loop is

$$I = \frac{\mathcal{E}}{R} = \frac{0.01 \text{ V}}{0.01 \text{ Ω}} = 1 \text{ A}.$$

Because of this current, a force F acts on the loop, in the direction *opposite* to its motion, equal to

$$\begin{aligned} F = IBl &= (1 \text{ A})(1 \text{ T})(0.1 \text{ m}) \\ &= 0.1 \text{ N}. \end{aligned}$$

To make the loop move with constant velocity, despite this resisting force, requires an equal and opposite additional force. The rate at which this additional force does work is the *power* needed to move the loop,

$$P = Fv = (0.1 \text{ N})(0.1 \text{ m·s}^{-1}) = 0.01 \text{ W}.$$

The product $\mathcal{E}I$ is

$$\mathcal{E}I = (0.01 \text{ V})(1 \text{ A}) = 0.01 \text{ W}.$$

Thus the rate of energy conversion, $\mathcal{E}I$, equals the mechanical power input, Fv, to the system.

Energy conversion associated with induced current in a loop

An emf is induced in a loop that rotates in a magnetic field.

EXAMPLE 32–2   The rectangular loop in Fig. 32–4, of length a and width b, is rotating with uniform angular velocity ω about the y-axis. The entire loop lies in a uniform, constant $\boldsymbol{B}$ field, parallel to the z-axis. We wish to calculate the induced emf in the loop, from Eq. (32–2).

SOLUTION   The velocity v of either side of the loop of length a has magnitude

$$v = \omega\left(\frac{b}{2}\right).$$

The motional emf in each side of length a is

$$\mathcal{E} = vB \sin \theta \, a = \tfrac{1}{2}\omega Bab \sin \theta.$$

These two emf's add, so the total emf due to these two sides is

$$\mathcal{E} = \omega Bab \sin \theta.$$

The magnetic forces on the other two sides of the loop are transverse to these sides and so contribute nothing to the emf.

As an alternative approach, we may use Eq. (32–5). The direction of $v \times \boldsymbol{B}$ in each side of length a is shown in the diagram. Its magnitude is

$$|v \times \boldsymbol{B}| = vB \sin \theta = \omega \frac{b}{2} B \sin \theta.$$

The motional forces in the other two sides of the loop are transverse to these sides and contribute nothing to the emf. The line integral of $v \times \boldsymbol{B}$ around the loop reduces to $2|v \times \boldsymbol{B}|a$, so

$$\mathcal{E} = \int v \times \boldsymbol{B} \cdot dl = 2\left(\omega \frac{b}{2} B \sin \theta\right)a$$
$$= \omega Bab \sin \theta.$$

The product ab equals the area A of the loop, and if the loop lies in the xy-plane at $t = 0$, then $\theta = \omega t$. Hence

$$\mathcal{E} = \omega AB \sin \omega t. \tag{32–6}$$

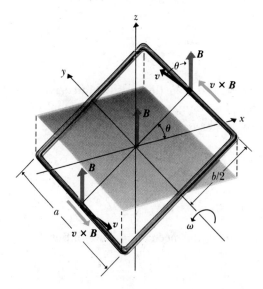

32–4 Rectangular loop rotating with constant angular velocity in a uniform magnetic field.

The emf therefore varies sinusoidally with time. The *maximum* emf $\mathcal{E}_m$, which occurs when $\sin \omega t = 1$, is

$$\mathcal{E}_m = \omega AB,$$

so we can write Eq. (32–6) as

$$\mathcal{E} = \mathcal{E}_m \sin \omega t. \tag{32–7}$$

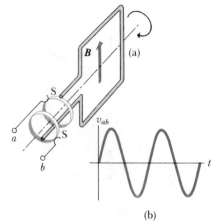

(b)

32–5 (a) Schematic diagram of an alternator, using a conducting loop rotating in a magnetic field. Connections to the external circuit are made by means of the slip rings S. (b) The resulting emf at terminals *ab*.

The rotating loop is the prototype of one type of alternating-current generator, or *alternator*; it develops a sinusoidally varying emf. The emf is maximum (in absolute value) when $\theta = 90°$ or $270°$ and the long sides are moving at right angles to the field. The emf is zero when $\theta = 0$ or $180°$ and the long sides are moving parallel or antiparallel to the field. We will show in the next section that the emf depends only on the *area A* of the loop and not on its shape.

The rotating loop in Fig. 32–4 can serve as the source in an external circuit by use of *slip rings* S, S, which rotate with the loop, as shown in Fig. 32–5a. Stationary brushes sliding on the rings are connected to the output terminals *a* and *b*. The instantaneous terminal voltage v_{ab}, on open circuit, equals the instantaneous emf. Figure 32–5b is a graph of v_{ab} as a function of time.

A similar scheme may be used to obtain an emf that, though varying with time, always has the same sign. We connect the loop to a split ring, or *commutator*, as in Fig. 32–6a. At angular positions where the emf reverses, the connections to the external circuit are interchanged. The resulting emf is shown in Fig. 32–6b. This device is the prototype of a dc generator. Commercial dc generators have a large number of coils and commutator segments; their terminal voltage is not only unidirectional but also practically constant.

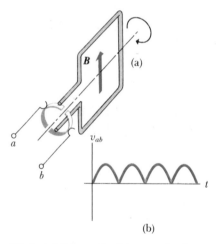

(b)

32–6 (a) Schematic diagram of a dc generator, using a split-ring commutator. (b) The resulting induced emf at terminals *ab*.

A rotating disk as a dc generator: principle of the Faraday disk dynamo

EXAMPLE 32–3 A disk of radius R, shown in Fig. 32–7, lies in the *xy*-plane and rotates with uniform angular velocity ω about the *z*-axis. The disk is in a uniform, constant B field parallel to the *z*-axis. Consider a short portion of a narrow radial segment of the disk, of length dr. Its velocity is $v = \omega r$, and since v is at right angles to B, the magnitude of $v \times B$ in the segment is

$$|v \times B| = vB = \omega r B.$$

The direction of $v \times B$ is radially outward. The motional emf due to this segment is

$$d\mathcal{E} = v \times B \cdot dl = vB \, dr = \omega r B \, dr.$$

The total emf between center and rim is

$$\mathcal{E} = \int_0^R v \times B \cdot dl = \omega B \int_0^R r \, dr = \tfrac{1}{2}\omega B R^2. \tag{32–8}$$

All the radial segments of the disk are in *parallel*, so the emf between center and rim is the same as that in any radial segment. Thus the entire disk is a *source*,

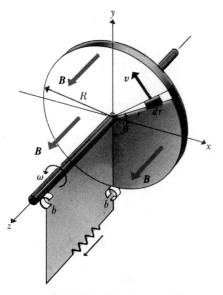

32-7 A Faraday disk dynamo. The emf is induced along radial lines of the rotating disk and is connected to an external circuit through the sliding contacts *bb*.

and the emf between center and rim equals $\omega BR^2/2$. The source can be included in a closed circuit by completing the circuit through sliding contacts of brushes b, b.

The emf in such a disk was studied by Faraday; the device is a called a *Faraday disk dynamo*, or a *homopolar generator*.

We pointed out in Section 30–4 that the magnetic-field force on a moving charged particle never does any *work* on the particle because $\boldsymbol{F}$ and $\boldsymbol{v}$ are always perpendicular. Yet in defining motional emf, we have spoken about the work done by the equivalent non-electrostatic force $q\boldsymbol{v} \times \boldsymbol{B}$. How is this apparent paradox resolved?

The resolution is somewhat subtle, and we cannot discuss it in detail here. Briefly, the vertical motion of the charges in the moving conductor in Fig. 32–3 causes a transverse (horizontal) magnetic-field force and thus a transverse displacement of charge, corresponding to the Hall effect discussed in Section 30–10. Hence there is a transverse *electrostatic* field in the conductor, from left to right in the example of Fig. 32–3. As the conductor moves to the right, it is this electrostatic force that actually does work on the moving charges. Detailed analysis shows that this work is the same *as though* it had been done by the vertical non-electrostatic or motional force $q\boldsymbol{v} \times \boldsymbol{B}$ during the vertical displacement of the moving charges.

32–3 FARADAY'S LAW

Relating induced emf to changing magnetic flux in a circuit

We analyzed the induced emf in the circuit of Fig. 32–3 by computing the magnetic-field forces on the mobile charges in the moving conductor. We can also consider this situation from another viewpoint based on the changing *magnetic flux* through the circuit. When the conductor moves toward the right a distance ds, the *area* enclosed by the circuit *abcd* increases by

$$dA = l\,ds,$$

and the change in magnetic *flux* through the circuit is

$$d\Phi = B\,dA = Bl\,ds.$$

The *rate of change* of flux is therefore

$$\frac{d\Phi}{dt} = \left(\frac{ds}{dt}\right)Bl = vBl. \tag{32–9}$$

But the product vBl equals the induced emf $\mathcal{E}$, so this equation states that *the induced emf in a circuit is numerically equal to the rate of change of the magnetic flux through it.*

To state this relationship in equation form, we need some sign rules. When we defined magnetic flux in Section 30–3 as the integral $\int \boldsymbol{B} \cdot d\boldsymbol{A}$, we pointed out the ambiguity in defining the direction of $d\boldsymbol{A}$ for an open surface. (We resolved that ambiguity for a *closed* surface with Gauss's law by making $d\boldsymbol{A}$ point always *out of* the surface.) For an open surface with a boundary curve, we can arbitrarily pick the direction for $d\boldsymbol{A}$. Then if the magnetic-field lines pass through the surface in the same direction as $d\boldsymbol{A}$, we count the corresponding flux as positive (because $\boldsymbol{B} \cdot d\boldsymbol{A}$ is positive); when they pass through in the direction opposite to $d\boldsymbol{A}$, the flux is negative.

That is half of our sign convention; the other half has to do with the sign of $\mathcal{E}$. We define a positive sense of circulation around the boundary curve by curling the fingers of the right hand around the $d\boldsymbol{A}$ vector with the thumb in the direction of $d\boldsymbol{A}$. The direction the fingers curl then defines the positive sense of going around the boundary. If an emf causes a current in this direction in a conductor lying along the boundary, it is positive; if in the opposite direction, negative.

Comparing these definitions with Fig. 32–3, suppose we choose $d\boldsymbol{A}$ to point out the plane, toward the reader. Then with the given direction of $\boldsymbol{B}$, the *flux* is negative. As the bar moves, the flux becomes larger in magnitude, and hence more negative, and $d\Phi/dt$ is a *negative* quantity. But as our discussion has shown, the induced current is counterclockwise, and according to the above definition, the associated $\mathcal{E}$ is *positive*. So, at least in this situation, $\mathcal{E}$ and $d\Phi/dt$ have opposite signs, and we write

$$\mathcal{E} = -\frac{d\Phi}{dt}. \qquad (32\text{--}10)$$

This sign rule will become clearer with continued study and additional examples. We must be very precise about signs right from the beginning, in order to relate this equation to other phenomena that we will study later.

Equation (32–10) is called **Faraday's law,** or *Faraday's law of induction.* It may appear to be just an alternative form of Eq. (32–2) for the emf in a moving conductor. It turns out, though, that it has much broader significance than we might expect from our derivation. It is applicable to *any* circuit through which there is a varying flux, even when no part of the circuit is moving and thus no emf appears that we can attribute directly to a $\boldsymbol{v} \times \boldsymbol{B}$ force.

Suppose, for example, that two loops of wire are located as in Fig. 32–8. A current in circuit 1 sets up a magnetic field whose magnitude at every point is proportional to this current. Part of this flux passes through circuit 2, and if the current in circuit 1 increases or decreases, the flux through circuit 2 will also vary. Circuit 2 is not moving in a magnetic field, so no "motional" emf is induced in it. However, a change does occur in the flux through the circuit, and it is found experimentally that an emf appears in circuit 2, given by $\mathcal{E} = -d\Phi/dt$. In such a situation no one portion of circuit 2 can be considered the source of emf; the *entire circuit* constitutes the source.

Here is another example. Suppose we set up a magnetic field within the toroidal winding of Fig. 32–9, link the toroid with a conducting ring, and vary the current in the winding of the toroid. We have shown that the flux lines set up by a steady current in a toroidal winding are confined to the space enclosed by the winding; not only is the ring not *moving* in a magnetic field, but if the current were steady, the ring would not even be *in* a magnetic field! Field lines do pass through the area bounded by the ring, however, and the flux changes as the current in the windings changes. Equation (32–10) predicts an induced emf in the ring, and we find by experiment that the emf actually exists. In case you have not recognized it, the apparatus in Fig. 32–9 is a *transformer* with a one-turn secondary; the phenomenon we are now discussing is the basis of operation for every transformer.

A slightly complicated sign convention for changing flux and emf

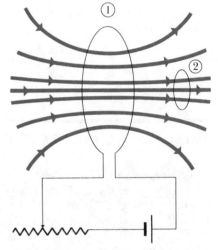

32–8 As the current in circuit 1 is varied, the magnetic flux through circuit 2 changes.

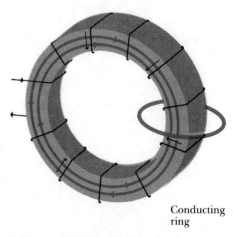

Conducting ring

32–9 An emf is induced in the ring when the flux in the toroid varies.

Faraday's law works both for moving conductors and for stationary conductors in changing magnetic fields.

To sum up, then, an emf is induced in a circuit whenever the magnetic flux through the circuit varies with time. The flux may change in various ways: For example, (1) a conductor may move through a stationary magnetic field, as in Figs. 32–2 and 32–3, or (2) the magnetic field through a stationary conducting loop may change with time, as in Figs. 32–8 and 32–9. For case (1), the emf may be computed *either* from

$$\mathcal{E} = vBl$$

or from

$$\mathcal{E} = -\frac{d\Phi}{dt}.$$

For case (2), the emf may be computed *only* from

$$\mathcal{E} = -\frac{d\Phi}{dt}.$$

If a coil has N turns, the emf is multiplied by N.

If we have a coil of N turns and the flux varies at the same rate through each turn, the induced emf's in the turns are in *series* and must be added; the total emf is

$$\mathcal{E} = -N\frac{d\Phi}{dt}. \tag{32–11}$$

PROBLEM-SOLVING STRATEGY: *Faraday's law*

1. First, be sure you understand what is making the flux change. Is the conductor moving? Is the magnetic field changing? Or both?

2. Define a positive direction for your area, and use the sign rule introduced above to write an expression for $d\Phi/dt$ that is consistent with that rule. If your conductor has N turns in a coil, don't forget to multiply by N.

3. Use Faraday's law to get the emf. Interpret the sign of your result with reference to the sign rules to determine the direction of the induced current. If the circuit resistance is known, the current can be calculated.

EXAMPLE 32–4   A certain coil of wire consists of 500 circular loops of radius 4 cm. It is placed between the poles of a large electromagnet, where the magnetic field is uniform, perpendicular to the plane of the coil, and increasing at a rate of 0.2 T·s⁻¹. What is the magnitude of the resulting induced emf?

Emf induced in a coil between the poles of an electromagnet

SOLUTION   The flux Φ at any time is given by $\Phi = BA$, and the rate of change of flux by $d\Phi/dt = (dB/dt)A$. In our problem, $A = \pi(0.04\ \text{m})^2 = 0.00503\ \text{m}^2$ and

$$\frac{d\Phi}{dt} = (0.2\ \text{T·s}^{-1})(0.00503\ \text{m}^2)$$

$$= 0.00101\ \text{T·m}^2\text{·s}^{-1} = 0.00101\ \text{Wb·s}^{-1}.$$

From Eq. (32–11), the magnitude of the induced emf is

$$\mathcal{E} = \left| N\frac{d\Phi}{dt} \right| = (500)(0.00101\ \text{Wb·s}^{-1}) = 0.503\ \text{V}.$$

If the coil is tilted so that a line perpendicular to its plane makes an angle of 30° with B, then only the component $B \cos 30°$ contributes to the flux through the coil. In that case the induced emf has magnitude $\mathcal{E} = (0.503 \text{ V})(\cos 30°) = 0.435 \text{ V}$.

EXAMPLE 32–5 Consider again the rotating rectangular loop in Fig. 32–4, discussed in Example 32–2 (Section 32–1). This time we compute the emf from Eq. (32–10). The flux through the loop equals its area A multiplied by the component of B perpendicular to the area; that is, $B \cos \theta$. This is also equal to the magnitude B multiplied by the projected area, shaded in Fig. 32–4, on a plane (the xy-plane) perpendicular to B:

$$\Phi = \boldsymbol{B} \cdot \boldsymbol{A} = BA \cos \theta = BA \cos \omega t.$$

Then

Emf induced in a rotating loop (again)

$$\mathcal{E} = -\frac{d\Phi}{dt} = \omega BA \sin \omega t,$$

in agreement with Eq. (32–6).

Note that the maximum value of $\mathcal{E}$ occurs when $\theta = 90°$ and the flux through the loop is zero, and that $\mathcal{E} = 0$ when $\theta = 0$ and the flux is maximum. The emf depends not on the flux through the loop, but on its *rate of change*.

EXAMPLE 32–6 Consider again the rotating disk in Fig. 32–7, discussed in Example 32–3 (Section 32–2). To compute the emf from Faraday's law, Eq. (32–10), consider the circuit to be the periphery of the shaded areas in Fig. 32–7. The rectangular portion in the yz-plane is fixed. The area of the shaded section in the xy-plane is $\frac{1}{2}R^2\theta$, and the flux through it is

Faraday's law applied to Faraday's disk: an alternative derivation of the emf

$$\Phi = \tfrac{1}{2}BR^2\theta.$$

As the disk rotates, the shaded area increases. In a time dt, the angle θ increases by $d\theta = \omega dt$. The flux increases by

$$d\Phi = \tfrac{1}{2}BR^2 \, d\theta = \tfrac{1}{2}BR^2\omega \, dt,$$

and the induced emf has magnitude

$$\mathcal{E} = \left|\frac{d\Phi}{dt}\right| = \tfrac{1}{2}BR^2\omega,$$

in agreement with the previous result.

EXAMPLE 32–7 *The search coil.* A useful experimental method for measuring magnetic-field strength uses a small, closely wound coil of N turns called a *search coil.* Suppose first that the plane of the windings of the search coil is initially perpendicular to a magnetic field B. If the area enclosed by the coil is A, the flux Φ through it is $\Phi = BA$. Now if the coil is quickly given a quarter-turn about one of its diameters so that its plane becomes parallel to the field, or if it is quickly snatched from its position to another where the field is known to be zero, the flux through it decreases rapidly from BA to zero. During the time that the flux is decreasing, an emf of short duration is induced in the coil, and a momentary induced current occurs in the external circuit to which the coil is connected. As we will see, the total flux change is proportional to the total *charge* that flows around the circuit.

The search coil: a way to measure magnetic fields

The current at any instant is

$$i = \frac{\mathcal{E}}{R},$$

where R is the combined resistance of external circuit and search coil, $\mathcal{E}$ is the instantaneous induced emf, and i is the instantaneous current. From Faraday's law, disregarding the negative sign,

$$\mathcal{E} = N \frac{d\Phi}{dt}, \qquad i = \frac{N}{R} \frac{d\Phi}{dt}.$$

The total charge Q that flows through the circuit is given by

$$Q = \int_0^t i \, dt = \frac{N}{R} \int d\Phi = \frac{N\Phi}{R}.$$

The field is proportional to the total charge that flows in the circuit.

Thus Φ and B are given by

$$\Phi = \frac{RQ}{N} \quad \text{and} \quad B = \frac{\Phi}{A} = \frac{RQ}{NA}. \tag{32–12}$$

It is not difficult to construct an instrument that measures the total charge passing through it. If the external circuit contains such an instrument, Q may be measured directly. We may then use Eq. (32–12) to compute B.

Strictly speaking, this method gives only the *average* field over the area of the coil. But if the area is small, this value is very nearly equal to the field at the center of the coil.

32–4 INDUCED ELECTRIC FIELDS

Faraday's law applied to changing fields and stationary conductors

The examples of induced emf analyzed in Section 32–3 were of two types, involving either a conductor moving in a magnetic field or changing flux through a stationary conductor. Another example of the second type is shown in Fig. 32–10, where a solenoid is encircled by a conducting loop with a galvanometer in it. A current I in the winding of the solenoid sets up a magnetic field $\boldsymbol{B}$ along the solenoid axis, as shown, and there is a magnetic flux $\Phi = BA$ through the surface bounded by the loop. When the current I changes, causing the flux to change, the galvanometer indicates a current in the wire *during the time that the flux is changing*. The associated emf is equal in magnitude to the *time rate of change of flux* through the surface bounded by the wire. Furthermore, the directions of the induced emf and current are given by the same sign rule described in Section 32–3. If we take the surface element dA to point in the direction of $\boldsymbol{B}$, and if I is increasing, then $d\Phi/dt$ is positive. The induced current is in the counterclockwise sense around the loop, as shown in Fig. 32–10. Thus we have

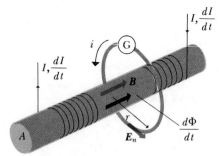

32–10 The windings of a long solenoid carry a current I that is increasing at a rate dI/dt. The magnetic flux in the solenoid is increasing at a rate $d\Phi/dt$, and this changing flux passes through a wire loop of arbitrary size and shape. An emf $\mathcal{E}$ is induced in the loop, given by $\mathcal{E} = d\Phi/dt$.

$$\mathcal{E} = -\frac{d\Phi}{dt}. \tag{32–13}$$

Equation (32–13) is again Faraday's law. We emphasize once more that it is valid for flux changes caused by motion of a conductor in a magnetic field and by magnetic-field changes in the area bounded by a stationary conductor.

EXAMPLE 32–8 Suppose the long solenoid in Fig. 32–10 is wound with 1000 turns per meter, and the current in its windings is increasing at the rate of 100 A·s^{-1}. The cross-sectional area of the solenoid is 4 cm^2 = 4×10^{-4} m^2.

The magnetic-field magnitude inside the solenoid, at points not too near its ends, is given by Eq. (31–28): $B = \mu_0 nI$. The flux Φ in the solenoid is

$$\Phi = BA = \mu_0 nIA.$$

The rate of change of flux is

$$\frac{d\Phi}{dt} = \mu_0 nA \frac{dI}{dt}$$
$$= (4\pi \times 10^{-7} \text{ Wb·A}^{-1}\text{·m}^{-1})(1000 \text{ turns·m}^{-1})$$
$$\times (4 \times 10^{-4} \text{ m}^2)(100 \text{ A·s}^{-1})$$
$$= 16\pi \times 10^{-6} \text{ Wb·s}^{-1}.$$

The magnitude of the induced emf is

$$|\mathcal{E}| = \frac{d\Phi}{dt} = 16\pi \times 10^{-6} \text{ V} = 16\pi \text{ }\mu\text{V}.$$

For conductors moving in magnetic fields, we derived Faraday's law for induced emf from the line integral of $v \times B$, the magnetic force per unit charge. But when the conductors are not moving, we must conclude that the induced current in the loop is caused by an **induced electric field.** This field is not caused by a distribution of electric charge but rather is associated with the changing magnetic field. To emphasize this difference, we call it a **non-electrostatic field,** and denote it by E_n. Then the induced emf in this case is the line integral of E_n around the loop, and we can restate Faraday's law as

An induced electric field is associated with an induced emf.

$$\oint E_n \cdot dl = -\frac{d\Phi}{dt}. \qquad (32\text{–}14)$$

As an example, suppose the loop in Fig. 32–10 is a circle of radius r. Because of axial symmetry, the non-electrostatic field E_n has the same magnitude at every point on the circle and is tangent to it at each point. The line integral in Eq. (32–14) becomes simply the magnitude E_n times the circumference $2\pi r$ of the circle. Thus at a distance r from the axis, the magnitude of the induced electric field is given by

$$E_n = \frac{1}{2\pi r} \frac{d\Phi}{dt}. \qquad (32\text{–}15)$$

The direction of E_n is as shown in the figure. We know that E_n has to have this direction because the integral of $E_n \cdot dl$ has to be negative when $d\Phi/dt$ is positive.

In summary, Eq. (32–10) is valid for two rather different situations. In one, an emf is induced by the magnetic forces on charges in a conductor moving through a magnetic field. In the other, a time-varying magnetic field induces an electric field of nonelectrostatic nature in a stationary conductor and hence induces an emf. The E_n field in the latter case differs from an *electrostatic* field in an important way. Its line integral around a closed path is

The analogy between magnetically induced electric fields and magnetic fields caused by displacement current

not zero, so it is not a *conservative* field. In contrast, an electrostatic field is *always* conservative, as discussed in Section 26–1. Despite this difference, however, the effect of an electric field in exerting a force $F = qE$ on a charged particle is the same, whether the field is produced by a charge distribution or by a changing magnetic field.

Electric field produced by a changing magnetic field is analogous to the role of *displacement current,* discussed in Section 29–5, where a changing *electric* field acts as a source of *magnetic* field. These relations thus exhibit a symmetry in the behavior of the two fields. We will return to this relationship in Section 32–7, and again in Chapter 35 in connection with electromagnetic waves.

32–5 LENZ'S LAW

Lenz's law: an alternative way to determine directions of induced emf's and currents

Lenz's law is a convenient alternative method for determining the sign or direction of an induced emf or current, or the direction of the associated non-electrostatic field. It always gives the same results as the sign rules we have introduced for $\mathcal{E}$ and $d\Phi/dt$ in connection with Faraday's law, but it is often easier to use. It also helps us gain intuitive understanding of various induction phenomena. H. F. E. Lenz (1804–1864) was a German scientist who, without knowing about the work of Faraday and Henry, duplicated many of their discoveries at about the same time. **Lenz's law** states:

> The direction of an induced current is such as to oppose the cause producing it.

The "cause" of the current may be the motion of a conductor in a magnetic field or the change of flux through a stationary circuit. In the first case, the direction of the induced current in the moving conductor is such that the direction of the sideways force exerted on the conductor by the magnetic field is opposite in direction to its motion. The motion of the conductor is therefore "opposed."

The effect always opposes its cause.

In the second case, the induced current sets up a magnetic field of its own that, within the area bounded by the circuit, is (a) *opposite* to the original field if this is *increasing,* but (b) is in the *same* direction as the original field if the latter is *decreasing.* Thus it is the *change in flux* through the circuit (not the flux itself) that is "opposed" by the induced current.

In order to have an induced current, we must have a closed circuit. If a conductor does not form a closed circuit, then we mentally complete the circuit between the ends of the conductor and use Lenz's law to determine the direction of the current. The polarity of the ends of the open-circuit conductor may then be deduced.

EXAMPLE 32–9   In Fig. 32–3, when the conductor moves to the right, a counterclockwise current is induced in the loop. The force exerted by the field on the moving conductor as a result of this current is to the left, *opposing* the conductor's motion.

EXAMPLE 32–10   In Fig. 32–4, the direction of the induced current in the loop is the same as that of the non-electrostatic field E_n. The magnetic field exerts forces on the loop as a result of this current; the force on the right side of the loop (of length a) is in the $+x$-direction, and that on the left side is in the $-x$-direction. The resulting torque is opposite in direction to ω and thus opposes the motion.

EXAMPLE 32–11 In Fig. 32–10, when the solenoid current is increasing, the induced current in the loop is counterclockwise. The additional field caused by this current, at points inside the loop, is opposite in direction to that of the solenoid; hence the induced current tends to oppose the increase in flux through the loop by causing flux in the opposite sense.

Lenz's law is also directly related to energy conservation. For instance, in Example 32–9 above, the induced current in the loop dissipates energy at the rate I^2R, and this energy must be supplied by the force that makes the conductor move despite the magnetic force opposing its motion. The work done by this applied force, in fact, must equal the energy dissipated in the circuit resistance. If the induced current were to have the opposite direction, the resulting force on the moving conductor would make it move faster and faster, violating energy conservation.

Lenz's law is a consequence of energy conservation.

The content of Lenz's law is contained in the sign rules introduced for use with Faraday's law. Let us review those briefly. We designate a positive direction for the surface through which we calculate the flux; this is the direction of the area element dA in the flux integral. This direction defines positive and negative flux and its rate of change. So, for example, if B has the same direction as dA and is increasing in magnitude, or if the area is increasing, then $d\Phi/dt$ is positive. If B has the same direction as dA but is decreasing in magnitude, or if the area is decreasing, $d\Phi/dt$ is negative. We invite you to think of other possibilities. Finally, in evaluating $\oint E_n \cdot dl$, we apply the right-hand rule to dA. For example, if the area element lies in the plane of this page and we define dA to point toward you, then we go counterclockwise around the circuit in evaluating the line integral of E_n.

Relation of Lenz's law to sign conventions in Faraday's law

EXAMPLE 32–12 In Fig. 32–3 let us take dA as parallel to B, away from the reader. Then Φ and $d\Phi/dt$ are both positive. Here $v \times B$ plays the role of E_n. The integral $\oint E_n \cdot dl$ must be taken clockwise around the loop. E_n is different from zero only along the line ab, where its direction is upward. Thus

Some exercise in using Lenz's law

$$\oint E_n \cdot dl = -E_n l.$$

This quantity is negative because E_n and dl have opposite directions, and it is consistent with Eq. (32–10), in which both sides are positive. Alternatively, we could have taken dA as toward the reader. In that case Φ and $d\Phi/dt$ are both negative, the loop is traversed in the counterclockwise sense, and $\oint E_n \cdot dl$ is positive. Either way, consistency with Eq. (32–10) requires that the direction of E_n must be from b to a, which is the direction of current in the moving conductor.

EXAMPLE 32–13 In Fig. 32–10, suppose the current in the solenoid has the direction shown and is decreasing. What is the direction of the induced current in the loop?

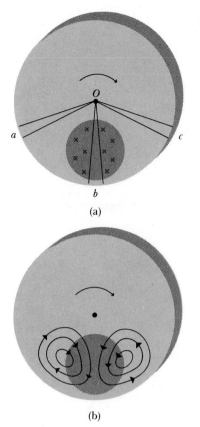

(a)

(b)

32–11 Eddy currents in a rotating disk.

Eddy currents dissipate energy: eddy-current brakes.

Eddy currents in transformer cores are a nuisance: How can we minimize them?

SOLUTION Take the area elements $d\mathbf{A}$ as pointing to the right, parallel to $\mathbf{B}$. Then the flux is positive but decreasing, and $d\Phi/dt$ is negative. Hence, according to Eq. (32–10), when we calculate $\mathcal{E} = \oint \mathbf{E}_n \cdot d\mathbf{l}$ by traversing the loop clockwise (looking in the direction of $\mathbf{B}$) the result must be positive. Hence $\mathbf{E}_n$ points clockwise around the loop, and the induced current must have this direction. Lenz's law gives the same result.

32–6 EDDY CURRENTS

In the examples of induction phenomena considered thus far, the induced currents resulting from induced emf's have been confined to well-defined paths in wires and other components forming a *circuit*. However, many pieces of electrical equipment contain masses of metal moving in magnetic fields or located in changing magnetic fields. In such situations it is possible to have induced currents that circulate throughout the volume of the material; because of their circulating nature, we call them **eddy currents.**

As an example, consider a disk that rotates in a magnetic field perpendicular to the plane of the disk but confined to a limited portion of the disk's area, as shown in Fig. 32–11. Element Ob is moving across the field and has an emf induced in it. Elements Oa and Oc are not in the field but, in common with all other elements located outside the field, provide return conducting paths along which charges displaced along Ob can return from b to O. A general eddy-current circulation is therefore set up in the disk, somewhat as sketched in Fig. 32–11b.

The currents in the neighborhood of radius Ob experience a side thrust that *opposes* the motion of the disk, while the return currents, since they lie outside the field, do not experience such a thrust. The interaction between the eddy currents and the field therefore results in a braking action on the disk. This apparatus finds some technical applications and is known as an eddy-current brake.

As a second example of eddy currents, consider the core of an alternating-current transformer, shown in Fig. 32–12a. The alternating current in the primary winding P sets up an alternating flux within the core, and an induced emf develops in the secondary winding S because of the continual change in flux through it. The iron core, however, is also a conductor, and any section such as AA can be thought of as a number of closed conducting circuits, one within the other (Fig. 32–12b). The flux through each of these circuits is

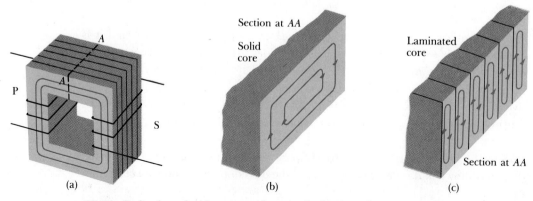

(a)   (b)   (c)

32–12 Reduction of eddy currents by use of a laminated core.

continually changing, so eddy currents circulate in the entire volume of the core, the lines of flow lying in planes perpendicular to the flux. These eddy currents are very undesirable both because of the energy they dissipate and because of the flux they themselves set up.

In all actual transformers, the eddy currents are greatly reduced by the use of a *laminated* core, that is, one built up of thin sheets, or laminae. The electrical resistance between the surfaces of the laminations (due either to a natural coating of oxide or to an insulating varnish) effectively confines the eddy currents to individual laminae (Fig. 32–12c). The flux through each loop and the resulting emf are small, and the currents and their heating effects are minimized.

In small transformers where eddy-current losses must be kept to an absolute minimum, the cores are sometimes made of *ferrites*, which are complex oxides of iron and other metals. These materials are ferromagnetic but have relatively high resistivity.

Ferrites: nonconducting ferromagnetic materials

32–7 MAXWELL'S EQUATIONS

We are now in a position to wrap up in a wonderfully neat package all the relationships between electric and magnetic fields and their sources that we have studied in the past several chapters. The package consists of four equations, called **Maxwell's equations.** You may remember Maxwell as the discoverer of the concept of displacement current, which we studied in Sections 29–5 and 31–9.

Two of Maxwell's equations involve integrals of E and B over closed surfaces. The first is simply Gauss's law, stating that the surface integral of $E_\perp$ over any closed surface equals $1/\epsilon_0$ times the total charge Q enclosed within the surface:

Gauss's law is one of Maxwell's equations.

$$\oint E \cdot dA = \frac{Q}{\epsilon_0}. \qquad (32\text{–}16)$$

The second is the analogous relation for *magnetic* fields, stating that the surface integral of $B_\perp$ over any closed surface is always zero:

$$\oint B \cdot dA = 0. \qquad (32\text{–}17)$$

This statement means, among other things, that there is no such thing as magnetic charge as a source of magnetic field.

There are no isolated magnetic poles: the basis of another of Maxwell's equations.

The third equation is Ampere's law, including displacement current, and the fourth is Faraday's law. Ampere's law, as stated in Section 31–9, states that both conduction current I_C and displacement current $\epsilon_0 \, d\Psi/dt$, where Ψ is electric flux, act as sources of magnetic field:

Ampere's law, as generalized by Maxwell to include displacement current as a source of magnetic field

$$\oint B \cdot dl = \mu_0 \left(I_C + \epsilon_0 \frac{d\Psi}{dt} \right). \qquad (32\text{–}18)$$

Faraday's law, which we have studied in this chapter, states that a changing magnetic field or magnetic flux induces an electric field:

Faraday's law: the fourth Maxwell equation

$$\oint E \cdot dl = -\frac{d\Phi}{dt}. \qquad (32\text{–}19)$$

In our discussions we have been careful to describe this induced electric field as a non-electrostatic field and to denote it as E_n. But if an *electrostatic* field, produced by electric charges, is also present in the same region, it is always *conservative,* and so its line integral around any closed path is always zero. Hence in Eq. (32–19) **E** is the *total* electric field, including both electrostatic and non-electrostatic contributions.

Comparing Eqs. (32–18) and (32–19), we see a remarkable symmetry. Equation (32–18) says that a changing electric field or electric flux creates a magnetic field, and Eq. (32–19) says that a changing magnetic field or magnetic flux creates an electric field. Indeed, in empty space, where there is no conduction current and $I_C = 0$, the two equations have the same form, apart from a numerical constant and a negative sign, with the roles of **E** and **B** reversed in the two equations. This same symmetry is also evident in the first two equations; in empty space, where there is no charge, the equations are identical in form, one containing **E,** the other **B.**

But the most remarkable thing about the third and fourth equations is this: A time-varying field of *either* kind induces a field of the other kind in neighboring regions of space. As Maxwell recognized, these relationships predict the existence of electromagnetic disturbances consisting of time-varying electric and magnetic fields that travel or *propagate* from one region of space to another, even if no matter is present in the intervening space. Such disturbances are called *electromagnetic waves,* and we now know that they provide the physical basis for light, radio and television waves, infrared, ultraviolet, x-rays, and the rest of the electromagnetic spectrum. We will return to this vitally important topic for further study in Chapter 35.

We can rewrite Eqs. (32–18) and (32–19) in a different but equivalent form by introducing the definitions of electric and magnetic flux, $\Psi = \int \mathbf{E} \cdot d\mathbf{A}$ and $\Phi = \int \mathbf{B} \cdot d\mathbf{A},$ respectively. In empty space, where $I_C = 0$, we obtain

$$\oint \mathbf{B} \cdot d\mathbf{l} = \mu_0 \epsilon_0 \frac{d}{dt} \oint \mathbf{E} \cdot d\mathbf{A}, \tag{32–20}$$

$$\oint \mathbf{E} \cdot d\mathbf{l} = -\frac{d}{dt} \oint \mathbf{B} \cdot d\mathbf{A}. \tag{32–21}$$

Again we notice the symmetry between the roles of the **E** and **B** fields in these expressions.

Although it may not be obvious, *all* the basic relations between fields and their sources are contained in Maxwell's equations. We can derive Coulomb's law from Gauss's law, we can derive the law of Biot and Savart from Ampere's law, and so on. When we add the equation that gives the force on a charged particle in **E** and **B** fields, namely,

$$F = q(\mathbf{E} + v \times \mathbf{B}),$$

we have all the fundamental relations of electromagnetism!

The fact that electromagnetism can be wrapped up so neatly and elegantly is a very satisfying discovery. In conciseness and generality, Maxwell's equations are comparable to Newton's law of motion and to the laws of thermodynamics. Indeed, that is really what science is all about—learning how to express very broad and general physical laws in a concise and compact form. Maxwell's synthesis of electromagnetism stands as a towering intellectual achievement, comparable to the Newtonian synthesis described at the end of Section 6–5 and to the development of relativity, quantum mechanics, and the understanding of DNA in our own century. All beautiful, and all monuments to the achievements of which the human intellect is capable!

A changing electric or magnetic field induces a field of the other type: the basis of electromagnetic waves.

Maxwell's equations display similarities in the behavior of electric and magnetic fields.

Maxwell's equations: wrapping up all the principles of electromagnetism in one beautiful, compact package

SUMMARY

When a conducting loop moves in a magnetic field or is stationary in a region of changing magnetic field, there is an induced current with a corresponding induced emf, whether the magnetic field is caused by a permanent magnet or by another current loop. The induced current depends on the rate of change of magnetic flux through the loop.

When a conductor of length l moves with velocity v in a uniform magnetic field $\boldsymbol{B}$, the induced emf is given by

$$\mathcal{E} = vBl. \tag{32-2}$$

More generally, the induced emf in a conductor moving in a $\boldsymbol{B}$ field is

$$\mathcal{E} = \int_b^a v \times \boldsymbol{B} \cdot d\boldsymbol{l}. \tag{32-5}$$

When a conducting loop of area A rotates with angular velocity ω in a uniform magnetic field of magnitude B, the induced emf is

$$\mathcal{E} = \omega AB \sin \omega t. \tag{32-6}$$

When a disk of radius R rotates with angular velocity ω in a magnetic field of magnitude B, with the disk axis parallel to B, the induced emf is

$$\mathcal{E} = \frac{1}{2} \omega BR^2. \tag{32-8}$$

Faraday's law of induction states that the induced emf in a conducting loop through which the magnetic flux Φ is changing is given by

$$\mathcal{E} = -\frac{d\Phi}{dt}. \tag{32-10}$$

This equation is valid whether the changing flux is caused by motion of the coil or by time variation of the magnetic field. If the conductor is a coil of N turns, the emf is multiplied by N. When a stationary conductor is in a changing magnetic field, the induced emf is associated with a non-electrostatic field $\boldsymbol{E}_n$, such that

$$\oint \boldsymbol{E}_n \cdot d\boldsymbol{l} = -\frac{d\Phi}{dt}. \tag{32-14}$$

Lenz's law, a useful rule to find the directions of induced currents and emf's, states that the induced current or emf always acts to oppose or cancel the change that caused it. Lenz's law can be derived from Faraday's law but is sometimes easier to use.

When a bulk piece of conducting material, such as a metal, is in a changing magnetic field or moves through a field, currents called eddy currents are induced in the volume of the material.

The relationships between electric and magnetic fields and their sources can be stated compactly in four equations called Maxwell's equations. They are

$$\oint \boldsymbol{E} \cdot d\boldsymbol{A} = \frac{Q}{\epsilon_0}, \tag{32-16}$$

$$\oint \boldsymbol{B} \cdot d\boldsymbol{A} = 0, \tag{32-17}$$

KEY TERMS

induced current
induced emf
motional electromotive force
Faraday's law
induced electric field
nonelectrostatic field
Lenz's law
eddy currents
Maxwell's equations

$$\oint \boldsymbol{B} \cdot d\boldsymbol{l} = \mu_0 \left(I_C + \epsilon_0 \frac{d\Psi}{dt}\right), \qquad (32\text{--}18)$$

$$\oint \boldsymbol{E} \cdot d\boldsymbol{l} = -\frac{d\Phi}{dt}. \qquad (32\text{--}19)$$

The first equation is Gauss's law of electrostatics; the second states the absence of magnetic charge; the third is Ampere's law, generalized by Maxwell to include displacement current; and the fourth is Faraday's law. Together they form a complete basis for the relation of $\boldsymbol{E}$ and $\boldsymbol{B}$ fields to their sources. These equations enabled Maxwell to predict the existence of electromagnetic waves, an outstanding achievement in the development of physical science.

QUESTIONS

32–1 In most parts of the northern hemisphere the earth's magnetic field has a vertical component directed *into* the earth. An airplane flying east generates an emf between its wingtips. Which wingtip acquires an excess of electrons and which a deficiency?

32–2 A sheet of copper is placed between the poles of an electromagnet, so the magnetic field is perpendicular to the sheet. When it is pulled out, a considerable force is required, and the force required increases with speed. What is happening?

32–3 In Fig. 32–4, if the angular velocity ω of the loop is doubled, then the frequency with which the induced current changes direction doubles, and the maximum emf also doubles. Why? Does the torque required to turn the loop change?

32–4 If we compare the conventional dc generator (Fig. 32–6) and the Faraday disk dynamo (Fig. 32–7), what are some advantages and disadvantages of each?

32–5 Some alternating-current generators use a rotating permanent magnet and stationary coils. What advantages does this scheme have? What disadvantages?

32–6 When a conductor moves through a magnetic field, the magnetic forces on the charges in the conductor cause an emf. But if this phenomenon is viewed in a frame of reference moving with the conductor, there is no motion, yet there is still an emf. How is this paradox resolved?

32–7 Two circular loops lie adjacent to each other. One is connected to a source that supplies an increasing current; the other is a simple closed ring. Is the induced current in the ring in the same direction as that in the ring connected to the source, or opposite? What if the current in the first ring is decreasing?

32–8 A farmer claimed that the high-voltage transmission lines running parallel to his fence induced dangerously large voltages on the fence. Is this within the realm of possibility?

32–9 Small one-cylinder gasoline engines sometimes use a device called a *magneto* to supply current to the spark plug. A permanent magnet is attached to the flywheel, and a stationary coil is mounted adjacent to it. What happens when the magnet passes the coil?

32–10 A current-carrying conductor passes through the center of a metal ring, perpendicular to its plane. If the current in the conductor increases, is a current induced in the ring?

32–11 A student asserted that if a permanent magnet is dropped down a vertical copper pipe, it eventually reaches a terminal velocity, even if there is no air resistance. Why should this be? Or should it?

EXERCISES

Section 32–2 Motional Electromotive Force

32–1 In Fig. 32–3 a rod of length $l = 0.40$ m moves in a magnetic field of magnitude $B = 1.2$ T. The emf induced in the moving rod is found to be 2.40 V.

a) What is the speed of the rod?

b) If the total circuit resistance is 1.2 Ω, what is the induced current?

c) What force (magnitude and direction) does the field exert on the rod as a result of this current?

32–2 In Fig. 32–2 a rod of length $l = 0.25$ m moves with constant speed of 6.0 m·s⁻¹ in the direction shown. The induced emf is found to be 3.0 V.

a) What is the magnitude of the magnetic field?

b) Which point is at higher potential, a or b?

32–3 In Fig. 32–13 a rod of length $l = 0.15$ m moves in a magnetic field $\boldsymbol{B}$ directed into the plane of the figure. If $B = 0.5$ T and the rod moves with velocity $v = 4$ m·s⁻¹ in the direction shown,

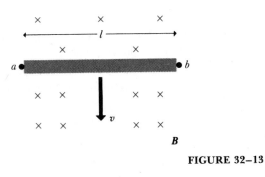

FIGURE 32–13

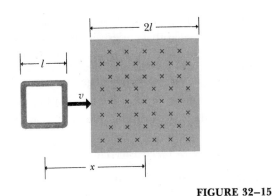

FIGURE 32–15

a) What is the motional emf induced in the rod?

b) What is the potential difference between the ends of the rod?

c) Which point, a or b, is at higher potential?

32–4 A conducting rod AB in Fig. 32–14 makes contact with metal rails CA and DB. The apparatus is in a uniform magnetic field 0.5 T, perpendicular to the plane of the diagram.

a) Find the magnitude and direction of the emf induced in the rod when it is moving toward the right with a speed 4 m·s^{-1}.

b) If the resistance of circuit $ABDC$ is 0.2 Ω (assumed constant), find the force required to maintain the rod in motion. Neglect friction.

c) Compare the rate at which mechanical work is done by the force (Fv) with the rate of development of heat in the circuit (i^2R).

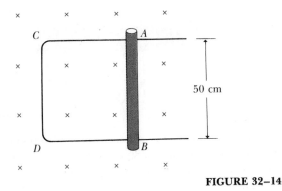

FIGURE 32–14

32–5 A square loop of wire with resistance R is moved at constant velocity v across a uniform magnetic field confined to a square region whose sides are twice the length of those of the square loop. (See Fig. 32–15.)

a) Sketch a graph of the external force F needed to move the loop at constant velocity, as a function of the distance x, from $x = -2l$ to $x = +2l$. (The distance x is negative when the center of the loop is to the left of the center of the magnetic-field region. Take positive force to be to the right.)

b) Sketch a graph of the induced current in the loop as a function of x. Take clockwise currents to be positive.

Section 32–3 Faraday's Law

32–6 A coil of 1000 turns enclosing an area of 20 cm² is rotated from a position where its plane is perpendicular to the earth's magnetic field to one where its plane is parallel to the field, in 0.02 s. What average emf is induced if the earth's magnetic field is 6×10^{-5} T?

32–7 A closely wound rectangular coil of 50 turns has dimensions of 12 cm × 25 cm. The plane of the coil is rotated from a position where it makes an angle of 45° with a magnetic field 2 T to a position perpendicular to the field in time $t = 0.1$ s. What is the average emf induced in the coil?

32–8 A flat, square coil of ten turns has sides of length 0.12 m. The coil rotates in a magnetic field of 0.025 T.

a) What is the angular velocity of the coil if the maximum emf produced is 20 mV?

b) What is the average emf at this velocity?

32–9 A Faraday disk dynamo is to be used to supply current to a large electromagnet that requires 20,000 A at 1.0 V. The disk is to be 0.6 m in radius, and it turns in a magnetic field of 1.2 T supplied by a smaller electromagnet.

a) How many revolutions per second must the disk turn?

b) What torque is required to turn the disk, assuming that all the mechanical energy is dissipated as heat in the large electromagnet?

32–10 The cross-sectional area of a closely wound search coil having 20 turns is 1.5 cm² and its resistance is 4 Ω. The coil is connected through leads of negligible resistance to a charge-measuring instrument having internal resistance 16 Ω. Find the quantity of charge displaced when the coil is pulled quickly out of a region where $B = 1.8$ T to a point where the magnetic field is zero. The plane of the coil, when in the field, makes an angle of 90° with the magnetic field.

32–11 A closely wound search coil has an area of 4 cm², 160 turns, and a resistance of 50 Ω. It is connected to a charge-measuring instrument whose resistance is 30 Ω. When the coil is rotated quickly from a position parallel to a uniform magnetic field to one perpendicular to the field, the instrument indicates a charge of 4×10^{-5} C. What is the magnitude of the field?

32–12 A coil 4 cm in radius, containing 500 turns, is placed in a magnetic field that varies with time according to $B = 0.01t + (2 \times 10^{-4})t^3$, where B is in teslas and t is in seconds. The coil is connected to a 500-Ω resistor, and its plane is perpendicular to the magnetic field.

a) Find the induced emf in the coil as a function of time.

b) What is the current in the resistor at time $t = 10$ s?

Section 32–4 Induced Electric Fields

32–13 A long, straight solenoid of cross-sectional area 6 cm^2 is wound with ten turns of wire per centimeter, and the windings carry a current of 0.25 A. A secondary winding of two turns encircles the solenoid. When the primary circuit is opened, the magnetic field of the solenoid becomes zero in 0.05 s. What is the average induced emf in the secondary?

32–14 The magnetic field within a long, straight solenoid of circular cross section and radius R is increasing at a rate of dB/dt.

a) What is the rate of change of flux through a circle of radius r_1 inside the solenoid, normal to the axis of the solenoid, and with center on the solenoid axis?

b) Find the induced electric field E_n inside the solenoid, at a distance r_1 from its axis. Show the direction of this field in a diagram.

c) What is the induced electric field *outside* the solenoid, at a distance r_2 from the axis?

d) Sketch a graph of the magnitude of E_n as a function of the distance r from the axis, from $r = 0$ to $r = 2R$.

e) What is the induced emf in a circular turn of radius $R/2$?

f) Of radius R?

g) Of radius $2R$?

32–15 The magnetic field B at all points within the colored circle of Fig. 32–16 has an initial magnitude of 0.5 T. It is directed into the plane of the diagram and is decreasing at the rate of -0.1 T·s^{-1}.

a) What is the shape of the field lines of the induced E_n field in Fig. 32–16, within the colored circle?

b) What are the magnitude and direction of this field at any point of the circular conducting ring of radius 0.10 m, and what is the emf in the ring?

c) What is the current in the ring if its resistance is 2 Ω?

d) What is the potential difference between points a and b of the ring?

Section 32–5 Lenz's Law

32–16 A circular loop of wire is in a region of spatially uniform magnetic field, as shown in Fig. 32–16. The magnetic field is directed into the plane of the figure. Calculate the direction (clockwise or counterclockwise) of the induced current in the loop when

a) B is increasing,

b) B is decreasing,

c) B is constant with value B_0.

32–17 A cardboard tube is wound with two windings of insulated wire, as in Fig. 32–17. Terminals a and b of winding A may be connected to a battery through a reversing switch. State whether the induced current in the resistor R is from left to right, or from right to left, in the following circumstances:

a) The current in winding A is from a to b and is increasing;

b) the current is from b to a and is decreasing;

c) the current is from b to a and is increasing.

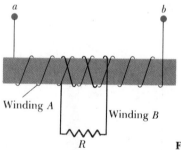

Winding A   Winding B

R   **FIGURE 32–17**

32–18 Using Lenz's law, determine the direction of the current in resistor ab of Fig. 32–18 when

a) switch S is opened after having been closed for several minutes;

b) coil B is brought closer to coil A, with the switch closed;

c) the resistance of R is decreased while the switch remains closed.

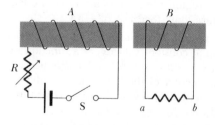

FIGURE 32–18

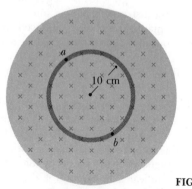

FIGURE 32–16

PROBLEMS

32–19 The long rectangular loop in Fig. 32–19, of width l, mass m, and resistance R, starts from rest in the position shown and is acted on by a constant force $\boldsymbol{F}$. At all points in the colored area there is a uniform magnetic field $\boldsymbol{B}$ perpendicular to the plane of the diagram.

a) Sketch a graph of the velocity of the loop as a function of time.

b) Find the terminal velocity.

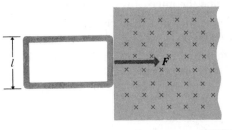

FIGURE 32–19

32–20 The cube in Fig. 32–20, 1 m on a side, is in a uniform magnetic field of 0.2 T, directed along the positive y-axis. Wires A, C, and D move in the directions indicated, each with a speed of 0.5 m·s^{-1}.

a) What is the motional emf along the length of each wire?

b) What is the potential difference between the ends of each wire?

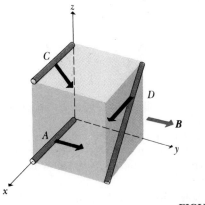

FIGURE 32–20

32–21 A slender rod 1 m long rotates about an axis through one end and perpendicular to the rod, with an angular velocity of 2 rev·s^{-1}. The plane of rotation of the rod is perpendicular to a uniform magnetic field of 0.5 T.

a) What is the induced emf in the rod?

b) What is the potential difference between its ends?

32–22 The rectangular loop in Fig. 32–21, of area A and resistance R, rotates at uniform angular velocity ω about the y-axis. The loop lies in a uniform magnetic field $\boldsymbol{B}$ in the direction of the x-axis. Sketch the following graphs:

a) the flux Φ through the loop as a function of time (let $t = 0$ in the position shown in Fig. 32–21);

b) the rate of change of flux $d\Phi/dt$;

c) the induced emf in the loop;

d) the torque Γ needed to keep the loop rotating at constant angular velocity;

e) the induced emf if the angular velocity is doubled.

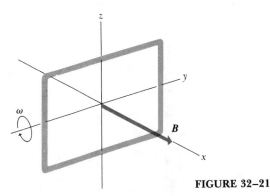

FIGURE 32–21

32–23 In Problem 32–22 and Fig. 32–21, let $A = 400$ cm^2, $R = 2\ \Omega$, $\omega = 10$ rad·s^{-1}, $B = 0.5$ T. Find

a) the maximum flux through the loop,

b) the maximum induced emf,

c) the maximum torque.

d) Show that the work of the external torque in one revolution is equal to the energy dissipated in the loop during one revolution.

32–24 Suppose the loop in Fig. 32–21 is

a) rotated about the z-axis;

b) rotated about the x-axis;

c) rotated about an edge parallel to the y-axis.

What is the maximum induced emf in each case, if the numerical parameters are as given in Problem 32–23?

32–25 A flexible circular loop 0.10 m in diameter lies in a magnetic field of magnitude 1.2 T, directed into the plane of the diagram in Fig. 32–22. The loop is pulled at the points indicated by the arrows, forming a loop of zero area in 0.2 s.

a) Find the average induced emf in the circuit.

b) What is the direction of the current in R?

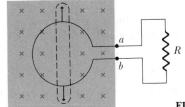

FIGURE 32–22

32–26 The current in the wire AB of Fig. 32–23 is upward and increasing steadily at a rate di/dt.

a) At an instant when the current is i, what are the magnitude and direction of the field B at a distance r from the wire?

b) What is the flux $d\Phi$ through the narrow shaded strip?

c) What is the total flux through the loop?

d) What is the induced emf in the loop?

e) Evaluate the numerical value of the induced emf if $a = 0.10$ m, $b = 0.30$ m, $l = 0.20$ m, and $di/dt = 2$ A·s⁻¹.

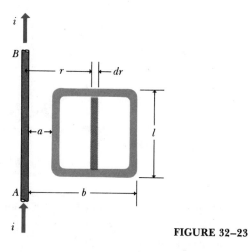

FIGURE 32–23

32–27 The magnetic field B at all points within a circular region of radius R is uniform in space and directed into the plane of Fig. 32–24. If the magnetic field is increasing at a rate dB/dt, what are the magnitude and direction of the force on a stationary point charge of positive charge q located at points a, b, and c? Point a is a distance r above the center of the region, point b is a distance r to the right of the center, and point c is at the center of the region.

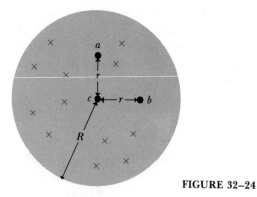

FIGURE 32–24

32–28 A circular conducting ring of radius $r_0 = 0.5$ m is oriented in the xy-plane and is in a region of magnetic field B, where

$$B = B_0[1 - 3(t/t_0)^2 + 2(t/t_0)^3]k.$$

Assume $t_0 = 0.01$ s and is constant, r is the distance of an arbitrary point from the center of the ring, t is the time,

and k is a unit vector in the positive z-direction. Also, $B_0 = 8 \times 10^{-2}$ T and is constant. At points a and b (see Fig. 32–25) there is a small gap in the ring with wires leading to an external circuit of resistance $R = 12\ \Omega$. There is no magnetic field at the location of the external circuits.

a) Derive an expression, as a function of time, for the total magnetic flux Φ enclosed by the ring.

b) Determine the emf induced in the ring at time $t = 0.005$ s. What is the polarity?

c) Because of the internal resistance of the ring, the current through R at the time given in part (b) is only 0.3 A. Determine the internal resistance of the ring.

d) Determine the emf in the ring at a time $t = 0.0121$ s. What is the polarity?

e) Determine the time at which the current through R reverses its direction.

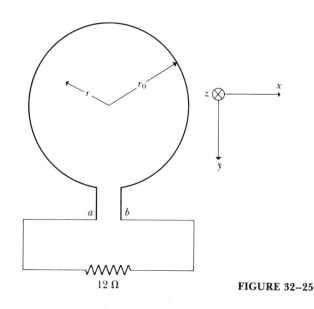

FIGURE 32–25

32–29 A search coil used to measure magnetic fields is to be made with a radius of 2 cm. It is to be designed so that flipping it 180° in a field of 0.1 T causes a total charge of 1×10^{-4} C to flow in a charge-measuring instrument when the total circuit resistance is 50 Ω. How many turns should the coil have?

32–30 A solenoid 0.50 m long and 0.08 m in diameter is wound with 500 turns. A closely wound coil of 20 turns of insulated wire surrounds the solenoid at its midpoint, and the terminals of the coil are connected to a charge-measuring instrument. The total circuit resistance is 25 Ω.

a) Find the quantity of charge displaced through the instrument when the current in the solenoid is quickly decreased from 3 A to 1 A.

b) Draw a sketch of the apparatus, showing clearly the directions of the windings of the solenoid and coil, and of the current in the solenoid. What is the direction of the current in the coil when the solenoid current is decreasing?

32–31 The long, straight wire in Fig. 32–26a carries constant current I. A metal bar of length l is moving at constant velocity v, as shown in the figure. Point a is a distance d from the wire.

a) Calculate the emf induced in the rod.

b) Which point, a or b, is at higher potential?

c) If the bar is replaced by a rectangular wire loop of resistance R, as shown in Fig. 32–26b, what is the magnitude of the current induced in the loop?

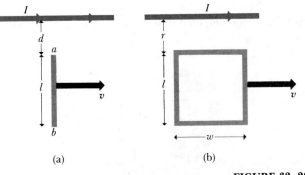

(a)   (b)

FIGURE 32–26

32–32 A rectangular loop of width a and slide wire of mass m are as shown in Fig. 32–27. A uniform magnetic field $\mathbf{B}$ is directed perpendicular to the plane of the loop, into the plane of the figure. The slide wire is given an initial velocity of v_0 and then released. Assume that there is no friction between the slide wire and the loop and that the resistance of the loop is negligible compared to the resistance R of the slide wire.

a) Obtain an expression for F, the magnitude of the force exerted on the wire while it is moving at speed v.

b) Show that the distance x which the wire moves before coming to rest is given by $x = mv_0R/(a^2B^2)$.

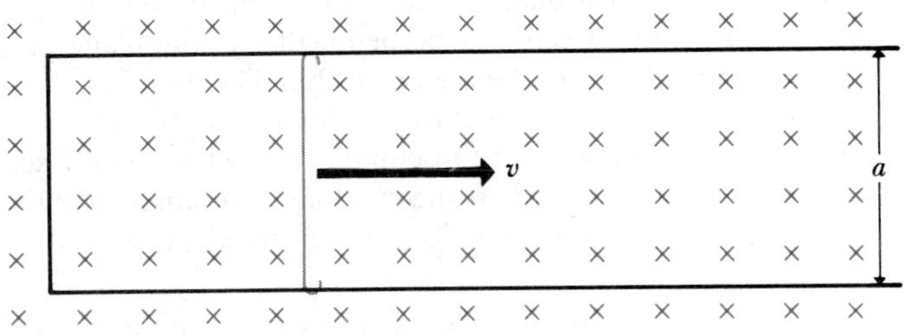

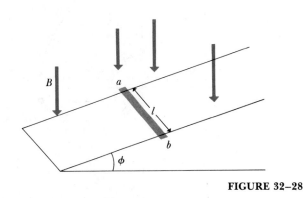

FIGURE 32–27

32–33 A circular ring of radius $a = 20$ cm is rotated about its vertical diameter with a constant angular velocity $\omega = 1000$ rad·s^{-1}. At time $t = 0$ the plane of the ring is perpendicular to a magnetic field that has a fixed direction into the page and whose magnitude varies with time according to the relation

$$B = B_0e^{-t/\tau},$$

where t is the time, $B_0 = 0.1$ T and $\tau = 0.02$ s. The resistance R of the ring is 0.01 Ω.

a) Obtain an expression for the magnetic flux Φ enclosed by the ring as an explicit function of time.

b) Using Faraday's law, derive an expression for the induced emf $\mathcal{E}$ as an explicit function of time.

c) Determine the current I at $t = 0$.

d) Determine the first time t_0 at which $\mathcal{E} = 0$. (Be careful; the answer is not π/ω.)

e) Determine the first time t_m at which $\mathcal{E}$ becomes a maximum and determine its maximum value.

CHALLENGE PROBLEMS

32–34 A metal bar of length l, mass m, and resistance R is placed on frictionless metal rails inclined at an angle ϕ above the horizontal. The rails have negligible resistance. There is a uniform magnetic field of magnitude B directed downward in Fig. 32–28. The bar is released from rest and slides down the rails.

a) Is the direction of the current induced in the bar from a to b, or from b to a?

b) What is the terminal velocity of the bar?

c) What is the induced current in the bar when the terminal velocity has been reached?

d) After the terminal velocity has been reached, at what rate is electrical energy being converted to heat in the resistance of the bar?

e) After the terminal velocity has been reached, at what rate is work being done on the bar by gravity? Compare your answer to that in (d).

FIGURE 32–28

32–35 A thin rod of length $l = 1$ m and mass $m = 0.1$ kg is free to rotate in a horizontal plane about one of its ends. The other end rests on a frictionless circular rail whose resistance per unit length λ is 2×10^{-2} Ω·m^{-1}. There is a narrow gap in the rail at point a; this gap has infinite resistance. A flexible wire of negligible resistance connects the end of the rod at the pivot and point a of the rail. The rod has negligible resistance. The rod and rail are in the region of a uniform magnetic field whose magnitude is 0.3 T and whose direction is perpendicular to the plane of rotation, into the page. (See Fig. 32–29.) The angular position θ of the rod is measured counterclockwise from a line drawn from the pivot to point a. The rod is initially set into motion such that when the rod makes an angle of 90°, the middle of the rod is moving with an initial velocity $v_0 = 20 \text{ m·s}^{-1}$.

a) Obtain an expression for the current I induced in the rod when its midpoint is moving with a velocity v and its angular position is θ.

b) Obtain an expression for the rate P at which electrical energy is converted into thermal energy when the angular position of the rod is θ.

c) Show that the velocity v of the midpoint of the rod is related to the angular position θ of the rod by the relation

$$v = v_0 - \left[\frac{3B^2l^2}{8m\lambda}\right] \ln (\theta/\theta_0),$$

where θ_0 is the initial angular position of the rod. [*Hint:* Equate the answer from part [b] to the time rate of decrease of the kinetic energy of the rod.]

d) Calculate θ_f, the angular position of the rod after it comes to rest. (Obtain a numerical value.)

e) Repeat the problem, but this time for a circular rail that is continuous, with no gap. When you redo part (c), you should obtain the relation

$$v = v_0 - \left[\frac{3B^2l^2}{8m\lambda}\right] \ln \left[\frac{\theta(2\pi - \theta_0)}{\theta_0(2\pi - \theta)}\right].$$

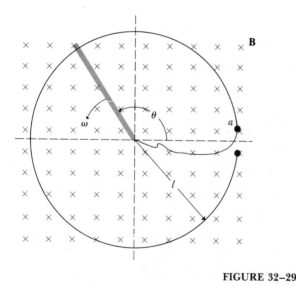

FIGURE 32–29

32–36 A square conducting loop, 20 cm on a side, is placed in the same magnetic field as in Exercise 32–15. (See Fig. 32–30.)

a) Copy Fig. 32–30, and show by vectors the directions and relative magnitudes of the induced electric field E_n at points a, b, and c.

b) Prove that the component of E_n along the loop has the same value at every point of the loop and is equal to that of the ring of Fig. 32–16 (Exercise 32–15).

c) What is the current induced in the loop if its resistance is 2 Ω?

d) What is the potential difference between points a and b?

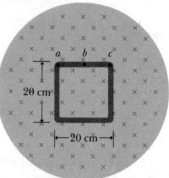

FIGURE 32–30

32–37 A uniform square conducting loop, 20 cm on a side, is placed in the same magnetic field as in Exercise 32–15, with side ac along a diameter and with point b at the center of the field. (See Fig. 32–31.)

a) Copy Fig. 32–31, and show by vectors the directions and relative magnitudes of the induced electric field E_n at the lettered points.

b) What is the induced emf in side ac?

c) What is the induced emf in the loop?

d) What is the current in the loop if its resistance is 2 Ω?

e) What is the potential difference between points a and c? Which is at higher potential?

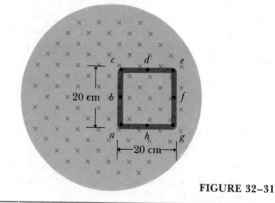

FIGURE 32–31

33

INDUCTANCE

THE PHENOMENA OF INDUCTION AND INDUCED EMF, WHICH WE STUDIED IN Chapter 32, have immediate practical applications in a variety of electric-circuit devices, including transformers and inductors. When two coils are adjacent, a changing current in one induces an emf in the other; this is the operating principle of the *transformer,* and the coupling between the coils is characterized by their *mutual inductance.* A changing current in a single coil also causes an induced emf in that same coil, and the relationship of current to emf is described by the *self-inductance* of the coil. A study of energy relationships in inductors leads to the concept of energy stored in magnetic fields. We study several simple circuits containing inductors, including one that can undergo electrical oscillations analogous to those of a mechanical harmonic oscillator. This circuit analysis is fundamental to our study of alternating-current circuits in Chapter 34.

33–1 MUTUAL INDUCTANCE

An emf is induced in a stationary circuit whenever the magnetic flux through the circuit varies with time. If the variation in flux is brought about by a varying current in a second circuit, it is convenient to express the induced emf in terms of the varying *current,* rather than in terms of the varying *flux.* We will use the symbol i to represent the instantaneous value of a varying current.

Figure 33–1 is a cross-sectional view of two closely wound coils of wire. A current i_1 in coil 1 sets up a magnetic field as indicated by the color lines, and some of these lines pass through coil 2. Let the resulting flux through coil 2 be Φ_2. The magnetic field is proportional to i_1, so Φ_2 is also proportional to i_1. When i_1 changes, Φ_2 changes; this changing flux induces an emf $\mathcal{E}_2$ in coil 2, given by

$$\mathcal{E}_2 = N_2 \frac{d\Phi_2}{dt}. \tag{33–1}$$

The proportionality of Φ_2 and i_1 could be represented in the form $\Phi_2 =$ (constant) i_1, but it is more convenient to include the number of turns N_2 in

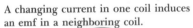

A changing current in one coil induces an emf in a neighboring coil.

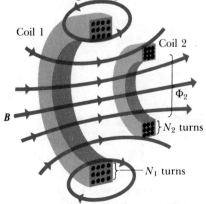

33–1 A portion of the flux set up by a current in coil 1 links with coil 2.

767

the relation. Introducing a proportionality constant M, we write

$$N_2 \Phi_2 = M i_1. \qquad (33\text{--}2)$$

From this,

$$N_2 \frac{d\Phi_2}{dt} = M \frac{di_1}{dt},$$

and we can rewrite Eq. (33–1) as

$$\mathcal{E}_2 = M \frac{di_1}{dt}. \qquad (33\text{--}3)$$

Mutual inductance: the relation between changing current and induced emf

The constant M, which depends only on the geometry of the two coils, is called their **mutual inductance.** It is defined by Eq. (33–2), which may be written

$$M = \frac{N_2 \Phi_2}{i_1}. \qquad (33\text{--}4)$$

The entire discussion can be repeated for the case where a changing current i_2 in coil 2 causes a changing flux Φ_1 and hence an emf $\mathcal{E}_1$ in coil 1. We might expect that the constant M would be different in this case, since the two coils are not, in general, symmetric. It turns out, however, that M is always the same in this case as in the case considered above, so *a single value of the mutual inductance characterizes completely the induced-emf interaction of two coils.*

The henry: the SI unit of inductance

The SI unit of mutual inductance, from Eq. (33–4), is one *weber per ampere.* An equivalent unit, obtained by reference to Eq. (33–3), is one *volt-second per ampere.* These two equivalent units are called one **henry** (1 H), in honor of Joseph Henry (1797–1878), one of the discoverers of electromagnetic induction. Thus the unit of mutual inductance is

$$1 \text{ H} = 1 \text{ Wb·A}^{-1} = 1 \text{ V·s·A}^{-1}.$$

A simple example of a mutual-inductance calculation

EXAMPLE 33–1   A long solenoid of length l and cross-sectional area A is closely wound with N_1 turns of wire. A small coil of N_2 turns surrounds it at its center, as in Fig. 33–2. A current i_1 in the solenoid sets up a $\boldsymbol{B}$ field at its center; from Eq. (31–28), the magnitude of $\boldsymbol{B}$ is

$$B = \mu_0 n i_1 = \frac{\mu_0 N_1 i_1}{l}.$$

The flux through the central section equals BA, and since all of this flux links with the small coil, the mutual inductance is

$$M = \frac{N_2 \Phi_2}{i_1} = \frac{N_2}{i_1} \left(\frac{\mu_0 N_1 i_1}{l} \right) A = \frac{\mu_0 A N_1 N_2}{l}.$$

If $l = 0.50$ m, $A = 10 \text{ cm}^2 = 10^{-3} \text{ m}^2$, $N_1 = 1000$ turns, $N_2 = 10$ turns,

$$M = \frac{(4\pi \times 10^{-7} \text{ Wb·A}^{-1} \text{·m}^{-1})(10^{-3} \text{ m}^2)(1000)(10)}{0.5 \text{ m}}$$

$$= 25.1 \times 10^{-6} \text{ Wb·A}^{-1}$$

$$= 25.1 \times 10^{-6} \text{ H} = 25.1 \ \mu\text{H}.$$

33–2 A long solenoid with cross-sectional area A and N_1 turns, surrounded by a small coil with N_2 turns.

33–2 SELF-INDUCTANCE

In discussing mutual inductance, we assumed that one circuit acted as the source of magnetic field and that the emf under consideration was induced in a separate independent circuit linking some of the flux created by the first circuit. But whenever a current is present in any circuit, this current sets up a magnetic field that links with the *same* circuit and varies when the current varies. Hence any circuit with a varying current has an induced emf in it resulting from the variation in *its own* magnetic field. Such an emf is called a **self-induced electromotive force (emf).**

As an example, suppose a circuit contains a coil with N turns of wire, as in Fig. 33–3, and that a flux Φ passes through each turn. In analogy to Eq. (33–4), we define the **self-inductance** L of the circuit, or simply its **inductance**:

$$L = \frac{N\Phi}{i}. \qquad (33\text{–}5)$$

This relation can be written as

$$N\Phi = Li.$$

If Φ and i change with time, then

$$N\frac{d\Phi}{dt} = L\frac{di}{dt},$$

and, since the self-induced emf $\mathcal{E}$ has magnitude

$$\mathcal{E} = N\frac{d\Phi}{dt},$$

it follows that

$$\mathcal{E} = L\frac{di}{dt}. \qquad (33\text{–}6)$$

The self-inductance of a circuit is therefore *the self-induced emf per unit rate of change of current.* The SI unit of self-inductance is one henry.

A circuit, or part of a circuit, that has inductance is called an **inductor.** An inductor is represented by the symbol

The direction of a self-induced (non-electrostatic) field can be found from Lenz's law. We consider the "cause" of this field, and hence of the emf associated with it, to be the *changing current* in the conductor. If the current is increasing, the direction of the induced field is *opposite* to that of the current. If the current is *decreasing,* the induced field is in the *same* direction as the current. Thus it is the *change* in current, not the current itself, that is "opposed" by the induced field.

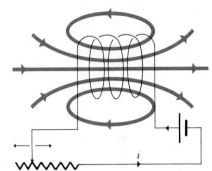

33–3 A flux Φ linking a coil of N turns. When the current in the circuit changes, the flux changes also, and a self-induced emf appears in the circuit.

A changing current in a coil induces an emf in that coil: pulling yourself up by your bootstraps.

Self-inductance: the ratio of induced emf to rate of change of current

The sign of the induced emf in an inductor

EXAMPLE 33–2 An air-core toroidal solenoid of cross-sectional area A and mean radius r is closely wound with N turns of wire. We neglect the variation of B across the section, assuming its average value to be very nearly equal to the value at the center of the cross section, as given by Eq. (31–29). Then the flux in the

Self-inductance in a toroidal solenoid

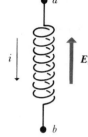

33–4 When di/dt is positive, the non-electrostatic induced field $\mathbf{E}_n$ is in the direction shown, and the emf is treated as a potential difference with V_{ab} positive; that is, a is at higher potential than b. When di/dt is negative, $\mathbf{E}_n$ has the opposite direction, and V_{ab} is negative.

toroid is

$$\Phi = BA = \frac{\mu_0 NiA}{2\pi r}.$$

Since all of the flux links with each turn, the self-inductance is

$$L = \frac{N\Phi}{i} = \frac{\mu_0 N^2 A}{2\pi r}.$$

Thus if $N = 100$ turns, $A = 10 \text{ cm}^2 = 10^{-3} \text{m}^2$, $r = 0.10$ m,

$$L = \frac{(4\pi \times 10^{-7} \text{ Wb·A}^{-1}\text{·m}^{-1})(100)^2(10^{-3}\text{m}^2)}{2\pi(0.10 \text{ m})}$$

$$= 20 \times 10^{-6} \text{ H} = 20 \ \mu\text{H}.$$

EXAMPLE 33–3 If the current in the coil above increases uniformly from zero to 1 A in 0.1 s, find the magnitude and direction of the self-induced emf.

SOLUTION

$$\mathcal{E} = L\frac{di}{dt} = (20 \times 10^{-6} \text{ H})\frac{1 \text{ A}}{0.1 \text{ s}}$$

$$= 2.0 \times 10^{-4} \text{ V}.$$

The current is increasing, so according to Lenz's law the direction of the emf is opposite to that of the current. In Fig. 33–4, suppose the inductor terminals are a and b and there is an *increasing* current from a to b in the inductor. Then the induced field $\mathbf{E}_n$ and the emf are in the direction from b to a, like a battery with a as the (+) terminal and b as the (−) terminal.

Sign rules for using Kirchhoff's rules with inductors

When Kirchhoff's loop rule (Section 29–2) is used with circuits containing inductors, this self-induced emf is treated as though it were a potential difference, with a at higher potential than b. Thus if we define the positive direction of the current i as from a to b in the inductor, then

$$V_{ab} = L\frac{di}{dt}. \tag{33–7}$$

The self-inductance of a circuit depends on its size, shape, number of turns, and so on. It also depends on the magnetic properties of the material enclosed by the circuit.

PROBLEM-SOLVING STRATEGY: Self-inductance

1. In this chapter we view the inductor primarily as a *circuit* device, so circuit analysis is of primary importance. In general all the voltages, currents, and capacitor charges are now functions of time, not constants as they have been in most of our previous analyses. But Kirchhoff's rules, which we studied in Chapter 29, are still valid. When the voltages and currents vary with time, Kirchhoff's loop rule is a relationship that holds at each instant of time, and it often gives us *differential equations* to solve, rather than just algebraic equations. This means that the solutions

will depend on initial conditions as well as on the constants that describe the circuits.

2. As in all circuit analysis, getting the signs right is often more challenging than understanding the principles. We suggest you review the strategy in Section 29–2 as preparation for study of the circuits in Sections 33–4 through 33–6. In addition, give close attention to the sign rule described with Eq. (33–7). Then look for the applications of Kirchhoff's loop rule in these discussions.

In the examples above we calculated the magnetic field by assuming that the conductor was surrounded by vacuum. If matter is present, then when we calculate the field we must replace the constant μ_0, the permeability of vacuum, by the permeability of the material, $\mu = K_m\mu_0$, as discussed in Section 31–8. If the material is diamagnetic or paramagnetic, this makes very little difference. If the material is *ferromagnetic*, however, the difference is of crucial importance. An inductor wound on a soft iron core having $K_m = 5000$ has an inductance approximately 5000 times as great as the same coil with an air core. Iron-core inductors are very widely used in a variety of electronic and electric-power applications. An added complication is that with ferromagnetic materials, the magnetization is not always a linear function of magnetizing current, especially as saturation is approached. As a result, the inductance can depend on current in a fairly complicated way. In our discussion we will ignore this complication and assume the inductance to be constant. This is a reasonable assumption even for a ferromagnetic material if the magnetization remains well below the saturation level.

A magnetic material multiplies the inductance by K_m.

33–3 ENERGY IN AN INDUCTOR

A changing current in an inductor causes an emf; while the current is changing, the source supplying the current must maintain a potential difference between its terminals and therefore must supply energy to the inductor. Let us calculate the total energy input needed to establish a final current I in an inductor of inductance L if the initial current is zero.

If the current at some instant is i and is changing at the rate di/dt, the induced emf at that instant is $\mathcal{E} = L\, di/dt$, and the instantaneous power P supplied by the current source is

To establish a current in an inductor, the circuit must supply energy to it.

$$P = \mathcal{E}i = Li\frac{di}{dt}.$$

The energy dW supplied in time dt is $P\, dt$, or

$$dW = Li\, di,$$

and the total energy supplied while the current increases from zero to a final value I is

$$W = L\int_0^I i\, di = \frac{1}{2}LI^2. \tag{33–8}$$

After the current has reached its final steady value, $di/dt = 0$ and the power input is zero. The energy that has been supplied to the inductor is needed to establish the magnetic field in and around the inductor. We can think of this as a *potential energy* associated with the current. We define this energy to be zero when there is no current; then when the current is I, the potential energy has the value $\frac{1}{2}LI^2$. In the discussion of *mechanical* potential energy in Chapter 7, we defined the potential energy of a system in a certain position as the work done by the forces of the system when it returns from this position to the position where the potential energy is defined to be zero. Our inductor's energy is consistent with this viewpoint; when the inductor current decreases from an initial value I to the reference value $I = 0$, the inductor does an amount of electrical work on the external circuit equal to $\frac{1}{2}LI^2$. If the current in the circuit is interrupted suddenly by opening a switch, the energy may be dissipated in an arc across the switch contacts.

The energy stored in an inductor is proportional to the square of the current.

Stored energy in an inductor is associated with the magnetic field produced by the current.

We can also consider the energy to be associated with the magnetic field itself, and we can develop a relation analogous to the one obtained for electric-field energy in Section 27–4, Eqs. (27–9) and (27–18). As we did then, we consider here only one simple case, the toroidal solenoid; this system has the advantage that its magnetic field is confined completely to a finite region of space in its interior. As in Example 33–2 (Section 33–2), we assume the cross-sectional area A is small enough that we can consider the magnetic field to be uniform over the area. The volume in the toroid is then approximately equal to the circumferential length $l = 2\pi r$ multiplied by the area A. From the preceding example, the self-inductance of the toroidal solenoid is

$$L = \frac{\mu_0 N^2 A}{l}.$$

When the current in the windings is I, the energy stored in the toroid is

$$W = \frac{1}{2}LI^2 = \frac{1}{2}\left(\frac{\mu_0 N^2 A}{l}\right)I^2.$$

We can think of this energy as localized in the volume enclosed by the windings, equal to lA. The energy *per unit volume*, or **energy density** u, is then

$$u = \frac{W}{lA} = \frac{1}{2}\mu_0\left(\frac{N^2 I^2}{l^2}\right).$$

We can express this relation in terms of the magnetic field B inside the toroid. From Eq. (31–29),

$$B = \frac{\mu_0 NI}{2\pi r} = \frac{\mu_0 NI}{l}. \tag{33–9}$$

Energy density in a magnetic field

Squaring this and rearranging, we find

$$\frac{N^2 I^2}{l^2} = \frac{B^2}{\mu_0^2}.$$

When we substitute this expression into the above equation, we finally find

$$u = \frac{B^2}{2\mu_0}. \tag{33–10}$$

This is the analog of the expression for the energy per unit volume in the electric field of an air capacitor, $\frac{1}{2}\epsilon_0 E^2$, derived in Section 27–4. We will need these expressions for electric- and magnetic-field energy when we study energy associated with electromagnetic waves in Chapter 35.

When a magnetic material is present, the energy density is smaller by a factor of K_m.

When the material inside the toroid is not vacuum but a material having magnetic permeability $\mu = K_m\mu_0$, we have to replace μ_0 by μ in Eq. (31–29). The energy per unit volume in the magnetic field is then

$$u = \frac{B^2}{2\mu}. \tag{33–11}$$

33–4 THE *R–L* CIRCUIT

Time-varying current in a circuit containing a resistor and an inductor

In our discussion of inductors thus far we have neglected the *resistance* of the windings. But every inductor found in the real world must have some resistance, unless its windings are superconducting. We can represent a real-life inductor as an ideal zero-resistance inductor with self-inductance *L*, in series with a resistance *R*, as shown in Fig. 33–5. This diagram can also represent a resistor in series with an inductor; in that case *R* is the *total* resistance of the combination.

The circuit of Fig. 33–5 shows some interesting behavior. By means of switch S_1, the *R–L* combination can be connected to a source with constant terminal voltage *V*, or it may be "shorted" by closing switch S_2. Suppose both switches are initially open, and then at some initial time $t = 0$ switch S_1 is closed. Because of the self-induced emf in the inductor, the current does not immediately rise to its final value at the instant S_1 is closed. Instead, it grows at a rate that depends on the values of *R* and *L* in the circuit.

Let *i* represent the instantaneous current at some time *t* after the switch is closed; then di/dt is its rate of increase at that time. The potential difference across the inductor is

$$v_{cb} = L\frac{di}{dt},$$

and the potential difference across the resistor is

$$v_{ac} = iR.$$

Since $V = v_{ac} + v_{cb}$, it follows that

$$V = L\frac{di}{dt} + iR. \tag{33–12}$$

The rate of increase of current is therefore

$$\frac{di}{dt} = \frac{V - iR}{L} = \frac{V}{L} - \frac{R}{L}i. \tag{33–13}$$

At the instant the circuit is first closed, $i = 0$ and the current starts to grow at the rate

$$\left(\frac{di}{dt}\right)_{\text{initial}} = \frac{V}{L}.$$

The greater the self-inductance *L*, the more slowly the current increases.

As the current increases, the term Ri/L increases also, and hence the *rate* of increase of current becomes smaller and smaller, as Eq. (33–13) shows. When the current reaches its final *steady-state* value *I*, its rate of increase is zero. Then

$$0 = \frac{V}{L} - \left(\frac{R}{L}\right)I$$

and

$$I = \frac{V}{R}.$$

That is, the *final* current does not depend on the self-inductance; it is the same as it would be in a pure resistance *R* connected to a source having emf *V*.

To obtain an expression for the current as a function of time, we proceed just as we did for the problem of the charging capacitor in Section 29–4. We

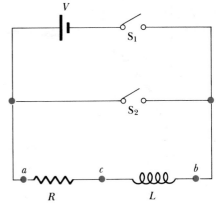

33–5 An *R–L* series circuit.

Applying Kirchhoff's loop rule to the circuit

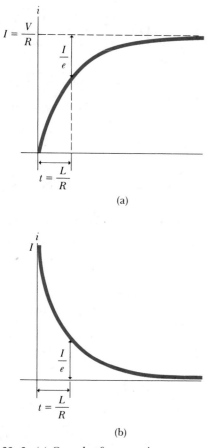

33–6 (a) Growth of current in a circuit containing inductance and resistance. (b) Decay of current in a circuit containing inductance and resistance.

When the inductance is large, the current in the circuit grows slowly.

Energy relations in an R–L circuit

first rearrange Eq. (33–13):

$$\frac{di}{(V/R) - i} = \frac{R}{L}dt.$$

Integrating both sides, we find

$$-\ln\left(\frac{V}{R} - i\right) = \left(\frac{R}{L}\right)t + \text{constant}.$$

The integration constant is evaluated by noting that the initial current is zero, so $i = 0$ at time $t = 0$:

$$\text{constant} = -\ln\frac{V}{R}.$$

Rearranging again, we obtain

$$\ln\left(\frac{V}{R} - i\right) - \ln\frac{V}{R} = \ln\left(1 - \frac{Ri}{V}\right) = -\left(\frac{R}{L}\right)t, \qquad (33\text{–}14)$$

$$i = \frac{V}{R}(1 - e^{-(R/L)t}).$$

From this,

$$\frac{di}{dt} = \frac{V}{L}\,e^{-(R/L)t}. \qquad (33\text{–}15)$$

We note that at time $t = 0$, $i = 0$ and $di/dt = V/L$, and that as $t \to \infty$, $i \to V/R$ and $di/dt \to 0$, as predicted above.

Figure 33–6a is a graph of Eq. (33–14) and shows the variation of current with time. The instantaneous current i first rises rapidly, then increases more slowly and approaches asymptotically the final value $I = V/R$. At a time τ equal to L/R, the current has risen to $(1 - 1/e)$ or about 0.63 of its final value. This time is called the **time constant** or the *decay constant* for the circuit:

$$\tau = \frac{L}{R}. \qquad (33\text{–}16)$$

For a given value of R, the time constant increases when the inductance L increases. Thus, although the graph of i versus t has the same general shape whatever the inductance, the current rises rapidly to its final value if L is small and slowly if L is large. For example, if $R = 100\ \Omega$ and $L = 10\ H$,

$$\frac{L}{R} = \frac{10\ H}{100\Omega} = 0.1\ s,$$

and the current increases to about 63% of its final value in 0.1 s. But if $L = 0.01\ H$,

$$\frac{L}{R} = \frac{0.01\ H}{100\Omega} = 10^{-4}\ s,$$

and only 10^{-4} s is required for the current to increase to 63% of its final value.

Energy considerations offer us additional insight into the behavior of an R–L circuit. The instantaneous rate at which the source delivers energy to the circuit is $P = Vi$. The instantaneous rate at which energy is dissipated in the resistor is i^2R, and the rate at which energy is stored in the inductor is $iv_{cb} = Li\,di/dt$. Multiplying Eq. (33–12) by i, we find

$$Vi = Li\frac{di}{dt} + i^2R. \tag{33-17}$$

This means that part of the power Vi supplied by the source is dissipated (i^2R) in the resistor, and part is stored ($Li\,di/dt$) in the inductor. This discussion is completely analogous to our power analysis for a charging capacitor at the end of Section 29–4.

Now suppose switch S_1 in the circuit of Fig. 33–5 has been closed for a long time, so that the final current $I = V/R$ has been reached. Redefining our initial time, we close switch S_2 at time $t = 0$, short-circuiting the battery. (We can then open S_1 to save the battery from ruin.) The current through R and L does not instantaneously go to zero but decays smoothly, as shown in Fig. 33–6b. We challenge you to follow the same pattern as in the analysis above to show that the current varies with time according to

$$i = Ie^{-Rt/L}. \tag{33-18}$$

The time constant, $\tau = L/R$, is the time for the current to decrease to $1/e$, or about 37% of its original value. The energy needed to maintain the current during this decay is provided by the energy stored in the magnetic field of the inductor. The detailed energy analysis is simpler this time. In place of Eq. (33–12) we have

$$0 = L\frac{di}{dt} + iR.$$

Multiplying through by i, we find

$$0 = Li\frac{di}{dt} + i^2R. \tag{33-19}$$

In this case, of course, $Li\,di/dt$ is negative, and this equation shows that the rate of *decrease* of energy stored in the inductor is equal to the rate of dissipation of energy, i^2R, in the resistor.

33–5 THE *L–C* CIRCUIT

We studied the behavior of an $R–C$ circuit in Section 29–4 and that of an $R–L$ circuit in Section 33–4. In both cases the behavior is characterized by an exponential approach to some steady-state situation. When a circuit contains *both* inductance L and capacitance C, entirely new modes of behavior appear, characterized by *oscillating* current and charge.

Consider first the $L–C$ circuit in Fig. 33–7, containing an ideal resistanceless inductor and a capacitor. We charge the capacitor to a potential difference V_m, as shown in Fig. 33–7a, and then close the switch. The capacitor immediately starts to discharge through the inductor. Figure 33–7b shows the situation later when the capacitor has completely discharged and the potential difference between its terminals (and those of the inductor) has decreased to zero. During this discharge, the current in the inductor has established a magnetic field in the space around it. This magnetic field now decreases, inducing an emf in the inductor in the same direction as the cur-

When the battery is short-circuited, the current doesn't suddenly drop to zero.

Energy relations for decaying current in an R–L circuit

The L–C circuit: an electrical system that vibrates

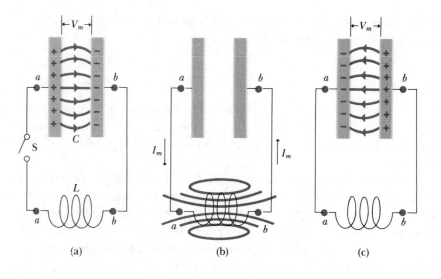

33–7 Energy transfer between electric and magnetic fields in an oscillating L–C circuit.

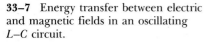

An electrical version of the harmonic oscillator

rent. The current therefore persists, although with decreasing magnitude, until the magnetic field has disappeared and the capacitor has been charged in the *opposite* sense to its initial polarity, as in Fig. 33–7c. The process now repeats itself in the reverse direction. In the absence of energy losses, the charges on the capacitor surge back and forth indefinitely. This process is called an **electrical oscillation.**

From the energy standpoint, the oscillations of an electrical circuit consist of a transfer of energy back and forth from the electric field of the capacitor to the magnetic field of the inductor. The *total* energy associated with the circuit remains constant. This is analogous to the transfer of energy in an oscillating mechanical system from kinetic to potential, and vice versa, with the total energy remaining constant.

The frequency of the electrical oscillations of a circuit containing inductance and capacitance only (a so-called L–C circuit) may be calculated in exactly the same way as the frequency of oscillation of a body attached to a spring, discussed in Chapter 11. We suggest you review that discussion before going on. In the mechanical problem, a body of mass m is attached to a spring of force constant k. Suppose we displace the body a distance A from its equilibrium position and release it from rest at time $t = 0$. Then, as shown in the left column of Table 33–1, the kinetic energy of the system at any later time is

TABLE 33–1 Oscillation of a Mass on a Spring Compared with the Electrical Oscillation in an L–C Circuit

Mass on a Spring	Circuit Containing Inductance and Capacitance
Kinetic energy = $\frac{1}{2}mv^2$	Magnetic energy = $\frac{1}{2}Li^2$
Potential energy = $\frac{1}{2}kx^2$	Electrical energy $\dfrac{q^2}{2C}$
$\frac{1}{2}mv^2 + \frac{1}{2}kx^2 = \frac{1}{2}kA^2$	$\dfrac{1}{2}Li^2 + \dfrac{q^2}{2C} = \dfrac{Q^2}{2C}$
$v = \pm\sqrt{k/m}\,\sqrt{A^2 - x^2}$	$i = \pm\sqrt{1/LC}\,\sqrt{Q^2 - q^2}$
$v = \dfrac{dx}{dt}$	$i = \dfrac{dq}{dt}$
$x = A\cos\sqrt{k/m}\,t = A\cos\omega t$	$q = Q\cos\sqrt{1/LC}\,t = Q\cos\omega t$
$v = -\omega A\sin\omega t = -v_{max}\sin\omega t$	$i = -\omega Q\sin\omega t = -I\sin\omega t$

$\frac{1}{2}mv^2$, and its elastic potential energy is $\frac{1}{2}kx^2$. Because the system is conservative, the sum of these values equals the initial energy of the system, $\frac{1}{2}kA^2$. The velocity v at any position is obtained just as in Section 11–2, Eq. (11–4):

$$v = \pm\sqrt{\frac{k}{m}}\sqrt{A^2-x^2}. \tag{33–20}$$

The velocity v equals dx/dt, and we found in Section 11–2 that the coordinate x as a function of t is

$$x = A\cos\left(\sqrt{\frac{k}{m}}\right)t = A\cos\omega t, \tag{33–21}$$

where the angular frequency ω is

$$\omega = \sqrt{\frac{k}{m}}.$$

We recall also that the ordinary frequency f, the number of cycles per unit time, is given by $f = \omega/2\pi$.

In the electrical problem, also a conservative system, a capacitor of capacitance C is given an initial charge Q and, at time $t = 0$, is connected to the terminals of an inductor of self-inductance L. The magnetic energy of the inductor at any later time is analogous to the kinetic energy of the vibrating body and is given by $\frac{1}{2}Li^2$. The electrical energy of the capacitor corresponds to the elastic potential energy of the spring and is given by $q^2/2C$, where q is the charge on the capacitor. The sum of these equals the initial energy of the system, $Q^2/2C$. That is,

$$\frac{1}{2}Li^2 + \frac{q^2}{2C} = \frac{Q^2}{2C}.$$

Solving for i, we find that when the charge on the capacitor is q, the current i is

$$i = \pm\sqrt{\frac{1}{LC}}\sqrt{Q^2 - q^2}. \tag{33–22}$$

Comparing this expression with Eq. (33–20), we see that the current $i = dq/dt$ varies with time in the same way as the velocity $v = dx/dt$ in the mechanical problem. Continuing the analogy, we conclude that q is given as a function of time by

$$q = Q\cos\left(\sqrt{\frac{1}{LC}}\right)t = Q\cos\omega t. \tag{33–23}$$

The angular frequency ω of the electrical oscillations is therefore

$$\omega = \sqrt{\frac{1}{LC}}. \tag{33–24}$$

This frequency is called the *natural frequency* of the L–C circuit. As Table 33–1 shows, it is analogous to the equation $\omega = \sqrt{k/m}$ for the angular frequency of a harmonic oscillator.

These results may also be derived directly from an analysis of the L–C circuit. At each instant the capacitor voltage $v_{ab} = q/C$ must equal that of the inductor, $v_{ab} = L\,di/dt$. Also, because of the choice of positive direction for i,

Energy relations in the L–C circuit

How to find the frequency of oscillation of an L–C circuit

we have $i = -dq/dt$. Combining these relations, we find

$$\frac{d^2q}{dt^2} + \frac{1}{LC} q = 0, \qquad (33\text{--}25)$$

which is analogous to Eq. (11–15) for the harmonic oscillator. The solutions of this differential equation, functions whose second derivative is equal to $-1/LC$ times the original function, are

$$q = Q \cos \omega t, \qquad (33\text{--}26a)$$

$$q = Q \sin \omega t, \qquad (33\text{--}26b)$$

$$q = Q \cos (\omega t + \phi), \qquad (33\text{--}26c)$$

where $\omega = 1/\sqrt{LC}$, and Q and ϕ are constants. Just as with the harmonic oscillator, the choice of one of these functions is determined by the initial conditions. If at time $t = 0$, the capacitor has maximum charge and $i = 0$, as in the discussion above, then we use Eq. (33–26a). If at $t = 0$, $q = 0$ but i is different from zero, we use Eq. (33–26b). And if both q and i are different from zero at time $t = 0$, the more general form, Eq. (33–26c), must be used.

The striking parallel between the mechanical and electrical systems shown in Table 33–1 is only one of many such examples in physics. So close is the parallel between electrical and mechanical (and acoustical) systems that it is possible to solve complicated mechanical and acoustical problems by setting up analogous electrical circuits and measuring the currents and voltages that correspond to the mechanical and acoustical quantities to be determined. This is the basic principle of one kind of *analog computer*. We see that in our comparison between the harmonic oscillator and the *L–C* circuit, m corresponds to L, k to $(1/C)$, x to q, and v to i. This analogy can be extended to *damped oscillations*, which we consider in the next section.

Analogs between mechanical and electrical systems are used in analog computers.

33–6 THE *L–R–C* CIRCUIT

In our discussion of the *L–C* circuit we did not include any *resistance*. This omission is an idealization, of course; every real inductor has resistance in its windings, and there may also be resistance in the connecting wires. The effect of resistance is to dissipate the electromagnetic energy in the circuit and convert it to heat; thus resistance in an electric circuit plays a role analogous to that of friction in a mechanical system.

The *L–R–C* circuit: the electrical analog of damped harmonic motion

To study this situation in greater detail, we consider an inductor with inductance L and a resistor of resistance R connected in series across the terminals of a charged capacitor. As before, the capacitor starts to discharge as soon as the circuit is completed, but because of i^2R losses in the resistor, the energy of the inductor when the capacitor is completely discharged is *less* than the original energy of the capacitor. In the same way, the energy of the capacitor when the magnetic field has collapsed is still smaller, and so on.

The oscillations die out because the resistor dissipates energy.

If the resistance R is relatively small, the circuit still oscillates, but with **damped harmonic motion,** as shown in Fig. 33–8a. If we increase R, the oscillations die out more rapidly. When R reaches a certain value, the circuit no longer oscillates, and we say that it is **critically damped,** as in Fig. 33–8b. For still larger values of R the circuit is **overdamped,** as in Fig. 33–8c. You

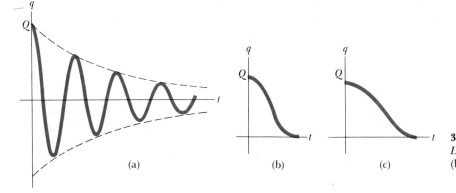

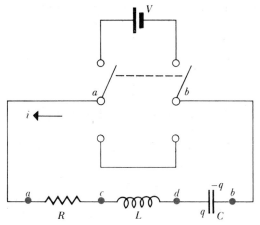

33–8 Graphs of q versus t in an *L–R–C* circuit. (a) Small damping. (b) Critically damped. (c) Overdamped.

may recall that we used these same terms in Section 11–8 for the analogous mechanical situation, the damped harmonic oscillator.

To analyze this behavior in detail, consider the circuit shown in Fig. 33–9. Let us find the current i and the capacitor charge q as functions of time; the analysis is sketched below. First we close the switch in the upward position for a long enough time so that the capacitor acquires its final charge $Q = CV$ and any initial oscillations have died out. Then at time $t = 0$ we flip the switch to the downward position. To find how q and i vary with time, we apply Kirchhoff's loop rule. Starting at point a and going around the loop in the direction *acdba*, we obtain the equation

$$iR + L\frac{di}{dt} + \frac{q}{C} = 0.$$

Replacing i with dq/dt and rearranging, we obtain

$$\frac{d^2q}{dt^2} + \frac{R}{L}\frac{dq}{dt} + \frac{1}{LC}q = 0. \qquad (33\text{–}27)$$

Note that when $R = 0$, this equation reduces to Eq. (33–25).

Solutions of Eq. (33–27) can be obtained by general methods of differential equations. The form of this solution depends on whether R is large or small. When R is less than $2(L/C)^{1/2}$, the solution has the form

$$q = VCe^{-(R/2L)t} \cos\left(\sqrt{\frac{1}{LC} - \left(\frac{R}{2L}\right)^2}\,\right)t. \qquad (33\text{–}28)$$

33–9 An *L–R–C* series circuit.

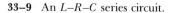

 When R is greater than $2(L/C)^{1/2}$, the solution is

$$q = e^{-(R/2L)t}\left[Ae^{\left(\sqrt{\left(\frac{R}{2L}\right)^2 - \frac{1}{LC}}\right)t} + Be^{\left(-\sqrt{\left(\frac{R}{2L}\right)^2 - \frac{1}{LC}}\right)t}\right], \quad (33\text{--}29)$$

where A and B are constants determined by V, R, L, and C.

Equation (33–28) corresponds to the **underdamped** behavior shown in Fig. 33–8a; the function represents a sinusoidal oscillation with an exponentially decaying amplitude. Note that the angular frequency of the oscillation is no longer $1/(LC)^{1/2}$ but is *less* than this because of the term containing R. The frequency ω' of the damped oscillations is thus given by

$$\omega' = \sqrt{\frac{1}{LC} - \frac{R^2}{4L^2}}. \quad (33\text{--}30)$$

As R increases, ω' becomes smaller and smaller. When $R^2 = 4L/C$, the quantity under the radical becomes zero and the case of *critical damping* has been reached (Fig. 33–8b). For still larger values of R, the behavior is no longer oscillatory but is described as the sum of two exponential functions, as in Fig. 33–8c; the circuit is then *overdamped*.

We emphasize once more that this behavior is completely analogous to that of the damped harmonic oscillator studied in Section 11–8. We invite you to verify, for example, that if you start with Eq. (11–44) and substitute L for m, $1/C$ for k, and R for b, the result is Eq. (33–30). Similarly, the crossover point between underdamping and overdamping occurs at $b^2 = 4km$ for the mechanical system and at $R^2 = 4L/C$ for the electrical one. Can you find still other aspects of this analogy?

It is possible, with appropriate electronic circuitry, to feed energy *into* an L–R–C circuit at the same rate as the rate of dissipation by i^2R losses. It is as though we had inserted a *negative resistance* into the circuit to make the *total* circuit resistance exactly zero. The circuit then oscillates with sustained oscillations of constant amplitude, just as the idealized L–C circuit with no resistance does.

Additional interesting aspects of the behavior of this circuit emerge when a sinusoidally varying emf is included in the circuit. This leads us to the study of alternating-current (ac) circuit analysis, the principal topic of Chapter 34.

SUMMARY

A changing current i_1 in one circuit induces an emf $\mathcal{E}$ in a second circuit if some of the magnetic flux created by the current links the second circuit. The induced emf is given by

$$\mathcal{E}_2 = M\frac{di_1}{dt}, \quad (33\text{--}3)$$

where M is a constant called the mutual inductance. It depends on the geometry of the two coils and on the material between them. The SI unit of mutual inductance, the weber per ampere, is called the henry, abbreviated H.

A changing current i in any circuit induces an emf $\mathcal{E}$ in that same circuit, called a self-induced emf. The self-induced emf is given by

$$\mathcal{E} = L\frac{di}{dt}, \quad (33\text{--}6)$$

The frequency of the damped oscillator is smaller than it would be with no resistance.

If the resistance is too large, there is no oscillation at all.

There are ways to overcome energy losses due to resistance, and to produce sustained oscillations.

KEY TERMS
mutual inductance
henry
self-induced electromotive force (emf)
inductance (self-inductance)
inductor
energy density
time constant
electrical oscillation
damped harmonic motion
critically damped
overdamped
underdamped

where L is a constant depending on the geometry of the circuit and the material surrounding it, called the self-inductance, or simply inductance. A circuit device, usually including a coil of wire, intended to have a substantial inductance is called an inductor.

An inductor with inductance L carrying current I has energy $\frac{1}{2}LI^2$. This energy was fed into the inductor when the current was established, and it is fed back into the circuit when the current decreases to zero again. This energy is associated with the magnetic field of the inductor, with an energy density u (energy per unit volume) given by

$$u = \frac{B^2}{2\mu_0} \qquad (33-10)$$

if the field is in vacuum, or

$$u = \frac{B^2}{2\mu} \qquad (33-11)$$

if it is in a material having magnetic permeability μ.

In a circuit containing a resistor R, an inductor L, and a source of emf V, the growth and decay of current are exponential, with a characteristic time τ called the time constant, given by

$$\tau = \frac{L}{R}. \qquad (33-16)$$

This is the time required for the current to approach within a fraction $1/e$ of its final value.

A circuit containing an inductance L and a capacitance C undergoes electrical oscillations, with angular frequency ω given by

$$\omega = \sqrt{\frac{1}{LC}}. \qquad (33-24)$$

Such a circuit is analogous to a mechanical harmonic oscillator, with the mass m analogous to inductance L, the force constant k to the reciprocal of capacitance $1/C$, the displacement x to charge q, and the velocity v to current i.

A series circuit containing inductance, resistance, and capacitance undergoes damped oscillations for sufficiently small resistance. As R increases, the damping increases; at a certain value of R the behavior becomes overdamped and no longer oscillates. The crossover between underdamping and overdamping occurs when

$$R^2 = 4L/C,$$

and the frequency ω' of damped oscillations when R is less than this critical value is

$$\omega' = \sqrt{\frac{1}{LC} - \frac{R^2}{4L^2}}. \qquad (33-30)$$

There is a direct analogy between every aspect of the behavior of the $L-R-C$ circuit and the mechanical damped harmonic oscillator. This analogy and similar ones are widely used in analog computers.

QUESTIONS

33–1 A resistor is to be made by winding a wire around a cylindrical form. In order to make the inductance as small as possible, it is proposed that we wind half the wire in one direction and the other half in the opposite direction. Would this achieve the desired result? Why or why not?

33–2 In Fig. 33–1, if coil 2 is turned 90° so that its axis is vertical, does the mutual inductance increase or decrease?

33–3 The toroidal solenoid is one of the few configurations for which it is easy to calculate self-inductance. What features of the toroidal solenoid give it this simplicity?

33–4 Two identical closely wound circular coils, each having self-inductance L, are placed side by side, close together. If they are connected in series, what is the self-inductance of the combination? What if they are connected in parallel? Can they be connected so that the total inductance is zero?

33–5 If two inductors are separated enough so that practically no flux from either links the coils of the other, show that the equivalent inductance of two inductors in series or parallel is obtained by the same rules for combining resistance.

33–6 Two closely wound circular coils have the same number of turns, but one has twice the radius of the other. How are the self-inductances of the two coils related?

33–7 One of the great problems in the field of energy resources and utilization is the difficulty of storing electrical energy in large quantities economically. Discuss the possibility of storing large amounts of energy by means of currents in large inductors.

33–8 In what regions in a toroidal solenoid is the energy density greatest? Least?

33–9 Suppose there is a steady current in an inductor. If one attempts to reduce the current to zero instantaneously by opening a switch, a big fat arc appears at the switch contacts. Why? What happens to the induced emf in this situation? Is it physically possible to stop the current instantaneously?

33–10 In the $R–L$ circuit of Fig. 33–5, is the current in the resistor always the same as that in the inductor? How do you know?

33–11 In the $R–L$ circuit of Fig. 33–5, when switch S_1 is closed, the potential V_{ab} changes suddenly and discontinuously, but the current does not. Why can the voltage change suddenly but not the current?

33–12 In the $R–L–C$ circuit, what criteria could be used to decide whether the system is overdamped or underdamped? For example, could one compare the maximum energy stored during one cycle to the energy dissipated during one cycle?

EXERCISES

Section 33–1 Mutual Inductance

33–1 A solenoid of length 10 cm and radius 2 cm is wound uniformly with 1000 turns. A second coil of 50 turns is wound around the solenoid at its center. What is the mutual inductance of the two coils?

33–2 A toroidal solenoid (cf. Section 31–7) has a radius of 10 cm and a cross-sectional area of 5 cm², and is wound uniformly with 1000 turns. A second coil with 500 turns is wound uniformly on top of the first. What is the mutual inductance?

33–3 Two coils have mutual inductance $M = 0.01$ H. The current i_1 in the first coil increases at a uniform rate of 0.05 A·s⁻¹.

a) What is the induced emf in the second coil? Is it constant?

b) Suppose that the current described is in the second coil rather than the first; what is the induced emf in the first coil?

Section 33–2 Self-Inductance

33–4

a) Show that the two expressions for self-inductance, namely,

$$\frac{N\Phi}{i} \quad \text{and} \quad \frac{\mathcal{E}}{di/dt},$$

have the same units.

b) Show that L/R and RC both have units of time.

c) Show that 1 Wb·s⁻¹ equals 1 V.

33–5 An inductor of inductance 5 H carries a current that decreases at a uniform rate, $di/dt = -0.02$ A·s⁻¹. Find the self-induced emf; what is its polarity?

33–6 An inductor with $L = 40$ H carries a current i that varies with time according to $i = (0.1$ A$) \sin ([120\pi$ s⁻¹$]t)$. Find an expression for the induced emf. What is the phase of $\mathcal{E}$ relative to i?

33–7

a) Find the self-inductance of the toroidal solenoid in Exercise 33–2 if only the 1000-turn coil is used.

b) If both coils are used, connected in series with each other, what is the self-inductance of the combination?

33–8 A solenoid has length 15 cm, radius 4 cm, and 2000 turns. It is filled with a core of relative permeability 600. Calculate the self-inductance of the solenoid.

Section 33–3 Energy in an Inductor

33–9 A toroidal solenoid has a mean radius of 0.12 m and a cross-sectional area of 20×10^{-4} m². It is found that when the current is 20 A, the energy stored is 0.1 J. How many turns does the winding have?

33–10 An inductor used in a dc power supply has an inductance of 20 H and a resistance of 200 Ω and carries a current of 0.1 A.

a) What is the energy stored in the magnetic field?

b) At what rate is energy dissipated in the resistor?

33–11 Derive in detail Eq. 33–11 for the energy density in a toroidal solenoid filled with a magnetic material.

33–12 A magnetic field of magnitude $B = 0.4$ T is uniform across a volume of 0.02 m^3. Calculate the total magnetic energy in the volume if

a) the volume is free space;

b) the volume is filled with material of relative permeability 600.

Section 33–4 The R–L Circuit

33–13 An inductor of inductance 3 H and resistance 6 Ω is connected to the terminals of a battery of emf 12 V and of negligible internal resistance. Find

a) the initial rate of increase of current in the circuit,

b) the rate of increase of current at the instant when the current is 1 A,

c) the current 0.2 s after the circuit is closed,

d) the final steady-state current.

33–14 The resistance of a 10-H inductor is 200 Ω. The inductor is suddenly connected across a potential difference of 10 V.

a) What is the final steady current in the inductor?

b) What is the initial rate of increase of current?

c) At what rate is the current increasing when its value is one-half the final current?

d) At what time after the circuit is closed does the current equal 99% of its final value?

e) Compute the current at the following times after the circuit is closed: 0, 0.025 s, 0.05 s, 0.075 s, 0.10 s, and 0.20 s. Show the results in a graph.

33–15 Write an equation corresponding to Eq. (33–12) for the current in Fig. 33–5 just after switch S$_2$ is closed and S$_1$ is opened, if the initial current is I. Solve the resulting differential equation and verify Eq. (33–18).

33–16 In Fig. 33–5 let $V = 200$ V, $R = 500$ Ω, and $L = 0.1$ H. With switch S$_2$ open, switch S$_1$ is closed and left until a constant current is established. Then S$_2$ is closed and S$_1$ opened, so the battery is taken out of the circuit.

a) What is the initial current in the resistor?

b) What is the current in the resistor at $t = 0.2 \times 10^{-3}$ s?

c) What is the potential difference between points b and c at $t = 0.2 \times 10^{-3}$ s? Which point is at higher potential?

d) How long does it take the current to decrease to half its initial value?

Section 33–5 The L–C Circuit

33–17 An inductor having $L = 40$ mH is to be combined with a capacitor to make an L–C circuit with natural frequency 2×10^6 Hz. What value of capacitance should be used?

33–18 The maximum capacitance of a variable air capacitor is 35 pF.

a) What should be the self-inductance of a coil connected to this capacitor if the natural frequency of the L–C circuit is to be 550×10^3 Hz, corresponding to one end of the AM radio broadcast band, when the capacitor is set to its maximum capacitance?

b) The frequency at the other end of the broadcast band is 1550×10^3 Hz. What must be the minimum capacitance of the capacitor if the natural frequency is to be adjustable over the range of the broadcast band?

33–19 Show that differential Eq. (33–25) is satisfied by the function $q = Q \cos \omega t$, with ω given by $1/(LC)^{1/2}$.

Section 33–6 The L–R–C Circuit

33–20 Show that the quantity $(L/C)^{1/2}$ has units of resistance (ohms).

33–21 An L–R–C circuit has $L = 0.5$ H, $C = 0.1 \times 10^{-3}$ F, and resistance R.

a) What is the natural angular frequency of the circuit when $R = 0$?

b) What value must R have to give a 10% decrease in natural frequency compared to the value calculated in part (a)?

PROBLEMS

33–22 A solenoid has length l_1, radius r_1, and number of turns N_1. A second, smaller solenoid of length l_2, radius r_2, and number of turns N_2 is placed at the center of the first solenoid, such that their axes coincide.

a) What is the mutual inductance of the pair of solenoids?

b) If the current in the larger solenoid is increasing at the rate di_1/dt, what is the emf induced in the small solenoid?

c) If the current in the smaller solenoid is increasing at the rate di_2/dt, what is the emf induced in the larger solenoid?

33–23 A coil of wire, initially carrying no current, has a current of 50 A established in 10 s. The current increases at a steady (constant) rate during this time, and the charging current induces an emf of 25 V in the wire.

a) Determine the self-inductance of the coil

b) Determine the total magnetic flux through the coil when the current is 50 A.

c) If the resistance of the coil is 25 Ω, determine the ratio of the rate at which energy is being stored in the magnetic field to the rate at which energy is being dissipated by the resistance at the instant when the current is 50 A.

33–24 A toroidal solenoid has two coils with N_1 and N_2 turns, respectively; it has radius r and cross-sectional area A.

a) Derive an expression for the self-inductance L_1 when only the first coil is used, and that for L_2 when only the second coil is used.

b) Derive an expression for the mutual inductance of the two coils.

c) Show that $M^2 = L_1L_2$. This result is valid whenever all the flux linked by one coil is also linked by the other.

33–25 The current in a resistanceless inductor is caused to vary with time as in the graph of Fig. 33–10.

a) Sketch the pattern that would be observed on the screen of an oscilloscope connected to the terminals of the inductor. (The oscilloscope spot sweeps horizontally across the screen at a constant speed, and its vertical deflection is proportional to the potential difference between the inductor terminals.)

b) Explain why the inductor can be described as a "differentiating circuit."

FIGURE 33–10

33–26 A coaxial cable consists of a small solid conductor of radius r_a supported by insulating disks on the axis of a thin-walled tube of inner radius r_b. Show that the self-inductance of a length l of the cable is

$$L = l\frac{\mu_0}{2\pi}\ln\left[\frac{r_b}{r_a}\right].$$

Assume the inner and outer conductors carry equal currents in opposite directions. (*Hint:* Use Ampere's law to find the magnetic field at any point in the space between the conductors. Write the expression for the flux $d\Phi$ through a narrow strip of length l parallel to the axis, of width dr, at a distance r from the axis of the cable and lying in a plane containing the axis. Integrate to find the total flux linking a current i in the central conductor.)

33–27 Uniform electric and magnetic fields **E** and **B** occupy the same region of free space. If $E = 400$ V·m^{-1}, what is B if the energy densities in the electric and magnetic fields are equal?

33–28 The 1000-turn toroidal solenoid described in Exercise 33–2 carries a current of 5.0 A.

a) What is the energy density in the magnetic field?

b) What is the total magnetic-field energy? Find the answer by using Eq. (33–8) and also by multiplying the energy density from (a) by the volume of the toroid, which is $2\pi rA$; compare the two results.

33–29 An inductor having inductance L and resistance R carries a current I. Show that the time constant is equal to *twice the ratio* of the energy stored in the magnetic field to the rate of dissipation of energy in the resistance.

33–30 Refer to Exercise 33–13.

a) What is the power input to the inductor from the battery, as a function of time, if the circuit is completed at $t = 0$?

b) What is the rate of dissipation of energy in the resistance of the inductor, as a function of time?

c) What is the rate at which the energy of the magnetic field in the inductor is increasing, as a function of time?

d) Compare the results of (a), (b), and (c).

33–31 Refer to Exercise 33–13.

a) How much energy is stored in the magnetic field of the inductor one time constant after the battery has been connected? Compute this value both by integrating the expression in Problem 33–30(c) and by using Eq. (33–8), and compare the results.

b) Integrate the expression obtained in Problem 33–30(a) to find the *total* energy supplied by the battery during the time interval considered in (a).

c) Integrate the expression obtained in Problem 33–30(b) to find the *total* energy dissipated in the resistance of the inductor during the same time period.

d) Compare the results obtained in (a), (b), and (c).

33–32 Refer to Exercise 33–16.

a) What is the total energy initially stored in the inductor?

b) At $t = 0.2 \times 10^{-3}$ s, at what rate is the energy stored in the inductor decreasing?

c) At $t = 0.2 \times 10^{-3}$ s, at what rate is electrical energy being converted into heat in the resistor?

d) Obtain an expression for the rate at which electrical energy is being converted into heat in the resistor, as a function of time. Integrate this expression from $t = 0$ to $t = \infty$ to obtain the total electrical energy dissipated in the resistor. Compare your result to that of part (a).

33–33 An inductor is made with two coils wound close together on a form, so all the flux linking one coil also links the other. The number of turns is the same in each. If the inductance of one coil is L, what is the inductance when the two coils are connected

a) in series?

b) in parallel?

In each case the current travels in the same sense around each coil.

c) If an L–C circuit with this inductor has natural frequency ω using one coil, what is the natural frequency when the two coils are used in series?

33–34 An L–C circuit consists of an inductor of $L = 0.8$ H and a capacitor of $C = 0.4 \times 10^{-3}$ F. The initial charge on the capacitor is 5 μC.

a) What is the maximum voltage across the capacitor?

b) What is the maximum current in the inductor?

c) What is the maximum energy stored in the inductor?

d) When the current in the inductor has half its maximum value, what is the charge on the capacitor and what is the energy stored in the inductor?

33–35 The equation preceding Eq. (33–27) may be converted into an energy relation. Multiply both sides of this equation by $i = dq/dt$. The first term then becomes i^2R. Show that the second can be written $d(\frac{1}{2}Li^2)/dt$, and the third can be written $d(q^2/2C)/dt$. What does the resulting equation say about energy conservation in the circuit?

33–36 An inductor of resistance R and self-inductance L is connected in series with a noninductive resistor of resistance R_0 to a constant potential difference V (Fig. 33–11).

a) Find the expression for the potential difference v_{cb} across the inductor at any time t after switch S_1 is closed.

b) Let $V = 20$ V, $R_0 = 50$ Ω, $R = 150$ Ω, $L = 5$ H. Compute v_{ac} and v_{cb} for $t = 0$, 0.5τ, τ, 1.5τ, and 2.5τ, where τ is the time constant of the circuit. Also calculate v_{ac} and v_{cb} for $t \to \infty$. Sketch graphs of v_{ac} and v_{cb} versus time from time zero to infinity.

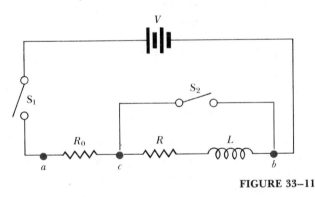

FIGURE 33–11

33–37 After the current in the circuit of Fig. 33–11 has reached its final steady value, the switch S_2 is closed, thus short-circuiting the inductor. (Switch S_1 remains closed.)

a) Derive an expression for the currents through R, R_0, and S_2 as functions of time.

b) What will be the magnitude and direction of the current in S_2, 0.01 s after S_2 is closed? Use the numerical values given in Problem 33–36.

33–38 A popular demonstration of self-inductance often used in a physics course employs a circuit such as the one

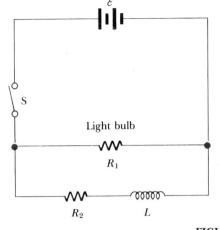

FIGURE 33–12

shown in Fig. 33–12. Switch S is closed and the light bulb, represented by resistance R_1, is seen to just barely glow. After a period of time switch S is opened and the bulb is then seen to light up brightly for a short time. This effect is easy to understand if one thinks of an inductor as a device that imparts an "inertia" to the current, preventing a discontinuous change in the current through it.

a) Derive, as explicit functions of time, expressions for i_1 and i_2, the currents through the light bulb and inductor, after switch S is closed.

b) After a long time, steady-state conditions can be assumed. Obtain expressions for the steady-state currents in the bulb and the inductor.

c) Switch S is now opened. Obtain an expression for the current through the inductor and light bulb as an explicit function of time.

d) You have been asked to design demonstration apparatus using the circuit shown in Fig. 33–12 with a 50-H inductor and a 60-W light bulb. You are to connect a resistor in series with the inductor, and R_2 represents the sum of that resistance plus the internal resistance of the inductor. When switch S is opened, a transient current is to be set up that starts at 0.9 A and is not to fall below 0.3 A until after 0.2 s. For simplicity assume the resistance of the light bulb is constant and equals the resistance the bulb must have to dissipate 60 W at 120 V. Determine R_2 and $\mathcal{E}$ for the given design considerations.

e) With the numerical values determined in (d), what is the current through the light bulb just before the switch is opened? Does the result confirm the qualitative description of what one observes in the demonstration?

CHALLENGE PROBLEMS

33–39 A certain toroidal solenoid has a rectangular cross section, as shown in Fig. 33–13. It has N uniformly spaced turns, with air inside. The magnetic field at a point inside the toroid is given by Eq. (31–29). *Do not* assume the field to be uniform over the cross section.

a) Show that the magnetic flux through a cross section of

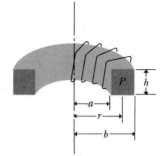

FIGURE 33-13

the toroid is

$$\Phi = \frac{\mu_0 N I h}{2\pi} \ln\left[\frac{b}{a}\right].$$

b) Show that the self-inductance of the toroidal solenoid is given by

$$L = \frac{\mu_0 N^2 h}{2\pi} \ln\left[\frac{b}{a}\right].$$

c) The fraction b/a may be written as

$$\frac{b}{a} = \frac{a + b - a}{a} = 1 + \frac{b - a}{a}.$$

The power series expansion for $\ln(1 + x)$ is $\ln(1 + x) = x + x^2/2 + \cdots$. Hence show that when $b - a$ is much less than a, the self-inductance is approximately equal to

$$L = \frac{\mu_0 N^2 h (b - a)}{2\pi a}.$$

Compare this result with that obtained in Example 33-2.

33-40 For the toroidal solenoid described in Challenge Problem 33-39 calculate the total energy when the current is I. *Do not* assume the field to be uniform over the cross section. Obtain the total energy in two ways:

a) Use Eq. (33-8) with the expression for L obtained in Problem 33-39.

b) Integrate Eq. (33-10) through the volume of the toroid. Compare the results of parts (a) and (b).

33-41 Consider the circuit shown in Fig. 33-14. The values of the circuit elements are as follows: $\mathcal{E} = 50$ V, $L = 10$ H, $C = 20$ μF, $R_1 = 25$ Ω, and $R_2 = 5000$ Ω. Switch S is closed at time $t = 0$, causing a current i_1 through the inductive branch and a current i_2 through the capacitive branch. The initial charge on the capacitor is zero, and the charge at time t is q_2.

a) Derive expressions for i_1, i_2, and q_2 as explicit functions of time.

b) What is the initial current through the inductive branch? What is the initial current through the capacitive branch?

c) What will be the currents through the inductive and capacitive branches a long time after the switch has been closed? How long is a "long time"? Explain.

d) At what time t_1 will the currents i_1 and i_2 be equal? (*Hint:* Consider using series expansions for the exponentials.)

e) For the conditions given in part (d), determine i_1.

f) The total current through the battery is $i = i_1 + i_2$. At what time t_2 will i equal one-half its final value? (*Hint:* As in (d), the numerical work is greatly simplified if one makes suitable approximations.)

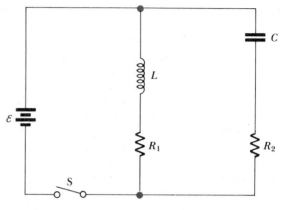

FIGURE 33-14

33-42 Consider the circuit shown in Fig. 33-15. The circuit elements are as follows: $\mathcal{E} = 50$ V, $L = 20$ H, $C = 6.25$ μF, and $R = 1000$ Ω. At time $t = 0$, switch S is closed. The current through the inductance is i_1, the current through the capacitor branch is i_2, and the charge on the capacitor is q_2.

a) Using Kirchhoff's rules, verify the circuit equations

$$R(i_1 + i_2) + L\left(\frac{di_1}{dt}\right) = \mathcal{E},$$

$$R(i_1 + i_2) + \frac{q_2}{C} = \mathcal{E}.$$

b) What are the initial values of i_1, i_2, and q_2?

c) Show by direct substitution that the following solutions for i_1 and q_2 satisfy the circuit equations from part (a). Also show that they satisfy the initial conditions.

$$i_1 = (\mathcal{E}/R)(1 - [(2\omega RC)^{-1}\sin(\omega t) + \cos(\omega t)]e^{-\beta t}),$$

$$q_2 = \left(\frac{\mathcal{E}}{\omega R}\right)e^{-\beta t}\sin(\omega t),$$

where $\beta = (2RC)^{-1}$ and $\omega = [(LC)^{-1} - (2RC)^{-2}]^{\frac{1}{2}}$.

d) Determine the time t_1 at which i_2 first becomes zero.

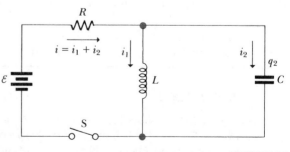

FIGURE 33-15

33–43 A tank containing a liquid has turns of wire wrapped around it, causing it to act as an inductor. The tank can be used to measure the liquid content by using its inductance to determine the height of the liquid. The inductance of the tank changes from a value of L_0, corresponding to a relative permeability of 1, when the tank is empty to a value of L_f, corresponding to a relative permeability of K_m (the relative permeability of the liquid) when the tank is full. The appropriate electronic circuitry can determine the inductance to five significant figures, and hence the effective relative permeability of the combined air and liquid within the rectangular cavity of the tank. Each of the four sides of the tank has a width W and a height D. (See Fig. 33–16.) The height of the liquid in the tank is d. Neglect any fringing effects, and also assume that the relative permeability of the tank material can be neglected.

a) Derive an expression for d as a function of L, the inductance corresponding to a certain fluid height, and L_0 and L_f.

b) What would be the inductance (to five significant figures) for a tank one-quarter full, one-half full, and three-quarters full, if the tank contains liquid oxygen? Take $L_0 = 1.25$ H. The magnetic susceptibility of liquid oxygen is $\chi_m = 1.52 \times 10^{-3}$.

c) Repeat part (b) for mercury. The magnetic susceptibility of mercury is given in Table 31–1.

d) For which material would this gauge be more practical?

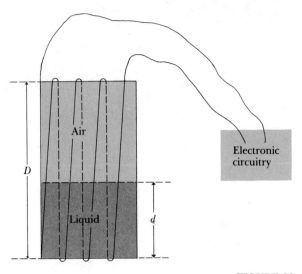

FIGURE 33–16

34

ALTERNATING CURRENTS

In many practical circuits, the currents are sinusoidal functions of time.

IN OUR STUDY OF CIRCUITS IN CHAPTERS 28 AND 29, WE CONCENTRATED primarily on *direct-current* circuits, in which all the currents, voltages, and emf's are *constant,* that is, not functions of time. However, many electric circuits of practical importance, including most household and industrial power-distribution systems, use **alternating current,** in which the voltages and currents vary with time, often in a *sinusoidal* manner. Alternating currents are of utmost importance in technology and industry. Transmission of power over long distances is much more economical with alternating than with direct current. Circuits used in modern communication equipment, including radio and television, make extensive use of alternating current. In this chapter we study the behavior of circuits with sinusoidally varying voltages and currents. Many of the same principles used in Chapters 28, 29, and 33 are applicable here as well, and we also introduce several new concepts related to the circuit behavior of inductors and capacitors.

34–1 AC SOURCES AND PHASORS

An alternator is a source of sinusoidally varying voltage and current.

We have already studied several sources of alternating emf or voltage. A coil of wire, rotating with constant angular velocity in a magnetic field, develops a sinusoidal alternating emf, as discussed in Section 32–2. This simple device is the prototype of the commercial alternating-current generator, or **alternator.** An L–C circuit, as discussed in Section 33–5, oscillates sinusoidally; and with the proper circuitry it provides an alternating potential difference having a frequency, depending on its purpose, that may range from a few hertz to many millions of hertz.

We first consider several circuits connected to an alternator or oscillator that maintains between its terminals a sinusoidal alternating potential difference

$$v = V \cos \omega t. \tag{34–1}$$

Sinusoidally varying voltages and currents are described in terms of their amplitude, frequency, and phase.

Here V is the maximum potential difference, or the **voltage amplitude;** v is the *instantaneous* potential difference; and ω is the *angular frequency,* equal to 2π times the frequency f. In the United States, commercial electric-power distribution systems always use a frequency of $f = 60$ Hz, corresponding to $\omega = 377 \text{ s}^{-1}$. For brevity, the alternator or oscillator will be referred to as an **ac source.** The circuit-diagram symbol for an ac source is $\otimes$.

Analysis of alternating-current circuits is facilitated by use of rotating vector diagrams similar to those used in the study of harmonic motion in Section 11–3. In such diagrams the instantaneous value of a quantity that varies sinusoidally with time is represented by the *projection*, onto a horizontal axis, of a vector whose length corresponds to the amplitude of the quantity; the vector rotates counterclockwise with constant angular velocity ω. These rotating vectors are called **phasors,** and diagrams containing them are called **phasor diagrams.** In Section 11–3 we used a single phasor to represent the position and motion of a point mass undergoing simple harmonic motion. In this chapter we will find phasors convenient for *adding* sinusoidal voltages and currents; with them we can apply the method of vector addition to combine sinusoidal quantities with phase differences. We will find a similar use for phasors in Chapter 39, in our study of interference phenomena in optics.

Phasor diagrams: a handy way to handle phase relationships in ac circuits

34–2 RESISTANCE, INDUCTANCE, AND CAPACITANCE

The simplest problem in ac-circuit analysis consists of a resistor of resistance R connected between the terminals of an ac source, as in Fig. 34–1a. Suppose the instantaneous potential of point a with respect to point b is given by $v = V \cos \omega t$. (We could equally well have chosen the sine function instead of the cosine.) The instantaneous current i in the resistor is

In a resistor, the voltage and current are in phase.

$$i = \frac{v}{R} = \frac{V}{R} \cos \omega t.$$

The maximum current I, or the **current amplitude,** is

$$I = \frac{V}{R}, \tag{34–2}$$

and we can therefore write

$$i = I \cos \omega t. \tag{34–3}$$

The current and voltage are both proportional to $\cos \omega t$, so the current is *in phase* with the voltage. The current and voltage amplitudes, from Eq. (34–2), are related in the same way as in a dc circuit.

Figure 34–1b shows graphs of i and v as functions of time. The fact that the curve representing current has the greater amplitude in the diagram is of no significance, because the choice of vertical scales for i and v is arbitrary. The corresponding phasor diagram is given in Fig. 34–1c. Because i and v are in phase and have the same frequency, the current and voltage phasors rotate together. Their projections on the horizontal axis represent the instantaneous current and voltage, respectively.

Next, suppose that a capacitor of capacitance C is connected across the source, as in Fig. 34–2a. The instantaneous charge q on the capacitor is

$$q = Cv = CV \cos \omega t. \tag{34–4}$$

In this case the instantaneous current is equal to the *rate of change* of the capacitor charge and so is proportional to the rate of change of voltage. Tak-

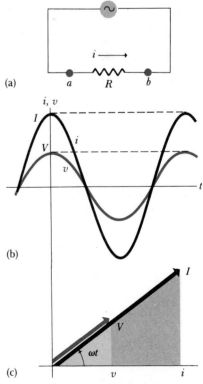

34–1 (a) Resistance R connected across an ac source. (b) Graphs of instantaneous voltage and current. (c) Phasor diagram; current and voltage are in phase.

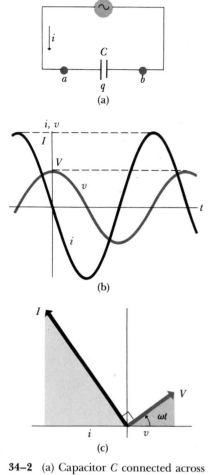

34–2 (a) Capacitor C connected across an ac source. (b) Graphs of instantaneous voltage and current. (c) Phasor diagram; voltage *lags* current by 90°.

Capacitive reactance: the ratio of voltage amplitude to current amplitude for a capacitor

The ohm: the unit of reactance as well as resistance

ing the derivative of Eq. (34–4), we find

$$i = \frac{dq}{dt} = -\omega CV \sin \omega t. \qquad (34\text{–}5)$$

Thus if the voltage is represented by a *cosine* function, the current is represented by a *negative sine* function, as shown in Fig. 34–2b. The current is *not* in phase with the voltage; the current is greatest at times when the v curve is rising or falling most steeply, and is zero at times when the v curve "levels off"; that is, when it reaches its maximum and minimum values.

The voltage and current are "out of step" or *out of phase* by a quarter-cycle, with the current a quarter-cycle ahead. The peaks of current occur a quarter-cycle *before* the corresponding voltage peaks. This result is also represented by the diagram of Fig. 34–2c, which shows a current phasor ahead of the voltage phasor by a quarter-cycle, or 90°. This *phase difference* between current and voltage can also be obtained by rewriting Eq. (34–5), using the trigonometric identity $\cos(A + 90°) = -\sin A$:

$$i = \omega CV \cos(\omega t + 90°). \qquad (34\text{–}6)$$

We can think of the expression for i in a capacitor as a cosine function with a "head start" of 90° compared with that for the voltage. We say that the current *leads* the voltage by 90° or that the voltage *lags* the current by 90°. Alternatively, if i is given by $i = \omega CV \cos \omega t$, then $v = V \cos(\omega t - 90°)$.

For uniformity of language in later discussions, we will usually describe the phase of the voltage relative to the current. Thus if the current i in a circuit is given by

$$i = I \cos \omega t$$

and the voltage v between two points is

$$v = V \cos(\omega t + \phi),$$

we call ϕ the **phase angle,** understanding that it gives the phase of the voltage relative to the current. It is always an angle between $-90°$ and $+90°$. Hence for the capacitor we have $\phi = -90°$.

Equations (34–5) and (34–6) show that the maximum current I is given by

$$I = \omega CV. \qquad (34\text{–}7)$$

This expression can be put in the same form as that for the maximum current in a resistor ($I = V/R$) if we write Eq. (34–7) as

$$I = \frac{V}{1/\omega C}$$

and define a quantity X_C, called the **capacitive reactance** of the capacitor, as

$$X_C = \frac{1}{\omega C}. \qquad (34\text{–}8)$$

Then

$$I = \frac{V}{X_C}. \qquad (34\text{–}9)$$

From Eq. (34–9) the SI unit of capacitive reactance is one *volt per ampere* (1 V·A^{-1}), or one *ohm* (1 Ω).

The reactance of a capacitor is inversely proportional both to the capacitance C and to the angular frequency ω; the greater the capacitance and the higher the frequency, the *smaller* the reactance X_C.

EXAMPLE 34–1 At an angular frequency of 1000 rad·s^{-1}, the reactance of a 1-μF capacitor is

$$X_C = \frac{1}{\omega C} = \frac{1}{(10^3 \text{ rad·s}^{-1})(10^{-6} \text{ F})} = 1000 \ \Omega.$$

At a frequency of 10,000 rad·s^{-1}, the reactance of the same capacitor is only 100 Ω, and at a frequency of 100 rad·s^{-1}, it is 10,000 Ω.

Finally, suppose we connect a pure inductor having a self-inductance L and zero resistance to an ac source, as in Fig. 34–3. The potential v of point a with respect to point b is given by $v = L \, di/dt$, and we have

$$L\frac{di}{dt} = V \cos \omega t \qquad (34\text{–}10)$$

and

$$di = \frac{V}{L} \cos \omega t \, dt.$$

Integrating both sides gives

$$i = \frac{V}{\omega L} \sin \omega t + \text{constant}.$$

If $i = 0$ at time $t = 0$, then constant = 0 and

$$i = \frac{V}{\omega L} \sin \omega t. \qquad (34\text{–}11)$$

Again, the voltage and current are a quarter-cycle out of phase, but this time the voltage *leads* the current by 90°. This may also be seen by rewriting Eq. (34–11) using the trigonometric identity $\cos(A - 90°) = \sin A$:

$$i = \frac{V}{\omega L} \cos(\omega t - 90°). \qquad (34\text{–}12)$$

This result and the phasor diagram of Fig. 34–3c show that the voltage can be viewed as a cosine function with a "head start" of 90°. Alternatively, if i is given by

$$i = \frac{V}{\omega L} \cos \omega t,$$

then the voltage v is

$$v = V \cos(\omega t + 90°).$$

The phase angle ϕ of the voltage with respect to the current is $\phi = +90°$.

In Fig. 34–3b the points of maximum voltage correspond on the graphs to points of maximum steepness of the current curve, and the points of zero

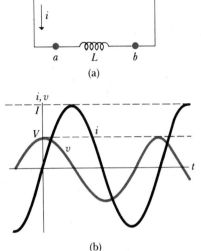

(a)

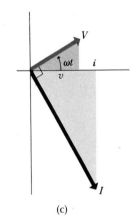

(b)

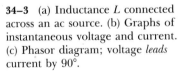

(c)

34–3 (a) Inductance L connected across an ac source. (b) Graphs of instantaneous voltage and current. (c) Phasor diagram; voltage *leads* current by 90°.

voltage correspond to the points where the current curve levels off at its maximum and minimum values.

From Eq. (34–11) or (34–12) the maximum current I is

$$I = \frac{V}{\omega L},\tag{34-13}$$

so we can also write

$$i = I \sin \omega t = I \cos (\omega t - 90°).\tag{34-14}$$

Inductive reactance: the ratio of voltage amplitude to current amplitude for an inductor

We define the **inductive reactance** X_L of an inductor as

$$X_L = \omega L,\tag{34-15}$$

and we can write Eq. (34–13) in the same form as for a resistor:

$$I = \frac{V}{X_L}.\tag{34-16}$$

The SI unit of inductive reactance is also one *ohm*.

Inductive reactance is directly proportional to frequency.

The reactance of an inductor is directly proportional both to its inductance L and to the angular frequency ω; the greater the inductance and the higher the frequency, the *larger* the reactance.

EXAMPLE 34–2   At an angular frequency of 1000 rad·s^{-1}, the reactance of a 1-H inductor is

$$X_L = \omega L = (10^3 \text{ rad·s}^{-1})(1 \text{ H}) = 1000 \ \Omega.$$

At a frequency of 10,000 rad·s^{-1} the reactance of the same inductor is 10,000 Ω, while at a frequency of 100 rad·s^{-1} it is only 100 Ω.

The graphs in Fig. 34–4 summarize the variations with frequency of the resistance of a resistor and the reactances of an inductor and of a capacitor. As the frequency increases, the reactance of the inductor approaches infinity, and that of the capacitor approaches zero. As the frequency decreases, the inductive reactance approaches zero, and the capacitive reactance approaches infinity. The limiting case of zero frequency corresponds to a dc circuit.

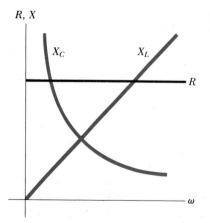

34–4 Graphs of R, X_L, and X_C as functions of frequency.

PROBLEM-SOLVING STRATEGY: *Alternating-current circuits*

1. In ac-circuit problems it is nearly always easiest to work with angular frequency ω. But in problems you may be given the ordinary frequency f, expressed in Hz. Don't forget to make the conversion, using $\omega = 2\pi f$.

2. Keep in mind a few basic facts about phase relationships. For a resistor, voltage and current are always *in phase*, and the two corresponding phasors in a phasor diagram always have the same direction. For a capacitor, the voltage always *lags* the current by 90° (i.e., $\phi = -90°$), and the voltage phasor is always turned 90° clockwise from the current phasor. For an inductor, the voltage always *leads* the current by 90° (i.e., $\phi = +90°$), and the voltage phasor is always turned 90° counterclockwise from the current phasor.

3. Remember that Kirchhoff's rules are just as applicable for ac circuits as for dc. All the voltages and currents are sinusoidal functions of time instead of being constant, but Kirchhoff's rules hold at each instant. Thus in a series circuit, the instantaneous current is the same in each circuit element; in a parallel circuit, the instantaneous potential difference is the same for each circuit element.

4. Reactance and impedance (coming in the next section) are a lot like resistance; each represents the ratio of voltage amplitude V to current amplitude I in a circuit element or combination of elements. But keep in mind that phase relations are always lurking in the shadows; resistance, reactance, and impedance have to be combined by *vector* addition of the corresponding phasors. When you have several circuit elements in series, for example, it isn't correct simply to *add* all the numerical values of resistance and reactance; that would ignore the phase relations.

34–3 THE *L–R–C* SERIES CIRCUIT

Many ac circuits of great usefulness include resistance, inductive reactance, and capacitive reactance. A simple series circuit is shown in Fig. 34–5a. In analyzing this and similar circuits, we will use a phasor diagram that includes the voltage and current phasors for each of the components. Here, because of Kirchhoff's loop rule, the instantaneous *total* voltage across all three components is equal to the source voltage at that instant, and its phasor is the *vector sum* of the phasors for the individual voltages. The complete phasor diagram for this circuit is shown in Fig. 34–5b. This may appear complex, but it is not as bad as it looks; we will explain it step by step.

The instantaneous current i has the same value at all points of the circuit. Thus a *single phasor I*, of length proportional to the current amplitude, represents the current in each circuit element.

Let us use the symbols v_R, v_L, and v_C for the instantaneous voltages across R, L, and C, and V_R, V_L, and V_C for their maximum values. We denote the instantaneous and maximum source voltages by v and V. Then $v = v_{ab}$, $v_R = v_{ac}$, $v_L = v_{cd}$, and $v_C = v_{db}$.

We have shown that the potential difference between the terminals of a resistor is *in phase* with the current in the resistor, and that its maximum value V_R is

$$V_R = IR.$$

Thus the phasor V_R in Fig. 34–5b, in phase with the current vector, represents the voltage across the resistor. Its projection on the horizontal axis, at any instant, gives the instantaneous potential difference v_R.

The voltage across an inductor *leads* the current by 90°. The voltage amplitude is

$$V_L = IX_L.$$

Using phasor diagrams to analyze voltage and current relations in an *L–R–C* circuit

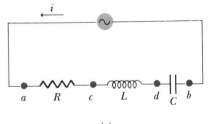

(a)

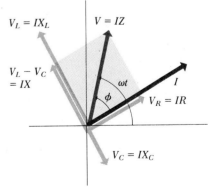

(b)

34–5 (a) A series *L–R–C* circuit. (b) Phasor diagram.

The phasor V_L in Fig. 34–5b represents the voltage across the inductor, and its projection at any instant onto the horizontal axis equals v_L.

The voltage across a capacitor *lags* the current by 90°. The voltage amplitude is

$$V_C = IX_C.$$

The phasor V_C in Fig. 34–5b represents the voltage across the capacitor, and its projection at any instant onto the horizontal axis equals v_C.

The instantaneous potential difference v between terminals a and b equals at every instant the (algebraic) sum of the potential differences v_R, v_L, and v_C; that is, it equals the sum of the projections of the phasors V_R, V_L, and V_C. But the *projection* of the *vector sum* of these phasors is equal to the *sum* of their *projections*, so this vector sum V must be the phasor that represents the source voltage. To form the vector sum, we first subtract the phasor V_C from the phasor V_L (since these always lie along a straight line), giving the phasor $V_L - V_C$. Since this is at right angles to the phasor V_R, the magnitude of the phasor V is

$$V = \sqrt{V_R{}^2 + (V_L - V_C)^2} = \sqrt{(IR)^2 + (IX_L - IX_C)^2}$$

$$= I\sqrt{R^2 + (X_L - X_C)^2}.$$

The quantity $X_L - X_C$ is called the **reactance** of the circuit, denoted by X:

$$X = X_L - X_C. \qquad (34\text{–}17)$$

Finally, we define the **impedance** Z of the circuit as

$$Z = \sqrt{R^2 + (X_L - X_C)^2} = \sqrt{R^2 + X^2}, \qquad (34\text{–}18)$$

so we can write

$$V = IZ \quad \text{or} \quad I = \frac{V}{Z}. \qquad (34\text{–}19)$$

Thus the *form* of the equation relating current and voltage *amplitudes* is the same as that for a dc circuit; impedance Z plays the same role as resistance R in a dc circuit. Note, however, that the impedance is actually a function of R, L, and C, as well as of the frequency ω. The complete expression for Z, for a series circuit, is

$$Z = \sqrt{R^2 + X^2}$$

$$= \sqrt{R^2 + (X_L - X_C)^2}$$

$$= \sqrt{R^2 + [\omega L - (1/\omega C)]^2}. \qquad (34\text{–}20)$$

The unit of impedance, from Eq. (34–19) or (34–20), is one *volt per ampere* (1 V·A^{-1}) or one *ohm*.

The expressions for the impedance Z of (a) an R–L series circuit, (b) an R–C series circuit, and (c) an L–C series circuit can be obtained from Eq. (34–20) by letting (a) $X_C = 0$, (b) $X_L = 0$, and (c) $R = 0$. Can you verify this statement?

Equation (34–20) gives the impedance Z only for a *series* L–R–C circuit. But the impedance of *any* network can be *defined* by Eq. (34–19) as the ratio of the voltage amplitude to the current amplitude.

The angle ϕ shown in Fig. 34–5b is the phase angle of the source voltage v with respect to the current i. From the diagram,

$$\tan \phi = \frac{V_L - V_C}{V_R} = \frac{I(X_L - X_C)}{IR} = \frac{X}{R}. \qquad (34–21)$$

How do we find the phase of voltage with respect to current for an *L–R–C* circuit?

Hence if the current is represented by a cosine function,

$$i = I \cos \omega t,$$

the source voltage *leads* the current by an angle ϕ between 0 and 90°, and its equation is

$$v = V \cos (\omega t + \phi).$$

Figure 34–5b has been constructed for a circuit in which $X_L > X_C$. If $X_L < X_C$, vector V lies on the opposite side of the current vector I and the voltage *lags* the current. In this case $X = X_L - X_C$ is a *negative* quantity, $\tan \phi$ is negative, and ϕ is a negative angle between 0 and −90°.

To summarize, we can say that the *instantaneous* potential differences in an ac-series circuit add *algebraically*, just as in a dc circuit, while the voltage *amplitudes* add *vectorially*.

Combining voltages in the series *L–R–C* circuit requires vector addition.

EXAMPLE 34–3 In the series circuit of Fig. 34–5, let $R = 300~\Omega$, $L = 0.9$ H, $C = 2.0~\mu$F, and $\omega = 1000$ rad·s^{-1}. Then

$$X_L = \omega L = 900~\Omega,$$

$$X_C = \frac{1}{\omega C} = 500~\Omega.$$

The reactance X of the circuit is

$$X = X_L - X_C = 400~\Omega,$$

and the impedance Z is

$$Z = \sqrt{R^2 + X^2} = 500~\Omega.$$

If the circuit is connected across an ac source of voltage amplitude 50 V, the current amplitude is

A numerical example of an *L–R–C* circuit

$$I = \frac{V}{Z} = 0.10 \text{ A}.$$

The phase angle ϕ is

$$\phi = \arctan \frac{X}{R} = 53°.$$

The voltage amplitude across the resistor is

$$V_R = IR = 30 \text{ V}.$$

The voltage amplitudes across the inductor and capacitor are, respectively,

$$V_L = IX_L = 90 \text{ V}, \qquad V_C = IX_C = 50 \text{ V}.$$

Transients: additional voltages and
currents that die out with time

The analysis above describes the *steady-state* condition of a circuit, the condition that prevails after the circuit has been connected to the source for a long time. When the source is first connected, there may be additional voltages and currents, called *transients,* whose nature depends on the time in the cycle when the circuit is initially completed. A detailed analysis of transients is beyond our scope. They always die out after a sufficiently long time and therefore do not affect the steady-state behavior of the circuit.

34–4 AVERAGE AND ROOT–MEAN–SQUARE VALUES

In this section we consider various ways of *measuring* ac quantities. The *instantaneous* potential difference between any two points in an ac circuit can be measured by connecting a calibrated oscilloscope between the points. The instantaneous current can be measured by connecting an oscilloscope across a resistor in the circuit. But the usual d'Arsonval galvanometer, described in Section 29–3, has too large a moment of inertia to follow the instantaneous values of an alternating current, except possibly at extremely low frequencies. At ordinary frequencies, such as 60-Hz household power, it averages out the sinusoidally varying torque on its coil; the meter deflection is thus proportional to the *average* current.

How do we calculate the average value of
a quantity that varies with time?

The **average value** f_{av} of any quantity $f(t)$ that varies with time, over a time interval from t_1 to t_2, is defined as

$$f_{av} = \frac{1}{t_2 - t_1} \int_{t_1}^{t_2} f(t)\, dt. \qquad (34\text{–}22)$$

This equation may be interpreted graphically. The integral represents the *area* bounded by the curve of $f(t)$ and by the two vertical lines t_1 and t_2. The product $f_{av}(t_2 - t_1)$ is the area of a rectangle of height f_{av} and base $(t_2 - t_1)$. Thus f_{av} is the height of a rectangle having the same area as the area under the curve between the same two values of t.

Let us apply this definition to a sinusoidally varying quantity—for example, a current given by

$$i = I \sin \omega t.$$

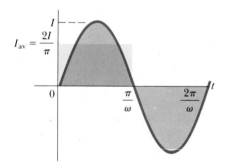

The period τ of this sinusoidal current is given by $\tau = 1/f = 2\pi/\omega$. From Eq. (34–22) the average value of i for a *half-period* or half-cycle, from $t = 0$ to $t = \pi/\omega$, is

34–6 The average value of a sinusoidal current over a half-cycle is $2I/\pi$. The average over a complete cycle is zero.

$$I_{av} = \frac{\omega}{\pi} \int_0^{\pi/\omega} I \sin \omega t\, dt = \frac{2}{\pi} I. \qquad (34\text{–}23)$$

That is, the average current is $2/\pi$ (about $\frac{2}{3}$) times the maximum current, and the area under the rectangle in Fig. 34–6 equals the area under one loop of the sine curve.

The average value of any sinusoidally
varying quantity is zero.

The average current for a *complete cycle* (or any number of complete cycles) is

$$I_{av} = \frac{\omega}{2\pi} \int_0^{2\pi/\omega} I \sin \omega t\, dt = 0,$$

as we should expect, since the *positive* area of the loop between 0 and π/ω is equal to the *negative* area of the loop between π/ω and $2\pi/\omega$. Hence if a sinusoidal current is sent through a d'Arsonval meter, the meter reads zero!

Such a meter can be used in an ac circuit, however, if it is connected in the *full-wave rectifier circuit* shown in Fig. 34–7a. As explained in Section 28–5, an ideal rectifier, or diode, offers zero resistance to current in the forward direction and infinite resistance to current in the opposite direction. Study of Fig. 34–7a shows that the current through the galvanometer is always upward, for either polarity of the source. Thus for a sinusoidal source, the current through the meter has the waveform shown as a solid curve in Fig. 34–7b. Although pulsating, it is always in the same direction, and its average value is *not* zero. Then, when provided with the necessary series resistance or shunt, as for a dc voltmeter or ammeter, the meter can serve as an ac voltmeter or ammeter.

The average value of the rectified current is the same as the average current in a half-cycle in Fig. 34–6, or $2/\pi$ times the maximum current I. Hence if the meter deflects full scale with a steady current I_0, it also deflects full scale when the *average value* of the rectified current, $2I/\pi$, is equal to I_0. The current amplitude I, when the meter deflects full scale, is then

$$I = \frac{\pi I_0}{2}.$$

For example, if $I_0 = 1$ A, $I = 1.57$ A.

Most ac meters are calibrated to read not the maximum value of the current or voltage, but the **root–mean–square (rms) value,** that is, the *square root* of the *average* value of the *square* of the current or voltage. Because i^2 is always positive, the average of i^2 is never zero, even when the average of i itself is zero. Also, rms values of voltage and current are useful in relationships for *power* in ac circuits; we return to this point in Section 34–5.

Figure 34–8 shows graphs of a sinusoidally varying current and of its square. If $i = I \sin \omega t$, then

$$i^2 = I^2 \sin^2 \omega t = I^2 [\tfrac{1}{2}(1 - \cos 2\omega t)]$$

$$= \tfrac{1}{2}I^2 - \tfrac{1}{2}I^2 \cos 2\omega t.$$

The average value of i^2, or the *mean square current*, is equal to the constant term $\tfrac{1}{2}I^2$, since the average value of $\cos 2\omega t$, over any number of complete cycles, is zero:

$$(i^2)_{av} = \frac{I^2}{2}.$$

The root–mean–square current is the square root of this quantity, or

$$I_{rms} = \sqrt{(i^2)_{av}} = \frac{I}{\sqrt{2}}. \tag{34–24}$$

In the same way, the root–mean–square value of a sinusoidal voltage is

$$V_{rms} = \frac{V}{\sqrt{2}}. \tag{34–25}$$

Voltages and currents in power-distribution systems are always referred to in terms of their rms values. Thus when we speak of our household power supply as "120-volt ac," we mean that the rms voltage is 120 V. The voltage

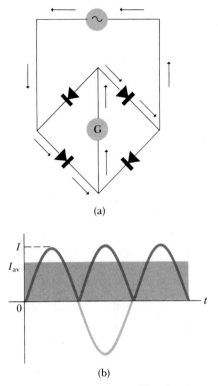

(a)

(b)

34–7 (a) A full-wave rectifier circuit. (b) Graph of a full-wave rectified current and its average value.

The root–mean–square value of a sinusoidally varying quantity is proportional to the amplitude.

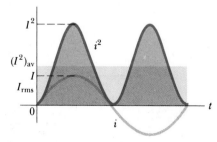

34–8 The average value of the square of a sinusoidally varying current, over any number of half-cycles, is $I^2/2$. The root–mean–square value is $I/\sqrt{2}$.

Voltages and currents in ac circuits are usually stated in terms of rms values, not amplitudes.

amplitude is

$$V = \sqrt{2}\,V_{\text{rms}} = 170 \text{ V.}$$

All the voltage-current relations developed in preceding sections have been stated in terms of amplitudes (maximum values) of voltage and current, but they remain valid when rms values are used; in each case the difference is simply a multiplicative factor of $1/\sqrt{2}$.

34–5 POWER IN AC CIRCUITS

How to calculate the electrical power input to a circuit element

Because electric power plays a central role in our systems for distributing, converting, and using energy, it is a concept of utmost practical importance. When a source with an instantaneous potential difference v supplies an instantaneous current i to an ac circuit, the instantaneous power p it supplies is given by

$$p = vi.$$

First we consider power relations for individual circuit elements.

If the circuit consists of a pure resistance R, as in Fig. 34–1, i and v are *in phase*. The graph representing p is obtained by multiplying together at every instant the ordinates of the graphs of v and i in Fig. 34–1b, and it is shown in Fig. 34–9a. The product vi is positive when v and i are both positive or both negative. That is, energy is supplied *to* the resistor at every instant, for either direction of instantaneous current, although the *rate* at which it is supplied is not constant.

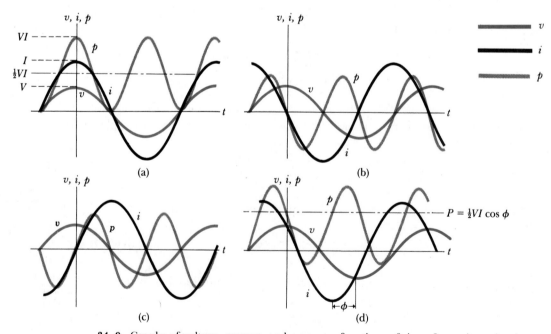

34–9 Graphs of voltage, current, and power as functions of time, for various circuits. (a) Instantaneous power input to a resistor. The average power is $\frac{1}{2}VI$. (b) Instantaneous power input to a capacitor. The average power is zero. (c) Instantaneous power input to a pure inductor. The average power is zero. (d) Instantaneous power input to an arbitrary ac circuit. The average power is $\frac{1}{2}VI \cos \phi = V_{\text{rms}}I_{\text{rms}} \cos \phi$.

The power curve is symmetrical about a value equal to one-half its maximum ordinate VI, so the *average power* P is

$$P = \tfrac{1}{2}VI. \tag{34–26}$$

The average power can also be written

$$P = \frac{V}{\sqrt{2}} \frac{I}{\sqrt{2}} = V_{\mathrm{rms}}I_{\mathrm{rms}}. \tag{34–27}$$

Furthermore, since $V_{\mathrm{rms}} = I_{\mathrm{rms}}R$, we have

$$P = I_{\mathrm{rms}}{}^2R. \tag{34–28}$$

The average power input to a resistor depends on the rms current and the resistance.

Note that Eqs. (34–27) and (34–28) have the same form as those for a dc circuit.

Suppose next that the circuit consists of a capacitor, as in Fig. 34–2. The current and voltage are then 90° out of phase. When the curves of v and i are multiplied together (the product vi is *negative* when v and i have *opposite* signs), we get the power curve in Fig. 34–9b, which is symmetrical about the horizontal axis. The average power is therefore *zero!*

The average power input to a pure inductor or capacitor is zero.

To see why this is so, recall that positive power means that energy is supplied *to* a device, and negative power means that energy is supplied *by* a device. The process we are considering is merely the charge and discharge of a capacitor. During the intervals when p is positive, energy is supplied to charge the capacitor; when p is negative, the capacitor is discharging and returning energy to the source.

Figure 34–9c is the power curve for a pure inductor. As with a capacitor, the current and voltage are out of phase by 90°, and the average power is zero. Energy is supplied to establish a magnetic field around the inductor and is returned to the source when the field collapses.

In a circuit containing any combination of resistors, capacitors, and inductors, the current and voltage differ in phase by an angle ϕ, and

The power input to a circuit element or a circuit depends on the phase of the voltage with respect to the current.

$$p = [V \cos (\omega t + \phi)][I \cos \omega t]. \tag{34–29}$$

The instantaneous power curve has the form shown in Fig. 34–9d. The area under the positive loops is greater than that under the negative loops, and the net average power is positive.

The preceding analyses have shown that when v and i are *in phase*, the average power equals $\tfrac{1}{2}VI$; when v and i are 90° *out of phase*, the average power is zero. In the general case, when v and i differ by an angle ϕ, the average power equals $\tfrac{1}{2}V$, multiplied by $I \cos \phi$, the component of I that is *in phase* with V. That is,

$$P = \tfrac{1}{2}VI \cos \phi = V_{\mathrm{rms}}I_{\mathrm{rms}} \cos \phi. \tag{34–30}$$

This equation is the general expression for the power input to *any* ac circuit. The factor $\cos \phi$ is called the **power factor** of the circuit. For a pure resistance, $\phi = 0$, $\cos \phi = 1$, and $P = V_{\mathrm{rms}}I_{\mathrm{rms}}$. For a pure (resistanceless) capacitor or inductor, $\phi = \pm 90°$, $\cos \phi = 0$, and $P = 0$.

A low power factor is often a liability.

A low power factor (large angle of lag or lead) is usually undesirable in power circuits because, for a given potential difference, a large current is needed to supply a given amount of power, resulting in large heat losses in the transmission lines. Many types of ac machinery draw a lagging current; the

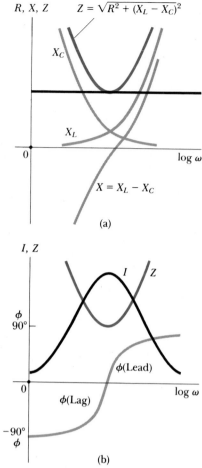

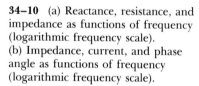

34–10 (a) Reactance, resistance, and impedance as functions of frequency (logarithmic frequency scale).
(b) Impedance, current, and phase angle as functions of frequency (logarithmic frequency scale).

At resonance, the inductor and capacitor voltages cancel exactly.

At resonance, the voltages across individual circuit elements can be larger than the source voltage.

The phase of voltage relative to current changes sign as the frequency goes through resonance.

power factor can be corrected by connecting a capacitor in parallel with the load. The leading current drawn by the capacitor compensates for the lagging current in the other branch of the circuit. The capacitor itself takes no net power from the line.

34–6 SERIES RESONANCE

The impedance of an L–R–C series circuit depends on the frequency, since inductive reactance is directly proportional to frequency and capacitive reactance is inversely proportional to frequency. Equation (34–20) shows explicitly the frequency dependence of the impedance of such a circuit. Figure 34–10 shows graphs of R, X_L, X_C, and Z as functions of ω. We have used a logarithmic frequency scale because of the wide range of frequencies covered. Note that at one particular frequency, X_L and X_C are numerically equal and $X = X_L - X_C$ is zero. Hence at this frequency the impedance Z has its *minimum* value, equal simply to the resistance R.

If we connect an ac source having constant voltage amplitude but variable frequency across the circuit, the current amplitude I varies with frequency as shown in Fig. 34–10b; it has its *maximum* value at the frequency where the impedance Z is *minimum*. This peaking of the current amplitude at a certain frequency is called **resonance.** It is easy to find the frequency ω_0 at which the resonance peak occurs; when the inductive and capacitive reactances are equal, we have

$$X_L = X_C, \qquad \omega_0 L = \frac{1}{\omega_0 C}, \qquad \omega_0 = \frac{1}{\sqrt{LC}}. \qquad (34\text{–}31)$$

Note that this frequency is equal to the natural frequency of oscillation of an L–C circuit, derived in Section 33–5, Eq. (33–24).

To understand the resonance peak in current amplitude more fully, consider the voltages in the circuit of Fig. 34–5. The current at any instant is the same in L and C. As we have learned, the voltage across an inductor always *leads* the current by 90°, or a quarter-cycle, and the voltage across a capacitor always *lags* the current by 90°. Thus, comparing the phase of the instantaneous voltage v_{cd} across L and the voltage v_{db} across C, we find that these two voltages always differ in phase by 180°, or a half-cycle. Therefore they have opposite signs at each instant. If, in addition, the *amplitudes* of the two voltages are equal, then they add to zero at each instant, and the *total* voltage v_{cb} across the L–C combination is exactly zero! As we found above, this phenomenon occurs only at one particular frequency, which we call the **resonant frequency.** Depending on the numerical values of R, L, and C, the voltages across L and C individually can be larger than that across R, so at frequencies sufficiently close to the resonant frequency, the voltages across L and C individually can be much *larger* than the source voltage!

The *phase* of the voltage in this circuit, relative to the current, also varies with frequency in an interesting way. At frequencies below resonance, X_C is greater than X_L; the phase relation is dominated by the capacitive reactance. The voltage *lags* the current, and the phase angle ϕ is between zero and −90°. At frequencies above resonance, X_L is greater than X_C, and the inductive reactance dominates. The voltage *leads* the current, and the phase angle is between zero and +90°. This variation of ϕ with frequency is shown in Fig. 36–10b. We can obtain an expression for ϕ as a function of ω from the phasor diagram in Fig. 34–5b and from Eq. (34–21):

$$\phi = \arctan \frac{X}{R} = \arctan \frac{\omega L - (1/\omega C)}{R}. \qquad (34\text{--}32)$$

If the inductance L or the capacitance C of a circuit can be varied, the resonant frequency can also be varied. This is the procedure by which a radio or television receiving set may be "tuned" to receive the signal from the desired station. In the early days of radio this was accomplished by use of capacitors with movable metal plates whose overlap could be varied to change C. Nowadays it is more common to use a variable inductor with a ferrite core that slides in and out of a coil to vary L.

EXAMPLE 34–4 The series circuit in Fig. 34–11 is connected to the terminals of an ac source whose frequency is variable but whose rms terminal voltage is constant and equal to 100 V. Find (a) the resonant frequency, (b) the inductive and capacitive reactances and the impedance at the resonant frequency, (c) the rms current at resonance, and (d) the rms potential difference across each circuit element at resonance.

A numerical example of series resonance in an ac circuit

SOLUTION a) The resonant frequency is

$$\omega_0 = \frac{1}{\sqrt{LC}} = \frac{1}{\sqrt{(2 \text{ H})(0.5 \times 10^{-6} \text{ F})}} = 1000 \text{ rad·s}^{-1}.$$

b) At this frequency,

$$X_L = (1000 \text{ rad·s}^{-1})(2 \text{ H}) = 2000 \ \Omega,$$

$$X_C = \frac{1}{(1000 \text{ rad·s}^{-1})(0.5 \times 10^{-6} \text{ F})} = 2000 \ \Omega,$$

$$X = X_L - X_C = 0.$$

From Eq. (34–20), the impedance Z at resonance is equal to the resistance: $Z = R = 500 \ \Omega$.

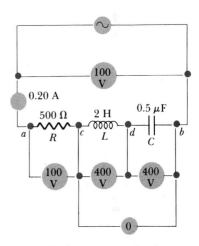

34–11 Series resonant circuit.

c) At resonance, the rms current is

$$I = \frac{V}{Z} = \frac{V}{R} = 0.20 \text{ A}.$$

d) The rms potential difference across the resistor is

$$V_R = IR = 100 \text{ V}.$$

The rms potential differences across the inductor and capacitor are, respectively,

$$V_L = IX_L = (0.20 \text{ A})(2000 \ \Omega) = 400 \text{ V},$$

$$V_C = IX_C = (0.20 \text{ A})(2000 \ \Omega) = 400 \text{ V}.$$

The rms potential difference across the inductor-capacitor combination (V_{cb}) is

$$V = IX = I(X_L - X_C) = 0.$$

This is true because the instantaneous potential differences across the inductor and the capacitor have equal amplitudes but are 180° out of phase, and so add to zero at each instant. Note also that at resonance, V_R is equal to the source voltage V, while V_L and V_C are both considerably *larger* than V.

The current in a series L–R–C circuit reaches its maximum value at the resonant frequency. Figure 34–12 shows a graph of current as a function of frequency for the circuit in Fig. 34–11, discussed in the example above. This curve is often called a *response curve* or a *resonance curve*. As expected, the curve has a peak at $\omega = 1000 \text{ s}^{-1}$, the resonant frequency.

What happens when we change R? The figure also shows graphs of I as a function of ω for $R = 200 \ \Omega$ and for $R = 2000 \ \Omega$. Note that the smaller R is, the taller and sharper the peak is, and the more rapidly it drops off as the frequency ω of the source moves away from the resonant frequency ω_0. Looking back at the expression for the impedance Z of the circuit, Eq. (34–20), we can see the reason for this behavior. When ω is far away from ω_0, either greater or less, Z is dominated by one of the reactance terms; ω_L becomes large at high frequencies, $1/\omega_C$ at low frequencies. This behavior does not change when we change R, nor does the value of ω_0 change. But at resonance, where the two reactances cancel, $Z = R$; the height of the peak is determined by R, and the behavior near resonance is also dominated by this value. Thus a small R gives a sharply peaked response curve, and a large R gives a broad, flat curve.

This distinction is crucially important in the design of radio and television receiving circuits. The sharply peaked curve is what makes it possible to discriminate between two stations broadcasting on adjacent frequency bands. But if the peak is *too* sharp, some of the information in the received signal is lost. Finally, note that the shape of the resonance curve is related to the overdamped and underdamped oscillations studied in Section 33–6. Reviewing that discussion, we see that a sharply peaked resonance curve corresponds to a small value of R and a lightly damped oscillating system; a broad, flat curve goes with a large value of R and a heavily damped system.

Resonance phenomena occur in all areas of physics; we have already seen one example in the forced oscillation of the harmonic oscillator, studied in Section 11–8. In that case the amplitude of a mechanical oscillation peaked at a driving-force frequency close to the natural frequency of the system, and the

The resonance curve is tall and sharp when R is small, smaller and broader when R is larger.

Practical applications of resonance in tuning radio and TV receivers

Resonance behavior occurs with both electrical and mechanical oscillating systems.

34–12 Graph of rms current I as a function of frequency ω (the dark color curve) for the circuit of Example 34–4. The other curves show the relationship for different values of the circuit resistance R, 2000 Ω (black curve) and 200 Ω (light color curve).

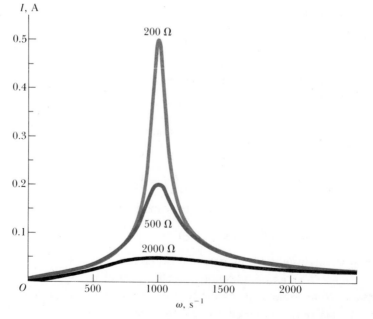

behavior of the L–R–C circuit is analogous to this. We suggest you review that discussion now, looking for the analogies. Other important examples of resonance occur in acoustics, in atomic and nuclear physics, and in the study of fundamental particles (high-energy physics).

34–7 PARALLEL RESONANCE

A different kind of resonance occurs when L, R, and C are connected in *parallel*, as shown in Fig. 34–13a. We can analyze this circuit by using the same procedure as for the series circuit. In this case the instantaneous potential difference v is the same for all elements and is equal to the source voltage. Figure 34–13b shows a phasor diagram; the single phasor V represents this common voltage. There are three separate currents, one in each branch, and the three corresponding current phasors are also shown. The phasor I_R, with amplitude V/R and in phase with V, represents the current in the resistor. Phasor I_L, with amplitude V/X_L and lagging V by 90°, represents the current in the inductor. Phasor I_C, with amplitude V/X_C and leading V by 90°, represents the current in the capacitor.

The instantaneous current i, by Kirchhoff's point rule, equals the (algebraic) sum of the instantaneous currents i_R, i_L, and i_C and is represented by the phasor I, the vector sum of phasors I_R, I_L, and I_C. Angle ϕ is the phase angle of current with respect to source voltage (the negative of the phase angle of voltage with respect to current).

From Fig. 34–13,

$$I = \sqrt{I_R{}^2 + (I_C - I_L)^2} = \sqrt{\left(\frac{V}{R}\right)^2 + \left(\omega C V - \frac{V}{\omega L}\right)^2}$$

$$= V\sqrt{\frac{1}{R^2} + \left(\omega C - \frac{1}{\omega L}\right)^2}. \tag{34–33}$$

The maximum current I is frequency-dependent, as expected. It is *minimum* when the second factor in the radical is zero; this occurs when the two reactances have equal magnitudes, at the resonant frequency ω_0 given by Eq. (34–31).

Comparing this equation with Eq. (34–19), we see that the *impedance* Z of this parallel combination is given by

$$\frac{1}{Z} = \sqrt{\frac{1}{R^2} + \left(\omega C - \frac{1}{\omega L}\right)^2}. \tag{34–34}$$

Impedance of a parallel L–R–C circuit

At resonance $1/Z$ is minimum, so Z itself has its *maximum* value at $\omega = \omega_0 = (1/LC)^{1/2}$.

Thus at resonance the total current in the parallel L–R–C circuit is *minimum*, in contrast to the L–R–C series circuit, which has *maximum* current at resonance. This distinction can be understood by noting that, in the parallel circuit, the currents in L and C are *always* exactly a half-cycle out of phase. When they also have equal *magnitudes*, they cancel each other completely, and the total current is simply the current through R. Indeed, when $\omega C = 1/\omega L$, Eq. (34–33) becomes simply $I = V/R$. This does *not* mean that there is *no* current in L or C at resonance, but only that the two currents cancel. If R is large, the equivalent impedance of the circuit near resonance is much *larger* that the individual reactances X_L and X_C.

At resonance, the inductor and capacitor currents in a parallel circuit cancel exactly.

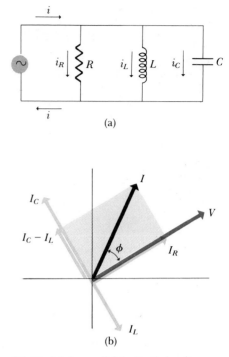

34–13 (a) A parallel L–R–C circuit. (b) Phasor diagram showing current phasors for the three branches. The single voltage phasor V represents the voltage across all three branches.

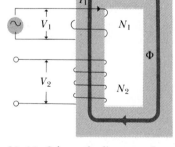

Transformers are used in power-distribution systems to "step down" the voltage from the large values (typically 500 kV) used for long-distance transmission to values used for local distribution and consumption (typically 120 V to 2400 V). (Courtesy of Pacific Gas and Electric.)

Transformers: the key to efficient long-distance power transmission

34–8 TRANSFORMERS

One of the great advantages of ac over dc for electric-power distribution is that it is much easier to step voltage levels up and down with ac than with dc. For long-distance power transmission it is desirable to use as high a voltage and as small a current as possible; this reduces I^2R heating losses in the transmission lines, and smaller wires can be used, saving on material costs. Present-day transmission lines routinely operate at rms voltages of the order of 500 kV. Yet safety considerations and insulation requirements dictate relatively low voltages in generating equipment and in household and industrial power distribution. The standard voltage for household wiring is 120 V in the United States and Canada, 240 V in most of Western Europe. The necessary voltage conversion is accomplished by use of **transformers.**

In principle, the transformer consists of two coils, electrically insulated from each other but wound on the same core. Transformers used in power-distribution systems have soft iron cores. An alternating current in one winding sets up an alternating flux in the core, and the induced electric field produced by this varying flux induces an emf in the other winding, as we learned in Sections 32–3 and 33–1. Energy is thus transferred from one winding to another via the core flux and its associated induced electric field. The winding to which power is supplied is called the **primary;** the winding from which power is delivered is called the **secondary.** The circuit symbol for an iron core transformer is

Small but inevitable energy losses in a transformer

The power output of a transformer is necessarily less than the power input because of unavoidable losses. These losses include resistance (I^2R) losses in the primary and secondary windings, and losses in the core due to hysteresis (Section 31–8) and eddy currents (Section 32–6). Hysteresis losses are minimized by the use of soft iron having a narrow hysteresis loop; and eddy currents are minimized by laminating the core. In spite of these losses, transformer efficiencies are usually well over 90%; in large installations they may reach 99%.

An idealized transformer has no energy losses at all.

We will consider only an idealized transformer in which there are *no* losses and in which *all* of the flux is confined to the iron core, so that the same flux links both primary and secondary. The transformer is shown schematically in Fig. 34–14. A primary winding of N_1 turns and a secondary winding of N_2 turns both encircle the core in the same sense. An ac source with voltage amplitude V_1 is connected to the primary; we first assume the secondary to be open, so that there is no secondary current. The primary winding then functions merely as an inductor.

The primary current, which is small, lags the primary voltage by 90° and is called the *magnetizing* current. Because the primary voltage is 90° out of phase with the current, the power input to the transformer is zero. The core flux is in phase with the primary current. Since the same flux links both primary and secondary, the induced emf *per turn* is the same in each. The ratio of primary

34–14 Schematic diagram of a transformer with secondary open.

to secondary induced emf is therefore equal to the ratio of primary to secondary turns, or

$$\frac{\mathcal{E}_2}{\mathcal{E}_1} = \frac{N_2}{N_1}.$$

Since the windings are assumed to have zero resistance, the induced emf's $\mathcal{E}_1$ and $\mathcal{E}_2$ are numerically equal to the corresponding terminal voltages V_1 and V_2, and

$$\frac{V_2}{V_1} = \frac{N_2}{N_1}. \tag{34–35}$$

A transformer functions as a voltage-conversion device.

Hence by properly choosing the turn ratio N_2/N_1, we may obtain any desired secondary voltage from a given primary voltage. If $V_2 > V_1$, we have a *step-up* transformer; if $V_2 < V_1$, we have a *step-down* transformer.

Consider next the effect of closing the secondary circuit. The secondary current I_2 and its phase angle ϕ_2 will, of course, depend on the nature of the secondary circuit. As soon as the latter is closed, some power must be delivered by the secondary (except when $\phi_2 = 90°$); and, from energy considerations, an equal amount of power must be supplied to the primary. The process by which the transformer is able to draw the necessary power is as follows. When the secondary circuit is open, the core flux is produced by the primary current only. But when the secondary circuit is closed, *both* primary and secondary currents set up flux in the core. The secondary current, by Lenz's law, tends to weaken the core flux and therefore to decrease the induced emf in the primary. But (in the absence of losses) the induced emf in the primary must equal the primary terminal voltage, which is assumed to have constant amplitude. The primary *current* therefore increases until the core flux is restored to its original no-load magnitude.

Faraday's law in a transformer core: flux relations under various operating conditions

If the secondary circuit is completed by a resistance R, $I_2 = V_2/R$. From energy considerations, the power delivered to the primary equals that taken out of the secondary (neglecting losses), so

$$V_1 I_1 = V_2 I_2. \tag{34–36}$$

Combining this equation with the above expression for I_2 and Eq. (34–35) to eliminate V_2 and I_2, we find

$$I_1 = \frac{V_1}{(N_1/N_2)^2 R}. \tag{34–37}$$

Hence, when the secondary circuit is completed through a resistance R, the result is the same as if the *source* had been connected directly to a resistance equal to R multiplied by the reciprocal of the *square* of the turns ratio. In other words, the transformer "transforms" not only voltages and currents but resistances (more generally, impedances) as well. It can be shown that maximum power is supplied by a source to a resistor when its resistance equals the internal resistance of the source. (See Challenge Problem 28–40.) The same principle applies in both dc and ac circuits. When a high-impedance ac source must be connected to a low-impedance circuit, as when an audio amplifier is connected to a loudspeaker, the impedance of the source can be *matched* to that of the circuit by use of a transformer having the correct turns ratio.

A transformer changes the effective impedance of a circuit.

SUMMARY

KEY TERMS
alternating current
alternator
voltage amplitude
ac source
phasors
phasor diagrams
current amplitude
phase angle
capacitive reactance
inductive reactance
reactance
impedance
average value
root–mean–square (rms) value
power factor
resonance
resonant frequency
transformers
primary
secondary

An alternator or ac source produces an emf that varies sinusoidally with time. Voltages and currents that vary sinusoidally with time can be represented by vectors called phasors. A phasor rotates counterclockwise with constant angular velocity ω equal to the angular frequency of the sinusoidal quantity, and its projection on the horizontal axis at any instant represents the instantaneous value of the quantity.

If the current i in an ac circuit is given by

$$i = I \cos \omega t$$

and the voltage v between two points by

$$v = V \cos (\omega t + \phi),$$

then ϕ is called the phase angle of the voltage relative to the current.

The voltage across a resistor R is in phase with the current, and the amplitudes are related by

$$V = IR. \tag{34–2}$$

The voltage across a capacitor C lags the current by 90°; the amplitudes are related by

$$V = IX_C, \tag{34–9}$$

where $X_C = 1/\omega C$ is the capacitive reactance of the capacitor.

The voltage across an inductor L leads the current by 90°; the amplitudes are related by

$$V = IX_L, \tag{34–16}$$

where $X_L = \omega L$ is the inductive reactance of the inductor.

In an $L–R–C$ series circuit, the voltage and current amplitudes are related by

$$V = IZ, \tag{34–19}$$

where Z is the impedance of the circuit, given by

$$Z = \sqrt{R^2 + X^2}$$

$$= \sqrt{R^2 + (X_L - X_C)^2}$$

$$= \sqrt{R^2 + [\omega L - (1/\omega C)]^2}. \tag{34–20}$$

The phase angle ϕ of the voltage relative to the current is given by

$$\tan \phi = \frac{V_L - V_C}{V_R} = \frac{I(X_L - X_C)}{IR} = \frac{X}{R}. \tag{34–21}$$

The SI unit of capacitive or inductive reactance or impedance is the ohm.

The average value of a sinusoidal current with amplitude I over a half-cycle is given by $I_{av} = 2I/\pi$. This is also equal to the rectified average for a full cycle. The root–mean–square (rms) value of a sinusoidally varying quantity is $1/\sqrt{2}$ times the amplitude; thus $I_{rms} = I/\sqrt{2}$ and $V_{rms} = V/\sqrt{2}$.

The average power input P to an ac circuit is given by

$$P = V_{rms}I_{rms} \cos \phi, \tag{34–30}$$

where ϕ is the phase angle of voltage with respect to current. The quantity $\cos \phi$ is called the power factor.

The current in an L–R–C series circuit becomes maximum, and the impedance minimum, at a frequency $\omega_0 = 1/(LC)^{1/2}$ called the resonant frequency. This phenomenon is called resonance. At resonance the voltage and current are in phase, and the impedance Z is equal to the resistance R.

The current in an L–R–C parallel circuit becomes minimum, and the impedance maximum, at this same resonant frequency ω_0. The impedance Z of this circuit at any frequency is given by

$$\frac{1}{Z} = \sqrt{\frac{1}{R^2} + \left(\omega C - \frac{1}{\omega L}\right)^2}. \tag{34–34}$$

At resonance, $Z = R$.

A transformer is used to transform the voltage and current levels in an ac circuit. If the primary winding has N_1 turns and the secondary N_2, the two voltages are related by

$$\frac{V_2}{V_1} = \frac{N_2}{N_1}. \tag{34–35}$$

If there are no energy losses in the transformer, then the primary and secondary voltages and currents are related by

$$V_1 I_1 = V_2 I_2. \tag{34–36}$$

QUESTIONS

34–1 Some electric-power systems formerly used 25-Hz alternating current instead of the 60 Hz that is now standard. The lights flickered noticeably. Why is this not a problem with 60 Hz?

34–2 Power-distribution systems in airplanes sometimes use 400-Hz ac. What advantages and disadvantages does this have compared to the standard 60 Hz?

34–3 Fluorescent lights often use an inductor, called a "ballast," to limit the current through the tubes. Why is it better to use an inductor than a resistor for this purpose?

34–4 At high frequencies a capacitor becomes a short circuit. Discuss.

34–5 At high frequencies an inductor becomes an open circuit. Discuss.

34–6 Household electric power in most of Western Europe is 240 V, rather than the 120 V that is standard in the United States and Canada. What advantages and disadvantages does each system have?

34–7 The current in an ac-power line changes direction 120 times per second, and its average value is zero. So how is it possible for power to be transmitted in such a system?

34–8 Electric-power connecting cords, such as lamp cords, always have two conductors that carry equal currents in opposite directions. How might one determine, by using measurements only at the midpoints along the lengths of the wires, the direction of power transmission in the cord?

34–9 Are the equations for the average and rms values of current, Eqs. (34–23) and (34–24), correct when the variation with time is not sinusoidal? Explain.

34–10 Electric-power companies like to have their power factors (cf. Section 34–5) as close to unity as possible. Why?

34–11 Some electrical appliances operate equally well on ac or dc, while others work only on ac or only on dc. Give examples of each, and explain the differences.

34–12 When a series-resonant circuit is connected across a 120-V ac line, the voltage rating of the capacitor may be exceeded even if it is rated at 200 or 400 V. How can this be?

34–13 In a parallel-resonant circuit connected across a 120-V line, it is possible for the maximum current rating in the inductor to be exceeded even if the total current through the circuit is very small. How can this be?

34–14 Can a transformer be used with dc? What happens if a transformer designed for 120-V ac is connected to a 120-V dc line?

34–15 During the last quarter of the nineteenth century there was great and acrimonious controversy over whether ac or dc should be used for power transmission. Edison favored dc, George Westinghouse ac. What arguments might each proponent have used to promote his scheme?

EXERCISES

$$\text{Angular frequency} = 2\pi \text{ (frequency)},$$
$$\omega(\text{rad·s}^{-1}) = 2\pi f(\text{Hz}).$$

Section 34–2 Resistance, Inductance, and Capacitance

34–1

a) What is the reactance of a 1-H inductor at a frequency of 60 Hz?

b) What is the inductance of an inductor whose reactance is 1 Ω at 60 Hz?

c) What is the reactance of a 1-μF capacitor at a frequency of 60 Hz?

d) What is the capacitance of a capacitor whose resistance is 1 Ω at 60 Hz?

34–2

a) Compute the reactance of a 10-H inductor at frequencies of 60 Hz and 600 Hz.

b) Compute the reactance of a 10-μF capacitor at the same frequencies.

c) At what frequency is the reactance of a 10-H inductor equal to that of a 10-μF capacitor?

34–3 A 1-μF capacitor is connected across an ac source whose voltage amplitude is kept constant at 50 V but whose frequency can be varied. Find the current amplitude when the angular frequency is

a) 100 rad·s^{-1},   b) 1000 rad·s^{-1},   c) 10,000 rad·s^{-1}.

34–4 The voltage amplitude of an ac source is 50 V, and its angular frequency is 1000 rad·s^{-1}. Find the current amplitude if the capacitance of a capacitor connected across the source is

a) 0.01 μF,   b) 1.0 μF,   c) 100 μF.

d) Construct a log-log plot of current amplitude versus capacitance.

34–5 An inductor of self-inductance 10 H and of negligible resistance is connected across the source in Exercise 34–3. Find the current amplitude when the angular frequency is

a) 100 rad·s^{-1},   b) 1000 rad·s^{-1},   c) 10,000 rad·s^{-1}.

d) Construct a log-log plot of current amplitude versus frequency.

34–6 Find the current amplitude if the self-inductance of a resistanceless inductor connected across the source of Exercise 34–4 is

a) 0.01 H,   b) 1.0 H,   c) 100 H.

d) Construct a log-log plot of current amplitude versus self-inductance.

Section 34–3 The L–R–C Series Circuit

34–7 The expression for the impedance Z of an R–L series circuit can be obtained from Eq. (34–20) by setting $X_C = 0$,

which corresponds to $C = \infty$, whereas for an R–C series circuit one obtains the impedance Z from Eq. (34–20) by setting $L = 0$. Explain.

34–8 In an L–R–C series circuit, the source has a constant voltage amplitude of 50 V and a frequency of 1000 rad·s^{-1}. $R = 300$ Ω, $L = 0.9$ H, and $C = 2.0$ μF. Suppose a series circuit contains only the resistor and the inductor in series.

a) What is the impedance of the circuit?

b) What is the current amplitude?

c) What are the voltage amplitudes across the resistor and across the inductor?

d) What is the phase angle ϕ of the source voltage with respect to the current? Does the source voltage lag or lead the current?

e) Construct the phasor diagram.

34–9 Same as Exercise 34–8, except that the circuit consists of only the resistor and the capacitor in series. For part (c) calculate the voltage amplitudes across the resistor and across the capacitor.

34–10 Same as Exercise 34–8, except that the circuit consists of only the capacitor and the inductor in series. For part (c), calculate the voltage amplitudes across the capacitor and across the inductor.

34–11 A 400-Ω resistor is in series with a 0.1-H inductor and a 0.5-μF capacitor. Compute the impedance of the circuit and draw the phasor diagram

a) at a frequency of 500 Hz;

b) at a frequency of 1000 Hz.

Compute, in each case, the phase angle of the source voltage with respect to the current, and state whether the source voltage lags or leads the current.

34–12

a) Compute the impedance of an L–R–C series circuit at angular frequencies of 1000, 750, and 500 rad·s^{-1}. Take $R = 300$ Ω, $L = 0.9$ H, and $C = 2.0$ μF.

b) Describe how the current amplitude varies as the frequency of the source is slowly reduced from 1000 rad·s^{-1} to 500 rad·s^{-1}.

c) What is the phase angle of the source voltage with respect to the current when $\omega = 1000$ rad·s^{-1}? Construct the phasor diagram when $\omega = 1000$ rad·s^{-1}.

d) Repeat part (c) for $\omega = 500$ rad·s^{-1}.

Section 34–5 Power in AC Circuits

34–13 The circuit in Exercise 34–11 carries an rms current of 0.25 A with a frequency of 100 Hz.

a) What average power is delivered by the source?

b) What average power is consumed in the resistor?

c) In the capacitor?

d) In the inductor?

e) Compare your answers in (a) to the sum of (b), (c), and (d).

34–14 A series circuit has a resistance of 75 Ω and an impedance of 150 Ω. What power is consumed in the circuit when a voltage of 120 V (rms) is impressed across it?

Section 34–6 Series Resonance

34–15 Consider the $L–R–C$ series circuit of Exercise 34–12. The voltage amplitude of the source is 50 V.

a) At what angular frequency is the circuit in resonance?

b) What is the power factor at resonance? Sketch the phasor diagram at the resonant frequency.

c) What is the reading of each voltmeter in Fig. 34–15 when the source frequency equals the resonant frequency? The voltmeters are calibrated to read rms voltages.

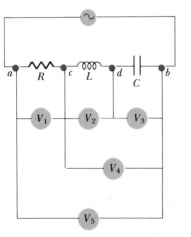

FIGURE 34–15

d) What would be the resonant angular frequency if the resistance were reduced to 100 Ω?

e) What would then be the rms current at resonance?

34–16 In an $L–R–C$ series circuit, $R = 250$ Ω, $L = 0.5$ H, and $C = 0.02$ μF.

a) What is the resonant angular frequency of the circuit?

b) The capacitor can withstand a peak voltage of 350 V. If the voltage source operates at the resonant frequency, what maximum voltage amplitude can it have if the maximum capacitor voltage is not exceeded?

Section 34–7 Parallel Resonance

34–17 For the circuit of Fig. 34–13a, let $V = 120$ V, $R = 200$ Ω, $L = 0.5$ H, and $C = 0.2$ μF.

a) What is the resonant angular frequency of the circuit?

b) Sketch the phasor diagram at the resonant frequency.

c) At the resonant frequency, what is the current through the source?

d) At the resonant frequency, what is the current through the resistor?

34–18 Consider the circuit of Fig. 34–13a, with the same numerical values as in Exercise 34–17. At resonance, what is

a) the average rate at which electrical energy is being delivered by the source?

b) the average rate at which electrical energy is being dissipated in the resistor? Compare to the result of (a).

c) Is the current through the inductor, and hence the energy stored in its magnetic field, zero at all times? If not, how can the result obtained in (b) be explained?

d) Calculate the maximum energy stored in the inductor.

e) Calculate the maximum energy stored in the capacitor.

Section 34–8 Transformers

34–19 A transformer connected to a 120-V ac line is to supply 12 V to a low-voltage lighting system for a model-railroad village. The total equivalent resistance of the system is 2 Ω.

a) What should be the ratio of primary to secondary turns of the transformer?

b) What current must the secondary supply?

c) What power is delivered to the load?

d) What resistance connected directly across the 120-V line would draw the same power as the transformer? Show that this quantity is equal to 2 Ω times the square of the ratio of primary to secondary turns.

34–20 A step-up transformer connected to a 120-V ac line is to supply 18,000 V for a neon sign. To reduce shock hazard, a fuse is to be inserted in the primary circuit; the fuse is to blow when the current in the secondary circuit exceeds 10 mA.

a) What is the ratio of secondary to primary turns of the transformer?

b) What power must be supplied to the transformer when the secondary current is 10 mA?

c) What current rating should the fuse in the primary circuit have?

34–21 The internal resistance of an ac source is 10,000 Ω.

a) What should be the ratio of primary to secondary turns of a transformer to match the source to a load of resistance of 10 Ω? ("Matching" means that the effective load resistance equals the internal resistance of the source. Refer to Exercise 34–19, part d.)

b) If the voltage amplitude of the source is 100 V, what is the voltage amplitude in the secondary circuit, under open-circuit conditions?

PROBLEMS

34–22 At a certain frequency ω_1, the reactance of a certain capacitor equals that of a certain inductor.

a) If the frequency is changed to $\omega_2 = 2\omega_1$, what is the ratio of the reactance of the inductor to that of the capacitor, and which reactance is larger?

b) If the frequency is changed to $\omega_3 = \omega_1/3$, what is the ratio of the reactance of the inductor to that of the capacitor, and which reactance is larger?

34–23 A coil has a resistance of 20 Ω. At a frequency of 100 Hz, the voltage across the coil leads the current in it by 30°. Determine the inductance of the coil.

34–24 Five infinite-impedance voltmeters, calibrated to read rms values, are connected as shown in Fig. 34–15. Take R, L, C, and V as given in Exercise 34–8. What is the reading of each voltmeter

a) if $\omega = 500$ rad·s^{-1}?

b) if $\omega = 1000$ rad·s^{-1}?

34–25 An inductor having a reactance of 25 Ω and a resistance R gives off heat at the rate of 10 J·s^{-1} when it carries a current of 0.5 A (rms). What is the impedance of the inductor?

34–26 A circuit draws 330 W from a 110-V (rms), 60-Hz ac line. The power factor is 0.6, and the current lags the source voltage.

a) What is the net resistance R of the circuit?

b) Find the capacitance of the series capacitor that will result in a power factor of unity when it is added to the original circuit.

c) What power will then be drawn from the supply line?

34–27 A series circuit has an impedance of 50 Ω and a power factor of 0.6 at 60 Hz, the voltage lagging the current.

a) Should an inductor or a capacitor be placed in series with the circuit to raise its power factor?

b) What size element will raise the power factor to unity?

34–28 A resistor of 500 Ω and a capacitor of 2 μF are connected in parallel to an ac generator supplying a constant voltage amplitude of 282 V and having an angular frequency of 377 rad·s^{-1}. Find

a) the current amplitude in the resistor,

b) the current amplitude in the capacitor,

c) the phase angle (sketch the phasor diagram for the current phasors),

d) the line-current amplitude.

34–29 A 100-Ω resistor, a 0.1-μF capacitor, and a 0.1-H inductor are connected in parallel to a voltage source with an amplitude of 100 V.

a) What is the resonant frequency? The resonant angular frequency?

b) What is the maximum total current through the parallel combination at the resonant frequency?

c) What is the maximum current in the resistor at resonance?

d) What is the maximum current in the inductor at resonance?

e) What is the maximum current in the branch containing the capacitor at resonance?

f) What is the maximum energy stored in the inductor at resonance? In the capacitor?

34–30 The same three components as in Problem 34–29 are connected in *series* to a voltage source with an amplitude of 100 V.

a) What is the resonant frequency? The resonant angular frequency?

b) What is the maximum current in the resistor at resonance?

c) What is the maximum voltage across the capacitor at resonance?

d) What is the maximum voltage across the inductor at resonance?

e) What is the maximum energy stored in the capacitor at resonance? In the inductor?

34–31 The current in a certain circuit varies with time as shown in Fig. 34–16. Find the average current and the rms current, in terms of I_0.

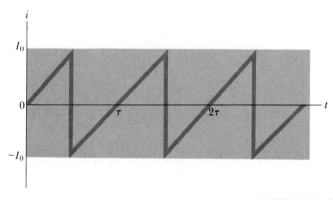

FIGURE 34–16

34–32 Consider a series circuit with a source of terminal voltage amplitude $V = 120$ V and frequency ω. The inductor inductance is $L = 1.2$ H, the capacitor capacitance is $C = 0.8$ μF, and the resistor resistance is $R = 400$ Ω.

a) What is the resonant angular frequency ω_0 of the circuit?

b) What is the current through the source at resonance?

c) For what two values of the angular frequency, ω_1 and ω_2, will the current be half the resonant value? The quantity $|\omega_1 - \omega_2|$ defines the *width* of the resonance.

d) Calculate the resonance width for $R = 4$ Ω, 40 Ω, and 400 Ω.

CHALLENGE PROBLEMS

34–33

a) At what angular frequency is the voltage amplitude across the *resistor* in an L–R–C series circuit at maximum value?

b) At what angular frequency is the voltage amplitude across the *inductor* at maximum value?

c) At what angular frequency is the voltage amplitude across the *capacitor* at maximum value?

34–34 The voltage drop across a circuit element in an ac circuit is not necessarily in phase with the current through that circuit element; hence the voltage amplitudes across the circuit elements in a branch in an ac circuit do not add algebraically. A method commonly employed to simplify the analysis of an ac circuit driven by a sinusoidal source is to represent the impedance Z as a complex number rather than as a real number. The resistance R is taken to be the real part of the impedance, and the reactance $X = X_L - X_C$ is taken to be the imaginary part. Thus for a branch containing a resistor, inductor, and capacitor in series, the complex impedance would be given by $Z_{cpx} = R + iX$, where $i^2 = -1$. If the voltage amplitude across the branch is V_{cpx} and the current amplitude is I_{cpx}, these quantities are related by $I_{cpx} = V_{cpx}/Z_{cpx}$. The actual current amplitude would be computed by taking the absolute value of the complex representation of the current amplitude; that is, $I = (I^*_{cpx}I_{cpx})^{1/2}$. The phase angle ϕ of the current with respect to the source voltage is given by

$$\tan(\phi) = Im(I_{cpx})/Re(I_{cpx}).$$

The voltage amplitudes V_{Rcpx}, V_{Lcpx}, and V_{Ccpx} across the resistance, inductance, and capacitance, respectively, are found by multiplying I_{cpx} by R, iX_L, or $-iX_C$, respectively. If we use the complex representation for the voltage amplitudes, the voltage drop across a branch can be found by algebraic addition of the voltage drops across each circuit element; that is,

$$V_{cpx} = V_{Rcpx} + V_{Lcpx} + V_{Ccpx}.$$

When we want to find the actual value of any current amplitude or voltage amplitude, the absolute value of the appropriate quantity is then used.

Consider the series L–R–C circuit shown in Fig. 34–17. The angular frequency ω is 1000 s^{-1}, and the voltage amplitude of the source is 100 V. The values of the circuit elements are as shown.

Use the phasor-diagram techniques presented in Section 34–3 to solve for

a) the current amplitude;

b) the phase angle ϕ of the current with respect to the source voltage. (Note that this angle is the negative of the phase angle defined in Fig. 34–5.)

Now apply the procedure outlined above to the same circuit.

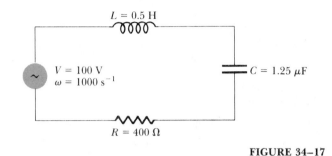

FIGURE 34–17

c) Determine the impedance of the circuit, Z_{cpx}, as represented by a complex number. Take the absolute value to obtain Z, the actual impedance of the circuit.

d) Take the voltage amplitude of the source, V_{cpx}, to be real, and find the complex current amplitude I_{cpx}. Find the actual current amplitude by taking the absolute value of I_{cpx}.

e) Find the phase angle ϕ of the current with respect to the source voltage by using the real and imaginary parts of I_{cpx}, as explained above.

f) Find the complex representations of the voltages across the resistance, the inductance, and the capacitance.

g) Add the answers found in part (f) and verify that the sum of these complex numbers is real and equal to 100 V, the voltage of the source.

34–35 The use of complex numbers to represent impedance, voltage amplitudes, and current amplitudes in an ac circuit is not restricted to series circuits but can be applied to parallel circuits as well. In the case of parallel circuits, the (complex) impedances are combined by using the same reciprocal combination rule used in a dc circuit for finding the equivalent resistance of resistors in parallel.

Consider the circuit shown in Fig. 34–18. The voltage amplitude of the source is 100 V, and its angular frequency is $\omega = 1000$ s^{-1}. The values of the passive circuit elements are as shown.

a) Determine the (complex) impedance of the branch containing the 0.3-H inductance and label it Z_{1cpx}.

b) Determine the (complex) impedance of the branch containing the 2.5-μF capacitor and label it Z_{2cpx}.

c) Determine the (complex) equivalent impedance Z_{cpx} of the circuit. Do this by using the relation $1/Z_{cpx} = 1/Z_{1cpx} + 1/Z_{2cpx}$.

d) Determine the current I_{cpx} that passes through the source by using the relation $I_{cpx} = V_{cpx}/Z_{cpx}$, where $V_{cpx} = 100$ V.

e) Determine the actual circuit current amplitude I and the phase angle ϕ of the current through the source relative to the source voltage.

f) Determine the (complex) current I_{1cpx} through the branch containing the 0.3-H inductor.

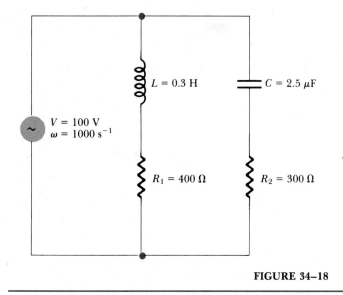

$L = 0.3$ H $C = 2.5\ \mu$F

$V = 100$ V
$\omega = 1000$ s^{-1}

$R_1 = 400\ \Omega$ $R_2 = 300\ \Omega$

FIGURE 34–18

g) Determine the actual current amplitude I_1 through this branch, as well as the phase angle ϕ_1 of the branch current with respect to the circuit voltage.

h) Determine the complex current I_{2cpx} through the branch containing the capacitor.

i) Determine the actual current I_2 through this branch, as well as the phase angle ϕ_2 of the branch current with respect to the circuit voltage.

j) Verify that $I_{cpx} = I_{1cpx} + I_{2cpx}$.

35

ELECTROMAGNETIC WAVES

AN ELECTROMAGNETIC WAVE CONSISTS OF TIME-VARYING ELECTRIC AND magnetic fields that travel or propagate through space with a definite speed. The theoretical basis for electromagnetic waves, discovered by Maxwell in 1865, rests on the ability of a time-varying *electric* field and its associated displacement current to act as a source of *magnetic* field, and on the ability of a time-varying *magnetic* field to act as a source of *electric* field, according to Faraday's law. Electromagnetic waves carry energy and momentum and show the property of polarization. The simplest electromagnetic waves are sinusoidal waves, in which electric and magnetic fields vary sinusoidally with time. The spectrum of electromagnetic waves covers an extremely broad range of frequency and wavelength, and these waves play a central role in a wide variety of physical phenomena and life processes. In particular, light consists of electromagnetic waves. This chapter forms a bridge between electromagnetism and optics; we can understand such basic optical concepts as reflection and refraction on the basis of the electromagnetic nature of light.

35–1 INTRODUCTION

In the last several chapters we studied various aspects of electric and magnetic fields, falling in two general categories. The first category includes fields that do not vary with time. The electrostatic field of a distribution of charges at rest and the magnetic field of a steady current in a conductor are examples of fields that do not vary with time at any individual point, although they may vary from point to point in space. For these situations we could treat the electric and magnetic fields independently, without worrying very much about interactions between them.

The second category includes fields that *do* vary with time, and in all such cases we found that it is *not* possible to treat the fields independently. Faraday's law tells us that a time-varying magnetic field acts as a source of electric field. This field is manifested in the induced emf's in inductances and transformers. Similarly, in developing the general formulation of Ampere's law (Section 31–9), which is valid for charging capacitors and similar situations as

A bootstrap effect: Each field acts as a source of the other field.

Electromagnetic waves: time-varying
electric and magnetic fields propagating
through space

well as for ordinary conductors, we found that a changing electric field acts as a source of magnetic field. This mutual interaction between the two fields is summarized neatly in Maxwell's equations, which we studied in Section 32–7.

Thus, when *either* an electric or a magnetic field is changing with time, a field of the other kind is induced in adjacent regions of space. We are led (as Maxwell was) to consider the possibility of an electromagnetic disturbance, consisting of time-varying electric and magnetic fields, that can propagate through space from one region to another even when no matter is present in the intervening region. Such a disturbance, if it exists, will have the properties of a *wave,* and an appropriate term is **electromagnetic wave.** Such waves do exist, of course, and it is a familiar fact that radio and television transmission, light, x-rays, and many other phenomena are examples of electromagnetic radiation. In this chapter we will show how the existence of such waves is related to the principles of electromagnetism we have studied thus far, and we will examine the properties of these waves.

Maxwell's prediction of the existence of
electromagnetic waves: an extraordinary
achievement in theoretical physics

As so often happens in the development of science, the theoretical understanding of electromagnetic waves originally took a considerably more devious path than the one just outlined. In the early days of electromagnetic theory (the early nineteenth century), two different units of electric charge were used, one for electrostatics, the other for magnetic phenomena involving currents. In the particular system of units in common use at the time, these two units of charge had different physical dimensions; their *ratio* turned out to have units of velocity. This in itself was not so astounding, but experimental measurements revealed that the ratio had a numerical value precisely equal to the speed of light, 3.00×10^8 m·s^{-1}. At the time, physicists regarded this as an extraordinary coincidence and had no idea how to explain it.

Maxwell's equations: all of
electromagnetism in one neat package

During his search for understanding of this result, Maxwell proved in 1865 that an electromagnetic disturbance should propagate in free space with a speed equal to that of light, and hence that light waves were very likely to be electromagnetic in nature. At the same time he discovered that the basic principles of electromagnetism could be expressed in terms of the four equations we now call **Maxwell's equations.** To review our discussion of Section 32–7, these four relationships are (1) Gauss's law, (2) the absence of magnetic charge, (3) Ampere's law, including displacement current, and (4) Faraday's law. Maxwell's equations, in the integral form used in this text, are

$$\oint \boldsymbol{E} \cdot d\boldsymbol{A} = \frac{Q}{\epsilon_0}, \qquad \text{(Gauss's law)} \qquad (25\text{–}17)$$

$$\oint \boldsymbol{B} \cdot d\boldsymbol{A} = 0, \qquad \text{(no magnetic charge)} \qquad (31\text{–}16)$$

$$\oint \boldsymbol{B} \cdot d\boldsymbol{l} = \mu_0 \Big(I_{\mathrm{C}} + \epsilon_0 \frac{d\Psi}{dt} \Big), \qquad \text{(Ampere's law)} \qquad (31\text{–}32)$$

$$\oint \boldsymbol{E} \cdot d\boldsymbol{l} = -\frac{d\Phi}{dt}. \qquad \text{(Faraday's law)} \qquad (32\text{–}14)$$

These equations apply to electric and magnetic fields *in vacuum.* If a material is present, the permittivity ϵ_0 and permeability μ_0 of free space are replaced by the permittivity ϵ and permeability μ of the material. We have already remarked that Maxwell's synthesis of electromagnetism in these four equations is one of the great milestones of theoretical physics, comparable to the Newtonian synthesis 150 years earlier.

In 1887 Heinrich Hertz actually *produced* electromagnetic waves, with the aid of oscillating circuits, and received and detected these waves with other circuits tuned to the same frequency. Hertz then produced electromagnetic standing waves and measured the distance between adjacent nodes (one-half wavelength), to determine the wavelength. Knowing the resonant frequency of his circuits, he then found the speed of the waves from the fundamental wave equation $c = \lambda f$ and verified Maxwell's theoretical value directly. The SI unit of frequency, one cycle per second, is named the *hertz* in honor of Hertz.

The possible use of electromagnetic waves for purposes of long-distance communication does not seem to have occurred to Hertz. It remained for the enthusiasm and energy of Marconi and others to make radio communication a familiar household phenomenon.

Hertz produced electromagnetic waves in his laboratory: a dramatic confirmation of Maxwell's theoretical predictions.

35–2 SPEED OF AN ELECTROMAGNETIC WAVE

In developing the relation of electromagnetic-wave propagation to familiar electromagnetic principles, we begin with a particularly simple and somewhat artificial example of an electromagnetic wave. We first postulate a particular configuration of fields, and then test whether it is consistent with the principles mentioned above, particularly Faraday's law and Ampere's law including displacement current.

Using an x–y–z-coordinate system, as shown in Fig. 35–1, we imagine that all space is divided into two regions by a plane perpendicular to the x-axis (parallel to the yz-plane). At all points to the left of this plane, there is a uniform electric field E in the $+y$-direction and a uniform magnetic field B in the $+z$-direction, as shown. Furthermore, we suppose that the boundary surface, which may also be called the *wave front,* moves to the right with a constant speed c, as yet unknown. Thus the E and B fields travel to the right into previously field-free regions, with a definite speed. The situation, in short, describes a rudimentary electromagnetic wave, provided that it is consistent with the laws of electromagnetism.

We will not worry about how such a field configuration might actually be produced. Instead, we simply ask whether it is consistent with the laws of electrodynamics, that is, with Maxwell's equations. First we apply Faraday's law to a rectangle in the xy-plane, as in Fig. 35–2. The rectangle is located so that at some instant the wave front has progressed partway through it, as shown. In a time dt, the boundary surface moves a distance $c\,dt$ to the right, sweeping out an area $ac\,dt$ of the rectangle. In this time the magnetic flux increases by $B(ac\,dt)$. We take the vector area element to point in the $+z$-direction, so this is a positive quantity. The rate of change of magnetic flux is given by

$$\frac{d\Phi}{dt} = Bac. \tag{35–1}$$

According to Faraday's law, this expression must equal the negative of the line integral $\oint E \cdot dl$, evaluated counterclockwise around the boundary of the area. The field at the right end is zero, and the top and bottom sides do not contribute because there the component of E along dl is zero. Thus only the left side contributes to the integral. On that side, E and dl have opposite directions, so the value of the integral is $-Ea$. When we divide out a, Faraday's law gives

$$E = cB. \tag{35–2}$$

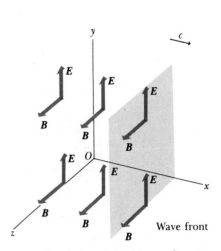

35–1 An electromagnetic wave front. The E and B fields are uniform over the region to the left of the plane, but are zero everywhere to the right of it. The plane representing the wave front moves to the right with speed c.

Checking for consistency with Faraday's law

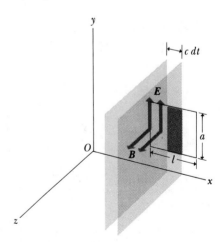

35–2 In time dt the wave front moves to the right a distance $c\,dt$. The magnetic flux through the rectangle in the xy-plane increases by an amount $d\Phi$ equal to the flux through the shaded rectangle of area $ac\,dt$; that is, $d\Phi = Bac\,dt$.

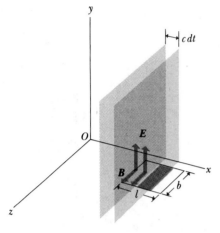

35–3 In time dt the electric flux through the rectangle in the xz-plane increases by an amount equal to E times the area $bc\,dt$ of the shaded rectangle; that is, $d\Psi = Ebc\,dt$. Thus $d\Psi/dt = Ebc$.

Thus the wave we have postulated is consistent with Faraday's law only if $\boldsymbol{E}$, $\boldsymbol{B}$, and c are related as in Eq. (35–2).

Next we apply Ampere's law to a rectangle in the xz-plane, as shown in Fig. 35–3. There is no conduction current, so $I_\mathrm{C} = 0$. In terms of electric flux Ψ, Ampere's law with displacement current but no conduction current is

$$\oint \boldsymbol{B} \cdot d\boldsymbol{l} = \mu_0\epsilon_0 \frac{d\Psi}{dt}. \tag{35–3}$$

In Fig. 35–3 we take the vector area to point in the $+y$-direction, so Ψ is positive and increasing. The line integral is then evaluated counterclockwise around the boundary of the area. The change in electric flux $d\Psi$ in time dt is the area $bc\,dt$ swept out by the wave front, multiplied by E. In evaluating the line integral of $\boldsymbol{B}$, we note that B is zero on the right end of the rectangle and is perpendicular to the boundary on the front and back sides. Thus only the left end, where $\boldsymbol{B}$ and $d\boldsymbol{l}$ are parallel, contributes to the integral, and we find $\oint \boldsymbol{B} \cdot d\boldsymbol{l} = Bb$. Combining these results with Eq. (35–3) and dividing out the common factor b, we obtain

$$B = \epsilon_0\mu_0 cE. \tag{35–4}$$

Thus Ampere's law is obeyed only if B, c, and E are related as in Eq. (35–4).

Since *both* Ampere's law and Faraday's law must be obeyed at the same time, Eqs. (35–2) and (35–4) must both be satisfied. This can happen only when $\epsilon_0\mu_0 c = 1/c$, or

$$c = \frac{1}{\sqrt{\epsilon_0\mu_0}}. \tag{35–5}$$

Inserting the numerical values of these quantities, we find

$$c = \frac{1}{\sqrt{(8.85 \times 10^{-12})(4\pi \times 10^{-7})}} = 3.00 \times 10^8 \text{ m·s}^{-1}.$$

The postulated field configuration *is* consistent with the laws of electrodynamics, provided the wave front moves with the speed given above, which of course is recognized as the speed of light.

We have chosen a particularly simple wave for study in order to avoid undue mathematical complexity, but this special case nevertheless illustrates several important features of *all* electromagnetic waves:

1. The wave is **transverse;** both $\boldsymbol{E}$ and $\boldsymbol{B}$ are perpendicular to the direction of propagation of the wave and to each other.

2. There is a definite ratio between the magnitudes of $\boldsymbol{E}$ and $\boldsymbol{B}$.

3. The wave travels in vacuum with a definite and unchanging speed.

It is not difficult to generalize this discussion to a more realistic situation. Suppose we have several wave fronts in the form of parallel planes perpendicular to the x-axis and all moving to the right with speed c. And suppose that, within a single region between two planes, the $\boldsymbol{E}$ and $\boldsymbol{B}$ fields are the same at all points in the region, but that they differ from region to region. An extension of the development above shows that such a situation is also consistent with Ampere's and Faraday's laws, provided the wave fronts all move with the speed c given by Eq. (35–5). From this picture it is only a short additional step to a wave picture in which the $\boldsymbol{E}$ and $\boldsymbol{B}$ fields at any instant vary smoothly rather than in steps, as we move along the x-axis, and the entire field pattern moves to the right with speed c. In Section 35–5 we consider waves in which the dependence of $\boldsymbol{E}$ and $\boldsymbol{B}$ on position and time is *sinusoidal;* first, however, we consider the *energy* associated with an electromagnetic wave.

35–3 ENERGY IN ELECTROMAGNETIC WAVES

It is a familiar fact that there is energy associated with electromagnetic waves. Two simple examples are the energy in the sun's radiation and cooking with microwave ovens. To derive detailed relationships for the energy in an electromagnetic wave, we begin with the expressions derived in Sections 27–4 and 33–3 for the **energy densities** associated with electric and magnetic fields; we suggest you review those derivations now. Specifically, Eqs. (27–9) and (33–10) show that the total energy density u in a region of space where $\boldsymbol{E}$ and $\boldsymbol{B}$ fields are present is given by

> The total energy density is the sum of the energy densities associated with the electric and magnetic fields.

$$u = \frac{1}{2}\epsilon_0 E^2 + \frac{1}{2\mu_0}B^2. \tag{35–6}$$

We found in Section 35–2 that $\boldsymbol{E}$ and $\boldsymbol{B}$ in an electromagnetic wave are not independent but are related by

$$B = \frac{E}{c} = \sqrt{\epsilon_0\mu_0}E. \tag{35–7}$$

Thus the energy density may also be expressed as

$$
\begin{aligned}
u &= \frac{1}{2}\epsilon_0 E^2 + \frac{1}{2\mu_0}\left(\sqrt{\epsilon_0\mu_0}E\right)^2 \\
&= \epsilon_0 E^2, \tag{35–8}
\end{aligned}
$$

which shows that, in a wave, the energy density associated with the $\boldsymbol{E}$ field is equal to that of the $\boldsymbol{B}$ field.

Because the $\boldsymbol{E}$ and $\boldsymbol{B}$ fields in the simple wave considered above advance with time into regions where originally no fields were present, it is clear that the wave transports energy from one region to another. We can describe this energy transfer in terms of energy transferred *per unit time, per unit cross-sectional area* for an area perpendicular to the direction of wave travel. This quantity will be denoted by S. This is analogous to the concept of current density, which is the charge per unit time transferred across unit area perpendicular to the direction of flow.

To see how the energy flow is related to the fields, consider a stationary plane, perpendicular to the x-axis, that coincides with the wave front at a certain time. In a time dt after this, the wave front moves a distance $c\,dt$ to the right. Considering an area A on the stationary plane, we note that the energy in the space to the right of this area must have passed through it to reach its new location. The volume dV of the relevant region is the base area A times the length $c\,dt$, and the total energy dU in this region is the energy density u times this volume:

$$dU = \epsilon_0 E^2 Ac\,dt. \tag{35–9}$$

Since this much energy passed through area A in time dt, the energy flow S

The comet Mrkos, photographed August 22 through 27, 1957. The tail of the comet is pushed away from the sun and split into two distinct parts by radiation pressure from the sun's electromagnetic radiation and by the "solar wind," a stream of particles emitted by the sun. (Courtesy of Hale Observatories.)

Energy flow can be expressed as power per unit time, per unit area, is
per unit area.

$$S = \frac{1}{A}\frac{dU}{dt} = \epsilon_0 c E^2.$$

Using Eq. (35–7), we obtain the alternative forms

$$S = \frac{\epsilon_0}{\sqrt{\epsilon_0 \mu_0}}E^2 = \sqrt{\frac{\epsilon_0}{\mu_0}}E^2 = \frac{EB}{\mu_0}. \qquad (35\text{–}10)$$

The unit of S is energy per unit time, per unit area. The SI unit of S is $1\ \text{J}\cdot\text{s}^{-1}\cdot\text{m}^{-2}$ or $1\ \text{W}\cdot\text{m}^{-2}$.

We can define a *vector* quantity that describes both the magnitude and the direction of the energy-flow rate:

$$\boldsymbol{S} = \frac{1}{\mu_0}\boldsymbol{E}\times\boldsymbol{B}. \qquad (35\text{–}11)$$

The Poynting vector: both magnitude and direction of energy flow

$\boldsymbol{S}$ is called the **Poynting vector;** its magnitude is given by Eq. (35–10), and its direction is the direction of propagation of the wave. The magnitude EB/μ_0 *gives the flow of energy through a cross section perpendicular to the propagation direction, per unit area and per unit time.* The total energy flow per unit time (power, P) through any surface is given by the integral

The average magnitude of the Poynting vector is called intensity.

$$P = \int \boldsymbol{S}\cdot d\boldsymbol{A}$$

over the surface.

Intensity is average power per unit area.

The electric and magnetic fields at any point in a wave vary with time, so the Poynting vector at any point is also a function of time. The *average* value of the magnitude of the Poynting vector at a point is called the **intensity** of the radiation at that point. As mentioned above, the SI unit of intensity is the watt per square meter ($\text{W}\cdot\text{m}^{-2}$). In Section 35–5 we consider the relation between the intensity of a sinusoidal wave and the amplitudes of the sinusoidally varying electric and magnetic fields.

EXAMPLE 35–1 For the wave described in Section 35–2, suppose

$$E = 100\ \text{V}\cdot\text{m}^{-1} = 100\ \text{N}\cdot\text{C}^{-1}.$$

Find the value of B, the energy density, and the rate of energy flow per unit area S.

SOLUTION From Eq. (35–7),

$$B = \frac{E}{c} = \frac{100\ \text{V}\cdot\text{m}^{-1}}{3.0\times10^8\ \text{m}\cdot\text{s}^{-1}} = 3.33\times10^{-7}\ \text{T}.$$

From Eq. (35–8),

$$\begin{aligned}
u &= \epsilon_0 E^2 = (8.85\times10^{-12}\ \text{C}^2\cdot\text{N}^{-1}\cdot\text{m}^{-2})(100\ \text{N}\cdot\text{C}^{-1})^2 \\
&= 8.85\times10^{-8}\ \text{N}\cdot\text{m}^{-2} = 8.85\times10^{-8}\ \text{J}\cdot\text{m}^{-3};
\end{aligned}$$

$$\begin{aligned}
S &= EB/\mu_0 = (100\ \text{V}\cdot\text{m}^{-1})(3.33\times10^{-7}\ \text{T})/(4\pi\times10^{-7}\ \text{Wb}\cdot\text{A}^{-1}\cdot\text{m}^{-1}) \\
&= 26.5\ \text{V}\cdot\text{A}\cdot\text{m}^{-2} = 26.5\ \text{W}\cdot\text{m}^{-2}.
\end{aligned}$$

Alternatively,

$$S = \epsilon_0 c E^2$$

$$= (8.85 \times 10^{-12}\,\text{C}^2\cdot\text{N}^{-1}\cdot\text{m}^{-2})(3.0 \times 10^8\,\text{m}\cdot\text{s}^{-1})(100\,\text{N}\cdot\text{C}^{-1})^2$$

$$= 26.5\,\text{W}\cdot\text{m}^{-2}.$$

The idea that energy can travel through empty space without the aid of any matter in motion may seem strange, yet this is the very mechanism by which energy reaches us from the sun. The conclusion that electromagnetic waves transport energy is as inescapable as the conclusion that energy is required to establish electric and magnetic fields. It is also possible to show that electromagnetic waves carry *momentum p*, with a corresponding momentum density (momentum p per unit volume V) of magnitude

Electromagnetic waves transport momentum as well as energy.

$$\frac{p}{V} = \frac{EB}{\mu_0 c^2} = \frac{S}{c^2}. \tag{35–12}$$

This momentum is a property of the field alone and is not associated with moving mass. There is also a corresponding momentum-flow rate; just as the energy density u corresponds to S, the rate of energy flow per unit area, the momentum density given by Eq. (35–12) corresponds to the momentum-flow rate per unit area

$$\frac{1}{A}\frac{dp}{dt} = \frac{S}{c} = \frac{EB}{\mu_0 c}, \tag{35–13}$$

which represents the momentum transferred per unit surface area, per unit time.

This momentum is responsible for the phenomenon of **radiation pressure.** When an electromagnetic wave is completely absorbed by a surface perpendicular to the propagation direction, the time rate of change of momentum equals the force on the surface. Thus the force per unit area, or pressure, is equal to S/c. If the wave is totally reflected, the momentum change is twice as great, and the pressure is $2S/c$. For example, the value of S for direct sunlight is about $1.4\,\text{kW}\cdot\text{m}^{-2}$, and the corresponding pressure on a completely absorbing surface is

Radiation pressure: a manifestation of momentum in electromagnetic waves

$$\frac{S}{c} = \frac{1.4 \times 10^3\,\text{W}\cdot\text{m}^{-2}}{3.0 \times 10^8\,\text{m}\cdot\text{s}^{-1}} = 4.7 \times 10^{-6}\,\text{Pa}$$

$$= 4.7 \times 10^{-6}\,\text{N}\cdot\text{m}^{-2}.$$

The pressure on a totally reflecting surface is $9.4 \times 10^{-6}\,\text{N}\cdot\text{m}^{-2}$.

Radiation pressure is important in the structure of stars. Gravitational attractions tend to shrink a star, but this tendency is balanced by radiation pressure in maintaining the size of the star through most stages of its evolution. The pressure of the sun's radiation is responsible for pushing the tail of a comet away from the sun.

The usefulness of the Poynting vector is not limited to wave situations; it also describes energy flow in situations where the fields are constant.

The Poynting vector is meaningful even when there is no wave motion.

EXAMPLE 35–2 A long cylindrical conductor of radius R and length l carries a steady current I, and the potential difference between its ends is V. Find the magnitude and direction of the Poynting vector at a point on the surface, and relate this to the energy input to the conductor.

SOLUTION The electric field is uniform inside the conductor, with magnitude $E = V/l$ and direction parallel to the cylinder's axis. The magnitude of the magnetic field at a point on the surface is given by Eq. (31–14), with $x = R$: $B = \mu_0 I / 2\pi R$. Its direction is tangent to the surface, perpendicular to the axis and to E. The magnitude of the Poynting vector is

$$S = \frac{EB}{\mu_0} = \frac{VI}{2\pi Rl}.$$

Equation (35–11) and the right-hand rule show that at each point on the surface, the direction of S is radially inward, toward the axis. Now VI is just the total *power* input P to the conductor, and $2\pi Rl$ is its surface area. Thus these quantities are related by

$$P = SA.$$

Although this situation is not a wave phenomenon, it does illustrate the usefulness of the Poynting vector in computing energy flow associated with electric and magnetic fields.

35–4 ELECTROMAGNETIC WAVES IN MATTER

Electromagnetic waves can travel through dielectrics.

Thus far we have talked only about electromagnetic waves in vacuum, but it is easy to extend our analysis to include electromagnetic waves in dielectrics. The wave speed is not the same as in vacuum, so we denote it by v instead of c. Faraday's law is unaltered, but Eq. (35–2) is replaced by $E = vB$. In Ampere's law the displacement-current density is given not by $\epsilon_0\, dE/dt$ but by $\epsilon\, dE/dt = K\epsilon_0\, dE/dt$. In addition, the constant μ_0 in Ampere's law must be replaced by $\mu = K_m \mu_0$. Thus Eq. (35–4) is replaced by

$$B = \mu \epsilon v E,$$

and the wave speed is given by

$$v = \frac{1}{\sqrt{\epsilon\mu}} = \frac{1}{\sqrt{KK_m}} \frac{1}{\sqrt{\epsilon_0\mu_0}}. \tag{35–14}$$

For many dielectrics the relative permeability K_m is very nearly equal to unity; in such cases we say

$$v \simeq \frac{1}{\sqrt{K}} \frac{1}{\sqrt{\epsilon_0\mu_0}} = \frac{c}{\sqrt{K}}.$$

What is the speed of electromagnetic waves in a dielectric?

The speed depends on the electric and magnetic properties of the material.

Because K is always greater than unity, the speed v of electromagnetic waves in a dielectric is always *less* than the speed c in vacuum, by a factor of $1/\sqrt{K}$. The ratio of the speed c in vacuum to the speed v in a material is known in optics as the **index of refraction** n of the material. For most dielectrics, where $K_m \simeq 1$, n is given by

$$\frac{c}{v} = n = \sqrt{KK_m} \simeq \sqrt{K}. \tag{35–15}$$

We also have to make corresponding modifications in the expressions for the energy density and the Poynting vector. The energy density is now given by

$$u = \frac{1}{2}\epsilon E^2 + \frac{1}{2\mu}B^2 = \epsilon E^2. \qquad (35-16)$$

The energy densities in the E and B fields are still equal. The Poynting vector is now

$$S = \frac{1}{\mu}E \times B. \qquad (35-17)$$

The waves described above cannot propagate any appreciable distance in a *conducting* material because the E field leads to currents that provide a mechanism for dissipating the energy of the wave. For an ideal conductor with zero resistivity, E must be zero everywhere inside the material. When an electromagnetic wave strikes such a material, it is totally reflected. Real conductors with finite resistivity permit some penetration of the wave into the material, with partial reflection. A polished metal surface is usually a good *reflector* of electromagnetic waves, but metals are not *transparent* to radiation.

Why can't electromagnetic waves travel through a conductor?

35–5 SINUSOIDAL WAVES

Sinusoidal electromagnetic waves are directly analogous to sinusoidal transverse mechancal waves on a stretched string. We studied these in Chapter 21; we suggest you review that discussion now, especially Section 21–3. In a sinusoidal electromagnetic wave, the E and B fields at any point in space are sinusoidal functions of time, and at any instant of time the spatial variation of the fields is also sinusoidal.

The simplest sinusoidal electromagnetic waves share with the waves described in Section 35–2 the property that at any instant the fields are uniform over any plane perpendicular to the direction of propagation. Such a wave is called a **plane wave.** The entire pattern travels in the direction of propagation with speed c. The directions of E and B are perpendicular to the direction of propagation (and to each other), so the wave is *transverse*.

Plane waves travel in one direction and are particularly simple to describe.

The frequency f, the wavelength λ, and the speed of propagation c are related by the equation applicable to all periodic-wave motion, namely, $c = f\lambda$. If the frequency f is the power-line frequency of 60 Hz, the wavelength is

$$\lambda = \frac{c}{f} = \frac{3 \times 10^8 \text{ m·s}^{-1}}{60 \text{ Hz}} = 5 \times 10^6 \text{ m} = 5000 \text{ km},$$

which is of the order of the earth's radius! Hence at this frequency even a distance of many miles includes only a small fraction of a wavelength. But if the frequency is 10^8 Hz (100 MHz), typical of commercial FM radio stations, the wavelength is

$$\lambda = \frac{3 \times 10^8 \text{ m·s}^{-1}}{10^8 \text{ Hz}} = 3 \text{ m},$$

and a moderate distance can include many complete waves.

Figure 35–4 shows a sinusoidal electromagnetic wave traveling in the $+x$-direction. The E and B vectors are shown only for a few points on the x-axis. If

An example of a plane sinusoidal electromagnetic wave

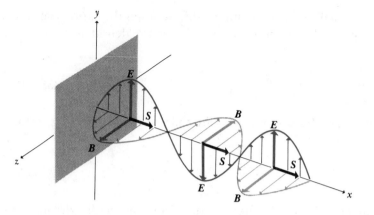

35–4 *E*, *B*, and *S* vectors in a sinusoidal electromagnetic wave traveling in the positive *x*-direction. The position of the wave at time $t = 0$ is shown.

The *E* and *B* fields are perpendicular to each other and to the direction of propagation.

we construct a plane perpendicular to the *x*-axis at a particular point and at a particular instant, the fields have the same values at all points in that plane. The values are, of course, different on different planes. In those planes in which the *E* vector is in the +*y*-direction, *B* is in the +*z*-direction; where *E* is in the −*y*-direction, *B* is in the −*z*-direction. In both cases the direction of the Poynting vector, given by Eq. (35–11), is along the +*x*-direction.

We can describe electromagnetic waves by means of *wave functions*, just as we did in Section 21–3 for waves on a string. One form of the equation of a transverse wave traveling to the right along a stretched string is Eq. (21–7):

$$y = A \sin (\omega t - kx),$$

where *y* is the transverse displacement from its equilibrium position, at time *t*, of a point of the string whose coordinate is *x*. The quantity *A* is the maximum displacement, or *amplitude*, of the wave; ω is its *angular frequency*, equal to 2π times the frequency *f*; and *k* is the *propagation constant*, equal to $2\pi/\lambda$, where λ is the wavelength.

Wave functions for a plane sinusoidal electromagnetic wave

Let *E* and *B* represent the instantaneous values, and E_{max} and B_{max} the maximum values, or *amplitudes*, of the electric and magnetic fields in Fig. 35–4. The equations of the traveling electromagnetic wave are then

$$E = E_{max} \sin (\omega t - kx), \qquad B = B_{max} \sin (\omega t - kx). \qquad (35–18)$$

The sine curves in Fig. 35–4 represent instantaneous values of *E* and *B*, as functions of *x*, at time $t = 0$. The wave travels to the right with speed *c*.

The instantaneous value *S* of the magnitude of the Poynting vector is

$$S = \frac{EB}{\mu_0} = \frac{E_{max}B_{max}}{\mu_0} \sin^2 (\omega t - kx)$$

$$= \frac{E_{max}B_{max}}{2\mu_0}[1 - \cos 2(\omega t - kx)].$$

The time-average value of $\cos 2(\omega t - kx)$ is zero, so the average value S_{av} of the Poynting vector magnitude is

$$S_{av} = \frac{E_{max}B_{max}}{2\mu_0}.$$

This is the *average power* transmitted per unit area; as noted in Section 35–3, it is called the *intensity* of the radiation, denoted by *I*. By using the relations

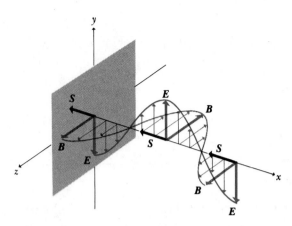

35-5 Electric and magnetic fields of a sinusoidal wave traveling in the negative x-direction. The position of the wave at time $t = 0$ is shown.

$E_{max} = B_{max}c$ and $\epsilon_0\mu_0 = 1/c^2$, we can express the intensity in several equivalent forms:

$$I = S_{av} = \frac{E_{max}B_{max}}{2\mu_0} = \frac{E_{max}^2}{2\mu_0 c} = \frac{1}{2}\sqrt{\frac{\epsilon_0}{\mu_0}}E_{max}^2 = \frac{1}{2}\epsilon_0 c E_{max}^2. \quad (35\text{--}19)$$

We invite you to verify that these expressions are all equivalent.

We can also obtain various expressions for the intensity of a wave in a material. In Eq. (35–19) we simply replace ϵ_0 by ϵ, μ_0 by μ, and c by v.

Figure 35–5 represents schematically the electric and magnetic fields of a wave traveling in the *negative* x-direction. At points where E is in the positive y-direction, B is in the *negative* z-direction, and where E is in the negative y-direction, B is in the *positive* z-direction. The Poynting vector is in the *negative* x-direction at all points. (Compare this with Fig. 35–4, which shows a wave traveling in the *positive* x-direction.) The equations of the wave are

$$E = -E_{max}\sin(\omega t + kx),$$
$$B = B_{max}\sin(\omega t + kx). \quad (35\text{--}20)$$

Finding the intensity of a sinusoidal electromagnetic wave

EXAMPLE 35–3 A radio station radiates a sinusoidal wave with an average total power of 50 kW. Assuming it radiates equally in all directions (which is unlikely in real-world situations), find the amplitudes E_{max} and B_{max} at a distance of 100 km from the antenna.

SOLUTION First we find the magnitude S of the Poynting vector. We surround the antenna with an imaginary sphere of radius 100 km $= 1.00 \times 10^5$ m. This sphere has area

$$A = 4\pi R^2 = 4\pi(1.00 \times 10^5 \text{ m})^2 = 12.6 \times 10^{10} \text{ m}^2.$$

All the power radiated passes through this surface, so the power per unit area is

$$S = \frac{P}{A} = \frac{P}{4\pi R^2} = \frac{5.00 \times 10^4 \text{ W}}{12.6 \times 10^{10} \text{ m}^2} = 3.98 \times 10^{-7} \text{ W·m}^{-2}.$$

But from Eq. (35–19),

$$S_{av} = \frac{E_{max}^2}{2\mu_0 c} \quad \text{and} \quad E_{max} = \sqrt{2\mu_0 c S_{av}},$$

A radio transmitter: an example of the relation of power to amplitude in a wave so

$$E_{max} = \sqrt{2(4\pi \times 10^{-7}\,\text{Wb·A}^{-1}\text{·m}^{-1})(3 \times 10^8\,\text{m·s}^{-1})(3.98 \times 10^{-7}\,\text{W·m}^{-2})}$$
$$= 1.73 \times 10^{-2}\,\text{V·m}^{-1},$$
$$B_{max} = E_{max}/c = 5.77 \times 10^{-11}\,\text{T}.$$

Note that E_{max} is of an order of magnitude comparable to many laboratory phenomena, but that B_{max} is extremely small compared to the **B** fields we have seen in previous chapters.

What is polarization?

Electromagnetic waves exhibit *polarization*. This concept was introduced at the end of Section 21–4 in the context of transverse waves on a string, and we suggest you review that discussion now. In the present context, note that for a wave traveling in the x-direction the choice of the y-direction for **E** was arbitrary. We could just as well have specified the z-axis for **E;** then when **E** is in the $+z$-direction, **B** is in the $-y$-direction, and so on.

A wave in which **E** always lies along a certain axis is said to be *linearly polarized* along that axis. More generally, we can think of *any* wave traveling in the x-direction as a superposition of waves linearly polarized in the y- and z-directions. A superposition of two linearly polarized waves with the same frequency and amplitude but a 90° phase difference yields a wave that is *circularly polarized*. We will study polarization phenomena in greater detail, with special emphasis on polarization of light, in Chapter 36.

35–6 STANDING WAVES

Electromagnetic waves can be reflected by a conducting surface.

Electromagnetic waves can be *reflected;* a conducting surface can serve as a reflector. The superposition principle holds for electromagnetic waves just as for all electric and magnetic fields, and the superposition of an incident wave and a reflected wave can form a **standing wave.** The situation is analogous to standing waves on a stretched string; we studied these in Section 22–2, and you should review that discussion now.

Suppose a sheet of an ideal conductor, having zero resistivity, is placed in the yz-plane of Fig. 35–5, and that the wave shown, traveling in the negative x-direction, is incident on it. The essential characteristic of an ideal conductor is that no electric field can ever exist within it; any attempt to establish a field is immediately canceled by rearrangement of the mobile charges in the conductor. Thus **E** must always be zero everywhere in this plane, and the **E** field of the incident wave induces sinusoidal currents in the conductor so that **E** is zero inside it.

Superposition of an incident and a reflected wave forms a standing wave.

These induced currents produce a reflected wave, traveling out from the plane, to the right. From the superposition principle, the total **E** field at any point to the right of the plane is the vector sum of the **E** fields of the incident and reflected waves; the same is true for the total **B** field.

Suppose the incident wave is described by the wave functions of Eqs. (35–20) and the reflected wave by the wave functions of Eqs. (35–18). (Compare these with Eqs. [21–7] and [21–8] for transverse waves on a string.) From the superposition principle, the total fields at any point are given by

$$E = E_{max}[-\sin(\omega t + kx) + \sin(\omega t - kx)],$$
$$B = B_{max}[\sin(\omega t + kx) + \sin(\omega t - kx)].$$

These expressions may be expanded and simplified by using the identity

Wave functions for a standing wave

$$\sin (A \pm B) = \sin A \cos B \pm \cos A \sin B.$$

The results are

$$E = -2E_{max} \cos \omega t \sin kx, \qquad (35\text{–}21a)$$

$$B = 2B_{max} \sin \omega t \cos kx. \qquad (35\text{–}21b)$$

The first of these is analogous to Eq. (22–1) for a stretched string. We see that at $x = 0$, E is *always* zero; this condition is required by the nature of the ideal conductor, which plays the same role as a fixed point at the end of the string. Furthermore, E is zero at all times in those planes for which $\sin kx = 0$; that is, $kx = 0, \pi, 2\pi, \ldots$, or

$$x = 0, \quad \frac{\lambda}{2}, \quad \lambda, \quad \frac{3\lambda}{2}, \quad \ldots .$$

These are called the **nodal planes** of the E field.

The total magnetic is zero at all times in those planes for which $\cos kx = 0$, or at which

The nodal planes of the B field do not coincide with those of the E field.

$$x = \frac{\lambda}{4}, \quad 3\frac{\lambda}{4}, \quad 5\frac{\lambda}{4}, \quad \ldots .$$

These are the nodal planes of the B field. The magnetic field is *not* zero at the conducting surface ($x = 0$), and there is no reason it should be. The nodal planes of one field are midway between those of the other, and the nodal planes of either field are separated by one-half wavelength. Figure 35–6 shows a standing-wave pattern at one instant of time.

The total electric field is a *cosine* function of t, and the total magnetic field is a *sine* function of t. The fields are therefore 90° out of phase. At times when $\cos \omega t = 0$, the electric field is zero *everywhere* and the magnetic field is maximum. When $\sin \omega t = 0$, the magnetic field is zero *everywhere* and the electric field is maximum.

Pursuing the stretched-string analogy, we may now insert a second conduction plane, parallel to the first and a distance L from it, along the $+x$-axis. This is analogous to the stretched string held at the points $x = 0$ and $x = L$. A standing wave can exist only when L is an integer multiple of $\lambda/2$. Hence the

When there are two reflecting surfaces, normal modes result.

35–6 E and B vectors in a standing wave. The pattern does not move along the x-axis, but the E and B vectors grow and diminish with time at each point. At each point E is maximum when B is minimum, and conversely. The position of the wave at time $t = 0$ is shown.

possible wavelengths are

$$\lambda_n = \frac{2L}{n}, \qquad n = 1, 2, 3, \ldots, \tag{35-22}$$

and the corresponding frequencies are

$$f_n = \frac{c}{\lambda_n} = n \frac{c}{2L}, \qquad n = 1, 2, 3, \ldots. \tag{35-23}$$

Thus there is a set of *normal modes*, each with a characteristic frequency, wave shape, and node pattern. Measurement of the node positions makes it possible to measure the wavelength. If the frequency is known, the wave speed can be determined. This technique was, in fact, used by Hertz in his pioneering investigations of electromagnetic waves.

Reflection and transmission at a dielectric interface

Conducting surfaces are not the only reflectors of electromagnetic waves; reflections also occur at an interface between two insulating materials having different dielectric or magnetic properties. The mechanical analog is a junction of two strings with equal tension but different linear mass density. In general, a wave incident on such a boundary surface is partly transmitted into the second material and partly reflected back into the first. The partial transmission and reflection of light at a glass surface is a familiar phenomenon; light is transmitted through a sheet of glass, but its surfaces also reflect light.

PROBLEM-SOLVING STRATEGY: *Electromagnetic waves*

1. In the problems of this chapter, the most important advice we can give is to make sure you know what the basic relationships are, such as the relation of **E** to **B**, how the wave speed is determined, the transverse nature of the waves, and so on.

2. In the discussion of sinusoidal waves in Sections 35–5 and 35–6 you will need to use the language of sinusoidal waves you learned in Chapters 21 and 22. Don't hesitate to go back and review that material, including the problem-solving strategies suggested in those chapters. In particular, keep in mind the basic

relationships for periodic waves, $c = \lambda f$ and $\omega = ck$. Be careful to distinguish between the ordinary frequency f, usually expressed in hertz, and the angular frequency $\omega = 2\pi f$, expressed in s^{-1}. Also remember that the wave number k is $k = 2\pi/\lambda$.

3. In the discussion of standing waves, make sure what you mean by nodes and antinodes, and which field you are talking about. Nodes of **E** coincide with antinodes of **B**, and conversely. Compare this situation to the distinction between pressure nodes and displacement nodes in Section 22–4.

35–7 THE ELECTROMAGNETIC SPECTRUM

Electromagnetic waves occur in nature with an enormous range of frequencies and wavelengths.

Electromagnetic waves cover an extremely broad spectrum of wavelength and frequency, as shown in Fig. 35–7. Radio and TV transmission, visible light, infrared and ultraviolet radiation, and gamma rays all form parts of the **electromagnetic spectrum.** It is well established that light consists of electromagnetic waves; it is a small part of a very broad class of electromagnetic radiations, all having the general characteristic described in preceding sections, including the propagation speed (in vacuum) $c = 3.00 \times 10^8$ m·s^{-1} but differing in frequency f and wavelength λ. The general wave relation $c = \lambda f$ holds for each.

Wavelengths of light are less than a thousandth of a millimeter.

The wavelengths of visible light (i.e., of electromagnetic waves that are perceived by the sense of sight) can be measured by methods to be explained

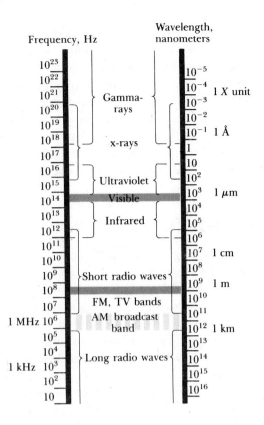

35–7 A chart of the electromagnetic spectrum.

in Chapter 39; they are in the range 4 to 7×10^{-7} m (400 to 700 nm). The corresponding range of frequencies is about 7.5 to 4.3×10^{14} Hz.

Because of the very small magnitudes of light wavelengths, it is convenient to measure them in small units of length. Three such units are commonly used: the micrometer (1 μm), the nanometer (1 nm) (both accented on the *first* syllable), and the angstrom (1 Å):

$$1 \ \mu\text{m} \ = \ 10^{-6} \ \text{m} = 10^{-4} \ \text{cm},$$

$$1 \ \text{nm} \ = \ 10^{-9} \ \text{m} = 10^{-7} \ \text{cm},$$

$$1 \ \text{Å} \ = \ 10^{-10} \ \text{m} = 10^{-8} \ \text{cm}.$$

In older literature, the micrometer is sometimes called the *micron,* and the nanometer is sometimes called the *millimicron;* these terms are now obsolete. Most workers in the fields of optical-instrument design, color, and physiological optics express wavelengths in *nanometers.* For example, the wavelength of the yellow light from a sodium-vapor lamp is 589 nm, but many spectroscopists would identify this same wavelength as 5890 Å.

Different parts of the visible spectrum evoke the sensations of different colors. Wavelengths for colors in the visible spectrum are (very approximately) as follows:

The color of light depends on its wavelength or frequency.

400 to 440 nm	Violet
440 to 480 nm	Blue
480 to 530 nm	Green
530 to 590 nm	Yellow
590 to 630 nm	Orange
630 to 700 nm	Red

Monochromatic (single-frequency) light: a useful idealization

By the use of special sources or special filters, it is possible to select a narrow band of wavelengths, with a range of, say, from 1 to 10 nm. Such light is approximately *monochromatic* (single-color) light. Light consisting of only one wavelength is an idealization that is impossible to attain in practice but is useful in theoretical calculations. When the expression "monochromatic light of wavelength 550 nm" is used in theoretical discussions, it refers to one wavelength, but in descriptions of laboratory experiments it means a small band of wavelengths *around* 550 nm. One distinguishing characteristic of light from a *laser* is that it is much more nearly monochromatic than light obtainable in any other way.

35–8 RADIATION FROM AN ANTENNA

Radiation from an antenna is not usually a plane wave.

The waves we have been discussing are called *plane waves;* this name refers to the fact that if we construct any plane perpendicular to the direction of propagation of the wave, then at any instant the *E* and *B* fields are uniform over that plane. Although plane waves are the simplest of all electromagnetic waves to describe and analyze, they are by no means the simplest to produce experimentally. Any charge or current distribution that oscillates sinusoidally with time produces sinusoidal electromagnetic waves, but in general there is no reason to expect them to be plane waves.

The simplest example of an oscillating charge distribution is an **oscillating dipole,** which is a pair of electric charges of equal magnitude and opposite sign, with the charge magnitude varying sinusoidally with time. Such an oscillating dipole can be constructed in various ways, but we need not be concerned with the details.

The radiation from an oscillating dipole is *not* a plane wave, but it travels out in all directions from the source. The wave fronts are not planes but expanding concentric spheres centered at the source. At points far from the source, the *E* and *B* fields are perpendicular to the direction from the source and to each other; in this sense the wave is still transverse. The value of *S* drops off as the square of the distance from the source. The intensity (the average value of *S*) depends on the direction from the source; it is greatest in directions perpendicular to the dipole axis, and *S* = 0 in directions parallel to the axis.

The radiation pattern from a dipole source is shown schematically in Fig. 35–8. The figure shows a cross section of the radiation pattern at one instant. The oscillating dipole *P* is located at the centers of the spheres. At all points in the plane of the figure, the *E* field lies in the plane and the *B* field is perpendicular to it. The *E* field is shown by colored arrows, and the direction of *B* by crosses (where it points into the plane) and circles with dots (where it points out of the plane). We invite you to verify that the direction of the Poynting vector *S* is radially outward from the source at every point.

Electromagnetic waves can be *reflected* by conducting surfaces. When the surface is large compared to the wavelength of the radiation, the reflection behaves like reflection of light rays from a mirror, which we will study in Chapter 37. Large parabolic mirrors several meters in diameter are used as both transmitting and receiving antennas for microwave communications signals; typical wavelengths are a few centimeters. The transmitting reflector produces a wave that radiates in a narrow, well-defined beam; the receiving reflector gathers wave energy over its whole area and reflects it to the focus of

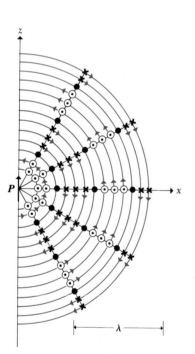

35–8 Cross section in the *xz*-plane of radiation from an oscillating electric dipole *P*. The wave fronts are expanding concentric spheres centered at *P*. The *E* field at every point lies in the plane; the colored lines show the directions of *E* at various points. The *B* field at every point is perpendicular to the plane. At points with circles, *B* comes out of the plane, and at points with crosses it is into the plane. The direction of the Poynting vector *S* is radially outward at every point.

35–9 A microwave-receiving antenna 64 m in diameter at Goldstone Tracking Station, California, the prime station of a worldwide network of stations used by NASA to monitor unmanned interplanetary spacecraft.

the parabola, where a detecting device is placed. Figure 35–9 shows an antenna installation using a large parabolic reflector.

SUMMARY

Maxwell's equations, which incorporate all the basic relationships of electric and magnetic fields and their sources (charges and currents), predict the existence of electromagnetic disturbances that can propagate through empty space and travel with a speed equal to the measured value of the speed of light. The simplest such wave is a plane wave, in which E and B are uniform over any plane perpendicular to the propagation direction, such that E and B are zero everywhere to the left of a certain plane and have constant values everywhere to the right. For such a wave disturbance to be consistent with Faraday's law, the two field magnitudes must be related by

$$E = cB, \tag{35–2}$$

and for consistency with Ampere's law they must be related by

$$B = \epsilon_0 \mu_0 cE, \tag{35–4}$$

where c is the propagation speed. For both of these requirements to be satisfied, c must be given by

$$c = \frac{1}{\sqrt{\epsilon_0 \mu_0}} = 3.00 \times 10^8 \text{ m·s}^{-1}. \tag{35–5}$$

Electromagnetic waves are transverse; the E and B fields are perpendicular to the direction of propagation and to each other. There is a definite ratio between E and B in a wave, and the waves travel in vacuum with a definite and unchanging speed c, given by Eq. (35–5).

The energy density in an electromagnetic wave can be expressed as

$$u = \frac{1}{2}\epsilon_0 E^2 + \frac{1}{2\mu_0}B^2 = \epsilon_0 E^2. \tag{35–8}$$

KEY TERMS

electromagnetic wave
Maxwell's equations
transverse wave
energy densities
Poynting vector
intensity
radiation pressure
index of refraction
plane wave
standing wave
nodal planes
electromagnetic spectrum
oscillating dipole

The energy-flow rate (power per unit area) is given by the Poynting vector

$$S = \frac{1}{\mu_0} E \times B. \tag{35-11}$$

The time-average value of the magnitude EB/μ_0 of the Poynting vector is called the intensity of the wave. These waves also carry momentum; the momentum per unit volume has magnitude

$$\frac{EB}{\mu_0 c^2} = \frac{S}{c^2}, \tag{35-12}$$

and the rate of transfer of momentum per unit cross-sectional area is

$$\frac{1}{A}\frac{dp}{dt} = \frac{S}{c} = \frac{EB}{\mu_0 c}. \tag{35-13}$$

When an electromagnetic wave travels through a dielectric, the wave speed v is given by

$$v = \frac{1}{\sqrt{\epsilon\mu}} = \frac{1}{\sqrt{KK_m}}\frac{1}{\sqrt{\epsilon_0\mu_0}}. \tag{35-14}$$

If the relative permeability K_m can be taken as unity, then

$$v \simeq \frac{1}{\sqrt{K}}\frac{1}{\sqrt{\epsilon_0\mu_0}} = \frac{c}{\sqrt{K}}.$$

In this case the index of refraction $n = c/v$ is given by

$$\frac{c}{v} = n = \sqrt{KK_m} \simeq \sqrt{K}. \tag{35-15}$$

For a sinusoidal electromagnetic wave traveling in the $+x$-direction, both E and B are sinusoidal functions of the quantity $(\omega t - kx)$, and at each point the sinusoidal variations of E and B are in phase. For a wave in the $-x$-direction, E and B are sinusoidal functions of $(\omega t + kx)$.

If a reflecting surface is placed at $x = 0$, it becomes a nodal plane for the E field; that is, E is zero everywhere in that plane. A wave traveling in the $-x$-direction is reflected, and the incident and reflected waves form a standing wave. There are nodal planes for E at $kx = 0$, π, 2π, and so on, and nodal planes for B at $kx = \pi/2, 3\pi/2, 5\pi/2$, and so on. At each point the sinusoidal variations of E and B are 90° out of phase; the nodes of B coincide with the antinodes of E, and conversely.

The electromagnetic spectrum covers a range of frequencies from at least 1 to 10^{24} Hz and a correspondingly broad range of wavelengths. Visible light is a very small part of this spectrum, with wavelengths of 4 to 7×10^{-7} m or 400 to 700 nm.

An oscillating dipole produces a wave that radiates out in all directions. At points far from the source, E and B are perpendicular to each other and to the radial direction, so this wave is transverse. The intensity depends on direction; it is zero along the dipole axis and greatest in the plane perpendicular to that axis.

QUESTIONS

35–1 In Ampere's law, is it possible to have both a conduction current and a displacement current at the same time? Is it possible for the effects of the two kinds of current to cancel each other exactly, so that *no* magnetic field is produced?

35–2 By measuring the electric and magnetic fields at a point in space where there is an electromagnetic wave, can one determine the direction from which the wave came?

35–3 Sometimes neon signs located near a powerful radio station are seen to glow faintly at night, even though they are not turned on. What is happening?

35–4 Light can be *polarized;* is this a property of all electromagnetic waves, or is it unique to light? What about polarization of sound waves? What fundamental distinction in wave properties is involved?

35–5 How does a microwave oven work? Why does it heat materials that are conductors of electricity, including most foods, but not insulators, such as glass or ceramic dishes? Why do the directions forbid putting anything metallic in the oven?

35–6 Electromagnetic waves can travel through vacuum, where there is no matter. We usually think of vacuum as empty space, but is it *really* empty if electric and magnetic fields are present? What *is* vacuum, anyway?

35–7 Give several examples of electromagnetic waves that are encountered in everyday life. How are they all alike? How do they differ?

35–8 We are surrounded by electromagnetic waves emitted by many radio and television stations. How is a radio or television receiver able to select a single station among all this mishmash of waves? What happens inside a radio receiver when the dial is turned to change stations?

35–9 The metal conducting rods on a television antenna are always in a horizontal plane. Would they work as well if they were vertical?

35–10 If a light beam carries momentum, should a person holding a flashlight feel a recoil analogous to the recoil of a rifle when it is fired? Why is this recoil not actually observed?

35–11 The nineteenth-century inventor Nikolai Tesla proposed to transmit large quantities of electrical energy across space by using electromagnetic waves instead of conventional transmission lines. What advantages would this scheme have? What disadvantages?

35–12 Does an electromagnetic *standing wave* have energy? Momentum? What distinction can be drawn between a standing wave and a propagating wave on this basis?

35–13 If light is an electromagnetic wave, what is its frequency? Is this a proper question to ask?

35–14 When an electromagnetic wave is reflected from a moving reflector, the frequency of the reflected wave is different from that of the initial wave. Explain physically how this can happen. (Some radar systems used for highway-speed control operate on this principle.)

35–15 The ionosphere is a layer of ionized air 100 km or so above the earth's surface. It acts as a reflector of radio waves of frequency less than about 30 MHz, but not of higher frequency. How does this reflection occur? Why does it work better for lower frequencies than for higher?

EXERCISES

Section 35–2 Speed of an Electromagnetic Wave

35–1 In a TV picture, ghost images are formed when the signal from the transmitter travels directly to the receiver and also indirectly after reflection from a building or other large metallic mass. In a 25-in. set the ghost is about 1 cm to the right of the principal image if the reflected signal arrives 1 μs after the principal signal. In this case, what is the difference in path length for the two signals?

35–2 The maximum electric field in the vicinity of a certain radio transmitter is 1.0×10^{-3} V·m^{-1}. What is the maximum magnitude of the B field? How does this compare in magnitude with the earth's field?

35–3 A certain radio station broadcasts at a frequency of 1020 kHz. At a point some distance from the transmitter, the maximum magnetic field of the electromagnetic wave it emits is found to be 1.6×10^{-11} T.

a) What is the wavelength of the wave?

b) What is the maximum electric field?

Section 35–3 Energy in Electromagnetic Waves

Section 35–5 Sinusoidal Waves

35–4 Consider each of the electric- and magnetic-field orientations given below. In each case what is the direction of propagation of the wave?

a) $E = Ei$, $B = Bj$;

b) $E = -Ej$, $B = Bi$;

c) $E = Ek$, $B = -Bi$;

d) $E = Ej$, $B = -Bk$.

35–5 Verify that all the expressions in Eq. (35–19) are equivalent.

35–6 A certain plane electromagnetic wave emitted by a microwave antenna has a wavelength of 3.0 cm and a maximum magnitude of electric field of 2.0×10^{-2} V·m^{-1}.

a) What is the frequency of the wave?

b) What is the maximum magnetic field?

c) What is the intensity (average power per unit area) of the wave, if the wave is sinusoidal?

35–7 At a distance of 50 km from a radio station antenna, the electric-field amplitude is measured to be $E_{max} = 2 \times 10^{-2}$ V·m^{-1}.

a) What is the magnetic-field amplitude B_{max} at this same point?

b) Assuming that the antenna radiates equally in all directions (which is probably not the case), what is the total power output of the station?

c) At what distance from the antenna would $E_{max} = 1 \times 10^{-2}$ V·m^{-1}, half the above value?

35–8 Assume that 10% of the power input to a 100-W lamp is radiated uniformly as light of wavelength 500 nm (1 nm = 10^{-9} m). At a distance of 2 m from the source, the electric and magnetic fields vary sinusoidally according to the equations $E = E_{max} \sin(\omega t + \phi)$ and $B = B_{max} \sin(\omega t + \phi)$. Calculate E_{max} and B_{max} for the 500-nm light.

35–9 If the intensity of direct sunlight is 1.4 kW·m^{-2}, find

a) the momentum density (momentum per unit volume);

b) the momentum flow rate (momentum carried through a surface area A in unit time) in the sunlight. (*Note.* This equals the radiation pressure.)

35–10 The intensity of a bright light source is 900 W·m^{-2}. Find the radiation pressure (in pascals) on

a) a totally absorbing surface;

b) a totally reflecting surface.

Also express your results in atmospheres.

Section 35–4 Electromagnetic Waves in Matter

35–11 An electromagnetic wave propagates in a ferrite material having $K = 10$ and $K_m = 1000$. Find

a) the speed of propagation;

b) the wavelength of a wave having a frequency of 100 MHz.

Section 35–6 Standing Waves

35–12 For the standing wave given by Eqs. (35–21):

a) Plot the energy density as a function of x, $0 < x < \pi/k$, for the times $t = 0$, $\pi/4\omega$, $\pi/2\omega$, $3\pi/4\omega$, and π/ω.

b) Find the direction of S in the regions $0 < x < \pi/2k$ and $\pi/2k < x < \pi/k$ at the times $t = \pi/4\omega$ and $3\pi/4\omega$.

c) Use your results in (b) to explain the plots obtained in (a).

Section 35–7 The Electromagnetic Spectrum

35–13 What is the wavelength in meters, microns, nanometers, and angstrom units of

a) soft x-rays of frequency 2×10^{17} Hz?

b) green light of frequency 5.6×10^{14} Hz?

PROBLEMS

35–14 The energy flow to the earth associated with sunlight is about 1.4 kW·m^{-2}.

a) Find the maximum values of E and B for a wave of this intensity.

b) The distance from the earth to the sun is about 1.5×10^{11} m. Find the total power radiated by the sun.

35–15 For a sinusoidal electromagnetic wave in vacuum, such as that described by Eqs. (35–18), show that the average density of energy in the electric field is the same as that in the magnetic field.

35–16 A plane sinusoidal electromagnetic wave has a wavelength of 3.0 cm and an E-field amplitude of 30 V·m^{-1}.

a) What is the frequency?

b) What is the B-field amplitude?

c) What is the intensity?

d) What average force does this radiation exert on a totally absorbing surface of area 0.5 m^2 perpendicular to the direction of propagation?

35–17 A very long solenoid of n turns per unit length and radius a carries a current i that is increasing at a constant rate di/dt.

a) Calculate the induced electric field at a point inside the solenoid at a distance r from the solenoid axis.

b) Compute the magnitude and direction of the Poynting vector at this point.

35–18 A capacitor consists of two circular plates of radius r separated by a distance l. Neglecting fringing, show that while the capacitor is being charged, the rate at which energy flows into the space between the plates is equal to the rate at which the electrostatic energy stored in the capacitor increases. (*Hint:* Integrate the Poynting vector over the surface of the space between the plates.)

35–19 A circular loop of wire can be used as a radio frequency antenna. If a 0.5-m diameter antenna is located 100 m from a 10-MHz source with a total power of 1 MW, what is the maximum emf induced in the loop? (Assume that the plane of the antenna loop is perpendicular to the direction of the magnetic field of the radiation, and that the source radiates uniformly in all directions.)

35–20 It has been proposed that to aid in meeting the energy needs of the United States, solar-power-generating satellites could be placed in earth orbit, and the power they generate could be beamed down to earth as microwave radiation. For a microwave beam with a cross-sectional area of 50 m^2 and a total power of 1 kW at the earth's surface, what is the magnitude of the electric field of the beam on the earth's surface?

35–21 The nineteenth-century inventor Nikolai Tesla proposed to transmit electric power via electromagnetic waves. Suppose power is to be transmitted in a beam of cross-sectional area 100 m^2; what **E** and **B** strengths are required to transmit an amount of power comparable to that handled by modern transmisson lines (of the order of 500 kV and 1000 A)?

35–22 A space-walking astronaut has run out of fuel for her jet-pack and is floating 20 m from the space shuttle with zero relative velocity. The astronaut and all her equipment have a total mass of 200 kg. If she uses her 100-W flashlight as a light rocket, how long will it take her to reach the shuttle?

CHALLENGE PROBLEMS

35–23 The concept of solar sailing has appeared in science fiction and proposals to NASA. A solar sailcraft uses a large low-mass sail and the energy and momentum of sunlight for propulsion. The total power output of the sun is 4×10^{26} W.

a) Should the sail be absorbing or reflective? Why?

b) How large a sail is necessary to propel a 10^5 kg spacecraft against the gravitational force of the sun? (Express your result in square miles.)

c) Explain why your answer to (b) is independent of the distance from the sun.

35–24 Electromagnetic radiation is emitted by accelerating charges. The rate of energy emission from an accelerating charge that has charge q and acceleration a is given by

$$\frac{dE}{dt} = \frac{q^2 a^2}{6\pi\epsilon_0 c^3},$$

where c is the speed of light.

a) Verify that this equation is dimensionally correct.

b) If a proton with a kinetic energy of 5 MeV is traveling in a particle accelerator in a circular orbit of radius 1 m, what fraction of its energy does it radiate per second?

c) Consider an electron orbiting with the same velocity and radius. What fraction of its energy does it radiate per second?

35–25 The electron in a hydrogen atom can be considered to be in a circular orbit with a radius of 0.053 nm and an energy of 13.6 eV. If the electron behaved classically, how much energy would it radiate per second? (See Challenge Problem 35–24). What does this tell you about the use of classical physics in describing the atom?

PART SEVEN

OPTICS

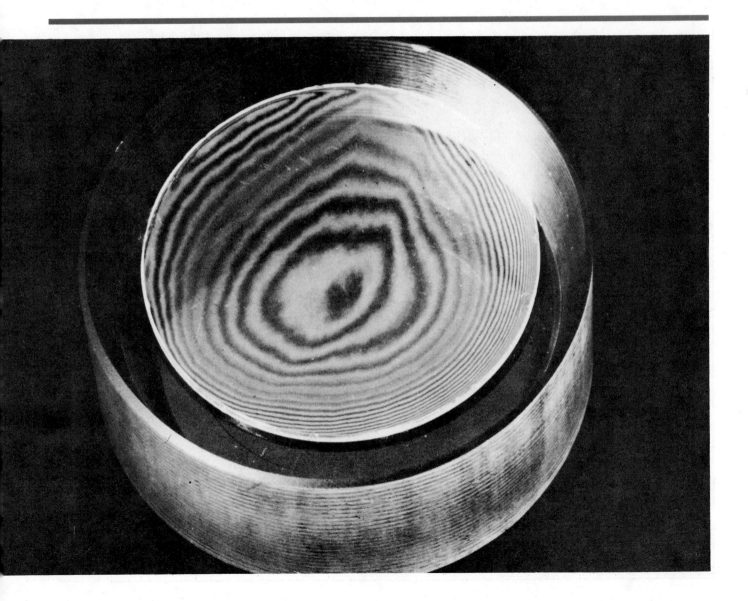

PERSPECTIVE

We have now completed our study of the principles of electromagnetism. We learned that the concept of magnetic field enables us to describe the interactions of moving charges and currents, including the action of these as *sources* of magnetic field and the *forces* that the field exerts on moving charges and currents. We also studied the interplay between electric and magnetic fields; a time-varying field of one kind acts as a source of the other kind of field in neighboring regions of space. This two-way interaction, which can take place in empty space even when there is no electric charge present, forms the basis of electromagnetic waves.

Our study of electromagnetism has had both fundamental and practical significance. On the practical side, we have learned how magnetic forces are used to deflect electron beams in TV picture tubes, how electric motors and generators work, and the operation of ac circuits, so vital in communications, electric-power distribution, and every area of contemporary electronics. From a fundamental point of view, we have seen how Maxwell's equations wrap up all of electromagnetism in a neat, concise package comparable to Newton's laws and the laws of thermodynamics. As we have remarked before, this is what science is all about—learning to generalize from experimental evidence and to express broad and general physical laws concisely and compactly. Maxwell's synthesis of electromagnetism stands as a towering intellectual achievement, comparable to the Newtonian synthesis in the preceding century, and to the development of relativity, quantum mechanics, and the understanding of DNA in our own time. All beautiful, and all monuments to the achievements of the human intellect!

Now we proceed to the subject of *optics*, originally the study of light but now broadened to include other electromagnetic radiation as well. Light is known to be electromagnetic radiation; it forms a small part of the electromagnetic spectrum, with frequencies ranging from less than 1 Hz to at least 10^{25} Hz. The wave model of light and other electromagnetic radiation provides the basis for understanding a variety of optical phenomena, including polarization, interference, and diffraction.

We begin with the special case where the wave picture can be simplified further by representing light in terms of *rays*. We study the reflection of rays and their bending or *refraction* when they pass from one material to another. The ray representation forms the basis of *geometrical optics* and is the model used to analyze mirrors, lenses, and common optical instruments. Geometrical optics is more limited in its scope than the general wave picture, but it is much simpler. We use it to study the optical behavior of several practical devices, including cameras, projectors, optical systems, the human eye, and various kinds of microscopes and telescopes.

Next we proceed to several optical phenomena that require the more general wave description of light for their understanding. Interference and diffraction phenomena show departures of light from the straight-line propagation that would occur if the ray picture were exactly correct. The corresponding branch of optics is called *physical optics,* to distinguish it from the more restricted geometrical optics. Interference effects enable us to measure wavelengths of light, design nonreflective coatings for lenses, measure atomic spacings in crystal lattices, and understand the fundamental limitations on the resolution of optical instruments. Finally, we study the principles of holography, one of the most exciting and useful developments in modern optics and a striking application of the wave nature of light.

36

THE NATURE AND PROPAGATION OF LIGHT

WE BEGIN THIS CHAPTER WITH A GENERAL DISCUSSION OF THE NATURE OF light and of light sources. Light is electromagnetic radiation and is thus a *wave* phenomenon. Many aspects of the propagation of light can be described by a *ray* model; rays travel in straight lines in homogeneous materials but can be reflected and also bent, or *refracted*, at interfaces between materials. In this chapter we study several basic phenomena associated with reflection and refraction of light. Like all transverse waves, light waves are *polarized;* we examine several aspects of polarization phenomena.

36–1 NATURE OF LIGHT

Until the time of Newton (1642–1727), most scientists (including Newton) thought that light consisted of streams of some sort of particles (which they called *corpuscles*) emanating from light sources. But at about this time, Huygens and others suggested that light might be a *wave* phenomenon. Indeed, diffraction effects that are now recognized as wave phenomena were observed by Grimaldi as early as 1665, but their significance was not understood then. By the early nineteenth century, evidence that light is a wave phenomenon grew more persuasive. The experiments of Fresnel, Thomas Young, and others revealed many phenomena that can be understood on the basis of a wave picture but not on the corpuscular model. We will study these phenomena in detail in Chapter 39.

What *is* light, anyway?

The next great step was taken in 1873 by Maxwell, who predicted the existence of electromagnetic waves and calculated their speed of propagation, as we learned in Chapter 35. In 1887 Hertz succeeded in producing short-wavelength electromagnetic waves and showing that they had all the properties of light waves. They could be reflected, refracted, focused by a lens, polarized, and so on. Thus the evidence grew more and more conclusive that light is indeed an electromagnetic-wave phenomenon.

Maxwell established that light is electromagnetic radiation with very short wavelengths.

Successful as the wave picture of light is, it is not the whole story. Several phenomena associated with emission and absorption of light reveal a particle

Particle aspects of light: twentieth-century concepts

aspect, in the sense that the energy carried by light waves is packaged in discrete bundles called *photons* or *quanta*. We will explore some of these phenomena in Chapter 41. These apparently contradictory wave and particle properties have been reconciled only since 1930 with the development of quantum electrodynamics, a comprehensive theory that includes *both* wave and particle properties. *Propagation* of light is best described by a wave model, but emission and absorption phenomena require a particle approach.

36–2 SOURCES OF LIGHT

Emission of electromagnetic radiation by hot bodies

The fundamental sources of all electromagnetic radiation are electric charges in motion. All bodies emit electromagnetic radiation as a result of thermal motion of their molecules; this radiation, called *thermal radiation*, is a mixture of different wavelengths. At a temperature of 300°C the most intense of these waves has a wavelength of 5000×10^{-9} m or 5000 nm, which is in the *infrared* region. At a temperature of 800°C a body emits enough visible radiant energy to be self-luminous and appears "red hot," although most of the energy emitted is still carried by infrared waves. At 3000°C, about the temperature of an incandescent lamp filament, the radiation contains enough of the visible wavelengths, between 400 nm and 700 nm, that the body appears "white hot." In modern incandescent lamps, the filament is a coil of fine tungsten wire. An inert gas such as argon is introduced to reduce evaporation of the filament. Incandescent lamps vary in size from one no larger than a grain of wheat to one with a power input of 5000 W, used for illuminating airfields.

Carbon-arc lamps are among the brightest light sources.

One of the brightest sources of light is the *carbon arc*. Two carbon rods, typically 10 cm to 20 cm long and 1 cm in diameter, are connected to a 120-V or 240-V dc source. They are touched together momentarily and then pulled apart a few millimeters. An arc forms, and the resulting intense electron bombardment of the positive rod causes an extremely hot crater to form at its end. This crater, whose temperature is typically 4000°C, is the source of light. Carbon-arc lights are used in most theater motion-picture projectors and in large searchlights and lighthouses.

Light produced by electrical discharge in a gas or vapor

Some light sources use an arc discharge through a conducting metal vapor, such as mercury or sodium. The vapor is contained in a sealed bulb with two electrodes, which are connected to a power source. Argon is sometimes added to permit a glow discharge that helps vaporize and ionize the metal. The bluish light of mercury-arc lamps and the bright orange-yellow of sodium-vapor lamps are familiar in highway and other outdoor lighting.

How do fluorescent lamps work?

An important variation of the mercury-arc lamp is the *fluorescent* lamp, consisting of a glass tube containing argon and mercury vapor, with tungsten electrodes. When an electric discharge takes place in the mercury-argon mixture, the emitted radiation is mostly in the ultraviolet region. The ultraviolet radiation is absorbed in a thin layer of material, called a *phosphor*, which is the white coating on the interior walls of the glass tube. The phosphor has the property of *fluorescence*, which means that it emits visible light when illuminated by ultraviolet radiation. Various phosphors can be used to obtain various colors of light. Fluorescent lamps have much higher efficiency of conversion of electrical energy to visible light than do incandescent lamps.

What is so special about laser light? A partial answer.

A special light source that has attained prominence in the last twenty years is the *laser*. It can produce a very narrow beam of enormously intense radiation. High-intensity lasers have been used to cut through steel, fuse high-melting-point materials, and bring about many other effects that are important in physics, chemistry, biology, and engineering. An equally significant characteristic of laser light is that it is much more nearly *monochromatic,* or single-frequency, than any other light source. We will study the operation of one type of laser in Chapter 41.

36–3 THE SPEED OF LIGHT

The speed of light in empty space is a fundamental constant of nature. The first successful measurement of the speed of light was made in 1676 by the Danish astronomer Olaf Roemer, who measured the period of revolution of one of Jupiter's satellites. The period (about 42 hr) appeared longer when Earth was moving in its orbit *away from* Jupiter than when it was moving *toward* Jupiter. Roemer correctly concluded that the difference was due to the displacement of the earth during one revolution of the satellite. When Earth is moving away from Jupiter, light leaving the satellite at the end of a particular revolution has to travel farther than light leaving at the beginning of the same revolution. In that case the apparent time for one revolution, measured on Earth, is a few seconds longer than the actual time. Six months later, when Earth is moving toward Jupiter, the situation is reversed, and the apparent period is a few seconds shorter than the actual value.

Measuring the speed of light three centuries ago

The first successful determination of the speed of light from purely *terrestrial* measurements was made by the French scientist Fizeau in 1849. A schematic diagram of his apparatus is shown in Fig. 36–1. The light source S directs light toward a toothed wheel T that can be rotated at high speed. G is an inclined plate of clear glass. When the wheel is stationary, light passes through one of the openings between the teeth. The light is reflected from M, retraces its path, and is in part reflected from the glass plate G into the eye of an observer at E. The role of the lenses is to concentrate the light on the toothed wheel and on mirror M.

Measuring the speed of light on Earth: a series of increasingly precise refinements

When the wheel T is rotating, the light from S is "chopped up" into a succession of wave pulses of limited length. A pulse travels through an opening between two teeth of the wheel, is reflected by the mirror M, and returns to the wheel. At a certain speed of rotation, the wheel turns just enough during this time so the return path is blocked by a tooth of the wheel, and *no* reflected light is seen by the observer at E. From a knowledge of (a) the angular velocity and radius of the wheel, (b) the distance between openings, and

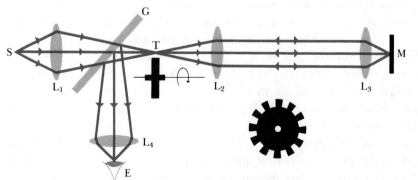

36–1 Fizeau's toothed-wheel method for measuring the speed of light. S is a light source; L_1, L_2, L_3, and L_4 are lenses. T is the toothed wheel, M is a mirror, and G is a glass plate.

(c) the distance from wheel to mirror, the speed of light may be computed. Fizeau obtained the value 3.15×10^8 m·s^{-1}.

Fizeau's apparatus was modified by Foucault, who replaced the toothed wheel with a rotating mirror. The most precise measurements by the Foucault method were made by the American physicist Albert A. Michelson (1852–1931). His first experiments were performed in 1878; the last, underway at the time of his death, were completed in 1935 by Pease and Pearson.

From analysis of all measurements up to 1983, the most probable value for the speed of light is

$$c = 2.99792458 \times 10^8 \text{ m·s}^{-1}.$$

This number is based on the definition of the meter in terms of the krypton wavelength and the definition of the second in terms of the cesium clock, as described in Section 1–2. The definition of the second is precise to within one part in 10 trillion (10^{13}), whereas the definition of the meter is much less precise, about four parts in a billion (10^9). Thus it has become advantageous to redefine the meter in terms of the unit of time by *defining* the speed of light to have a specific value and then defining the meter in terms of the distance traveled by light in one second.

Accordingly, in November 1983 the General Conference of Weights and Measures *defined* the speed of light in vacuum to be precisely 299,792,458 m·s^{-1} and defined one meter to be the distance traveled by light in a time of 1/299,792,458 s, with the second defined by the cesium clock.

As we found in Chapter 35, the speed of any electromagnetic wave in empty space is given by

$$c = \frac{1}{\sqrt{\epsilon_0 \mu_0}}.$$

In SI units, μ_0 is assigned the value of precisely $4\pi \times 10^{-7}$ N·s^2·C^{-2}. Thus the preceding equation provides a very precise means of evaluating the electrical constant ϵ_0:

$$\epsilon_0 = \frac{1}{\mu_0 c^2} = 8.85418782 \times 10^{-12} \text{ C}^2\text{·N}^{-1}\text{·m}^{-2}.$$

This value has much greater precision than could be obtained from direct measurements of forces between electric charges.

36–4 WAVES, WAVE FRONTS, AND RAYS

The concept of **wave front** provides a convenient language for describing the propagation of any kind of wave. We define a wave front as *the locus of all points at which the phase of vibration of a physical quantity associated with the wave is the same*. A familiar example is the crest of a water wave; when we drop a pebble in a calm pool, the expanding circles formed by the wave crests are wave fronts. When sound waves spread out in all directions from a pointlike source, any spherical surface concentric with the source is a wave front. The surfaces over which the pressure is maximum and those over which it is minimum form sets of expanding spheres as the wave travels outward from the source. The *phase* of the pressure variation is the same at all points on one of the spherical surfaces. In diagrams of wave motion we usually draw only a few wave fronts,

Does it make sense to *define* the speed of light to have a certain value?

The speed of light is closely related to the constant in Coulomb's law.

Describing waves in terms of wave fronts and rays: some basic language

often those that correspond to the maxima and minima of the disturbance, such as the crests and troughs of a water wave. For a sinusoidal wave, wave fronts corresponding to maximum displacements in opposite directions are separated from each other by one-half wavelength. Two consecutive wave fronts corresponding to maximum displacement in the same direction are separated by one wavelength.

For a light wave (or any other electromagnetic wave), the quantity that corresponds to the pressure in a sound wave is the electric or magnetic field. Often it is not necessary to indicate in a diagram either the magnitude or the direction of the field; instead we simply show the *shapes* of the wave fronts or their intersections with some reference plane. For example, the electromagnetic waves radiated by a small light source may be represented by *spherical* surfaces concentric with the source or, as in Fig. 36–2a, by the intersections of these surfaces with the plane of the diagram. At a sufficiently great distance from the source, where the radii of the spheres have become very large, the spherical surfaces can be considered planes and we have a *plane* wave, as in Fig. 36–2b.

For some phenomena, it is convenient to represent a light wave by **rays** rather than by wave fronts. The corresponding branch of optics is called *geometrical optics*. Indeed, rays were used to describe light long before its wave nature was firmly established and, in a particle theory of light, rays are merely the paths of the particles. From the wave viewpoint, *a ray is an imaginary line drawn in the direction in which the wave is traveling.* Thus in Fig. 36–2a the rays are the radii of the spherical wave fronts, and in Fig. 36–2b they are straight lines perpendicular to the wave fronts. In fact, in every case where waves travel in a homogeneous isotropic material, the rays are straight lines normal to the wave fronts. At a boundary surface between two materials, such as the surface between a glass plate and the air outside it, the direction of a ray changes, but the portions in the air and in the glass are straight lines.

Although the ray picture provides an adequate description of many reflection and refraction phenomena found in mirrors and lenses, several other optical phenomena, such as polarization and diffraction, require a more detailed wave theory for their understanding.

36–5 REFLECTION AND REFRACTION

In many familiar optical phenomena, a wave strikes an interface between two optical materials, such as air and glass or water and glass. When the interface is smooth (i.e., when its irregularities are small compared with the wavelength), the wave is in general partly reflected and partly transmitted into the second material, as shown in Fig. 36–3a. For example, when you look into a store window from the street, you see a reflection of the street scene, but a person in the store can look *through* the window at the same scene.

The segments of plane waves shown in Fig. 36–3a can be represented by bundles of rays forming *beams* of light, as in Fig. 36–3b. For simplicity in discussing the various angles, we often consider only one ray in each beam, as in Fig. 36–3c. Representing these waves in terms of rays is the basis of the branch of optics called **geometrical optics.** This chapter and the next two are concerned primarily with geometrical optics and with optical phenomena that can be understood on the basis of rays, without explicit use of the wave nature of light.

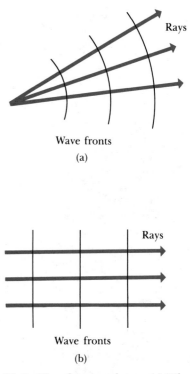

36–2 Wave fronts and rays. (a) When the wave fronts are spherical, the rays radiate out from the center of the spheres. (b) When the wave fronts are planes, the rays are parallel.

Some optical phenomena can't be analyzed by using rays.

Geometrical optics: a very simple model based on reflection and refraction of rays

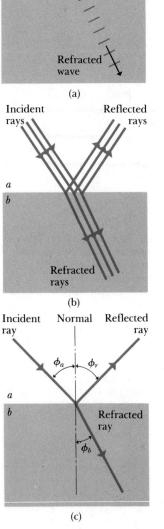

36–3 (a) A plane wave is in part reflected and in part refracted at the boundary between two media. (b) The waves in (a) are represented by rays. (c) For simplicity, only one example of incident, reflected, and refracted rays is drawn.

Index of refraction: describing the refractive properties of an optical material

At an interface between two optical materials, the directions of the incident, reflected, and refracted (transmitted) beams of light are described in terms of the angles they make with the *normal* to the surface at the point of incidence. For this purpose we need indicate only one ray, as in Fig. 36–3c, although a single ray of light is a geometrical abstraction. Experimental studies of the directions of the incident, reflected, and refracted beams lead to the following results

1. *The incident, reflected, and refracted rays, and the normal to the surface, all lie in the same plane.* Hence, if the incident ray is in the plane of the diagram, and the boundary surface between the two materials is perpendicular to this plane, the reflected and refracted rays are in the plane of the diagram.

2. *The angle of reflection ϕ_r is equal to the angle of incidence ϕ_a for all wavelengths and any pair of substances.* Thus

$$\phi_r = \phi_a. \qquad (36\text{–}1)$$

The experimental result that $\phi_r = \phi_a$, and that the incident and reflected rays and the normal all lie in the same plane, is known as the **law of reflection.**

3. For monochromatic light and for a given pair of substances, a and b, on opposite sides of the surface of separation, *the ratio of the sine of the angle ϕ_a* (between the ray in substance a and the normal) *and the sine of the angle ϕ_b* (between the ray in substance b and the normal) *is constant.* Thus

$$\frac{\sin \phi_a}{\sin \phi_b} = \text{constant}. \qquad (36\text{–}2)$$

This experimental result, together with the fact that the incident and refracted rays and the normal to the surface all lie in the same plane, is known as the **law of refraction** or **Snell's law,** after Willebrord Snell (1591–1626). There is some doubt that Snell actually discovered it.

The laws of reflection and refraction relate the *directions* of the rays; they say nothing about the *intensities* of the reflected and refracted rays. These depend on the angle of incidence; for the present we simply state that the fraction reflected is smallest at *normal* incidence, where it is a few percent, and that it increases with increasing angle of incidence to 100% at grazing incidence, when $\phi_a = 90°$.

When a ray of light approaches the interface from *below* in Fig. 36–3, there are again reflected and refracted rays; these, together with the incident ray and the normal, all lie in the same plane. The laws of reflection and refraction apply, whether the incident ray is in material a or b. The passage of a refracted ray, transmitted from one material to another, is reversible. It follows the same path in going from b to a as when going from a to b.

Let us now consider a beam of monochromatic light traveling *in vacuum*, making an angle of incidence ϕ_0 with the normal to the surface of a substance a, and let ϕ_a be the angle of refraction in the substance. The constant in Snell's law is then called the **index of refraction** of substance a and is denoted by n_a:

$$\frac{\sin \phi_0}{\sin \phi_a} = n_a. \qquad (36\text{–}3)$$

This definition of index of refraction may seem unrelated to the definition we encountered in Section 35–4. In fact, though, they are equivalent. We will return to this point in Section 36–12.

The index of refraction (also called *refractive index*) depends not only on the substance but also on the wavelength of the light. If no wavelength is stated, the index is often assumed to be that corresponding to the yellow light from a sodium lamp, of wavelength 589 nm. This wavelength is near the middle of the visible spectrum, and sodium lamps are simple, inexpensive, nearly monochromatic, and easy to use.

The index of refraction of most common glasses used in optical instruments lies between 1.46 and 1.96. A few substances have indexes larger than this value; diamond is one, with an index of 2.42, and rutile (crystalline titanium dioxide) is another, with an index of 2.62. Indexes of refraction for several solids and liquids are given in Table 36–1. The index of refraction of *air* at standard conditions is about 1.0003; for most purposes the *index of refraction of air can be assumed to be unity.* The index of refraction of a gas increases uniformly as the density of the gas increases.

For reasons to be discussed in Section 36–12, all materials have a refractive index greater than unity; for example, note the values in Table 36–1. Thus for a ray passing from vacuum into a material, the angle of refraction ϕ_a is always *less than* the angle of incidence ϕ_0. That is, the ray is bent *toward* the normal. When light travels in the opposite direction, from a material into vacuum, the reverse is true and the ray is bent *away from* the normal.

We can express the constant in Eq. (36–2) in terms of the indexes of refraction of the two materials. To develop the relation, we consider two parallel-sided plates of substances a and b placed parallel to each other with space between them, as in Fig. 36–4a. We assume that the surrounding medium is vacuum, although the behavior would differ only very slightly if it were air. A ray of monochromatic light starts at the lower left with an angle of incidence ϕ_0. The angle between the ray and the normal in substance a is ϕ_a, and the light emerges from substance a at an angle ϕ_0 equal to its incident angle. The light ray therefore enters plate b with an angle of incidence ϕ_0, makes an angle ϕ_b in substance b, and emerges again at an angle ϕ_0. Exactly the same path would be traversed if the same light ray were to start at the upper right and enter substance b at an angle ϕ_0. Moreover, *the angles are independent of the thickness of the space between the two plates* and are the same when the space shrinks to nothing, as in Fig. 36–4b.

Applying Snell's law to the refractions at the surface between vacuum and substance a, and at the surface between vacuum and substance b, we have

$$\frac{\sin \phi_0}{\sin \phi_a} = n_a,$$

$$\frac{\sin \phi_0}{\sin \phi_b} = n_b.$$

Dividing the second equation by the first, we obtain

$$\frac{\sin \phi_a}{\sin \phi_b} = \frac{n_b}{n_a}, \tag{36–4}$$

which shows that *the constant in Snell's law for the refraction between substances a and b is the ratio of the indices of refraction.* From Eq. (36–4) we see that the simplest and most symmetrical way of writing Snell's law for any two substances a and b, and for any direction, is

$$n_a \sin \phi_a = n_b \sin \phi_b. \tag{36–5}$$

TABLE 36–1 Index of Refraction for Yellow Sodium Light ($\lambda = 589$ nm)

Substance	Index of Refraction
Solids	
Ice (H_2O)	1.309
Fluorite (CaF_2)	1.434
Polystyrene	1.49
Rock salt (NaCl)	1.544
Quartz (SiO_2)	1.544
Zircon ($ZrO_2 \cdot SiO_2$)	1.923
Diamond (C)	2.417
Fabulite ($SrTiO_3$)	2.409
Rutile (TiO_2)	2.62
Glasses (typical values)	
Crown	1.52
Light flint	1.58
Medium flint	1.62
Dense flint	1.66
Lanthanum flint	1.80
Liquids at 20°C	
Methyl alcohol (CH_3OH)	1.329
Water (H_2O)	1.333
Ethyl alcohol (C_2H_5OH)	1.36
Carbon tetrachloride (CCl_4)	1.460
Turpentine	1.472
Glycerine	1.473
Benzene	1.501
Carbon disulfide (CS_2)	1.628

Photograph of a soda straw in a glass of water, showing refraction at the top and side surfaces of the water. (Nancy Rodger, Exploratorium.)

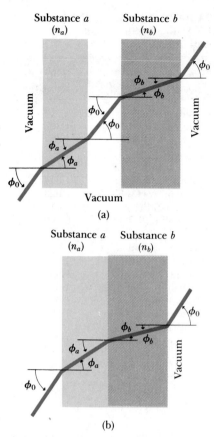

36-4 The transmission of light through parallel plates of different substances. The incident and emerging rays are parallel, regardless of direction and regardless of the thickness of the space between adjacent slabs.

The frequency of light doesn't change when it passes from one material to another.

The wavelength is smaller in a material than for the same light in vacuum.

EXAMPLE 36–1 In Fig. 36–3, material a is water and material b is a glass with index of refraction 1.50. If the incident ray makes an angle of 60° with the normal, find the directions of the reflected and refracted rays.

SOLUTION According to Eq. (36–1), the angle the reflected ray makes with the normal is the same as that of the incident ray. Hence $\phi_r = \phi_a = 60°$. To find the direction of the refracted ray we use Eq. (36–5), with $n_a = 1.33$, $n_b = 1.50$, and $\phi_a = 60°$. We find

$$n_a \sin \phi_a = n_b \sin \phi_b,$$

$$(1.33)(\sin 60°) = (1.50)(\sin \phi_b),$$

$$\phi_b = 50.2°.$$

The second material has a larger refractive index than the first, and the refracted ray is bent toward the normal. As Eq. (36–5) shows, this is always the case when the second index is larger than the first. In the opposite case, where the second index is *smaller* than the first, the ray is always bent *away from* the normal.

The index of refraction of a material is closely related to the speed of light in the material. We will develop this relationship in detail in Section 36–12, but we state the principal results here for reference. Light always travels *more slowly* in a material than in vacuum, and the ratio of the two speeds is equal to the index of refraction. Thus the speed of light v in a material having index of refraction n is given by

$$v = \frac{c}{n}, \quad \text{or} \quad n = \frac{c}{v}. \tag{36–6}$$

When light passes from one material to another, its frequency f does not change, for the following reason. When light interacts with matter, the electrons in the material absorb energy from the light and undergo vibration motion with the same frequency as the light. This motion causes reradiation of the energy *with the same frequency*. Thus, since $v = \lambda f$, when v is less than the wave speed c in vacuum, λ is also correspondingly reduced. Thus the wavelength λ of light in a material is less than the wavelength λ_0 of the same light in vacuum, by a factor n:

$$\lambda = \frac{\lambda_0}{n}. \tag{36–7}$$

Two final comments about reflection and refraction need to be made. First, reflection occurs at a highly polished surface of an *opaque* material such as a metal. There is *no* refracted ray, but the reflected ray still behaves according to Eq. (36–1). Second, if the reflecting surface of either a transparent or an opaque material is rough, with irregularities on a scale comparable to or larger than the wavelength of light, reflection occurs not in a single direction but in all directions; such reflection is called *diffuse* reflection. Conversely, reflections in a single direction from smooth surfaces are called *regular* reflections or *specular* reflections.

PROBLEM-SOLVING STRATEGY: *Reflection and refraction*

1. In geometrical optics problems involving rays and angles, *always* start by drawing a big, neat diagram. Use a ruler and protractor. Label all known angles and indexes of refraction.

2. Don't forget that by convention we always measure the angles of incidence, reflection, and refraction from the *normal* to the surface where the reflection and refraction occur, never from the surface itself.

3. You will often have to use some simple geometry or trigonometry in working out angular relations. The sum of the interior angles in a triangle is 180°, and so on. Often it helps to think through the problem, asking yourself, "What do I need to know in order to find this angle?" or "What other angles or other quantities can I compute using the information given in the problem?"

36–6 TOTAL INTERNAL REFLECTION

Figure 36–5a shows a number of rays diverging from a point source P in material a having index of refraction n_a. The rays strike the surface of a second material b having index n_b, where $n_a > n_b$. From Snell's law,

$$\sin \phi_b = \frac{n_a}{n_b} \sin \phi_a.$$

Because n_a/n_b is greater than unity, $\sin \phi_b$ is larger than $\sin \phi_a$. Thus there must be some value of ϕ_a *less than* 90° for which $\sin \phi_b = 90°$. This is illustrated by ray 3 in the diagram, which emerges just grazing the surface at an angle of refraction of 90°. The angle of incidence for which the refracted ray emerges tangent to the surface is called the **critical angle** and is denoted by ϕ_{crit} in the diagram. If the angle of incidence is *greater than* the critical angle, the sine of the angle of refraction, as computed by Snell's law, would have to be greater

> Total internal reflection: light can be trapped inside a material (but not outside).

> Total internal reflection occurs when the angle of incidence exceeds the critical value.

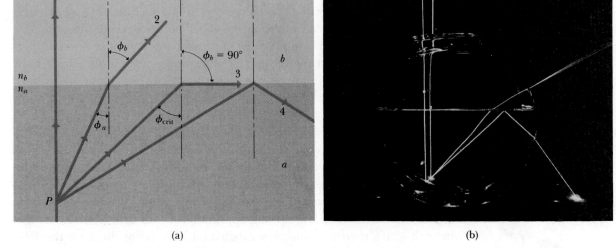

(a)   (b)

36–5 (a) Total internal reflection. The angle of incidence ϕ_a for which the angle of refraction is 90°, is called the critical angle. (b) Rays of laser light enter the water in the fishbowl from above; they are reflected at the bottom by mirrors tilted at slightly different angles, and one ray undergoes total internal reflection at the air–water interface. (The Exploratorium.)

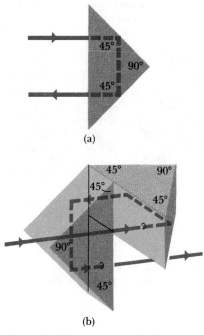

(a)

(b)

36–6 (a) A Porro prism. (b) A combination of two Porro prisms.

Total internal reflection is used in binocular prisms.

Fiber optics: a dramatic example of total internal reflection

Fiber optics for communication: the wave of the future

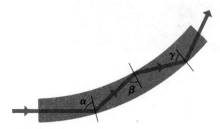

36–7 A light ray "trapped" by internal reflections.

than unity. This is impossible; beyond the critical angle, the ray *cannot pass* into the upper material but is completely reflected internally at the boundary surface. **Total internal reflection** occurs only when a ray is incident on the surface of a material whose index of refraction is *smaller than* that of the material in which the ray is traveling.

The critical angle for two given materials may be found by setting $\phi_b = 90°$ or $\sin \phi_b = 1$ in Snell's law. We then have

$$\sin \phi_{\text{crit}} = \frac{n_b}{n_a}. \qquad (36\text{–}8)$$

The critical angle of a glass–air surface, taking 1.50 as a typical index of refraction of glass, is

$$\sin \phi_{\text{crit}} = \frac{1}{1.50} = 0.67, \qquad \phi_{\text{crit}} = 42°.$$

The fact that this angle is slightly less than 45° makes it possible to use a prism with angles of 45°, 45°, and 90° as a totally reflecting surface. The advantages of totally reflecting prisms over metallic surfaces as reflectors are, first, that the light is *totally* reflected, while no metallic surface reflects 100% of the light incident on it, and second, the reflecting properties are permanent and not affected by tarnishing. Offsetting these is the fact that some loss of light occurs by reflection at the surfaces where light enters and leaves the prism, although coating the surfaces with so-called nonreflecting films can reduce this loss considerably.

A 45°–45°–90° prism, used as in Fig. 36–6a, is called a *Porro* prism. Light enters and leaves at right angles to the hypotenuse and is totally reflected at each of the shorter faces. The total deviation (change of direction of the rays) is 180°. Two Porro prisms are sometimes combined, as in Fig. 36–6b, an arrangement often found in binoculars.

If a beam of light enters at one end of a transparent rod, as in Fig. 36–7, the light is totally reflected internally and is "trapped" within the rod even if it is curved, provided the curvature is not too great. Such a rod is sometimes referred to as a *light pipe.* A bundle of fine glass fibers behaves in the same way and has the advantage of being flexible. A bundle may consist of thousands of individual fibers, each of the order of 0.002 mm to 0.01 mm in diameter. If the fibers are assembled in the bundle so that the relative positions of the ends are the same (or mirror images) at both ends, the bundle can transmit an image, as shown in Fig. 36–8.

Fiber-optic devices have found a wide range of medical applications in instruments called *endoscopes,* which can be inserted directly into the bronchial tubes, the bladder, the colon, and so on, for direct visual examination. A bundle of fibers can be enclosed in a hypodermic needle for study of tissues and blood vessels far beneath the skin.

Fiber optics are now also finding applications in communication systems, where they are used to transmit a modulated laser beam. Because the frequency of the modulated beam is very much higher than those used in wire or radio communication, an enormous amount of information can be transmitted through one fiber-optic cable. For example, the Carnegie-Mellon University computer system, consisting of several thousand microcomputers networked with mainframe computers, will be linked partly by fiber-optic ca-

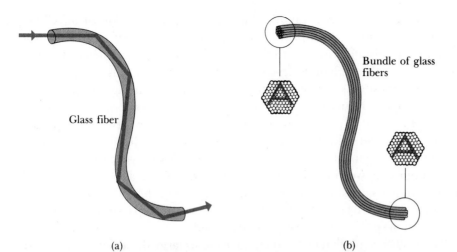

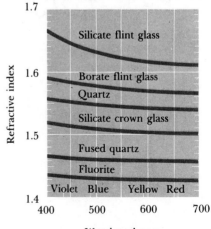

36–8 (a) Total internal reflection in a single fiber. (b) Image transmission by a bundle of fibers. (c) A fiber-optic cable used to transmit a modulated laser beam for communication purposes compared to a larger copper cable that has equal information-transmitting capacity. (Courtesy of Corning Glass Works.)

bles as shown in Fig. 36–8c. In the future, most telephone systems are likely to be connected by fiber optics.

36–7 DISPERSION

Most light beams are a superposition of waves with wavelengths extending throughout the visible spectrum. The speed of light *in vacuum* is the same for all wavelengths, but the speed in a material substance is different for different wavelengths. Hence the index of refraction of a material depends on wavelength. Any wave medium in which the speed of a wave varies with wavelength is said to show **dispersion.** Figure 36–9 shows the variation of index of refraction with wavelength for a few common optical materials. The value of n usually *decreases* with increasing wavelength and so *increases* with increasing frequency. Thus light of longer wavelength usually has greater speed in a material than does light of shorter wavelength.

36–9 Variation of index of refraction with wavelength.

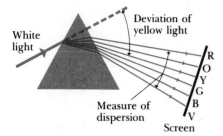

36–10 Dispersion by a prism. The band of colors on the screen is called a spectrum.

What makes a diamond sparkle? Do fake diamonds sparkle as much?

Polarization: a property of all transverse waves, electromagnetic and otherwise

Figure 36–10 shows a ray of white light (a superposition of all visible wavelengths) incident on a prism. The deviation (change of direction) produced by the prism increases with increasing index of refraction and decreasing wavelength. Violet light is deviated most and red least, with other colors in intermediate positions. When it comes out of the prism, the light is spread out into a fan-shaped beam, as shown. The light is said to be *dispersed* into a spectrum. The amount of dispersion depends on the *difference* between the refractive index for violet light and that for red light. From Fig. 36–9 we can see that for a substance such as fluorite, whose refractive index for yellow light is small, the difference between the indexes for red and violet is also small. But, for silicate flint glass, both the index for yellow light and the difference between extreme indexes are large. In other words, for most transparent materials, the greater the deviation, the greater the dispersion.

The brilliance of diamond is due in part to its large dispersion and in part to its unusually large refractive index. In recent years synthetic crystals of titanium dioxide and of strontium titanate, with about eight times the dispersion of diamond, have been produced.

36–8 POLARIZATION

Polarization phenomena occur with transverse waves. Our principal concern in this chapter is with electromagnetic waves, especially light, but to introduce basic concepts we first consider the mechanical example of transverse waves on a string, as discussed in Chapter 21. For a string whose equilibrium position is along the x-axis, the displacements may be along the y-direction, as in Fig. 36–11a. In this case the string always lies in the xy-plane. But the displacements might instead be along the z-axis, as in Fig. 36–11b, so that the string lies in the xz-plane.

36–11 (a) Transverse wave on a string, polarized in the y-direction. (b) Wave polarized in the z-direction. (c) Barrier with a frictionless vertical slot passes components polarized in the y-direction but blocks those polarized in the z-direction, acting as a polarizing filter.

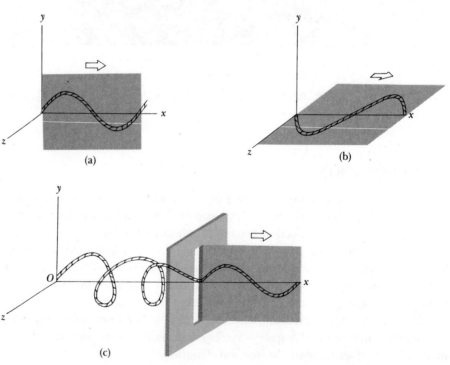

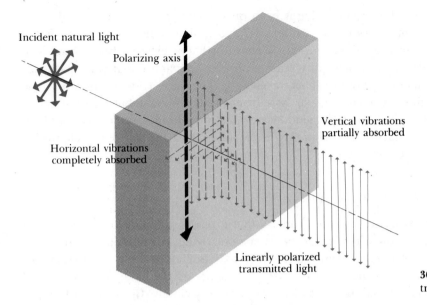

Incident natural light

Polarizing axis

Horizontal vibrations completely absorbed

Vertical vibrations partially absorbed

Linearly polarized transmitted light

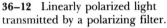

36–12 Linearly polarized light transmitted by a polarizing filter.

A wave having only y-displacements in the discussion above is said to be **linearly polarized** in the y-direction, and the one with only z-displacements is linearly polarized in the z-direction. It is easy in principle to construct a mechanical *filter* that permits only waves with a certain polarization direction to pass. An example is shown in Fig. 36–11c; the string can slide vertically in the slot without friction, but no horizontal motion is possible. This filter passes waves polarized in the y-direction but blocks those polarized in the z-direction.

This same language can be applied to light and other electromagnetic waves, which also exhibit polarization. As we learned in Chapter 35, an electromagnetic wave consists of fluctuating electric and magnetic fields, perpendicular to each other and to the direction of propagation. By convention, the direction of polarization is taken to be that of the *electric*-field vector, not the magnetic field, because most mechanisms for detecting electromagnetic waves employ principally the electric-field forces on electrons in materials. That is, the most common manifestations of electromagnetic radiation are due chiefly to the electric-field force, not the magnetic-field force.

Polarizing filters, or **polarizers,** can be made for electromagnetic waves; the details of construction depend on the wavelength. For microwaves having a wavelength of a few centimeters, a grid of closely spaced, parallel conducting wires insulated from each other will pass waves whose E fields are perpendicular to the wires but not those with E fields parallel to the wires. For light, the most common polarizing filter is a material known by the trade name Polaroid, widely used for sunglasses and polarizing filters for camera lenses. This material works on the principle of preferential absorption, passing waves polarized parallel to a certain axis in the material (called the **polarizing axis**) with 80% or more transmission, but offering only 1% or less transmission to waves with polarization perpendicular to this axis. The action of such a polarizing filter is shown schematically in Fig. 36–12.

Waves emitted by a radio transmitter are usually linearly polarized; a vertical rod antenna of the type widely used for CB radios emits waves that in a horizontal plane around the antenna are polarized in the vertical direction

Linear polarization: when the disturbance is always along one direction

A polarizing filter transmits only waves polarized along a certain direction.

Light emerging from a polarizing filter is always linearly polarized.

Ordinary light is a random mixture of all states of polarization.

(parallel to the antenna). Light from ordinary sources is *not* polarized, for a slightly subtle reason. The "antennas" that radiate light waves are the molecules of which the light sources are composed. The electrically charged particles in the molecules acquire energy in some way and radiate this energy as electromagnetic waves of short wavelength. The waves from any one molecule may be linearly polarized, like those from a radio antenna; but since any actual light source contains a tremendous number of molecules, oriented at random, the light emitted is a random mixture of waves linearly polarized in all possible transverse directions.

Behavior of an ideal polarizing filter (or polarizer).

36–9 POLARIZING FILTERS

An ideal polarizer has the property that it passes 100% of the incident light polarized in the direction of the filter's polarizing axis but blocks completely all light polarized perpendicular to this axis. Such a device is an unattainable idealization, but the concept is useful in clarifying the basic ideas. Some real polarizers approximate this ideal behavior very closely. In Fig. 36–13, unpolarized light is incident on a polarizer in the form of a flat plate. The polarizing axis is represented by the broken line. As the polarizer is rotated about an axis parallel to the incident ray, the intensity of the transmitted light (measured by the photocell) does not change. (Recall from Sections 35–3 and 35–5 that intensity, the average magnitude of the Poynting vector, is the power per unit area transmitted by the light.) The polarizer transmits the components of the

36–13 The intensity of the transmitted linearly polarized light, measured by the photocell, is the same for all orientations of the polarizing filter.

36–14 The analyzer transmits only the component parallel to its transmission direction or polarizing axis.

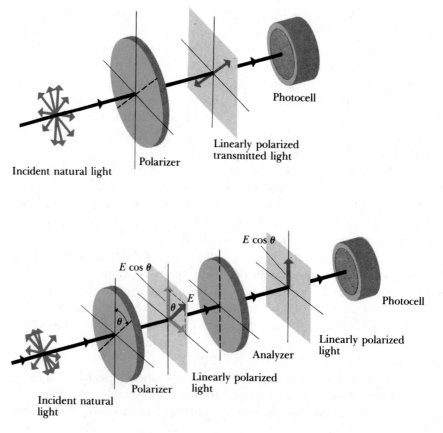

incident waves in which the **E** vector is parallel to the transmission direction of the polarizer, and by symmetry the components are equal for all angles perpendicular to the beam direction.

The intensity of the light transmitted through the polarizer can be measured with the photocell in Fig. 36–13; it is found to be exactly one-half that of the incident light. We can understand why this should be, as follows: The incident light can always be resolved into components polarized parallel to the polarizer axis and components polarized perpendicular to it. Because the incident light is a random mixture of all states of polarization, these two components are, on the average, equal. Thus (in the ideal polarizer) exactly half of the incident intensity, the part corresponding to the component parallel to the polarizer axis, is transmitted.

When unpolarized light strikes a polarizing filter, the transmitted light has half the incident intensity.

Suppose now that we insert a second polarizer between the first polarizer and the photocell, as in Fig. 36–14. Let the transmission direction of the second polarizer, or *analyzer*, be vertical, and let that of the first polarizer make an angle θ with the vertical. The linearly polarized light transmitted by the polarizer may be resolved into two components as shown, one parallel and the other perpendicular to the transmission direction of the analyzer. Only the parallel component, of amplitude $E \cos \theta$, is transmitted by the analyzer. The transmitted intensity is maximum when $\theta = 0$ and is zero when $\theta = 90°$, or when polarizer and analyzer are *crossed*.

To find the transmitted intensity at intermediate angles, we recall the discussion of energy in electromagnetic waves in Sections 35–3 and 35–5. In particular, Eq. (35–18) shows that the intensity is proportional to the *square* of the amplitude; thus we have

Using vector components to analyze transmission of polarized light by polarizing filters

$$I = I_{\max} \cos^2 \theta, \tag{36–9}$$

where $I_{\max}$ is the maximum intensity of light transmitted (at $\theta = 0$) and I is the amount transmitted at angle θ. This relation, discovered experimentally by Etienne Louis Malus in 1809, is called **Malus' law.**

The angle θ is the angle between the transmission directions of polarizer and analyzer. If either the analyzer or the polarizer is rotated, the amplitude of the transmitted beam varies with the angle between them according to Eq. (36–19).

PROBLEM-SOLVING STRATEGY: *Linear polarization*

1. Remember that in light or any electromagnetic wave the **E** field is perpendicular to the propagation direction and is the direction of polarization (or opposite to that direction). The polarization direction can be thought of as a two-headed arrow. When working with polarizing filters, you are really dealing with components of **E** parallel and perpendicular to the polarizing axis. Everything you know about components of vectors is applicable here.

2. The intensity (average power per unit area) of a wave is proportional to the *square* of its amplitude, as shown by Eq. (35–18). If you find that two waves differ in amplitude by a certain factor, their intensities differ by the square of that factor.

3. In Section 36–11 we will encounter problems involving superposition of two linearly polarized waves with perpendicular directions of polarization. Their relative phase is crucial; if they are in phase, the resultant is again linearly polarized, but if they are not, the resultant is circularly or elliptically polarized. In such cases, pay close attention to phase relationships.

EXAMPLE 36–2 In Fig. 36–14, the incident unpolarized light has intensity I_0. Find the intensity transmitted by the first polarizer and by the second, if the angle θ is 30°.

SOLUTION As explained above, the intensity after the first filter is $I_0/2$. According to Eq. (36–9), the second filter reduces the intensity by a factor of $\cos^2 30° = \frac{3}{4}$. Thus the intensity transmitted by the second polarizer is $(I_0/2)(\frac{3}{4}) = 3I_0/8$.

Reflected light is preferentially polarized.

Unpolarized light can be partially polarized by *reflection*. When unpolarized light strikes a reflecting surface between two optical materials, preferential reflection occurs for those waves in which the electric-field vector is perpendicular to the plane of incidence (the plane containing the incident ray and the normal to the surface). At one particular angle of incidence, called the **polarizing angle** ϕ_p, no light whatever is reflected except that in which the electric vector is perpendicular to the plane of incidence. This case is shown in Fig. 36–15.

When light is incident at the polarizing angle, *none* of the component parallel to the plane of incidence is reflected; this component is 100% transmitted in the *refracted* beam. For the component perpendicular to the plane of incidence, the fraction reflected depends on the index of the reflecting material. About 15% is reflected if the reflecting surface is glass. Hence the *reflected* light is weak and *completely* polarized. The *refracted* light is a mixture of the parallel component, all of which is refracted, and the remaining 85% of the perpendicular component. It is therefore strong but only *partially* polarized. At angles of incidence other than the polarizing angle, some of the component parallel to the plane of incidence is reflected, so that, except at the polarizing angle, the reflected light is not completely linearly polarized.

When the angle of incidence is Brewster's angle, the reflected light is completely polarized.

In 1812 Sir David Brewster noticed that when the angle of incidence is equal to the polarizing angle ϕ_p, the reflected ray and the refracted ray are

36–15 When light is incident at the polarizing angle, the reflected light is linearly polarized.

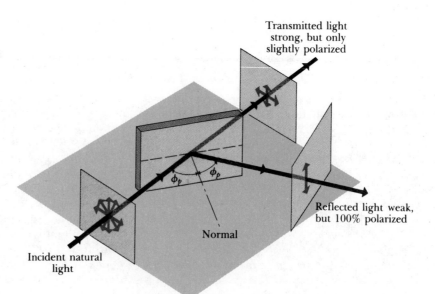

Transmitted light strong, but only slightly polarized

Reflected light weak, but 100% polarized

Normal

Incident natural light

perpendicular to each other, as shown in Fig. 36–16. When this is the case, the angle of refraction ϕ' becomes the complement of ϕ_p, so that $\sin \phi' = \cos \phi_p$. Since, according to the law of refraction,

$$n \sin \phi_p = n' \sin \phi', \qquad (36\text{–}10)$$

we find $n \sin \phi_p = n' \cos \phi_p$ and

$$\tan \phi_p = \frac{n'}{n}, \qquad (36\text{–}11)$$

a relation known as **Brewster's law.**

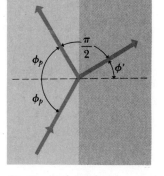

36–16 At the polarizing angle the reflected and transmitted rays are perpendicular to each other.

Some crystalline materials exhibit different refractive indexes for different directions of polarization. A common example is calcite; when a calcite crystal is oriented appropriately in a beam of unpolarized light, its refractive index (for $\lambda = 589$ nm) is 1.658 for one direction of polarization and 1.486 for the perpendicular direction. Such crystals are said to be *doubly refracting,* or **bire-fringent.** They can be used in arrangements or prisms that separate spatially the two polarized components of an unpolarized beam and reflect one compo-nent out of the beam direction. These polarizing filters are more efficient than Polaroid filters and also much more expensive.

Birefringence: different indexes of refraction for different polarizations

Some doubly refracting crystals exhibit **dichroism,** a selective absorption in which one of the polarized components is *absorbed* much more strongly than the other. If the crystal is cut of the proper thickness, one component is practi-cally extinguished by absorption, while the other is transmitted in appreciable amount, as indicated in Fig. 36–12. Tourmaline is one example of such a dichroic crystal.

Dichroism: preferential absorption of some polarization states

An early form of Polaroid, invented by Edwin H. Land in 1928, consisted of a thin layer of tiny needlelike dichroic crystals of herapathite (iodoquinine sulfate), in parallel orientation, embedded in a plastic matrix and enclosed for protection between two transparent plates. A more recent modification, devel-oped by Land in 1938 and known as an H-sheet, is a *molecular* polarizer. It consists of long polymeric molecules of polyvinyl alcohol (PVA) that have been given a preferred direction by stretching and have been stained with an ink containing iodine that causes the sheet to exhibit dichroism. The PVA sheet is laminated to a support sheet of cellulose acetate butyrate.

Dichroic materials can make polarizing filters.

Polaroid sheet is widely used in sunglasses, where, from the standpoint of its polarizing properties, it plays the role of the analyzer in Fig. 36–14. We have seen that when unpolarized light is reflected, there is a preferential re-flection for light polarized perpendicular to the plane of incidence. When sunlight is reflected from a horizontal surface, the plane of incidence is verti-cal. Hence the reflected light contains a preponderance of light polarized in the horizontal direction. When such reflection occurs at smooth asphalt road surfaces or the surfaces of a lake, it causes unwanted "glare," and vision is improved by eliminating it. The transmission direction of the Polaroid sheet in the sunglasses is vertical, so none of the horizontally polarized light is trans-mitted to the eyes.

What are the advantages of polarizing sunglasses?

Apart from this polarizing feature, these glasses serve the same purpose as any dark glasses, absorbing 50% of the incident light. In an unpolarized beam, half the light can be considered as polarized horizontally and half vertically, and only the vertically polarized light is transmitted. The sensitivity of the eye is independent of the state of polarization of the light.

36–10 SCATTERING OF LIGHT

The sky is blue. Sunsets are red. Skylight is partially polarized, as can readily be verified by looking at the sky directly overhead through a polarizing filter. It turns out that one phenomenon is responsible for all three of these effects.

In Fig. 36–17, sunlight (unpolarized) comes from the left along the z-axis and passes over an observer looking vertically upward along the y-axis. Molecules of the earth's atmosphere are located at point O. The electric field in the beam of sunlight sets the electric charges in the molecules in vibration. Since light is a transverse wave, the direction of the electric field in any component of the sunlight lies in the xy-plane, and the motion of the charges takes place in this plane. There is no field, and hence no vibration, in the direction of the x- and y-axes.

A component of the incident light at an angle θ with the x-axis sets the electric charges in the molecules vibrating in the same direction, as indicated by the heavy line through point O. We can resolve this vibration into two components, one along the x-axis and the other along the y-axis. Each component in the incident light produces the equivalent of two molecular "antennas," oscillating with the same frequency as the incident light, and lying along the x- and y-axes.

We mentioned in Section 35–8 that an antenna does not radiate in the direction of its own length. Hence the antenna along the y-axis does not send any light to the observer directly below it. It does, of course, send out light in other directions. The only light reaching the observer comes from the component of vibration along the x-axis, and, as is the case with the waves from any antenna, this light is linearly polarized, with the electric field parallel to the antenna. The vectors on the y-axis below point O show the direction of polarization of the light reaching the observer.

The process just described is called **scattering.** The energy of the scattered light is removed from the original beam, which becomes weakened in the process. Detailed analysis of the scattering processes shows that the intensity

Scattering of light leads to preferential polarization.

Blue light is scattered more than red.

Why is the sky blue?
Why are sunsets red?

36–17 Scattered light is linearly polarized.

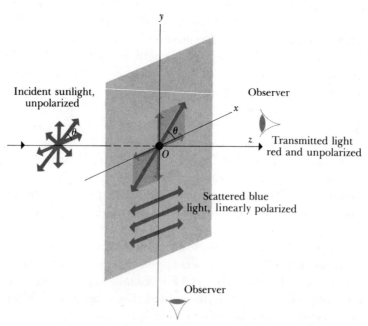

Incident sunlight, unpolarized

Observer

Transmitted light red and unpolarized

Scattered blue light, linearly polarized

Observer

of the scattered light increases with increasing frequency; blue light is scattered more than red, with the result that the hue of the scattered light is blue.

Because skylight is partially polarized, polarizers are useful in photography. The sky can be darkened in a photograph by appropriate orientation of the polarizer axis. The effect of atmospheric haze can be reduced in exactly the same way, and unwanted reflections can be controlled just as with polarizing sunglasses, discussed in Section 36–9.

Toward evening, when sunlight has to travel a large distance through the earth's atmosphere to reach a point over or nearly over an observer, a large proportion of the blue light in sunlight is removed from it by scattering. White light minus blue light is yellow or red in hue. Thus when sunlight, with the blue component removed, is incident on a cloud, the light reflected from the cloud to the observer has the yellow or red hue so commonly seen at sunset. If the earth had no atmosphere, we would receive *no* skylight at the earth's surface, and the sky would appear as black in the daytime as it does at night. To an astronaut in a spaceship or on the moon, the sky appears black, not blue.

In outer space, the sky is black, not blue.

36–11 CIRCULAR AND ELLIPTICAL POLARIZATION

Up to this point we have discussed polarization phenomena in terms of *linearly polarized* light. Light (and all other electromagnetic radiation) may also have *circular* or *elliptical* polarization. To introduce these new concepts, we return again to mechanical waves on a stretched string. In Fig. 36–11, suppose the two linearly polarized waves in parts (a) and (b) are in phase and have equal amplitude. When they are *superposed*, each point in the string has simultaneously y- and z-displacements of equal magnitude, and a little thought shows that the resultant wave lies in a plane oriented at 45° to the x- and z-axes (i.e., in a plane making a 45° angle with the xy- and xz-planes). The amplitude of the resultant wave is larger by a factor of $\sqrt{2}$ than that of either component wave, and the resultant wave is again linearly polarized.

Superposing two linearly polarized waves with phase differences

But now suppose one of the equal-amplitude component waves differs in phase by a quarter-cycle from the other. Then the resultant motion of each point corresponds to a superposition of two simple harmonic motions at right angles with a quarter-cycle phase difference. The motion is then no longer confined to a single plane, and it can be shown that each point on the rope moves in a *circle* in a plane parallel to the yz-plane. Successive points on the rope have successive phase differences, and the overall motion of the string then has the appearance of a rotating helix. This particular superposition of two linearly polarized waves is called **circular polarization.** By convention, the wave is said to be *right circularly polarized* when, as in the present instance, the sense of motion of a particle of the string, to an observer looking *backward* along the direction of propagation, is *clockwise.* The wave is *left circularly polarized* when it appears counterclockwise to that observer. Left circular polarization would be the result if the phase difference between y- and z-components were opposite to that in our example.

*In circularly polarized light, the **E** and **B** vectors rotate around the direction of propagation.*

If the phase difference between the two component waves is something other than a quarter-cycle, or if the two component waves have different amplitudes, then each point on the string traces out not a circle but an *ellipse.* The resulting wave is said to be **elliptically polarized.**

How to make circularly polarized light: a recipe

For electromagnetic waves of radio frequencies, circular or elliptical polarization can be produced by using two antennas at right angles, fed from the same transmitter but with a phase-shifting network that introduces the appropriate phase difference. For light, the phase shift can be introduced by use of a birefringent material. If two waves with perpendicular directions of polarization are in phase as they enter the material, they travel with different speeds because the refractive index is different for the two waves. In general, they are no longer in phase when they emerge from the crystal. If the thickness of the crystal is such as to introduce a quarter-cycle phase difference, then the crystal converts linearly polarized light to circularly polarized light. Such a crystal is called a **quarter-wave plate.** We challenge you to show that such a plate also converts circularly polarized light to linearly polarized light!

When a polarizer and an analyzer are mounted in the "crossed" position, that is, with their transmission directions at right angles to each other, no light is transmitted through the combination. But if a doubly refracting crystal is inserted between polarizer and analyzer, the light transmitted through the crystal is, in general, elliptically polarized, and some light will be transmitted by the analyzer. Thus the field of view, dark in the absence of the crystal, becomes light when the crystal is inserted.

Photoelasticity: using polarized light to measure mechanical stresses in materials

Some substances, such as glass and various plastics, though not normally doubly refracting, become so when subjected to mechanical stress. This is the basis of the science of **photoelasticity.** Stresses in opaque engineering materials such as girders, boiler plates, and gear teeth can be analyzed by constructing a transparent model of the object, usually of a plastic, subjecting it to stress and examining it between a polarizer and an analyzer in the crossed position. Very complicated stress distributions, such as those around a hole or gear tooth, that are practically impossible to analyze mathematically, may thus be studied by optical methods. Figure 36–18 shows two photographs of photoelastic models under stress.

Liquids are not normally doubly refracting, but some become so when an electric field is established within them. This phenomenon is known as the *Kerr effect*. The existence of the Kerr effect makes it possible to construct an electrically controlled "light valve." A cell with transparent walls contains the liquid between a pair of parallel plates. The cell is inserted between crossed Polaroid disks. Light is transmitted when an electric field is set up between the plates and is cut off when the field is removed.

Optical activity: one of the differences between dextrose and levulose

When a beam of linearly polarized light is sent through certain types of crystals and certain liquids, the direction of polarization of the emerging linearly polarized light is found to be different from the original direction. This phenomenon is called *rotation of the direction of polarization,* and substance that exhibit this effect are said to be *optically active*. Those that rotate the direction of polarization to the right, looking along the advancing beam, are called *dextrorotatory*, or right-handed; those that rotate it to the left are *levorotatory*, or left-handed.

Optical activity in crystals that do not have mirror-image symmetry

Optical activity may be due to an asymmetry of the molecules of a substance, or it may be a property of a crystal as a whole. For example, solutions of cane sugar are dextrorotatory, indicating that the optical activity is a property of the sugar molecule. The molecules of the sugars dextrose and levulose are mirror images, and their optical activities are opposite. The rotation of the direction of polarization by a sugar solution is used commercially as a method

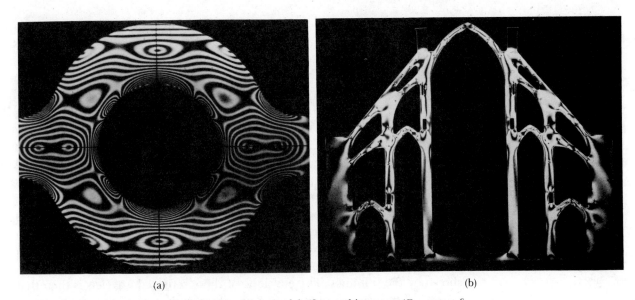

(a)   (b)

36–18 (a) Photoelastic stress analysis of a plastic model of a machine part. (Courtesy of Dr. W. M. Murray, Massachusetts Institute of Technology.) (b) Stress analysis of a model of a cross section of a Gothic cathedral. The masonry construction used for this kind of building had great strength in compression but very little in tension. Inadequate buttressing and high winds sometimes caused tensile stresses in normally compressed structural elements, leading to some spectacular collapses. (Sepp Seitz/Woodfin Camp.)

of determining the proportion of cane sugar in a given sample. Crystalline quartz is also optically active; some natural crystals are right-handed and others left-handed. Here the optical activity is a consequence of the crystalline structure, since the activity disappears when the quartz is melted and allowed to resolidify into a glassy, noncrystalline state called fused quartz.

36–12 HUYGENS' PRINCIPLE

The principles of reflection and refraction of light rays that we introduced in Section 36–5 were discovered experimentally long before the wave nature of light was firmly established. These principles may, however, be derived from *wave* considerations and thus shown to be consistent with the wave nature of light. To establish this connection we use a principle called **Huygens' principle.** This principle, stated originally by Christian Huygens in 1678, is a geometrical method for finding, from the known shape of a wave front at some instant, the shape of the wave front at some later time. Huygens assumed that *every point of a wave front may be considered the source of secondary wavelets that spread out in all directions with a speed equal to the speed of propagation of the wave.* The new wave front is then found by constructing a surface *tangent* to the secondary wavelets or, as it is called, the *envelope* of the wavelets.

Huygens' principle is illustrated in Fig. 36–19. The original wave front AA' is traveling as indicated by the small arrows. We wish to find the shape of the wave front after a time interval t. Let v represent the speed of propagation. We construct several circles (traces of spherical wavelets) of radius $r = vt$, with centers along AA'. The trace of the envelope of these wavelets, which is the new wave front, is the curve BB'. The speed v is assumed to be the same at all points and in all directions.

Huygens' principle: an important concept in wave propagation

36–19 Geometric construction illustrating Huygens' principle.

Deriving the law of reflection from Huygens' principle

Deriving the law of refraction from Huygens' principle

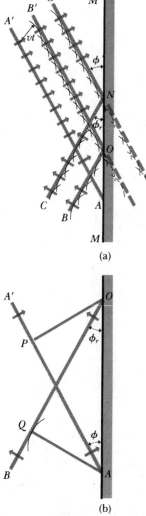

36–20 (a) Successive positions of a plane wave AA' as it is reflected from a plane surface. (b) A portion of (a).

To derive the law of reflection from Huygens' principle, we consider a plane wave approaching a plane reflecting surface. In Fig. 36–20a, the lines AA', BB', and CC' represent successive positions of a wave front approaching the surface MM'. The actual planes are perpendicular to the plane of the figure. Point A on the wave front AA' has just arrived at the reflecting surface. The position of the wave front after a time interval t may be found by applying Huygens' principle. With points on AA' as centers, we draw several secondary wavelets of radius vt, where v is the speed of propagation of the wave. Those wavelets originating near the upper end of AA' spread out unhindered, and their envelope gives that portion of the new wave surface OB'. The wavelets originating near the lower end of AA', however, strike the reflecting surface. If the surface had not been there, they would have reached the positions shown by the broken circular arcs. The effect of the reflecting surface is to *reverse the direction* of travel of those wavelets that strike it, so that part of a wavelet that would have penetrated the surface actually lies to the left of it, as shown by the full lines. The envelope of these reflected wavelets is then that portion of the wave front OB. The trace of the entire wave front at this instant is the broken line BOB'. A similar construction gives the line CNC' for the wave front after another time interval t.

The angle ϕ between the incident *wave front* and the *surface* is the same as that between the incident *ray* and the *normal* to the surface and is therefore the angle of incidence. Similarly, ϕ_r is the angle of reflection. To find the relation between these angles, we consider Fig. 36–20b. From O, we draw $OP = vt$, perpendicular to AA'. Now OB, by construction, is tangent to a circle of radius vt with center at A. Hence, if we draw AQ from A to the point of tangency, the triangles APO and OQA are equal (right triangles with the side AO in common and with $AQ = OP$). The angle ϕ therefore equals the angle ϕ_r, and we have the law of reflection.

We can derive the law of *refraction* by a similar procedure. In Fig. 36–21a, we consider a wave front, represented by line AA', for which point A has just arrived at the boundary surface SS' between two transparent materials a and b, with indexes of refraction n_a and n_b. (The *reflected* waves are not shown in the figure; they proceed exactly as in Fig. 36–20.) We can apply Huygens' principle to find the position of the refracted wave fronts after a time t.

With points on AA' as centers, we draw a number of secondary wavelets. Those originating near the upper end of AA' travel with speed v_a and, after a time interval t, are spherical surfaces of radius $v_a t$. The wavelet originating at point A, however, is traveling in the second material b with speed v_b and at time t is a spherical surface of radius $v_b t$. The envelope of the wavelets from the original wave front is the plane whose trace is the broken line BOB'. A similar construction leads to the trace CPC' after a second interval t.

The angles ϕ_a and ϕ_b between the surface and the incident and refracted wave fronts are, respectively, the angle of incidence and the angle of refraction. To find the relation between these angles, refer to Fig. 36–21b. Draw $OQ = v_a t$, perpendicular to AQ, and draw $AB = v_b t$, perpendicular to BO. From the right triangle AOQ,

$$\sin \phi_a = \frac{v_a t}{AO},$$

and from the right triangle AOB,

$$\sin \phi_b = \frac{v_b t}{AO}.$$

Hence

$$\frac{\sin \phi_a}{\sin \phi_b} = \frac{v_a}{v_b}.$$ (36–12)

Since v_a/v_b is a constant, Eq. (36–12) expresses Snell's law, and we have derived Snell's law from a wave theory! This result emphasizes that the general wave theory gives the same results as the more specialized ray picture in cases where the ray picture is applicable.

The most general form of Snell's law is given by Eq. (36–4), namely,

$$\frac{\sin \phi_a}{\sin \phi_b} = \frac{n_b}{n_a}.$$

Comparing this with Eq. (36–12), we see that

$$\frac{v_a}{v_b} = \frac{n_b}{n_a},$$

and

$$n_a v_a = n_b v_b.$$

When either material is vacuum, $n = 1$ and the speed is c. Hence

$$n_a = \frac{c}{v_a}, \qquad n_b = \frac{c}{v_b},$$ (36–13)

showing that *the index of refraction of any material is the ratio of the speed of light in vacuum to the speed in the material.* We stated Eq. (36–13) without proof at the end of Section 36–5. The speed of light in a material is always *less* than in vacuum; hence for any material n is always greater than unity.

In Fig. 36–21, if t is chosen to be the period τ of the wave, the spacing is $v\tau$, which is the wavelength λ. The figure shows that when v_b is less than v_a, the wavelength in the second material is smaller than in the first. When a light wave proceeds from one material to another, where the speed is different, the wavelength changes *but not the frequency*, as we discussed in Section 36–5. Since $v_a = f\lambda_a$ and $v_b = f\lambda_b$,

$$\frac{\lambda_a}{v_a} = \frac{\lambda_b}{v_b}$$

and

$$\lambda_a \frac{c}{v_a} = \lambda_b \frac{c}{v_b}.$$

Therefore,

$$\lambda_a n_a = \lambda_b n_b.$$

If either material is vacuum, the index is 1 and the wavelength in vacuum is represented by λ_0. Hence

$$\lambda_a = \frac{\lambda_0}{n_a}, \qquad \lambda_b = \frac{\lambda_0}{n_b},$$ (36–14)

showing that *the wavelength in any material is the wavelength in vacuum divided by the index of refraction of the medium.* We stated Eq. (36–14) without proof in Section 36–5.

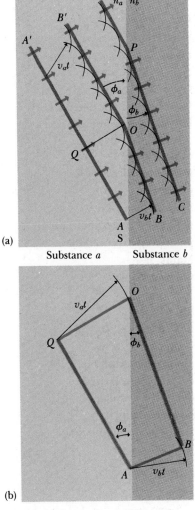

(a)

Substance *a*   Substance *b*

(b)

36–21 (a) Successive positions of a plane wave front *AA'* as it is refracted by a plane surface. (b) A portion of (a). The case $v_b < v_a$ is shown.

The wavelength is smaller in a material than for the same wave in vacuum.

SUMMARY

KEY TERMS

wave front

rays

geometrical optics

law of reflection

law of refraction (Snell's law)

**index of refraction
   (refractive index)**

critical angle

total internal reflection

dispersion

linearly polarized

polarizing filters (polarizers)

polarizing axis

Malus' law

polarizing angle

Brewster's law

birefringent

dichroism

scattering

circular polarization

elliptically polarized

quarter-wave plate

photoelasticity

Huygens' principle

Light is electromagnetic radiation with wavelengths in the range 400 to 700 nm. It is emitted by moving electric charges within atoms and molecules that have been given excess energy by heating or by electrical discharge. Commonly used light sources include incandescent lamps, carbon arcs, sodium- and mercury-arc lamps, and fluorescent lamps. The light emitted by a laser is much more nearly monochromatic (single-frequency) than that of any other source.

The speed of light is a fundamental physical constant. It is defined to have a certain precise value; this, along with the unit of time defined by an atomic clock, defines the unit of length.

A wave front is a line or surface of constant phase; wave fronts move with a speed equal to the propagation speed of the wave. The propagation of light can also be represented by rays, which are lines along the direction of propagation, perpendicular to the wave fronts. Representation of light by rays is the basis of geometrical optics.

The law of reflection and the law of refraction (Snell's law) govern the behavior of light rays at an interface between two optical materials. The incident, reflected, and refracted rays and the normal to the plane of interface all lie in a single plane. The law of reflection states that the angles of incidence and reflection are equal; the law of refraction states that the ratio of the sines of the angles of incidence and refraction is constant for a given pair of materials. Angles of incidence, reflection, and refraction are always measured from the normal to the surface.

When a ray is incident in vacuum on material a, with incident and refracted angles ϕ_0 and ϕ_a, respectively, the index of refraction n_a of the material is defined as

$$\frac{\sin \phi_0}{\sin \phi_a} = n_a. \qquad (36\text{--}3)$$

The general statement of the law of refraction, in terms of indexes of refraction, is

$$n_a \sin \phi_a = n_b \sin \phi_b. \qquad (36\text{--}5)$$

The speed of light is less in a material than in vacuum by a factor n, and the wavelength in a material is smaller than in vacuum by the same factor n.

When a ray travels from a material of greater index of refraction toward one of smaller index, total internal reflection occurs when the angle of incidence exceeds a critical value ϕ_{crit} given by

$$\sin \phi_{\text{crit}} = \frac{n_b}{n_a}. \qquad (36\text{--}8)$$

The index of refraction of a material depends on the wavelength of light, usually decreasing with increasing wavelength. This is called dispersion.

Electromagnetic waves, like all transverse waves, exhibit polarization. The direction of polarization of a linearly polarized wave is defined as the direction of the E field. A polarizing filter passes radiation that is linearly polarized in the direction of its polarizing axis, and blocks radiation polarized perpendicularly to that axis. When linearly polarized light is incident on a polarizing filter with its axis at an angle θ to the direction of polarization, the transmitted intensity I is given by

$$I = I_{max} \cos^2 \theta, \qquad\qquad (36\text{--}9)$$

where I_{max} is the intensity when $\theta = 0$. This relation is called Malus' law.

When unpolarized light strikes an interface between two materials, the reflected light is completely polarized perpendicular to the plane of incidence if the angle of incidence ϕ_p is given by

$$\tan \phi_p = \frac{n'}{n}. \qquad\qquad (36\text{--}11)$$

This relation is called Brewster's law.

Materials having different indexes of refraction for two perpendicular directions of polarization are said to be birefringent. Those that show preferential absorption for one polarization direction are dichroic; they are used in polarizing filters for light. Some materials become birefringent under mechanical stress; these form the basis of photoelastic stress analysis.

Light can be scattered by air molecules. The scattered light is preferentially polarized. When two linearly polarized waves with a phase difference are superposed, the result is circularly or elliptically polarized light. In this case the E vector is not confined to a plane containing the direction of propagation but describes a circle or ellipse in the planes perpendicular to the direction of propagation.

Huygens' principle states that if the position of a wave front at one instant is known, the position of the front at a later time can be constructed by imagining the front as a source of secondary wavelets. Huygens' principle can be used to derive the laws of reflection and refraction.

QUESTIONS

36–1 During a thunderstorm one always sees the flash of lightning before hearing the accompanying thunder. Discuss this in terms of the various wave speeds. Can this phenomenon be used to determine how far away the storm is?

36–2 When hot air rises around a radiator or from a heating duct, objects behind it appear to shimmer or waver. What is happening?

36–3 Light requires about 8 min to travel from the sun to the earth. Is it delayed appreciably by the earth's atmosphere?

36–4 Sometimes when looking at a window one sees two reflected images, slightly displaced from each other. What causes this?

36–5 An object submerged in water appears to be closer to the surface than it actually is. Why? Swimming pools are always deeper than they look; is this the same phenomenon?

36–6 A ray of light in air strikes a glass surface. Is there a range of angles for which total reflection occurs?

36–7 As shown in Table 36–1, diamond has a much larger refractive index than glass. Is there a larger or smaller range of angles for which total internal reflection occurs for

diamond, than for glass? Does this have anything to do with the fact that a real diamond has more sparkle than a glass imitation?

36–8 Sunlight or starlight passing through the earth's atmosphere is always bent toward the vertical. Why? Does this mean that a star is not really where it appears to be?

36–9 The sun or moon usually appears flattened just before it sets. Is this related to refraction in the earth's atmosphere, mentioned in Question 36–8?

36–10 A student claimed that, because of atmospheric refraction (cf. Question 36–8), the sun can be seen after it has set, and that the day is therefore longer than it would be if the earth had no atmosphere. First, what does he mean by saying the sun can be seen after it has set? Second, comment on the validity of his conclusion.

36–11 It has been proposed that automobile windshields and headlights should have polarizing filters to reduce the glare of oncoming lights during night driving. Would this work? How should the polarizing axes be arranged? What advantages would this scheme have? What disadvantages?

36–12 A salesperson at a bargain counter claims that a certain pair of sunglasses has Polaroid filters; you suspect they are just tinted plastic. How could you find out for sure?

36–13 When unpolarized light is incident on two crossed polarizers, no light is transmitted. A student asserted that if a third polarizer is inserted between the other two, some transmission may occur. Does this make sense? How can adding a third filter *increase* transmission?

36–14 How could you determine the direction of the polarizing axis of a single polarizer?

36–15 In three-dimensional movies, two images are projected on the screen, and the viewers wear special glasses to sort them out. How does this work?

36–16 In Fig. 36–17, since the light scattered out of the incident beam is polarized, why is the transmitted beam not also partially polarized?

36–17 Light from blue sky is strongly polarized because of the nature of the scattering process described in Section 36–10. But light scattered from white clouds is usually *not* polarized. Why not?

36–18 When a sheet of plastic food wrap is placed between two crossed polarizers, no light is transmitted. When the sheet is stretched in one direction, some light passes through. What is happening?

36–19 Television transmission usually uses plane-polarized waves. It has been proposed to use circularly polarized waves to improve reception. Why? (*Hint:* See Problem 36–24.)

36–20 Can sound waves be reflected? Refracted? Give examples. Does Huygens' principle apply to sound waves?

36–21 Why should the wavelength of light change, but not its frequency, in passing from one material to another?

36–22 When light is incident on an interface between two materials, the angle of the refracted ray depends on the wavelength, but the angle of the reflected ray does not. Why should this be?

EXERCISES

Section 36–3 The Speed of Light

36–1 The orbital period of the moon of Jupiter studied by Roemer is 42 hr.

a) Using data from Appendix F, how far along its orbit around the sun does the earth travel in 42 hr?

b) How much time does it take light to travel the distance calculated in (a)?

36–2 Fizeau's measurements of the speed of light were continued by Cornu, using Fizeau's apparatus but with the distance from the toothed wheel to the mirror increased to 22.9 km. One of the toothed wheels used was 40 mm in diameter and had 180 teeth. Find the angular velocity at which it should rotate so that light transmitted through one opening will return through the next.

Section 36–5 Reflection and Refraction

36–3 A ray of light is incident on a plane surface separating two transparent substances of refractive indices 1.60 and 1.40. The angle of incidence is 30°, and the ray originates in the medium of higher index. Compute the angle of refraction.

36–4 A parallel-sided plate of glass having a refractive index of 1.60 is held on the surface of water in a tank. A ray coming from above makes an angle of incidence of 45° with the top surface of the glass.

a) What angle does the ray make with the normal in the water?

b) How does this angle vary with the refractive index of the glass?

36–5 A parallel beam of light makes an angle of 30° with the surface of a glass plate having a refractive index of 1.50.

a) What is the angle between the refracted beam and the surface of the glass?

b) What should be the angle of incidence ϕ with this plate for the angle of refraction to be $\phi/2$, where both angles are measured relative to the normal?

36–6 The density of the earth's atmosphere increases as the surface of the earth is approached. This increase in density is accompanied by a corresponding increase in refractive index.

a) Draw a diagram showing how the light from a star or planet is bent as it goes through the atmosphere. Indicate the apparent position of the light source.

b) Explain how one can see the sun after it has set.

c) Explain why the setting sun appears flattened.

36–7 The speed of light of wavelength 656 nm in heavy flint glass is 1.60×10^8 m·s^{-1}. What is the index of refraction of this glass?

36–8 Light of a certain frequency has a wavelength in water of 442 nm. What is the wavelength of this light when it passes into carbon disulfide?

36–9

a) What is the speed of light of wavelength 500 nm (in vacuum) in glass whose index of refraction at this wavelength is 1.50?

b) What is the wavelength of these waves in the glass?

36–10 Prove that a ray of light reflected from a plane mirror rotates through an angle of 2θ when the mirror rotates through an angle θ about an axis perpendicular to the plane of incidence.

36–11 A parallel beam of light is incident on a prism, as shown in Fig. 36–22. Part of the light is reflected from one face and part from another. Show that the angle θ between the two reflected beams is twice the angle A between the two reflecting surfaces.

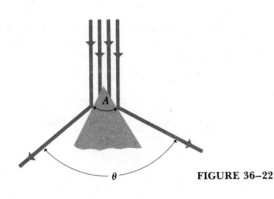

FIGURE 36–22

Section 36–6 Total Internal Reflection

36–12 A ray of light in glass of index of refraction 1.50 is incident on an interface with air. What is the *largest* angle the ray can make with the normal and not be totally reflected back into the glass?

36–13 The speed of a sound wave is 330 m·s^{-1} in air and 1320 m·s^{-1} in water.

a) What is the critical angle for a sound wave incident on the surface between air and water?

b) Which medium has the higher "index of refraction" for sound?

36–14 A point source of light is 20 cm below the surface of a body of water. Find the diameter of the largest circle at the surface through which light can emerge from the water.

Section 36–9 Polarizing Filters

36–15 Unpolarized light of intensity I_0 is incident on a polarizing filter, and the emerging light strikes a second polarizing filter with its axis 45° to that of the first. Determine

a) the intensity of the emerging beam;

b) its state of polarization.

36–16 A polarizer and an analyzer are oriented so that the maximum amount of light is transmitted. To what fraction of its maximum value is the intensity of the transmitted light reduced when the analyzer is rotated through

a) 30°?

b) 45°?

c) 60°?

36–17 Three polarizing filters are stacked, with the polarizing axes of the second and third at 45° and 90°, respectively, with that of the first.

a) If unpolarized light of intensity I_0 is incident on the stack, find the intensity and state of polarization of light emerging from each filter.

b) If the second filter is removed, how does the situation change?

36–18 Light traveling in air strikes a glass plate at an angle of incidence of 60°; part of the beam is reflected and part

refracted. It is observed that the reflected and refracted portions make an angle of 90° with each other. What is the index of refraction of the glass?

36–19 The critical angle of light in a certain substance is 45°. What is the polarizing angle?

36–20

a) At what angle above the horizontal must the sun be for sunlight reflected from the surface of a calm body of water to be completely polarized?

b) What is the plane of the **E** vector in the reflected light?

36–21 A parallel beam of "natural" light is incident at an angle of 58° (with respect to the normal) on a plane glass surface. The reflected beam is completely linearly polarized.

a) What is the refractive index of the glass?

b) What is the angle of refraction of the transmitted beam?

Section 36–10 Scattering of Light

36–22 A beam of light, after passing through the Polaroid disk P_1 in Fig. 36–23, traverses a cell containing a scattering medium. The cell is observed at right angles through another Polaroid disk P_2. Originally the disks are oriented so that the brightness of the field as seen by the observer is a maximum.

a) Disk P_2 is now rotated through 90°. Is extinction produced?

b) Disk P_1 is now rotated through 90°. Is the field bright or dark?

c) Disk P_2 is then restored to its original position. Is the field bright or dark?

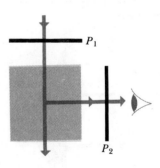

FIGURE 36–23

Section 36–11 Circular and Elliptical Polarization

36–23 What is the state of polarization of the light transmitted by a quarter-wave plate when the electric vector of the incident linearly polarized light makes an angle of 30° with the optic axis?

36–24 A beam of right circularly polarized light is reflected at normal incidence from a reflecting surface. Is the reflected beam right or left circularly polarized? Explain.

PROBLEMS

36-25 An inside corner of a cube is lined with mirrors. A ray of light is reflected successively from each of three mutually perpendicular mirrors; show that its final direction is always exactly opposite to its initial direction. This principle is used in tail-light lenses and reflecting highway signs.

36-26 High frequency ($f = 1$ to 5 MHz) soundwaves, called ultrasound, are now being used by physicians to image internal organs. The speed of these ultrasound waves is 1480 m·s⁻¹ in muscle and 350 m·s⁻¹ in air.

a) At what angle would an ultrasound beam enter the heart if it left the lungs at an angle of 8° from the normal to the heart wall?

b) What is the critical angle for sound waves in air incident on muscle?

36-27 Old photographic plates were made of glass with a light-sensitive emulsion on the front surface. This emulsion was somewhat transparent. When a bright point source is focused on the front of the plate, the developed photograph will show a halo around the image of the spot. If the glass plate is 3 mm thick and the halos have a radius of 2.2 mm, what is the index of refraction of the glass? (*Hint:* Consider total internal reflection of the light from the focused spot at the back surface of the glass.)

36-28 A glass plate 3 mm thick, of index of refraction 1.50, is placed between a point source of light of wavelength 600 nm (in vacuum) and a screen. The distance from source to screen is 3 cm. How many waves are there between source and screen?

36-29 The prism of Fig. 36-24 has a refractive index of 1.414, and the angles A are 30°. Two light rays m and n are parallel as they enter the prism. What is the angle between them after they emerge?

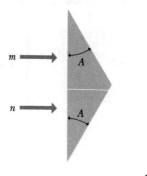

FIGURE 36-24

36-30 A 45°-45°-90° prism is immersed in water. A ray of light is incident normally on one of its shorter faces. What is the minimum index of refraction the prism must have if this ray is to be totally reflected within the glass, at the long face of the prism?

36-31 Light is incident normally on the short face of a 30°-60°-90° prism, as in Fig. 36-25. A drop of liquid is placed on the hypotenuse of the prism. If the index of the prism is 1.50, find the maximum index the liquid may have if the light is to be totally reflected.

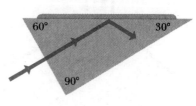

FIGURE 36-25

36-32 The glass vessel shown in Fig. 36-26a contains a large number of small, irregular pieces of glass and a liquid. The dispersion curves of the glass and of the liquid are shown in Fig. 36-26b. Explain the behavior of a parallel beam of white light as it traverses the vessel. (This is known as a *Christiansen filter*.)

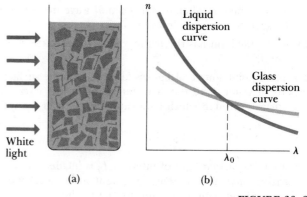

FIGURE 36-26

36-33 Three polarizing filters are stacked, with the polarizing axes of the second and third at angles θ and 90°, respectively, with that of the first. Unpolarized light of intensity I_0 is incident on the stack.

a) Derive an expression for the intensity of light transmitted through the stack, as a function of I_0 and θ.

b) For what value of θ does the maximum transmission occur?

36-34 It is desired to rotate the direction of polarization of linearly polarized light 90°, by using two Polaroid filters. Explain how this can be done, and find the final intensity in terms of the incident intensity.

36-35 A light source is made of an unpolarized component of intensity I_0 and a polarized component of intensity I_p. The plane of polarization of the polarized component is oriented at an angle of θ with respect to the vertical. The following data give the intensity measured through a polarizer with an orientation of ϕ with respect to the vertical.

$\phi(°)$	$I_{total}(W \cdot m^{-2})$
0	18.4
10	21.4
20	23.7
30	24.8
40	24.8
50	23.7
60	21.4
70	18.4
80	15.0
90	11.6
100	8.6
110	6.3
120	5.2
130	5.2
140	6.3
150	8.6
160	11.6
170	15.0
180	18.4

Rotation (°)		Concentration (g/100 ml)
l-leucine	l-glutamic acid	
− 0.11	0.124	1.0
− 0.22	0.248	2.0
− 0.55	0.620	5.0
− 1.10	1.24	10.0
− 2.20	2.48	20.0
− 5.50	6.20	50.0
−11.0	12.4	100.0

36–38 In Fig. 36–27, A and C are Polaroid sheets whose transmission directions are as indicated. B is a sheet of doubly refractive material whose optic axis is vertical. All three sheets are parallel. Unpolarized light enters from the left. Discuss the state of polarization of the light at points 2, 3, and 4.

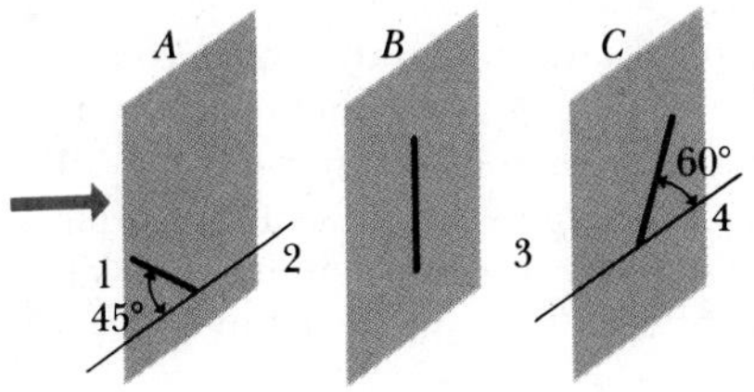

FIGURE 36–27

a) What is the orientation of the polarized component? (That is, what is the angle θ?)

b) What are the values of I_0 and I_p?

36–36 The refractive index of a certain flint glass is 1.65. For what incident angle is light reflected from the surface of this glass completely polarized if the glass is immersed in

a) air?

b) water?

36–37 Many biologically active molecules are also optically active. Plane polarized light on traversing a solution of these compounds has its plane of polarization rotated. Some compounds rotate the polarization clockwise, while others rotate the polarization counterclockwise. The amount of rotation depends on the amount of material in the light path. The following data give the amount of rotation through two amino acids over a path length of 100 cm. From these data find the relationship between the concentration and polarization rotation. Find both the functional form valid for both materials and the value of any compound-specific parameters.

36–39 A certain birefringent material has indexes of refraction n_1 and n_2 for the two perpendicular components of linearly polarized light passing through it. The corresponding wavelengths are $\lambda_1 = \lambda_0/n_1$ and $\lambda_2 = \lambda_0/n_2$, where λ_0 is the wavelength in vacuum.

a) If the crystal is to function as a quarter-wave plate, the number of wavelengths of each component within the material must differ by $\frac{1}{4}$. Thus show that the minimum thickness for a quarter-wave plate is

$$d = \frac{\lambda_0}{4(n_1 - n_2)}.$$

b) Find the minimum thickness of a quarter-wave plate made of calcite, if the indexes of refraction are 1.658 and 1.486 and the wavelength is $\lambda_0 = 589$ nm.

CHALLENGE PROBLEMS

36–40 A ray of light traveling with speed c leaves point 1 of Fig. 36–28 and is reflected to point 2. The ray strikes the reflecting surface a horizontal distance x from point 1.

a) Show that time t required for the light to travel from 1 to 2 is

$$t = \frac{\sqrt{y_1^2 + x^2} + \sqrt{y_2^2 + (l - x)^2}}{c}.$$

b) Take the derivative of t with respect to x. Set the

derivative equal to zero to show that this time reaches its *minimum* value when $\theta_1 = \theta_2$, which is the law of reflection and corresponds to the actual path taken by the light. This is an example of Fermat's *principle of least time*, which states that among all possible paths between two points, the one actually taken by a ray of light is that for which the time of travel is a *minimum*. (In fact, there are some cases where the time is a maximum rather than a minimum.)

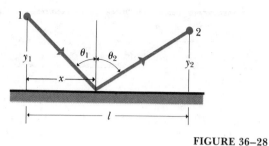

FIGURE 36-28

36-41 A ray of light goes from point A in a medium where the velocity of light is v_1 to point B in a medium where the velocity is v_2, as in Fig. 36-29. The ray strikes the interface a horizontal distance x to the right of point A.

a) Show that the time required for the light to go from A to B is
$$t = \frac{\sqrt{h_1{}^2 + x^2}}{v_1} + \frac{\sqrt{h_2{}^2 + (l - x)^2}}{v_2}.$$

b) Take the derivative of t with respect to x. Set this derivative equal to zero to show that this time reaches its *minimum* value when $n_1 \sin \theta_1 = n_2 \sin \theta_2$. This is Snell's law, and corresponds to the actual path taken by the light. This problem is a second example of Fermat's principle of least time, discussed in Challenge Problem 36-40.

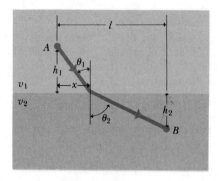

FIGURE 36-29

36-42 Light is incident at an angle ϕ_1 (as in Fig. 36-30) on the upper surface of a transparent plate, the surfaces of the plate being plane and parallel to each other.

a) Prove that $\phi_1 = \phi_2'$.

b) Show that this relation is true for any number of different parallel plates.

c) Prove that the lateral displacement d of the emergent beam is given by the relation
$$d = t \frac{\sin (\phi_1 - \phi_1')}{\cos \phi_1'},$$
where t is the thickness of the plate.

d) A ray of light is incident at an angle of 60° on one surface of a glass plate 2 cm thick, of index 1.50. The

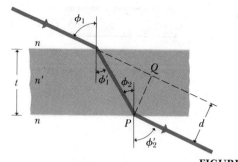

FIGURE 36-30

medium on either side of the plate is air. Find the lateral displacement between the incident and emergent rays.

36-43 Light passes symmetrically through a prism having apex angle A, as shown in Fig. 36-31.

a) Show that the angle of deviation δ (the angle between the initial and final directions of the ray) is given by
$$\sin \frac{A + \delta}{2} = n \sin \frac{A}{2}.$$

b) Use the result of (a) to find the angle of deviation for a ray of light passing symmetrically through a prism having three equal angles ($A = 60°$) and $n = 1.50$.

c) A certain glass has a refractive index of 1.50 for red light (700 nm) and 1.52 for violet light (400 nm). If both colors pass through symmetrically, as described in (a), and if $A = 60°$, find the difference between the angles of deviation for the two colors.

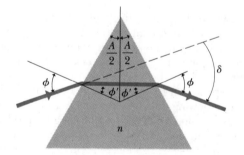

FIGURE 36-31

36-44 Consider two vibrations, one along the x-axis,
$$x = a \sin (\omega t - \alpha),$$
and the other along the y-axis, of equal amplitude and frequency, but differing in phase,
$$y = a \sin (\omega t - \beta).$$

Let us write them as follows:
$$\frac{x}{a} = \sin \omega t \cos \alpha - \cos \omega t \sin \alpha, \qquad (1)$$

$$\frac{y}{a} = \sin \omega t \cos \beta - \cos \omega t \sin \beta. \qquad (2)$$

a) Multiply Eq. (1) by $\sin \beta$ and Eq. (2) by $\sin \alpha$ and then subtract the resulting equations.

b) Multiply Eq. (1) by $\cos \beta$ and Eq. (2) by $\cos \alpha$ and then subtract the resulting equations.

c) Square and add the results of (a) and (b).

d) Derive the equation $x^2 + y^2 - 2xy \cos \delta = a^2 \sin^2 \delta$, where $\delta = \alpha - \beta$.

e) Use the above result to justify each of the diagrams in Fig. 36–32. In the figure the angle given is the phase difference between two simple harmonic motions, one horizontal (along the x-axis) and the other vertical (along the y-axis), and of the same frequency and amplitude. The figure thus shows the resultant motion from the superposition of the two perpendicular harmonic motions.

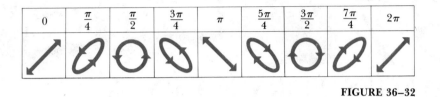

0	$\dfrac{\pi}{4}$	$\dfrac{\pi}{2}$	$\dfrac{3\pi}{4}$	π	$\dfrac{5\pi}{4}$	$\dfrac{3\pi}{2}$	$\dfrac{7\pi}{4}$	2π

FIGURE 36–32

37

IMAGES FORMED BY A SINGLE SURFACE

IN THE PRECEDING CHAPTER WE DISCUSSED THE PRINCIPLES THAT GOVERN reflection and refraction of a ray of light at a reflecting surface or an interface between two materials. We now consider the behavior of several rays that diverge from a common point and strike a reflecting or refracting surface. A central idea in this discussion is the concept of *image.* After reflection or refraction, the rays emerge with directions characteristic of having passed through some other common point called the *image point.* This discussion lays the foundation for analysis of many familiar optical instruments, including camera lenses, magnifiers, the human eye, microscopes, and telescopes. These instruments will be discussed in Chapter 38.

37–1 REFLECTION AT A PLANE SURFACE

Consider the situation shown in Fig. 37–1. Rays diverge from point P, called an **object point,** and are reflected or refracted (or both) at an interface between two transparent materials. The direction of each *reflected* ray is given by the law of reflection, and that of each transmitted or *refracted* ray by Snell's law. Here and in the following discussion we denote the two refractive indexes as n and n', without using the more elaborate subscript notation of Chapter 36.

Let us concentrate first on the *reflected* rays. Reflection can occur at an interface between two transparent materials or at a highly polished surface of an opaque material such as a metal, in which case the surface is usually called a *mirror.* We will use the term *mirror* to include all these possibilities.

As mentioned above, the key concept is that of **image.** After reflection the rays appear to have diverged from some common point, which we call the *image point.* In some cases, the rays coming from the surface after reflection really *do* meet at a common point and then diverge again after passing it; such a point is called a **real-image** point. In other cases the rays diverge *as though* they had passed through such a point, which is then called a **virtual-image** point. In many situations the image point exists only in an approximate sense, that is, when certain approximations are used in the calculations. This chapter and the next are devoted primarily to a study of the formation and properties of images.

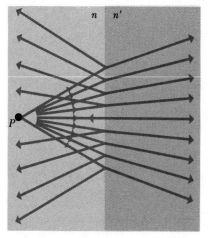

37–1 Reflection and refraction of rays at a plane interface between two transparent materials.

What's the difference between a real image and a virtual image?

868

Figure 37–2a shows two rays diverging from a point P at a distance s to the left of a plane mirror. The ray PV, incident normally on the mirror, returns along its original path. The ray PB, making an angle u with PV, strikes the mirror at an angle of incidence $\phi = u$ and is reflected at an angle $r = \phi = u$. When extended backward, the reflected ray intersects the normal to the surface at point P'. The angle u' is equal to r and hence to u.

Figure 37–2b shows several rays diverging from P. The construction of Fig. 37–2a can be repeated for each of them, and we see that the directions of the outgoing rays are the same as though they had originated at point P', which is therefore the *image* of P. The rays do not, of course, actually pass through this point; in fact, if the mirror is opaque, there is no light at all on the right side. Thus P' is a *virtual* image. Nevertheless, P' is a very real point in the sense that it describes the final directions of all the rays that originally diverged from P.

From the symmetry of the figure, we see that P' lies on a line perpendicular to the mirror passing through P, and that P and P' are equidistant from the mirror, on opposite sides. Thus *for a plane mirror, the image of an object point lies on the extension of the normal from the object point to the mirror, and the object and image points are equidistant from the mirror.*

Before proceeding further, we pause here to introduce some conventions that anticipate later situations where the object and image may be on either side of a reflecting or refracting surface. We adopt the following:

1. When the object is on the same side of the reflecting surface as the incoming light, the object distance s is positive; otherwise it is negative.

2. When the image is on the same side of the reflecting surface as the outgoing light, the image distance s is positive; otherwise it is negative.

For a mirror, the incoming and outgoing sides are always the same; in Fig. 37–2 they are both the left side. The rules above have been stated in this form so they may be applied also to *refracting* surfaces, where light comes in one side and goes out the opposite side.

In Fig. 37–2 the object distance s is *positive* because the object point P is on the incoming side—that is, the left side—of the reflecting surface. The image distance s' is *negative* because the image point P' is *not* on the outgoing side of the surface. Thus object and image distances are related simply by

$$s = -s'. \tag{37–1}$$

Next we consider an object of finite size, parallel to the mirror, represented by the arrow PQ in Fig. 37–3. Two of the rays from Q are shown, and *all* rays from Q diverge from its image Q' after reflection. Other points of the object PQ have image points between P' and Q'. As the figure shows, the object PQ and image $P'Q'$ have the same size and orientation: $y = y'$.

The ratio of image to object size, y'/y, is called the **lateral magnification** m; that is,

$$m = \frac{y'}{y}. \tag{37–2}$$

Thus we have found that *the lateral magnification for a plane mirror is unity*. When you look at yourself in a plane mirror, you do not look any larger or smaller than you really are.

Using the law of reflection to locate a virtual image

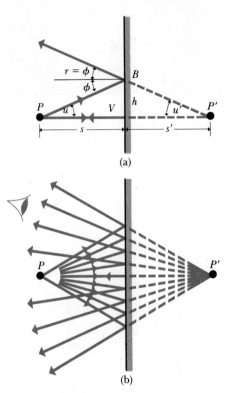

(a)

(b)

37–2 After reflection at a plane surface, all rays originally diverging from the object point P now diverge from the point P', although they do not *originate* at P'. Point P' is called the *virtual image* of point P. The eye sees some of the outgoing rays and perceives them as having come from point P'.

Lateral magnification is the ratio of image height to object height.

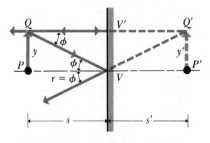

37–3 Construction for determining the height of an image formed by reflection at a plane surface.

The image formed by a plane mirror is reversed; the image of a right hand is a left hand, and so on. (Photo by Chip Clark.)

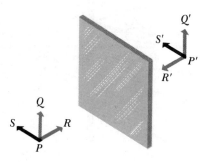

37–4 The image formed by a plane mirror is virtual, erect, and reversed, and is the same size as the object.

Some terms to describe spherical reflecting surfaces

In general, if an object transverse to the optic axis is represented by an arrow, its image may point in the same direction as the object or in the opposite direction. When, as in Fig. 37–3, the directions are the same, the image is called **erect;** if they are opposite, the image is **inverted.** The image formed by a plane mirror is always erect.

The three-dimensional virtual image of a three-dimensional object, formed by a plane mirror, is shown in Fig. 37–4. The images $P'Q'$ and $P'S'$ are parallel to their objects, but $P'R'$ is reversed relative to PR. The image of a three-dimensional object formed by a plane mirror is the same size as the object in both its longitudinal and transverse dimensions. The image and object are not identical in all respects, however, but are related in the same way as are a right hand and a left hand. To verify this statement, point your thumbs along PR and $P'R'$, your forefingers along PQ and $P'Q'$, and your middle fingers along PS and $P'S'$. When an object and its image are related in this way, the image is said to be **reversed.** When the transverse dimensions of object and image are in the same direction, the image is erect. Thus a plane mirror forms an erect but reversed image.

37–2 REFLECTION AT A SPHERICAL SURFACE

Next we consider the formation of an image of a *spherical* mirror. Figure 37–5a shows a spherical mirror of radius of curvature R, with its concave side facing the incident light. The **center of curvature** of the surface is at C. Point P is an object point; for the moment, assume that P is farther from V than is the center of curvature. The ray PV, passing through C, strikes the mirror normally and is reflected back on itself. Point V is called the **vertex** of the mirror, and the line PCV is the **optic axis.**

Ray PB, at an angle u with the axis, strikes the mirror at B, where the angle of incidence is ϕ and the angle of reflection is $r = \phi$. The reflected ray intersects the axis at point P'. We will show that *all* rays from P intersect the axis at the *same* point P', as in Fig. 37–5b, no matter what u is, provided that u is a *small* angle. Point P' is therefore the *image* of object point P. The object distance, measured from the vertex V, is s, and the image distance is s'. The object point P is on the same side as the incident light, so the object distance s is positive. The image point P' is on the same side as the reflected light, so the image distance s' is also positive.

Unlike the reflected rays in Fig. 37–2, the reflected rays in Fig. 37–5b actually intersect at point P' and then diverge from P' *as if* they had originated at this point. The image P' is called *real*, and a real image corresponds to a *positive* image distance.

Making use of the fact that an exterior angle of a triangle equals the sum of the two opposite interior angles, and considering the triangles PBC and $P'BC$ in Fig. 37–5a, we have

$$\theta = u + \phi, \qquad u' = \theta + \phi.$$

Eliminating ϕ between these equations gives

$$u + u' = 2\theta. \tag{37–3}$$

We may now introduce a sign convention for radii of curvature of spherical surfaces:

> When the center of curvature C is on the same side as the outgoing (reflected) light, the radius of curvature is positive; otherwise it is negative.

In Fig. 37–5, R is positive because the center of curvature C is on the same side of the mirror as the reflected light. This is always the case for reflection from the concave side of a surface; for a convex surface, the center of curvature is on the opposite end from the reflected light, and R is negative.

We may now compute the image distance s'. Let h represent the height of point B above the axis, and δ the short distance from V to the foot of this vertical line. We now write expressions for the tangents of u, u', and θ, remembering that s, s', and R are all positive quantities:

$$\tan u = \frac{h}{s - \delta}, \qquad \tan u' = \frac{h}{s' - \delta}, \qquad \tan \theta = \frac{h}{R - \delta}.$$

These trigonometric equations cannot be solved as simply as the corresponding algebraic equations for a plane mirror. *If the angle u is small,* however, the angles u' and θ will be small also. Since the tangent of a small angle is nearly equal to the angle itself (in radians), we can replace $\tan u'$ by u', and so on, in the equations above. Also if u is small, the distance δ can be neglected compared with s', s, and R. Hence, approximately, for small angles,

$$u = \frac{h}{s}, \qquad u' = \frac{h}{s'}, \qquad \theta = \frac{h}{R}.$$

Substituting in Eq. (37–3) and canceling h, we obtain

$$\frac{1}{s} + \frac{1}{s'} = \frac{2}{R} \qquad\qquad (37\text{–}4)$$

as a general relation among the three quantities s, s', and R. Since this equation does not contain the angle u, *all* rays from P making sufficiently small angles with the axis will, after reflection, intersect at P'. Such rays, nearly parallel to the axis, are called *paraxial rays.*

We must understand that Eq. (37–4), as well as many similar relations to be derived later in this chapter and the next, is the result of a calculation containing approximations and is valid only for paraxial rays. (The term **paraxial approximation** is often used for the approximations just described.) As the angle increases, the point P' moves closer to the vertex; a spherical mirror, unlike a plane mirror, does not form precisely a point image of a point object. This property of a spherical mirror is called **spherical aberration.**

If $R = \infty$, the mirror becomes *plane,* and Eq. (37–4) reduces to Eq. (37–1), previously derived for this special case.

Now suppose we have an object of finite size, represented by the arrow PQ in Fig. 37–6, perpendicular to the axis PV. The image of P formed by paraxial rays is at P'. Since the object distance for point Q is very nearly equal to that for the point P, the image of $P'Q'$ is nearly straight and is perpendicular to the axis. Note that object and image have different sizes, y and y', respectively, and that they have opposite orientation. We have defined the *lateral magnification m*

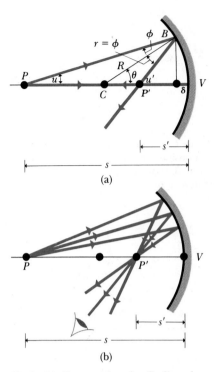

37–5 (a) Construction for finding the position of the image P' of a point object P, formed by a concave spherical mirror. (b) If the angle u is small, *all* rays from P intersect at P'. The eye sees some of the outgoing rays and perceives them as having come from P'.

A fundamental relation for spherical mirrors

What is the paraxial approximation?

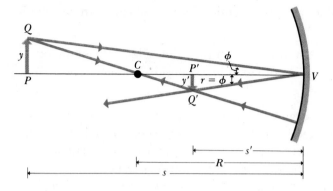

37–6 Construction for determining the position, orientation, and height of an image formed by a concave spherical mirror.

as the ratio of image to object size:

$$m = \frac{y'}{y}.$$

Because triangles PVQ and $P'VQ'$ in Fig. 37–6 are *similar*, we also have the relation $y/s = -y'/s'$. The negative sign is needed because object and image are on opposite sides of the optic axis; if y is positive, y' is negative. Hence

$$m = \frac{y'}{y} = -\frac{s'}{s}. \qquad (37\text{–}5)$$

A negative value of m indicates that the image is *inverted* relative to the object, as the figure shows. In cases to be considered later, where m may be either positive or negative, a positive value always corresponds to an erect image, a negative value to an inverted one. For a *plane* mirror, $s = -s'$, and hence $y' = y$, as we have already shown.

Although the ratio of image size to object size is called the *magnification*, the image formed by a mirror or lens may be *smaller* than the object. The magnification is then less than unity. The image formed by an astronomical telescope mirror or by a camera lens is much smaller than the object. It is interesting to note that for three-dimensional objects the ratio of image-to-object distances measured *along* the optic axis is different from the ratio of *lateral* distances (which we have called the lateral magnification). In particular, if m is a small fraction, the three-dimensional image of a three-dimensional object is reduced *longitudinally* much more than it is reduced *transversely*. Figure 37–7 illustrates this effect. Also, the image formed by a spherical mirror, like that of a plane mirror, is always reversed.

37–7 Schematic diagram of an object and its real, inverted, reduced image formed by a concave mirror.

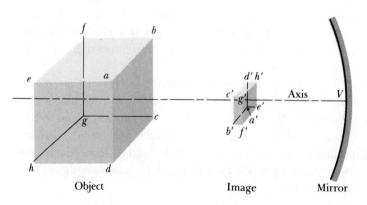

EXAMPLE 37-1 A concave mirror forms an image, on a wall 3 m from the mirror, of the filament of a headlight lamp 10 cm in front of the mirror. (a) What is the radius of curvature of the mirror? (b) What is the height of the image if the height of the object is 5 mm?

SOLUTION

a) Both object distance and image distance are positive, so from Eq. (37-4),

$$s = 10 \text{ cm}, \qquad s' = 300 \text{ cm},$$

$$\frac{1}{10 \text{ cm}} + \frac{1}{300 \text{ cm}} = \frac{2}{R},$$

$$R = 19.4 \text{ cm}.$$

The mirror in a headlight lamp forms a real image of the lamp filament.

Since the radius is positive, a concave mirror is required.

b) From Eq. (37-5),

$$m = \frac{y'}{y} = -\frac{s'}{s} = -\frac{300 \text{ cm}}{10 \text{ cm}} = -30.$$

The image is therefore inverted (m is negative) and is 30 times the height of the object, or (30)(5 mm) = 150 mm.

In Fig. 37-8a the *convex* side of a spherical mirror faces the incident light, so R is negative. Ray PB is reflected with the angle of reflection r equal to the angle of incidence ϕ, and the reflected ray, projected backward, intersects the axis at P'. As in the case of a concave mirror, *all* rays from P will, after reflection, diverge from the same point P', provided that the angle u is small; so P' is the image of P. The object distance s is positive, the image distance s' is negative, and the radius of curvature R is negative.

Figure 37-8b shows two rays diverging from the head of the arrow PQ, and the virtual image $P'Q'$ of this arrow. We leave it to you to show, by the same procedure used for a concave mirror, that

$$\frac{1}{s} + \frac{1}{s'} = \frac{2}{R},$$

Convex mirrors: The same formulas can be used, but the results are different.

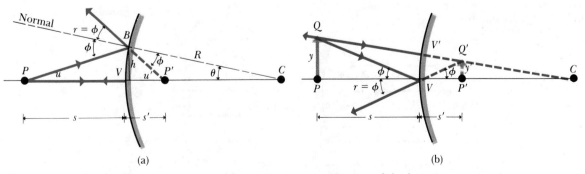

37-8 Construction for finding (a) the position and (b) the magnification of the image formed by a convex mirror.

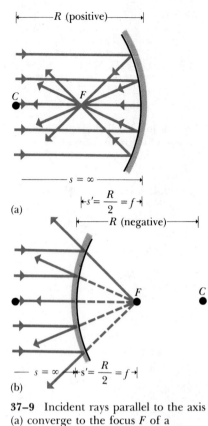

(a)

(b)

37–9 Incident rays parallel to the axis (a) converge to the focus F of a concave mirror, (b) diverge as though coming from the focus F of a convex mirror.

and that the lateral magnification is

$$m = \frac{y'}{y} = -\frac{s'}{s}.$$

These expressions are exactly the same as those for a concave mirror, as they must be when a consistent sign convention is adopted.

37–3 FOCUS AND FOCAL LENGTH

When an object point is at a very large distance from a spherical mirror, all rays from that point that strike the mirror are (in the paraxial approximation) parallel to one another. The object distance is $s = \infty$ and, from Eq. (37–4),

$$\frac{1}{\infty} + \frac{1}{s'} = \frac{2}{R}, \qquad s' = \frac{R}{2}.$$

When R is positive (concave mirror), the situation is as shown in Fig. 37–9a. A beam of incident parallel rays converges after reflection at a point F a distance $R/2$ from the vertex of the mirror. Point F is called the *focal point* or simply the **focus,** and its distance from the vertex, denoted by f, is called the **focal length.**

When R is negative (convex mirror), as in Fig. 37–9b, the image point is behind the mirror and f is negative. In that case the outgoing rays do not converge at a point but instead diverge as though they had come from the point F behind the mirror. In this case F is called a *virtual focus.*

The entire discussion may be reversed, as shown in Fig. 37–10. When the image distance s' is very large, the outgoing rays are parallel to the optic axis. The object distance s is then given by

$$\frac{1}{s} + \frac{1}{\infty} = \frac{2}{R}, \qquad s = \frac{R}{2}.$$

In Fig. 37–10a, rays coming in toward the mirror pass through the focus and diverge from it, and after reflection they are parallel to the optic axis. In Fig. 37–10b, the incoming rays are converging as though they would meet at the virtual focus F, and they are reflected parallel to the optic axis.

Thus for both concave and convex mirrors, the focal length f is related to the radius of curvature R by

$$f = \frac{R}{2}. \tag{37–6}$$

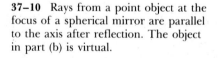

How is the focal length of a mirror related to its radius of curvature?

37–10 Rays from a point object at the focus of a spherical mirror are parallel to the axis after reflection. The object in part (b) is virtual.

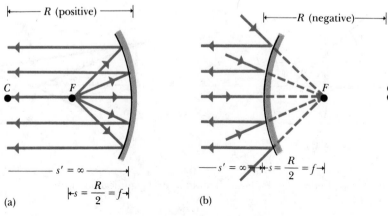

For a concave mirror both f and R are positive, and for a convex mirror both are negative.

The relation between object and image distances for a mirror, Eq. (37–4), may now be written as

$$\frac{1}{s} + \frac{1}{s'} = \frac{1}{f}. \tag{37–7}$$

37–4 GRAPHICAL METHODS

The position and size of the image formed by a mirror may be found by a simple graphical method. This method consists of finding the point of intersection, after reflection from the mirror, of a few particular rays diverging from some point of the object *not* on the mirror axis, such as point Q in Fig. 37–11. Then (neglecting aberrations) *all* rays from this point that strike the mirror will intersect at the same point. Four rays that we can always draw easily are shown in Fig. 37–11. These are called **principal rays.**

Principal rays: four rays you can always draw without guessing

1. *A ray parallel to the axis,* after reflection, passes through the focus of a concave mirror or appears to come from the focus of a convex mirror.

2. *A ray through (or proceeding toward) the focus* is reflected parallel to the axis.

3. *A ray along the radius* through C (extended if necessary) intersects the surface normally and is reflected back along its original path.

4. *A ray to the vertex* is reflected and forms equal angles with the optic axis.

Once we have found the position of the image point by means of the intersection of any two of these principal rays (1, 2, 3, 4), we can draw the path of any other ray from the same object point.

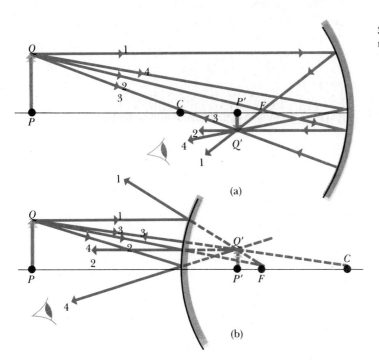

(a)

(b)

37–11 Rays used in the graphical method of locating an image.

Several examples of principal-ray diagrams for spherical mirrors, along with the corresponding calculations

EXAMPLE 37–2   A concave mirror has a radius of curvature of magnitude 20 cm. Find graphically the image of an object in the form of an arrow perpendicular to the axis of the mirror at each of the following object distances: 30 cm, 20 cm, 10 cm, and 5 cm. Check the construction by computing the size and magnification of the image.

SOLUTION   The graphical constructions are shown in the four parts of Fig. 37–12. Study each of these diagrams carefully, comparing each numbered ray with the description above. Several points are worth noting. First, in (b) the object and image distances are equal; ray 3 cannot be drawn in this case because a ray from Q through the center of curvature C does not strike the mirror. For the same reason, ray 2 cannot be drawn in (c); in this case the outgoing rays are parallel, corresponding to an infinite image distance. In (d) the outgoing rays have no real intersection point; they must be extended backward to find the point from which they appear to diverge, that is, from the *virtual-image point Q'*.

Measurement of the figures, with appropriate scaling, gives the following approximate image distances: (a) 15 cm, (b) 20 cm, (c) ∞ or $-\infty$, (d) -10 cm. To *compute* these distances, we first note that $f = R/2 = 10$ cm; then we use Eq. (37–7):

a)
$$\frac{1}{30 \text{ cm}} + \frac{1}{s'} = \frac{1}{10 \text{ cm}}, \qquad s' = 15 \text{ cm},$$

b)
$$\frac{1}{20 \text{ cm}} + \frac{1}{s'} = \frac{1}{10 \text{ cm}}, \qquad s' = 20 \text{ cm},$$

37–12   Image of an object at various distances from a concave mirror, showing principal-ray construction.

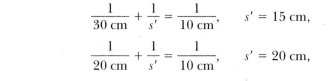

(a)

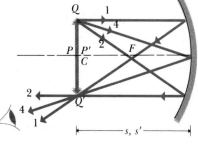

(b)

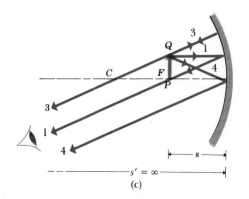

(c)

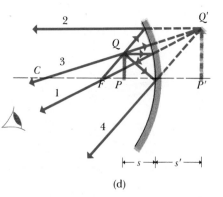

(d)

c)
$$\frac{1}{10 \text{ cm}} + \frac{1}{s'} = \frac{1}{10 \text{ cm}}, \qquad s' = \infty \text{ (or } -\infty\text{)},$$

d)
$$\frac{1}{5 \text{ cm}} + \frac{1}{s'} = \frac{1}{10 \text{ cm}}, \qquad s' = -10 \text{ cm}.$$

The lateral magnifications measured from the figures are approximately (a) $-\frac{1}{2}$, (b) -1, (c) ∞ or $-\infty$, and (d) $+2$. *Computing* the magnifications from Eq. (37–4), we find

An image may be at infinity; if so, it is infinitely large.

a)
$$m = -(15 \text{ cm})/(30 \text{ cm}) = -\tfrac{1}{2},$$

b)
$$m = -(20 \text{ cm})/(20 \text{ cm}) = -1,$$

c)
$$m = -(\pm\infty \text{ cm})/(10 \text{ cm}) = \mp\infty,$$

d)
$$m = -(-10 \text{ cm})/(5 \text{ cm}) = +2.$$

In (a) and (b) the image is inverted; in (d) it is erect.

PROBLEM-SOLVING STRATEGY: *Image formation by mirrors*

1. The principal-ray diagram is to geometrical optics what the free-body diagram is to mechanics! When the problem concerns image formation by a mirror, *always* draw a principal-ray diagram first. The same advice should be applied to lenses in the next chapter. It is usually best to orient your diagrams consistently, with the incoming rays traveling from left to right. Don't draw a million other rays at random; stick with the principal rays, the ones you know something about. If your principal rays don't converge at a real-image point, you may have to extend them backward to locate a virtual-image point. We recommend drawing the extensions with broken lines. Another useful aid is to color-code your principal rays, using red for (1), green for (2), black for (3), and blue for (4), in the list above; or something like that.

2. Pay careful attention to signs on object and image distances, radii of curvature, and object and image heights. Make certain you understand that the same set of sign rules works for all four cases in this chapter: reflection and refraction from plane and spherical surfaces. A negative sign on one of the quantities mentioned above *always* has significance; apply the equations carefully and with consistent use of the sign rules, and they will give you correct results!

3. Later in the chapter we will get into refraction as well as reflection. Remember that when a ray passes from a material of smaller index of refraction to one of larger index, it is always bent *toward* the normal; when going from larger to smaller, it is bent *away from* the normal.

37–5 REFRACTION AT A PLANE SURFACE

Suppose we want to find the image of a point object formed by rays *refracted* at a plane or spherical surface. The method is essentially the same as for reflection; the only difference is that Snell's law replaces the law of reflection. We let n represent the refractive index of the material on the "incoming" side of the surface and n' that of the material on the "outgoing" side, and we use the same convention of signs as in reflection.

Consider first a plane surface, shown in Fig. 37–13, and assume $n' > n$. This is not a necessary restriction, but the picture looks a little different if $n' < n$. A ray from the object point P toward the vertex V is incident normally and passes into the second material without deviation. A ray making an angle u with the axis is incident at B with an angle of incidence $\phi = u$. We find the

Applying the law of refraction to a ray bent at a plane interface between two materials

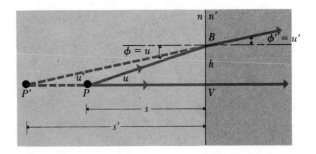

37–13 Construction for finding the position of the image P' of a point object P, formed by a refraction at a plane surface.

angle of refraction, ϕ', from Snell's law:

$$n \sin \phi = n' \sin \phi'.$$

The two rays both appear to come from the image point P' after refraction. From the triangles PVB and $P'VB$,

$$\tan \phi = \frac{h}{s}, \qquad \tan \phi' = \frac{h}{-s'}. \tag{37–8}$$

We must write $-s'$, since the image point is on the side *opposite* that of the refracted (outgoing) light.

Again, small-angle approximations simplify things greatly.

If the angle u is small, the angles ϕ, u', and ϕ' are small also and therefore approximately

$$\tan \phi = \sin \phi, \qquad \tan \phi' = \sin \phi'.$$

Then Snell's law can be written as

$$n \tan \phi = n' \tan \phi',$$

Image position and size for a plane refracting surface

and from Eq. (37–8), after canceling h, we have

$$\frac{n}{s} = -\frac{n'}{s'},$$

or

$$\frac{s'}{s} = -\frac{n'}{n}. \tag{37–9}$$

This is an *approximate* relation, *valid for paraxial rays only*. That is, a plane refracting surface does *not* image all rays from a point object at the same image point, but only those rays that are nearly perpendicular to the refracting surface.

Consider next the image of a finite object, as in Fig. 37–14. The two rays diverging from point Q appear to diverge from its image Q' after refraction,

37–14 Construction for determining the height of an image formed by refraction at a plane surface.

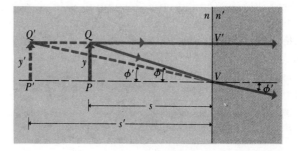

and $P'Q'$ is the image of the object PQ. As the figure shows, the object and the image are the same size, and so the lateral magnification is unity:

$$m = \frac{y'}{y} = 1. \qquad (37\text{–}10)$$

The image *distance* is greater than the object distance, but image and object are the same height.

Here is a familiar example of refraction at a plane surface. When you look vertically downward into the quiet water of a pond or swimming pool, the apparent depth is less than the actual depth. Figure 37–15 illustrates this situation. Two rays are shown diverging from a point Q at a distance s below the surface. Here, n' (air) is less than n (water), and the ray through V is deviated *away from* the normal. The rays after refraction appear to diverge from Q', and the arrow PQ, to an observer looking vertically downward, appears lifted to the position $P'Q'$. From Eq. (37–9),

$$s' = -\frac{n'}{n}s = -\frac{1.00}{1.33}s = -0.75\,s.$$

The apparent depth s' is therefore only three-fourths of the actual depth s. The same phenomenon accounts for the apparent sharp bend in an oar when a portion of it extends below a water surface. The submerged portion appears lifted above its actual position.

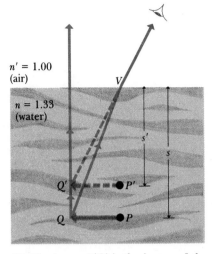

37–15 Arrow $P'Q'$ is the image of the underwater object PQ. The angles of the rays with the vertical are exaggerated for clarity.

Water is deeper than it appears to be.

37–6 REFRACTION AT A SPHERICAL SURFACE

Finally, we consider refraction at a spherical surface. In Fig. 37–16, P is an object point at a distance s to the left of a spherical surface of radius R, with center of curvature C. The refractive indexes at the left and right of the surface are n and n', respectively. Ray PV, incident normally at V, passes into the second material without deviation. Ray PB, making an angle u with the axis, is incident at an angle ϕ with the normal and is refracted at an angle ϕ'. These rays intersect at P' at a distance s' to the right of the vertex. The figure is drawn for the case $n' > n$.

We can prove that if the angle u is small, *all* rays from P intersect at the same point P', so P' is the *real image* of P. The object and image distances are both positive. The radius of curvature is positive also, because the center of curvature is on the side of the outgoing or refracted light.

Applying the law of refraction to a ray bent at a spherical surface

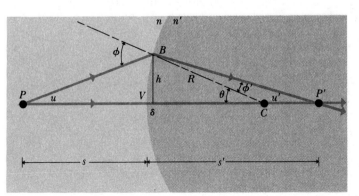

37–16 Construction for finding the position of the image P' of a point object P, formed by refraction at a spherical surface.

When we considered rays from P that are *reflected* at the surface, as in Fig. 37–8, the radius of curvature was negative. Thus the radius of curvature of a given surface has one sign for reflected light and the opposite sign for refracted light. This apparent inconsistency is resolved by noting that in both cases R is positive when the center of curvature is on the "outgoing" side of the reflecting or refracting surface and negative when it is not on the "outgoing" side.

From the triangles PBC and $P'BC$, we have

$$\phi = \theta + u, \qquad \theta = u' + \phi'. \tag{37-11}$$

From Snell's law,

$$n \sin \phi = n' \sin \phi'.$$

Small-angle approximations, one more time

Also, the tangents of u, u', and θ are

$$\tan u = \frac{h}{s + \delta}, \qquad \tan u' = \frac{h}{s' - \delta}, \qquad \tan \theta = \frac{h}{R - \delta}.$$

For paraxial rays, we may approximate both the sine and tangent of an angle by the angle itself and neglect the small distance δ. Snell's law then becomes

$$n\phi = n'\phi',$$

and, combining with the first of Eqs. (37–11), we obtain

$$\phi' = \frac{n}{n'}(u + \theta).$$

The basic formula for image formation by a spherical refracting surface

Substituting this in the second of Eqs. (37–11) gives

$$nu + n'u' = (n' - n)\theta.$$

Using the small-angle approximations for the tangents of u, u', and θ, and canceling h, we obtain

$$\frac{n}{s} + \frac{n'}{s'} = \frac{n' - n}{R}. \tag{37-12}$$

This equation does not contain the angle u, so the image distance is the same for all paraxial rays from P.

If the surface is plane, $R = \infty$ and this equation reduces to Eq. (37–9), already derived for the special case of a plane surface.

The magnification is found from the construction in Fig. 37–17. We draw two rays from point Q, one through the center of curvature C and the other incident at the vertex V. From the triangles PVQ and $P'Q'V$,

$$\tan \phi = \frac{y}{s}, \qquad \tan \phi' = \frac{-y'}{s'},$$

and from Snell's law

$$n \sin \phi = n' \sin \phi'.$$

Magnification of a spherical refracting surface

For small angles,

$$\tan \phi = \sin \phi, \qquad \tan \phi' = \sin \phi',$$

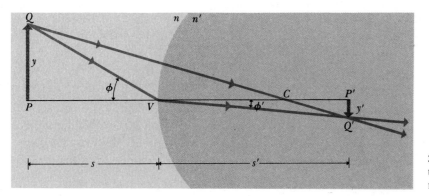

37–17 Construction for determining the height of an image formed by refraction at a spherical surface.

and hence

$$\frac{ny}{s} = -\frac{n'y'}{s'},$$

or

$$m = \frac{y'}{y} = -\frac{ns'}{n's}. \qquad (37\text{–}13)$$

This is the same relation previously derived for a *plane* surface. In that case, from Eq. (37–9), $ns' = -n's$, and $m = 1$.

EXAMPLE 37–3   One end of a cylindrical glass rod (Fig. 37–18) is ground to a hemispherical surface of radius $R = 20$ mm. Find the image distance of a point object on the axis of the rod, 80 mm to the left of the vertex. The rod is in air.

SOLUTION   We are given

$$n = 1, \quad n' = 1.5,$$
$$R = +20 \text{ mm}, \quad s = +80 \text{ mm}.$$

From Eq. (37–12),

$$\frac{1}{80 \text{ mm}} + \frac{1.5}{s'} = \frac{1.5 - 1}{+20 \text{ mm}},$$
$$s' = +120 \text{ mm}.$$

The image is therefore formed at the right of the vertex (s' is positive) and at a distance of 120 mm from it.

Suppose that the object is an arrow 1 mm high, perpendicular to the axis. Then, from Eq. (37–13),

$$m = -\frac{ns'}{n's} = -\frac{(1)(120 \text{ mm})}{(1.5)(80 \text{ mm})} = -1.$$

That is, the image is the same height as the object but is inverted.

EXAMPLE 37–4   The rod in Example 37–3 is immersed in water of index 1.33; the other quantities have the same values as before. Find the image distance (Fig. 37–19).

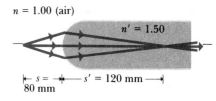

FIGURE 37–18

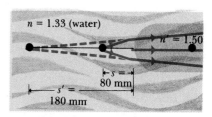

FIGURE 37–19

SOLUTION

$$\frac{1.33}{80 \text{ mm}} + \frac{1.5}{s'} = \frac{1.5 - 1.33}{+20 \text{ mm}},$$

$$s' = -185 \text{ mm}.$$

The fact that s' is negative means that the rays, after refraction by the surface, are not converging but *appear* to diverge from a point 180 mm to the *left* of the vertex. We have met a similar case before in the refraction of spherical waves by a plane surface and have called the point a *virtual image*. In this example, then, the surface forms a virtual image 180 mm to the left of the vertex.

Equations (37–12) and (37–13) can be applied to both convex and concave refracting surfaces when the appropriate sign convention is used, and they apply whether n' is greater or less than n. You should construct diagrams like Figs. 37–16 and 37–17, when R is negative and $n' < n$, and use them to derive Eqs. (37–12) and (37–13) for these cases.

SUMMARY

KEY TERMS

object point

image

real image

virtual image

lateral magnification

erect image

inverted image

reversed image

center of curvature

vertex

optic axis

paraxial approximation

focus

focal length

principal rays

When rays diverge from an object point P and are reflected or refracted, their directions after reflection or refraction are as though they had diverged from a point P' called the image point. If they really do converge at P' and diverge again beyond it, P' is a real image of P; if they only appear to have come from P', then it is a virtual image. Images can be either erect or inverted. An image formed by a plane or spherical mirror is always reversed; the image of a right hand is a left hand.

The lateral magnification in any reflecting or refracting situation is defined as the ratio of image height to object height:

$$m = \frac{y'}{y}. \tag{37–2}$$

When m is positive, the image is erect; when negative, inverted.

Relations derived in this chapter for the object and image positions in reflection or refraction with a spherical surface are valid only for rays nearly parallel to the optic axis; these are called paraxial rays, and the corresponding approximation is the paraxial approximation. The failure of nonparaxial rays to converge precisely at an image point is called spherical aberration.

The focus of a mirror is the point at which parallel rays converge after reflection from a concave mirror, or the point from which they appear to diverge after reflection from a convex mirror. Conversely, rays diverging from the focus of a concave mirror are parallel after reflection, and rays converging toward the focus of a convex mirror are parallel after reflection. The distance from the focus to the vertex is called the focal length, denoted as f.

All the formulas for object distance s and image distance s' for plane and spherical mirrors and refracting surfaces are summarized in Table 37–1. The equation for a plane surface can be obtained from the corresponding equation for a spherical surface by setting $R = \infty$.

TABLE 37–1

	Plane Mirror	Spherical Mirror	Plane Refracting Surface	Spherical Refracting Surface
Object and image distances	$\dfrac{1}{s} + \dfrac{1}{s'} = 0$	$\dfrac{1}{s} + \dfrac{1}{s'} = \dfrac{2}{R} = \dfrac{1}{f}$	$\dfrac{n}{s} + \dfrac{n'}{s'} = 0$	$\dfrac{n}{s} + \dfrac{n'}{s'} = \dfrac{n' - n}{R}$
Lateral magnification	$m = -\dfrac{s'}{s} = 1$	$m = -\dfrac{s'}{s}$	$m = -\dfrac{ns'}{n's} = 1$	$m = -\dfrac{ns'}{n's}$

The following set of sign rules can be applied uniformly to plane and spherical reflecting and refracting surfaces:

s is positive when the object is on the incoming side of surface, negative otherwise.

s' is positive when the image is on the outgoing side of surface, negative otherwise.

R is positive when the center of curvature is on the outgoing side of surface, negative otherwise.

m is positive when the image is erect, negative when inverted.

QUESTIONS

37–1 Can a person see a real image by looking backward along the direction from which the rays come? A virtual image? Can you tell by looking whether an image is real or virtual? How *can* the two be distinguished?

37–2 Why does a plane mirror reverse left and right but not top and bottom?

37–3 For a spherical mirror, if $s = f$, then $s' = \infty$, and the lateral magnification m is infinite. Does this make sense? If so, what does it mean?

37–4 According to the discussion of the preceding chapter, light rays are reversible. Are the formulas in Table 37–1 still valid if object and image are interchanged? What does reversibility imply with respect to the *forms* of the various formulas?

37–5 If a spherical mirror is immersed in water, does its focal length change?

37–6 For what range of object positions does a concave spherical mirror form a real image? What about a convex spherical mirror?

37–7 If a piece of photographic film is placed at the location of a real image, the film will record the image. Can this be done with a virtual image? How might one record a virtual image?

37–8 When a room has mirrors on two opposite walls, an infinite series of reflections can be seen. Discuss this phenomenon in terms of images. Why do the distant images appear darker?

37–9 When observing fish in an aquarium filled with water, one can see clearly only when looking nearly

perpendicularly to the glass wall; objects viewed at an oblique angle always appear blurred. Why? Do the fish have the same problem when looking at you?

37–10 Can an image formed by one reflecting or refracting surface serve as an object for a second reflection or refraction? Does it matter whether the first image is real or virtual?

37–11 A concave mirror (sometimes surrounded by lights) is often used as an aid for applying cosmetics to the face. Why is such a mirror always concave rather than convex? What considerations determine its radius of curvature?

37–12 A student claimed that one can start a fire on a sunny day by use of the sun's rays and a concave mirror. How is this done? Is the concept of image relevant? Could one do the same thing with a convex mirror?

37–13 A person looks at his reflection in the concave side of a shiny spoon. Is it right side up or inverted? What if he looks in the convex side?

37–14 In Example 37–2 (Section 37–4), there appears to be an ambiguity for the case $s = 10$ cm, as to whether s' is ∞ or $-\infty$, and as to whether the image is erect or inverted. How is this resolved? Or is it?

37–15 "See yourself as others see you." Can you do this with an ordinary plane mirror? If not, how *can* you do it?

37–16 The shadow formed under a tree by sunlight partially blocked by leaves ordinarily is sprinkled with small circles of light and irregular patterns. During a solar eclipse, however, the shadow is sprinkled instead with a pattern of overlapping crescents that copy the shape of the occluded sun. Why?

EXERCISES

Section 37–1 Reflection at a Plane Surface

37–1 A candle 6 cm tall is 80 cm to the left of a plane mirror. Where is the image formed by the mirror, and what is the height of this image?

37–2 The image of a tree just covers the length of a 5-cm plane mirror when the mirror is held 30 cm from the eye. The tree is 100 m from the mirror. What is its height?

Section 37–2 Reflection at a Spherical Surface

Section 37–3 Focus and Focal Length

Section 37–4 Graphical Methods

37–3 The diameter of the moon is 3480 km, and its distance from the earth is 386,000 km. Find the diameter of the image of the moon formed by a spherical concave telescope mirror of focal length 4 m.

37–4 A spherical concave shaving mirror has a radius of curvature of 30 cm.

a) What is the magnification when the face is 10 cm from the vertex of the mirror?

b) Where is the image?

37–5 An object 2 cm high is placed 12 cm away from a concave spherical mirror having radius of curvature of 20 cm.

a) Draw a principal-ray diagram showing formation of the image.

b) Determine the position, size, orientation, and nature of the image.

37–6 A concave mirror has a radius of curvature of 20 cm.

a) What is its focal length?

b) If the mirror is immersed in water (refractive index 1.33), what is its focal length?

37–7 Prove that the image formed of a real object by a convex mirror is always virtual, no matter what the object position.

37–8 Repeat Exercise 37–5 for the case where the mirror is convex.

37–9 An object 1.5 cm tall is placed 8 cm away from the vertex of a convex spherical mirror whose radius of curvature has magnitude 20 cm.

a) Draw a principal-ray diagram showing formation of the image.

b) Determine the position, size, orientation, and nature of the image.

Section 37–5 Refraction at a Plane Surface

37–10 A ray of light in air makes an angle of incidence of 45° at the surface of a sheet of ice. The ray is refracted within the ice at an angle of 30°.

a) What is the critical angle for the ice?

b) A speck of dirt is embedded 3 cm below the surface of the ice. What is its apparent depth when viewed at normal incidence?

37–11 A skin diver is 2 m below the surface of a lake. A bird flies overhead 3 m above the surface of the lake. When the bird is directly overhead, how far above the diver does it appear to be?

37–12 A tank whose bottom is a mirror is filled with water to a depth of 20 cm. A small object hangs motionless 8 cm under the surface of the water. What is the apparent depth of its *image* when viewed at normal incidence?

Section 37–6 Refraction at a Spherical Surface

37–13 Equations (37–12) and (37–13) were derived in the text for the case where R is positive and $n < n'$. (See Figs. 37–16 and 37–17.)

a) Carry through the derivation of these two equations for the case where $R > 0$ and $n > n'$.

b) Carry through the derivation for $R < 0$ and $n < n'$.

37–14 The end of a long glass rod 8 cm in diameter has a convex hemispherical surface 4 cm in radius. The refractive index of the glass is 1.50. Determine the position of the image if an object is placed on the axis of the rod at the following distances to the left of its end:

a) infinitely far,   b) 16 cm,

c) 4 cm.

37–15 The rod of Exercise 37–14 is immersed in a liquid. An object 60 cm from the end of the rod and on its axis is imaged at a point 100 cm inside the rod. What is the refractive index of the liquid?

37–16 The left end of a long glass rod 10 cm in diameter, of index 1.50, is ground and polished to a convex hemispherical surface of radius 5 cm. An object in the form of an arrow 1 mm long, at right angles to the axis of the rod, is located on the axis 20 cm to the left of the vertex of the convex surface. Find the position and magnification of the image of the arrow formed by paraxial rays incident on the convex surface.

37–17 Repeat Exercise 37–16 for the case where the end of the rod is ground to a *concave* hemispherical surface of radius 5 cm.

37–18 A small tropical fish is at the center of a spherical fish bowl 30 cm in diameter. Find its apparent position and magnification to an observer outside the bowl. The effect of the thin walls of the bowl may be neglected.

PROBLEMS

37–19 What is the size of the smallest vertical plane mirror in which an observer standing erect can see his full-length image?

37–20 An object is placed between two mirrors arranged at right angles to each other.

a) Locate all the images of the object.

b) Draw the paths of rays from the object to the eye of an observer.

37–21 A concave mirror is to form an image of the filament of a headlight lamp on a screen 4 m from the mirror. The filament is 5 mm high, and the image is to be 40 cm high.

a) How far in front of the vertex of the mirror should the filament be placed?

b) What should be the radius of curvature of the mirror?

37–22 An object is 16 cm from the center of a silvered spherical glass Christmas tree ornament 8 cm in diameter. What are the position and magnification of its image?

37–23 If the light striking a convex mirror does not diverge from an object point but instead converges toward a point at a (negative) distance s to the right of the mirror, this point is called a *virtual object*.

a) For a convex mirror having radius of curvature 10 cm, for what range of virtual-object positions is a real image formed?

b) What is the orientation of this real image?

c) Draw a principal-ray diagram showing formation of such an image.

37–24 A microscope is focused on the upper surface of a glass plate. A second plate is then placed over the first. In order to focus on the bottom surface of the second plate, the microscope must be raised 1 mm. In order to focus on the upper surface, it must be raised 2 mm *farther*. Find the index of refraction of the second plate. (This problem illustrates one method of measuring index of refraction.)

37–25 What should be the index of refraction of a transparent sphere so that paraxial rays from an infinitely distant object will focus at the vertex of the surface opposite the point of incidence?

37–26 A transparent rod 40 cm long is cut flat at one end and rounded to a hemispherical surface of 12-cm radius at the other end. A small object is embedded within the rod along its axis and halfway between its ends. When viewed from the flat end of the rod, the apparent depth of the object is 12.5 cm. What is its apparent depth when viewed from the curved end?

37–27 A solid glass hemisphere having a radius of 10 cm and a refractive index of 1.50 is placed with its flat face downward on a table. A parallel beam of light of circular cross section 1 cm in diameter travels directly downward and enters the hemisphere along its diameter.

a) What is the diameter of the circle of light formed on the table?

b) How does your result depend on the radius of the hemisphere?

37–28 In Fig. 37–20a the first focal length f is the value of s corresponding to $s' = \infty$; in Fig. 37–20b the second focal length f' is the value of s' when $s = \infty$.

a) Prove that $n/n' = f/f'$.

b) Prove that the general relation between object and image distance is

$$\frac{f}{s} + \frac{f'}{s'} = 1.$$

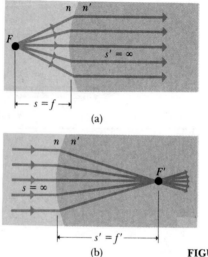

(a)

(b)   **FIGURE 37–20**

37–29 The longitudinal magnification is defined as $m' = ds'/ds$. It relates the longitudinal dimension of a small object to the longitudinal dimension of its image.

a) For a spherical mirror show that $m' = -m^2$. What is the significance of the fact that m' is *always* negative?

b) A wire frame in the form of a small cube 1 cm on a side is placed with its center on the axis of a concave mirror of radius of curvature 30 cm. The sides of the cube are all either parallel or perpendicular to the axis. The cube face toward the mirror is 60 cm to the left of the mirror vertex. Find

i) the location of the image of this face and of the opposite face of the cube;

ii) the lateral and longitudinal magnifications;

iii) the dimensions of each of the six cube faces of the image.

37–30 Refer to Problem 37–29. Show that the longitudinal magnification m' for refraction at a spherical surface is given by

$$m' = -\frac{n'}{n}m^2.$$

CHALLENGE PROBLEMS

37–31 You are sitting in your parked car and notice a jogger approach in the convex side mirror, which has a radius of curvature 2 m. If the jogger is running at a speed of 4 m·s^{-1}, how fast does the runner appear to be moving when she is

a) 10 m away?   b) 2 m away?

37–32 Two mirrors are placed together as shown in Fig. 37–21.

a) Show that a source in front of these mirrors and its two images lie on a circle.

b) Find the center of the circle.

c) Where should an observer stand to be able to see both images?

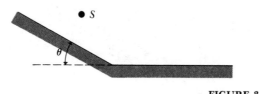

FIGURE 37–21

37–33 Spherical aberration is a blurring of the image formed by a mirror or lens because parallel rays striking the mirror far from the optic axis are focused at a different point than are rays near the axis. This problem is usually minimized by using only the center of a spherical mirror.

a) Show that for a spherical concave mirror the focus moves toward the mirror as the parallel rays move toward the outer edge of the mirror. (*Hint:* Derive an analytic expression for the distance from the vertex to the focus of the ray for a particular parallel ray, in terms of the radius of curvature R of the mirror and the angle θ. θ is the angle between the incident ray and the line connecting the center of curvature of the mirror and the point where the ray strikes the mirror.)

b) What value of θ produces a 2% change in the location of the focus, compared to the $\theta \approx 0$ value?

37–34 Parabolic mirrors are preferred in critical applications because they do not show any spherical aberration. (See Challenge Problem 37–33.) Show that for a mirror with a concave parabolic surface ($y = ax^2$), parallel rays are focused at $f = \frac{1}{4}a$ no matter how far from the axis they strike the mirror.

38

LENSES AND OPTICAL INSTRUMENTS

THE MOST FAMILIAR AND SIGNIFICANT SIMPLE OPTICAL DEVICE (EXCEPT for the plane mirror) is the *lens*. In this chapter we introduce the concepts of focus and focal length for a lens and then study the formation of images by lenses; we also develop graphical methods for analyzing image formation. Lenses are central to the operation of many familiar optical devices, including the human eye, magnifiers, projectors, and cameras. Other devices, such as telescopes and compound microscopes, use combinations of lenses or of lenses and mirrors. Here, as in the preceding chapter, the concept of *image* provides the key to analyzing and understanding these optical devices. We again base our analysis on the ray model of light, an adequate representation of many aspects of lens and mirror behavior.

38–1 THE THIN LENS

A lens is an optical system that includes two refracting surfaces. For now we concentrate primarily on lenses with two *spherical* surfaces sufficiently close together that the distance between them (the thickness of the lens) can be neglected. Such a device is called a **thin lens.** The behavior of a lens can be analyzed in detail by repeated application of the results of Section 37–6 for refraction by a single spherical surface. We postpone this derivation until Section 38–4 in order to move immediately into a discussion of the properties of thin lenses.

A lens of the type shown in Fig. 38–1 has the property that a beam of parallel rays converges, after passing through the lens, at a point F', as in Fig. 38–1a. Similarly, rays passing through point F emerge from the lens as a beam of parallel rays, as in Fig. 38–1b. The points F and F' are called the first and second *focal points*, or *foci* (plural form of **focus**), and the distance f is called the **focal length.** The concepts of focus and focal lengths are directly analogous to the same concepts used for spherical mirrors in Section 37–3. The two focal lengths in Fig. 38–1, both labeled f, are always equal for a thin lens, even when the two sides have different curvatures. We will derive this somewhat surprising result in Section 38–4. The central horizontal line is called the

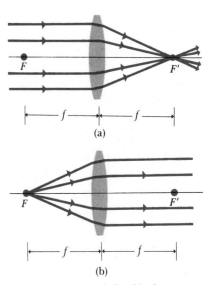

38–1 Focal points of a thin lens.

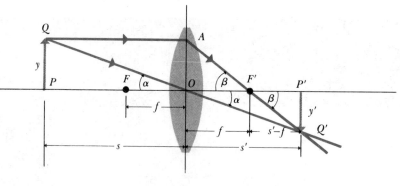

38–2 Construction used to find image position for a thin lens. The ray QAQ' is shown as bent at the midplane of the lens rather than at the two surfaces, to emphasize that the thickness of the lens is assumed to be very small.

optic axis, as with spherical mirrors. The centers of curvature of the two spherical surfaces lie on (and in fact define) the optic axis.

The focal length f depends on the refractive index of the lens material (and on that of the surrounding matter if it is not vacuum) and on the radii of curvature of the spherical surfaces. We will derive this relationship, called the *lensmaker's equation,* in Section 38–4.

Using the properties of the foci to derive the equation relating object and image positions for a thin lens

A thin lens forms an image of an object of finite size. Figure 38–2 shows how the position of the image may be calculated. Let s and s' be the object and image distances, respectively, and let y and y' be the object and image heights. This is the same notation we used in Chapter 37. Ray QA, parallel to the optic axis before refraction, passes through the second focus F' after refraction. Ray QOQ' passes undeflected straight through the center of the lens, because at the center the two surfaces are parallel and (we have assumed) very close together.

The angles labeled α in Fig. 38–2 are vertical angles and are equal. Therefore the two right triangles PQO and $P'Q'O$ are *similar,* and the ratios of corresponding sides are equal. Thus

$$\frac{y}{s} = -\frac{y'}{s'}, \quad \text{or} \quad \frac{y'}{y} = -\frac{s'}{s}. \tag{38–1}$$

(The negative sign arises because y' is negative, according to the convention used in Chapter 37.) Also, the two angles labeled β are equal, and the two right triangles OAF' and $P'Q'F'$ are similar, so

$$\frac{y}{f} = -\frac{y'}{s'-f}, \quad \text{or} \quad \frac{y'}{y} = -\frac{s'-f}{f}. \tag{38–2}$$

The basic object–image relation for a thin lens

We now equate Eqs. (38–1) and (38–2), divide by s', and rearrange to obtain

$$\frac{1}{s} + \frac{1}{s'} = \frac{1}{f}. \tag{38–3}$$

Magnification of a thin lens: a simple relationship

This analysis also gives us immediately the magnification $m = y'/y$ of the system; from Eq. (38–1),

$$m = -\frac{s'}{s}. \tag{38–4}$$

The negative sign tells us that when s and s' are both positive, the image is *inverted,* and y and y' have opposite signs.

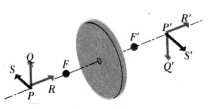

Equations (38–3) and (38–4) are the basic equations for thin lenses; their *form* is exactly the same as the corresponding equations for spherical mirrors, Eqs. (37–5) and (37–7). As we will see, the same sign conventions we used for spherical mirrors are also applicable to lenses.

The three-dimensional image of a three-dimensional object, formed by a lens, is shown in Fig. 38–3. Since point R is nearer the lens than point P, its image, from Eq. (38–3), is farther from the lens than is point P', and the image $P'R'$ points in the same direction as the object PR. Arrows $P'S'$ and $P'Q'$ are reversed in space, relative to PS and PQ. Although we speak of the image as "inverted," only its transverse dimensions are reversed.

We invite you to compare Fig. 38–3 with Fig. 37–4, showing the image formed by a plane mirror. Note that the image formed by a lens, although inverted, is *not* reversed. That is, if the object is a left hand, its image is also a left hand. This fact may be verified by pointing the left thumb along PR, the left forefinger along PQ, and the left middle finger along PS. A rotation of 180° about the thumb as an axis then brings the fingers into coincidence with $P'Q'$ and $P'S'$. In other words, *inversion* of an image is equivalent to a rotation of 180° about the lens axis.

38–3 A lens forms a three-dimensional image of a three-dimensional object.

A lens doesn't turn a right hand into a left hand.

38–2 DIVERGING LENSES

A bundle of parallel rays incident on the lens shown in Fig. 38–1 *converges* to a real image after passing through the lens. Thus the lens is called a **converging lens.** Because its focal length is a positive quantity, it is also called a *positive lens.*

A bundle of parallel rays incident on the lens shown in Fig. 38–4 *diverges* after refraction, and so the lens is called a **diverging lens.** Because its focal length is a negative quantity, it is also called a *negative lens.* The foci of a negative lens are reversed relative to those of a positive lens. The second focus F' of a negative lens is the point from which rays, originally parallel to the axis, *appear to diverge* after refraction, as in Fig. 38–4a. Incident rays converging toward the first focus F, as in Fig. 38–4b, emerge from the lens parallel to its axis.

Equations (38–3) and (38–4) apply both to negative and to positive lenses. Various types of lenses, both converging and diverging, are shown in Fig. 38–5. In Section 38–4 we will see that any lens thicker in the center than at the edges is a converging lens with positive f, and any lens thicker at the edges than in the center is a diverging lens with negative f.

What's the difference between a converging lens and a diverging lens?

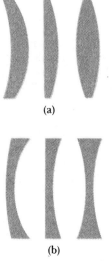

38–5 (a) Meniscus, plano-convex, and double-convex converging lenses. (b) Meniscus, plano-concave, and double concave diverging lenses. A converging lens is always thicker at its center than at its edges, and the reverse is true for all diverging lenses.

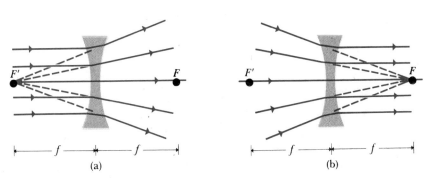

38–4 Focal points and focal length of a diverging lens.

38–3 GRAPHICAL METHODS

We can determine the position and size of an image formed by a thin lens with a graphical method very similar to the one used in Section 37–4 for spherical mirrors. Again we draw a few special rays, called **principal rays,** diverging from a point of the object that is *not* on the optic axis. The intersection of these rays, after passing through the lens, determines the position and size of the image. In using this graphical method, we will consider the entire deviation of a ray as occurring at the midplane of the lens, as shown in Fig. 38–6; this is consistent with the assumption that the distance between the lens surfaces is negligible.

The three principal rays whose paths can be traced easily are shown in Fig. 38–6:

1. *A ray parallel to the axis,* after refraction by the lens, passes through the second focal point of a converging lens or appears to come from the second focal point of a diverging lens.

2. *A ray through the center of the lens* is not appreciably deviated, since the two lens surfaces through which the central ray passes are very nearly parallel and close together if the lens is thin.

3. *A ray through (or proceeding toward) the first focal point* emerges parallel to the axis.

Once the position of the image point has been found by means of the intersection of any two rays 1, 2, and 3, the path of any other ray from the same point, such as ray 4 in Fig. 38–6, may be drawn. A few examples of this procedure are given in Fig. 38–7. The unnumbered rays in Fig. 38–7 are not principal rays.

We urge you to study each of these diagrams (drawn for several different object distances) very carefully, comparing each numbered ray with the above description. Several points are worth noting. In (d) the object is at the focus; ray 3 cannot be drawn because it does not pass through the lens. In (e) the

38–6 Principal-ray diagram, showing graphical method of locating an image. (a) A converging lens; (b) a diverging lens.

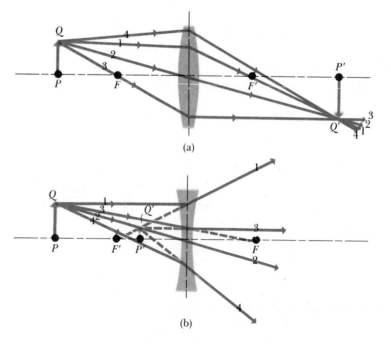

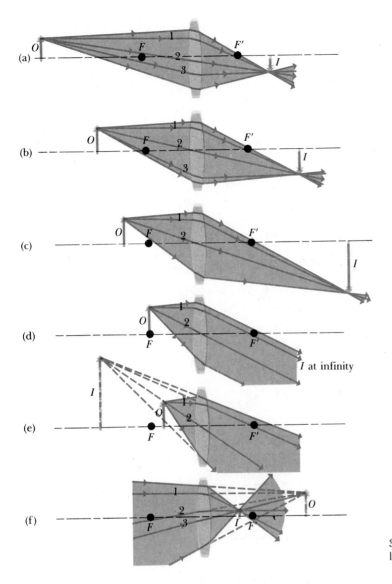

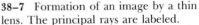

38-7 Formation of an image by a thin lens. The principal rays are labeled.

object distance is less than the focal length. The outgoing rays are divergent, and the *virtual image* is located by extending the outgoing rays backward. In this case the image distance s' is negative. Part (f) corresponds to a *virtual object*. The incoming rays are not diverging from a real-object point but are converging as though they would meet at the virtual-object point O on the right side. The object distance s is negative in this case. The image is real, and the image distance s' is positive and less than f.

EXAMPLE 38-1 A converging lens has a focal length of 20 cm. Find graphically the image location for an object at each of the following distances from the lens: 50 cm, 20 cm, 15 cm, and −40 cm. Determine the magnification in each case. Check your results by calculating the image position and magnification from Eqs. (38–3) and (38–4).

SOLUTION The appropriate principal-ray diagrams are shown in Fig. 38–7, parts (a), (d), (e), and (f). The approximate image distances, from measurements of these diagrams, are 35 cm, ∞, −40 cm, and 15 cm, and the approximate magnifications are $-\frac{2}{3}$, ∞, +3, and $+\frac{1}{3}$.

Calculating the image positions from Eq. (38–3), we find

$$\frac{1}{s} + \frac{1}{s'} = \frac{1}{f},$$

a) $\qquad \dfrac{1}{50 \text{ cm}} + \dfrac{1}{s'} = \dfrac{1}{20 \text{ cm}}, \qquad s' = 33.3 \text{ cm},$

d) $\qquad \dfrac{1}{20 \text{ cm}} + \dfrac{1}{s'} = \dfrac{1}{20 \text{ cm}}, \qquad s' = \infty,$

e) $\qquad \dfrac{1}{15 \text{ cm}} + \dfrac{1}{s'} = \dfrac{1}{20 \text{ cm}}, \qquad s' = -60 \text{ cm},$

f) $\qquad \dfrac{1}{-40 \text{ cm}} + \dfrac{1}{s'} = \dfrac{1}{20 \text{ cm}}, \qquad s' = 13.3 \text{ cm}.$

The graphical results are fairly close to these except for (e), where the precision of the diagram is limited by the fact that the rays extended backward have nearly the same direction.

From Eq. (38–4), the magnifications are

$$m = -\frac{s'}{s},$$

a) $\qquad m = -\dfrac{33.3 \text{ cm}}{30 \text{ cm}} = -\dfrac{2}{3},$

d) $\qquad m = -\dfrac{\infty}{20 \text{ cm}} = -\infty,$

e) $\qquad m = -\dfrac{-60 \text{ cm}}{15 \text{ cm}} = +4,$

f) $\qquad m = -\dfrac{13.3 \text{ cm}}{-40 \text{ cm}} = +\dfrac{1}{3}.$

Several principal-ray diagrams for a converging lens, along with the calculated image positions and magnifications

---

PROBLEM-SOLVING STRATEGY: *Image formation by a thin lens*

1. The strategy outlined at the end of Section 37–4 is equally applicable here, and we suggest you review it now. Always begin with a principal-ray diagram; orient your diagrams consistently so that light travels from left to right. For a lens there are only three principal rays, compared to four for a mirror. Don't just sketch these diagrams; draw the rays with a ruler, measuring the distances carefully. Draw them so they bend at the midplane of the lens, as shown in Fig. 38–6. Be sure to draw *all three* principal rays. The intersection of any two determines the image, but if the third doesn't pass through the same intersection point, you know you have made a mistake. Some redundancy can be a virtue! When there is a virtual image, you will have to extend the outgoing rays backward; in that case the image lies on the incoming side of the lens. Don't let that alarm you.

2. The same set of sign rules used in Chapter 37 is still applicable for thin lenses, and we will extend it in the next section to include radii of curvature of lens surfaces. Be extremely careful to get your signs right and to interpret the signs of results correctly.

3. Always determine the image position and size *both* graphically and by calculating. This procedure gives an extremely useful consistency check.

4. An additional point we will encounter in the next section is that the *image* from one lens or mirror may serve as the *object* for another. In that case, be careful in finding the object and image *distances* for this intermediate image; be sure you include the correct distance between the two elements (lenses and/or mirrors).

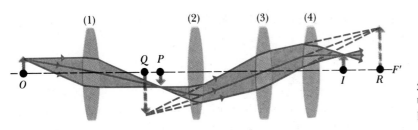

38–8 The object for each lens, after the first, is the image formed by the preceding lens.

38–4 IMAGES AS OBJECTS

An image formed by one lens or refracting surface can serve as the object for a second lens or refracting surface. Many optical systems, such as camera lenses, microscopes, and telescopes, use more than one lens. In each case the image formed by any one lens serves as the object for the next lens. Figure 38–8 shows various possibilities. Lens 1 forms a real image at P of a real object at O. This real image serves as a real object for lens 2. The virtual image at Q formed by lens 2 is a real object for lens 3. If lens 4 were not present, lens 3 would form a real image at R. Although this image is never formed, it serves as a **virtual object** for lens 4, which forms a real image at I.

These considerations can also be applied to images formed by individual refracting surfaces. In particular, we can derive the thin-lens equation, Eq. (38–3), by repeated application of the single-surface equation, Eq. (37–12). For this derivation, consider the situation of Fig. 38–9. The figure shows several rays diverging from point Q of an object PQ. The first surface forms a virtual image of Q at Q'. This virtual image serves as a real object for the second surface of the lens, which forms a real image of Q' at Q''. Distance s_1 is the object distance for the first surface; s_1' is the corresponding image distance. The object distance for the second surface is s_2, equal to the sum of s_1' and the lens thickness t, and s_2' is the image distance for the second surface.

If the lens is thin enough so that its thickness t is negligible in comparison with the distances s_1, s_1', s_2, and s_2', we may assume that s_1' equals $-s_2$ and measure object and image distances from either vertex of the lens. We will also assume that the medium on both sides of the lens is air, with index of refraction 1.00. For the first refraction, Eq. (37–12) becomes

$$\frac{1}{s_1} + \frac{n}{s_1'} = \frac{n-1}{R_1}.$$

Refraction at the second surface yields the equation

$$\frac{n}{s_2} + \frac{1}{s_2'} = \frac{1-n}{R_2}.$$

What is a virtual object?

The object–image relation for a lens, in terms of the properties of the lens

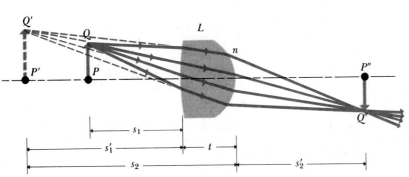

38–9 The image formed by the first surface of a lens serves as the object for the second surface.

Adding these two equations, and remembering that the lens is so thin that $s_2 = -s_1'$, we find

$$\frac{1}{s_1} + \frac{1}{s_2'} = (n - 1)\left(\frac{1}{R_1} - \frac{1}{R_2}\right).$$

Since s_1 is the object distance for the thin lens and s_2' is the image distance, the subscripts may be omitted, and we finally obtain

$$\frac{1}{s} + \frac{1}{s'} = (n - 1)\left(\frac{1}{R_1} - \frac{1}{R_2}\right). \tag{38-5}$$

The usual sign conventions apply to this equation. In Fig. 38–10, s, s', and R_1 are positive quantities, but R_2 is negative.

Comparing Eq. (38–5) with our previous form of the thin-lens equation, Eq. (38–3), we see that the focal length is given by

$$\frac{1}{f} = (n - 1)\left(\frac{1}{R_1} - \frac{1}{R_2}\right), \tag{38-6}$$

which is known as the **lensmaker's equation.** Thus in this derivation we obtain the thin-lens equation and also find how the focal length is related to the refractive index and the radii of curvature of the surfaces.

We can also obtain the focal length directly from Eq. (38–5) by recalling that the focus is the image of an infinitely distant object; when $s = \infty$, $s' = f$. Inserting these values into Eq. (38–5) yields Eq. (38–6) immediately.

How to calculate the focal length of a lens from its shape and material

---

EXAMPLE 38–2 In Fig. 38–10, let the absolute magnitudes of the radii of curvature of the lens surfaces be 20 cm and 5 cm, respectively. Since the center of curvature of the first surface is on the side of the outgoing light, $R_1 = +20$ cm, and since that of the second surface is not, $R_2 = -5$ cm. Let $n = 1.50$. Then

$$\frac{1}{f} = (1.50 - 1)\left(\frac{1}{20 \text{ cm}} - \frac{1}{-5 \text{ cm}}\right),$$

$$f = +8 \text{ cm}.$$

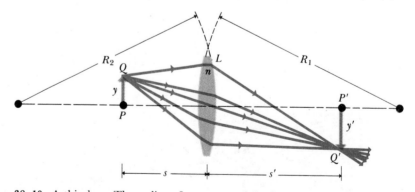

38–10 A thin lens. The radius of curvature of the first surface is R_1, and for the second surface R_2. In this case R_1 is positive but R_2 is negative. The focal length f is positive, and the lens is a converging lens.

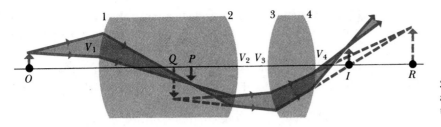

38–11 The object for each surface, after the first, is the image formed by the preceding surface.

Equation (38–6) can be generalized to the situation where the lens is immersed in a material having a refractive index greater than unity. We invite you to work out the lensmaker's equation for this more general situation.

An example of formation of successive images: The image from each surface is the object for the next.

We can extend the same method of analysis to systems consisting of several *thick* lenses; an example is shown in Fig. 38–11. The arrow at point O represents a small object at right angles to the optic axis. A narrow cone of rays diverging from the head of the arrow is traced through the system. Surface 1 forms a real image of the arrow at point P. Distance OV_1 is the object distance for the first surface, and distance V_1P is the image distance. Both distances are positive.

The image at P, formed by surface 1, serves as the object for surface 2. The object distance is PV_2 and is positive, since the direction from P to V_2 is the same as that of the incident light. The second surface forms a virtual image at point Q. The image distance is V_2Q and is negative because the direction from V_2 to Q is opposite to that of the refracted light.

The image at Q, formed by surface 2, serves as the object for surface 3. The object distance is QV_3 and is positive. The image at Q, although virtual, constitutes a *real object* so far as surface 3 is concerned. The rays incident on surface 3 are rendered converging and, except for the interposition of surface 4, would converge to a real image at point R. Even though this image is never formed, distance V_3R is the image distance for surface 3 and is positive.

The rays incident on surfaces 1, 2, and 3 have all been diverging, and the object distance has been the distance from the surface to the point from which the rays were actually or apparently diverging. The rays incident on surface 4, however, are *converging*, and there is no point at the left of the vertex from which they diverge or appear to diverge. The *image at R, toward which the rays are converging,* is the object for surface 4, and since the direction from R to V_4 is opposite to that of the incident light, the object distance RV_4 is negative. The image at R is called a *virtual object* for surface 4. In general, whenever a *converging* cone of rays is incident on (and interrupted by) a surface, the point toward which the rays are converging serves as the object, the object distance is negative, and the point is called a virtual object.

Finally, surface 4 forms a real image at I; the image distance is V_4I and is positive.

In our analysis of compound microscopes and telescopes in Sections 38–11 and 38–12, we will see examples where the image formed by a lens or mirror serves as the object for another lens.

38–5 LENS ABERRATIONS

The relatively simple equations we have derived for object and image distances, focal lengths, radii of curvature, and so on, have been based on the *paraxial approximation;* all rays were assumed to be *paraxial,* that is, to make

Lens aberrations: What happens when the paraxial approximation isn't good enough?

small angles with the axis. In general, however, a lens must form images not only of points on its axis but also of points that lie off the axis. Furthermore, because of the finite size of the lens, the cone of rays that forms the image of any point is of finite size. Nonparaxial rays proceeding from a given object point *do not*, in general, all intersect at precisely the same point after refraction by a lens, and the image formed by these rays is never a perfectly sharp one. Furthermore, the focal length of a lens depends on its index of refraction, which varies with wavelength. Therefore, if the light proceeding from an object is a mixture of wavelengths, different wavelengths are imaged at different points.

The departures of an actual image from the predictions of simple theory are called **aberrations.** Those caused by the variation of index with wavelength are the **chromatic aberrations.** Others, which would arise even if the light were monochromatic, are **monochromatic aberrations.** Lens aberrations are not caused by faulty construction of the lens, such as the failure of its surfaces to conform to a truly spherical shape, but are inevitable consequences of the laws of refraction at spherical surfaces.

Monochromatic aberrations are all related to the limitations of the paraxial approximation. *Spherical aberration* is the failure of rays from a point object on the optic axis to converge to a point image. Instead, the rays converge within a circle of minimum radius, called *the circle of least confusion*, and then diverge again, as shown in Fig. 38–12. The corresponding effect for points off the axis produces images that are comet-shaped figures rather than circles; this effect is called *coma*.

Astigmatism is the imaging of a point off the axis as two perpendicular *lines*. In this aberration the rays from a point object converge, at a certain distance from the lens, to a line in the plane defined by the optic axis and the object point. At a somewhat different distance from the lens, they converge to a second line *perpendicular* to this plane. The circle of least confusion appears between these two positions at a location that depends on the object point's distance from the axis as well as its distance from the lens. As a result, object points lying in a plane are, in general, imaged not in a plane but in some curved surface; this effect is called *curvature of field*. Finally, the image of a straight line that does not pass through the optic axis may be curved. As a result, the image of a square with the axis through its center may resemble a barrel (sides bent outward) or a pincushion (sides bent inward). This effect, called *distortion*, is not related to lack of sharpness of the image but results from a change in lateral magnification with distance from the axis.

Chromatic aberrations result directly from the variation of index of refraction with wavelength. When an object is illuminated with white light containing a mixture of wavelengths, different wavelengths are imaged at different points. The magnification of a lens also varies with wavelength; this effect is

Lens aberrations show themselves in several interesting ways.

Chromatic aberrations: What happens when the index of refraction varies with wavelength?

38–12 Spherical aberration. The circle of least confusion is shown by C–C.

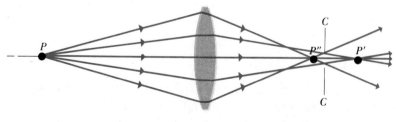

responsible for the rainbow-fringed images seen with inexpensive binoculars or telescopes.

It is impossible to eliminate all these aberrations from a single lens, but in a compound lens of several elements, the aberrations of one element may partially cancel those of another element. Design of such lenses is an extremely complex problem, which has been aided greatly in recent years by the use of computers. It is still impossible to eliminate all aberrations, but it *is* possible to decide which ones are most troublesome for a particular application and to design accordingly.

38–6 THE EYE

The essential parts of the human eye, considered as an optical system, are shown in Fig. 38–13. The eye is nearly spherical in shape and about 2.5 cm in diameter. The front portion is somewhat more sharply curved and is covered by a tough, transparent membrane C, called the *cornea*. The region behind the cornea contains a liquid A called the *aqueous humor*. Next comes the *crystalline lens L*, which is a capsule containing a fibrous jelly, hard at the center and progressively softer at the outer portions. The crystalline lens is held in place by ligaments that attach it to the ciliary muscle M. Behind the lens, the eye is filled with a thin, watery jelly V, called the *vitreous humor*. The indices of refraction of both the aqueous humor and the vitreous humor are nearly equal to that of water, about 1.336. The crystalline lens, while not homogeneous, has an average index of 1.437. This is not very different from the indices of the aqueous and vitreous humors; most of the refraction of light entering the eye occurs at the cornea.

Refraction at the cornea and the surfaces of the lens produces a *real image* of the object being viewed; the image is formed on the light-sensitive *retina R*, lining the rear inner surface of the eye. The *rods* and *cones* in the retina act like an array of miniature photocells; they sense the image and transmit it via the *optic nerve O* to the brain. Vision is most acute in a small central region called the *fovea centralis Y*, about 0.25 mm in diameter.

In front of the lens is the *iris*, containing an aperture of variable diameter called the *pupil P*, which opens and closes to adapt to changing light intensity. The receptors of the retina also have intensity-adaptation mechanisms.

For an object to be seen sharply, the image must be formed exactly at the location of the retina. The lens-to-retina distance, corresponding to s', does not change, but the eye accommodates to different object distances s by changing the focal length of its lens. When the ciliary muscle surrounding the lens contracts, the lens bulges and the radii of curvature of its surfaces decrease, decreasing the focal length. For the normal eye, an object at infinity is sharply focused when the ciliary muscle is relaxed, and maximum tension in this muscle provides the appropriate focal length for an object at a distance of about 25 cm. This process is called *accommodation*.

The extremes of the range over which distinct vision is possible are known as the *far point* and the *near point* of the eye. The far point of a normal eye is at infinity. The position of the near point depends on the amount by which the curvature of the crystalline lens may be increased. The range of accommodation gradually diminishes with age as the crystalline lens loses its flexibility. For this reason, the near point gradually recedes as one grows older. This recession of the near point is called *presbyopia*. Following is a table of the

The eye is an optical instrument: using geometrical optics to analyze its image formation.

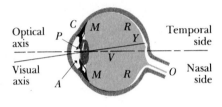

38–13   The eye.

How can the eye focus on objects at varying distances?

Why can't older people focus on close objects?

approximate average position of the near point at various ages:

Age (years)	Near Point (cm)
10	7
20	10
30	14
40	22
50	40
60	200

For example, an average person 50 years of age cannot focus on an object closer than about 40 cm.

38–7 DEFECTS OF VISION

Several common defects of vision result from incorrect distance relations between the parts of the optical system of the eye. A normal eye forms an image on the retina of an object at infinity when the eye is relaxed, as in Fig. 38–14a. In the *myopic* (nearsighted) eye, the eyeball is too long from front to back in comparison with the radius of curvature of the cornea, and rays from an object at infinity are focused in front of the retina. The most distant object for which an image can be formed on the retina is then nearer than infinity. In the *hyperopic* (farsighted) eye, the eyeball is too short, and the image of an infinitely distant object is behind the retina. The myopic eye produces too much convergence in a parallel bundle of rays for an image to be formed on the retina; the hyperopic eye, not enough convergence.

Astigmatism refers to a defect in which the surface of the cornea is not spherical but is more sharply curved in one plane than another. Astigmatism makes it impossible, for example, to focus clearly on the horizontal and vertical bars of a window at the same time.

These defects can be corrected by the use of corrective lenses ("glasses" or contact lenses). The near point of either a presbyopic or a hyperopic eye is farther from the eye than normal. To see clearly an object at normal reading distance (usually assumed to be 25 cm), we must place in front of the eye a lens of such focal length that it forms a virtual image of the object at or beyond the near point. Thus the function of the lens is not to make the object appear larger, but in effect to move the object farther away from the eye to a point where a sharp retinal image can be formed.

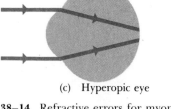

(a) Normal eye

(b) Myopic eye

(c) Hyperopic eye

38–14 Refractive errors for myopic (nearsighted) and hyperopic (farsighted) eye viewing a very distant object.

Using lenses to correct for farsightedness

EXAMPLE 38–3   The near point of a certain hyperopic eye is 100 cm in front of the eye. What lens should be used to see clearly an object 25 cm in front of the eye?

SOLUTION   We want the lens to form an image of the object at a location corresponding to the near point of the eye, 100 cm from it. Thus we have

$$s = +25 \text{ cm}, \qquad s' = -100 \text{ cm},$$
$$\frac{1}{f} = \frac{1}{s} + \frac{1}{s'} = \frac{1}{+25 \text{ cm}} + \frac{1}{-100 \text{ cm}},$$
$$f = +33 \text{ cm}.$$

That is, a converging lens of focal length 33 cm is required.

The far point of a *myopic* eye is nearer than infinity. To see clearly objects beyond the far point, a lens must be used that will form an image of such objects not farther from the eye than the far point.

EXAMPLE 38-4 The far point of a certain myopic eye is 1 m in front of the eye. What lens should be used to see clearly an object at infinity?

SOLUTION Assume the image to be formed at the far point. Then

$$s = \infty, \qquad s' = -100 \text{ cm},$$

$$\frac{1}{f} = \frac{1}{s} + \frac{1}{s'} = \frac{1}{\infty} + \frac{1}{-100 \text{ cm}},$$

$$f = -100 \text{ cm}.$$

A *diverging* lens of focal length 100 cm is required.

Astigmatism is corrected by means of a *cylindrical* lens. The curvature of the cornea in a horizontal plane may have the proper value such that rays from infinity are focused on the retina. In the vertical plane, however, the curvature may not be sufficient to form a sharp retinal image. When a cylindrical lens with its axis horizontal is placed before the eye, the rays in a horizontal plane are unaffected, while the additional convergence of the rays in a vertical plane now causes these to be sharply imaged on the retina.

Using cylindrical surfaces to correct astigmatism

Optometrists describe the converging or diverging effect of lenses in terms not of the focal length, but of its *reciprocal*. The reciprocal of the focal length of a lens is called its *power*. If the focal length is in meters, the power is in *diopters*. Thus the power of a positive lens whose focal length is 1 m is 1 diopter; if the focal length is negative, the power is negative also. In the two examples above, the required powers are +3.0 diopter and −1.0 diopter, respectively.

A diopter is not an ancient flying dinosaur.

38-8 THE MAGNIFIER

The apparent size of an object is determined by the size of its retinal image. If the eye is unaided, this size depends on the *angle* subtended by the object at the eye, called its *angular size*. To examine a small object in detail, we bring it close to the eye in order to make the subtended angle and the retinal image as large as possible. Since the eye cannot focus sharply on objects closer than the near point, a particular object subtends the maximum possible viewing angle at an unaided eye when placed at this point. (We assume here that the near point is 25 cm from the eye.)

The magnifying glass: getting the image in the right place

When a converging lens is placed in front of the eye, it forms a virtual image farther from the eye than the object. Thus the object may be moved closer to the eye, and the angular size of the image may be substantially larger than the angular size of the object at 25 cm without the lens. A lens used in this way is called a *magnifying glass*, or simply a *magnifier*. Viewing of the virtual image is most comfortable when it is placed at infinity, and in the following discussion we assume this is done.

The principle of the magnifier is shown in Fig. 38-15. In (a) the object is at the near point, where it subtends an angle *u* at the eye. In (b) a magnifier in

Angular magnification: making things look bigger than they are

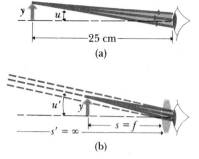

(a)

(b)

38–15 A simple magnifier.

front of the eye forms an image at infinity, and the angle subtended at the magnifier is u'. The **angular magnification** M (not to be confused with the *lateral magnification m*) is defined as the ratio of the angle u' to the angle u. We can find the value of M as follows. From Fig. 38–15, u and u' are given (in radians) by

$$u = \frac{y}{25 \text{ cm}} \quad \text{(approximately)},$$

$$u' = \frac{y}{f} \quad \text{(approximately)}.$$

Hence

$$M = \frac{u'}{u} = \frac{y/f}{y/25} = \frac{25}{f} \quad (f \text{ in centimeters}). \tag{38–7}$$

It appears at first that we can make the angular magnification as large·as we like by decreasing the focal length f. In fact, the aberrations of a simple double-convex lens set a limit to M of about 2X or 3X. If these aberrations are corrected, the magnification may be carried as high as 20X. It is not ordinarily practical to correct aberrations adequately for greater magnification; instead, greater magnification is attained by use of a compound microscope, discussed in Section 38–11.

38–9 THE CAMERA

The essential elements of a camera are a lens equipped with a shutter, a light-tight enclosure, and a light-sensitive film to record an image. An example is shown in Fig. 38–16. The lens forms a real image, on the film, of the object being photographed, just as the lens of the human eye forms a real image on the retina, the eye's "film." The lens may be moved closer to or farther from the film to provide proper image distances for various object distances. All but the most inexpensive lenses have several elements to permit partial correction of various aberrations. A typical example is the Zeiss "Tessar" design, shown in Fig. 38–17.

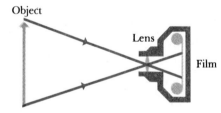

38–16 Essential elements of a camera.

Controlling the amount of light that strikes the film

In order for the image to be recorded properly on the film, the total light energy per unit area reaching the film (the "exposure") must fall within certain limits; this is controlled by the *shutter* and the *lens aperture*. The shutter controls the time during which light enters the lens, typically adjustable in steps corresponding to factors of about two, from 1 s to $\frac{1}{1000}$ s or thereabouts. The light-gathering capacity of the lens is proportional to its effective area; this may be varied by means of an adjustable aperture, or *diaphragm*, which is a nearly circular hole of variable diameter. The aperture size is usually described in terms of its "f-number," which is the number by which the focal length of the lens must be divided to obtain the diameter of the aperture. Hence a lens having $f = 50$ mm and an aperture diameter of 25 mm is said to have an aperture of $f/2$, or an f-number of 2.

Because the light-gathering capacity of a lens is proportional to its area and thus to the *square* of its diameter, changing the diameter by a factor of $\sqrt{2}$ corresponds to a factor of two in exposure. Thus adjustable apertures usually have scales labeled with successive numbers related by factors of $\sqrt{2}$, such as

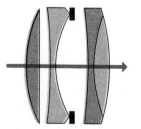

38–17 Zeiss "Tessar" lens design.

$$f/2, \quad f/2.8, \quad f/4, \quad f/5.6, \quad f/8, \quad f/11, \quad f/16,$$

and so on. The larger numbers represent smaller apertures and exposures, and each step corresponds to a factor of two in exposure.

The choice of focal length for a camera lens depends on the film size and the desired angle of view, or *field*. For the popular 35-mm cameras, with image size of 24×36 mm, the normal lens is usually about 50 mm in focal length and permits an angle of about 45°. A longer focal-length lens, used with the same film size, provides a smaller angle of view and a larger image of part of the object, compared with a normal lens. This gives the impression that the camera is closer than it really is; such a lens is called a *telephoto* lens. At the other extreme, a lens of shorter focal length, such as 35 mm or 28 mm, permits a wider angle of view and is called a *wide-angle* lens.

The optical system for a television camera is the same in principle as for an ordinary camera. The film is replaced by an electronic system that scans the image with a series of 525 parallel lines. The image brightness at points along these lines is translated into electrical impulses that can be broadcast by using electromagnetic waves with frequencies of the order of 100 to 400 MHz. The entire picture is scanned 30 times each second, so 30×525, or 15,750, lines are scanned each second. Some TV receivers emit a faint high-pitched sound at this frequency.

> The focal length of the lens determines the angle of view.

> The optical system of a television camera

38–10 THE PROJECTOR

A projector for slides or motion pictures operates very much like a camera in reverse. The essential elements are shown in Fig. 38–18. Light from the source (an incandescent lamp bulb or, in large motion-picture projectors, a carbon-arc lamp) shines through the film, and the projection lens forms a real, enlarged, inverted image of the film on the projection screen. Additional lenses called *condenser* lenses are placed between lamp and film. Their function is to direct the light from the source so that most of it enters the projection lens after passing through the film. A concave mirror behind the lamp also helps direct the light. The condenser lenses must be large enough to cover the entire area of the film. The image on the screen is always inverted; this is why slides have to be put into a projector upside down.

The position and size of the image projected on the screen are determined by the position and focal length of the projection lens.

> A projector is like a camera in reverse: making an enlarged real image on a projection screen.

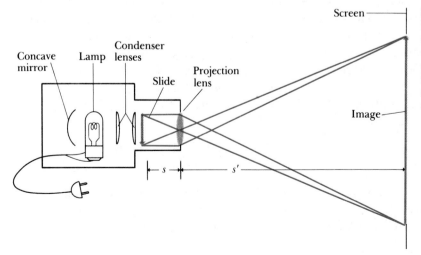

38–18 A slide projector. The concave mirror and condenser lenses gather and direct the light from the lamp so it will enter the projection lens after passing through the slide. The cooling fan, usually needed to take away excessive heat from the lamp, is not shown.

A 35-mm slide projector: geometrical optics at home

EXAMPLE 38–5 An ordinary 35-mm color slide has a picture area 24 × 36 mm. What focal-length projection lens would be needed to project an image 1.2 m × 1.8 m on a screen 5 m from the lens?

SOLUTION We need a lateral magnification (apart from sign) of 1.2 m/24 mm = 50. Thus, from Eq. (38–4), the ratio s'/s must also be 50. We are given $s' = 5$ m, so $s = 5$ m/50 = 0.1 m. Then from Eq. (38–3),

$$\frac{1}{f} = \frac{1}{0.1 \text{ m}} + \frac{1}{5 \text{ m}},$$

$$f = 0.098 \text{ m} = 98 \text{ mm}.$$

A commercially available lens would probably have $f = 100$ mm; this is a popular focal length for home slide projectors.

38–11 THE COMPOUND MICROSCOPE

The compound microscope uses two lens systems; the second looks at the image formed by the first.

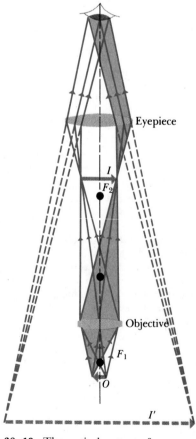

38–19 The optical system of a microscope.

When an angular magnification larger than that attainable with a simple magnifier is desired, it is necessary to use a *compound microscope,* usually called merely a *microscope.* The essential elements of a microscope are illustrated in Fig. 38–19. The object O to be examined is placed just beyond the first focal point F_1 of the **objective** lens, which forms a real and enlarged image I. This image lies just within the first focal point F_2 of the **eyepiece,** which forms a final virtual image of I at I'. As we stated earlier, the position of I' may be anywhere between the near and far points of the eye. Although both the objective and eyepiece of an actual microscope are highly corrected compound lenses, they are shown as simple thin lenses for simplicity.

Since the objective lens forms an enlarged real image that is viewed through the eyepiece, the overall angular magnification M of the compound microscope is the product of the *lateral* magnification m_1 of the objective and the *angular* magnification M_2 of the eyepiece. The former is given by

$$m_1 = -\frac{s_1'}{s_1},$$

where s_1 and s_1' are the object and image distances, respectively, for the objective lens. Ordinarily the object is very close to the focus, resulting in an image whose distance from the objective is much larger than its focal length f_1. Thus s_1 is approximately equal to f_1, and $m_1 = -s_1'/f_1$, approximately. The angular magnification of the eyepiece, from Eq. (38–7), is $M_2 = (25 \text{ cm})/f_2$, where f_2 is the focal length of the eyepiece, considered as a simple lens. Hence the overall magnification M of the compound microscope is, apart from a negative sign, which is customarily ignored,

$$M = m_1 M_2 = \frac{(25 \text{ cm})s_1'}{f_1 f_2}, \tag{38–8}$$

where s_1', f_1, and f_2 are measured in centimeters. Microscope manufacturers customarily specify the values of m_1 and M_2 for microscope components rather than the focal lengths of the objective and eyepiece.

38–12 TELESCOPES

The optical system of a refracting telescope is similar to that of a compound microscope. In both instruments the image formed by an objective is viewed through an eyepiece. The difference is that the telescope is used to examine large objects at large distances and the microscope is used to examine small objects close at hand.

The *astronomical* telescope is illustrated in Fig. 38–20. The objective lens forms a real, reduced image I of the object, and a virtual image of I is formed by the eyepiece. As with the microscope, the image I' may be formed anywhere between the near and far points of the eye. In practice, the objects examined by a telescope are at such large distances from the instrument that the image I is formed very nearly at the second focal point of the objective. Furthermore, if the image I' is at infinity, the image I is at the first focal point of the eyepiece. The distance between objective and eyepiece, or the length of the telescope, is therefore the *sum* of the focal lengths of objective and eyepiece, $f_1 + f_2$.

The angular magnification M of a telescope is defined as the ratio of the angle subtended at the eye by the final image I', to the angle subtended at the (unaided) eye by the object. This ratio may be expressed in terms of the focal lengths of objective and eyepiece as follows. In Fig. 38–20 the ray passing through F_1, the first focal point of the objective, and through F_2', the second focal point of the eyepiece, has been emphasized. The object (not shown) subtends an angle u at the objective and would subtend essentially the same angle at the unaided eye. Also, since the observer's eye is placed just to the right of the focal point F_2', the angle subtended at the eye by the final image is very nearly equal to the angle u'. Because bd is parallel to the optic axis, the distances ab and cd are equal to each other and to the height y' of the image I. Since u and u' are small, they may be approximated by their tangents. From the right triangles F_1ab and $F_2'cd$,

$$u = \frac{-y'}{f_1}, \qquad u' = \frac{y'}{f_2}.$$

Hence

$$M = \frac{u'}{u} = -\frac{y'/f_2}{y'/f_1} = -\frac{f_1}{f_2}. \tag{38–9}$$

The angular magnification M of a telescope is therefore equal to the ratio of the focal length of the objective to that of the eyepiece. The negative sign denotes an inverted image.

The astronomical telescope uses two lens systems; the second looks at the image formed by the first.

The overall magnification of a telescope.

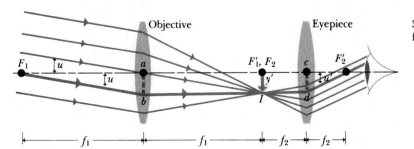

38–20 Optical system of a telescope; final image at infinity.

38–21 The prism binocular. (Courtesy of Bushnell Division of Bausch & Lomb Optical Co.)

Using prisms to turn an inverted image right side up

An inverted image is not a disadvantage if the instrument is to be used for astronomical observations, but it is desirable for a terrestrial telescope to form an erect image. This is accomplished in the *prism binocular* by a pair of 45°–45°–90° totally reflecting prisms inserted between objective and eyepiece, as shown in Fig. 38–21. The image is inverted by the four reflections from the inclined faces of the prisms. It is customary to stamp on a flat metal surface of a binocular two numbers separated by a multiplication sign, such as 7 × 50. The first number is the magnification and the second is the diameter of the objective lenses in millimeters, which determines the brightness of the image.

The reflecting telescope: a mirror and a lens

In the *reflecting telescope*, the objective lens is replaced by a concave mirror, as shown in Fig. 38–22. In large telescopes this scheme has many advantages, both theoretical and practical. The mirror is intrinsically free of chromatic aberrations, and spherical aberrations are much easier to correct than with a lens. The material need not be transparent, and the reflector can be made more rigid than a lens, which has to be supported only at its edges. The largest reflecting telescope in the world has a mirror over 5 m in diameter.

Huge mirrors are a lot easier to make than huge lenses.

Because the image is formed in a region traversed by incoming rays, this image can be observed directly with an eyepiece only by blocking off part of the incoming beam; this is practical only for the largest telescopes. Alternative schemes use a mirror to reflect the image out the side or through a hole in the mirror, as shown in Figs. 38–22b and 38–22c. This scheme is also used in some long-focal-length telephoto lenses for cameras. In the context of photography, such a system is called a *catadioptric lens,* which is a fancy name for an

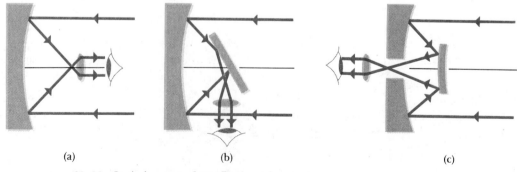

(a)   (b)   (c)

38–22 Optical systems for reflecting telescopes.

38–23 The reflector of the 200-inch Hale telescope at Palomar Observatory, being cleaned in preparation for the application of a new aluminum reflective surface. (Palomar Observatory Photograph.)

optical system containing both reflecting and refracting elements. The reflector for a large astronomical telescope is shown in Fig. 38–23.

SUMMARY

A thin lens has two focal points, or foci, at equal distances from the lens; this distance is called the focal length, denoted by f. An incoming parallel light beam converges to the focus for a converging lens (positive f) or diverges as though the rays had come from the focus for a diverging lens (negative f). Rays diverging from the focus of a converging lens emerge parallel, and rays directed toward the focus of a diverging lens (on the far side) emerge parallel, after refraction.

For an object at distance s from a thin lens, the lens forms an image at distance s'. The two distances are related by

$$\frac{1}{s} + \frac{1}{s'} = \frac{1}{f}, \qquad (38\text{--}3)$$

and the magnification is given by

$$m = -\frac{s'}{s}. \qquad (38\text{--}4)$$

The distances and focal length follow the same sign rules as in Chapter 37: s is positive if the object is on the incoming side, negative otherwise. The image distance s' is positive when the image is on the outgoing side, negative otherwise. The focal length f is positive for a converging lens, negative for a diverging lens.

A principal-ray diagram for a thin lens contains three rays that can be easily drawn; two are related to the foci, and the third passes through the center undeflected.

The focal length f of a thin lens is related to its index of refraction n and the radii of curvature R_1 and R_2 of its surfaces by the lensmaker's equation:

$$\frac{1}{f} = (n - 1)\left(\frac{1}{R_1} - \frac{1}{R_2}\right). \qquad (38\text{--}6)$$

KEY TERMS
thin lens
focus
focal length
optic axis
converging lens
diverging lens
principal rays
virtual object
lensmaker's equation
aberrations
chromatic aberrations
monochromatic aberrations
angular magnification
objective
eyepiece

A radius is positive if its center of curvature is on the outgoing side, negative otherwise. The same method of analysis used to derive this result can also be used for thick lenses, by successive application of the single-surface equation, Eq. (37–12).

Lens aberrations describe the failure of a lens to form a perfectly sharp image of an object. Monochromatic aberrations occur because of limitations of the paraxial approximation; chromatic aberrations result from the dependence of index of refraction on wavelength.

In the eye, a real image is formed on the retina and transmitted to the optic nerve. Adjustment for various object distances is made by the ciliary muscle, which squeezes the lens, making it bulge and decreasing its focal length. A nearsighted eye is too long for its lens, a farsighted eye too short. Focal lengths of corrective lenses are described in terms of their power; the power of a lens, in diopters, is the reciprocal of the focal length, in meters.

The simple magnifier creates a virtual image whose angular size is larger than that of the object itself at a distance of 25 cm, the nominal closest distance for comfortable viewing. The angular magnification is the ratio of the angular size of the virtual image to that of the object, at this distance.

The camera forms a real, inverted, reduced image (on film) of the object being photographed. The amount of light striking the film is controlled by the shutter speed and the aperture. The projector is essentially a camera in reverse; a lens forms an enlarged, real, inverted image on a screen of the slide or motion-picture film. Other lenses, called condenser lenses, concentrate the light from the lamp on the slide or film.

In the compound microscope, the objective lens forms a first image in the barrel of the instrument, and the eyepiece forms a final virtual image at infinity of this first image. The telescope operates on the same principle, but the object is far away. In a reflecting telescope the objective lens is replaced by a concave mirror, which eliminates chromatic aberrations.

QUESTIONS

38–1 Sometimes a wine glass filled with white wine forms an image of an overhead light on a white tablecloth. Would the same image be formed with an empty glass? With a glass of gin? Gasoline?

38–2 How could one very quickly make an approximate measurement of the focal length of a converging lens? Could the same method be applied if we wished to use a diverging lens?

38–3 If you look closely at a shiny Christmas tree ball, you can see nearly the entire room. Does the room appear right side up or upside down? Discuss your observations in terms of images.

38–4 A student asserted that any lens with spherical surfaces has a positive focal length if it is thicker at the center than at the edge, and negative if it is thicker at the edge. Do you agree?

38–5 The focal length of a simple lens depends on the color (wavelength) of light passing through it. Why? Is it possible for a lens to have a positive focal length for some colors and negative for others?

38–6 The human eye is often compared to a camera. In what ways is it similar to a camera? In what ways does it differ?

38–7 How could one make a lens for sound waves?

38–8 A student proposed to use a plastic bag full of air, immersed in water, as an underwater lens. Is this possible? If the lens is to be a converging lens, what shape should the air pocket have?

38–9 When a converging lens is immersed in water, does its focal length increase or decrease, compared with the value in air?

38–10 You are marooned on a desert island and want to use your eyeglasses to start a fire. Can this be done if you are nearsighted? If you are farsighted?

38–11 While lost in the mountains, a person who was nearsighted in one eye and farsighted in the other made a crude emergency telescope from the two lenses of his eyeglasses. How did he do this?

38–12 In using a magnifying glass, is the magnification greater when the glass is close to the object or when it is close to the eye?

38–13 When a slide projector is turned on without a slide in it, and the focus adjustment is moved far enough in one direction, a gigantic image of the light-bulb filament can be seen on the screen. Explain how this happens.

38–14 A spherical air bubble in water can function as a lens. Is it a converging or diverging lens? How is its focal length related to its radius?

38–15 As discussed in the text, some binoculars use prisms to invert the final image. Why are prisms better than ordinary mirrors for this purpose?

38–16 There have been reports of round fishbowls starting fires by focusing the sun's rays coming in a window. Is this possible?

38–17 How does a person judge distance? Can a person with vision in only one eye judge distance? What is meant by "binocular vision"?

38–18 Zoom lenses are widely used in television cameras and conventional photography. Such a lens has, effectively, a variable focal length; changes in focal length are accomplished by moving some lens elements relative to others. Try to devise a scheme to accomplish this effect.

38–19 Why can't you see clearly when your head is under water? Could you wear glasses so you *could* see under water? Would the lenses be converging or diverging?

EXERCISES

Section 38–1 The Thin Lens

Section 38–2 Diverging Lenses

Section 38–3 Graphical Methods

38–1 A converging lens has a focal length of 10 cm. For object distances of 30 cm, 20 cm, 15 cm, and 5 cm, determine

a) image position,

b) magnification,

c) whether the image is real or virtual,

d) whether the image is erect or inverted.

Be sure to draw a principal-ray diagram in each case.

38–2 A converging lens of focal length 10 cm forms a real image 1 cm high, 12 cm to the right of the lens. Determine the position and size of the object. Is the image erect or inverted? Draw a principal-ray diagram for this situation.

38–3 Repeat Exercise 38–1 for the case where the lens is diverging, with focal length −10 cm.

38–4 An object is 10 cm from a lens that forms an image 15 cm from the lens, on the side opposite to the object.

a) What is the focal length of the lens? Is the lens converging or diverging?

b) If the object is 1 cm high, how high is the image? Is it erect or inverted?

c) Draw a principal-ray diagram.

38–5 A lens forms an image of an object 20 cm from it. The image is 4 cm from the lens on the same side as the object.

a) What is the focal length of the lens? Is the lens converging or diverging?

b) If the object is 2 cm high, how high is the image? Is it erect or inverted?

c) Draw a principal-ray diagram.

38–6 Prove that the image of a real object formed by a diverging lens is *always* virtual.

Section 38–4 Images as Objects

38–7 A diverging meniscus lens (see Fig. 38–5b) of refractive index 1.48 has spherical surfaces whose radii are 4.0 cm and 2.5 cm. What would be the position of the image if an object were placed 15 cm in front of the lens?

38–8 Sketch the various possible thin lenses obtainable by combining two surfaces whose radii of curvature are, in absolute magnitude, 10 cm and 20 cm. Which are converging and which are diverging? Find the focal length of each if made of glass of index 1.50.

38–9 A layer of benzene ($n = 1.50$) 2 cm deep floats on water ($n = 1.33$) that is 4 cm deep. What is the apparent distance from the upper benzene surface to the bottom of the water layer, when viewed at normal incidence?

38–10 A transparent rod 40 cm long and of refractive index 1.50 is cut flat at one end and rounded to a hemispherical surface of 12-cm radius at the other end. An object is placed on the axis of the rod 10 cm from the hemispherical end.

a) What is the position of the final image?

b) What is its magnification?

38–11 Both ends of a glass rod 10 cm in diameter, of index 1.50, are ground and polished to convex hemispherical surfaces of radius 5 cm at the left end and radius 10 cm at the right end. The length of the rod between vertexes is 60 cm. An arrow 1 mm long, at right angles to the axis and 20 cm to the left of the first vertex, constitutes the object for the first surface.

a) What constitutes the object for the second surface?

b) What is the object distance for the second surface?

c) Is this object real or virtual?

d) What is the position of the image formed by the second surface?

38–12 The same rod as in Exercise 38–11 is now shortened to a distance of 10 cm between its vertices, the curvatures of its ends remaining the same.

a) What is the object distance for the second surface?

b) Is the object real or virtual?

c) What is the position of the image formed by the second surface?

d) Is the image real or virtual and is it erect or inverted, with respect to the original object?

e) What is the height of the final image?

38–13 A narrow beam of parallel rays enters a solid glass sphere in a radial direction. At what point outside the sphere are these rays brought to a focus? The radius of the sphere is 3 cm, and its index is 1.50.

38–14 The radii of curvature of the surfaces of a thin lens are $R_1 = +10$ cm and $R_2 = +30$ cm. The index is 1.50.

a) Compute the position and size of the image of an object in the form of an arrow 1 cm high, perpendicular to the lens axis, 40 cm to the left of the lens.

b) A second similar lens is placed 160 cm to the right of the first. Find the position and size of the final image.

c) Same as (b), except the second lens is 40 cm to the right of the first.

d) Same as (c), except the second lens, of focal length −40 cm, is diverging.

38–15 Three thin lenses, each of focal length 20 cm, are aligned on a common axis, and adjacent lenses are separated by 30 cm. Find the position of the image of a small object on the axis, 60 cm to the left of the first lens.

38–16 Two thin lenses with focal length of magnitude 10 cm, the first converging and the second diverging, are placed 5 cm apart. An object 2 mm tall is placed 20 cm in front of the first (converging) lens.

a) How far from this lens will the final image be formed?

b) Is the final image real or virtual?

c) What is the height of the final image?

38–17 An eyepiece consists of two similar positive thin lenses having focal lengths of 6 cm, separated by a distance of 3 cm. Where are the focal points of the eyepiece?

Section 38–6 The Eye

Section 38–7 Defects of Vision

38–18 What is the power of the lens required

a) by a hyperopic eye whose near point is at 125 cm?

b) by a myopic eye whose far point is at 50 cm?

38–19

a) Where is the near point of an eye for which a lens of power +2 diopters is prescribed?

b) Where is the far point of an eye for which a lens of power −0.5 diopter is prescribed for distant vision?

38–20 In a simplified model of the human eye, the aqueous and vitreous humors and the lens all have a refractive index of 1.40, and all the refraction occurs at the cornea, about 2.50 cm from the retina. What should be the radius of curvature of the cornea in order to focus an infinitely distant object on the retina?

38–21 For the model of the eye described in Exercise 38–20, what should be the radius of curvature of the cornea in order to focus an object 25 cm from the cornea?

Section 38–8 The Magnifier

38–22 A thin lens of focal length 10 cm is used as a simple magnifier.

a) What angular magnification is obtainable with the lens?

b) When an object is examined through the lens, how close can it be brought to the eye? Assume that the image viewed by the eye is at infinity.

38–23 The focal length of a simple magnifier is 10 cm.

a) How far in front of the magnifier should an object be placed if the image is formed at the observer's near point, 25 cm in front of her eye?

b) If the object is 1 mm high, what is the height of its image formed by the magnifier?

Assume the magnifier to be a thin lens placed very close to the eye.

Section 38–9 The Camera

38–24 During a lunar eclipse, a picture of the moon (diameter 3.48×10^6 m, distance from earth 3.8×10^8 m) is taken with a camera whose lens has a focal length of 50 mm. What is the diameter of the image on the film?

38–25 The picture size on ordinary 35-mm camera film is 24×36 mm. Focal lengths of lenses available for 35-mm cameras typically include 28, 35, 50 (the "standard" lens), 85, 100, 135, 200, and 300 mm, among others. Which of these lenses should be used to photograph the following objects, assuming the object is to fill most of the picture area?

a) A cathedral 100 m high and 150 m long, at a distance of 150 m.

b) An eagle with a wingspan of 2.0 m, at a distance of 15 m.

38–26 Camera A, having an $f/8$ lens 2.5 cm in diameter, photographs an object with the correct exposure of $\frac{1}{100}$ s. What exposure should camera B use in photographing the same object if it has an $f/4$ lens 5 cm in diameter?

38–27 The focal length of an $f/2.8$ camera lens is 8 cm.

a) What is the aperature diameter of the lens?

b) If the correct exposure of a certain scene is $\frac{1}{200}$ s at $f/2.8$, what would be the correct exposure at $f/5.6$?

Section 38–10 The Projector

38–28 The dimensions of the picture on a 35-mm color slide are 24×36 mm. An image of the slide is projected on a screen 10 m from the projector lens. The focal length of the projector lens is 12 cm.

a) How far should the slide be from the lens?

b) What will be the dimensions of the image on the screen?

38–29 In Example 38–5, for a 98-mm focal-length lens, it was found that the slide needs to be placed 10 cm in front of the lens to focus the image on a screen 5 m from the lens. Now assume instead that a 100-mm-focal-length lens is used.

a) If the screen remains 5 m from the lens, how far should the slide be in front of the lens?

b) Could the distance between the slide and lens be kept at 10 cm and the screen moved to achieve focus? Is so, how far and in what direction would it have to be moved?

Section 38–11 The Compound Microscope

38–30 A certain microscope is provided with objectives of focal lengths 16 mm, 4 mm, and 1.9 mm, and with eyepieces of angular magnification 5X and 10X. What is

a) the largest overall magnification obtainable?

b) the least?

Each objective forms an image 160 mm beyond its second focal point.

38–31 The focal length of the eyepiece of a certain microscope is 2.5 cm. The focal length of the objective is 16 mm. The distance between objective and eyepiece is 22.1 cm. The final image formed by the eyepiece is at infinity. Treat all lenses as thin.

a) What should be the distance from the objective to the object viewed?

b) What is the linear magnification produced by the objective?

c) What is the overall magnification of the microscope?

38–32 The image formed by a microscope objective of focal length 4 mm is 180 mm from its second focal point. The eyepiece has a focal length of 31.25 mm.

a) What is the magnification of the microscope?

b) The unaided eye can distinguish two points as separate if they are about 0.1 mm apart. What is the minimum separation that can be resolved with this microscope?

Section 38–12 Telescopes

38–33 A crude telescope is constructed of two eyeglass lenses of focal lengths 100 cm and 20 cm, respectively.

a) Find its angular magnification.

b) Find the height of the image formed by the objective of a building 80 m high at a distance of 2 km.

38–34 The eyepiece of a telescope has a focal length of 10 cm. The distance between objective and eyepiece is 2.1 m. What is the angular magnification of the telescope?

38–35 The moon subtends an angle at the earth of approximately $\frac{1}{2}°$. What is the diameter of the image of the moon produced by the objective of the Lick Observatory telescope, a refractor with a focal length of 18 m?

38–36 A reflecting telescope is made of a mirror of radius of curvature 0.50 m and an eyepiece of focal length 1.0 cm.

a) What is the angular magnification?

b) What should be the position of the eyepiece if both the object and final image are at infinity?

38–37 The new space telescope uses an optical system similar to that shown in Fig. 38–24 (Cassegrain system). The image of a far distant object is focused on the detector through a hole in the large (primary) mirror. The primary mirror has a focal length of 2.5 m, the distance between the vertexes of the two mirrors is 1.5 m, and the distance from the vertex of the primary mirror to the detector is 0.25 m. Should the smaller mirror (the secondary) be concave or convex, and what is its radius of curvature?

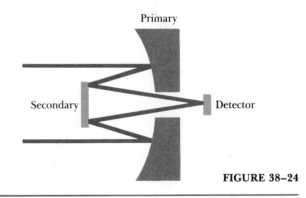

FIGURE 38–24

PROBLEMS

38–38 An object is placed 18 cm from a screen.

a) At what points between object and screen may a converging lens of 4-cm focal length be placed to obtain an image on the screen?

b) What is the magnification of the image for these positions of the lens?

38–39 An object is imaged by a lens on a screen placed 12 cm from the lens. When the lens is moved 2 cm farther from the object, the screen must be moved 2 cm closer to the object to refocus it. Determine the focal length of the lens.

38–40

a) Prove that when two thin lenses of focal lengths f_1 and f_2 are placed *in contact,* the focal length f of the combination is given by the relation

$$\frac{1}{f} = \frac{1}{f_1} + \frac{1}{f_2}.$$

b) A converging meniscus lens (Fig. 38–5a) has an index of refraction of 1.50, and the radii of its surfaces are 5 cm and 10 cm. The concave surface is placed upward and filled with water. What is the focal length of the water–glass combination?

38–41 When an object is placed at the proper distance in front of a converging lens, the image falls on a screen 20 cm from the lens. A diverging lens is now placed halfway between the converging lens and the screen, and it is found that the screen must be moved 20 cm farther away from the converging lens to obtain a sharp image. What is the focal length of the diverging lens?

38–42 A lens operates via Snell's law, bending light rays at each surface an amount determined by the index of the lens and the index of the medium in which the lens is located.

a) Equation (38–6) assumes that the lens is surrounded by air. Consider instead a thin lens immersed in a liquid of refractive index n_{liq}. Derive the equation analogous to Eq. (38–6) for the focal length f' of the lens.

b) A thin lens of index n has a focal length in vacuum of f. Use the result of (a) to show that when this lens is immersed in a liquid of index n_{liq}, it will have a new focal length of

$$f' = \left[\frac{n_{liq}(n-1)}{n - n_{liq}}\right]f.$$

38–43 A convex mirror and a concave mirror are placed on the same optic axis separated by a distance $L = 0.8$ m. The radius of curvature of each mirror has a magnitude of 0.4 m. A light source is located a distance x from the concave mirror, as shown in Fig. 38–25.

a) What distance x will result in the rays from the source returning to the source after reflecting first from the convex mirror and then from the concave mirror?

b) Repeat part (a), but now let the rays reflect first from the concave mirror and then from the convex one.

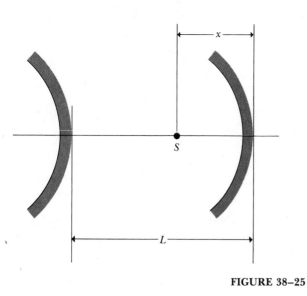

FIGURE 38–25

38–44 In the arrangement shown in Fig. 38–26, the candle is at the center of curvature of the concave mirror, whose focal length is 10 cm. The converging lens has a focal length of 35 cm and is 85 cm to the right of the candle. The candle is viewed by looking through the lens from the right.

a) Draw a principal-ray diagram that locates the final image.

b) Where is the final image?

c) Is the final image real or virtual?

d) Is the final image erect or inverted? (*Hint:* Are one or two images formed?)

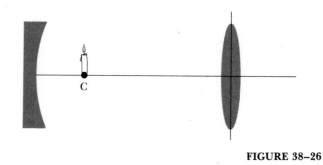

FIGURE 38–26

38–45 Your camera has a 50-mm-focal-length lens and a viewfinder that is 20 mm (high) by 30 mm (long). In taking a picture of a 3-m-long automobile, you find that the image of the auto fills only two-thirds of the viewfinder.

a) How far are you from the auto?

b) How close should you stand if you want to fill the viewfinder frame with the auto's image?

38–46 A glass rod of refractive index 1.50 is ground and polished at both ends to hemispherical surfaces of 5-cm radius. When an object is placed on the axis of the rod, 20 cm from one end, the final image is formed 40 cm from the opposite end. What is the length of the rod?

38–47 A thin-walled glass sphere of radius R is filled with water. An object is placed a distance $3R$ from the surface of the sphere. Determine the position of the final image. The effect of the glass wall may be neglected.

38–48 A thick-walled wine goblet can be considered a glass sphere of radius 5 cm with a spherical cavity of radius 4 cm. The index of refraction of the goblet glass is 1.50.

a) A beam of parallel light rays enters the empty goblet along a radius. Where, if anywhere, will an image be formed?

b) The goblet is filled with white wine ($n = 1.37$). Where is the image now formed?

38–49 A glass plate 2 cm thick, of index 1.50, having plane parallel faces, is held with its faces horizontal and its lower face 8 cm above a printed page. Find the position of the image of the page formed by rays making a small angle with the normal to the plate.

38–50 Rays from a lens are converging toward a point image P, as in Fig. 38–27. What thickness t of glass of index 1.50 must be interposed, as in the figure, for the image to be formed at P'?

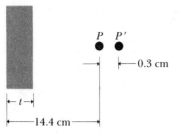

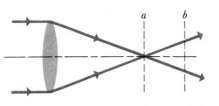

FIGURE 38–28

FIGURE 38–27

38–51 The *resolution* of a camera lens can be defined as the maximum number of lines per millimeter in the image that can barely be distinguished as separate lines. A certain lens has a focal length of 50 mm and resolution of 100 lines·mm^{-1}. What is the minimum separation of two lines in an object 100 m away if they are to be visible in the image as separate lines?

38–52

a) Show that when two thin lenses are placed in contact, the *power* of the combination in diopters, as defined in Section 38–7, is the sum of the powers of the separate lenses. Is this relation valid even when one lens has positive power and the other negative?

b) Two thin lenses of 25-cm and 40-cm focal lengths are in contact. What is the power of the combination?

38–53 A certain very nearsighted person cannot focus anything farther than 10 cm from the eye. If the radius of curvature of the cornea is 0.70 cm when the eye is focusing on an object 10 cm from it, and the indexes of refraction are as described in Exercise 38–20, what is the cornea-to-retina distance? What does this tell you about the shape of the nearsighted eye?

38–54 A camera lens is focused on a distant point source of light; the image is formed on a screen at a (Fig. 38–28). When the screen is moved backward a distance of 2 cm to b,

the circle of light on the screen has a diameter of 4 mm. What is the $f/$number of the lens?

38–55 A microscope with an objective of focal length 9 mm and an eyepiece of focal length 5 cm is used to project an image on a screen 1 m from the eyepiece. Let the image distance of the objective be 18 cm.

a) What is the lateral magnification of the image?

b) What is the distance between the objective and the eyepiece?

38–56 A certain reflecting telescope, constructed as in Fig. 38–22a, has a mirror 10 cm in diameter with radius of curvature 1.0 m and an eyepiece of focal length 1.0 cm. If the angular magnification has a magnitude of 48 and the object is at infinity, find the position of the lens and the position and nature of the final image. (*Note:* $|M|$ is *not* equal to $|f_1/f_2|$, so the image formed by the eyepiece is *not* at infinity.)

38–57 Figure 38–29 is a diagram of a *Galilean telescope*, or *opera glass*, with both object and final image at infinity. The image I serves as a virtual object for the eyepiece. The final image is virtual and erect.

a) Prove that the angular magnification is $M = -f_1/f_2$.

b) A Galilean telescope is to be constructed with the same objective lens as in Exercise 38–33. What focal length should the eyepiece have if this telescope is to have the same magnification as the one in Exercise 38–33?

c) Compare the lengths of the telescopes.

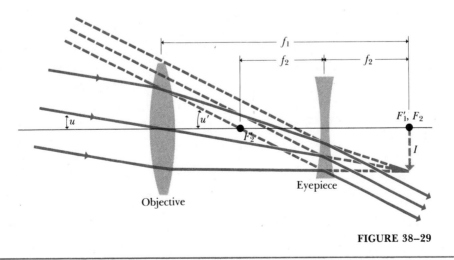

FIGURE 38–29

CHALLENGE PROBLEMS

38–58 A 20-cm long pencil is placed at a 45° angle, with its center 15 cm above the optic axis and 45 cm from a 25-cm-focal-length lens, as shown in Fig. 38–30.

a) Where is the image of the pencil? (Give the location of the images of the points A, B, and C on the object, which are located at the eraser, point, and center of the pencil, respectively.)

b) What is the length of the image?

c) How is the image oriented?

Neglect aberration effects.

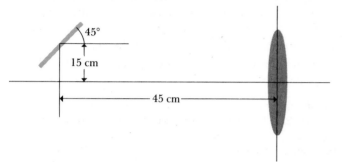

FIGURE 38–30

38–59

a) For a lens of focal length f, find the smallest distance possible between the object and its real image.

b) Sketch a graph of the distance between the object and real image as a function of the distance of the object from the lens. Does your sketch agree with the result you found in (a)?

38–60 A solid glass sphere of radius R and index 1.50 is silvered over one hemisphere, as in Fig. 38–31. A small object is located on the axis of the sphere at a distance $2R$ from the pole of the unsilvered hemisphere. Find the position of the final image after all refractions and reflections have taken place.

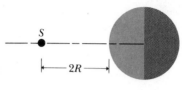

FIGURE 38–31

38–61 A symmetric double-convex thin lens made of glass of index 1.50 has a focal length in air of 30 cm. The lens is sealed into an opening in one end of a tank filled with water (index = 1.33). At the end of the tank opposite the lens is a plane mirror 80 cm from the lens.

a) Find the position of the image formed by the lens–water–mirror system, of a small object outside the tank on the lens axis and 90 cm to the left of the lens.

b) Is the image real or virtual?

c) Is it erect or inverted?

d) If the object has a height of 4 mm, what is the height of the image?

38–62 A person with normal vision cannot focus his or her eyes underwater.

a) Why not?

b) With the simplified model of the eye described in Exercise 38–20, what corrective lens (specified by focal length, as measured in air) would be needed to enable a person underwater to focus an infinitely distant object? (Be careful—the focal length of a lens underwater is not the same as in air! See Problem 38–42. Assume the corrective lens has a refractive index in air of 1.60.)

39

INTERFERENCE AND DIFFRACTION

IN OUR ANALYSIS OF IMAGE FORMATION BY LENSES AND MIRRORS, WE HAVE
represented light as *rays* that travel in straight lines in a homogeneous material
and that are bent according to simple laws at a reflecting or refracting surface.
This simple model forms the basis of *geometrical optics;* as we have seen, this
model is an adequate basis for analyzing a wide variety of phenomena involving lenses and mirrors.

In this chapter we will study the phenomena of *interference* and *diffraction*.
These are inherently *wave* phenomena, and the principles of geometrical optics are *not* an adequate basis for this study. Instead, we must return to the
more general view that light is a *wave motion*. When several waves overlap at a
point, their total effect depends on the *phases* of the waves as well as their
amplitudes. This part of the subject is called *physical optics*. When light passes
through apertures or around obstacles, patterns are formed that could not be
predicted on the basis of a ray model, but instead depend directly on the wave
nature of light. In several simple situations we can predict the characteristics
of these patterns in detail. We will look at several practical applications of
physical optics, including diffraction gratings, x-ray diffraction, and holography.

Light doesn't always behave like rays:
Interference and diffraction phenomena
show the wave nature of light.

39–1 INTERFERENCE AND COHERENT SOURCES

In our discussions of mechanical waves in Chapter 21 and electromagnetic
waves in Chapter 35, we often considered *sinusoidal* waves having a single
frequency and a single wavelength. In the context of optics, such a wave is
called a **monochromatic** (single-color) **wave.** Common sources of light, such as
an incandescent light bulb or a flame, *do not* emit monochromatic light but
rather emit a continuous distribution of wavelengths. A strictly monochromatic light wave is an unattainable idealization.

Monochromatic light has only one
wavelength and frequency: a useful
though unattainable idealization.

Monochromatic light can be *approximated* in the laboratory. For example,
we can pass continuous-spectrum light through a filter that blocks all but a
narrow range of wavelengths. Gas-discharge lamps, such as the mercury-arc
lamp, emit line spectra in which the light consists of a discrete set of colors,

Laser light is very nearly monochromatic.

each having a narrow band of wavelengths called a *spectrum line*. For example, the bright green line in the mercury spectrum has an average wavelength of 546.1 nm, with a spread of wavelength of the order of ± 0.001 nm, depending on the pressure and temperature of the mercury vapor in the lamp. By far the most nearly monochromatic source available at present is the *laser*, to be discussed in Chapter 41. The familiar helium-neon laser, inexpensive and readily available, emits visible light at 632.8 nm with a line width (wavelength range) of the order of ± 0.000001 nm, or about one part in 10^9. In the following discussion of interference and diffraction phenomena, we will nearly always assume that we are dealing with monochromatic waves.

The term **interference** refers to any situation in which two or more waves overlap in space. We introduced this term in Section 22–2 in connection with standing waves on a stretched string, formed by the superposition of two sinusoidal waves traveling in opposite directions. In such cases the total displacement at any point at any instant of time is governed by the **principle of linear superposition,** also introduced in Section 22–2. This principle, the most important in all of physical optics, states that *when two or more waves overlap, the resultant displacement at any point and at any instant may be found by adding the instantaneous displacements that would be produced at the point by the individual waves if each were present alone.* The term *displacement,* as used here, is a general one. If we are considering surface ripples on a liquid, the displacement means the actual displacement of the surface above or below its normal level. If the waves are sound waves, the term refers to the excess or deficiency of pressure. If the waves are electromagnetic, the displacement means the magnitude of the electric or magnetic field. When light of extremely high intensity passes through matter, the principle of linear superposition is *not* precisely obeyed, and the resulting phenomena are classified under the heading *nonlinear optics.* (These effects are beyond the scope of this book.)

To introduce the essential ideas of interference, consider first the problem of two identical sources of monochromatic waves, S_1 and S_2, separated in space by a certain distance. The two sources are permanently *in phase,* so that at every point in space there is a definite and unchanging phase relation between waves from the two sources. They might be, for example, two agitators in a ripple tank, two loudspeakers driven by the same amplifier, two radio antennas powered by the same transmitter, or two small apertures in an opaque screen, illuminated by the same monochromatic light source.

We position the sources S_1 and S_2 along the y-axis, equidistant from the origin, as shown in Fig. 39–1. Let P_0 be any point on the x-axis. From symmetry, the two distances S_1P_0 and S_2P_0 are equal; waves from the two sources thus require equal times to travel to P_0, and having left S_1 and S_2 in phase, they arrive at P_0 in phase. The total amplitude at P_0 is thus twice the amplitude of each individual wave.

Next we consider a point P_1, located so that its distance from S_2 is exactly one wavelength greater than its distance from S_1. That is,

$$S_2P_1 - S_1P_1 = \lambda.$$

Then any given wave crest from S_1 arrives at P_1 exactly one cycle earlier than the crest emitted at the same time from S_2, and again the two waves arrive *in phase.* Similarly, waves arrive in phase at all points P_2 for which the path difference is *two* wavelengths ($S_2P_2 - S_1P_2 = 2\lambda$), or indeed for *any* positive or negative integer number of wavelengths.

Superposition: the underlying principle in all interference and diffraction calculations

For two-source interference, the sources must have a definite phase relationship.

The two waves arriving at a point are not in phase because they travel different distances from the sources.

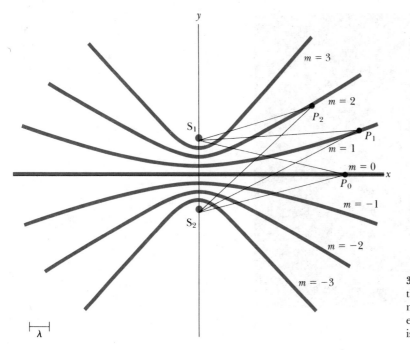

39–1 Curves of maximum intensity in the interference pattern of two monochromatic point sources. In this example the distance between sources is four times the wavelength.

When two waves arrive in phase, they add.

The addition of amplitudes that results when waves from two or more sources arrive at a point *in phase* is often called **constructive interference** or *reinforcement,* and our discussion shows that constructive interference occurs whenever the path difference for the two sources is an integral multiple of the wavelength:

$$S_2P - S_1P = m\lambda \qquad (m = 0, \pm 1, \pm 2, \pm 3, \ldots). \qquad (39\text{–}1)$$

In our example, the points satisfying this condition lie on the set of curves shown in Fig. 39–1.

Intermediate between these lines is a set of other lines for which the path difference for the two sources is a *half-integral* number of wavelengths. Waves from the two sources arrive at a point on one of these lines exactly a half-cycle out of phase, and the resultant amplitude is the *difference* between the two individual amplitudes. If the amplitudes are equal, which is approximately true when the distance from either source to P is much greater than the distance between sources, then the total amplitude at such a point is zero! This condition is called **destructive interference** or *cancellation.* In our example, the condition for destructive interference is

When two waves arrive a half-cycle out of phase, they subtract.

$$S_2P - S_1P = (m + \tfrac{1}{2})\lambda \qquad (m = 0, \pm 1, \pm 2, \pm 3, \ldots). \qquad (39\text{–}2)$$

An example of this interference pattern is the familiar ripple-tank pattern shown in Fig. 39–2. The two wave sources are two agitators driven by the same vibrating mechanism. The regions of both maximum and zero amplitude are clearly visible. The superposed color lines, corresponding to those in Fig. 39–1, show the lines of maximum amplitude.

A ripple tank is an inherently two-dimensional situation, but in some cases, such as two loudspeakers or two radio-transmitter antennas, the pattern is three-dimensional. We may think of rotating the color curves of Fig. 39–1

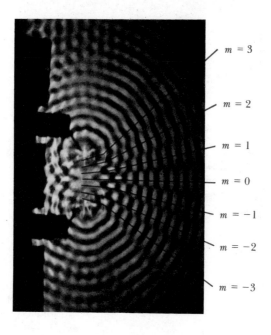

$m = 3$

$m = 2$

$m = 1$

$m = 0$

$m = -1$

$m = -2$

$m = -3$

39–2 Photograph of an interference pattern produced by water waves in a shallow ripple tank. The two wave sources are small balls moved up and down by the same vibrating mechanism. As waves move outward from the sources, they overlap and produce an interference pattern. The lines of maximum amplitude in the pattern are shown by the superposed color lines. (*PSSC Physics,* second edition, 1965; D. C. Heath and Co., with Educational Development Center, Inc., Newton, Mass.)

A definite phase relation between the sources is essential.

Coherent sources have a definite phase relationship.

about the horizontal line through the center; the resulting surfaces are the points where maximum constructive interference occurs.

In this discussion, the constant *phase* relationship between the sources is an essential requirement. If the relative phase of the sources changes, the positions of the maxima and minima in the resulting interference pattern also change. When the radiation is light, two sources can have a definite and constant phase relation *only when they both emit light coming from a single primary source.* There is no practical way to achieve such a relationship with two separate sources. The reason is a fundamental one associated with the mechanisms of light emission. In ordinary light sources, atoms of the material of the source are given excess energy by thermal agitation or by impact with accelerated electrons. An atom thus "excited" begins to radiate energy and continues until it has lost all the energy it can, typically in a time of the order of 10^{-8} s. A source ordinarily contains a very large number of atoms, and they radiate in an unsynchronized and random phase relationship. Thus emission from two such sources has a rapidly varying phase relation. The result is an interference pattern that constantly changes in a random manner, and ordinary observation does not show a visible interference pattern at all.

However, if the light from a single source is split so that parts of it emerge from two or more regions of space, forming two or more *secondary sources,* any random phase change in the source affects these secondary sources equally and does not change their *relative* phase. Two such sources derived from a single source and having a definite phase relation are said to be **coherent.**

The distinguishing feature of light from a *laser* is that the emission of light from many atoms is *synchronized* in frequency and phase, by mechanisms to be discussed in Chapter 41. As a result, the random phase changes mentioned above occur *much* less frequently. Definite phase relations are preserved over correspondingly much greater lengths in the beam. Accordingly, laser light is said to be much more *coherent* than ordinary light.

39–2 TWO-SOURCE INTERFERENCE

One of the earliest experimental demonstrations of the fact that light can produce interference effects was performed in 1800 by the English scientist Thomas Young. This was a crucial experiment because it added significantly to the evidence for the wave nature of light. His experiment involved the interference of light from two sources, as described in the preceding section.

A landmark experiment to demonstrate the wave nature of light and measure its wavelengths

Young's apparatus is shown schematically in Fig. 39–3a. Monochromatic light emerging from a narrow slit S_0 falls on a screen having two other narrow slits S_1 and S_2, 0.1 mm or so wide and 1.0 mm or less apart. According to Huygens' principle (Section 36–12), cylindrical wavelets spread out from slit S_0 and reach slits S_1 and S_2 in phase because they travel equal distances from S_0. A succession of Huygens wavelets then emerges from each slit; the two sets leave in phase, and therefore the two slits act as two *coherent* sources. But the waves do not necessarily arrive at point P in phase because of the path difference $(r_1 - r_2)$ for the two waves.

Coherent sources can be obtained by dividing the light from a single source.

In the following analysis we will assume that the distance R from the slits to the screen is so large compared to the slit spacing d that the lines from the slits to P are very nearly parallel, as shown in Fig. 39–3c. This is not an essential assumption, but it simplifies the following analysis considerably and is usually valid for the experiments we will discuss. In this case the path difference is given by

$$r_1 - r_2 = d \sin \theta. \qquad (39\text{--}3)$$

According to the discussion in the preceding section, constructive interference (reinforcement) occurs at point P when the path difference $d \sin \theta$ is some integral number of wavelengths, say $m\lambda$ ($m = 0, \pm 1, \pm 2, \pm 3$, etc.). Thus,

$$d \sin \theta = m\lambda \quad \text{or} \quad \sin \theta = \frac{m\lambda}{d}. \qquad (39\text{--}4)$$

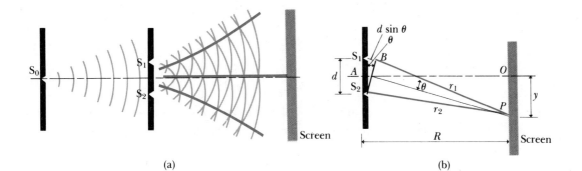

(a)

(b)

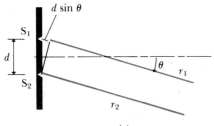

(c)

39–3   (a) Interference of light waves passing through two slits.
(b) Geometrical analysis of Young's experiment. (c) Approximate geometry when R is much larger than d.

Zeroth fringe

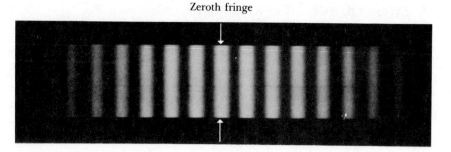

39-4 Interference fringes produced by Young's double-slit experiment.

Similarly, complete cancellation or destructive interference occurs when the path difference is a half-integral number of wavelengths, $(m + \frac{1}{2})\lambda$, where again $m = 0, \pm 1, \pm 2$, and so on:

$$d \sin \theta = \left(m + \frac{1}{2}\right)\lambda \quad \text{or} \quad \sin \theta = \frac{(m + \frac{1}{2})\lambda}{d}. \tag{39-5}$$

An interference pattern is a series of alternating bright and dark bands or fringes.

Thus the pattern on the screen of Fig. 39-3b is a succession of bright and dark bands; an actual photograph of such a pattern is shown in Fig. 39-4.

Let us derive an expression for the positions of the centers of the bright bands. Let y_m be the distance of the center of the mth bright band from the center of the central band at $\theta = 0$. The corresponding value of θ is θ_m, and

$$y_m = R \tan \theta_m.$$

In optical interference experiments, the y_m's are nearly always much smaller than R. In that case $\tan \theta_m$ is very nearly equal to $\sin \theta_m$, and

$$y_m = R \sin \theta_m.$$

Combining this with Eq. (39-4), we find

$$y_m = R\frac{m\lambda}{d} \tag{39-6}$$

and

$$\lambda = \frac{y_m d}{mR}. \tag{39-7}$$

A simple and direct way to measure the wavelength of light

We can measure R and d, as well as the positions y_m of the bright fringes. Thus this experiment provides a direct measurement of the wavelength λ.

EXAMPLE 39-1 In a two-slit interference experiment with two slits 0.2 mm apart, and a screen at a distance of 1 m, the third bright fringe is found to be displaced 7.5 mm from the central fringe. Find the wavelength of the light used.

SOLUTION Let λ be the unknown wavelength. Then

$$\lambda = \frac{y_m d}{mR} = \frac{(0.75 \text{ cm})(0.02 \text{ cm})}{(3)(100 \text{ cm})} = 5 \times 10^{-5} \text{ cm}$$

$$= 500 \times 10^{-9} \text{ m} = 500 \text{ nm}.$$

Although we have described this experiment as Young performed it with visible light, a completely analogous experiment could be done with any other electromagnetic waves, such as radio waves.

EXAMPLE 39–2   A radio station operating at a frequency of 1500 kHz = 1.5×10^6 Hz has two identical vertical dipole antennas spaced 400 m apart. Where are the intensity maxima in the resulting radiation pattern?

Interference with radio antennas: how to make a directional antenna

SOLUTION   The wavelength is $\lambda = c/f = 200$ m. The directions of the intensity *maxima* are those for which the path difference is zero or an integer number of wavelengths, as given by Eq. (39–4). Inserting the numerical values, we find

$$\sin \theta = \frac{m\lambda}{d} = \frac{m}{2}, \qquad \theta = 0, \ \pm 30°, \ \pm 90°.$$

In this example, values of m greater than 2 give values of $\sin \theta$ greater than unity and thus have no meaning; there is *no* direction for which the path difference is three or more wavelengths. Similarly, the directions having zero intensity (complete destructive interference) are given by Eq. (39–5):

$$\sin \theta = \frac{(m + \frac{1}{2})\lambda}{d} = \frac{m + \frac{1}{2}}{2}, \qquad \theta = \pm 14.5°, \ \pm 48.6°.$$

In this case values of m greater than 1 have no meaning, for the reason just mentioned.

39–3 INTENSITY DISTRIBUTION IN INTERFERENCE PATTERNS

In Section 39–2 we computed the directions of maximum and minimum intensity in the two-source interference pattern. We may also find the intensity at *any* point in the pattern. To do this we have to combine the two sinusoidally varying fields at a point P in the radiation pattern caused by the two sources, taking proper account of their phase difference. The intensity is then proportional to the square of the resultant electric-field amplitude, as we learned in Chapter 35.

Calculating the details of the intensity distribution in an interference pattern

Here is our program. The phase difference δ between the two sinusoidally varying fields at point P depends on the difference in path length. We have to relate δ to the location of P, and then we have to learn how to *add* the two sinusoidal functions with a phase difference of δ. From the amplitude E_P of the resultant sinusoidal wave at point P, we compute the **intensity** I. To obtain I we recall from Section 35–3 that I is equal to the average magnitude of the Poynting vector, S_{av}. For a sinusoidal wave with E-field amplitude E_P, this is given by Eq. (35–18) with E_{max} replaced by E_P. By using the relation $\epsilon_0\mu_0 = c^2$, we can express this in the following equivalent forms:

$$S_{av} = I = \frac{1}{2}\frac{E_P{}^2}{\mu_0 c} = \frac{1}{2}\sqrt{\frac{\epsilon_0}{\mu_0}}E_P{}^2 = \frac{1}{2}\epsilon_0 c E_P{}^2. \qquad (39–8)$$

We will assume the two sinusoidal functions have equal amplitude E and that the **E** fields lie along the same line; this assumes the sources are identical and neglects the slight amplitude difference caused by the unequal path lengths. If the sources are in phase, then the waves that arrive at P differ in phase by an amount proportional to the difference in their path lengths, $(r_1 - r_2)$. If the phase angle between these arriving waves is δ, then the two electric fields superposed at P might be

$$E_1 = E \cos \omega t,$$

$$E_2 = E \cos (\omega t + \delta).$$

The general relation between path-length difference and phase difference

To find δ, we note that if the path difference is one wavelength, then $\delta = 2\pi$ or $360°$. If it is equal to $\lambda/2$, the phase difference δ is π, and so on. That is, the ratio of δ to 2π is equal to the ratio of $(r_1 - r_2)$ to λ:

$$\frac{\delta}{2\pi} = \frac{r_1 - r_2}{\lambda}.$$

Thus a path difference $(r_1 - r_2)$ causes a phase difference δ given by

$$\delta = \frac{2\pi}{\lambda}(r_1 - r_2) = k(r_1 - r_2), \qquad (39\text{--}9)$$

where $k = 2\pi/\lambda$ is the *wave number* introduced in Section 21–3.

Using phasors to add sinusoidal functions with phase differences

If the material in the space between the sources and P is anything other than vacuum, the wavelength *in the material* must be used in Eq. (39–9). If the material has refractive index n, then

$$\lambda = \frac{\lambda_0}{n} \quad \text{and} \quad k = nk_0, \qquad (39\text{--}10)$$

where λ_0 and k_0 are the values of λ and k, respectively, in vacuum.

To add the two sinusoidal functions with a phase difference, we can use the *phasor* representation that we have previously used for simple harmonic motion (Section 11–3) and for voltages and currents in ac circuits (Section 34–1). We suggest you review these sections so that phasors are fresh in mind. Each sinusoidal function is represented by a rotating vector (phasor) whose projection on the horizontal axis at any instant represents the instantaneous value of the sinusoidal function. In Fig. 39–5, E_1 is the phasor representing the wave from source S_1, and E_2 is the phasor for S_2. E_2 is *ahead* of E_1 in phase by an angle δ, as shown in the diagram. The amplitude of the resultant sinusoidal wave at P is the magnitude of the phasor labeled E_P in the diagram, the *vector sum* of E_1 and E_2. To find E_P, we use the law of cosines:

Vector addition is used to combine phasors representing two sinusoidal functions.

$$E_P{}^2 = E_1{}^2 + E_2{}^2 - 2E_1E_2 \cos (\pi - \delta).$$

If the amplitudes of the two waves are equal, then $E_1 = E_2 = E$, and

$$E_P{}^2 = E^2 + E^2 - 2E^2 \cos (\pi - \delta)$$

$$= 2E^2 + 2E^2 \cos \delta = 2E^2(1 + \cos \delta)$$

$$= 4E^2 \cos^2 (\delta/2). \qquad (39\text{--}11)$$

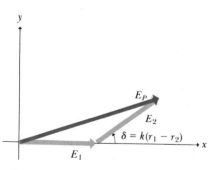

39–5 Phasor diagram of the variations of the electric field at a point P where two waves are superposed.

When the two waves are in phase, $\delta = 0$ and $E_p = 2E$. When they are exactly a half-cycle out of phase, $\delta = \pi$ (180°), $\delta/2 = \pi/2$, $\cos^2 (\delta/2) = 0$, and $E = 0$.

Thus the superposition of two sinusoidal waves with the same amplitude and frequency but in a phase difference yields a sinusoidal wave with amplitude between zero and twice the individual amplitudes, depending on the phase difference.

Next we relate δ to the geometry of the situation. The path difference is given by Eq. (39–3):

$$r_1 - r_2 = d \sin \theta.$$

Combining this with Eq. (39–9), we find

$$\delta = k(r_1 - r_2) = kd \sin \theta = \frac{2\pi d}{\lambda} \sin \theta. \qquad (39\text{–}12)$$

When we substitute this into Eq. (39–11), we find

$$E_P{}^2 = 4E^2 \cos^2 \left(\frac{1}{2} kd \sin \theta \right) = 4E^2 \cos^2 \left(\frac{\pi d}{\lambda} \sin \theta \right). \qquad (39\text{–}13)$$

Finally, the intensity I is given by Eq. (39–8):

$$I = \frac{1}{2} \epsilon_0 c E_P{}^2$$

$$= 2\epsilon_0 c E^2 \cos^2 \left(\frac{1}{2} kd \sin \theta \right)$$

$$= 2\epsilon_0 c E^2 \cos^2 \left(\frac{\pi d}{\lambda} \sin \theta \right). \qquad (39\text{–}14)$$

The intensity in a two-source interference pattern depends on the direction from the source.

We can simplify this result by expressing it in terms of the intensity I_0 at points where $\delta = 0$, such as $\theta = 0$. From Eq. (39–14),

$$I_0 = 2\epsilon_0 c E^2,$$

and finally

$$I = I_0 \cos^2 \left(\frac{1}{2} kd \sin \theta \right) \qquad (39\text{–}15)$$

$$= I_0 \cos^2 \left(\frac{\pi d}{\lambda} \sin \theta \right).$$

Note that at points of maximum intensity (maximum constructive interference), the intensity is *four times* (not twice) as great as it would be from either source alone. Of course, there are other points where it is *less* than the intensity from either source by itself.

In Section 39–2 we noted that in experiments with light we can describe the positions of the bright fringes with the coordinate y, as in Eq. (39–6), and that ordinarily $y \ll R$. In that case $\sin \theta$ is approximately equal to y/R, and we obtain the even simpler expressions

$$I = I_0 \cos^2 \left(\frac{kdy}{2R} \right) = I_0 \cos^2 \left(\frac{\pi dy}{\lambda R} \right). \qquad (39\text{–}16)$$

We can use these to calculate the intensity of light at any point and to compare the results with the photographically recorded pattern of Fig. 39–4.

Intensity distribution in an interference pattern from two radio antennas

EXAMPLE 39-3 In Fig. 39-1, suppose the two sources are identical radio antennas 10 m apart, radiating waves in all directions with a frequency $f = 30$ MHz. If the intensity in the $+x$-direction (corresponding to $\theta = 0$ in Fig. 39-2) is $I_0 = 0.02$ W·m^{-2}, what is the intensity in the direction $\theta = 45°$? In what direction is the intensity zero?

SOLUTION We want to use Eq. (39-15); the approximate relation of Eq. (39-16) cannot be used in this case because θ is not small. First we must find the wavelength, using the relation $c = \lambda f$. We find

$$\lambda = \frac{c}{f} = \frac{3.0 \times 10^8 \text{ m·s}^{-1}}{30 \times 10^6 \text{ s}^{-1}} = 10 \text{ m}.$$

The spacing between sources is $d = 10$ m, and Eq. (39-15) becomes

$$I = (0.02 \text{ W·m}^{-2}) \cos^2 \left[\frac{\pi(10 \text{ m})}{(10 \text{ m})} \sin \theta \right]$$
$$= (0.02 \text{ W·m}^{-2}) \cos^2 (\pi \sin \theta).$$

When $\theta = 45°$,

$$I = (0.02 \text{ W·m}^{-2}) \cos^2 (\pi \sin 45°) = 0.0073 \text{ W·m}^{-2}.$$

This is about 37% of the intensity at $\theta = 0$.

The intensity is zero when $\cos(\pi \sin \theta) = 0$; this occurs when $\pi \sin \theta = \pi/2$, $\sin \theta = \frac{1}{2}$, and $\theta = 30°$. By symmetry, the intensity is also zero when $\theta = 150°, 210°$, and $330°$ (or $-150°, -30°$).

This is not just a hypothetical problem. It is often desirable to beam most of the radiated energy from a radio transmitter in particular directions rather than uniformly in all directions. Pairs or rows of antennas are often used to produce the desired radiation pattern.

39-4 INTERFERENCE IN THIN FILMS

Interference in light reflected from two closely spaced surfaces.

The bright bands of color that you often see when light is reflected from a soap bubble or from a thin layer of oil floating on water are the result of an interference effect. Light waves are reflected from opposite surfaces of the thin films, and constructive interference between the two reflected waves occurs in different places for different wavelengths. The situation is represented schematically in Fig. 39-6. The line ab is a ray in a beam of light shining on the upper surface of a thin film. Part of the light is reflected at the first surface (ray bc), and part is transmitted (bd). At the second surface another partial reflection occurs, and some of the reflected light emerges (ray ef). The rays bc and ef come together at a point on the retina of the eye. Depending on the phase relationship, they may interfere constructively or destructively. Because different colors have different wavelengths, the interference may be constructive for some colors and destructive for others; hence the appearance of colored rings or fringes.

To keep things as simple as possible, let us consider interference of *monochromatic* light reflected from two nearly parallel surfaces. Figure 39-7 shows two plates of glass separated by a thin wedge of air; we want to consider interference between the two light waves reflected from the surfaces adjacent

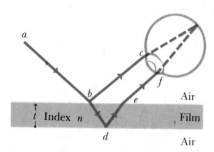

39-6 Interference between rays reflected from the upper and lower surfaces of a thin film.

to the air wedge, as shown. The situation is the same as in Fig. 39–6 except that the film thickness is not uniform. The refractions at the upper surface of the top plate in Fig. 39–7 do not change the situation in any essential way. If the observer is at a great distance compared to the other dimensions of the experiment and the rays are nearly perpendicular to the reflecting surfaces, then the path difference between the two waves (corresponding to $r_1 - r_2$ in the preceding section) is just twice the thickness d of the air wedge at each point. At points for which this path difference is an integer number of wavelengths, we expect to see constructive interference and a bright area, and where it is a half-integer number of wavelengths, destructive interference and a dark area. Along the line where the plates are in contact, there is *no* path difference and we expect a bright area.

It is easy enough to carry out this experiment; the bright and dark fringes appear as expected, but they are interchanged! Along the line of contact a *dark* fringe, not a bright one, is found. What happened? The inescapable conclusion is that one of the waves has undergone a half-cycle phase shift during its reflection, so that the two reflected waves are a half-cycle out of phase even though they have the same path length.

Further experiments show that a half-cycle phase change occurs whenever the material in which the wave is initially traveling before reflection has a *smaller* refractive index than the second material forming the interface. But when the first material has a *greater* refractive index than the second, such as a wave in glass reflected internally at a glass–air interface, *no* phase change occurs. Thus in Fig. 39–6 the wave reflected at point b undergoes the half-cycle phase shift, while the wave reflected at d does not. Similarly, in Fig. 39–7 the wave reflected from the upper surface of the lower plate has a half-cycle phase shift, while the other wave has none.

The path difference for the two waves is twice the thickness of the film.

39–7 Interference between two light waves reflected from the two sides of an air wedge separating two glass plates. The path difference is $2d$.

In some cases reflection causes an additional half-cycle phase shift.

EXAMPLE 39–4 Suppose the two glass plates in Fig. 39–7 are two microscope slides 10 cm long. At one end they are in contact, and at the other end they are separated by a thin piece of tissue paper 0.02 mm thick. What is the spacing of the resulting interference fringes? Is the fringe adjacent to the line of contact bright or dark? Assume monochromatic light with $\lambda = 500$ nm.

SOLUTION To answer the second question first, the fringe at the line of contact is dark because the wave reflected from the lower surface of the air wedge has undergone a half-cycle phase shift, while that from the upper surface has not. For this reason, the condition for *destructive* interference (a dark fringe) is that the path difference ($2d$) be an integer number of wavelengths:

$$2d = m\lambda \qquad (m = 0, 1, 2, 3, \ldots). \qquad (39\text{–}17)$$

From similar triangles in Fig. 39–7, d is proportional to the distance x from the line of contact:

$$\frac{d}{x} = \frac{h}{l}.$$

Combining this with Eq. (39–17), we find

$$\frac{2xh}{l} = m\lambda,$$

A simple example of interference from an air space between two glass plates

or

$$x = m\frac{l\lambda}{2h} = m\frac{(0.1 \text{ m})(500 \times 10^{-9} \text{ m})}{(2)(0.02 \times 10^{-3} \text{ m})} = m(1.25 \text{ mm}).$$

If the film has an index of refraction different from unity, the wavelength in the material is less than in vacuum.

Thus successive dark fringes, corresponding to successive integer values of m, are spaced 1.25 mm apart.

In this example, if the space between plates contains water ($n = 1.33$) instead of air, the phase changes are the same but the wavelength is $\lambda = \lambda_0/n = 376$ nm, and the fringe spacing is 0.94 mm. But now suppose the top plate is a glass with $n = 1.4$, the wedge is filled with a silicone grease having $n = 1.5$, and the bottom plate has $n = 1.6$. In this case there are half-cycle phase shifts at *both* surfaces bounding the wedge, and the line of contact corresponds to a *bright* fringe, not a dark one. The fringe spacing is again obtained by using the wavelength in the wedge (i.e., in the silicone grease), $\lambda = 500$ nm$/1.5 = 333$ nm; we invite you to show that the fringe spacing is 0.83 mm.

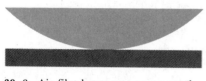

39–8 Air film between a convex and a plane surface.

Colored interference fringes formed by reflection of white light.

If we illuminate the arrangement in Fig. 39–7 first with blue, then with red light, the spacing of the red fringes is greater than that of the blue, as we should expect from the greater wavelength of the red light. The fringes produced by intermediate wavelengths occupy intermediate positions. If it is illuminated by white light, the color at any point is that due to the mixture of those colors that may be reflected at that point, while the colors for which the thickness is such as to result in destructive interference are absent. Just those colors that are absent in the reflected light, however, are found to predominate in the transmitted light. At any point, the color of the wedge as viewed by reflected light is *complementary* to its color as seen by transmitted light! Roughly speaking, the complement to any color is obtained by removing that color from white light. For example, the complement of blue is yellow, and the complement of green is magenta.

If the convex surface of a lens is placed in contact with a plane glass plate, as in Fig. 39–8, a thin film of air is formed between the two surfaces. The resulting circular interference fringes are shown in Fig. 39–9; they were studied by Newton and are called *Newton's rings*. When viewed by reflected light, the center of the pattern is black. Can you use the discussion of this section to show why this should be expected?

The surface of an optical part that is being ground to some desired curvature may be compared with that of another surface, known to be correct, by bringing the two in contact and observing the interference fringes. Figure 39–10 is a photograph made during the grinding of a telescope objective lens. The larger-diameter, thicker lower disk is the master, and the smaller upper disk is the lens under test. The "contour lines" are Newton's interference fringes; each one indicates an additional departure of the specimen from the master of $\frac{1}{2}$ wavelength of light. At ten lines from the center spot the distance between the two surfaces is five wavelengths, or about 0.0001 inch. This specimen is very poor; high-quality lenses are routinely ground with a precision of less than one wavelength!

Thin-film interference occurs in nonreflective coatings for lenses. A thin layer or film of hard transparent material with an index of refraction smaller

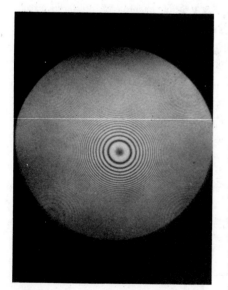

39–9 Newton's rings formed by interference in the air film between a convex and a plane surface. (Courtesy of Bausch & Lomb Optical Co.)

39–10 The surface of a telescope objective under inspection during manufacture. (Courtesy of Bausch & Lomb Optical Co.)

than that of the glass is deposited on the surface of the glass, as in Fig. 39–11. If the coating has the proper index of refraction, equal quantities of light will be reflected from its outer surface and from the boundary surface between it and the glass. Furthermore, since in both reflections the light is reflected from a medium of greater index than that in which it is traveling, the same phase change occurs in each reflection. It follows that if the film thickness is $\frac{1}{4}$ wavelength *in the film* (normal incidence is assumed), the light reflected from the first surface will be 180° out of phase with that reflected from the second, and complete destructive interference will result.

Of course, the thickness can be $\frac{1}{4}$ wavelength for only one particular wavelength. This is usually chosen in the yellow-green portion of the spectrum (about 550 mm), where the eye is most sensitive. Some reflection then takes place at both longer and shorter wavelengths, and the reflected light has a purple hue. The overall reflection from a lens or prism surface can be reduced in this way from 4% to 5% to a fraction of 1%. The treatment is extremely effective in eliminating stray reflected light and increasing the contrast in an image formed by highly corrected lenses having a large number of air–glass surfaces. A commonly used coating material is magnesium fluoride, MgF_2, with an index of 1.38. With this coating, the wavelength of green light in the coating is

$$\lambda = \frac{\lambda_0}{n} = \frac{550 \times 10^{-9}\,\text{m}}{1.38} = 4 \times 10^{-5}\,\text{cm},$$

and the thickness of a "nonreflecting" film of MgF_2 is 10^{-5} cm.

If a material whose index of refraction is *greater* than that of glass is deposited on glass to a thickness of $\frac{1}{4}$ wavelength, then the reflectivity is *increased*. For example, a coating of index 2.5 will allow 38% of the incident energy to be reflected, instead of the usual 4% when there is no coating. By use of multilayer coatings, it is possible to achieve reflectivity for a particular wavelength of almost 100%. These coatings are used for "one-way" windows and reflective sunglasses.

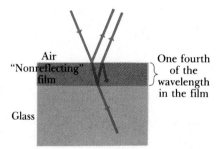

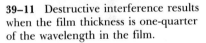

39–11 Destructive interference results when the film thickness is one-quarter of the wavelength in the film.

A film can't be perfectly nonreflecting for all wavelengths; some compromises are necessary.

39–5 THE MICHELSON INTERFEROMETER

A slightly complicated instrument of
great historical importance

The **Michelson interferometer** played an interesting role in the history of
science during the latter part of the nineteenth century and has had an equally
important role in establishing high-precision standards of the unit of length.
In contrast to the Young two-slit experiment, which uses light from two very
narrow sources, the Michelson interferometer uses light from a broad, spread-
out source. Figure 39–12 is a diagram of its principal features. The figure
shows the path of one ray from a point A of an extended but monochromatic
source. This ray strikes a glass plate C, the right side of which has a thin
coating of silver. Part of the light is reflected from the silvered surface at point
P to the mirror M_2 and back through C to the observer's eye. The remainder
of the light passes through the silvered surface and the compensator plate D
and is reflected from mirror M_1. It then returns through D and is reflected
from the silvered surface of C to the observer. The compensator plate D is cut
from the same piece of glass as plate C, so that its thickness will not differ from
that of C by more than a fraction of a wavelength. Its purpose is to ensure that
rays 1 and 2 pass through the same thickness of glass. Plate C is called a *beam
splitter*.

The whole apparatus is mounted on a very rigid frame, and a fine, very
accurate screw thread is used to move the mirror M_2. A common commercial
model of the interferometer is shown in Fig. 39–13. The light source is placed
to the left, and the observer is directly in front of the handle that turns the
screw.

An interference pattern is formed by
superposition of two waves traveling
different paths.

If the distances L_1 and L_2 in Fig. 39–12 are exactly equal, and the mirrors
M_1 and M_2 are exactly at right angles, the virtual image of M_1 formed by
reflection at the silvered surface of plate C coincides with mirror M_2. If L_1 and
L_2 are not exactly equal, the image of M_1 is displaced slightly from M_2; and if
the angle between the mirrors is not exactly a right angle, the image of M_1
makes a slight angle with M_2. Then the mirror M_2 and the virtual image of M_1
play the same roles as the two surfaces of a thin film, discussed in Section
39–3, and the same sort of interference fringes result from the light reflected
from these surfaces.

A technique for high-precision
wavelength measurements.

Suppose that the extended source in Fig. 39–12 is monochromatic with
wavelength λ, and that the angle between mirror M_2 and the virtual image of

39–12 The Michelson interferometer.

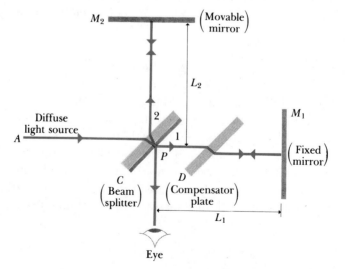

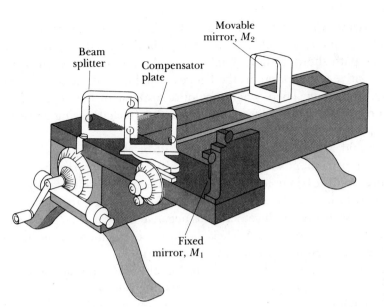

Beam
splitter

Compensator
plate

Movable
mirror, M_2

Fixed
mirror, M_1

39–13 A common type of Michelson interferometer.

M_1 is such that five or six vertical fringes are present in the field of view. If the mirror M_2 is now moved slowly either backward or forward by a distance $\lambda/2$, the effective film thickness changes by λ and each of the fringes moves either to the right or to the left through a distance equal to the spacing of the fringes. If the fringes are observed through a telescope whose eyepiece is equipped with a cross hair, and m fringes cross the cross hair when the mirror is moved a distance x, then

$$x = m\frac{\lambda}{2} \quad \text{or} \quad \lambda = \frac{2x}{m}. \qquad (39\text{–}18)$$

If m is several thousand, the distance x is large enough so that it can be measured with good precision, and hence a precise value of the wavelength λ can be obtained.

Until recently the meter was defined as a length equal to a specified number of wavelengths of the orange-red light of krypton-86. Before this standard could be established, it was necessary to measure as accurately as possible the number of these wavelengths in the *former* standard meter, defined as the distance between two scratches on a bar of platinum-iridium. The measurement was made with a modified Michelson interferometer many times and under very carefully controlled conditions. The number of wavelengths in a distance equal to the old standard meter was found to be 1,650,763.73 wavelengths. The meter was then defined as *exactly* this number of wavelengths. As we mentioned in Section 1–2, this definition has recently been superseded by a new length standard based on the unit of *time*.

Another application of the Michelson interferometer with considerable historical interest is the Michelson-Morley experiment. To understand the purpose of this experiment, we must recall that before the electromagnetic theory of light and Einstein's special theory of relativity became established, physicists believed that the propagation of light waves occurred in a medium called the **ether,** which was believed to permeate all space. In 1887 Michelson and Morley used the Michelson interferometer in an attempt to detect the motion of the earth through the ether. Suppose the interferometer in Fig.

The role of interferometry in establishing a standard of length

The Michelson-Morley experiment: a death-blow to the ether theory of light

39–12 were moving from left to right relative to the ether. According to nineteenth-century theory, this would lead to changes in the speed of light in the portions of the path shown as horizontal lines in the figure. There would be fringe shifts relative to the positions the fringes would have if the instrument were at rest in the ether. Then, when the entire instrument was rotated 90°, the other portions of the paths would be similarly affected, giving a fringe shift in the opposite direction.

Michelson and Morley expected a fringe shift of about four-tenths of a fringe when the instrument was rotated. The shift actually observed was less than a hundredth of a fringe and, within the limits of experimental uncertainty, appeared to be exactly zero. Despite its orbital velocity, the earth appeared to be at rest relative to the ether. This negative result baffled physicists of the time, and to this day the Michelson-Morley experiment is the most significant "negative-result" experiment ever performed.

Understanding of this result had to wait for Einstein's special theory of relativity, published in 1905. Einstein realized that the velocity of a light wave has the same magnitude c relative to *all* reference frames, whatever their velocity may be relative to each other. The presumed ether then plays no role, and the very concept of an ether has been given up. The theory of relativity is a well-established cornerstone of modern physics, and we will study it in detail in Chapter 40.

Evidence for the correctness of the special theory of relativity

39–6 FRESNEL DIFFRACTION

According to *geometrical* optics, if an opaque object is placed between a point light source and a screen, as in Fig. 39–14, the edges of the object cast a sharp shadow on the screen. No light reaches the screen at points within the geometrical shadow, while outside the shadow the screen is uniformly illuminated. But as we have seen, geometrical optics is an idealized model of the behavior of light, and there are many situations where the ray model of light is inadequate. Another important class of such phenomena is grouped under the heading **diffraction.**

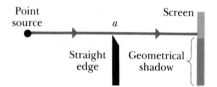

39–14 Geometrical shadow of a straight edge.

The photograph in Fig. 39–15 was made by placing a razor blade halfway between a pinhole illuminated by monochromatic light and a photographic film, so that the film made a record of the shadow cast by the blade. Figure 39–16 is an enlargement of a region near the shadow of an edge of the blade. The boundary of the *geometrical* shadow is indicated by the short arrows. The geometrical shadow is bordered by alternating bright and dark bands. Some light has "bent" around the edge into the shadow region, although this is not visible in the figure. Note also that the first bright band, just outside the geometrical shadow, is actually *brighter* than in the region of uniform illumination to the extreme left. This simple experimental setup serves to give some idea of the true complexity of what is often considered the most elementary of optical phenomena, the shadow cast by a small source of light.

Diffraction phenomena: When is a shadow not really a shadow?

Diffraction patterns such as that in Fig. 39–15 are not commonly observed in everyday life because most ordinary light sources are not point sources of monochromatic light. If the shadow of a razor blade is cast by a frosted-bulb incandescent lamp, for example, the light from every point of the surface of the lamp forms its own diffraction pattern, but these overlap to such an extent that no individual pattern can be observed.

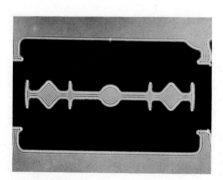

39–15 Shadow of a razor blade.

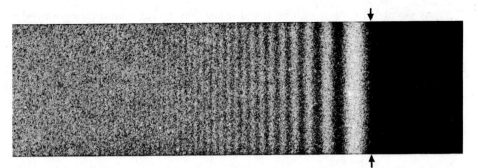

39–16 Shadow of a straight edge. The arrows show the position of the *geometric* shadow.

The term *diffraction* is applied to problems involving *the resultant effect produced by a limited portion of a wave front.* Since in most diffraction problems some light is found within the region of geometrical shadow, diffraction is sometimes defined as "the bending of light around an obstacle." It should be emphasized, however, that the process by which diffraction effects are produced is going on continuously in the propagation of *every* wave. Only if part of the wave is cut off by some obstacle do we observe diffraction effects. But every optical instrument uses only a limited portion of a wave; for example, a telescope uses only that portion of a wave admitted by the objective lens. Thus diffraction plays a role in practically all optical phenomena.

Figure 39–17 shows a diffraction pattern formed by a steel ball about 3 mm in diameter. Note the rings in the pattern, both outside and inside the geometrical shadow area, and the bright spot at the very center of the shadow. The existence of this spot was predicted in 1818 by the French mathematician Poisson. Ironically, Poisson himself was not a believer in the wave theory of light, and he published this *apparently* absurd result as a final ridicule of the wave theory. But almost immediately, the bright spot was observed experimentally by Arago. It had, in fact, been observed much earlier, in 1723 by Maraldi, but its significance was not recognized then.

Do you believe that the shadow of a circular object can have a bright spot at its center?

The essential features observed in diffraction effects can be predicted with the help of Huygens' principle: Every point of a wave surface can be considered the source of a secondary wavelet that spreads out in all directions. At

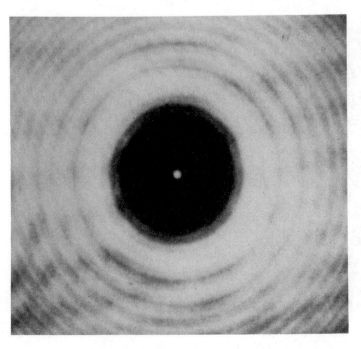

39–17 Fresnel diffraction pattern formed by a steel ball 3 mm in diameter. The Poisson bright spot is seen at the center of the shadow area. (Courtesy Prof. E. Hecht.)

every point we must combine the displacements that would be produced by the secondary wavelets, taking into account their amplitudes and relative phases. The mathematical operations are often quite complicated.

In Fig. 39–14 both the point source and the screen are at finite distances from the obstacle forming the diffraction pattern. This situation is described as **Fresnel diffraction** (after Augustin Jean Fresnel, 1788–1827), and the resulting pattern on the screen is called a *Fresnel diffraction pattern*. If the source, obstacle, and screen are far enough away so that the lines from the source to the obstacle and from the obstacle to a point in the pattern formed on the screen can be considered to be parallel, the phenomenon is called **Fraunhofer diffraction** (after Joseph von Fraunhofer, 1787–1826). The latter situation is simpler to treat in detail, and our analysis will be restricted to Fraunhofer diffraction.

39–7 FRAUNHOFER DIFFRACTION FROM A SINGLE SLIT

Suppose a monochromatic plane wave (in the ray picture, a beam of parallel rays) is incident on an opaque plate having a narrow slit. According to geometrical optics, the transmitted beam should have the same cross section as the slit, and a screen in the path of the beam would be illuminated uniformly over an area of the same size and shape as the slit, as in Fig. 39–18a. What is *actually* observed is the pattern shown in Fig. 39–18b. The beam spreads out horizontally after passing through the slit, and the diffraction pattern consists of a central bright band, which may be much wider than the slit width, bordered by alternating dark bands and bright bands of decreasing intensity. You can easily observe a diffraction pattern of this sort by looking at a point source such as a distant street light through a narrow slit formed between two fingers in front of your eye. The retina of your eye then corresponds to the screen.

Figure 39–19 shows a side view of the setup of Fig. 39–18, with the slit turned to a horizontal position. That is, in Fig. 39–19 we are looking along the

The analysis is simpler when all the distances are much larger than the wavelength of light.

Light emerging from an aperture spreads out beyond the area predicted by geometrical optics.

39–18 (a) Geometrical "shadow" of a slit. (b) Diffraction pattern of a slit. The slit width has been greatly exaggerated.

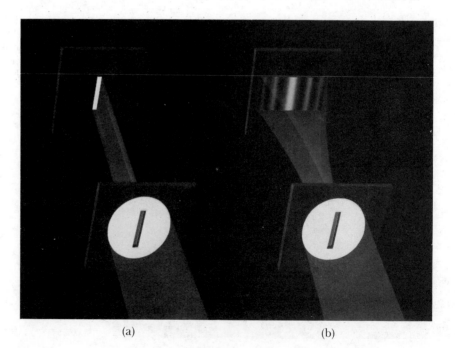

(a)   (b)

length of the slit. According to Huygens' principle, each element of area of the slit opening can be considered a source of secondary wavelets. In particular, we may imagine elements of area formed by subdividing the slit into narrow strips parallel to the long edges, perpendicular to the page in Fig. 39–19a. From each strip, cylindrical secondary wavelets spread out, as shown in cross section.

In Fig. 39–19b a screen is placed to the right of the slit. The resultant intensity at a point P is calculated by adding the contributions from the individual wavelets, taking proper account of their various phases and amplitudes. The problem becomes much simpler if the screen is far away, so the rays from the slit to the screen are parallel, as in Fig. 39–19c, or when a lens is placed as in Fig. 39–19d, in which case the rays to the lens are parallel and the lens forms in its focal plane a reduced *image* of the pattern that would be formed on an infinitely distant screen without the lens. You might ask whether the lens introduces additional phase shifts that are different for different parts of the wave front; it can be shown quite generally that the lens *does not* cause any additional phase shifts.

The situation of Fig. 39–19b is Fresnel diffraction; those in Figs. 39–19c and 39–19d, where the outgoing rays can be considered parallel, are Fraunhofer diffraction. Some aspects of Fraunhofer diffraction from a single slit can be deduced easily. First consider two narrow strips, one just below the top edge of the slit and one at its center, as in Fig. 39–20. The difference in path length to point P is $(a/2) \sin \theta$, where a is the slit width. Suppose this path difference happens to be equal to $\lambda/2$; then light from these two strips arrives at point P with a half-cycle phase difference, and cancellation occurs. Similarly, light from two strips just below these two will also arrive a half-cycle out of phase; and, in fact, light from *every* strip in the top half cancels out that from a corresponding strip in the bottom half, resulting in complete cancellation and giving a dark fringe in the interference pattern. Thus a dark fringe occurs whenever

$$\frac{a}{2} \sin \theta = \pm \frac{\lambda}{2} \quad \text{or} \quad \sin \theta = \pm \frac{\lambda}{a}. \qquad (39\text{–}19)$$

We may also divide the screen into quarters, sixths, and so on, and use the argument above to show that a dark fringe occurs whenever $\sin \theta = \pm 2\lambda/a$,

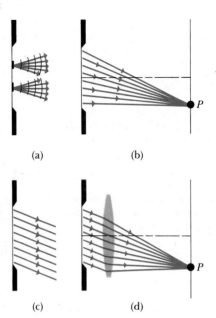

(a)   **(b)**

(c)   **(d)**

39–19 Diffraction by a single slit. Relation of Fresnel diffraction to Fraunhofer diffraction by a single slit.

Using the superposition principle to find the dark areas in the diffraction pattern

39–20 The wavefront is divided into a large number of narrow strips.

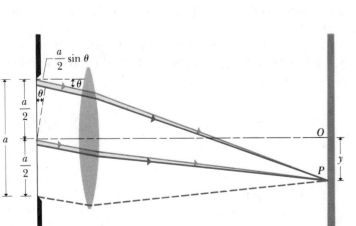

$\pm 3\lambda/a$, and so on. Hence the condition for a *dark* fringe is

$$\sin \theta = \frac{n\lambda}{a} \qquad (n = \pm 1, \pm 2, \pm 3, \ldots). \qquad (39\text{-}20)$$

An important difference between the single-slit pattern and the two-slit interference pattern

Using phasor addition to find the intensity distribution in a single-slit diffraction pattern

For example, if the slit width is equal to ten wavelengths, dark fringes occur at $\sin \theta = \pm\frac{1}{10}, \pm\frac{2}{10}, \pm\frac{3}{10}, \ldots$. Between the dark fringes are bright fringes. We also note that $\sin \theta = 0$ is a *bright* band, since then light from the entire slit arrives at P in phase. The central bright fringe is therefore twice as wide as the others, as Fig. 39–18 shows.

We can derive an expression for the intensity distribution for the single-slit pattern by the same method we used to obtain Eq. (39–15) for the two-slit pattern. We again imagine a plane wavefront at the slit subdivided into a large number of strips, each of which sends out rays in *all* directions toward the lens, as shown in Fig. 39–20. If we choose an arbitrary point P on a screen in the focal plane of the lens, only those rays making an angle θ with the axis will arrive at P. Point O on the screen is the special point at which all rays making the angle $\theta = 0$ arrive. Figure 39–21a is a phasor diagram for the situation, showing that when the slit is subdivided into 14 sections and each section emits a Huygens wavelet at the angle $\theta = 0$, all wavelets arrive in phase. The resultant amplitude at O is denoted by S.

With the same subdivision of the wavefront into 14 strips, the wavelets that make the angle θ and arrive at P have a slight phase difference between succeeding wavelets, and the corresponding phasor diagram is shown in Fig. 39–21b. The sum S is now the perimeter of a portion of a many-sided polygon and E_P, the amplitude of the resultant electric field at P, is the *chord*. The angle δ is the total phase difference between the wave from the bottom strip of Fig. 39–20 and the wave from the top strip.

In the limit, as the number of strips into which the slit is subdivided is increased indefinitely, the phasor diagram becomes an *arc of a circle*, as shown in Fig. 39–21c, with the arc length S equal to the length S in Fig. 39–21a. We can find the center C of this arc by constructing perpendiculars at A and B. From the definition of radian measure of angles, the radius of the arc is S/δ, and the resultant amplitude E_P (distance AB) is $2(S/\delta) \sin (\delta/2)$. We then have

$$E_P = S\frac{\sin \delta/2}{\delta/2},$$

where δ is the phase difference between the two wavelets at the extreme top and bottom edges of the slit.

From Eq. (39–9), the phase difference is $2\pi/\lambda$ times the path difference. From Fig. 39–20, the path difference between the ray from the top of the slit and that from the bottom is $a \sin \theta$. Therefore

$$\delta = \frac{2\pi}{\lambda} a \sin \theta, \qquad (39\text{-}21)$$

and

$$E_P = S\frac{\sin[\pi a(\sin \theta)/\lambda]}{\pi a(\sin \theta)/\lambda}.$$

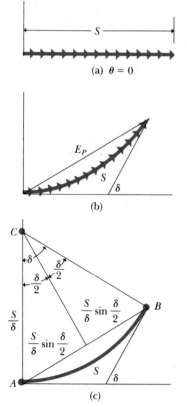

39–21 (a) Phasor diagram when all elementary electric fields are in phase ($\theta = 0$, $\delta = 0$). (b) Phasor diagram when each elementary electric field differs in phase slightly from the preceding one. (c) Limit reached by the phasor diagram when the slit is subdivided infinitely.

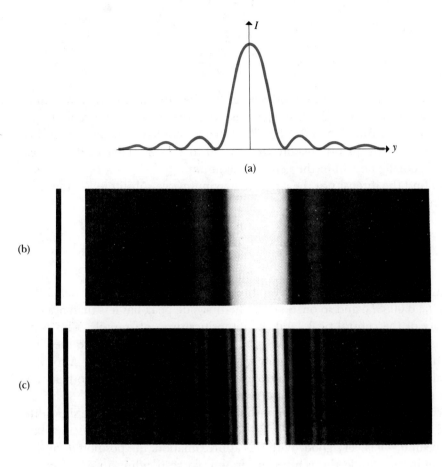

(a)

(b)

(c)

39-22 (a) Intensity distribution. (b) Photograph of the Fraunhofer diffraction pattern of a single slit. (c) Fraunhofer diffraction pattern of a double slit.

Since the intensity I is proportional to the *square* of the amplitude,

$$I = I_0 \left(\frac{\sin \delta/2}{\delta/2} \right)^2 = I_0 \left\{ \frac{\sin[\pi a (\sin \theta)/\lambda]}{\pi a (\sin \theta)/\lambda} \right\}^2, \qquad (39\text{-}22)$$

where I_0 is the intensity at O in Fig. 39-20 where $\theta = 0$.

Equation (39-22) is plotted in Fig. 39-22a, and a photograph of an actual diffraction pattern is shown directly under it in Fig. 39-22b. Note that the intensity of the central maximum is much greater than at any of the others, and that the peak intensities drop off rapidly as we go away from the center of the pattern.

We can calculate the positions of the peaks and the peak intensities from Eq. (39-22). The central peak occurs when $\delta = 0$. Substituting this value into Eq. (39-22) gives an indeterminate form (0/0), but by taking the limit as $\delta \to 0$, we find that when $\delta = 0$, $I = I_0$. The side maxima occur approximately where $\sin(\delta/2) = \pm 1$, or

$$\frac{\delta}{2} = \frac{3\pi}{2}, \frac{5\pi}{2}, \cdots,$$

or in general

$$\frac{\delta}{2} = \left(m + \frac{1}{2} \right) \pi \qquad (m = 1, 2, 3, \ldots), \qquad (39\text{-}23)$$

or, using Eq. (39–21),

$$\sin \theta = \left(m + \frac{1}{2}\right)\frac{\lambda}{a}. \tag{39–24}$$

Note that these maximum intensities occur between the minima that we found earlier, as given by Eq. (39–20). Note also that there is *no* side maximum at $\delta/2 = \pi/2$. The function of δ in Eq. (39–22) does not have a maximum at this value of δ but drops off steadily from $\delta = 0$ to $\delta = 2\pi$. That is, the values $\delta = \pm\pi$ actually lie within the central maximum.

The intensities of the side maxima are much less than for the central maximum.

To find the intensity at the *m*th side maximum, we substitute Eq. (39–23) back into Eq. (39–22) to obtain

$$I = \frac{I_0}{(m + \frac{1}{2})^2 \pi^2}. \tag{39–25}$$

Putting in $m = 1$, $m = 2$, . . . , we find the series of intensities $0.0450I_0$, $0.0162I_0$, $0.0083I_0$, and so on. So even the first side peak has less than 5% the intensity of the central peak, and the intensities drop off rapidly. This is in contrast to the two-slit pattern we studied in Section 39–3, where the side maxima were approximately as intense as the central maximum. Also, the central maximum in the single-slit pattern is twice as wide as the others, an effect not seen in the two-slit pattern.

With light, the wavelength λ is ordinarily much smaller than the slit width a, and as a result the values of θ in Eq. (39–22) are so small that the approximation $\sin \theta = \theta$ is very good. With this approximation, the position θ_1 of the first minimum beside the central maximum, corresponding to $\delta/2 = \pi$, is, from Eq. (39–21),

$$\theta_1 = \frac{\lambda}{a}. \tag{39–26}$$

Superposition of interference and diffraction patterns resulting from two slits of finite width

When a is of the order of a centimeter or more, θ_1 is so small that we can consider practically all the light to be concentrated at the geometrical focus.

The photograph in Fig. 39–22c shows the Fraunhofer diffraction pattern of *two* slits, each with the same width a as for Fig. 39–22b but separated by a distance $d = 4a$. The two-slit interference pattern, represented by the narrow fringes, is modified by the diffraction curve of part (a) because of the finite widths of the slits. Thus the first minimum in the single-slit pattern in Fig. 39–21c completely extinguishes the fourth bright fringe of the two-slit pattern.

39–8 THE DIFFRACTION GRATING

The diffraction grating: many equally spaced sources

Suppose that instead of a single slit or two slits side by side, we have a very large number of parallel slits, all with the same width and spaced equal distances apart. Such an arrangement is called a **diffraction grating;** the first one was constructed by Fraunhofer with fine wires. Gratings are now made by using a diamond point to scratch a large number of equally spaced grooves on a glass or metal surface, or by photographic reduction of a black-and-white pattern drawn with a pen.

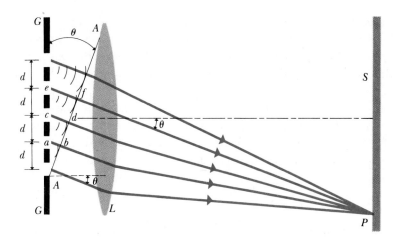

39–23 The plane diffraction grating.

In Fig. 39–23, *GG* represents the grating; the slits are perpendicular to the plane of the page. Only five slits are shown in the diagram, but an actual grating may contain several thousand, with spacing *d* of the order of 0.002 mm. A plane monochromatic wave is incident normally on the grating from the left side. The problem of finding the intensity pattern in the light transmitted by the grating then combines the principles of interference and diffraction. Each slit creates a diffraction pattern, and these then interfere with each other to produce the final pattern. The lens is included so we can view the pattern on a screen at a finite distance from the grating and still meet the conditions for Fraunhofer diffraction, that is, parallel rays emerging from the grating.

Let us assume that the slits are very narrow, so the diffracted beam from each spreads out over a wide enough angle for it to interfere with all the other diffracted beams. Consider first the light proceeding from elements of infinitesimal width at the lower edges of each opening and traveling in a direction making an angle θ with that of the incident beam, as in Fig. 39–23. A lens at the right of the grating forms in its focal plane a diffraction pattern similar to that which would appear on a screen at infinity.

Suppose the angle θ in Fig. 39–23 is taken so that the distance *ab* equals λ, the wavelength of the incident light. Then $cd = 2\lambda$, $ef = 3\lambda$, and so on. The waves from all these elements, since they are in phase at the plane of the grating, are also in phase along the plane *AA* and therefore reach the point *P* in phase. The same holds true for any set of elements in corresponding positions in the various slits.

If the angle θ is increased slightly, the waves from the various grating slits no longer arrive at *AA* in phase, and even an extremely small change in angle results in almost complete destructive interference among them, provided there are a large number of slits in the grating. Roughly speaking, for each slit in the grating we can find another slit whose wave arrives a half-cycle out of phase with the first one and cancels it. Hence the maximum at the angle θ is an extremely sharp one, differing from the rather broad maxima that result from interference or diffraction effects with a small number of openings.

As the angle θ is increased still further, a position is eventually reached in which the distance *ab* in Fig. 39–23 becomes equal to 2λ. Then *cd* equals 4λ, *cf* equals 6λ, and so on. The waves at *AA* are again all in phase; the path differ-

A crucial difference between the grating pattern and a two-slit pattern

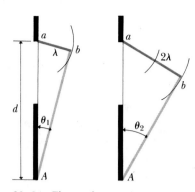

39–24 First-order maximum when $ab = \lambda$; second-order maximum when $ab = 2\lambda$.

Use of diffraction gratings in spectrometry

ence between adjacent waves is now 2λ, and another maximum results. Still others appear when $ab = 3\lambda, 4\lambda, \ldots$. We also find maxima at corresponding angles on the opposite side of the grating normal, as well as along the normal itself, since in the latter position the phase difference between waves reaching AA is zero.

From Fig. 39–24 we can find the angles of deviation at which the maxima occur. Consider the right triangle Aba. Let d be the distance between successive grating elements, called the *grating spacing*. The necessary condition for a maximum is that $ab = m\lambda$, where $m = 0, \pm1, \pm2, \pm3$, and so on. It follows that

$$\sin \theta = m\frac{\lambda}{d} \qquad (m = 0, \pm1, \pm2, \ldots) \qquad (39–27)$$

is the necessary condition for a maximum. The angle θ is also the angle by which the rays corresponding to the maxima have been *deviated* from the direction of the incident light.

In practice, the parallel beam of rays incident on the grating is usually produced by a lens with a narrow illuminated slit at its first focal point. These are not shown in Fig. 39–23. Each of the maxima in the diffraction pattern formed by the grating is then a sharp image of the slit. If the slit is illuminated by light consisting of a mixture of two or more wavelengths, the grating forms two or more series of images of the slit in different positions; each wavelength in the original light gives rise to a set of slit images deviated by the appropriate angles. If the slit is illuminated with white light, the diffraction grating forms a continuous group of images side by side. That is, white light is dispersed into continuous spectra. In contrast with the single spectrum produced by a prism, a grating forms several spectra on either side of the normal. Those that correspond to $m = \pm1$ in Eq. (39–27) are called *first order;* those that correspond to $m = \pm2$ are called *second order;* and so on. Since for $m = 0$ the deviation is zero, all colors combine to produce a white image of the slit in the direction of the incident beam.

As Eq. (39–27) shows, the sines of the deviation angles of the maxima are proportional to the ratio λ/d. Thus for substantial deviation to occur, the grating spacing d should be of the same order of magnitude as the wavelength λ. Gratings for use in or near the visible spectrum are ruled with from about 500 to 1500 lines per millimeter.

The diffraction grating is widely used in spectrometry, instead of a prism, as a means of dispersing a light beam into spectra. If the grating spacing is known, then from a measurement of the angle of deviation of any wavelength, the value of this wavelength may be computed. This is not true for a prism; the angles of deviation are not related in any simple way to the wavelengths but depend on the characteristics of the prism material. Since the index of refraction of optical glass varies more rapidly at the violet end than at the red end of the spectrum, the spectrum formed by a prism is always spread out more at the violet end than it is at the red. Also, while a prism deviates red light the least and violet the most, the reverse is true of a grating.

Two examples of diffraction-grating calculations

EXAMPLE 39–5 The wavelengths of the visible spectrum are approximately 400 nm to 700 nm. Find the angular breadth of the first-order visible spectrum produced by a plane grating having 6000 lines per centimeter, when white light is incident normally on the grating.

SOLUTION The grating spacing d is

$$d = \frac{1}{6000 \text{ lines·cm}^{-1}} = 1.67 \times 10^{-6} \text{ m}.$$

The angular deviation of the violet is

$$\sin \theta = \frac{400 \times 10^{-9} \text{ m}}{1.67 \times 10^{-6} \text{ m}} = 0.240,$$

$$\theta = 13.9°.$$

The angular deviation of the red is

$$\sin \theta = \frac{700 \times 10^{-9} \text{ m}}{1.67 \times 10^{-6} \text{ m}} = 0.420,$$

$$\theta = 24.8°.$$

Hence the first-order visible spectrum includes an angle of

$$24.8° - 13.9° = 10.9°.$$

EXAMPLE 39–6 Show that the violet of the third-order spectrum overlaps the red of the second-order spectrum.

SOLUTION The angular deviation of the third-order violet is

$$\sin \theta = \frac{(3)(400 \times 10^{-9} \text{ m})}{d}$$

and of the second-order red it is

$$\sin \theta = \frac{(2)(700 \times 10^{-9} \text{ m})}{d}.$$

Since the first angle is smaller than the second, whatever the grating spacing, the third order will *always* overlap the second.

PROBLEM-SOLVING STRATEGY: *Diffraction*

Here are several general comments about the kinds of problems you will encounter in this chapter. You may want to refer back to earlier sections and look for applications of these ideas.

1. Always draw a diagram; label all the distances and angles clearly. Several of the equations derived in this chapter contain an angle θ. Make absolutely sure you know where θ is on your diagram.

2. The equations for an intensity *maximum* in the two-source or grating pattern, Eqs. (39–4) and (39–27), look just like the equation for an intensity *minimum* in the single-slit pattern, Eq. (39–20). Be careful not to get these confused, and be sure you understand why the same equation gives intensity maxima in some situations and minima in others.

3. In x-ray diffraction, discussed in the next section, the angle θ is defined differently from the angles of incidence and reflection in geometrical optics. In this case, θ is the angle of the incoming beam relative to the crystal planes, not their normals. Be careful! Also notice that there are *two* conditions for an intensity maximum; the incident and scattering angles are equal, and the Bragg condition, Eq. (39–28), must be satisfied.

4. One more time we caution you about units. Light wavelengths are usually in nanometers (1 nm = 10^{-9} m), but if you refer to other books you may also find Ångstrom units; 1 Å = 10^{-10} m = $\frac{1}{10}$ nm. Dimensions of slits and apertures may be expressed in millimeters, and lens-to-screen distances in centimeters or meters. Lens focal lengths are traditionally expressed in millimeters. In coping with all this, be very careful not to drop any powers of ten!

39–9 X-RAY DIFFRACTION

X-rays have much shorter wavelengths than visible light.

X-rays were discovered by Röntgen in 1895, and early experiments suggested that they were electromagnetic waves with wavelengths of the order of 10^{-10} m. It was also strongly suspected at this time that in a crystalline solid the atoms are arranged in a lattice in a regular repeating pattern, with spacing between adjacent atoms also of the order of 10^{-10} m. Putting these two ideas together, Max von Laue (1879–1960) suggested in 1913 that a crystal might serve as a kind of three-dimensional diffraction grating for x-rays. That is, a beam of x-rays might be scattered by the individual atoms in a crystal, and the scattered waves might interfere just as the individual waves from a diffraction grating interfere.

Using a crystal as a diffraction grating for x-rays.

The first **x-ray diffraction** experiments were performed by Friederich and Knipping, and interference effects *were* observed. These experiments thus verified in a single stroke the hypothesis that x-rays *are* waves, or at least have wavelike properties, and that the atoms in a crystal *are* arranged in a regular pattern. Since that time, the phenomenon of x-ray diffraction by a crystal has proved an invaluable research tool, both as a way to measure x-ray wavelengths and as a method of studying the structure of crystals. Figure 39–25 is a diagram of the structure of a familiar crystal, sodium chloride.

To introduce the basic idea in a simple context, we consider first a two-dimensional scattering situation, as shown in Fig. 39–26a, where a plane wave is incident on a square array of scattering centers. The situation might be a ripple tank with an array of small posts, or 3-cm microwaves with an array of small conducting spheres, or x-rays with an array of atoms. In the case of electromagnetic waves, the wave induces an oscillating electric dipole moment in each scatterer, and each one emits a scattered wave. The total interference pattern is the superposition of all these scattered waves. To compute its nature we have to consider the total path differences for the various scattered waves, including the distances both from source to scatterer and from scatterer to observer.

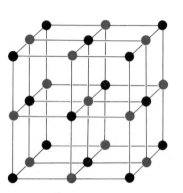

39–25 Model of arrangement of ions in a crystal of NaCl. Black circles, Na; color circles, Cl. The spacing of adjacent atom is 0.282 nm.

As Fig. 39–26a shows, the path length from source to observer is the same for all the scatterers in a single row if the two angles θ are equal, as shown. Scattered radiation from adjacent rows is *also* in phase if the path difference is an integer number of wavelengths. Figure 39–26b shows that the path difference for adjacent rows is $2d \sin \theta$. Thus the conditions for radiation from the *entire array* to reach the observer in phase are that (1) the angle of incidence must equal the angle of scattering, and (2) the path difference for adjacent

A crystal is analogous to a three-dimensional diffraction grating for x-rays.

39–26 Scattering of radiation from a square array. Interference from successive scatterers in a row is constructive when the angles of incidence and reflection are equal. Interference from adjacent rows is also constructive when Eq. (39–28) is satisfied.

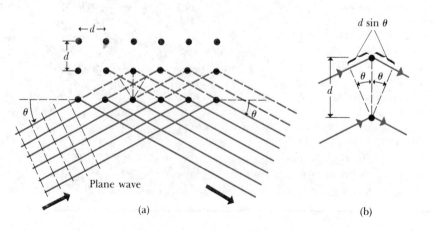

Plane wave

(a)

(b)

rows must equal $n\lambda$, where n is an integer. We can express the second condition as

$$2d \sin \theta = n\lambda \qquad (n = 1, 2, 3, \ldots). \qquad (39\text{–}28)$$

At points where this condition is satisfied, a strong maximum in the interference pattern is observed; otherwise no strong maximum is seen because there is no point where the radiation from all scatterers arrives in phase. We can describe this interference in terms of *reflections* of the wave from the horizontal rows of scatterers in Fig. 39–26a. Strong interference occurs at angles such that the incident and scattered angles are equal and Eq. (39–28) is satisfied.

We can extend this discussion to a three-dimensional array. Instead of *rows,* we consider *planes* of scatterers. Figure 39–27 shows that we can construct several different sets of parallel planes that pass through all the scatterers. Waves from all the scatterers in a given plane interfere constructively if the angles of incidence and scattering are equal. There is also constructive interference between planes when Eq. (39–28) is satisfied, where d is now the distance between adjacent planes. Because there are many different sets of parallel planes, there are also many values of d and many sets of angles corresponding to constructive interference for the whole crystal lattice. This phenomenon is called **Bragg reflection,** and Eq. (39–28) is called the **Bragg condition,** in honor of Sir William Bragg and his son Laurence Bragg, two pioneers in x-ray analysis. We must not let the term *reflection* obscure the fact that we are dealing with an *interference* effect.

Figure 39–28 is a photograph made by directing a narrow beam of x-rays at a thin section of a quartz crystal and allowing the scattered beam to strike a photographic film. As we predicted, nearly complete cancellation occurs for all but certain very specific directions, where constructive interference occurs and forms bright spots. Such a pattern is usually called an x-ray *diffraction* pattern, although as we have seen there is no fundamental distinction between the phenomena we call *interference* and those we call *diffraction*. Such patterns are also called Laue patterns.

If the crystal lattice spacing is known, we can determine the wavelength from the diffraction pattern, just as we determined wavelengths of visible light from measurements on diffraction patterns from slits or gratings. For example, we can determine the crystal lattice spacing for sodium chloride from its density and Avogadro's number. Conversely, once we know the x-ray wavelength, we can use x-ray diffraction to explore the structure and lattice spacing of crystals of unknown structure.

Indeed, x-ray diffraction has been by far the most important experimental tool in the investigation of crystal structure of solids. Atomic spacings in crystals can be measured precisely, and the details of the lattice arrangement of complex crystals can be determined. More recently, x-ray diffraction has played an important role in studies of the structures of liquids and of organic molecules. But the basic principles, superposition and interference, are exactly the same as for all the other phenomena we have studied in this chapter.

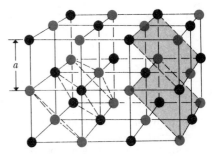

39–27 Cubic crystal lattice, showing two different families of crystal planes. The spacing of the planes on the left is $a/\sqrt{3}$; that of the planes on the right is $a/\sqrt{2}$. There are also three sets of planes parallel to the cube faces, with spacing a.

X-ray diffraction can be described in terms of reflection from crystal planes. Waves reflected from adjacent planes interfere with each other.

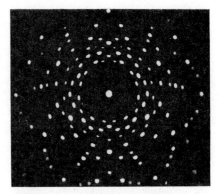

39–28 Laue diffraction pattern formed by directing a beam of x-rays at a thin section of quartz crystal. (Courtesy of Dr. B. E. Warren.)

X-ray diffraction: an important tool for investigating crystal structure

39–10 CIRCULAR APERTURES AND RESOLVING POWER

We have studied in detail the diffraction patterns formed by long, thin slits or arrays of slits. We also know from our introductory discussion that an aperture of *any* shape forms a diffraction pattern. The pattern formed by a *circular*

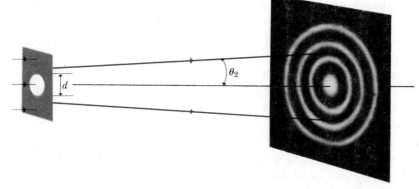

39–29 Diffraction pattern formed by a circular aperture of diameter d, consisting of a central bright spot and alternating dark and bright rings. The angular size θ_2 of the second dark ring is shown.

aperture is of special interest because of its role in limiting the resolving power of optical instruments. In principle we could compute the intensity at any point P in the diffraction pattern by dividing the area of the aperture into small elements, finding the resulting wave amplitude and phase at P, and then integrating over the aperture area to find the resultant amplitude and intensity at that point. In practice, the integration cannot be carried out in terms of elementary functions and has to be done by numerical approximation. We will simply describe the pattern and quote some relevant numbers.

The diffraction pattern formed by a circular aperture consists of a central bright spot surrounded by a series of bright and dark rings. Considering the axial symmetry of the situation, this is not surprising. In Fig. 39–29 we can characterize the size of the pattern in terms of the angle θ, representing the angular size of each ring. If the diameter of the aperture is d and the wavelength is λ, the angular size θ_1 of the first dark ring is found to be given by

$$\sin \theta_1 = 1.22\frac{\lambda}{d}; \tag{39–29}$$

that of the second dark ring is given by

$$\sin \theta_2 = 2.23\frac{\lambda}{d}; \tag{39–30}$$

the third by

$$\sin \theta_3 = 3.24\frac{\lambda}{d}; \tag{39–31}$$

and so on. Between the dark rings are bright rings with angular size given by

$$\sin \theta = 1.63\frac{\lambda}{d}, \quad 2.68\frac{\lambda}{d}, \quad 3.70\frac{\lambda}{d}, \tag{39–32}$$

and so on. The central bright spot is called the **Airy disk,** in honor of Sir George Airy (1801–1892), Astronomer Royal of England, who first derived the complete expression for the intensity in the pattern.

The *intensities* of the bright rings drop off very quickly; the peak intensity in the center of the first ring is only 1.7% of the value at the center of the Airy disk, and at the center of the second ring it is only 0.4%. Thus nearly all the radiant power (about 85%) is concentrated in the Airy disk. The angular size of the Airy disk can be taken to be that of the first dark ring, given by Eq.

(39–29). Figure 39–30 is a photograph of a diffraction pattern from a circular aperture 1.0 mm in diameter.

All of this has far-reaching implications for image formation of lenses and mirrors. In our study of optical instruments in Chapter 38, we assumed that a lens of focal length f focuses a parallel beam (plane wave) to a *point* at a distance f from the lens. This assumption ignored diffraction effects. We now see that what we get is not a point but the diffraction pattern just described. To put it another way, if we have two point objects, their images are not two points but two diffraction patterns. If the objects are close together, their diffraction patterns overlap; if they are close enough, their patterns overlap almost completely and cannot be distinguished. The effect is shown in Fig. 39–31, which shows the patterns for four (very small) "point" objects. In (a) the two images on the right have merged together; in (b), where we use a

39–30 Diffraction pattern formed by a circular aperture 1.0 mm in diameter.

39–31 Diffraction patterns of four "point" sources, with a circular opening in front of the lens. In (a) the opening is so small that the patterns at the right are just resolved, by Rayleigh's criterion. Increasing the aperture decreases the size of the diffraction patterns, as in (b) and (c).

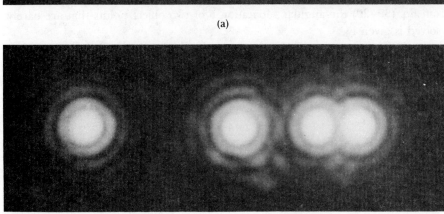

(a)

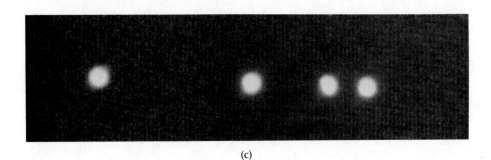

(b)

(c)

larger aperture diameter d with resulting smaller angular size of the Airy disk, the two right images are barely resolved. In (c), with a still larger aperture, they are well resolved.

A widely used criterion for resolution of two point objects, proposed by Lord Rayleigh and called **Rayleigh's criterion,** is that the points are just barely resolved (i.e., distinguishable) if the center of one diffraction pattern coincides with the first minimum of the other. In that case the angular separation of the image centers is given by Eq. (39–29). Because the angular separation of the *objects* is the same as that of the *images*, this means that two point objects are barely resolved, according to Rayleigh's criterion, if their angular separation is given by Eq. (39–29).

The minimum separation of two points that can just be resolved by an optical instrument is called the **limit of resolution** of the instrument. The smaller the limit of resolution, the greater the **resolving power** of the instrument. Diffraction sets the ultimate limits on resolution of lenses. If we look only at the formulas of geometrical optics, it appears we can make images as large as we like. While this is true, we always reach a point eventually where the image becomes larger but does not gain in detail.

Diffraction limits the sharpness of photographic images. Sharpness decreases when the lens is stopped down.

EXAMPLE 39–7   A camera lens of focal length $f = 50$ mm and maximum aperture $f/2$ forms an image of an object 10 m away. (a) If the resolution is limited by diffraction, what is the minimum distance between two points on the object that are barely resolved, and what is the corresponding distance between image points? (b) How does the situation change if the lens is "stopped down" to $f/16$? Assume $\lambda = 500$ nm in both cases.

SOLUTION   (a) The aperture diameter is $d = (50 \text{ mm})/2 = 25$ mm $= 25 \times 10^{-3}$ m. From Eq. (39–29) the angular separation θ of two object points that are barely resolved is given by

$$\sin \theta \cong \theta = 1.22 \frac{\lambda}{d}$$
$$= 1.22 \frac{500 \times 10^{-9} \text{ m}}{25 \times 10^{-3} \text{ m}}$$
$$= 2.44 \times 10^{-5}.$$

Let y be the separation of the object points and y' the separation of the corresponding image points. We know from our thin-lens analysis in Section 38–1 that, apart from sign, $y/s = y'/s'$. Thus the angular separations of the object points and the corresponding image points are both equal to θ. Thus

$$\frac{y}{10 \text{ m}} = 2.44 \times 10^{-5}, \qquad y = 2.44 \times 10^{-4} \text{ m} = 0.244 \text{ mm};$$

$$\frac{y'}{25 \text{ mm}} = 2.44 \times 10^{-5}, \qquad y' = 6.10 \times 10^{-4} \text{ mm} = 0.00061 \text{ mm}.$$

(b) The aperture is now $(50 \text{ mm})/16$, or one-eighth as large as before. The angular separation is eight times as great, and the values of y and y' are also eight times as great as before:

$$y = 1.95 \text{ mm}, \qquad y' = 0.00488 \text{ mm}.$$

Because setups to test the resolution of lenses often use series of parallel lines with varying spacing, the resolution in the image is often described in "lines per millimeter." Our lens would be described as having a resolution of about 1600 lines per mm when "wide open" and about 200 lines per mm when stopped down to $f/16$. Only the best-quality camera lenses approach this resolution. Photographers who always use the smallest possible aperture for maximum depth of focus and (presumably) maximum sharpness should be aware that diffraction effects become more significant at small apertures. Thus there is an optimization problem, balancing one cause of fuzzy images against another.

Another lesson to be learned is that resolution improves with shorter wavelengths. Ultraviolet microscopes have higher resolution than visible-light microscopes. In electron microscopes, which we will study in Chapter 42, the resolution is limited by the wavelengths associated with wavelike behavior of electrons. As we will see, electron wavelengths can be made 100,000 times smaller than wavelengths of visible light, with a corresponding gain in resolution. Finally, we note that part of the motivation for building very large reflecting telescopes is to increase the aperture diameter and thus minimize diffraction effects. The other reason, of course, is to provide greater light-gathering area for viewing very faint stars.

Why can electron microscopes have greater magnification than optical microscopes?

39–11 HOLOGRAPHY

Holography is a technique for recording and reproducing an image of an object without the use of lenses. Unlike the two-dimensional images recorded by an ordinary photograph or television system, a holographic image is truly three-dimensional. Such an image can be viewed from different directions to reveal different sides, and from various distances to reveal changing perspective.

The basic procedure for making a hologram is very simple in principle. A possible arrangement is shown in Fig. 39–32a. We illuminate the object to be holographed with monochromatic light, and we place a photographic film so

Holography: using interference to create three-dimensional images

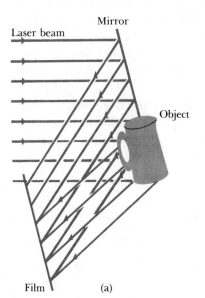

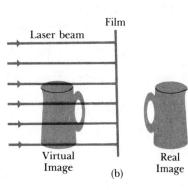

39–32 (a) The hologram is the record on film of the interference pattern formed with light directly from the source and light scattered from the object. (b) Images are formed when light is projected through the hologram.

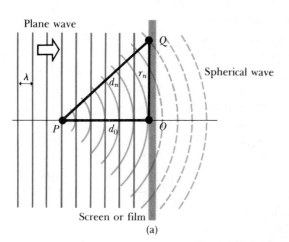

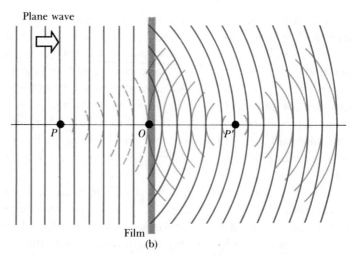

39–33 (a) Constructive interference of the plane and spherical waves occurs in the plane of the film at every point Q for which the distance d_n from P is greater than the distance d_0 from P to O by an integer number of wavelengths $n\lambda$. For the point shown, $n = 2$. (b) When a plane wave strikes the developed film, the diffracted wave consists of a wave converging to P' and then diverging again, and a diverging wave that appears to originate at P. These waves form the real and virtual images, respectively.

that it is struck by scattered light from the object and also by direct light from the source. In practice, the source must be a laser, for reasons to be discussed later. Interference between the direct and scattered light leads to the formation and recording of a complex interference pattern on the film.

To form the images, we simply project laser light through the developed film, as shown in Fig. 39–32b. Two images are formed, a virtual image on the side of the film nearer the source, and a real image on the opposite side.

A complete analysis of holography is beyond our scope, but we can gain some insight into the process by examining how a single point is holographed and imaged. Consider the interference pattern formed on a photographic film by the superposition of an incident plane wave and a spherical wave, as shown in Fig. 39–33a. The spherical wave originates at a point source P a distance d_0 from the film; P may in fact be a small object that scatters part of the incident plane wave. In any event, we assume that the two waves are monochromatic and coherent, and that the phase relation is such that constructive interference occurs at point O on the diagram. Then constructive interference will *also* occur at any point Q on the film that is farther from P than O is, by an integer number of wavelengths. That is, if $d_n - d_0 = n\lambda$, where n is an integer, then constructive interference occurs. The points where this condition is satisfied form circles centered at O, with radii r_n given by

$$d_n - d_0 = \sqrt{d_0{}^2 + r_n{}^2} - d_0 = n\lambda \qquad (n = 1, 2, 3, \ldots). \qquad (39\text{–}33)$$

Solving this equation for $r_n{}^2$, we find

$$r_n{}^2 = \lambda(2nd_0 + n^2\lambda).$$

How to make a hologram

Ordinarily d_0 is very much larger than λ, so we neglect the second term in parentheses, obtaining

$$r_n = \sqrt{2n\lambda d_0}. \qquad (39\text{–}34)$$

Since n must be an integer, the interference pattern consists of a series of concentric bright circular fringes, with the radii of the brightest regions given by Eq. (39–34). Between these bright fringes are darker fringes.

Now we develop the film and make a transparent positive print, so the bright-fringe areas have the greatest transparency on the film. It is then illuminated with monochromatic plane-wave light of the same wavelength as that used initially. In Fig. 39–33b, consider a point P' at a distance d_0 along the axis from the film. The centers of successive bright fringes differ in their distances from P' by an integer number of wavelengths, and therefore a strong *maximum* in the diffracted wave occurs at P'. That is, light converges to P' and then diverges from it on the opposite side, and P' is therefore a *real image* of point P.

This is not the entire diffracted wave, however; there is also a diverging spherical wave, which would represent a continuation of the wave originally emanating from P if the film had not been present. Thus the *total* diffracted wave is a superposition of a converging spherical wave forming a real image at P' and a diverging spherical wave shaped as though it had originated at P, forming a virtual image at P.

Because of the principle of linear superposition, what is true for the imaging of a single point is also true for the imaging of any number of points. The film records the superposed interference pattern from the various points, and when light is projected through the film the various image points are reproduced simultaneously. Thus the images of an extended object can be recorded and reproduced just as for a single point object.

In making a hologram, several practical problems must be overcome. First, the light used must be *coherent* over distances that are large compared to the dimensions of the object and its distance from the film. Ordinary light sources *do not* satisfy this requirement, for reasons discussed in Section 39–1, and laser light is essential. Second, extreme mechanical stability is needed. If any relative motion of source, object, or film occurs during exposure, even by as much as a wavelength, the interference pattern on the film is blurred enough to prevent satisfactory image formation. These obstacles are not insurmountable, however, and holography promises to become increasingly important in research, entertainment, and a wide variety of technological applications.

Making a hologram isn't as easy as it sounds.

SUMMARY

Monochromatic light is light having a single definite frequency. Coherence is a definite, unchanging phase relationship between two waves. When two coherent sources emit monochromatic light, their waves overlap, causing an interference pattern. The principle of linear superposition states that the total wave disturbance at a point at any instant is the sum of the disturbances from the separate waves. When the sources are in phase, constructive interference at a point occurs when the difference in path length from the two sources is zero or an integer number of wavelengths; destructive interference occurs when the path difference is a half-integer number of wavelengths. When the line from the sources to a point P makes an angle θ with the line perpendicular to the line of the sources, and when the distance between sources is d, the

KEY TERMS
monochromatic wave
interference
principle of linear superposition
constructive interference
destructive interference
coherent light
intensity
Michelson interferometer
ether
diffraction

Fresnel diffraction

Fraunhofer diffraction

diffraction grating

x-ray diffraction

Bragg reflection

Bragg condition

Airy disk

Rayleigh's criterion

limit of resolution

resolving power

holography

condition for constructive interference is

$$d \sin \theta = m\lambda \qquad (m = 0, \pm 1, \pm 2, \pm 3, \ldots), \qquad (39\text{–}4)$$

and the condition for destructive interference is

$$d \sin \theta = (m + \tfrac{1}{2})\lambda \qquad (m = 0, \pm 1, \pm 2, \pm 3, \ldots). \qquad (39\text{–}5)$$

When θ is very small, the position y_m of the mth bright fringe is given by

$$\lambda = \frac{y_m d}{mR}. \qquad (39\text{–}7)$$

When two sources emit waves in phase, the phase difference δ of the waves arriving at point P is related to the difference in path length $(r_1 - r_2)$ by

$$\delta = \frac{2\pi}{\lambda}(r_1 - r_2) = k(r_1 - r_2). \qquad (39\text{–}9)$$

When two sinusoidal waves of amplitude E and phase difference δ are superposed, the resultant amplitude E_P is

$$E_P{}^2 = 4E^2 \cos^2(\delta/2), \qquad (39\text{–}11)$$

and the intensity I is given by

$$I = 2\epsilon_0 c E^2 \cos^2\left(\frac{\pi d}{\lambda} \sin \theta\right). \qquad (39\text{–}14)$$

When light is reflected from both sides of a thin film of thickness d, constructive interference between the reflected waves occurs when

$$2d = m\lambda \qquad (m = 0, 1, 2, 3, \ldots) \qquad (39\text{–}17)$$

unless a half-cycle phase shift occurs at one surface; then this is the condition for destructive interference. A half-cycle phase shift occurs during reflection whenever the index of refraction in the second material is greater than that in the first.

The Michelson interferometer uses an extended monochromatic source and can be used for high-precision measurements of wavelengths. Its original purpose was to detect motion of the earth relative to a hypothetical ether, the supposed medium for electromagnetic waves. The concept of ether has been abandoned; the speed of light is the same relative to all observers. This is part of the foundation of the special theory of relativity.

Diffraction occurs when light passes through an aperture or around an edge. When source and observer are so far away from the obstructing surface that the outgoing rays can be considered parallel, it is called Fraunhofer diffraction, and when source or observer is at a finite distance it is Fresnel diffraction. For a single narrow slit of width a, the condition for destructive interference at a point P at an angle θ from the perpendicular to the surface of the slit is

$$a \sin \theta = n\lambda \qquad (n = 0, \pm 1, \pm 2, \pm 3, \ldots). \qquad (39\text{–}20)$$

The complete expression for the intensity I at any angle θ, in terms of the intensity I_0 at $\theta = 0$, is

$$I = I_0 \left(\frac{\sin \delta/2}{\delta/2}\right)^2 = I_0 \left\{\frac{\sin[\pi a(\sin \theta)/\lambda]}{\pi a(\sin \theta)/\lambda}\right\}^2. \qquad (39\text{–}22)$$

A diffraction grating consists of a large number of thin parallel slits, spaced a distance d apart. The condition for maximum intensity in the interference pattern is

$$d \sin \theta = m\lambda \qquad (m = 0, \pm 1, \pm 2, \pm 3, \ldots). \qquad (39\text{--}27)$$

This is the same condition for the two-source pattern, but for the grating the maxima are very sharp and narrow.

A crystal serves as a three-dimensional diffraction grating for waves having wavelengths of the order of magnitude of the lattice spacing; namely, x-rays. For a set of crystal planes spaced a distance d apart, constructive interference occurs when the angles of incidence and scattering are equal and when

$$2d \sin \theta = n\lambda \qquad (n = 0, 1, 2, 3, \ldots). \qquad (39\text{--}28)$$

This is called the Bragg condition.

The diffraction pattern from a circular aperture of diameter d consists of a central bright spot, called the Airy disk, and a series of concentric dark and bright rings. The angular size θ_1 of the first dark ring, which is also the angular size of the Airy disk, is given by

$$\sin \theta_1 = 1.22\lambda/d. \qquad (39\text{--}29)$$

Diffraction sets the ultimate limit on resolution (image sharpness) of optical instruments. According to Rayleigh's criterion, two object points are just barely resolved when their angular separation θ is given by Eq. (39–29).

A hologram is photographic record of an interference pattern formed by light scattered from an object and light coming directly from the source. It can be used to form three-dimensional images of the object.

QUESTIONS

39–1 Could an experiment similar to Young's two-slit experiment be performed with sound? How might this be carried out? Does it matter that sound waves are longitudinal and electromagnetic waves transverse?

39–2 At points of constructive interference between waves of equal amplitude, the intensity is four times that of either individual wave. Does this violate energy conservation? If not, why not?

39–3 In using the superposition principle to calculate intensities in interference and diffraction patterns, could one add the intensities of the waves instead of their amplitudes? What is the difference?

39–4 A two-slit interference experiment is set up and the fringes displayed on a screen. Then the whole apparatus is immersed in the nearest swimming pool. How does the fringe pattern change?

39–5 Would the headlights of a distant car form a two-source interference pattern? If so, how might it be observed? If not, why not?

39–6 A student asserted that it is impossible to observe interference fringes in a two-source experiment if the distance between sources is less than half the wavelength of the wave. Do you agree? Explain.

39–7 An amateur scientist proposed to record a two-source interference pattern by using only one source, placing it first in position S_1 in Fig. 39–1 and turning it on for a certain time, then placing it at S_2 and turning it on for an equal time. Does this work?

39–8 When a thin oil film spreads out on a puddle of water, the thinnest part of the film looks lightest in the resulting interference pattern. What does this tell you about the relative magnitudes of the refractive indexes of oil and water?

39–9 A glass windowpane with a thin film of water on it reflects less than when it is perfectly dry. Why?

39–10 In high-quality camera lenses, the resolution in the image is determined by diffraction effects. Is the resolution best when the lens is "wide open" or when it is "stopped down" to a smaller aperture? How does this behavior compare with the effect of aperture size on the depth of focus, that is, on the limit of resolution due to imprecise focusing?

39–11 If a two-slit interference experiment were done with white light, what would be seen?

39–12 Why is a diffraction grating better than a two-slit setup for measuring wavelengths of light?

39–13 Would the interference and diffraction effects described in this chapter still be seen if light were a longitudinal wave instead of transverse?

39–14 One sometimes sees rows of evenly spaced radio antenna towers. A student remarked that these act like diffraction gratings. What did he mean? Why would one *want* them to act like a diffraction grating?

39–15 Could x-ray diffraction effects with crystals be observed by using visible light instead of x-rays? Why or why not?

39–16 Does a microscope have better resolution with red light or blue light?

39–17 How could an interference experiment, such as one using a Michelson interferometer or fringes caused by a thin air space between glass plates, be used to measure the refractive index of air?

EXERCISES

Section 39–2 Two-Source Interference

39–1 Two slits are spaced 0.3 mm apart and are placed 50 cm from a screen. What is the distance between the second and third dark lines of the interference pattern when the slits are illuminated with light of 600-nm wavelength?

39–2 Young's experiment is performed with sodium light ($\lambda = 589$ nm). Fringes are measured carefully on a screen 100 cm away from the double slit, and the center of the twentieth fringe is found to be 11.78 mm from the center of the zeroth fringe. What is the separation of the two slits?

39–3 Light from a mercury-arc lamp is passed through a filter that blocks everything except for one spectrum line in the green region of the spectrum. It then falls on two slits separated by 0.6 mm. In the resulting interference pattern on a screen 2.5 m away, adjacent bright fringes are separated by 2.27 mm. What is the wavelength?

Section 39–3 Intensity Distribution in Interference Patterns

39–4 An AM radio station has a frequency of 1000 kHz; it uses two identical antennas at the same elevation, 150 m apart.

a) In what direction is the intensity maximum, considering points in a horizontal plane?

b) Calling the maximum intensity in (a) I_0, determine in terms of I_0 the intensity in directions making angles of 30°, 45°, 60°, and 90° to the direction of maximum intensity.

39–5 Consider a two-slit interference pattern, for which the intensity distribution is given by Eq. (39–15). Let θ_m be the angular position of the mth bright fringe, where the intensity is I_0. Assume that θ_m is small, so that $\sin \theta_m \approx \theta_m$. Let θ_m^+ and θ_m^- be the two angles on either side of θ_m for which $I = \frac{1}{2} I_0$. The quantity $\Delta\theta_m = |\theta_m^+ - \theta_m^-|$ is the half-width of the mth fringe. Calculate $\Delta\theta_m$. How does $\Delta\theta_m$ depend on m?

Section 39–4 Interference in Thin Films

39–6 Light of wavelength 500 nm is incident perpendicularly from air on a film 1×10^{-4} cm thick and of refractive index 1.375. Part of the light is reflected from the first surface of the film, and part enters the film and is reflected back at the second surface.

a) How many waves are contained along the path of this second part of the light in the film?

b) What is the phase difference between these two parts of the light as they leave the film?

39–7 In Example 39–4 suppose the top plate is glass with $n = 1.4$, the wedge is filled with silicone grease having $n = 1.5$, and the bottom plate is glass with $n = 1.6$. Calculate the spacing between the dark fringes.

39–8 A sheet of glass 10 cm long is placed in contact with a second sheet and is held at a small angle with it by a metal strip 0.1 mm thick placed under one end. The glass is illuminated from above with light of 546-nm wavelength. How many interference fringes are observed per centimeter in the reflected light?

39–9 Two rectangular pieces of plane glass are laid one upon the other on a table. A thin strip of paper is placed between them at one edge so that a very thin wedge of air is formed. The plates are illuminated by a beam of sodium light at normal incidence ($\lambda = 589$ nm). Interference fringes are formed, with ten fringes per centimeter length of wedge measured normal to the edges in contact. Find the angle of the wedge.

39–10

a) Is a thin film of quartz suitable as a nonreflecting coating for fabulite? (See Table 36–1.)

b) If so, what is the minimum thickness of the film required?

39–11

a) What is the thinnest film of a 1.40 refractive index coating on glass ($n = 1.50$) for which destructive interference of the violet component (400 nm) of an incident white light beam in air can take place by reflection?

b) What then is the residual color of the beam?

Section 39–5 The Michelson Interferometer

39–12 How far must the mirror M_2 (Fig. 39–12) of the Michelson interferometer be moved so that 3000 fringes of krypton-86 light ($\lambda = 606$ nm) will move across a line in the field of view?

39–13 A Mach-Zehnder interferometer is shown in Fig. 39–34. The mirrors at B and C are totally reflecting, while the mirrors at A and D are half-silvered, so they reflect $\frac{1}{2}$ the light and transmit $\frac{1}{2}$ the light. Light polarized in the plane of the mirrors is used so that the light waves will undergo a 90° phase shift at each mirror. Show that if the optical paths are such that the output at F is zero (total destructive interference), then the output at E will show total constructive interference.

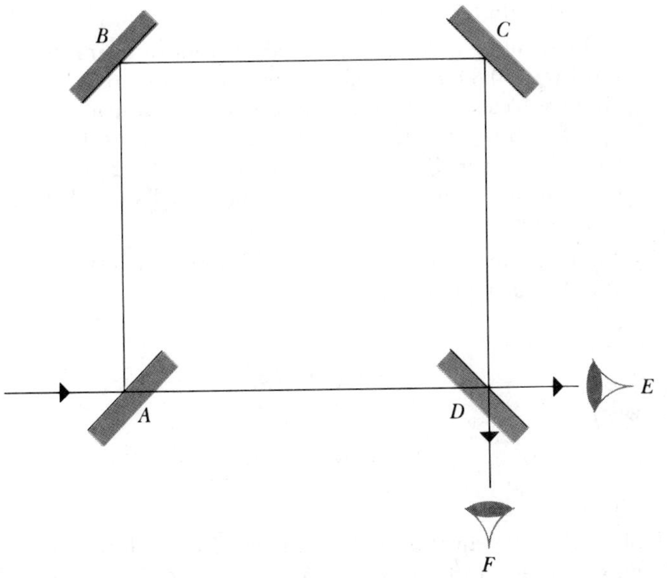

FIGURE 39–34

Section 39–7 Fraunhofer Diffraction from a Single Slit

39–14 Parallel rays of green mercury light of wavelength 546 nm pass through a slit of width 0.437 mm covering a lens of focal length 40 cm. In the focal plane of the lens, what is the distance from the central maximum to the first minimum?

39–15 Monochromatic light from a distant source is incident on a slit 0.8 mm wide. On a screen 3.0 m away, the distance from the central maximum of the diffraction pattern to the first minimum is measured to be 2 mm. Calculate the wavelength of the light.

39–16 Light of wavelength 589 nm from a distant source is incident on a slit 1.0 mm wide, and the resulting diffraction pattern is observed on a screen 2.0 m away. What is the distance between the two dark fringes on either side of the central bright fringe?

Section 39–8 The Diffraction Grating

39–17 Plane monochromatic waves of wavelength 600 nm are incident normally on a plane transmission grating having 500 lines·mm^{-1}. Find the angles of deviation in the first, second, and third orders.

39–18

a) What is the wavelength of light that is deviated in the first order through an angle of 20° by a transmission grating having 6000 lines·cm^{-1}?

b) What is the second-order deviation of this wavelength? Assume normal incidence.

39–19 A plane transmission grating is ruled with 4000 lines·cm^{-1}. Compute the angular separation in degrees between the α and δ lines of atomic hydrogen in the second-order spectrum. The wavelengths of these lines are, respectively, 656 nm and 410 nm. Assume normal incidence.

Section 39–10 Circular Apertures and Resolving Power

39–20 In Fig. 39–35, two point sources of light, a and b, at a distance of 50 m from lens L and 6 mm apart, produce images at c that are just resolved by Rayleigh's criterion. The focal length of the lens is 20 cm. What is the diameter of the diffraction circles at c?

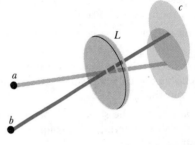

FIGURE 39–35

39–21 A telescope is used to observe two distant point sources 1 m apart. ($\lambda = 500$ nm.) The objective of the telescope is covered with a slit of width 1 mm. What is the maximum distance in meters at which the two sources may be distinguished?

39–22 A converging lens 8 cm in diameter has a focal length of 40 cm. If the resolution is diffraction limited, how far away can an object be if points on it 4 mm apart are to be resolved (by Rayleigh's criterion)? (Use $\lambda = 550$ nm.)

Section 39–11 Holography

39–23 If a hologram is made by using 600-nm light and then viewed with 500-nm light, how will the images look compared to those observed with 600-nm light?

39–24 A hologram is made by using 600-nm light and is then viewed with continuous-spectrum white light from an incandescent bulb. What will be seen?

39–25 Ordinary photographic film reverses black and white, in the sense that the most brightly illuminated areas become blackest upon development (hence the term *negative*). Suppose a hologram negative is viewed directly, without making a positive transparency. How will the resulting images differ from those obtained with the positive hologram?

PROBLEMS

39-26 Parallel light rays of wavelength $\lambda = 600$ nm fall on a single slit. On a screen 3 m away, the distance from the center of the central maximum to the center of the next maximum is 4 mm. What is the width of the slit?

39-27 In Exercise 39-1, suppose the entire apparatus is immersed in water. Then what is the distance between the second and third dark lines?

39-28 An FM radio station has a frequency of 100 MHz and uses two identical antennas mounted at the same elevation, 12 m apart. The resulting radiation pattern has a maximum intensity along a horizontal line perpendicular to the line joining the antennas.

a) At what other angles (measured from the line of maximum intensity) is the intensity maximum?

b) At what angles is it zero?

39-29 A glass plate 0.40 μm thick is illuminated by a beam of white light normal to the plate. The index of refraction of the glass is 1.50. What wavelengths within the limits of the visible spectrum ($\lambda = 400$ nm to $\lambda = 700$ nm) will be intensified in the reflected beam?

39-30 An oil tanker spills a large amount of oil ($n = 1.4$) into the sea.

a) If you are overhead and look down onto the oil spill, what predominant color do you see at a point where the oil is 410 nm thick?

b) If you swam under the slick and looked up at the same place in the slick as in (a), what color would predominate in the transmitted light?

39-31 The radius of curvature of the convex surface of a plano-convex lens is 1.20 m. The lens is placed convex side down on a plane glass plate and illuminated from above with red light of wavelength 650 nm. Find the diameter of the third bright ring in the interference pattern.

39-32 Newton's rings can be seen when a plano-convex lens is placed on a flat glass surface (Problem 39-31). If the lens has an index of refraction of $n = 1.50$ and the glass plate an index of $n = 1.80$, the rings are seen with a spacing of 0.50 mm. If a liquid of index 1.65 is added to the space between the lens and the plate, what will be the new spacing between the interference rings?

39-33 In a Young's two-slit experiment, a piece of glass with an index of refraction n and a thickness L is placed in front of the upper slit.

a) Describe qualitatively what happens to the interference pattern.

b) Derive an expression for the intensity I of the light at points on a screen as a function of n, L, and θ. Here θ is

the usual angle measured from the center of the two slits. That is, determine the equation analogous to Eq. (39-15).

c) From your result in (b) derive an expression for the values of θ that locate the maxima in the interference pattern. That is, derive an equation analogous to Eq. (39-4).

39-34 In Exercise 39-16, suppose the entire apparatus is immersed in water. Then what is the distance between the two dark fringes?

39-35 A slit of width a was placed in front of a lens of focal length 0.80 m. The slit was illuminated by parallel light of wavelength 600 nm, and the diffraction pattern of Fig. 39-22b was formed on a screen in the second focal plane of the lens. If the photograph of Fig. 39-22b represents an enlargement to twice the actual size, what was the slit width?

39-36 The intensity of light in the Fraunhofer diffraction pattern of a single slit is

$$I = I_0 (\sin \beta / \beta)^2,$$

where

$$\beta = (\pi a \sin \theta)/\lambda.$$

Show that the equation for the values of β at which I is a maximum is $\tan \beta = \beta$. How can you solve such an equation graphically?

39-37 What is the longest wavelength that can be observed in the fourth order for a transmission grating having 5000 lines per centimeter? Assume normal incidence.

39-38 A Michelson interferometer can be used to measure the index of refraction of gases by placing an initially evacuated tube in one arm of the interferometer. The gas is then slowly added to the tube, and the number of fringes that cross the telescope cross hairs are counted. If the length of the tube is 4 cm and the light source is a sodium lamp (589 nm), what is the index of refraction of the gas if 35 fringes are seen to pass the view of the telescope? (*Note.* For gases it is convenient to give the value of $n - 1$ rather than n itself, as the index differs little from unity.)

39-39 An astronaut in the space shuttle can just resolve two point sources on earth that are 20 m apart. Assume that the resolution is diffraction limited and use Rayleigh's criterion. What is his altitude above the earth? Treat his eye as a circular aperture of diameter 4.0 mm (the diameter of his pupil), and take 550 nm as the wavelength of the light.

CHALLENGE PROBLEMS

39-40 During the Battle of Britain (WW II), it was found that aircraft flying at certain low altitudes over the English Channel could not receive radio signals from transmission towers located on the cliffs of Dover, 200 m above the

water. At other altitudes the signals received were very strong. For an airplane flying 10 km from England and a radio signal with wavelength 4 m, at what altitudes above the water will the received signal be strongest? (Consider only altitudes much less than 10 km. Interference occurs between radio waves that travel directly to the plane and those that first reflect off the water.)

39–41 Consider a single-slit diffraction pattern. The center of the central maximum, where the intensity is I_0, is located at $\theta = 0$. Let θ_+ and θ_- be the two angles on either side of $\theta = 0$ for which $I = \frac{1}{2}I_0$. $\Delta\theta = |\theta_+ - \theta_-|$ is called the half-width of the central diffraction maximum. Solve for $\Delta\theta$, when the ratio between the slit width a and wavelength λ is

a) $a/\lambda = 2$,   b) $a/\lambda = 5$,

c) $a/\lambda = 10$.

(*Hint:* Your equation for θ_+ or θ_- cannot be solved analytically. You must use iteration—see Challenge Problem 27–35—or solve it graphically.)

39–42 Figure 39–36 shows an interferometer known as *Fresnel's biprism.* The magnitude of the prism angle A is extremely small.

a) If S_0 is a very narrow source slit, show that the separation of the two virtual coherent sources S_1 and S_2 is given by $d = 2aA(n - 1)$, where n is the index of refraction of the material of the prism.

b) Calculate the spacing of the fringes of green light of wavelength 500 nm on a screen 2 m from the biprism. Take $a = 0.20$ m, $A = 0.005$ rad, and $n = 1.5$.

39–43 The index of refraction of a glass rod is 1.48 at $T = 20°C$ and varies linearly with temperature, with a coefficient of $2.0 \times 10^{-5}(C°)^{-1}$. The coefficient of linear expansion of the glass is $5.0 \times 10^{-6}(C°)^{-1}$. At 20°C the length of the rod is 2.00 cm. Derive a relationship for the speed with which fringes will cross the field of view of a Michelson interferometer that has this glass rod in one arm, if the rod is being heated at a rate of $5C°\cdot min^{-1}$. The light source has wavelength $\lambda = 589$ nm, and the rod initially is at $T = 20°C$.

39–44 The yellow sodium D lines are a doublet with wavelengths of 589.0 and 589.6 nm and equal intensities.

a) How many lines/cm are required for a diffraction grating to resolve these two lines in the first-order spectrum? Assume the grating is placed 0.5 m from a screen and that the source image is 0.1 mm wide, so that resolution of the lines means their centers are 0.1 mm apart on the screen.

b) The sodium D lines are used as a source in a Michelson interferometer. As the mirror is moved, it is noted that in addition to moving across the field of view, the interference fringes periodically appear and disappear. Why? How far is the mirror moved between disappearances of the fringes?

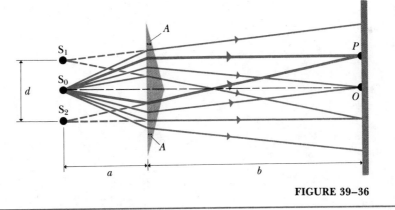

FIGURE 39–36

PART EIGHT

MODERN PHYSICS

PERSPECTIVE

In the past several chapters we have seen how the subject of optics grows out of electromagnetic theory. Maxwell's equations predict the existence of electromagnetic waves. Many optical phenomena, including polarization, interference, and diffraction, can be understood on the basis of an electromagnetic wave model of light. The simpler ray model of geometrical optics can be used to analyze mirrors, lenses, and optical instruments.

We have now arrived at a very significant milepost in our study of physics. We have stressed the roles of Newton's laws, the laws of thermodynamics, and Maxwell's equations in summarizing the most important laws of physics. All these principles were formulated in their present form by the year 1900. It would have been tempting to speculate then that the laws of physics were essentially complete, that all that remained was to work out the details of some complicated problems using well-established principles. Nothing would have been further from the truth! The revolution in scientific thought that has occurred since 1900 is at least as significant as the sum of all developments up to that year. Part of this revolution was the theory of relativity. Starting with the innocent-sounding premise that the laws of physics should be the same in all inertial frames of reference, Einstein showed that consistency with this premise required sweeping revisions of our concepts of space and time.

Optics, far from being a closed, nineteenth-century subject, has formed a significant gateway into twentieth-century physics. Ray and wave optics are an adequate description of the *propagation* of light. But for many other phenomena, especially the *emission* and *absorption* of light, classical optics is *not* adequate. A variety of experimental evidence shows that the energy in light is emitted and absorbed only in discrete packages

called *quanta* or *photons*. These packages, in turn, are related to the existence of characteristic spectra of elements and hence to the internal structure of atoms.

Understanding of atomic structure and the emission and absorption of light requires a new theory called *quantum mechanics* that involves fundamental changes in the language we use to describe mechanical systems. We abandon the description of a particle as a geometric point and regard it instead as an inherently spread-out entity that can have wave-like properties. Thus both particles (such as electrons) and radiation (photons) exhibit both particle and wave aspects. Like relativity, quantum mechanics requires sweeping revisions of our fundamental notions of space, time, and the description of motion. Quantum mechanics also provides the foundation for understanding many macroscopic (bulk) properties of matter on the basis of its microscopic (molecular) structure. Thus quantum mechanics is the starting point for modern materials science.

Another area of study is the atomic nucleus. Nuclear structure involves a kind of interaction distinct from gravitational and electromagnetic interactions. Some nuclei are unstable, leading to radioactivity. Some nuclear reactions, such as fission and fusion, have practical applications as sources of energy and as weapons.

We end the final chapter with a brief overview of high-energy physics, one of the present-day frontiers of physics. Many new particles have been discovered during the past forty years; the familiar particles, such as protons and neutrons, are not the most fundamental level of the structure of matter. There are still many unanswered questions in this vital area of present-day research, and our understanding of the physical world is far from complete!

40

RELATIVISTIC MECHANICS

IN EARLIER CHAPTERS OF THIS BOOK, ESPECIALLY THOSE ON MECHANICS, we have stressed the importance of inertial frames of reference. Newton's laws of motion are valid only in inertial frames, but they are valid in *all* inertial frames. Any frame moving with constant velocity with respect to an inertial frame is itself an inertial frame, and all such frames are equivalent with respect to expressing the basic principles of mechanics. *The laws of mechanics are the same in every inertial frame of reference.*

In 1905 Einstein proposed that this principle be extended to include *all* the basic laws of physics. This innocent-sounding proposition, often called the *principle of relativity,* has far-reaching and startling consequences. We will find, for example, that if the principle of conservation of momentum is to be valid in all inertial systems, the definition of momentum for particles moving at speeds comparable to the speed of light must be modified, and kinetic energy must also be redefined. Equally fundamental are the modifications needed in the *kinematic* aspects of motion. The resulting generalizations of the laws of mechanics are part of the **special theory of relativity,** the subject of this chapter. In studying this material you must be prepared to confront some ideas that at first sight will seem too strange to be believed, and you must learn to mistrust intuition when dealing with phenomena far removed from everyday experience!

Our discussion will center mostly on *mechanical* concepts, but the theory of relativity has significant consequences in *all* areas of physics, including thermodynamics, electromagnetism, optics, atomic and nuclear physics, and high-energy physics.

40–1 INVARIANCE OF PHYSICAL LAWS

Einstein's **principle of relativity** states that *the laws of physics are the same in every inertial frame of reference.* A familiar example of this principle in electromagnetism is the electromotive force induced in a coil of wire as a result of motion of a nearby permanent magnet. In the frame of reference where the *coil* is sta-

Describing physical phenomena in various frames of reference

955

tionary, the moving magnet causes a change of magnetic flux through the coil and hence an induced emf. In a different frame, where the *magnet* is stationary, the motion of the coil through a magnetic field causes magnetic-field forces on the mobile charges in the conductor, inducing an emf. According to the principle of relativity, both points of view have equal validity and both must predict the same result for the induced emf. As we have seen in Chapter 32, Faraday's law of electromagnetic induction can be applied to either description, so it does indeed satisfy this requirement.

Of equal significance is the prediction of the speed of light and other electromagnetic radiation, emerging from the development in Chapter 35. According to the principle of relativity, the speed of propagation of this radiation must be independent of the frame of reference and must be the same for all inertial frames.

Indeed, the speed of light plays a special role in the theory of relativity. According to the principle of relativity, light travels in vacuum with speed c, independent of the motion of the source. As we mentioned in Section 1–2, the numerical value of c is *defined* to be exactly 299,792,458 m·s^{-1}, and this number is used to define the unit of length in terms of the unit of time. The approximate value $c = 3.00 \times 10^8$ m·s^{-1} is within one part in 1000 of the exact value, and we will often use it when we do not need greater precision.

At one time it was thought that light traveled through a hypothetical medium called the *ether*, just as sound waves travel through air. In that case, its speed would depend on the motion of the observer relative to the ether and hence would be different in different directions. During the late nineteenth and early twentieth centuries, intensive efforts were made to find experimental evidence for the existence of the ether. The Michelson-Morley experiment, described in Section 39–5, was an effort to detect motion of the earth relative to the ether. This and all similar experiments yielded consistently negative results, and the ether concept has been discarded. Thus this experimental evidence confirms the prediction of the principle of relativity that the speed of light is the same in all frames of reference. With this result in mind, let us suppose the speed of light is measured by two observers, one at rest with respect to the light source, the other moving away from it. Both are in inertial frames of reference, and according to Einstein's principle of relativity, the laws of physics—in particular the speed of light—must be the same in both frames.

If this does not impress you, consider the following situation. A spaceship moving away from Earth at 1000 m·s^{-1} fires a missile with a speed of 2000 m·s^{-1} in a direction directly away from Earth. What is the missile's speed relative to Earth? Simple, you say. An elementary problem in relative velocity. The correct answer is 3000 m·s^{-1}. But now suppose there is a searchlight in the spaceship, pointing in the same direction that the missile was fired. An observer on the spaceship measures the speed of light emitted by the searchlight and obtains the value c. But, according to our previous discussion, the motion of the light after it has left the source cannot depend on the motion of the source. So when an observer on Earth measures the speed of this same light, he must also obtain the value c. This contradicts our elementary notion of relative velocities and may not appear to agree with common sense. But we have to recognize that "common sense" is intuition based on everyday experience, and this does not usually include measurements of the speed of light. Sometimes we must be prepared to accept results that seem not to make sense when they involve realms far removed from everyday observation.

Does light need a mechanical medium in which to travel?

The ether is a mythical beast, like the unicorn and phlogiston.

Some strange conclusions about relative velocities

Thus the principle of relativity, supported by experimental evidence, requires that the speed of light (in vacuum) is independent of the motion of the source and is the same in all frames of reference. To explore the consequences of this statement, consider first the *Newtonian* relationship between two inertial frames, labeled S and S' in Fig. 40–1. To keep things as simple as possible, we have not shown the z-axis on the figure. Let the x-axes of the two frames lie along the same line, but let the origin O' of S' move relative to the origin O of S with constant velocity u along the common x-axis. If the two origins coincide at time $t = 0$, then their separation at a later time t is ut.

We can describe a point P by coordinates (x, y, z) in S or by coordinates (x', y', z') in S'. Reference to the figure shows that these are related by

$$x = x' + ut, \qquad y = y', \qquad z = z'. \tag{40–1}$$

These equations are called the **Galilean coordinate transformation.**

If point P moves in the x-direction, its velocity v relative to S is given by $v = \Delta x / \Delta t$, and its velocity v' relative to S' is $v' = \Delta x' / \Delta t$. Intuitively it is clear that these are related by

$$v = v' + u. \tag{40–2}$$

This relation may also be obtained formally from Eqs. (40–1). Suppose the particle is at a point described by coordinate x_1 or x_1' at time t_1, and at x_2 or x_2' at time t_2. Then $\Delta t = t_2 - t_1$, and, from Eq. (40–1),

$$\Delta x = x_2 - x_1 = (x_2' - x_1') + u(t_2 - t_1)$$
$$= \Delta x' + u\,\Delta t,$$
$$\frac{\Delta x}{\Delta t} = \frac{\Delta x'}{\Delta t} + u,$$

and, in the limit as $\Delta t \to 0$,

$$v = v' - u,$$

in agreement with Eq. (40–2).

A fundamental problem now appears. Applied to the speed of light, Eq. (40–2) says $c = c' + u$. Einstein's principle of relativity, supported by experimental observations, says $c = c'$. This is a genuine inconsistency, not an illusion, and it demands resolution. If we accept the principle of relativity, we are forced to conclude that Eqs. (40–1) and (40–2), intuitively appealing as they are, *cannot* be correct but need to be modified to bring them into harmony with this principle.

The resolution involves modifying our basic kinematic concepts. The first modification involves an assumption so fundamental that it might seem unnecessary, namely, the assumption that the same *time scale* is used in frames S and S'. This may be stated formally by adding to Eqs. (40–1) a fourth equation:

$$t = t'.$$

Obvious though this assumption may seem, it is not correct when the relative speed u of the two frames of reference is comparable to the speed of light. The difficulty lies in the concept of *simultaneity*, which we examine next.

Transforming coordinates from one system to another

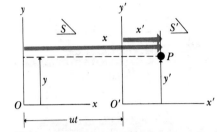

40–1 The position of point P can be described by the coordinates x and y in frame of reference S, or by x' and y' in S'. S' moves relative to S with constant velocity u along the common x–x' axis. The two origins O and O' coincide at time $t = t' = 0$.

Addition of velocities, according to the Galilean coordinate transformation

The assumption that all frames of reference have the same time scale needs to be reexamined.

40-2 RELATIVE NATURE OF SIMULTANEITY

To measure time intervals we need to use the concept of simultaneity.

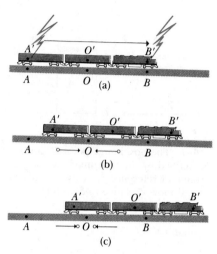

40-2 (a) To the stationary observer at point O, two lightning bolts appear to strike simultaneously. (b) The moving observer at point O' sees the light from the front of the train first and thinks that the bolt at the front struck first. (c) The two light pulses arrive at O simultaneously.

Simultaneity is not an absolute concept; events that seem simultaneous to one observer do not to another.

Comparing a time interval between two events in two coordinate systems

Measuring times and time intervals involves the concept of **simultaneity.** When a person says he awoke at seven o'clock, he means that two *events* (his awakening and the arrival of the hour hand of his clock at the number seven) occurred *simultaneously.* The fundamental problem in measuring time intervals is that, in general, two events that appear simultaneous in one frame of reference *do not* appear simultaneous in a second frame that is moving relative to the first, even if both are inertial frames.

The following thought experiment, devised by Einstein, illustrates this point. Consider a long train moving with uniform velocity, as shown in Fig. 40-2a. Two lightning bolts strike the train, one at each end. Each bolt leaves a mark on the train and one on the ground at the same instant. The points on the ground are labeled A and B in the figure, and the corresponding points on the train are A' and B'. An observer on the ground is located at O, midway between A and B; another observer is at O', moving with the train and midway between A' and B'. Both observers observe the lightning bolts by means of the light signals they emit.

Suppose the two light signals reach the observer at O simultaneously. He concludes that the two events took place at A and B simultaneously. But the observer at O' is moving with the train, and the light pulse from B' reaches him before the light pulse from A' does. He concludes that the lightning bolt at the front of the train happened *earlier* than the one at the rear. Thus the two events appear simultaneous to one observer but not to the other. *Whether two events at different space points are simultaneous depends on the state of motion of the observer.* It follows that *the time interval between two events at different space points is, in general, different for two observers in relative motion.*

You may want to argue that in this example the lightning bolts really *are* simultaneous, and that if the observer at O' could communicate with the distant points without time delay, he would realize this. But that would be erroneous; the finite speed of information transmission is not the real issue. If O' is really midway between A' and B', then, in his frame of reference, the time for a signal to travel from A' to O' is the same as from B' to O'. Two signals arrive simultaneously at O' only if they were emitted simultaneously at A' and B'. In this example they *do not* arrive simultaneously at O', and so O' must conclude that the events at A' and B' were *not* simultaneous.

Furthermore, there is no basis for saying either that O is right and O' is wrong, or the reverse, since, according to the principle of relativity, no inertial frame of reference is preferred over any other in the formulation of physical laws. Each observer is correct *in his own frame of reference.* In other words, simultaneity is not an absolute concept. Whether two events are simultaneous depends on the frame of reference, and the time interval between two events also depends on the frame of reference.

40-3 RELATIVITY OF TIME

To derive a quantitative relation between time intervals in different coordinate systems, we consider another thought experiment. As before, a frame of reference S' moves along the common x–x' axis with constant speed u relative to a frame S. For reasons that will become clear later, we assume that u is always less than the speed of light c. An observer O' in S' directs a source of light at a mirror a distance d away, as shown in Fig. 40–3a, and measures the time interval $\Delta t'$ for light to make the "round trip" to the mirror and back. The total distance is $2d$, so the time interval is

$$\Delta t' = \frac{2d}{c}. \qquad (40\text{–}3)$$

As measured in frame S, the time for the round trip is a different interval Δt. During this time the source moves relative to S a distance $u\,\Delta t$, and the total round-trip distance is not just $2d$ but is $2l$, where

$$l = \sqrt{d^2 + \left(\frac{u\,\Delta t}{2}\right)^2}.$$

In writing this expression, we have used the fact that the distance d looks the same to both observers. This can (and indeed must) be justified by other thought experiments, but we will not go into this matter now. The speed of light is the same for both observers, so the relation in S analogous to Eq. (40–3) is

$$\Delta t = \frac{2l}{c} = \frac{2}{c}\sqrt{d^2 + \left(\frac{u\,\Delta t}{2}\right)^2}. \qquad (40\text{–}4)$$

To obtain a relation between Δt and $\Delta t'$ that does not contain d, we solve Eq. (40–3) for d and substitute the result into Eq. (40–4), obtaining

$$\Delta t = \frac{2}{c}\sqrt{\left(\frac{c\,\Delta t'}{2}\right)^2 + \left(\frac{u\,\Delta t}{2}\right)^2}.$$

This may now be squared and solved for Δt; the result is

$$\Delta t = \frac{\Delta t'}{\sqrt{1 - u^2/c^2}}. \qquad (40\text{–}5)$$

We may generalize this important result: If two events (in our case, the departure and arrival of the light signal at O') occur at the same space point in a frame of reference S' and are separated in time by an interval $\Delta t'$, then the time interval Δt between these two events as observed in S is given by Eq. (40–5). Because the denominator is always smaller than unity, Δt is always *larger* than $\Delta t'$. Thus when the rate of a clock at rest in S' is measured by an observer in S, the rate measured in S is *slower* than the rate observed in S'. This effect is called **time dilation.** We note that Eq. (40–5) makes sense only when $u < c$; otherwise the denominator is imaginary.

It is important to note that the observer in S measuring the time interval Δt cannot do so with a single clock. In Fig. 40–3 the points of departure and return of the light pulse are different space points in S, although they are the same point in S'. If S tries to use a single clock, the finite time of communication between two points will cloud the issue. To avoid this, S may use two assistants with two clocks at the two relevant points. There is no difficulty in synchronizing two clocks in the same frame of reference; one procedure is to send a light pulse simultaneously to two clocks from a point midway between them, with the two assistants setting their clocks to a predetermined time when the pulses arrive. In thought experiments it is often helpful to imagine a large number of observers with synchronized clocks distributed conveniently in a single frame of reference. Only when a clock is moving relative to a given frame of reference do ambiguities of synchronization or simultaneity arise.

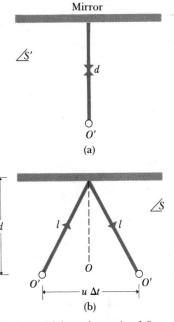

40–3 (a) Light pulse emitted from source at O' and reflected back along the same line, as observed in S'. (b) Path of the same light pulse, as observed in S. The positions of O' at the times of departure and return of the pulse are shown. The speed of the pulse is the same in S as in S', but the path is longer in S.

The relation between time intervals in two coordinate systems: Moving clocks run slower.

For some measurements we have to imagine many synchronized clocks distributed around a particular frame of reference.

EXAMPLE 40–1   A spaceship flies past earth with a speed of 0.99c (about 2.97×10^8 m·s⁻¹). A high-intensity signal light (perhaps a pulsed laser) blinks on and off, each pulse lasting 2×10^{-6} s as measured on the spaceship. At a certain instant the ship appears to an observer on earth to be directly overhead at an altitude of 1000 km and to be traveling perpendicular to the line of sight. What is the duration of each light pulse, as measured by this observer, and how far does the ship travel relative to the earth during each pulse?

An example of transformation of time intervals

SOLUTION   The observer does not see the pulse at the instant it is emitted, because the light signal requires a time equal to $(1000 \times 10^3 \text{ m})/(3 \times 10^8 \text{ m·s}^{-1})$, or $(\frac{1}{300})$ s, to travel from the ship to earth. But if the distance from the spaceship to the observer is essentially constant during the emission of a pulse, the time delays at the beginning and end of the pulse are equal, and the time *interval* is not affected.

Let S be the earth's frame of reference, S' that of the spaceship. Then, in the notation of Eq. (40–5), $\Delta t' = 2 \times 10^{-6}$ s. This interval refers to two events occurring at the same point relative to S', namely, the starting and stopping of the pulse. The corresponding interval in S is given by Eq. (40–5):

$$\Delta t = \frac{\Delta t'}{\sqrt{1 - u^2/c^2}} = \frac{2 \times 10^{-6} \text{ s}}{\sqrt{1 - (0.99)^2}} = 14.2 \times 10^{-6} \text{ s}.$$

Thus the time dilation in S is about a factor of seven. The distance d traveled in S during this interval is

$$d = u \, \Delta t = (0.99)(3 \times 10^8 \text{ m·s}^{-1})(14.2 \times 10^{-6} \text{ s})$$

$$= 4210 \text{ m} = 4.21 \text{ km}.$$

If the spaceship is traveling directly *toward* the observer, the time interval cannot be measured directly by a single observer because the time delay is not the same at the beginning and end of the pulse. One possible scheme, at least in principle, is to use *two* observers at rest in S, with synchronized clocks, one at the position of the ship when the pulse starts, the other at its position at the end of the pulse. These observers will again measure a time interval in S of 14.2×10^{-6} s.

Using an atomic clock for direct confirmation of time dilation

Time-dilation effects are not observed in everyday life because the speeds of all practical modes of transportation are much less than the speed of light. For example, for a jet airplane flying at 270 m·s⁻¹ (about 600 mi·hr⁻¹),

$$\frac{u^2}{c^2} = \left(\frac{270 \text{ m·s}^{-1}}{3 \times 10^8 \text{ m·s}^{-1}}\right)^2 = 8.1 \times 10^{-13},$$

and the time-dilation factor in Eq. (40–5) is approximately $1 + (4 \times 10^{-13})$. Thus to observe time dilation in this situation requires a clock with a precision of the order of one part in 10^{13}. However, as noted in Section 1–2, atomic clocks capable of this precision have recently been developed, and in the past few years experiments with such clocks in jet airplanes have verified Eq. (40–5) directly.

From the derivation of Eq. (40–5) and the spaceship example, we can see that a time interval between two events occurring at *the same point* in a given frame of reference is a more fundamental quantity than an interval between events at different points. The term *proper time* is used to denote an interval

A time interval between two events at the same space point in a frame of reference is called a proper time in that frame.

between two events occurring at the same space point. Hence Eq. (40–5) may be used *only* when $\Delta t'$ is a proper time interval in S'; in that case Δt is *not* a proper time interval in S. If, instead, Δt is proper in S, then Δt and $\Delta t'$ must be interchanged in Eq. (40–5).

When the relative velocity u of S and S' is very small, the factor $(1 - u^2/c^2)$ is very nearly equal to unity, and Eq. (40–5) approaches the Newtonian relation $\Delta t = \Delta t'$ (i.e., the same time scale for all frames of reference).

Equation (40–5) suggests an apparent paradox called the **twin paradox.** Consider identical-twin astronauts named Eartha and Astro. Eartha remains on earth while Astro takes off on a high-speed trip through the galaxy. Because of time dilation, Eartha sees Astro's heartbeat and all other life processes proceeding more slowly than his own. Therefore Eartha thinks Astro ages more slowly, so when Astro returns to earth he is younger than Eartha.

Now here is the paradox: Since all inertial frames are equivalent, cannot Astro make exactly the same arguments, to conclude that Eartha is in fact the younger? Thus each twin would think the other is younger, which is a paradox.

The resolution of this paradox is the recognition that the twins are *not* identical in all respects. If Eartha remains in an inertial frame at all times, Astro must at times have an acceleration with respect to inertial frames in order to turn around and come back. Eartha remains always at rest in the same inertial frame; Astro does not. There is a real physical difference between the circumstances of the twins. Careful analysis shows that Eartha is correct: When Astro returns, he *is* younger than Eartha.

The twin paradox: How can one twin grow old faster than the other?

The twin paradox isn't actually a paradox if you understand relativity.

40–4 RELATIVITY OF LENGTH

Just as the time interval between two events depends on the observer's frame of reference, the *distance* between two points may also depend on the observer's frame of reference. To measure a distance one must, in principle, observe the positions of two points, such as the two ends of a ruler, simultaneously; but what is simultaneous in one reference frame is not simultaneous in another.

To develop a relation between lengths in various coordinate systems, we consider another thought experiment. We attach a source of light pulses to one end of a ruler and a mirror to the other end, as shown in Fig. 40–4. Let the ruler be at rest in reference frame S' and its length in this frame be l'. Then the time $\Delta t'$ required for a light pulse to make the round trip from source to mirror and back is given by

$$\Delta t' = \frac{2l'}{c}. \tag{40–6}$$

This is a proper time interval, since departure and return occur at the same point in S'.

In S the ruler is moving with speed u, and it is displaced during this travel of the light pulse. Let the length of the ruler in S be l, and let the time of travel from source to mirror, as measured in S, be Δt_1. During this interval the ruler, with source and mirror attached, moves a distance $u \, \Delta t_1$, and the total length of path d from source to mirror is not l but

$$d = l + u \, \Delta t_1. \tag{40–7}$$

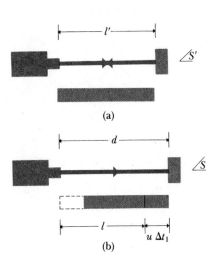

40–4 (a) A light pulse is emitted from a source at one end of a ruler, reflected from a mirror at the opposite end, and returned to the source position. (b) Motion of the light pulse as seen by an observer in S. The distance traveled from source to mirror is greater than the length l measured in S, by the amount $u \, \Delta t_1$, as shown.

But since the pulse travels with speed c, it is also true that

$$d = c\,\Delta t_1. \tag{40-8}$$

Combining Eqs. (40-7) and (40-8) to eliminate d, we find

$$c\,\Delta t_1 = l + u\,\Delta t_1,$$

or

$$\Delta t_1 = \frac{l}{c - u}. \tag{40-9}$$

In the same way we can show that the time Δt_2 for the return trip from mirror to source is

$$\Delta t_2 = \frac{l}{c + u}. \tag{40-10}$$

The *total* time $\Delta t = \Delta t_1 + \Delta t_2$ for the round trip, as measured in S, is

$$\Delta t = \frac{l}{c - u} + \frac{l}{c + u} = \frac{2l}{c(1 - u^2/c^2)}. \tag{40-11}$$

We also know that Δt and $\Delta t'$ are related by Eq. (40-5), since $\Delta t'$ is proper in S'. Thus Eq. (40-6) becomes

$$\Delta t \sqrt{1 - \frac{u^2}{c^2}} = \frac{2l'}{c}. \tag{40-12}$$

Moving objects appear shorter.

Finally, combining this expression with Eq. (40-11) to eliminate Δt, and simplifying, we obtain

$$l = l' \sqrt{1 - \frac{u^2}{c^2}}. \tag{40-13}$$

Thus the length l measured in S, in which the ruler is moving, is *shorter* than the length l' in S', where it is at rest. A length measured in the rest frame of the body is called a proper length; thus l' above is a proper length in S', and the length measured in any other frame is less than l'. This effect is called **length contraction.**

EXAMPLE 40-2 In Example 40-1 (Section 40-3), what distance does the spaceship travel during emission of a pulse, as measured in its rest frame?

SOLUTION The question is somewhat ambiguous since, of course, in its own frame of reference the ship is at rest. But suppose it leaves markers in space (such as small smoke bombs) that remain at rest relative to the earth, at the instants when the pulse starts and stops. Then an observer on the ship measures the distance between these markers, with the aid of observers behind the ship but moving with it, each with a clock synchronized with that of the ship. The distance d between the markers is a proper length in the earth's frame S. In the spaceship's frame S', the distance d' is contracted by the factor given in Eq. (40-13):

$$d' = d \sqrt{1 - \frac{u^2}{c^2}} = (4210 \text{ m}) \sqrt{1 - (0.99)^2}$$

$$= 594 \text{ m}.$$

(Note that because d, not d', is a proper length, we must reverse the roles of l and l'.) An observer in the spaceship can calculate its speed relative to earth from this set of data:

$$u = \frac{d'}{\Delta t'} = \frac{594 \text{ m}}{2 \times 10^{-6} \text{ s}} = 2.97 \times 10^8 \text{ m·s}^{-1},$$

which agrees with the initial data.

When u is very small compared to c, the contraction factor in Eq. (40–13) approaches unity, and in the limit of small speeds we recover the Newtonian relation $l = l'$. This and the corresponding result for time dilation shows that Eqs. (40–1) retain their validity in the limit of speeds much smaller than c; only at speeds comparable to c are modifications needed.

We have derived Eq. (40–13) for lengths measured in the direction *parallel* to the relative motion of the two frames of reference. Lengths measured *perpendicular* to the direction of motion are *not* contracted. To prove this, consider two identical rulers; one ruler lies along the y-axis with one end at the origin O of frame of reference S, and the other lies along the y'-axis with one end at the origin O' of S'. At the instant the two origins coincide, observers in the two frames of reference S and S' observe the positions of the upper ends of the rulers. If the observer in S thinks his ruler is shorter, the observer in S' must think his ruler is longer. But this would mean that there is some distinction between the two frames of reference, which violates our basic premise that all inertial frames of reference are equivalent. Thus both observers must conclude that the two rulers have the same length, despite the fact that to each observer, one is stationary and one is moving. Hence there is no length contraction perpendicular to the direction of relative motion of the coordinate systems.

Lengths perpendicular to the observer's motion do not contract.

40–5 THE LORENTZ TRANSFORMATION

The Galilean coordinate transformation, given by Eqs. (40–1) and the relation $t = t'$, is valid only in the limit when u is much smaller than c. We are now ready to derive a more general transformation that is not subject to this limitation. The more general relations are called the **Lorentz transformation.** In the limit of very small u, they reduce to the Galilean transformation, but they may also be used when u is comparable to c.

The Lorentz transformation: a coordinate transformation consistent with requirements of the principle of relativity

The basic problem is this: When an event occurs at point (x, y, z) at time t, as observed in a frame of reference S, what are the coordinates (x', y', z') and time t' of the event as observed in a second frame S' moving relative to S with constant velocity u along the x-direction?

To derive the transformation equations, we refer to Fig. 40–5, which is the same as Fig. 40–1. As before, we assume that the origins coincide at the initial time $t = t' = 0$. Then in S the distance from O to O' at time t is still ut. The coordinate x' is a proper length in S', so in S it appears contracted by the factor given in Eq. (40–13). Thus the distance x from O to P in S is given not simply by $x = ut + x'$ as in the Galilean transformation, but by

$$x = ut + x'\sqrt{1 - \frac{u^2}{c^2}}. \tag{40–14}$$

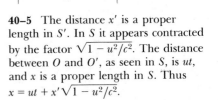

40–5 The distance x' is a proper length in S'. In S it appears contracted by the factor $\sqrt{1 - u^2/c^2}$. The distance between O and O', as seen in S, is ut, and x is a proper length in S. Thus $x = ut + x'\sqrt{1 - u^2/c^2}$.

Solving this equation for x', we obtain

$$x' = \frac{x - ut}{\sqrt{1 - u^2/c^2}}. \tag{40-15}$$

This equation is half of the Lorentz transformation; the other half is the equation giving t' in terms of x and t. To obtain it we note that the principle of relativity requires that the *form* of the transformation from S to S' be identical to that from S' to S, the only difference being a change in the sign of the relative velocity u. Thus, from Eq. (40-14), it must be true that

$$x' = -ut' + x\sqrt{1 - \frac{u^2}{c^2}}. \tag{40-16}$$

We may now equate Eqs. (40-15) and (40-16) to eliminate x' from this expression, obtaining the desired relation for t' in terms of x and t. We will leave the algebraic details for you to work out; the result is

$$t' = \frac{t - ux/c^2}{\sqrt{1 - u^2/c^2}}. \tag{40-17}$$

As we discussed previously, lengths perpendicular to the direction of relative motion are not affected by the motion, so $y' = y$ and $z' = z$.

Collecting all the transformation equations, we have

$$x' = \frac{x - ut}{\sqrt{1 - u^2/c^2}}, \qquad y' = y, \qquad z' = z, \qquad t' = \frac{t - ux/c^2}{\sqrt{1 - u^2/c^2}}. \tag{40-18}$$

These are the *Lorentz transformation* equations, the relativistic generalization of the Galilean transformation, Eqs. (40-1). When u is much smaller than c, the two transformations become identical, and $t = t'$. In both cases we assume that the relative motion of the two coordinate systems is along the common xx'-axis.

Using the Lorentz transformation to derive a relativistically correct velocity transformation

Next we consider the relativistic generalization of the Galilean velocity-transformation equations, Eq. (40-2). As we have noted, the Galilean relation is valid only in the limit when u is very small. The relativistic expression can easily be obtained from the Lorentz transformation. In the following discussion, we continue to use u as the velocity of frame of reference S' relative to S, and we use v and v' for the x-components of velocity of a particle as measured in S and S', respectively. Suppose a particle observed in S' is at point x_1' at time t_1' and point x_2' at time t_2'. Then its x-component of velocity v' is given by

$$v' = \frac{x_2' - x_1'}{t_2' - t_1'} = \frac{\Delta x'}{\Delta t'}. \tag{40-19}$$

To obtain the x-component of velocity in S, we use Eqs. (40-18) to translate this expression into terms of the corresponding positions x_1 and x_2 and times t_1 and t_2 observed in S. We find

$$x_2' - x_1' = \frac{x_2 - x_1 - u(t_2 - t_1)}{\sqrt{1 - u^2/c^2}} = \frac{\Delta x - u\,\Delta t}{\sqrt{1 - u^2/c^2}},$$

$$t_2' - t_1' = \frac{t_2 - t_1 - u(x_2 - x_1)/c^2}{\sqrt{1 - u^2/c^2}} = \frac{\Delta t - u\,\Delta x/c^2}{\sqrt{1 - u^2/c^2}}.$$

Using these results in Eq. (40–19), we find

$$v' = \frac{\Delta x - u\,\Delta t}{\Delta t - u\,\Delta x/c^2}$$

$$= \frac{(\Delta x/\Delta t) - u}{1 - (u/c^2)(\Delta x/\Delta t)}.$$

In the limit $\Delta t \to 0$, $\Delta x/\Delta t$ is just the x-component of velocity v measured in S, so we finally obtain

$$v' = \frac{v - u}{1 - uv/c^2}. \qquad (40\text{–}20)$$

Note that when u and v are much smaller than c, the denominator approaches unity, and we obtain the nonrelativistic result $v' = v - u$. The opposite extreme is the case $v = c$; then we find

$$v' = \frac{c - u}{1 - uc/c^2} = c.$$

That is, anything moving with speed c relative to S also has speed c relative to S', despite the relative motion of the two frames. This result demonstrates the consistency of Eq. (40–20) with our initial assumption that the speed of light is the same in all frames of reference.

We can also rearrange Eq. (40–20) to give v in terms of v'. We leave the algebraic details as a problem; the result is

$$v = \frac{v' + u}{1 + uv'/c^2}. \qquad (40\text{–}21)$$

> Anything moving with speed c relative to one observer also moves with speed c relative to any other observer.

EXAMPLE 40–3   A spaceship moving away from earth with a speed $0.9c$ fires a missile in the same direction as its motion, with a speed $0.9c$ relative to the spaceship. What is the missile's speed relative to the earth?

SOLUTION   Let the earth's frame of reference be S, the spaceship S'. Then $v' = 0.9c$ and $u = 0.9c$. The nonrelativistic velocity addition formula would give a velocity relative to earth of $1.8c$. The correct relativistic result, obtained from Eq. (40–21), is

$$v = \frac{0.9c + 0.9c}{1 + (0.9c)(0.9c)/c^2}$$

$$= 0.994c.$$

When u is less than c, a body moving with a speed less than c in one frame of reference also has a speed less than c in *every other* frame of reference. This is one reason for thinking that no material body can travel with a speed greater than that of light, relative to *any* frame of reference. The relativistic generalizations of energy and momentum, to be considered in the following sections, give further support to this hypothesis.

> Nothing can travel faster than light.

PROBLEM-SOLVING STRATEGY: Lorentz transformation

1. Try hard to understand the concepts of proper time and proper length. A time interval between two events that happen at the same point in a particular frame of reference is a proper time in that frame, and the time interval between the same two events is longer in any other frame. A length measured between two points at rest in a frame of reference is a proper length in that frame, and the distance between the points is shorter in any other frame.

2. You can get a lot of mileage out of the Lorentz transformation equations. It helps to make a list of what you know and don't know. Do you know the coordinates in one frame? The time of an event in one frame? What are your knowns and unknowns?

3. In velocity-transformation problems, if you have two observers measuring the motion of a body, decide which you want to call S and which S', identify the velocities v and v' clearly, and make sure you know the velocity u of S' relative to S. Use either form of the velocity-transformation equation, Eq. (40–20) or (40–21), whichever is more convenient.

4. Don't be discouraged if some of your results don't seem to make sense or if they disagree with your intuition. Reliable intuition about relativity takes time to develop. Keep trying to develop intuitive understanding; it will come.

40–6 MOMENTUM

To save conservation of momentum we have to revise our definition of momentum.

We have discussed the fact that Newton's laws of motion are *invariant* under the Galilean coordinate transformation, but that to satisfy the principle of relativity, this transformation must be replaced by the more general Lorentz transformation. This requires corresponding generalizations in the laws of motion and the definitions of momentum and energy.

The principle of conservation of momentum states that *when two bodies collide, the total momentum is constant*, provided there is no interaction except that of the two bodies with each other. However, when we consider a collision in one coordinate system S, in which momentum is conserved, and then use the Lorentz transformation to obtain the velocities in a second system S', we find that if the Newtonian definition of momentum ($p = mv$) is used, momentum is *not* conserved in the second system. According to the basic principle of relativity, if momentum conservation is a valid physical law, it must hold *exactly* in all inertial frames of reference and for all particle speeds, whether or not they are small compared to c. Thus if the Lorentz transformation is correct, we must revise the *definition* of momentum.

Deriving the correct relativistic generalization is beyond our scope, and we simply quote the result. The **relativistic momentum p** of a particle of mass m moving with velocity v is given by

$$p = \frac{mv}{\sqrt{1 - v^2/c^2}}. \tag{40–22}$$

Note that, as usual when the particle speed v is much less than c, this is approximately equal to the Newtonian expression $p = mv$, but that in general the momentum is greater in magnitude than mv.

In these momentum expressions, m is a *constant* characterizing the inertial properties of a particle. Because $p = mv$ is still valid in the limit of very small velocities, we see that m must be the same quantity we used (and learned to measure) in our study of Newtonian mechanics early in the book. In relativistic mechanics it is often called the **rest mass** of a particle.

In Newtonian mechanics, the second law of motion can be stated in the form

What is the relativistic relation between force and motion?

$$F = \frac{d\boldsymbol{p}}{dt}. \tag{40–23}$$

That is, force equals time rate of change of momentum. Experiments show that this result is still valid in relativistic mechanics, provided we use the relativistic momentum given by Eq. (40–22). That is, the relativistically correct generalization of Newton's second law is found to be

$$\boldsymbol{F} = \frac{d}{dt} \frac{m\boldsymbol{v}}{\sqrt{1 - v^2/c^2}}. \tag{40–24}$$

This relation has the consequence that constant force no longer causes constant acceleration. For example, when force and velocity are both along a single line, causing straight-line motion, it is not difficult to show that Eq. (40–24) becomes

$$F = \frac{m}{(1 - v^2/c^2)^{3/2}} \frac{dv}{dt},$$

or

$$\frac{dv}{dt} = a = \frac{F}{m}(1 - v^2/c^2)^{3/2}. \tag{40–25}$$

We leave the derivation of Eq. (40–25) as a problem. This equation shows that as a particle's speed increases, the acceleration caused by a given force continuously *decreases*. As the speed approaches c, the acceleration approaches zero, no matter how great a force is applied. Thus it is impossible to accelerate a particle from a state of rest to a speed equal to or greater than c, and the speed of light is sometimes referred to as "the ultimate speed."

Why can't we accelerate an object up to the speed of light?

Equation (40–22) is sometimes interpreted to mean that a rapidly moving particle undergoes an increase in mass. If the mass at zero velocity (the rest mass) is denoted by m, then the "relativistic mass" m_{rel} is given by

$$m_{\text{rel}} = \frac{m}{\sqrt{1 - v^2/c^2}}.$$

Indeed, when we consider the motion of a system of particles (such as gas molecules in a moving container), the total mass of the system is the sum of the relativistic masses of the particles rather than the sum of their rest masses.

Relativistic mass: a concept with uses and pitfalls

The concept of relativistic mass also has its pitfalls, however. It is *not* correct to say that the relativistic generalization of Newton's second law is $\boldsymbol{F} = m_{\text{rel}}\boldsymbol{a},$ and it is *not* correct that the relativistic kinetic energy of a particle is $K = \frac{1}{2}m_{\text{rel}}v^2$. So this concept must be approached with great caution. For the present discussion it is better to regard Eq. (40–22) as a generalized definition of momentum and to retain the meaning of m as an inertial quantity characteristic of each particle and independent of its state of motion.

40–7 WORK AND ENERGY

When we developed the relationship between work and kinetic energy in Chapter 7, we used Newton's laws of motion. As we have just seen, Newton's laws have to be generalized to bring them into harmony with the principle of

Deriving a relativistically correct work–energy relation

relativity; hence it is not surprising that we also need to generalize the definition of kinetic energy and the work–energy relation.

Because a constant force on a body does not cause a constant acceleration (except when the speed of the body is very small), even the simplest dynamics problems require the use of calculus. But we can derive fairly easily the relativistic generalization of the work–energy principle. We begin with the definition of work: $W = \int F\,dx$. According to Eq. (40–23), $F = dp/dt$. Combining these, we write

$$W_{12} = \int_{x_1}^{x_2} F\,dx = \int_{x_1}^{x_2} \frac{dp}{dt}dx. \tag{40–26}$$

In order to derive the generalized expression for kinetic energy K as a function of speed v, we would like to convert this to an integral on v. To do so, we first note that dp, dx, and dv are the infinitesimal increments in p, x, and v, respectively, in the time interval dt. Thus it is permissible to interchange dp and dx in the right side of Eq. (40–26) and rewrite it as

$$\int_{x_1}^{x_2} F\,dx = \int_{x_1}^{x_2} \frac{dp}{dt}\,dx = \int_{x_1}^{x_2} \frac{dx}{dt}dp. \tag{40–27}$$

Now $dx/dt = v$, and dp may be expressed in terms of dv as $dp = (dp/dv)dv$. Thus Eq. (40–27) becomes

$$\int_{x_1}^{x_2} F\,dx = \int_{v_1}^{v_2} v\frac{dp}{dv}dv. \tag{40–28}$$

Note that the limits on the integral on the right are the speeds v_1 and v_2 at positions x_1 and x_2, respectively. To find dp/dv in terms of v, we take the derivative of Eq. (40–22). We will let you work out the details; the result is

$$\frac{dp}{dv} = \frac{m}{(1 - v^2/c^2)^{3/2}}.$$

We substitute this into Eq. (40–28) to obtain, finally, an expression that we can integrate easily

$$W_{12} = \int_{x_1}^{x_2} F\,dx = \int_{v_1}^{v_2} \frac{mv\,dv}{(1 - v^2/c^2)^{3/2}}$$

$$= \frac{mc^2}{\sqrt{1 - v_2^2/c^2}} - \frac{mc^2}{\sqrt{1 - v_1^2/c^2}}, \tag{40–29}$$

where v_1 and v_2 are the initial and final speeds of the particle, respectively.

This result suggests defining kinetic energy as

$$\frac{mc^2}{\sqrt{1 - v^2/c^2}}. \tag{40–30}$$

The relativistic expression for kinetic energy

But this expression is not zero when $v = 0$; instead it becomes equal to mc^2. Thus the correct relativistic generalization of kinetic energy K is

$$K = \frac{mc^2}{\sqrt{1 - v^2/c^2}} - mc^2. \tag{40–31}$$

This expression, if correct, must reduce to the Newtonian expression $K = \frac{1}{2}mv^2$ when v is much smaller than c. It is not obvious that this is the case; to

A nuclear-fission explosion, a dramatic and frightening example of conversion of nuclear mass into energy. (Courtesy of U.S. Dept. of Energy.)

demonstrate that it is so, we can expand the radical by using the binomial theorem:

$$\left(1 - \frac{v^2}{c^2}\right)^{-1/2} = 1 + \frac{1}{2}\frac{v^2}{c^2} + \frac{3}{8}\frac{v^4}{c^4} + \frac{5}{16}\frac{v^6}{c^6} + \cdots.$$

Combining this with Eq. (40–31), we find

At slow speeds the relativistic kinetic energy expression reduces to the familiar Newtonian relation.

$$K = mc^2\left(1 + \frac{1}{2}\frac{v^2}{c^2} + \frac{3}{8}\frac{v^4}{c^4} + \cdots\right) - mc^2$$

$$= \frac{1}{2}mv^2 + \frac{3}{8}m\frac{v^4}{c^2} + \cdots. \qquad (40\text{–}32)$$

In each expression the dots stand for omitted terms. When v is much smaller than c, all the terms in the series except the first are negligibly small, and we obtain the classical $\frac{1}{2}mv^2$.

But what is the significance of the term mc^2 that had to be subtracted in Eq. (40–31)? Although Eq. (40–30) does not give the *kinetic energy* of the particle, perhaps it represents some kind of *total* energy, including both the kinetic energy and an additional energy mc^2, which the particle possesses even when it is not moving. Calling this total energy E and using Eq. (40–31), we find

$$E = K + mc^2 = \frac{mc^2}{\sqrt{1 - v^2/c^2}}. \qquad (40\text{–}33)$$

The energy mc^2 associated with mass rather than motion may be called the **rest energy** of the particle. This speculation does not prove that the concept of rest energy is meaningful, but it points the way toward further investigation.

Rest energy: energy associated with mass rather than motion

There is in fact direct experimental evidence of the existence of rest energy. The simplest example is the decay of the π° meson, an unstable particle that "decays"; in the decay process, the particle disappears and electromagnetic radiation appears. When the particle is at rest (and therefore has no

Conversion of mass into energy: direct confirmation of the validity of the concept of rest mass

Mass and energy: a more general
conservation principle

kinetic energy) before its decay, the total energy of the radiation produced is found to be exactly equal to mc^2. There are many other examples of fundamental particle transformations in which the total mass of the system changes; in every case, a corresponding energy change occurs, consistent with the assumption of a rest energy mc^2 associated with a rest mass m.

Although the principles of conservation of mass and of energy originally were developed quite independently, the theory of relativity shows that they are but two special cases of a single broader conservation principle, the *principle of conservation of mass and energy*. In some physical phenomena, neither mass nor energy is separately conserved, but the changes in these quantities are governed by a more general conservation principle: When a change in rest mass m occurs in an isolated system, an opposite change mc^2 in the total energy of other types must accompany this change.

The conversion of mass into energy is the fundamental principle involved in the generation of power through nuclear reactions, a subject we will discuss in Chapter 44. When a uranium nucleus undergoes fission in a nuclear reactor, the total mass of the resulting fragments is less than that of the parent nucleus, and the total kinetic energy of the fragments is equal to this mass deficit times c^2. This kinetic energy can be used to produce steam to operate turbines for electric-power generators or in a variety of other ways.

We can relate the total energy E (kinetic plus rest) of a particle directly to its momentum by combining Eqs. (40–22) and (40–33) to eliminate the particle's velocity. The simplest procedure is to rewrite these equations in the following forms:

$$\left(\frac{E}{mc^2}\right)^2 = \frac{1}{1 - v^2/c^2}; \qquad \left(\frac{p}{mc}\right)^2 = \frac{v^2/c^2}{1 - v^2/c^2}.$$

Subtracting the second of these equations from the first and rearranging, we find

$$E^2 = (mc^2)^2 + (pc)^2. \tag{40–34}$$

Again we see that for a particle at rest ($p = 0$), $E = mc^2$.

Can a particle have zero rest mass?

Equation (40–34) also suggests that a particle may have energy and momentum even when it has no rest mass. In such a case, $m = 0$ and

$$E = pc. \tag{40–35}$$

In fact, massless particles do exist. The most familiar examples are photons, the quanta of electromagnetic radiation. The existence of these particles is well established, and we will study them in greater detail in later chapters. They always travel with the speed of light; they are emitted and absorbed during changes of state of atomic systems, accompanied by corresponding changes in the energy and momentum of these systems.

40–8 INVARIANCE

Invariance: things that don't change
when you go from one coordinate system
to another

We have used the term **invariance** to describe the idea that, according to the principle of relativity, all the fundamental laws of physics must keep the same form, that is, must be *invariant,* when we shift our description from one inertial frame of reference to another. This word is also used in a more specific sense to describe *physical quantities* that have the same value in all inertial frames of reference. The speed of light is one obvious example, but there are others that are important.

For example, we can describe the position (x, y, z) and time t of an event in a frame of reference S, and the position (x', y', z') and time t' of that same event in a frame S' moving with constant velocity relative to S. In general, of course, x is not equal to x', nor y to y', nor t to t'. But we can prove that the quantity $(x^2 + y^2 + z^2 - c^2t^2)$ *does* have the same value in both systems, no matter what the event. That is, it is always true that the position and time in S are related to the position and time in S' by

$$x^2 + y^2 + z^2 - c^2t^2 = x'^2 + y'^2 + z'^2 - c^2t'^2. \qquad (40\text{–}36)$$

We leave the detailed proof of this as a problem; just substitute Eqs. (40–18) into the right side of Eq. (40–36) and simplify the result to show that it equals the left side.

We say that the quantity $(x^2 + y^2 + z^2 - c^2t^2)$ is *invariant* under the Lorentz transformation. This invariance is directly related to the invariance of the speed of light. To see this, suppose a light pulse starts from the origin O of frame S at time $t = 0$; the square of its distance from the origin at any later time t is $(x^2 + y^2 + z^2)$ and must equal $(ct)^2$. But when the same light pulse is observed in frame S', an exactly similar equation must hold for the coordinates and time in S'. Thus Eq. (40–36) is not so surprising.

We can obtain another useful invariant quantity from Eq. (40–34). The rest mass m of a particle is an inherent property of the particle, independent of the frame of reference. Thus the energy–momentum relation of Eq. (40–34) tells us that the quantity $(E^2 - p^2c^2)$ is *invariant*. That is, if a particle has energy E and momentum of magnitude p, computed from the velocity v measured in frame of reference S, and energy E' and momentum of magnitude p' computed from the velocity v' in S', then

Invariance of a quantity involving energy and momentum

$$E^2 - p^2c^2 = E'^2 - p'^2c^2. \qquad (40\text{–}37)$$

It is not very hard to show that this relation is also valid for a collection of particles. That is, if as usual we define the total energy (rest plus kinetic) of a set of particles as the sum of the energies of the individual particles, and the total momentum as the vector sum of momenta of particles, then Eq. (40–37) also holds if we interpret E as the *total* energy and p as the magnitude of the *total* momentum. These relations are useful in analyzing relativistic collision problems.

40–9 THE DOPPLER EFFECT

An additional important consequence of relativistic kinematics is the Doppler effect for electromagnetic radiation. In our previous discussion of the Doppler effect in Section 23–5, we quoted without proof the formula, Eq. (23–18), for the frequency shift that results from motion of a source relative to an observer. We can now derive that result.

The Doppler effect: The frequency seems different when the source and observer are in relative motion.

Suppose a light source at rest in a frame of reference S' emits light pulses with frequency f', as measured by an observer in that frame. The period, or time between pulses, is $\tau' = 1/f'$ as seen in S'. To an observer in frame S, the source is moving with speed u; the frequency of the pulses as seen by this observer is f, and the time between successive pulses measured by this observer

Virgo   78,000,000 light years distant
receding at 1200 km/sec

Corona
borealis   1,400,000,000 light years distant
receding at 22,000 km/sec

Hydra   3,960,000,000 light years distant
receding at 61,000 km/sec

The Doppler effect for light. Each photograph shows two dark bands characteristic of the absorption spectrum of calcium in the spectra of three galaxies. The recession of the sources from Earth causes Doppler "redshifts" to smaller frequencies and longer wavelengths. For a nearby galaxy in Virgo the redshift is small, but in the Hydra galaxy, nearly 50 times as far away, the shift is much greater. (Palomar Observatory Photograph.)

is $\tau = 1/f$. Because τ' is a proper time interval in S', τ and τ' are related by

$$\tau = \frac{\tau'}{\sqrt{1 - u^2/c^2}}. \tag{40–38}$$

During the interval τ between pulses, as seen in S, the source moves to the right a distance of $u\tau$ and the first pulse moves to the left a distance $c\tau$. Hence the distance between successive pulses approaching the observer in S is $(c + u)\tau$. But this is also equal to the wavelength λ, which in turn is equal to c/f. Thus we have

$$\lambda = (c + u)\tau = \frac{c}{f}. \tag{40–39}$$

We now use Eq. (40–38) to express τ in terms of τ' and then use the relation $\tau' = 1/f'$ to obtain

$$(c + u)\frac{\tau'}{\sqrt{1 - u^2/c^2}} = \frac{1}{f'}\frac{c + u}{\sqrt{1 - u^2/c^2}} = \frac{c}{f}. \tag{40–40}$$

Finally, we rearrange this to obtain

$$f = \sqrt{\frac{c - u}{c + u}}f'. \tag{40–41}$$

This relation has the same form as Eq. (23–18), with minor differences in notation.

The relativistic Doppler effect is different from the effect with sound.

Unlike the situation with sound, with light there is no distinction between motion of source and motion of observer; only the *relative* velocity of the two is significant. Equation (40–41) shows that when the source is moving away from the observer, the observed frequency is less than the frequency measured in the rest frame of the source. If the source is moving *toward* the observer, we change the sign of u in Eq. (40–41) to find the apparent *increase* in frequency. The last three paragraphs of Section 23–5 discuss several practical applications of the Doppler effect with light and other electromagnetic radiation; we suggest you review those paragraphs now.

40–10 RELATIVITY AND NEWTONIAN MECHANICS

The sweeping changes required by the principle of relativity go to the very roots of Newtonian mechanics, including the concepts of length and time, the equations of motion, and the conservation principles. Thus it may appear that the foundations on which Newton's mechanics are built have been destroyed. While this is true in one sense, it is essential to keep in mind that the Newtonian formulation still retains its validity whenever speeds are small compared with the speed of light. In such cases time dilation, length contraction, and the modifications of the laws of motion are so small that they are unobservable. In fact, every one of the principles of Newtonian mechanics survives as a special case of the more general relativistic formulation.

Newtonian mechanics is not *wrong;* it is *incomplete.* It is a limiting case of relativistic mechanics. It is approximately correct when all speeds are small compared to *c;* and in the limit when all speeds approach zero, it becomes exactly correct.

Thus relativity does not completely destroy Newtonian mechanics but *generalizes* it. After all, Newton's laws rest on a very solid base of experimental evidence, and it would be strange indeed to advance a new theory inconsistent with this evidence. So it always is with the development of physical theory. Whenever a new theory is in partial conflict with an older, established theory, it nevertheless must yield the same predictions as the old in areas where the old theory is supported by experimental evidence. Every new physical theory must pass this test, called the **correspondence principle,** which has come to be regarded as a fundamental procedural rule in all physical theory. There are many problems for which Newtonian mechanics is clearly inadequate, including all situations where particle speeds are comparable to that of light or where direct conversion of mass to energy occurs. But there is still a large area, including nearly all the behavior of macroscopic bodies in mechanical systems, in which Newtonian mechanics is still perfectly adequate.

At this point it is legitimate to ask whether the relativistic mechanics just discussed is the final word on this subject or whether *further* generalizations are possible or necessary. For example, inertial frames of reference have occupied a privileged position in all our discussions thus far. Should the principle of relativity be extended to noninertial frames as well?

Here is an example that illustrates some implications of this question. A student decides to go over Niagara Falls while enclosed in a large wooden box. During his free fall he can, in principle, perform experiments inside the box. An object released inside the box does not fall to the floor because both the box and the object are in free fall with a downward acceleration of 9.8 m·s^{-2}. But an alternative interpretation, from this man's point of view, is that the force of gravity has suddenly been turned off. Provided he remains in the box and it remains in free fall, he cannot tell whether he is indeed in free fall or whether the force of gravity has vanished. A similar problem appears in a space station in orbit around the earth. Objects in the spaceship appear weightless, but without going outside the ship there is no way to determine whether gravity has disappeared or the spaceship is in an accelerated (i.e., noninertial) frame of reference.

These considerations form the basis of Einstein's **general theory of relativity.** If we cannot distinguish experimentally between a gravitational field at a particular location and an accelerated reference system, then there cannot be any real distinction between the two. Pursuing this concept, we may try to

Is there anything left of Newtonian mechanics?

Newtonian mechanics is a special case of relativistic mechanics, when all the speeds are much smaller than *c.*

Generalizations to noninertial frames of reference

The general theory of relativity: What's the difference between gravitation and an accelerated frame of reference?

represent *any* gravitational field in terms of special characteristics of the coordinate system. This turns out to require even more sweeping revisions of our space–time concepts than did the special theory of relativity, and we find that, in general, the geometric properties of space are noneuclidean.

The basic ideas of the general theory of relativity are now well established, but some of the details remain speculative in nature. Its chief application is in cosmological investigations of the structure of the universe, the formation and evolution of stars, and related matters. It is not believed to have any relevance for atomic or nuclear phenomena or macroscopic mechanical problems of less than astronomical dimensions.

SUMMARY

KEY TERMS

special theory of relativity
principle of relativity
Galilean coordinate
  transformation
simultaneity
time dilation
twin paradox
length contraction
Lorentz transformation
relativistic momentum
rest mass
rest energy
invariance
correspondence principle
general theory of relativity

The principle of relativity states that all the fundamental laws of physics have the same form in all inertial frames of reference. In particular, the speed of light is the same in all inertial frames and is independent of the motion of the source. This is inconsistent with the Galilean coordinate transformation, which has to be modified. Simultaneity is not an absolute concept; two events that appear simultaneous in one frame of reference in general do not appear simultaneous in a second frame moving relative to the first.

If the time interval between two events occurring at the same space point in a frame S' is $\Delta t'$, and if S' moves with constant velocity u relative to S, the time interval Δt between the events, as observed in S, is given by

$$\Delta t = \frac{\Delta t'}{\sqrt{1 - u^2/c^2}}. \tag{40–5}$$

This effect is called time dilation, and $\Delta t'$ is called a proper time interval in S'.

If the distance between two points at rest in a frame S' is l', and if S' moves with constant velocity u relative to S, the distance l between the points as measured in S is

$$l = l' \sqrt{1 - \frac{u^2}{c^2}}. \tag{40–13}$$

This effect is called length contraction, and l' is called a proper length in S'.

The Lorentz transformation relates the coordinates and time of an event in one inertial coordinate system to the coordinates and time of the same event as observed in a second inertial coordinate system moving with constant velocity u relative to the first. The transformation equations are

$$x' = \frac{x - ut}{\sqrt{1 - u^2/c^2}},$$
$$y' = y,$$
$$z' = z,$$
$$t' = \frac{t - ux/c^2}{\sqrt{1 - u^2/c^2}}. \tag{40–18}$$

If a particle's velocity in S' is v', and S' moves relative to S with velocity u, then the particle's velocity v in S is given by

$$v = \frac{v' + u}{1 + uv'/c^2}. \tag{40–21}$$

In order for momentum conservation in collisions to hold in all coordinate systems, the definition of momentum must be generalized. For a particle of mass m moving with velocity v, the momentum p is defined as

$$p = \frac{mv}{\sqrt{1 - v^2/c^2}}. \qquad (40\text{-}22)$$

Generalizing the work–energy relation requires that we generalize the definition of kinetic energy K. For a particle of mass m moving with speed v,

$$K = \frac{mc^2}{\sqrt{1 - v^2/c^2}} - mc^2. \qquad (40\text{-}31)$$

This form suggests assigning a rest energy mc^2 to a particle, so that the total energy E, kinetic energy plus rest energy, is

$$E = K + mc^2 = \frac{mc^2}{\sqrt{1 - v^2/c^2}}. \qquad (40\text{-}33)$$

The total energy E and magnitude of momentum p for a particle of rest mass m are related by

$$E^2 = (mc^2)^2 + (pc)^2. \qquad (40\text{-}34)$$

The quantity

$$x^2 + y^2 + z^2 - c^2 t^2$$

is invariant; for a given event it has the same value in all inertial frames. Similarly, the quantity

$$E^2 - p^2 c^2$$

is invariant.

The special theory of relativity is a generalization of Newtonian mechanics. All the principles of Newtonian mechanics are present as limiting cases when all the speeds are small compared to c. Further generalization to include accelerated frames of reference and their relation to gravitational fields leads to the general theory of relativity.

QUESTIONS

40–1 What do you think would be different in everyday life if the speed of light were 10 m·s^{-1} instead of its actual value?

40–2 The average life span in the United States is about 70 years. Does this mean that it is impossible for an average person to travel a distance greater than 70 light-years away from the earth? (A light-year is the distance light travels in a year.)

40–3 What are the fundamental distinctions between an inertial frame of reference and a noninertial frame?

40–4 A physicist claimed that it is impossible to define what is meant by a rigid body in a relativistically correct way. Why?

40–5 Two events occur at the same space point in a particular frame of reference and appear simultaneous in that frame. Is it possible that they may not appear simultaneous in another frame?

40–6 Does the fact that simultaneity is not an absolute concept also destroy the concept of *causality*? If event A is to *cause* event B, A must occur first. Is is possible that in some frames A may appear to cause B, and in others B may appear to cause A?

40–7 A social scientist who has done distinguished work in fields far removed from physics has written a book purporting to refute the special theory of relativity. He begins with a premise that might be paraphrased as follows: "Either two events occur at the same time, or they don't; that's just common sense." How would you respond to this in the light of our discussion of the relative nature of simultaneity?

40–8 When an object travels across an observer's field of view at a relativistic speed, it appears not only foreshortened but also slightly rotated, with the side toward the observer shifted in the direction of motion relative to the side away from him. How does this come about?

40–9 According to the twin paradox mentioned in Section 40–3, if one twin stays on earth while the other takes off in a spaceship at relativistic speed and then returns, one will be older than the other. Can you think of a practical experiment, perhaps using two very precise atomic clocks, that would test this conclusion?

40–10 When a monochromatic light source moves toward an observer, its wavelength appears to be shorter than the value measured when the source is at rest. Does this contradict the hypothesis that the speed of light is the same for all observers? What about the apparent frequency of light from a moving source?

40–11 A student asserted that a massive particle must always have a speed less than that of light, while a massless particle must always travel at exactly the speed of light. Is he correct? If so, how do massless particles such as photons and neutrinos acquire this speed? Can they not start from rest and accelerate?

40–12 The theory of relativity sets an upper limit on the speed a particle can have. Are there also limits on its energy and momentum?

40–13 In principle, does a hot gas have more mass than the same gas when it is cold? Explain. In practice, would this be a measurable effect?

40–14 Why do you think the development of Newtonian mechanics preceded the more refined relativistic mechanics by so many years?

EXERCISES

Section 40–2 Relative Nature of Simultaneity

40–1 For the two trains discussed in Section 40–2, suppose the two lightning bolts appear simultaneous to an observer on the train. Show that they do not appear simultaneous to an observer on the ground. Which appears to come first?

Section 40–3 Relativity of Time

40–2 The π^+ meson, an unstable particle, lives on average about 2.6×10^{-8} s (measured in its own frame of reference) before decaying.

a) If such a particle is moving with respect to the laboratory with a speed of $0.8c$, what lifetime is measured in the laboratory?

b) What distance, measured in the laboratory, does the particle move before decaying?

40–3 The μ^+ meson (or positive muon) is an unstable particle with a lifetime of about 2.3×10^{-6} s (measured in the rest frame of the muon).

a) If the muon is made to travel at very high velocity relative to a laboratory, its lifetime is measured to be 1.6×10^{-5} s. Calculate the speed of the muon, expressed as a fraction of c.

b) What distance, measured in the laboratory, does the particle travel during its lifetime?

40–4 Two atomic clocks are carefully synchronized. One remains in New York while the other is loaded on a supersonic airplane that travels at an average speed of 400 m·s^{-1} and then returns to New York. When the plane returns, the elapsed time on the clock that stayed behind is 5 hr. By how much will the reading of the two clocks differ, and which clock will show the smaller elapsed time? (*Hint:* Use the fact that $v \ll c$ to simplify $\sqrt{1 - v^2/c^2}$ by making a binomial expansion.)

Section 40–4 Relativity of Length

40–5 In the year 2010 a spacecraft flies over Moon Station III at a speed of $0.8c$. A scientist on the moon measures the length of the moving spacecraft to be 200 m. The spacecraft later lands on the moon, and the same scientist measures the length of the now stationary spacecraft. What value does she get?

40–6 Two events are observed in a frame of reference S to occur at the same space point, the second occurring 2 s after the first. In a second frame S' moving relative to S, the second event is observed to occur 3 s after the first. What is the difference between the positions of the two events as measured in S'?

Section 40–5 The Lorentz Transformation

40–7 Show in detail the derivation of Eq. (40–17) from Eqs. (40–15) and (40–16).

40–8 Solve Eqs. (40–18) to obtain x and t in terms of x' and t', and show that the resulting transformation has the same form as the original one except for a change of sign for u.

40–9 Show the details of the derivation of Eq. (40–21) from Eq. (40–20).

40–10 Two particles emerge from a high-energy accelerator in opposite directions, each with a speed $0.6c$ as measured in the laboratory. What is the relative velocity of the particles?

Section 40–6 Momentum

40–11 At what speed is the momentum of a particle twice as great as the result obtained from the nonrelativistic expression mv?

40–12

a) At what speed does the momentum of a particle differ by 1% from the value obtained by using the nonrelativistic expression mv?

b) Is the correct relativistic value greater or less than that obtained from the nonrelativistic expression?

40–13 A particle moves along a straight line under the action of a force lying along the same line. Show that the acceleration $a = dv/dt$ of the particle is given by Eq. (40–25).

Section 40–7 Work and Energy

40–14 For a particle moving in the x-direction, use Eq. (40–22) to show that

$$\frac{dp}{dv} = \frac{m}{(1 - v^2/c^2)^{3/2}}.$$

40–15 What is the speed of a particle whose kinetic energy is equal to

a) its rest energy?   b) ten times its rest energy?

40–16

a) How much work must be done to accelerate a particle of mass m from rest to a speed of $0.1c$?

b) From a speed of $0.9c$ to a speed of $0.99c$?

40–17 Compute the kinetic energy of an electron (mass 9.11×10^{-31} kg) by using both the nonrelativistic and relativistic expressions, and compute the ratio of the two results, for speeds of

a) 1.0×10^8 m·s^{-1},   b) 2.8×10^8 m·s^{-1}.

40–18 In *positron annihilation*, an electron and a positron (a positively charged electron) collide and disappear, producing electromagnetic radiation. If each particle has a mass of 9.11×10^{-31} kg and they are at rest just before the annihilation, find the total energy of the radiation.

40–19 The total consumption of electrical energy per year in the United States is of the order of 10^{19} J. If matter could be converted completely into energy, how many kilograms

of matter would have to be converted to produce this much energy?

40–20 In a hypothetical nuclear-fusion reactor, two deuterium nuclei combine or "fuse" to form one helium nucleus. The mass of a deuterium nucleus, expressed in atomic mass units (u), is 2.0136 u; that of a helium nucleus is 4.0015 u. (1 u = 1.66×10^{-27} kg.)

a) How much energy is released when 1 kg of deuterium undergoes fusion?

b) The annual consumption of electrical energy in the United States is of the order of 10^{19} J. How much deuterium must react to produce this much energy?

40–21 Starting from Eq. (40–34), show that in the classical limit ($pc \ll mc^2$) the energy approaches the classical kinetic energy plus the rest mass energy.

40–22 Calculate, relativistically, the amount of work in MeV that must be done

a) to bring an electron from rest to a velocity of $0.4c$;

b) to increase its velocity from $0.4c$ to $0.8c$.

c) What is the ratio of the kinetic energy of the electron at the velocity of $0.8c$ to that of $0.4c$ when computed (1) from relativistic values and (2) from classical values?

Section 40–8 Invariance

40–23 Use the Lorentz transformation, Eq. (40–18), to prove Eq. (40–36).

Section 40–9 The Doppler Effect

40–24 In terms of c, what relative velocity u between source and observer produces

a) a 1% decrease in frequency?

b) a doubling of the frequency of the observed light?

40–25 How fast must you be approaching a red traffic light ($\lambda = 675$ nm) for it to appear green ($\lambda = 525$ nm)? (If you used this as an excuse for not getting a ticket for running a red light, do you think you might get a speeding ticket instead?)

PROBLEMS

40–26 A cube of metal with sides of length a and rest mass m sits at rest in a frame S_1. In S_1, therefore, the cube has density $\rho_1 = m/a^3$. Frame S_2 moves along the x-axis with a velocity v. To an observer in frame S_2, what is the density of the metal cube?

40–27 The Stanford Linear Accelerator Center (SLAC) uses a 3-km long tube in accelerating subatomic particles. The π^+ meson has a half-life of 2.6×10^{-8} s.

a) How fast must a π^+ meson travel if it is not to decay before it reaches the end of the tube? (Since v will be very close to c, write $v = [1 - \Delta]c$ and give your answer in terms of Δ rather than v.)

b) With a rest mass of 139.6 MeV, what is the π^+ meson's energy at the velocity calculated in (a)?

40–28 A photon of energy E is emitted by an atom of mass m, which recoils in the opposite direction.

a) Assuming that the atom can be treated nonrelativistically, compute the recoil velocity of the atom.

b) From the result of (a), show that the recoil velocity is much smaller than c whenever E is much smaller than the rest energy mc^2 of the atom.

40–29 A radioactive isotope of cobalt, ^{60}Co, emits an electromagnetic photon (γ ray) of energy 1.33 MeV. The cobalt nucleus contains 27 protons and 33 neutrons, each with a mass of about 1.66×10^{-27} kg.

a) If the nucleus is at rest before emission, what is the speed afterward?

b) Is it necessary to use the relativistic generalization of momentum?

40–30 In Problem 40–29, suppose the cobalt atom is in a metallic crystal containing 0.01 mol of cobalt (about 6.02×10^{21} atoms) and that the entire crystal recoils as a unit, rather than just the single nucleus. Find the recoil velocity. (This recoil of the entire crystal, rather than a single nucleus, is called the *Mössbauer effect*, in honor of its discoverer, who first observed it in 1958.)

40–31 A particle is said to be in the *extreme relativistic range* when its kinetic energy is much larger than its rest energy.

a) What is the speed of a particle (expressed as a fraction of c) such that the total energy is ten times the rest energy?

b) For such a particle, what percent error in the energy–momentum relation of Eq. (40–34) results if the term $(mc^2)^2$ is neglected? That is, what is the percentage difference between the left and right sides of Eq. (40–34) if $(mc^2)^2$ is neglected for a particle with the speed calculated in (a)?

40–32 A nuclear bomb containing 20 kg of plutonium explodes. The rest mass of the products of the explosion is less than the original rest mass by one part in 10^4.

a) How much energy is released in the explosion?

b) If the explosion takes place in $1 \mu s$, what is the average power developed by the bomb?

c) How much water could the released energy lift to a height of 1 km?

40–33 An electron in a certain x-ray tube is accelerated from rest through a potential difference of 180,000 V in going from the cathode to the anode. When it arrives at the anode, what is

a) its kinetic energy in eV?

b) its total energy?

c) its velocity?

d) What is the velocity of the electron, calculated classically?

40–34 A certain electron accelerator accelerates electrons through a potential difference of 6.5×10^7 V, so that their kinetic energy is 6.5×10^7 eV.

a) What is the ratio of the speed v of an electron having this energy to the speed of light c?

b) What would the speed be if computed from the principles of classical mechanics?

40–35 The Soviet physicist P. A. Cerenkov discovered that a charged particle traveling in a solid with a speed exceeding the speed of light in that material would radiate electromagnetic radiation. What is the minimum kinetic energy an electron must have while traveling inside a slab of crown glass ($n = 1.52$) in order to create Cerenkov radiation?

40–36 Construct a right triangle in which one of the angles is α, where $\sin \alpha = v/c$. (v is the speed of a particle, c the speed of light.) If the base of the triangle (the side adjacent to α) is the rest energy mc^2, show that

a) the hypotenuse is the total energy;

b) the side opposite α is c times the relativistic momentum.

c) Describe a simple graphical procedure for finding the kinetic energy K.

40–37 One of the wavelengths of light emitted by hydrogen atoms under normal laboratory conditions is $\lambda = 121.6$ nm. In the light emitted from a distant interstellar region, this same spectral line is observed to be Doppler-shifted to $\lambda = 430.4$ nm. How fast are the emitting atoms moving relative to the earth, and are they approaching the earth or receding from it?

40–38 A highway patrolman measures the speed of cars approaching him with a device that sends out electromagnetic waves of frequency f_0 and then measures the shift in frequency Δf of the waves reflected from the moving car. What fractional frequency shift $\Delta f/f_0$ is produced by a car speeding at 80 mph (35.8 m·s^{-1})? (See Problem 23–19.)

40–39 A particle moves along the x-axis with speed v. A force in the $+y$-direction, having magnitude F, is applied. Show that the magnitude of the acceleration initially is

$$a = \frac{F}{m}(1 - v^2/c^2)^{1/2}.$$

This result and that of Exercise 40–13 were sometimes interpreted in the early days of relativity as meaning that a particle has a "longitudinal mass" given by $m/(1 - v^2/c^2)^{3/2}$ and a "transverse mass" given by $m/(1 - v^2/c^2)^{1/2}$. These terms are no longer in common use.

40–40 An astronaut, her jet pack, and her spacesuit have a combined rest mass of 100 kg and a velocity of $1 \times 10^4 \text{ m·s}^{-1}$.

a) What is the difference between her Newtonian energy and her correct relativistic kinetic energy? Note that you cannot simply calculate both values and subtract, as the precision of most calculators is insufficient.

b) What fraction of the Newtonian energy is this difference?

40–41 A particle of mass m accelerated by a constant force F will, according to Newtonian mechanics, continue to accelerate without bound. That is, as $t \to \infty$, $v \to \infty$. Show that according to relativistic mechanics the particle's speed approaches c as $t \to \infty$ (*Note.* A standard integral is $\int [1 - x^2]^{-3/2} dx = x/\sqrt{1 - x^2}$.)

CHALLENGE PROBLEMS

40–42 The classical velocity of a particle of rest mass m and charge q in a circular orbit of radius R in a magnetic field is

$$v = (qBR)/m.$$

Find the corresponding velocity for a particle orbiting at relativistic speeds, and show that your result reduces to the classical result when v/c is small. (*Hint*: Use Eq. [40–24]. The relation between velocity and acceleration in circular motion, and between the velocity and the magnetic force, is the same relativistically as nonrelativistically.)

40–43 In high-energy-particle physics, new particles are created by colliding fast-moving projectile particles with stationary particles. Some of the kinetic energy of the incident particle is used to create mass of the new particle. A proton–proton collision can result in the creation of 2 pions (π^- and π^+):

$$p + p \rightarrow p + p + \pi^- + \pi^+.$$

a) Calculate the threshold kinetic energy of the incident proton that will allow this reaction to occur if the second proton is initially at rest. The rest mass of each π is 139.6 MeV. (*Hint*: Working in the center-of-mass frame is useful here. See Challenge Problem 8–57. But now the Lorentz transformation must be used to relate the velocities in the laboratory and the center-of-mass frames.)

b) How does this calculated threshold kinetic energy compare with the total rest mass energy of the created particles?

40–44 The French physicist Fizeau was the first to measure accurately the speed of light. (See Section 36–3.) He also found experimentally that the speed relative to the lab frame of light traveling in a tank of water that is itself moving at a speed V relative to the lab frame is

$$v = c/n + kV.$$

Fizeau called k the dragging coefficient and obtained an experimental value of $k = 0.44$. What is the value of k you would calculate from relativistic transformations?

40–45 Two events observed in a frame of reference S have positions and times given by (x_1, t_1) and (x_2, t_2), respectively.

a) Show that in a frame S' moving along the x-axis just fast enough so the two events occur at the same point in S', the time interval $\Delta t'$ between the two events is given by

$$\Delta t' = \sqrt{(\Delta t)^2 - \left(\frac{\Delta x}{c}\right)^2},$$

where $\Delta x = x_2 - x_1$, and $\Delta t = t_2 - t_1$. Hence show that, if $\Delta x \geq c\, \Delta t$, there is *no* frame S' in which the two events occur at the same point. The interval $\Delta t'$ is sometimes called the *proper-time interval* for the event. Is this term appropriate?

b) Show that if $\Delta x > c\, \Delta t$, there is a frame of reference S' in which the two events occur *simultaneously*. Find the distance between the two events in S'. This distance is sometimes called a *proper length*. Is this term appropriate?

c) Two events are observed in a frame of reference S' to occur simultaneously at points separated by a distance of 1 m. In a second frame S moving relative to S' along the line joining the two points in S', the two events appear to be separated by 2 m. What is the time interval between the events, as measured in S? (*Hint*: Apply the result obtained in [b].)

41
PHOTONS, ELECTRONS, AND ATOMS

IN THE PAST SEVERAL CHAPTERS WE HAVE STUDIED VARIOUS ASPECTS OF the propagation of light on the basis of an electromagnetic wave theory. The work of Maxwell, Hertz, and others has established firmly the electromagnetic nature of light. We understand interference, diffraction, and polarization on the basis of a *wave* model; when interference effects can be neglected, the further simplifications of *ray* optics permit us to analyze the behavior of lenses and mirrors. Together, these phenomena form the subject of *classical optics*.

Many other phenomena, however, show a different aspect of the nature of light, in which it seems to behave as a stream of *particles*. Among these are the photoelectric effect, the liberation of electrons from a surface by absorption of light energy; line spectra of elements; and the production and scattering of x-rays. All these point to a model in which the energy of light is carried in packages of a definite size, called *quanta* or *photons*. The energy of a single photon is proportional to the frequency of the radiation, and photons also carry momentum. Understanding the role of photons in emission and absorption phenomena requires some radical changes in our view of the nature of radiation and of matter itself. The new theory that emerges is called *quantum mechanics*. It provides the key to understanding many aspects of the structure and behavior of atoms and molecules that cannot be understood on the basis of the older "classical" theories of mechanics and electromagnetism.

41–1 EMISSION AND ABSORPTION OF LIGHT

The electromagnetic wave model of light provides an adequate basis for understanding many aspects of the propagation of light, including interference, diffraction, and polarization phenomena. It is considerably *less* successful with respect to the emission and absorption of light and other electromagnetic radiation from matter. In Section 35–1 we discussed the electromagnetic waves of Hertz; they were produced by oscillations with frequencies of the order of 10^8 Hz in a resonant L–C circuit similar to those studied in Chapter 33. Frequencies of visible light are much larger, of the order of 10^{15} Hz, far higher than the highest frequencies attainable with conventional electronic equipment.

Several mysteries about production of light

In the mid-nineteenth century theorists speculated that visible light might be produced by motion of electric charges within individual atoms rather than in macroscopic circuits. In fact, in 1862 Faraday placed a light source in a strong magnetic field in an attempt to determine whether the emitted radiation was changed by the effect of the field on the source. He was not able to detect any change, but when his experiments were repeated thirty years later by Zeeman with greatly improved equipment, changes *were* observed.

Each element has a characteristic line spectrum. How does it come about?

Particularly puzzling was the existence of *line spectra*. We have seen how a prism or grating spectrometer functions to disperse a beam of light into a *spectrum*. If the light source is an incandescent solid or liquid, the spectrum is *continuous;* that is, light of all wavelengths is present. But if the source is a gas through which an electrical discharge is passing, or a flame in which a volatile salt is present, the spectrum is of an entirely different character. Instead of a continuous band of color, only a few colors appear, in the form of isolated parallel lines. (Each "line" is an image of the spectrograph slit, deviated through an angle that depends on the frequency of the light forming the image.) A spectrum of this sort is termed a **line spectrum.**

The wavelengths of spectral lines are characteristic of the element emitting the light. That is, hydrogen always gives a set of lines in the same pattern, sodium produces another set, iron still another, and so on. The line structure of the spectrum extends into both the ultraviolet and infrared regions, where photographic or other means are required for its detection.

The existence of a characteristic spectrum for each element, discovered early in the nineteenth century, suggested a direct relation between the characteristics and internal structure of an atom and its spectrum. Attempts to understand this relation on the basis of Newtonian mechanics and classical electricity and magnetism were not successful, however.

The photoelectric effect: a puzzling aspect of the interaction of light and matter

There were other mysteries associated with the absorption of light. The *photoelectric effect*, discovered by Hertz in 1887 during his investigations of electromagnetic wave propagation, is the liberation of electrons from the surface of a conductor when light strikes the surface. When light is absorbed by the surface, energy is transferred to electrons near the surface, and some of the electrons acquire enough energy to surmount the potential-energy barrier at the surface and escape from the material into space. More detailed investigation revealed some puzzling features that could *not* be understood on the basis of classical optics. We will discuss these in the next section.

Still another area of unsolved problems centered around the production and scattering of *x-rays*, electromagnetic radiation with wavelengths shorter than those of visible light by a factor of the order of 10^4 and with correspondingly greater frequencies. These rays were produced in high-voltage glow discharge tubes, but the details of this process eluded understanding. Even worse, when these rays collided with matter, the scattered rays sometimes had a longer wavelength than the original ray. This is like directing a beam of blue light at a mirror and having it reflect as red!

All these phenomena, and several others, pointed forcefully to the conclusion that classical optics, successful though it was in explaining ray optics, interference, and polarization, nevertheless had its limitations. Understanding the phenomena cited above would require at least some generalization of classical theory. In fact, it has required something much more radical than that.

All these phenomena are concerned with the *quantum* theory of radiation, which includes the assumption that despite the *wave* nature of electromagnetic radiation, it has some properties akin to those of *particles*. In particular, the *energy* conveyed by an electromagnetic wave is always carried in units whose magnitude is proportional to the frequency of the wave. These units of energy are called photons or quanta.

Thus electromagnetic radiation emerges as an entity with a dual nature, having both wave and particle aspects. The remainder of this chapter will be devoted to the applications of this duality to some of the phenomena mentioned above, and to a study of this seemingly (but not actually) inconsistent nature of electromagnetic radiation.

41–2 THE PHOTOELECTRIC EFFECT

The **photoelectric effect** is the liberation of electrons from the surface of a conductor when light strikes the surface. The electrons absorb energy from the incident radiation and are thus able to overcome the potential-energy barrier that normally confines them in the material.

The photoelectric effect was first observed in 1887 by Hertz, quite by accident. He noticed that a spark would jump more readily between two spheres when their surfaces were illuminated by the light from another spark. Light seemed to facilitate the escape of charges from the surfaces. This idea in itself was not revolutionary. The existence of the surface potential-energy barrier was already known. In 1883 Edison had discovered **thermionic emission,** in which the escape energy is supplied by heating the material to a very high temperature, liberating electrons by a process analogous to boiling a liquid.

The photoelectric effect was investigated in detail by Hallwachs and Lenard, with quite unexpected results. We will describe their work in terms of more contemporary equipment.

A modern phototube is shown schematically in Fig. 41–1. A beam of light, indicated by the arrows, falls on a photosensitive surface K called the *cathode*. The battery or other source of potential difference creates an electric field in the direction from A (called the *collector* or *anode*) toward K, and electrons emitted from K are pushed by this field to the anode A. Anode and cathode are enclosed in a container that can be pumped down to a good vacuum; a residual pressure of 0.01 Pa (10^{-7} atm) or less is needed to avoid collisions of electrons with gas molecules. The photoelectric current is measured by the galvanometer G.

It is found that with a given material as emitter, no photoelectrons at all are emitted unless the wavelength of the light is *shorter* than some critical value. The corresponding *minimum* frequency is called the **threshold frequency** of the particular surface. The threshold frequency for most metals is in the ultraviolet region (corresponding to wavelengths of 200 to 300 nm), but for potassium and cesium oxide it lies in the visible spectrum (400 to 700 nm).

In energy relations involving electrons, it is often convenient to use the electronvolt. We introduced this unit in Section 26–7 during our study of electrical potential, and we suggest you review that section now. The volt, equal to one joule per coulomb, is a unit of *electrical potential*, but the electronvolt (1 eV) is a unit of *energy*. If we express a potential difference V in volts and the electron charge e in coulombs ($e = 1.602 \times 10^{-19}$ C), then the correspond-

Is light a wave or a particle? Yes!

Two mechanisms for electrons to be liberated from a surface

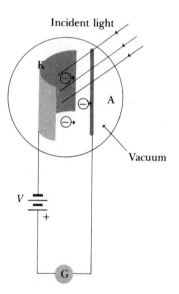

41–1 Schematic diagram of a photocell circuit.

Electrons are emitted only when the light frequency is less than some critical value.

ing potential energy eV is in joules. But if V is in volts and $e = 1$ electron charge, then the product eV is in electronvolts. Thus if $V = 5$ V, then $eV = 5$ eV. From the definition of the electronvolt, we see that the conversion factor is

$$1 \text{ eV} = 1.602 \times 10^{-19} \text{ J}.$$

Some electrons are emitted from the cathode with substantial initial speeds; this is shown by the fact that, even with *no* emf in the external circuit, a few electrons reach the collector, causing a small current in the external circuit. Indeed, even when the polarity of the potential difference V is reversed and the associated electric-field force on the electrons is back toward the cathode, some electrons still reach the anode. Only when the reversed potential V is made large enough so that the potential energy eV is greater than the maximum kinetic energy $\frac{1}{2}mv_{\max}^2$ with which the electrons leave the cathode does the electron flow stop completely. The critical reversed potential is called the *stopping potential*, denoted by V_0, and it provides a direct measurement of the maximum kinetic energy with which electrons leave the cathode, through the relation

$$\frac{1}{2}mv_{\max}^2 = eV_0. \qquad (41\text{–}1)$$

Surprisingly, it turns out that the maximum kinetic energy of an emitted electron *does not* depend on the *intensity* of the incident light, but *does* depend on its *wavelength* or *frequency*. When the light intensity increases, the photoelectric current also increases, but only because the *number* of emitted electrons increases, not the energy of an individual electron. This surprising result is hard to understand on the basis of classical theory, which would predict that greater intensity should yield some electrons with higher and higher energies.

Each electron absorbs the same amount of energy from the radiation.

The correct explanation of the photoelectric effect was given by Einstein in 1905. Extending a proposal made two years earlier by Planck, Einstein postulated that a beam of light consisted of small bundles of energy called **quanta** or **photons.** The energy E of a photon is proportional to its frequency f or is equal to its frequency multiplied by a constant. That is,

Photons: bundles of energy in electromagnetic radiation

$$E = hf, \qquad (41\text{–}2)$$

where h is a universal constant, called **Planck's constant,** whose numerical value is 6.626×10^{-34} J·s. When a photon collides with an electron at or just within the surface of a metal, the photon may transfer its energy to the electron. This transfer is an "all-or-nothing" process, the electron getting all the photon's energy or none at all. The photon then simply drops out of existence. The energy acquired by the electron may enable it to surmount the surface potential-energy barrier and escape from the surface of the metal.

Planck's constant: a new fundamental constant of nature

The height of the surface potential-energy barrier is called the **work function,** denoted by ϕ. That is, in leaving the surface of the metal, the electron loses an amount of kinetic energy ϕ. Some electrons may lose more than this if they start at some distance below the metal surface, but the *maximum* energy with which an electron can emerge is the energy gained from a photon minus the work function. Hence the maximum kinetic energy of the photoelectrons ejected by light of frequency f is

How much energy is needed to pull an electron away from a surface?

$$\frac{1}{2}mv_{\max}^2 = hf - \phi. \qquad (41\text{–}3)$$

Combining this with Eq. (41–1) leads to the relation

$$eV_0 = hf - \phi. \qquad (41\text{–}4)$$

Thus by measuring the stopping potential V_0 required for each of several values of frequency f for a given cathode material, we can determine both the work function ϕ for the material and the value of the quantity h/e. This experiment provides a direct measurement of the value of Planck's constant, in addition to a direct confirmation of Einstein's interpretation of photoelectric emission. Typical work functions are of the order of 1 to 5 eV.

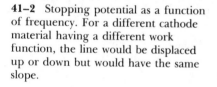

V_0 (volts)

3

2

1

0

0.25   0.50   0.75   1.0

$f(10^{15}$ Hz)

−1

41–2 Stopping potential as a function of frequency. For a different cathode material having a different work function, the line would be displaced up or down but would have the same slope.

EXAMPLE 41–1 For a certain cathode material used in a photoelectric-effect experiment, a stopping potential of 3.0 V was required for light of wavelength 300 nm, 2.0 V for 400 nm, and 1.0 V for 600 nm. Determine the work function for this material and the value of Planck's constant.

SOLUTION According to Eq. (41–4), a graph of V_0 as a function of f should be a straight line. We rewrite this equation as

$$V_0 = \frac{h}{e}f - \frac{\phi}{e}.$$

In this form we see that the *slope* of the line is h/e and the *intercept* on the vertical axis (corresponding to $f = 0$) is at $-\phi/e$. The frequencies, obtained from $f = c/\lambda$ and $c = 3.00 \times 10^8$ m·s^{-1}, are 1.0, 0.75, and 0.5×10^{15} s^{-1}, respectively. The graph is shown in Fig. 41–2. From it we find

$$-\frac{\phi}{e} = -1.0 \text{ V}, \qquad \phi = 1.0 \text{ eV} = 1.60 \times 10^{-19} \text{ J},$$

and

$$\frac{h}{e} = \frac{1.0 \text{ V}}{0.25 \times 10^{15} \text{ s}^{-1}} = 4.0 \times 10^{-15} \text{ J·C}^{-1}\text{·s},$$

$$h = (4.0 \times 10^{-15} \text{ J·C}^{-1}\text{·s})(1.60 \times 10^{-19} \text{ C})$$

$$= 6.4 \times 10^{-34} \text{ J·s}.$$

This experimental value differs by about 3.4% from the currently accepted value of 6.626×10^{-34} J·s.

The particlelike nature of electromagnetic radiation, used in our analysis of the photoelectric effect, has been established beyond any reasonable doubt. Figure 41–3 shows a very direct illustration of the particle aspects of light. A photon of electromagnetic radiation of frequency f and corresponding wavelength $\lambda = c/f$ has energy E given by

$$E = hf = \frac{hc}{\lambda}. \qquad (41\text{–}5)$$

Furthermore, according to relativity theory, every particle having energy must also have momentum, even if it has no rest mass. Photons have zero rest mass; according to Eq. (40–35), the momentum p of a photon of energy E has

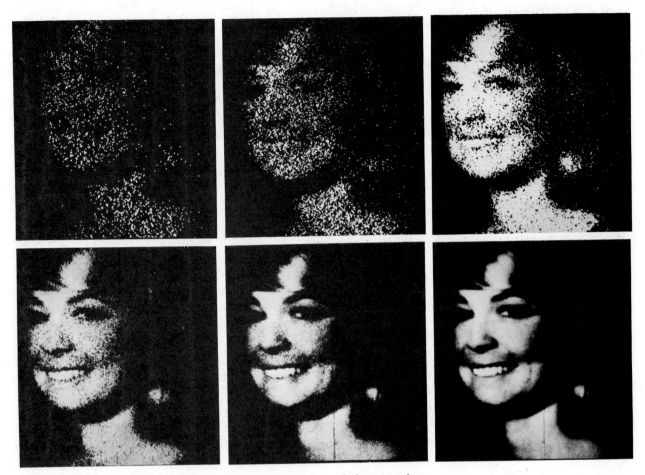

41–3 These photographs are images made by small numbers of photons, using electronic image amplification. In the upper pictures each spot corresponds to one photon; under extremely faint light (few photons) the pattern seems almost random, but it emerges more distinctly as the light level increases (lower pictures). (Used with permission of RCA Corporation.)

magnitude p given by

$$E = pc. \qquad (41-6)$$

Thus the wavelength of a photon and its momentum are related simply by

$$p = \frac{h}{\lambda}. \qquad (41-7)$$

We will use these relations frequently in the remainder of this chapter. We also note that Eq. (41–6) is consistent with the energy–momentum relation for electromagnetic waves that we derived in Section 35–3; this further confirms our view of the photon as a particle with zero rest mass.

Photons have zero rest mass and travel always with speed c.

41–3 LINE SPECTRA

The quantum hypothesis, used in the preceding section for the analysis of the photoelectric effect, also plays an important role in understanding atomic spectra, particularly the line spectra described in Section 41–1. We might

Trying to make sense out of the line spectrum of hydrogen

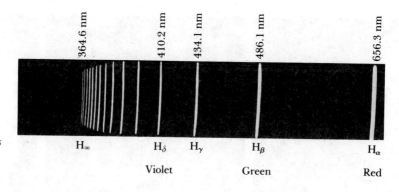

41–4 The Balmer series of atomic hydrogen. (Reproduced by permission from *Atomic Spectra and Atomic Structures* by Gerhard Herzberg. Copyright 1937 by Prentice-Hall, Inc.)

A numerical rule for the hydrogen-spectrum wavelengths

expect that the spectrum frequencies of the light emitted by a particular element would be arranged in some regular way. For instance, a radiating atom might be analogous to a vibrating string, emitting a fundamental frequency and its harmonics. At first sight, there does not seem to be any semblance of order or regularity in the lines of a typical spectrum. For many years unsuccessful attempts were made to correlate the observed frequencies with those of a fundamental and its overtones. Finally, in 1885 Johann Jakob Balmer (1825–1898) found by trial and error a simple formula that gives the frequencies of a group of lines emitted by atomic hydrogen. Since the spectrum of this element is relatively simple, we will consider it in more detail.

Under the proper conditions of excitation, atomic hydrogen may be made to emit the sequence of lines shown in Fig. 41–4. This sequence is called the **Balmer series.** A pattern is evident in this spectrum; the lines become closer and closer together as the end of the series is approached. The line of longest wavelength or lowest frequency, in the red, is known as H_α; the next line, in the blue-green, as H_β; the third is H_γ; and so on. Balmer found that the wavelengths of these lines were given accurately by the simple formula

$$\frac{1}{\lambda} = R\left(\frac{1}{2^2} - \frac{1}{n^2}\right), \tag{41–8}$$

where λ is the wavelength, R is a constant called the **Rydberg constant,** and n may have the integer values 3, 4, 5, and so on. If λ is in meters,

$$R = 1.097 \times 10^7 \text{ m}^{-1}.$$

Letting $n = 3$ in Eq. (41–8), we obtain the wavelength of the H_α-line:

$$\frac{1}{\lambda} = 1.097 \times 10^7 \text{ m}^{-1}\left(\frac{1}{4} - \frac{1}{9}\right)$$
$$= 1.524 \times 10^6 \text{ m}^{-1},$$

and

$$\lambda = 656.2 \text{ nm}.$$

For $n = 4$, we obtain the wavelength of the H_β-line, and so on. For $n = \infty$, we obtain the limit of the series, at $\lambda = 364.6$ nm. This is the *shortest* wavelength in the series.

Hydrogen has several spectral series in the ultraviolet, visible, and infrared regions of the spectrum.

Other series spectra for hydrogen have since been discovered. These are known, after their discoverers, as the Lyman, Paschen, Brackett, and Pfund series. The formulas for these follow.

Lyman series:

$$\frac{1}{\lambda} = R\left(\frac{1}{1^2} - \frac{1}{n^2}\right), \qquad n = 2, 3, \ldots ,$$

Paschen series:

$$\frac{1}{\lambda} = R\left(\frac{1}{3^2} - \frac{1}{n^2}\right), \qquad n = 4, 5, \ldots ,$$

Brackett series:

$$\frac{1}{\lambda} = R\left(\frac{1}{4^2} - \frac{1}{n^2}\right), \qquad n = 5, 6, \ldots ,$$

Pfund series:

$$\frac{1}{\lambda} = R\left(\frac{1}{5^2} - \frac{1}{n^2}\right), \qquad n = 6, 7, \ldots .$$

The Lyman series is in the ultraviolet region, and the Paschen, Brackett, and Pfund series are in the infrared. The Balmer series evidently fits into the scheme between the Lyman and Paschen series.

The Balmer formula, Eq. (41–8), may also be written in terms of the frequency of the light, by use of the relation

$$c = f\lambda, \quad \text{or} \quad \frac{1}{\lambda} = \frac{f}{c}.$$

Thus Eq. (41–8) becomes

Expressing frequencies of spectrum lines as differences of two terms

$$f = Rc\left(\frac{1}{2^2} - \frac{1}{n^2}\right), \tag{41–9}$$

or

$$f = \frac{Rc}{2^2} - \frac{Rc}{n^2}. \tag{41–10}$$

Each of the fractions on the right-hand side of Eq. (41–10) is called a **term,** and the frequency of every line in the series is given by the difference between two terms.

Only a few elements (hydrogen, singly ionized helium, doubly ionized lithium) have spectra that can be represented by a simple formula of the Balmer type. Nevertheless, it is possible to separate the more complicated spectra of other elements into series, and to express the frequency of each line in the series as the difference of two terms. The first term is constant for any one series, while the various values of the second term can be labeled by values of an integer index n analogous to the n appearing in Eq. (41–10). In a few simple cases the numerical values of the terms can be *calculated* from theoretical considerations, as we will see in Chapter 42. For complex atoms, however, the term values must be determined experimentally by analysis of spectra, often an extremely complex problem.

41–4 ENERGY LEVELS

Energy levels in atoms: the connection between spectra and photons

Every element has a characteristic line spectrum that must be related to the characteristics and structure of the *atoms* of that element. Part of the key to understanding the relation of atomic structure to atomic spectra was supplied in 1913 by the Danish physicist Niels Bohr, who applied to spectra the same concept of light quanta or *photons* that Einstein had used earlier in analyzing the photoelectric effect.

Bohr's hypothesis was as follows: Each atom, as a result of its internal structure (and, presumably, internal motion), can have a variable amount of *internal energy.* But the energy of an atom cannot change by any arbitrary amount; rather, each atom has a series of discrete **energy levels.** An atom can have an amount of internal energy corresponding to any one of these levels, but it cannot have an energy *intermediate* between two levels. All atoms of a given element have the same set of energy levels, but atoms of different elements have different sets.

An atom radiates only while going from one energy level to another.

While an atom is in one of these states corresponding to a definite energy, it does not radiate. However, an atom can make a transition from one energy level to a lower level by emitting a photon whose energy is equal to the energy difference between the initial and final states. If E_i is the initial energy of the atom, before such a transition, and E_f is its final energy, after the transition, then, since the photon's energy is hf, we have

$$hf = E_i - E_f. \tag{41–11}$$

For example, a photon of orange light of wavelength 600 nm has a frequency f given by

$$f = \frac{c}{\lambda} = \frac{3.00 \times 10^8 \ \mathrm{m \cdot s^{-1}}}{600 \times 10^{-9} \ \mathrm{m}} = 5.00 \times 10^{14} \ \mathrm{Hz}$$
$$= 5.00 \times 10^{14} \ \mathrm{s^{-1}}.$$

The corresponding photon energy is

$$E = hf = (6.63 \times 10^{-34} \ \mathrm{J \cdot s})(5.00 \times 10^{14} \ \mathrm{s^{-1}})$$
$$= 3.31 \times 10^{-19} \ \mathrm{J} = 2.07 \ \mathrm{eV}.$$

Energy levels provide a simple explanation for the Balmer formula and others.

Thus this photon must be emitted in a transition between two states of the atom differing in energy by 2.07 eV.

The Bohr hypothesis, if correct, would shed new light on the analysis of spectra on the basis of *terms*, as described in Section 41–3. For example, Eq. (41–10) gives the frequencies of the Balmer series in the hydrogen spectrum. Multiplied by Planck's constant h, this becomes

$$hf = \frac{Rch}{2^2} - \frac{Rch}{n^2}. \tag{41–12}$$

A formula for the energy levels of the hydrogen atom

If we now compare Eqs. (41–11) and (41–12), identifying $-Rch/n^2$ with the initial energy of the atom E_i and $-Rch/2^2$ with its final energy E_f, before and after a transition in which a photon of energy $hf = E_i - E_f$ is emitted, then Eq. (41–12) takes on the same form as Eq. (41–11). More generally, if we assume that the possible energy levels for the hydrogen atom are given by

$$E_n = -\frac{Rch}{n^2}, \qquad n = 1, 2, 3, \ldots, \tag{41–13}$$

then *all* the series spectra of hydrogen can be understood on the basis of transitions from one energy level to another. For the Lyman series the final state is always $n = 1$; for the Paschen series it is $n = 3$; and so on. Similarly, complex spectra of other elements, represented by terms, are understood on the basis that each term corresponds to an energy level; and a frequency, represented as a difference of two terms, corresponds to a transition between the two corresponding energy levels.

In 1914 a series of experiments by Franck and Hertz provided more direct experimental evidence for energy levels in atoms. In studying the motion of electrons through mercury vapor, under the action of an electric field, they found that a spectrum line at 254 nm was emitted by the vapor when the electron kinetic energy was greater than 4.9 eV, but not when it was less. This strongly suggests the existence of an energy level 4.9 eV above the ground state. A mercury atom is excited to this level by collision with an electron and subsequently decays to the ground state by emitting a photon. The energy of the photon, according to Eq. (41–2), should be

The Franck-Hertz experiment: independent confirmation of the existence of energy levels in atoms

$$E = hf = \frac{hc}{\lambda} = \frac{(6.63 \times 10^{-34}\,\text{J·s})(3.00 \times 10^8\,\text{m·s}^{-1})}{254 \times 10^{-9}\,\text{m}}$$

$$= 7.82 \times 10^{-19}\,\text{J} = \frac{7.82 \times 10^{-19}\,\text{J}}{1.60 \times 10^{-19}\,\text{J·eV}^{-1}}$$

$$= 4.9\,\text{eV},$$

in excellent agreement with the measured electron energy.

Although the Bohr hypothesis permits partial understanding of line spectra on the basis of energy levels of atoms, it is not yet complete because it provides no basis for *predicting* what the energy levels for any particular kind of atom should be. We will return to this problem in Chapter 42, where we will develop the principles of quantum mechanics needed for the understanding of the structure and energy levels of atoms.

PROBLEM-SOLVING STRATEGY: *Photons and energy levels*

1. Remember that with photons, as with any other periodic wave, the wavelength λ and frequency f are related by $f = c/\lambda$. The energy E of a photon can be expressed as hf or hc/λ, whichever is more convenient for the problem at hand. Be careful with units; if E is in joules, h must be in (J·s), λ in meters, and f in (seconds)$^{-1}$ or hertz. The magnitudes are in such an unfamiliar realm that common sense may not help if you goof by a factor of 10^{10}, so be careful with powers of ten.

2. It is often convenient to measure energy in electronvolts. The conversion $1\,\text{eV} = 1.602 \times 10^{-19}\,\text{J}$ comes in handy. When energies are in eV, you may want to express h in eV·s; in those units, $h = 4.136 \times 10^{-15}\,\text{eV·s}$. We invite you to verify this value.

3. Keep in mind that an electron moving through a potential difference of one volt gains or loses an amount of energy equal to one electronvolt. You will use the electronvolt a lot in this chapter and the next three, so it's worthwhile to become familiar with it now.

41–5 ATOMIC SPECTRA

As we have seen, the key to understanding atomic spectra is the concept of atomic *energy levels*. Every spectrum line corresponds to a specific transition between two energy levels of an atom, and the corresponding frequency is

Atomic spectra are related directly to atomic-energy levels.

given in each case by Eq. (41–11). Thus the fundamental problem for the spectroscopist is to determine the energy levels of an atom from the measured values of the wavelengths of the spectral lines that are emitted when the atom goes from one energy level to another. In the case of complicated spectra emitted by the heavier atoms, this is a complex task requiring tremendous ingenuity. Nevertheless, almost all atomic spectra have been analyzed, and the resulting energy levels have been tabulated with the aid of diagrams similar to the one shown for sodium in Fig. 41–5.

Every atom has a lowest energy level, representing the *minimum* energy the atom can have. This lowest energy level is called the **ground state,** and all higher levels are called **excited states.** As we have seen, a photon corresponding to a particular spectral line is emitted when an atom makes a transition from an excited state to a lower state. The only means discussed so far for raising the atom from the ground state to an excited state has been with the aid of an electric discharge. Let us consider now another method, involving *absorption* of radiant energy.

From Fig. 41–5 we can see that a sodium atom emits characteristic yellow light of wavelengths 589.0 and 589.6 nm when it makes transitions from the

An atom can gain or lose energy only in units determined by its energy levels.

41–5 Energy levels of the sodium atom. Numbers on the lines between levels are wavelengths. The column labels, such as $^2S_{1/2}$, refer to the quantum states of the electron, to be discussed in Chapter 43.

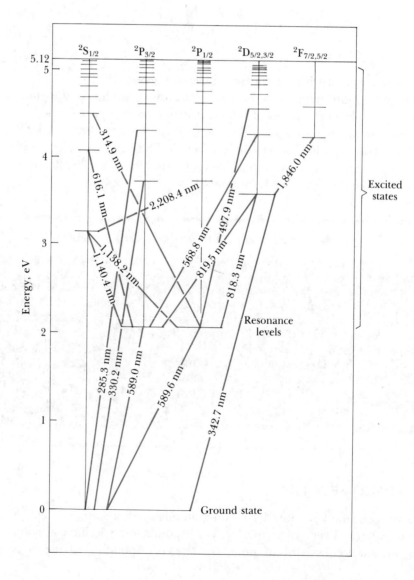

259.4 nm — 254.4 nm — 251.2 nm

41–6 Absorption spectrum of sodium. This is an image on photographic film. Black and white are reversed; the bright regions of the spectrum show as black, and the dark absorption lines show as light lines.

two closely spaced levels marked *resonance levels* to the ground state. Suppose a sodium atom in the ground state were to *absorb* a quantum of radiant energy of wavelength 589.0 or 589.6 nm. It would then undergo a transition in the opposite direction and be raised to one of the resonance levels. After a short time the atom returns to the ground state, emitting a photon. The average time spent in the excited state is called the *lifetime* of the state; for the resonance levels of the sodium atom, the lifetime is about 1.6×10^{-8} s.

This emission process is called **resonance radiation;** we can demonstrate it as follows. We pump the air out of a glass bulb and then introduce a small amount of pure metallic sodium. Then we concentrate a strong beam of the yellow light from a sodium-vapor lamp on the bulb. When we warm the bulb to evaporate some of the sodium, some of the atoms in the vapor absorb the 589-nm photons from the beam, then reemit them in all directions. This happens throughout the whole bulb, which glows with the yellow light characteristic of sodium.

A sodium atom in the ground state may absorb radiant energy of wavelengths other than the yellow resonance lines. All wavelengths corresponding to spectral lines *emitted* when the sodium atom returns to its ground state may also be *absorbed* by a sodium atom in the ground state. If, therefore, the continuous-spectrum light from a carbon arc is sent through an absorption tube containing sodium vapor and is then examined with a spectroscope, a series of dark lines appears, corresponding to the wavelengths absorbed, as shown in Fig. 41–6. This is called an **absorption spectrum.**

The sun's spectrum is an absorption spectrum. The main body of the sun emits a continuous spectrum, whereas the cooler vapors in the sun's atmosphere emit line spectra corresponding to all the elements present. When the intense light from the main body of the sun passes through the cooler vapor, the lines of these elements are absorbed. The light *emitted* by the cooler vapors is so small compared with the unabsorbed continuous spectrum that the continuous spectrum appears to be crossed by many faint *dark* lines. These were first observed by Fraunhofer and are called *Fraunhofer lines*. They may be observed with any student spectroscope pointed toward any part of the sky. Figure 41–7 shows Fraunhofer lines in a portion of the sun's spectrum.

Absorption spectra: An atom in its ground state can absorb only certain wavelengths, depending on its energy levels.

41–7 A portion of the solar spectrum, between 390 nm and 460 nm, showing the Fraunhofer lines corresponding to the absorption spectrum. (Courtesy of Mt. Wilson & Las Campanas Observatories, Carnegie Institute of Washington.)

41–6 THE LASER

If the energy difference between the ground state and the first excited state of an atom is E, the atom is capable of absorbing a photon whose frequency f is given by the Planck equation $E = hf$. The *absorption* of a photon by a normal atom A is depicted schematically in Fig. 41–8a. After absorbing the photon, the atom becomes an excited atom A*. A short time later, *spontaneous emission* takes place, and the excited atom returns to the ground state by emitting a photon of the same frequency as the one originally absorbed. The direction and phase of this photon are random, as shown in Fig. 41–8b. There is also a third process, first proposed by Einstein, called **stimulated emission,** shown schematically in Fig. 41–8c. Stimulated emission takes place when an incident photon encounters an excited atom and forces it to emit *another* photon with the same frequency, the same direction, the same phase, and the same polarization as the incident photon. The two photons thus have a definite phase relation and go off together as *coherent* radiation.

Suppose we have a large number of identical atoms in a gas or vapor, in a container with transparent walls, as in Fig. 41–8a. At moderate temperatures, if no radiation is incident on the container, most of the atoms are in the ground state, and only a few are in excited states. If there is an energy level E above the ground state, the ratio of the number of atoms n_E in this state (the *population* of the state) to the number of atoms n_0 in the ground state is very small.

Now suppose we send through the container a beam of radiation with frequency f corresponding to the energy difference E. Some of the atoms absorb photons of energy E and are raised to the excited state, and the population ratio n_E/n_0 increases. Because n_0 is originally so much larger than n_E, an enormously intense beam of light would be required to increase n_E to a value comparable to n_0. Therefore the rate at which energy is absorbed from the beam by the n_0 ground-state atoms far outweighs the rate at which energy is added to the beam by stimulated emission from the relatively rare (n_E) excited atoms.

But now suppose we can create a situation in which n_E is substantially increased compared to the normal equilibrium value; this condition is called **population inversion.** In this case the rate of energy radiation by stimulated emission may actually *exceed* the rate of absorption. The system then acts as a

41–8 Three interaction processes between an atom and radiation. The heavier waves on the right side of (c) show the presence of additional photons from stimulated emission.

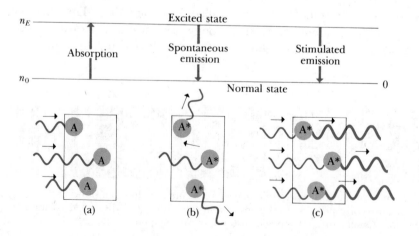

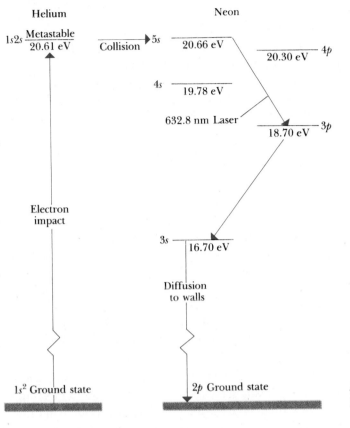

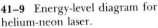

41–9 Energy-level diagram for helium-neon laser.

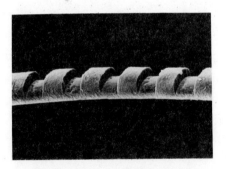

A human hair notched by an ultraviolet laser. The ultraviolet photons have enough energy to break molecular bonds; the fragments are swept away by remaining photon energy, causing very clean cuts with very little heating. Such lasers can be used for integrated-circuit fabrication and corneal surgery. (Courtesy of IBM Corp.)

Metastable states: atoms getting stuck in states from which radiative decay is impossible

source of radiation with photon energy E. Furthermore, since the photons are the result of stimulated emission, they all have the same frequency, phase, polarization, and direction. The resulting radiation is therefore very much more *coherent* than light from ordinary sources, where the emissions of individual atoms are *not* coordinated.

The necessary population inversion can be achieved in a variety of ways. As an example, consider the helium-neon laser, a simple, inexpensive laser available in many undergraduate laboratories. A mixture of helium and neon, each typically at a pressure of the order of 10^2 Pa (or 10^{-3} atm), is sealed in a glass enclosure provided with two electrodes. When a sufficiently high voltage is applied, a glow discharge occurs. Collisions between ionized atoms and electrons carrying the discharge current excite atoms to various energy states.

Figure 41–9 shows an energy-level diagram for the system. The notation used to label the various energy levels, such as $1s2s$ or $5s$, refers to the energy states of the electrons in the atoms and will be discussed in Section 43–1. A helium atom excited to the $1s2s$ state cannot return to the ground state by emitting a 20.61-eV photon, as might be expected. The reason, which we cannot discuss in detail here, is related to restrictions imposed by conservation of angular momentum. Such a state, in which single-photon radiative decay is impossible, is called a **metastable state.**

However, excited helium atoms *can* lose energy by energy-exchange collisions with neon atoms initially in the ground state. A $1s2s$ helium atom, with its internal energy of 20.61 eV and a little additional kinetic energy, can collide with a neon atom in the ground state, exciting the neon atom to the $5s$ excited

Induced emission produces a beam with many photons of the same energy, all in phase.

state at 20.66 eV and leaving the helium atom in the $1s^2$ ground state. Thus we have the necessary mechanism for a population inversion in neon, greatly enhancing the population in the 5s state. Stimulated emission from this state then results in the emission of highly coherent light at 632.8 nm, as shown on the diagram. In practice the beam is sent back and forth through the gas many times by a pair of parallel mirrors, so as to stimulate emission from as many excited atoms as possible. One of the mirrors is partially transparent, so a portion of the beam emerges as an external beam.

The net effect of all these processes taking place in a laser tube is a beam of radiation that is (1) very intense, (2) almost perfectly parallel, (3) almost monochromatic, and (4) spatially *coherent* at all points within a given cross section. To understand this fourth characteristic, recall the simple double-slit interference experiment. A mercury arc placed directly behind the double slit would not give rise to interference fringes because the light issuing from the two slits would come from different points of the arc and would not retain a constant phase relationship. When we use common laboratory arc-lamp sources, we must use the light from a very small portion of the source to illuminate the double slit. The slightly diverging beam from a laser, however, may be allowed to fall directly on a double slit (or other interferometer) because the light rays from any two points of a cross section are in phase and are said to exhibit "spatial coherence."

Lasers have many practical applications, some of them spectacular.

In recent years lasers have found a wide variety of practical applications. The high intensity of a laser beam makes it a convenient drill. For example, a laser beam can drill a very small hole in a diamond for use as a die in drawing very small-diameter wire. Because the photons in a laser beam are strongly correlated in their directions, a laser beam can travel long distances without appreciable spreading. This makes it a very useful tool for surveyors, especially in situations where great precision is required, such as a long tunnel drilled from both ends.

41–7 CONTINUOUS SPECTRA

Hot condensed matter emits a continuous spectrum, not a line spectrum.

Our discussion of the particle aspect of electromagnetic radiation has centered so far on its role in the understanding of line spectra emitted by elements in the gaseous state, where each atom behaves as an isolated system and interactions between atoms are negligible. Very hot materials in the solid and liquid states also emit radiation, but with a **continuous spectrum,** that is, continuous distribution of wavelengths rather than a line spectrum. By 1900 this radiation had been studied extensively, and several characteristics had been established.

The total rate of energy radiation depends on the fourth power of the absolute temperature.

First, the total rate of radiation of energy, per unit surface area, is proportional to the fourth power of the absolute temperature. We have seen this relationship before in Section 16–3, in our study of heat transfer mechanisms, and we suggest you review that section. The total heat current H (power, or energy per unit time) radiated from an ideal radiator of area A at absolute temperature T is given by the **Stefan-Boltzmann law:**

$$H = A\sigma T^4, \tag{41–14}$$

where σ is a fundamental physical constant called the **Stefan-Boltzmann constant.** In SI units,

$$\sigma = 5.6699 \times 10^{-8}\ \text{W·m}^{-2}\text{·K}^{-4}. \tag{41–15}$$

An alternative statement is that the *power emitted per unit area* for an ideal radiator is equal to σT^4.

Second, the power is not uniformly distributed over all wavelengths, but it can always be described by a power distribution function $F(\lambda)$ having the property that $F(\lambda)\,d\lambda$ is the power per unit area radiated with wavelengths in the interval $d\lambda$. The *integral* of $F(\lambda)$ over all wavelengths, represented by the area under the curve when $F(\lambda)$ is plotted as a function of λ, is the *total* power per unit area. The distribution functions for three different temperatures are shown in Fig. 41–10. Each has a peak wavelength λ_m, where the power is most concentrated. Experiment shows that λ_m is inversely proportional to T, and that in fact

$$\lambda_m T = \text{constant} = 2.90 \times 10^{-3}\ \text{m·K}. \qquad (41\text{–}16)$$

This rule is called the **Wien displacement law.** As the temperature rises, the peak of $F(\lambda)$ becomes higher and shifts to shorter wavelengths. This corresponds to the familiar fact that a body that glows yellow is hotter and brighter than one that glows red; yellow has a shorter wavelength than red. Finally, experiments show that the *shape* of the distribution function is the same for all temperatures; we can make a curve for one temperature fit any other temperature by simply changing the scales on the graph.

In the last decade of the nineteenth century, many attempts were made to *derive* these empirical results from basic principles. It was known by then that light is electromagnetic radiation, and it seemed natural to assume that light is produced by vibration of elementary charges in a material. Attempts were made to derive a function for the curves of Fig. 41–10 by treating the electrons in a material as harmonic oscillators with energies governed by the principle of equipartition of energy. We studied this principle in Section 20–5 in connection with heat capacities of gases. Finally, in 1900 Max Planck succeeded in deriving a function, now called the **Planck radiation law,** that agreed with experimentally obtained power-distribution curves. To do this he added to the classical equipartition theorem the additional assumption that a harmonic oscillator with frequency f can gain or lose energy only in discrete steps of magnitude hf, where h is the same constant that now bears his name.

Ironically, Planck himself originally regarded this *quantum hypothesis* as a calculational trick rather than a fundamental principle. But as we have seen, evidence for the quantum aspects of light accumulated, and by 1920 there was no longer any doubt about the validity of the concept. Indeed, the concept of discrete energy levels of microscopic systems really originated with Planck, not Bohr, and we have departed from the historical order of things by discussing atomic spectra before continuous spectra.

Planck's derivation of the power-distribution function is rather involved, and we quote his result here without derivation:

$$F(\lambda) = \frac{2\pi hc^2}{\lambda^5}\,\frac{1}{e^{hc/\lambda kT}-1}. \qquad (41\text{–}17)$$

where h is Planck's constant, c is the speed of light, k is Boltzmann's constant, T is the *absolute* temperature, and λ is the wavelength. This function turns out to agree well with experimentally determined power-distribution curves such as those in Fig. 41–10. It also contains the Wien displacement law and the Stefan-Boltzmann law as consequences.

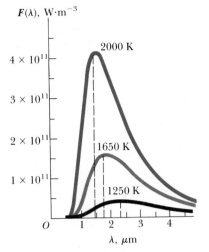

41–10 Power distribution function $F(\lambda)$ for continuous-spectrum radiation from an ideal radiator. The power per unit area in a wavelength interval $d\lambda$ at wavelength λ is $F(\lambda)\,d\lambda$. The total area under the curve at any temperature represents the total radiated power per unit area. Curves are plotted for three different temperatures, with darker color corresponding to higher temperature. The vertical broken lines show the value of λ_m in Eq. (41–18) for each temperature. As the temperature increases, the peak grows larger and shifts to shorter wavelengths; the total area under the curve is proportional to T^4.

The quantum hypothesis: calculational trick or fundamental truth?

The Wien displacement law and the Stefan-Boltzmann law can be derived from the Planck radiation formula.

To derive the Wien law we take the derivative of Eq. (41–17) and set it equal to zero to find the value of λ at which $F(\lambda)$ is maximum. We leave you the details as a problem; the result is

$$\lambda_{\mathrm{m}} = \frac{hc}{4.965kT}.$$ (41–18)

To obtain this result, you have to solve the equation

$$5 - x = 5e^{-x}.$$ (41–19)

The root of this equation is (approximately) 4.965. We invite you to evaluate the constant $hc/4.965k$ and show that it agrees with the empirical value given in Eq. (41–16).

We can obtain the Stefan-Boltzmann law by integrating Eq. (41–17) over all λ to find the *total* radiated power per unit area. This calculation involves looking up an integral in a table. Again we leave you the details as a problem; the result is

$$\int_0^\infty F(\lambda)\, d\lambda = \frac{2\pi^5 k^4}{15c^2 h^3} T^4 = \sigma T^4.$$ (41–20)

This is the Stefan-Boltzmann law, and it also shows that the constant σ in that law can be expressed in terms of a rather unlikely looking combination of other fundamental constants:

$$\sigma = \frac{2\pi^5 k^4}{15c^2 h^3}.$$ (41–21)

We invite you to plug in the values of k, c, and h, and verify that you get the value of σ quoted above.

The general form of Eq. (41–18) can be predicted from kinetic theory. If photons are emitted by microscopic oscillating charges, and if their energies are typically of the order of kT, as suggested by the equipartition theorem, then for a typical photon we would expect

$$E = \frac{hc}{\lambda} \quad \text{and} \quad \lambda = \frac{hc}{kT}.$$ (41–22)

Indeed, a photon with wavelength given by Eq. (41–18) has an energy of $4.965kT$.

41–8 X-RAY PRODUCTION AND SCATTERING

Production of x-rays: additional evidence for the validity of the photon concept

The production and scattering of **x-rays** gives additional support to the quantum view of electromagnetic radiation. X-rays are produced when rapidly moving electrons that have been accelerated through a potential difference of the order of 10^3 to 10^6 V strike a metal target. They were first produced by Wilhelm Röntgen (1845–1923) in 1895 with an apparatus similar in principle to that shown in Fig. 41–11a. Electrons are "boiled off" from the heated cathode by thermionic emission and are accelerated toward the anode (the target) by a large potential difference V. Most of the air is pumped out of the bulb so that electrons can travel from cathode to anode with only a small probability of collision with air molecules. The residual gas pressure is of the

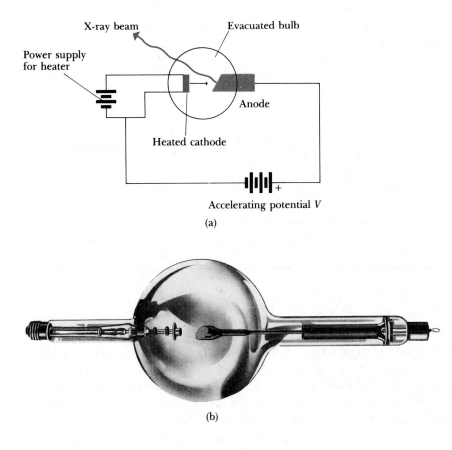

X-ray beam

Evacuated bulb

Power supply
for heater

Anode

Heated cathode

Accelerating potential V

(a)

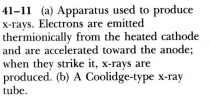

41–11 (a) Apparatus used to produce
x-rays. Electrons are emitted
thermionically from the heated cathode
and are accelerated toward the anode;
when they strike it, x-rays are
produced. (b) A Coolidge-type x-ray
tube.

(b)

order of 0.01 Pa or 10^{-7} atm. When V is a few thousand volts or more, a very
penetrating radiation is emitted from the anode surface. A common x-ray
tube of the type invented by Coolidge is shown in Fig. 41–11b.

Because of their origin, it is clear that x-rays are electromagnetic waves;
like light, they are governed by quantum relations in their interaction with
matter. Thus we can talk about x-ray photons or quanta, and the energy of an
x-ray photon is related to its frequency and wavelength just as for photons of
light, $E = hf = hc/\lambda$. Typical x-ray wavelengths are 0.001 to 1 nm (10^{-12} to
10^{-9} m). X-ray wavelengths can be measured quite precisely by crystal diffraction techniques, such as those we studied in Section 39–9.

Two distinct processes are involved in x-ray emission. Some electrons are
stopped by the target (anode), and all their kinetic energy is converted directly
to an x-ray photon. Other electrons transfer their energy partly or completely
to atoms in the target. These atoms are left in excited states, and when they
decay back to the ground state, they emit x-ray photons with energies characteristic of the element in the target.

X-ray photons have much higher energy
than visible-light photons.

EXAMPLE 41–2 Electrons are accelerated by a potential difference of 10 kV. If
an electron produces a photon on impact with the target, what is the wavelength
of the resulting x-rays?

SOLUTIONS   The photon energy $hf = hc/\lambda$ is equal to the kinetic energy eV of the electron just before impact; hence

$$eV = hf = \frac{hc}{\lambda}, \qquad\qquad (41-23)$$

and

$$\lambda = \frac{hc}{eV} = \frac{(6.626 \times 10^{-34} \text{ J·s})(3.00 \times 10^{8} \text{ m·s}^{-1})}{(1.602 \times 10^{-19} \text{ C})(1.00 \times 10^{4} \text{ V})}$$

$$= 1.24 \times 10^{-10} \text{ m}$$

$$= 0.124 \text{ nm}.$$

As we have mentioned, x-ray wavelengths can be measured directly by crystal diffraction, so this relation between accelerating potential and wavelength can be confirmed directly.

X-ray energy levels in atoms: the origin of x-ray line spectra

The atomic energy levels associated with excitation by x-rays are rather different in character from those associated with visible spectra. To understand them, we need some understanding of the arrangement of electrons in complex atoms, a topic we will study in detail in Chapter 43. For the present, we simply state that in a many-electron atom, the electrons are always arranged in concentric *shells* at increasing distances from the nucleus. These shells are labeled K, L, M, N, and so on. The K shell is closest to the nucleus, the L shell next, and so on. For any given atom in the ground state there is a definite number of electrons in each shell. Each shell has a maximum number of electrons it can accommodate, and we may speak of *filled shells* and *partially filled shells*. The K shell can contain at most 2 electrons. The next, the L shell, can contain 8. The third, the M shell, has a capacity for 18 electrons, while the N shell can hold 32. The sodium atom, for example, which contains 11 electrons, has 2 in the K shell, 8 in the L shell, and a single electron in the M shell. Molybdenum, with 42 electrons, has 2 in the K shell, 8 in the L shell, 18 in the M shell, 13 in the N shell, and 1 in the O shell.

Transitions involving the inner electrons in a many-electron atom produce x-ray spectra.

The *outer* electrons of an atom are responsible for the optical spectra of the elements. Relatively small amounts of energy (typically a few electronvolts) are needed to promote these electrons to excited states, and when they return to their normal states, wavelengths in or near the visible region are emitted. The inner electrons of a complex atom are closer to the nucleus and are more tightly bound; much more energy is required to displace them from their normal levels. As a result, we would expect a photon of much larger energy, and hence much higher frequency, to be emitted when the atom returns to its normal state after the displacement of an inner electron. The displacement of the *inner* electrons gives rise to the emission of x-rays.

What are x-ray energy levels?

If they have enough energy, some of the electrons accelerated in an x-ray tube will dislodge one of the inner electrons of a target atom, say one of the K electrons. This leaves a vacant space in the K shell, which is immediately filled by an electron from one of the outer shells, such as the L, M, N, O, . . . shell. The readjustment of the electrons is accompanied by a decrease in the energy of the atom, and an x-ray photon is emitted with energy just equal to this

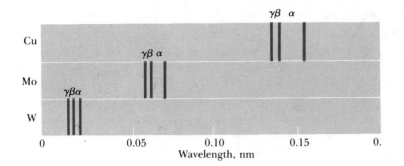

41–12 Wavelengths of the K_α-, K_β-, and K_γ-lines of copper, molybdenum, and tungsten.

decrease. Since the energy change is perfectly definite for atoms of a given element, the emitted x-rays should have definite frequencies. In other words, the **x-ray spectrum** should be a *line spectrum*. If the outermost electrons are in the N shell, there should be just three lines in the series, corresponding to the three possibilities that the vacant space may have been filled by an L, M, or N electron.

This is precisely what is observed. Figure 41–12 illustrates the so-called K series of the elements copper, molybdenum, and tungsten. Each series consists of three lines, known as the K_α-, K_β-, and K_γ-lines. The K_α-line is produced by the transition of an L electron to the vacated space in the K shell, the K_β-line by an M electron, and the K_γ-line by an N electron.

In addition to the K series, there are other series of x-ray lines, called the L, M, and N series, produced by the ejection of electrons from the L, M, and N shells rather than the K shell. Electrons in these outer shells are farther away from the nucleus and are not held as tightly as those in the K shell, so their removal requires less energy. The corresponding x-ray photons emitted when these vacancies are again filled have lower energy and longer wavelength.

Along with this x-ray *line* spectrum is a *continuous* spectrum of x-ray radiation, resulting from the stopping of electrons without exciting x-ray energy levels in the target. This spectrum extends indefinitely at the long-wavelength (low-energy) end but cuts off sharply at the short-wavelength end at a wavelength corresponding to a photon energy just equal to the electron energy before impact, as in Example 41–2.

Continuous x-ray spectra: What determines the short-wavelength limit?

This energy relationship is exactly the same as for the photoelectric effect (Section 41–2) except for the omission of the work-function term; this is negligible here because the x-ray energies are much larger than the work function. Indeed, x-ray emission can be thought of as an *inverse photoelectric effect*. In photoelectric emission the energy of a photon is transformed into kinetic energy of an electron; in x-ray production the kinetic energy of an electron is transformed into energy of a photon.

A phenomenon called **Compton scattering,** first observed in 1924 by A. H. Compton, provides additional direct confirmation of the quantum nature of electromagnetic radiation. When x-rays impinge on matter, some of the radiation is *scattered,* just as visible light falling on a rough surface undergoes diffuse reflection. Observations show that some of the scattered radiation has smaller frequency and longer wavelength than the incident radiation, and that the change in wavelength depends on the angle through which the radiation is scattered. Specifically, if the scattered radiation emerges at an angle ϕ

X-ray scattering: How can the wavelength change?

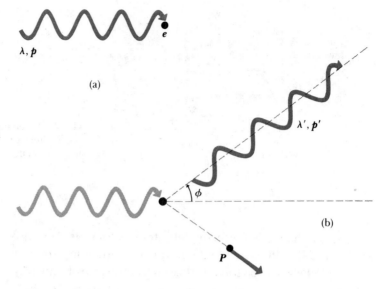

$\lambda, \boldsymbol{p}$

(a)

$\lambda', \boldsymbol{p}'$

ϕ

(b)

$\boldsymbol{P}$

41–13 Schematic diagram of Compton scattering showing (1) electron initially at rest with incident photon of wavelength λ and momentum $\boldsymbol{p}$; (b) scattered photon with longer wavelength λ' and momentum $\boldsymbol{p}'$ and recoiling electron with momentum $\boldsymbol{P}$. The direction of the scattered photon makes an angle ϕ with that of the incident photon, and the angle between $\boldsymbol{p}$ and $\boldsymbol{p}'$ is also ϕ.

A simple analysis of Compton scattering in terms of a collision of particles

with respect to the incident direction, as shown in Fig. 41–13, and if λ and λ' are the wavelengths of the incident and scattered radiation, respectively, we find that

$$\lambda' - \lambda = \frac{h}{mc}(1 - \cos \phi), \qquad (41\text{–}24)$$

where m is the electron mass.

Compton scattering cannot be understood on the basis of classical electromagnetic theory. On the basis of classical principles, the scattering mechanism is induced motion of electrons in the material, caused by the incident radiation. This motion must have the same frequency as that of the incident wave, and so the scattered wave radiated by the oscillating charges should have the same frequency. The frequency cannot be *shifted* by this mechanism.

In contrast, the quantum theory provides a beautifully simple explanation. We imagine the scattering process as a collision of two *particles,* the incident photon and an electron initially at rest, as in Fig. 41–13. The photon gives up some of its energy and momentum to the electron, which recoils as a result of this impact. The final scattered photon has less energy, smaller frequency, and longer wavelength than the initial one.

We can derive Eq. (41–24) from the principles of conservation of energy and momentum. We outline the derivation below and invite you to fill in the details. The electron energy may be in the relativistic range, so we have to use the relativistic energy–momentum relations, Eqs. (40–34) and (40–35). The initial photon has momentum p and energy pc; the final photon has momentum p' and energy $p'c$. The electron is initially at rest, so its initial momentum is zero and its initial energy is mc^2. The final electron momentum is P, and the final electron energy E is given by $E^2 = (mc^2)^2 + (Pc)^2$. Then energy conservation gives us the relation

$$pc + mc^2 = p'c + E,$$

or

$$(pc - p'c + mc^2)^2 = E^2 = (mc^2)^2 + (Pc)^2. \qquad (41\text{–}25)$$

We may eliminate the electron momentum P from this equation by using momentum conservation:

$$p = p' + P,$$

or

$$p - p' = P. \tag{41–26}$$

We take the scalar product of this quantity with itself, noting that $p \cdot p' = pp' \cos \phi$, and obtain

$$P^2 = p^2 + p'^2 - 2pp' \cos \phi. \tag{41–27}$$

This expression for P^2 may now be substituted into Eq. (41–25) and the left side multiplied out. A common factor c^2 is divided out; several terms cancel, and when the resulting equation is divided through by (pp'), the result is

$$\frac{mc}{p'} - \frac{mc}{p} = 1 - \cos \phi. \tag{41–28}$$

Finally, we substitute $p = h/\lambda$ and $p' = h/\lambda'$ and rearrange again to obtain Eq. (41–24).

X-rays have many practical applications in medicine and industry. Because they can penetrate several centimeters of solid matter, they can be used to visualize the interiors of materials opaque to ordinary light, such as broken bones or defects in structural steel. The object to be visualized is placed between an x-ray source and a large sheet of photographic film; the darkening of the film is proportional to the radiation exposure. A crack or air bubble allows greater transmission and shows as a dark area. Bones appear lighter than the surrounding flesh because they contain greater proportions of elements with high atomic number (and greater absorption) than flesh, where the light elements carbon, hydrogen, and oxygen predominate. This technique is not very effective in discriminating slightly different absorption characteristics, as found with many kinds of tumors.

Using x-rays to see into things and through things

In the past decade, several vastly improved x-ray techniques have been developed. One widely used system is *computerized axial tomography;* the corresponding instrument is called a CAT-scanner. The x-ray source produces a thin fan-shaped beam that is detected on the opposite side of the subject by an array of several hundred detectors in a line. Each detector measures absorption along a thin line through the subject. The entire apparatus is rotated around the subject in the plane of the beam during a few seconds. The changing reactions of the detectors are recorded digitally; a computer processes this information and reconstructs a picture of density over an entire cross section of the subject. Density differences as small as 1% can be detected with CAT-scans, and tumors and other anomalies much too small to be seen with older x-ray techniques can be detected.

More sophisticated x-ray imaging techniques

X-rays cause damage to living tissues. As x-ray photons are absorbed in tissues, they break molecular bonds and create highly reactive free radicals (such as neutral H and OH), which in turn can disturb the molecular structure of proteins and especially genetic material. Young and rapidly growing cells are particularly susceptible; hence x-rays are useful for selective destruction of

Biological effects of x-rays: some good, some bad

cancer cells. Conversely, however, a cell may be damaged by radiation but survive, continue dividing, and produce generations of defective cells; hence x-rays can *cause* cancer. Even when the organism itself shows no apparent damage, excessive radiation exposure can cause changes in the reproductive system that will affect the organism's offspring. The use of x-rays in medical diagnosis has become an area of great concern in recent years; a careful assessment of the balance between risks and benefits of radiation exposure is essential in each individual case.

SUMMARY

KEY TERMS

line spectrum

photoelectric effect

photons (quanta)

thermionic emission

threshold frequency

Planck's constant

work function

Balmer series

Rydberg constant

term

energy levels

ground state

excited states

resonance radiation

absorption spectrum

stimulated emission

population inversion

metastable state

continuous spectrum

Stefan-Boltzmann law

Stefan-Boltzmann constant

Wien displacement law

Planck radiation law

x-rays

x-ray spectrum

Compton scattering

The wave and ray pictures of electromagnetic radiation provide an adequate model to analyze the propagation of light. However, several phenomena involving emission and absorption point to a particle aspect of its behavior, in which the energy comes in quanta or photons. The energy E of a photon is proportional to its frequency f: $E = hf$, where h is Planck's constant, a fundamental constant of nature.

In the photoelectric effect, an electron is liberated from a conducting surface by absorption of a photon. The energy required to surmount the potential-energy barrier at the surface and escape is called the work function. The kinetic energy of the liberated electrons can be determined by measuring their stopping potential. Thus both the work function and Planck's constant can be determined.

Line spectra of atoms are associated with the existence of discrete energy levels. An atom emits a photon of a certain energy when it makes a transition between two energy levels differing in energy by that amount. The energy levels of the hydrogen atom are given by

$$E_n = -\frac{Rch}{n^2} \qquad (n = 1, 2, 3, \ldots). \qquad (41\text{--}13)$$

The lowest-energy state of an atom is called the ground state, and all higher-energy states are called excited states. An absorption spectrum results when continuous-spectrum radiation is partially absorbed while exciting atoms from the ground state.

The laser operates on the principle of stimulated emission, where an atom is stimulated to emit a photon by the presence of other photons of the same energy. For a laser to operate, there must be a means for creating a population inversion in which there are excited metastable states that cannot decay directly to the ground state.

Continuous spectra are emitted by very hot solid and liquid materials. The total rate of radiation is proportional to T^4, where T is the absolute temperature, and is given by

$$H = A\sigma T^4. \qquad (41\text{--}14)$$

This is the Stefan-Boltzmann radiation law. The distribution of the radiated power among various wavelengths is described by a distribution function $F(\lambda)$. At any temperature T, the wavelength λ_m at which this function has a maximum is given by

$$\lambda_m T = \text{constant} = 2.90 \times 10^{-3} \text{ m·K}. \qquad (41\text{--}16)$$

This relation is called the Wien displacement law. This and the Stefan-Boltzmann law can both be derived from the Planck radiation law. In deriving this law, Planck treated the electrons in a material as harmonic oscillators, with the added assumption that each electron could gain or lose energy only in increments hf, where f is the oscillator frequency. Historically, this was the first use of the quantum hypothesis. The Planck radiation law is

$$F(\lambda) = \frac{2\pi hc^2}{\lambda^5} \frac{1}{e^{hc/\lambda kT} - 1}. \tag{41–17}$$

X-rays are produced when rapidly moving electrons strike a target. The kinetic energy of an electron is converted directly to energy of a photon, or it excites an atom to an x-ray energy level; when the atom decays back to the ground state, it emits an x-ray photon. The first process gives a continuous spectrum of x-rays, with maximum energy (and minimum wavelength) determined by the electron's initial kinetic energy; the second process yields an x-ray line spectrum that depends on the element of the target.

In Compton scattering, an x-ray photon is scattered by an electron; the electron recoils, absorbing some of the photon's energy and momentum, and the scattered photon has lower energy and longer wavelength than the original one. This process can be analyzed as a collision between two particles; for a photon scattered through an angle ϕ, the increase in wavelength is given by

$$\lambda' - \lambda = \frac{h}{mc}(1 - \cos \phi). \tag{41–24}$$

QUESTIONS

41–1 In analyzing the photoelectric effect, how can we be sure that each electron absorbs only *one* photon?

41–2 In what ways do photons resemble other particles such as electrons? In what ways do they differ? Do they have mass? Electric charge? Can they be accelerated? What mechanical properties do they have?

41–3 Considering a two-slit interference experiment, if the photons are not synchronized with each other (i.e., are not coherent) and if half go through each slit, how can they possibly interfere with each other? Is there any way out of this paradox?

41–4 Can you devise an experiment to measure the work function of a material?

41–5 How might the energy levels of an atom be measured directly, that is, without recourse to analysis of spectra?

41–6 Would you expect quantum effects to be generally more important at the low-frequency end of the electromagnetic spectrum (radio waves) or at the high-frequency end (x-rays and gamma rays)? Why?

41–7 Most black-and-white photographic film (with the exception of some special-purpose films) is less sensitive at the far red end of the visible spectrum than at the blue end and has almost no sensitivity to infrared. How can these properties be understood on the basis of photons?

41–8 Human skin is relatively insensitive to visible light, but ultraviolet radiation can be quite destructive. Does this have anything to do with photon energies?

41–9 Does the concept of photon energy shed any light (no pun intended) on the question of why x-rays are so much more penetrating than visible light?

41–10 The phosphorescent materials that coat the inside of a fluorescent lamp tube convert ultraviolet radiation (from the mercury-vapor discharge inside the tube) to visible light. Could one also make a phosphor that converts visible light to ultraviolet?

41–11 As a body is heated to very high temperature and becomes self-luminous, the apparent color of the emitted radiation shifts from red to yellow and finally to blue as the temperature increases. Why the color shift?

41–12 Elements in the gaseous state emit line spectra with well-defined wavelengths; but hot solid bodies usually emit a continuous spectrum, that is, a continuous smear of wavelengths. Can you account for this difference?

41–13 Could Compton scattering occur with protons as well as electrons? Suppose, for example, one directed a beam of x-rays at a liquid-hydrogen target. What similarities and differences in behavior would be expected?

EXERCISES

$$e = 1.602 \times 10^{-19} \text{ C}$$
$$m = 9.110 \times 10^{-31} \text{ kg}$$
$$h = 6.626 \times 10^{-34} \text{ J·s}$$
$$N_A = 6.022 \times 10^{23} \text{ atoms·mol}^{-1}$$

Energy equivalent of 1 u = 931.5 MeV

$$\frac{e}{m} = 1.758 \times 10^{11} \text{ C·kg}^{-1}$$
$$k = 1.381 \times 10^{-23} \text{ J·K}^{-1}$$
$$1 \text{ eV} = 1.602 \times 10^{-19} \text{ J}$$
$$1 \text{ u} = 1.661 \times 10^{-27} \text{ kg}$$
$$\epsilon_0 = 8.854 \times 10^{-12} \text{ C}^2 \cdot \text{N}^{-1} \cdot \text{m}^{-2}$$

Section 41–2 The Photoelectric Effect

41–1 A nucleus in an excited state emits a γ-ray photon of energy 1 MeV.

a) What is the photon frequency?

b) What is the photon wavelength?

c) How does the wavelength compare with typical nuclear radii (of the order of 10^{-15} m)?

41–2 A sodium-vapor lamp emits light of wavelength 589 nm. If the total power of the emitted light is 6 W, how many photons are emitted per second?

41–3 A photon of orange light has a wavelength of 600 nm. Find the frequency, momentum, and energy of the photon; express the energy both in joules and in electronvolts.

41–4 A laser used to weld detached retinas emits light of wavelength 633 nm with a power of 0.5 W, in pulses 20 ms in duration.

a) How much energy is in each pulse, in joules? In electronvolts?

b) What is the energy of one photon, in joules? In electronvolts?

c) How many photons are in each pulse?

41–5 A radio station broadcasts at a frequency of 100 MHz, with a total power output of 50 kW.

a) What is the energy of the emitted photons, in joules? In electronvolts?

b) How many photons are emitted per second?

41–6 In the photoelectric effect, what is the relation between the threshold frequency f_0 and the work function ϕ?

41–7 A photoelectric surface has a work function of 4.00 eV. What is the maximum speed of the photoelectrons emitted by light of frequency 3×10^{15} Hz?

41–8 The photoelectric threshold wavelength of tungsten is 2.73×10^{-5} cm. Calculate the maximum kinetic energy of the electrons ejected from a tungsten surface by ultraviolet radiation of wavelength 1.80×10^{-5} cm. (Express the answer in electronvolts.)

41–9 When ultraviolet light of wavelength 2.54×10^{-5} cm from a mercury arc falls on a clean copper surface, the retarding potential necessary to stop emission of photoelectrons is 0.59 V. What is the photoelectric threshold wavelength for copper?

41–10 The photoelectric work function of potassium is 2.0 eV. If light having a wavelength of 360 nm falls on potassium, find

a) the stopping potential;

b) the kinetic energy in electronvolts of the most-energetic electrons ejected;

c) the speeds of these electrons.

Section 41–3 Line Spectra

Section 41–4 Energy Levels

Section 41–5 Atomic Spectra

41–11 Calculate (a) the frequency and (b) the wavelength of the H_β-line of the Balmer series for hydrogen. This line is emitted in the transition from $n = 4$ to $n = 2$.

41–12 Find the longest and shortest wavelengths in the Lyman, Balmer, and Paschen series for hydrogen. In what region of the electromagnetic spectrum does each series lie?

41–13 The silicon-silicon single bond that forms the basis of the (mythical) silicon-based creature the Horta has a bond strength of 3.2 eV. What wavelength photon would you need in a (mythical) phasor disintegration gun to destroy the Horta?

41–14 The energy-level scheme for the mythical one-electron element Searsium is shown in Fig. 41–14. The potential energy of an electron is taken to be zero at an infinite distance from the nucleus.

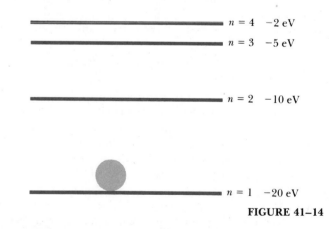

	$n = 4$	-2 eV
	$n = 3$	-5 eV
	$n = 2$	-10 eV
	$n = 1$	-20 eV

FIGURE 41–14

a) How much energy (in electronvolts) does it take to ionize an electron from the ground state?

b) A 15-eV photon is absorbed by a Searsium atom. When the atom returns to its ground state, what possible energies can the emitted photons have?

c) What will happen if a photon with an energy of 8 eV strikes a Searsium atom? Why?

d) If photons emitted from Searsium transitions $n = 4$ to $n = 2$ and from $n = 2$ to $n = 1$ will eject photoelectrons from an unknown metal, but the photon emitted from the transition $n = 3$ to $n = 2$ will not, what are the limits (maximum and minimum possible values) of the work function of the metal?

Section 41–6 The Laser

41–15 In Fig. 41–9, compute the energy difference for the $5s$–$3p$ transition in neon; express your result in electronvolts and in joules. Compute the wavelength of a photon having this energy, and compare your result with the observed wavelength of the laser light.

41–16 In the helium-neon laser, what wavelength corresponds to the $3p$–$3s$ transition in neon? Why is this not observed in the beam with the same intensity as the 632.8-nm laser line?

41–17 How many photons per second are emitted by a 0.5-mW He-Ne laser that has a wavelength of 633 nm?

Section 41–7 Continuous Spectra

41–18 What is λ_m, the wavelength at the peak of the Planck distribution, and the corresponding frequency f_m, at the following Kelvin temperatures:

a) 3 K?   **b)** 300 K?   **c)** 3000 K?

41–19

a) Show that the maximum in the Planck distribution, Eq. (41–17), occurs at a wavelength λ_m given by $\lambda_m = hc/4.965kT$ (Eq. [41–18]). As discussed in the text, 4.965 is the root of Eq. (41–19).

b) Evaluate the constants in the expression derived in (a) to show that $\lambda_m T$ has the numerical value given in the Wien displacement law, Eq. (41–16).

Section 41–8 X-Ray Production and Scattering

41–20

a) What is the minimum potential difference between the filament and the target of an x-ray tube if the tube is to produce x-rays of wavelength 0.05 nm?)

b) What is the shortest wavelength produced in an x-ray tube operated at 2×10^6 V?

41–21 Complete the derivation of the Compton-scattering formula, Eq. (41–24), following the outline given in Eqs. (41–25) through (41–28).

41–22 X-rays are produced in a tube operating at 50.0 kV. After emerging from the tube, some x-rays strike a target and are Compton-scattered through an angle of 60°.

a) What is the original x-ray wavelength?

b) What is the wavelength of the scattered x-rays?

c) What is the energy of the scattered x-rays (in electronvolts)?

41–23 X-rays with initial wavelength 0.5×10^{-10} m undergo Compton scattering. For what scattering angle is the wavelength of the scattered x-rays greater by 1% than that of the incident x-rays?

PROBLEMS

41–24

a) If the average wavelength emitted by a 100-W light bulb is 600 nm, and 10% of the input power is emitted as visible light, approximately how many visible light photons are emitted per second?

b) At what distance would this correspond to 100 photons per square centimeter, per second, if the light is emitted uniformly in all directions?

41–25 The light-sensitive compound on most photographic films is silver bromide, AgBr. A film is "exposed" when the light energy absorbed dissociates this molecule into its atoms. (The actual process is more complex, but the quantitative result does not differ greatly.) The energy of dissociation of AgBr is 1.00×10^5 J·mol^{-1}. Find

a) the energy in electronvolts,

b) the wavelength,

c) the frequency of the photon that is just able to dissociate a molecule of silver bromide.

d) What is the energy in electronvolts of a quantum of radiation having a frequency of 100 MHz?

e) Explain the fact that light from a firefly can expose a photographic film, whereas the radiation from a TV station transmitting 50,000 W at 100 MHz cannot.

f) Will photographic films stored in a light-tight container be ruined (exposed) by the radio waves passing through them? Explain.

41–26 The directions of emission of photons from a source of radiation are random. According to the wave theory, intensity of radiation from a point source varies inversely as the square of the distance from the source. Show that the number of photons from a point source passing out through a unit area is also given by an inverse-square law.

41–27 What will be the change in the stopping potential for photoelectrons emitted from a surface if the wavelength of the incident light is reduced from 400 nm to 360 nm?

41–28 The photoelectric work functions for particular samples of certain metals are as follows: cesium, 2.00 eV; copper, 4.00 eV; potassium, 2.25 eV; and zinc, 3.60 eV.

a) What is the threshold wavelength for each metal?

b) Which of these metals could not emit photoelectrons when irradiated with visible light?

41–29 When a certain photoelectric surface is illuminated with light of different wavelengths, the stopping potentials in the table below are observed.

Wavelength (nm)	Stopping Potential (V)
366	1.48
405	1.15
436	0.93
492	0.62
546	0.36
579	0.24

Plot the stopping potential as ordinate against the frequency of the light as abscissa. Determine

a) the threshold frequency,

b) the threshold wavelength,

c) the photoelectric work function of the material (in electronvolts),

d) the value of Planck's constant h (assuming the value of e is known).

41–30 An unknown element is found to have an absorption spectrum with lines at 4, 7, and 9 eV, and its ionization potential is 10 eV.

a) Draw an energy-level diagram for this element.

b) If a 9-eV photon is absorbed, what energies can the subsequently emitted photons have?

41–31

a) What is the least amount of energy in electronvolts that must be given to a hydrogen atom initially in its ground state so that it can emit the H_β-line (see Exercise 41–11) in the Balmer series?

b) How many different possibilities of spectral line emissions are there for this atom, when the electron starts in the $n = 4$ level and eventually ends up in the ground state? Calculate the wavelength of the emitted photon in each case.

41–32 If hydrogen were monatomic, at what kinetic temperature would the average translational kinetic energy be equal to the energy required to raise a hydrogen atom from the ground state to the $n = 2$ excited state?

41–33 If electrons in a metal had the same energy distribution as molecules in a gas at the same temperature (which is not actually the case), at what temperature would the average electron kinetic energy equal 1 eV, typical of work functions of metals?

41–34 An x-ray tube is operating at 150,000 V and 10 mA.

a) If only 1% of the electric power supplied is converted into x-rays, at what rate is the target being heated in joules per second?

b) If the target has a mass of 0.300 kg and a specific heat of 147 J·kg^{-1}·C°$^{-1}$, at what average rate would its temperature rise if there were no thermal losses?

c) What must be the physical properties of a practical target material? What would be some suitable target elements?

41–35

a) Calculate the maximum increase in x-ray wavelength that can occur during Compton scattering.

b) What is the energy (in electronvolts) of the smallest-energy x-ray photon for which Compton scattering could result in doubling the original wavelength?

41–36 A photon of wavelength 0.1200 nm is Compton-scattered through an angle of 180°.

a) What is the wavelength of the scattered photon?

b) How much energy is given to the electron?

c) What is the recoil speed of the electron? Is it necessary to use the relativistic kinetic-energy relationship?

41–37 A photon with $\lambda = 0.100$ nm collides with an electron at rest. After the collision the photon's wavelength is 0.110 nm.

a) What is the kinetic energy of the electron after the collision?

b) If the electron is suddenly stopped (for example, in a solid target), it emits a photon. What is the wavelength of this photon?

41–38 An electron with an energy of 1000 eV suddenly emits a photon and then continues on with an energy of 600 eV. This photon is then totally absorbed by another electron. How much momentum is transferred to the second electron by the photon?

CHALLENGE PROBLEMS

41–39

a) Write the Planck distribution law in terms of the frequency f rather than the wavelength λ, to obtain $F(f)$.

b) Show that

$$\int_0^\infty F(\lambda)\, d\lambda = \frac{2\pi^5 k^4}{15c^2 h^3}T^4,$$

where $F(\lambda)$ is the Planck distribution formula of Eq. (41–

17). *Hint:* Change the integration variable from λ to f. You will need to use the following tabulated integral:

$$\int_0^\infty \frac{x^3}{e^{\alpha x}-1}\, dx = \frac{1}{240}\left(\frac{2\pi}{\alpha}\right)^4.$$

c) The result of (b) is H/A and has the form of the Stefan-Boltzmann law, $H/A = \sigma T^4$ (Eq. 41–14). Evaluate the constants in (b) to show that σ has the value given in Eq. (41–15).

42

QUANTUM MECHANICS

WE HAVE SEEN IN THE PRECEDING CHAPTER THAT SOME ASPECTS OF emission and absorption of light, including atomic spectra, can be understood on the basis of the photon concept, together with the concept of discrete energy levels in atoms. But a complete theory should also offer some means of *predicting*, on theoretical grounds, the values of these energy levels for any particular atom. We begin this chapter with a discussion of a partially successful attempt to do so for the hydrogen atom. This attempt, the Bohr model of the hydrogen atom, cannot be generalized to atoms having more than one electron, and more drastic departures from nineteenth-century ideas are needed. In particular, we discuss the extension of the wave-particle duality, firmly established for electromagnetic radiation, to include particles as well as radiation. The entities (such as electrons) that we are accustomed to calling *particles* may in some situations exhibit *wavelike* behavior.

The new theory we will study in this chapter requires fundamental changes in the language we use to describe the state of a mechanical system. A particle can no longer be described as a single point moving in space but is an inherently spread-out entity. As the particle moves, the spread-out character has some of the properties of a *wave;* for example, particles can undergo *diffraction*. This new theory is called *quantum mechanics;* in it we find the key to understanding the structure of atoms and molecules, including their spectra, chemical behavior, and many other properties. Quantum mechanics has the happy effect of restoring unity to our description of both particles and radiation, and wave concepts are central to the entire theory.

42–1 THE BOHR ATOM

At the same time (1915) that Bohr advanced his hypothesis about the relation of spectrum-line frequencies to energy levels of atoms, he also proposed a mechanical model of the simplest atom, hydrogen. By combining some classical mechanics with a postulate that had no basis in previous physics, he was able to calculate the energy levels of hydrogen and obtain agreement with

The Bohr model: predicting the energy levels of the hydrogen atom

values determined from spectra. His description is called the **Bohr model** of the hydrogen atom.

Bohr's was not by any means the first attempt to understand the internal structure of atoms. Starting in 1906, Rutherford and his coworkers had performed experiments on the scattering of alpha particles (helium nuclei emitted from radioactive elements) by thin metallic foils. These experiments, which we will discuss in Chapter 44, showed that each atom contains a dense, compact nucleus whose size (of the order of 10^{-15} m) is very much smaller than the overall size of the atom (of the order of 10^{-10} m). The nucleus is surrounded by a swarm of electrons.

A dynamic picture of the atom: electrons in orbits

To account for the fact that the negatively charged electrons remain at relatively large distances from the positively charged nucleus despite the electrostatic attraction the nucleus exerts on the electrons, Rutherford postulated that the electrons *revolve* about the nucleus in orbits, more or less as the planets in the solar system revolve around the sun, but with the electrical attraction providing the necessary centripetal force.

Why doesn't an orbiting electron radiate energy continuously?

This assumption, however, has an unfortunate consequence. A body moving in a circle accelerates continuously toward the center of the circle. According to classical electromagnetic theory, an accelerating electron radiates energy. The total energy of the electrons would therefore decrease continuously, their orbits would become smaller and smaller, and eventually they would spiral into the nucleus and come to rest. Furthermore, according to classical theory the *frequency* of the electromagnetic waves emitted by a revolving electron should equal the frequency of revolution. As the electrons radiated energy, their angular velocities would change continuously and they would emit a *continuous* spectrum (a mixture of all frequencies), in contradiction to the *line* spectrum actually observed.

Electrons in stable nonradiating orbits

Faced with the dilemma that electromagnetic theory predicted an unstable atom emitting radiant energy of all frequencies, while observation showed stable atoms emitting only a few frequencies, Bohr concluded that, in spite of the success of electromagnetic theory in explaining large-scale phenomena, it could not be applied to processes on an atomic scale. He therefore postulated that an electron in an atom can revolve in certain **stable orbits,** each having a definite associated energy, *without* emitting radiation, contrary to the predictions of classical electromagnetic theory. According to Bohr, an atom radiates only when it makes a transition from one of these stable orbits to another, at the same time emitting (or absorbing) a photon of appropriate energy and frequency, given by Eq. (41–11).

The permitted orbits have definite values of angular momentum.

To determine the radii of the "permitted" orbits, Bohr introduced what must be regarded in hindsight as a brilliant intuitive guess. He noted that the *units* of Planck's constant h, usually written as J·s, are the same as the units of angular momentum, usually written as kg·m^2·s^{-1}, and he postulated that only those orbits are permitted for which the angular momentum is an integral multiple of $h/2\pi$. Recall from Section 9–12 that the angular momentum of a particle of mass m, moving with tangential speed v in a circle of radius r, is mvr. Hence the condition above may be stated as

$$mvr = n\frac{h}{2\pi},$$

where $n = 1, 2, 3$, and so on. Each value of n corresponds to a permitted value of the orbit radius, which we denote from now on by r_n, and a corresponding speed v_n. With this notation, the equation above becomes

$$mv_nr_n = n\frac{h}{2\pi}.$$

(42–1)

We now incorporate this condition into the analysis of the hydrogen atom. This atom consists of a single electron of charge $-e$, revolving about a single proton of charge $+e$. The proton is nearly 2000 times as massive as the electron, and we will assume the proton does not move. The electrostatic force of attraction between the charges,

$$F = \frac{1}{4\pi\epsilon_0}\frac{e^2}{r_n^2},$$

provides the centripetal force and, from Newton's second law,

$$\frac{1}{4\pi\epsilon_0}\frac{e^2}{r_n^2} = \frac{mv_n^2}{r_n}.$$

(42–2)

Calculating the energy corresponding to a particular electron orbit

When Eqs. (42–1) and (42–2) are solved simultaneously for r_n and v_n, we obtain

$$r_n = \epsilon_0\frac{n^2h^2}{\pi m e^2},$$

(42–3)

$$v_n = \frac{1}{\epsilon_0}\frac{e^2}{2nh}.$$

(42–4)

Let

$$\epsilon_0\frac{h^2}{\pi m e^2} = r_1.$$

(42–5)

Then Eq. (42–3) becomes

$$r_n = n^2r_1,$$

The permitted states are labeled by a value of an integer called a quantum number.

and the permitted, nonradiating orbits have radii r_1, $4r_1$, $9r_1$, and so on. The appropriate value of n is called the **quantum number** of the orbit.

The numerical values of the quantities on the left side of Eq. (42–5) are

$$\epsilon_0 = 8.854 \times 10^{-12}\ \mathrm{C^2 \cdot N^{-1} \cdot m^{-2}},$$

$$h = 6.626 \times 10^{-34}\ \mathrm{J \cdot s},$$

$$m = 9.110 \times 10^{-31}\ \mathrm{kg},$$

$$e = 1.602 \times 10^{-19}\ \mathrm{C}.$$

Hence r_1, the radius of the first Bohr orbit, is

$$r_1 = \frac{(8.854 \times 10^{-12}\ \mathrm{C^2 \cdot N^{-1} \cdot m^{-2}})(6.626 \times 10^{-34}\ \mathrm{J \cdot s})^2}{(3.14)(9.110 \times 10^{-31}\ \mathrm{kg})(1.602 \times 10^{-19}\ \mathrm{C})^2}$$

$$= 0.53 \times 10^{-10}\ \mathrm{m} = 0.53 \times 10^{-8}\ \mathrm{cm}.$$

This is in good agreement with atomic diameters as estimated by other methods, namely, about 10^{-8} cm.

The kinetic energy of the electron in any orbit is

The energy of each state depends on the value of its quantum number in a simple way.

$$K_n = \frac{1}{2}mv_n^2 = \frac{1}{\epsilon_0^2}\frac{me^4}{8n^2h^2},$$

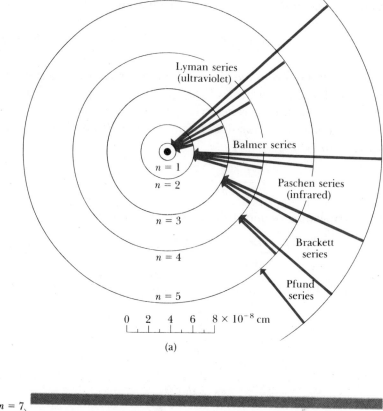

(a)

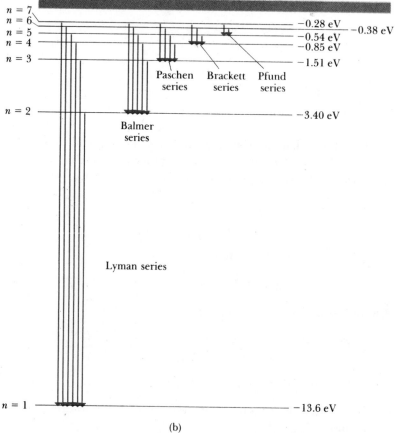

(b)

42–1 (a) "Permitted" orbits of an electron in the Bohr model of a hydrogen atom. The transitions responsible for some of the lines of the various series are indicated by arrows. (b) Energy-level diagram, showing transitions corresponding to the various series.

and the potential energy is

$$U_n = -\frac{1}{4\pi\epsilon_0}\frac{e^2}{r_n} = -\frac{1}{\epsilon_0^2}\frac{me^4}{4n^2h^2},$$

The total energy, E_n, is therefore

$$E_n = K_n + U_n = -\frac{1}{\epsilon_0^2}\frac{me^4}{8n^2h^2}. \qquad (42\text{–}6)$$

The total energy has a negative sign because the reference level of potential energy is taken to be zero with the electron at an infinite distance from the nucleus. Since we are interested only in energy *differences*, this is not of importance. The energy levels can be displayed graphically as in Fig. 42–1b.

Thus the possible states of the atoms are labeled by values of the integer n, which we call the *quantum number* for the system. For each value of n there are corresponding values of orbit radius, angular momentum, and total energy. The energy of the atom is least when $n = 1$, for then E_n has its largest negative value. This is the **ground state** of the atom; in this state the electron is in the smallest orbit, with radius r_1. For $n = 2, 3, \ldots$, the absolute value of E_n is smaller and the energy is progressively larger (less negative). The orbit radius increases as n^2, according to Eq. (42–3).

The ground state has the smallest energy and the smallest quantum number.

As a result of collisions with rapidly moving electrons in an electrical discharge, or by other means, the atom may temporarily acquire enough energy to raise the electron to some higher energy and larger orbit. The atom is then said to be in an **excited state.** This state is unstable; the electron eventually falls back to a state of lower energy, emitting a photon in the process.

Let n_1 be the quantum number of some excited state and n_2 be the quantum number of the lower state to which the electron returns after the emission process. Then E_i, the initial energy, is

$$E_i = -\frac{1}{\epsilon_0^2}\frac{me^4}{8n_1^2h^2},$$

and E_f, the final energy, is

$$E_f = -\frac{1}{\epsilon_0^2}\frac{me^4}{8n_2^2h^2}.$$

The decrease in energy, $E_i - E_f$, which we set equal to the energy hf of the emitted photon, is

Photon energies are differences between the initial and final energy states of an atom.

$$E_i - E_f = hf = -\frac{1}{\epsilon_0^2}\frac{me^4}{8n_1^2h^2} + \frac{1}{\epsilon_0^2}\frac{me^4}{8n_2^2h^2},$$

or

$$f = \frac{1}{\epsilon_0^2}\frac{me^4}{8h^3}\left(\frac{1}{n_2^2} - \frac{1}{n_1^2}\right). \qquad (42\text{–}7)$$

This equation has the same form as the Balmer formula, Eq. (41–9), for the frequencies in the hydrogen spectrum if we place

$$\frac{1}{\epsilon_0^2}\frac{me^4}{8h^3} = Rc, \qquad (42\text{–}8)$$

and let $n_2 = 1$ for the Lyman series, $n_2 = 2$ for the Balmer series, and so on.

The Lyman series is therefore the group of lines emitted by electrons returning from some excited state to the ground state. The Balmer series is the group of lines emitted by electrons returning from some higher excited state, but stopping in the *second orbit* ($n = 2$) instead of falling directly to the ground state. That is, an electron returning from the third orbit ($n = 3$) to the second orbit ($n = 2$) emits the H_α-line. One returning from the fourth orbit ($n = 4$) to the second ($n = 2$) emits the H_β-line, and so on. These transitions are shown in Fig. 42–1.

Evaluating the Rydberg constant in terms of other fundamental constants

Every quantity in Eq. (42–8) may be determined quite independently of the Bohr theory. Apart from this theory, we have no reason to expect these quantities to be related in this particular way. The quantities m and e, for instance, are found from experiments on free electrons; h may be found from the photoelectric effect; R is determined by measurements of wavelengths; and c is the speed of light. But if we substitute the values of these quantities, obtained by such diverse means, into Eq. (42–8), we find it *does* hold exactly, within the limits of experimental error, providing direct confirmation of Bohr's theory.

Ionization energy: How much energy is required to remove the electron completely from a hydrogen atom?

The ionization energy of the hydrogen atom (the energy required to remove the electron completely) can also be predicted from the Bohr theory. Ionization corresponds to a transition from the ground state ($n = 1$) to an infinitely large orbit radius ($n = \infty$). The predicted energy is 13.6 eV, and again this is in excellent agreement with the experimentally measured value.

The Bohr model can be extended easily to other one-electron atoms, such as the singly ionized helium atom, the doubly ionized lithium atom, and so on. If the nuclear charge is Ze (where Z is the atomic number) instead of just e, the effect in the analysis above is to replace e^2 everywhere by Ze^2. In particular, the orbits, described by Eq. (42–3), become smaller by a factor of Z, and the energy levels, given by Eq. (42–6), are all multiplied by Z^2. We invite you to verify these statements.

What is the mechanism by which radiation of a certain frequency is emitted by an atom?

Although the Bohr model was successful in predicting the energy levels of the hydrogen atom, it raised as many questions as it answered. It combined elements of classical physics with new postulates that were inconsistent with classical ideas. It provided no insight into what happens *during* a transition from one orbit to another, and the stability of certain orbits was achieved at the expense of discarding the only picture available at the time of the electromagnetic mechanism for the atom to radiate energy. There was no clear justification for restricting the angular momentum to multiples of $h/2\pi$, except that it led to the right answer. Furthermore, attempts to extend the model to atoms with two or more electrons were not successful. We will see in the next section that an even more radical departure from classical concepts was required before the understanding of atomic structure could progress further.

The Bohr model was a mix of classical ideas with new assumptions inconsistent with classical principles.

42–2 WAVE NATURE OF PARTICLES

Wave-particle duality: a property of electrons as well as photons

The next major advance in understanding atomic structure came in 1923, about ten years after the Bohr theory. This was a suggestion by de Broglie that since light is dualistic in nature, behaving in some situations like waves and in others like particles, the same might be true of matter. That is, electrons and protons, which until that time had been thought to be purely particlelike, might in some circumstances behave like *waves*. Specifically, de Broglie postulated that a free electron of mass m, moving with speed v, should have a wavelength λ related to its momentum $p = mv$ in exactly the same way as the wavelength and momentum of a photon are related, as expressed by Eq. (41–7), $\lambda = h/p$. Thus the **de Broglie wavelength** of an electron is given by

$$\lambda = \frac{h}{mv}, \qquad (42-9)$$

where h is the same Planck's constant that appears in the frequency–energy relation for photons.

This wave hypothesis, unorthodox though it seemed at the time, almost immediately received direct experimental confirmation. We described in Section 39–9 how the layers of atoms in a crystal can serve as a diffraction grating for x-rays. An x-ray beam is strongly reflected when it strikes a crystal at such an angle that the waves scattered from the atomic layers combine to reinforce one another. The essential point here is that the existence of these strong reflections is evidence of the *wave* nature of x-rays.

In 1927 Davisson and Germer, working in the Bell Telephone Laboratories, were studying the surface of a crystal of nickel by directing a beam of *electrons* at the surface and observing the electrons reflected at various angles. It might be expected that even the smoothest surface attainable would still look rough to an electron, and that the electron beam would therefore be diffusely reflected. But the **Davisson-Germer experiment** showed that the electrons were reflected in almost the same way that x-rays would be reflected from the same crystal; that is, they were being *diffracted*. The de Broglie wavelengths of the electrons in the beam were computed from their known speed, with the help of Eq. (42–9); and the angles at which strong reflection took place were found to be the same as those at which x-rays of the same wavelength would be reflected. The discovery of **electron diffraction** gave strong support to de Broglie's hypothesis.

A diffraction experiment with electrons: direct confirmation of the wave nature of particles

This wave hypothesis clearly required sweeping revisions of our fundamental concepts regarding the description of matter. What we are accustomed to calling a *particle* actually behaves like a particle only if we do not look too closely. In general, a particle has to be described as a spread-out entity that is not entirely localized in space; and at least in some cases this spreading out appears as a periodic pattern suggesting *wavelike* properties. The wave and particle aspects are not inconsistent; the particle model is an *approximation* of a more general wave picture. We are reminded of the ray picture of geometrical optics, a special case of the more general wave picture of physical optics. Indeed, there is a very close analogy between optics and the description of the motion of particles.

Within a few years after 1923, the wave hypothesis of de Broglie was developed by Heisenberg, Schrödinger, Dirac, Born, and many others, into a complete theory called **quantum mechanics.** In the following sections we will sketch the main lines of thought in a nonmathematical way and describe some of the experimental evidence for the wave nature of material particles. We will show how the quantum numbers that were introduced in such an artificial way by Bohr now enter naturally into the theory of atomic structure.

One of the essential features of quantum mechanics is that a particle is no longer described as located at a single point, but is described instead in terms of a *function* that has various values at various points in space. The spatial

Quantum mechanics: some drastic revisions in our ideas of how to describe position and motion of a particle

distribution describing a *free* electron may have a recurring pattern characteristic of a wave that propagates through space. Electrons *in atoms* can be visualized as diffuse clouds surrounding the nucleus. The idea that the electrons in an atom move in definite orbits such as those in Fig. 42–1 has been abandoned. The orbits themselves, however, were never an essential part of Bohr's theory, since the quantities that determine the frequencies of the emitted photons are the *energies* corresponding to the orbits. The new theory still assigns definite energy states to an atom. In the hydrogen atom the energies are the same as those given by Bohr's theory; in more complicated atoms, where the Bohr theory does not work, the quantum mechanical picture is in excellent agreement with observation.

To show how quantization arises in atomic structure, we can use an analogy with the classical mechanical problem of a vibrating string held at its ends. We worked out the normal modes of this system in Section 22–3, and we suggest that you review that discussion. When the string vibrates, the ends must be nodes, but nodes may occur at other points also. The general requirement is that the length of the string shall equal some *integral* number of half-wavelengths.

In a similar way, the principles of quantum mechanics lead to a differential equation (Schrödinger's equation) that must be satisfied by an electron in an atom, subject also to certain boundary conditions. Let us think of an electron as a wave extending in a circle around the nucleus. In order for the wave to "come out even" and join onto itself smoothly, the circumference of this circle must include some *integral number* of wavelengths, as suggested by Fig. 42–2. The wavelength of a particle of mass m, moving with speed v, is given, according to wave mechanics, by Eq. (42–9), $\lambda = h/mv$. Then if r is the radius and $2\pi r$ the circumference of the circle occupied by the wave, we must have $2\pi r = n\lambda$, where $n = 1, 2, 3$, and so on. Since $\lambda = h/mv$, this equation becomes

$$2\pi r = n\frac{h}{mv}, \qquad mvr = n\frac{h}{2\pi}. \qquad (42\text{--}10)$$

But mvr is the angular momentum of the electron, so we see that the wave-mechanical picture leads naturally to Bohr's postulate that the angular momentum equals some integral multiple of $h/2\pi$.

To be sure, the idea of wrapping a wave around in a circular orbit is a rather vague notion. But the agreement of Eq. (42–10) with Bohr's hypothesis is much too remarkable to be a coincidence; and it strongly suggests that the wave properties of electrons do indeed have something to do with atomic structure.

A wave picture of atomic structure: standing waves in an atom

The Bohr quantum condition as a consequence of the de Broglie wave hypothesis

42–2 Diagrams showing the idea of wrapping a standing wave around a circular orbit. For the wave to join onto itself smoothly, the circumference of the orbit must be an integral number n of wavelengths. Examples are shown for $n = 2$, 3, and 4.

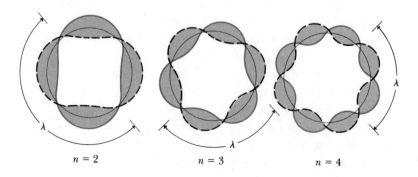

$n = 2$ $\qquad\qquad$ $n = 3$ $\qquad\qquad$ $n = 4$

PROBLEM-SOLVING STRATEGY: *Atomic physics*

1. In atomic physics, the orders of magnitude of physical quantities are so unfamiliar that often common sense isn't much help in judging the reasonableness of a result. It helps to remind yourself of some typical magnitudes of various quantities:

> Size of an atom, 10^{-10} m,
>
> Mass of an atom, 10^{-26} kg,
>
> Mass of an electron, 10^{-30} kg,
>
> Energy of an atomic state, 1 to 10 eV or 10^{-18} J (but some interaction energies are much *smaller* than this),
>
> Speed of an electron in the Bohr atom, 10^6 m·s^{-1},
>
> Electric charge, 10^{-19} C,
>
> kT at room temperature, $\frac{1}{40}$ eV.

You may want to add items to the list. This will also help you in Chapter 44, where we have to deal with magnitudes characteristic of nuclear rather than atomic structure, often different by factors of 10^6 or so. In working out problems, be very careful to handle powers of ten properly. A gross error may not be obvious.

2. As in the last chapter, energies may be expressed either in joules or in electronvolts. Be sure you use consistent units. Lengths, such as wavelengths, are always in meters if you use the other quantities consistently in SI units, such as $h = 6.626 \times 10^{-34}$ J·s. If you want nanometers or something else, don't forget to convert.

3. Aside from these calculational details, the main challenges of this chapter are conceptual, not computational. Try to keep an open mind when you encounter new and sometimes jarring ideas. Eventually you will come to appreciate the fact that a photon *can* have both wavelike and particlelike properties. Don't get discouraged; intuitive understanding of quantum mechanics takes some time to develop. Keep trying!

EXAMPLE 42–1   Find the speed and the kinetic energy of a neutron ($m = 1.675 \times 10^{-27}$ kg) having a de Broglie wavelength of 0.1 nm, typical of atomic spacing in crystals. Compare the energy with the average kinetic energy of a gas molecule at room temperature ($T = 20°C$).

An example of a wavelength calculation for a particle

SOLUTION   From Eq. (42–9),

$$v = \frac{h}{\lambda m} = \frac{6.626 \times 10^{-34} \text{ J·s}}{(0.1 \times 10^{-9} \text{ m})(1.675 \times 10^{-27} \text{ kg})}$$

$$= 3.96 \times 10^3 \text{ m·s}^{-1};$$

$$K = \tfrac{1}{2}mv^2 = \tfrac{1}{2}(1.675 \times 10^{-27} \text{ kg})(3.96 \times 10^3 \text{ m·s}^{-1})^2$$

$$= 1.31 \times 10^{-20} \text{ J} = 0.0818 \text{ eV}.$$

The average translational kinetic energy of a molecule of an ideal gas is given by Eq. (20–10):

$$K = \tfrac{3}{2}kT = (\tfrac{3}{2})(1.38 \times 10^{-23} \text{ J·K}^{-1})(293 \text{ K})$$

$$= 6.06 \times 10^{-21} \text{ J} = 0.0378 \text{ eV}.$$

Thus the two energies are of comparable magnitude, and indeed a neutron having an energy in this range is called a *thermal neutron*. Diffraction of thermal neutrons can be used to study crystal and molecular structure in the same way as x-ray diffraction; neutron diffraction has proved especially useful in the study of large organic molecules.

42–3 THE ELECTRON MICROSCOPE

Using an electron beam to form greatly enlarged images of microscopic objects

An electron beam can be used to form an image of an object in exactly the same way as a light beam. A ray of light is bent by reflection or refraction, and an electron trajectory is bent by an electric or magnetic field. Rays of light diverging from a point on an object can be brought to convergence by a converging lens, and electrons diverging from a small region can be brought to convergence by an electrostatic or magnetic lens. Figure 42–3 (parts a and b) shows the behavior of a simple type of electrostatic lens, and Figure 42–3c shows the analogous optical system. In each case the image can be made larger than the object by appropriate design; hence both devices can act as magnifiers.

The analogy between light rays and electrons goes deeper. The ray model of geometrical optics is an approximate representation of the more general wave picture, and geometrical optics is valid whenever interference and diffraction effects can be neglected. Similarly, we have seen in Section 42–2 that the model of an electron as a point particle following a line trajectory is an approximate description of the actual behavior of the electron, valid when effects associated with the wave nature of electrons can be neglected.

Resolution is better with short-wavelength electrons than with much longer wavelength visible light.

Herein lies the value of the **electron microscope.** The *resolution* of an optical microscope is limited by diffraction effects, as discussed in Section 39–10. Using a wavelength of 500 nm, typical of visible light, no optical microscope can resolve objects smaller than a few hundred nanometers, no matter how carefully its lenses are made. The resolution of an electron microscope is similarly limited by the wavelengths of the electrons, but these may be many thousands of times *smaller* than wavelengths of visible light. Hence the useful

42–3 (a) An electrostatic lens. The two cylinders are at different electrical potentials V_a and V_b; there is an electric field in the region between them, and the corresponding equipotential lines are shown in black. (b) Cross-sectional view from the side. The electron trajectories are shown in color; electrons diverging from point A are brought to a focus at point B. The behavior of magnetic lenses is similar. (c) Optical analog of the electrostatic lens in part (a).

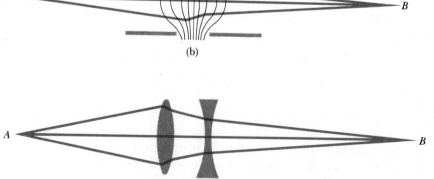

magnification of an electron microscope can be thousands of times as great as that of an optical microscope.

It is important to understand that the ability of the electron microscope to form an image *does not* depend on the wave properties of electrons. Their trajectories can be computed by treating them as charged particles under the action of electric- and magnetic-field forces. In the matter of *resolution*, however, their wave properties do become significant.

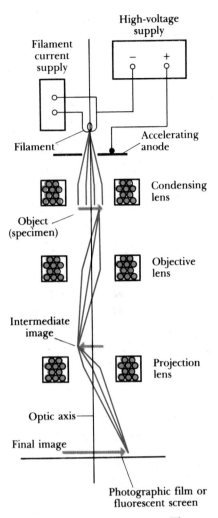

EXAMPLE 42–2 An electron beam is formed by a setup similar to that of the cathode-ray tube, discussed in Section 26–8. If the accelerating voltage is 10 kV, what is the wavelength of the electrons?

SOLUTION The wavelength is determined by Eq. (42–9). To find the speed we use conservation of energy; the kinetic energy $\frac{1}{2}mv^2$ of an electron equals the loss of potential energy eV. Thus

$$\frac{1}{2}mv^2 = eV, \qquad v = \sqrt{\frac{2eV}{m}}.$$

Inserting this result into Eq. (42–9), we find

$$\lambda = \frac{h}{m}\sqrt{\frac{m}{2eV}} = \frac{h}{\sqrt{2meV}}$$

$$= \frac{6.63 \times 10^{-34}\ \text{J·s}}{\sqrt{2(9.11 \times 10^{-31}\ \text{kg})(1.60 \times 10^{-19}\ \text{C})(10^4\ \text{V})}} \qquad (42\text{–}11)$$

$$= 1.23 \times 10^{-11}\ \text{m}$$

$$= 0.0123\ \text{nm}.$$

This value is smaller than typical wavelengths of visible light (around 500 nm) by a factor of about 40,000.

Most electron microscopes use magnetic rather than electrostatic lenses, for practical reasons. A common setup includes three lenses in a compound-microscope arrangement, as shown in Fig. 42–4. Electrons are emitted from a hot cathode and accelerated by a potential difference, typically 10 to 100 kV. The electrons pass through a condenser lens and are formed into a parallel beam before passing through the specimen or object to be viewed. The objective lens then forms an intermediate image of this object, and the projection lens produces a final real image of that image. These lenses play the roles of the objective and eyepiece lenses, respectively, of a compound optical microscope. The final image is recorded on photographic film or projected on a fluorescent screen for viewing or photographing. The entire apparatus, including the specimen, must be enclosed in a vacuum container, just as with the cathode-ray tube; otherwise electrons would collide with air molecules, muddling up the image. The specimen to be viewed is very thin, typically 10 nm to 100 nm, so the electrons are not slowed appreciably as they pass through.

We might think that with electrons of wavelength 0.01 nm, as in the example above, the resolution would be 0.01 nm or less. In fact, it is seldom better

42–4 An electron microscope. The magnetic lenses, consisting of coils of wire carrying currents, are shown in cross section. The condensing lens forms a parallel beam of electrons that strikes the object. The objective lens forms an intermediate image that serves as the object for the final image formed by the projection lens. All images are *real*. The final image is projected on photographic film or a fluorescent screen. The magnification of each lens may be of the order of 100X, and the overall magnification of the order of 10,000X. For greater magnification an additional intermediate lens may be used. The color lines are not continuous electron trajectories through the instrument but are drawn from lens to lens to show the formation of images. The angles of the electron paths with the optic axis are greatly exaggerated; in actual instruments these angles are usually less than 0.01 rad or 0.5°. The entire apparatus is enclosed in a vacuum chamber, not shown in the diagram.

than 0.5 nm, for several reasons. Large-aperture magnetic lenses have aberrations analogous to those of optical lenses, as discussed in Section 38–5. In addition, the focal length of a magnetic lens depends on the current in the coil, which must be controlled precisely, and on the electron speed, which is never a single precise value; the latter effect is the equivalent of chromatic aberration.

An improved electron microscope concept for studying surface details

A useful variation is the *scanning electron microscope*. The electron beam is focused to a very fine line and is swept across the specimen just as the electron beam in a TV picture tube traces out the picture. As the beam scans the specimen, electrons are knocked off; these are collected by a collecting anode kept at a potential a few hundred volts positive with respect to the specimen. The current in the collecting anode is amplified and used to modulate the electron beam in a cathode-ray tube, which is swept in synchronism with the microscope beam. Thus the cathode-ray tube traces out a greatly magnified image of the specimen. This scheme has the advantages that the beam need not pass through the specimen and that knock-off electron production depends on the angle at which the beam strikes the surface. Thus scanning electron micrographs have a much greater three-dimensional appearance than conventional ones. The resolution is not as great, typically of the order of 10 nm, but still much greater than the optical microscope. A scanning electron microscope is shown in Fig. 42–5, and a photograph made with such an instrument is shown in Fig. 42–6.

42–5 A scanning electron microscope, showing a vacuum chamber in the background and cathode-ray tube monitors in the foreground. (Courtesy of General Electric.)

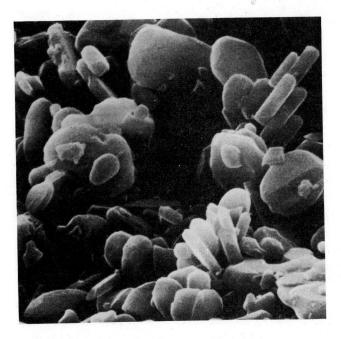

42–6 Photograph of powdered aluminum oxide. The overall magnification is about 4500X. (Manfred Kage-Peter Arnold, Inc.)

42–4 PROBABILITY AND UNCERTAINTY

We have learned that the entities we customarily call *particles* can, in some circumstances, behave like *waves*. Figure 42–7 shows a *wave packet* or *wave pulse* that has both wave and particle properties. The regular spacing λ_{av} between successive maxima is characteristic of a wave, but there is also a particlelike localization in space. To be sure, the wave pulse is not localized at a single point, but in any experiment that detects only dimensions much larger than Δx, the wave pulse appears to be a localized particle. Although it would be simplistic to say that Fig. 42–7 is a picture of an electron, the figure does suggest that wave and particle properties are not necessarily incompatible.

It would *not* be correct to regard the wave of Fig. 42–7 as having only a single wavelength. A sinusoidal wave with a definite wavelength has no beginning and no end; to make a wave *pulse* we must superpose many sinusoidal waves having various wavelengths. Thus Fig. 42–7 is a wave having such a distribution of wavelengths, and λ_{av} is an average value. This distribution has additional important implications, which we will explore at the end of this section.

The discovery of the dual wave-particle nature of matter has forced a drastic revision of the language used to describe the behavior of a particle. In classical Newtonian mechanics, we think of a particle as an idealized geometrical point that, at any instant of time, has a perfectly definite location in space and is moving with a definite velocity. As we will see, such a specific description is, in general, not possible. On a sufficiently small scale, there are fundamental limitations on the precision with which the position and velocity of a particle can be described; and some aspects of a particle's behavior can be stated only in terms of *probabilities*.

To illustrate the nature of the problem, let us consider again the single-slit diffraction experiment described in Section 39–7. We learned there that most (85%) of the intensity in the diffraction pattern is concentrated in the central maximum; the angular size of this maximum is determined by the positions of

Can something be a particle and a wave at the same time?

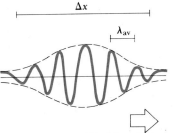

Direction of motion

42–7 A wave pulse or packet. There is an average wavelength λ_{av}, the distance between adjacent peaks, but the wave is localized at any instant in a region with length of the order of Δx. The broken lines are called the envelope of the pulse; all the peaks lie on the envelope curves, which approach zero at both ends of the pulse.

Particle and wave aspects of a single-slit diffraction experiment

the first intensity *minimum* on either side of the central maximum. Using Eq. (39–20), with $n = 1$, we find that the angle θ in Fig. 39–20 between the central peak and the minimum on either side is given by $\sin \theta = \lambda/a$, where a is the slit width. If λ is much smaller than a, then θ is very small, $\sin \theta$ is very nearly equal to θ, and the relation may be simplified further to

$$\theta = \frac{\lambda}{a}. \tag{41–12}$$

Now we perform the same experiment again but use a beam of *electrons* instead of a beam of monochromatic light. The apparatus must be evacuated to avoid collisions of electrons with air molecules; and there are other experimental details that need not concern us. The electron beam can be produced with a setup similar in principle to the electron gun in a cathode-ray tube, which produces a narrow beam of electrons all having the same direction and speed, and therefore also the same wavelength. Such an experiment is shown schematically in Fig. 42–8.

> *We cannot predict precisely the trajectory of an individual particle.*

The result of this experiment, as again recorded on photographic film or by means of more sophisticated detectors, is a diffraction pattern identical to that shown in Fig. 39–22b; this provides additional direct evidence of the wave nature of electrons. Most electrons strike the film in the vicinity of the central maximum, but a few strike farther from the center, near the edges of that maximum and also in the subsidiary maxima on both sides. Thus, if we believe that electrons are waves, the wave behavior in this experiment presents no surprises.

> *The diffraction pattern is a probability distribution.*

Interpreted in terms of *particles,* however, this experiment poses very serious problems. First, although the electrons all have the same initial state of motion, they do not all follow the same path. In fact, *the trajectory of an individual electron cannot be predicted from knowledge of its initial state.* The best we can do is to say that most of the electrons go to a certain region, fewer go to other regions, and so on; alternatively, we can describe the *probability* for an individual electron to strike each of various areas on the film. This fundamental indeterminacy has no counterpart in Newtonian mechanics, where the motion of a particle or a system is always predictable if the initial position and motion are known with sufficient precision.

Second, there are fundamental *uncertainties* in both position and momentum of an individual particle, and these two uncertainties are related inseparably. To illustrate this point, we note that, in Fig. 42–8, an electron striking the

42–8 An electron diffraction experiment. The graph at the right shows the degree of blackening of the film, which in any region is proportional to the number of electrons striking that region. The components of momentum of an electron striking the outer fringe of the central maximum are shown.

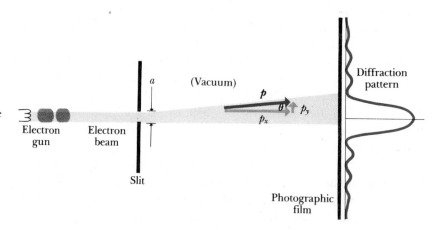

film at the outer edge of the central maximum, at angle θ, must have a component of momentum p_y in the y-direction, as well as a component p_x in the x-direction, despite the fact that initially the beam was directed along the x-axis. From the geometry of the situation, the two components are related by $p_y/p_x = \tan \theta$; and if θ is small, we may approximate $\tan \theta = \theta$, obtaining

$$p_y = p_x\theta. \qquad (42\text{--}13)$$

Neglecting any electrons striking the film outside the central maximum (that is, at angles greater than λ/a), we see that the y-component of momentum may be as large as

The uncertainties in the transverse position and momentum have a reciprocal relationship.

$$p_y = p_x\left(\frac{\lambda}{a}\right). \qquad (42\text{--}14)$$

Hence the *uncertainty* Δp_y in the y-component of momentum is at least as great as $p_x\lambda/a$:

$$\Delta p_y \geq p_x\frac{\lambda}{a}.$$

Thus the narrower the slit width a, the broader the diffraction pattern and the greater the uncertainty in the y-component of momentum.

Now the electron wavelength λ is related to the momentum $p_x = mv_x$ by the de Broglie relation, Eq. (42–9), which may be rewritten as $\lambda = h/p_x$. Using this result in the equation above and simplifying, we find

The more accurately the position is known, the less accurately the momentum can be known.

$$\Delta p_y \geq p_x\left(\frac{h}{ap_x}\right) = \frac{h}{a},$$

or

$$\Delta p_y a \geq h. \qquad (42\text{--}15)$$

To interpret this result, we note that the slit width a represents the uncertainty in *position* of an electron as it passes through the slit; we do not know through which particular part of the slit each particle passes. Thus the y-components of *both* position and momentum have uncertainties, and the two uncertainties are related by Eq. (42–15). We can reduce the *momentum* uncertainty only by increasing the slit width, which increases the *position* uncertainty; and conversely, when we decrease the position uncertainty by narrowing the slit, the diffraction pattern broadens and the corresponding momentum uncertainty increases.

All of this may be bitter medicine for a reader steeped in the tradition of nearly three centuries of Newtonian mechanics; but the weight of experimental evidence leaves us no alternative. To those who protest that the lack of a definite position and momentum is contrary to common sense, we reply that what we call common sense is based on familiarity gained through experience, and that our usual experience includes very little contact with the microscopic behavior of particles. Thus we must sometimes be prepared to accept conclusions that seem contrary to intuition when we are dealing with areas far removed from everyday experience.

Some rather jarring conclusions about what we can and cannot know about particle motion

In more general discussions of uncertainty relations, it is customary to describe the uncertainty of a quantity in terms of the statistical concept of *standard deviation*, a measure of the spread or dispersion of a set of numbers around their average value. If a coordinate x has an uncertainty Δx, defined in this way, and if the corresponding momentum component p_x has an uncertainty Δp_x, then the two uncertainties are found to be related in general by the inequality

$$\Delta x \, \Delta p_x \geq \frac{h}{2\pi}. \tag{42-16}$$

Uncertainty is a fundamental fact of nature, not a shortcoming of our experimental technique.

Equation (42–16) is one form of the **Heisenberg uncertainty principle;** it states that, in general, neither the momentum nor the position of a particle can be predicted with arbitrarily great precision, as classical physics would predict. Instead, the two quantities play complementary roles, as described above. One might protest that greater precision could be attained by using more sophisticated particle detectors in various areas of the slit or by other means; but this turns out to be impossible. To detect a particle the detector must *interact* with it, and this interaction unavoidably changes the state of motion of the particle. A more detailed analysis of such hypothetical experiments shows that the uncertainties we have described are fundamental and intrinsic; they cannot be circumvented even in principle by any experimental technique, no matter how sophisticated.

Constructing a wave pulse by superposing sinusoidal waves: another view of uncertainty

Additional insight into the uncertainty principle is provided by the wave packet shown in Fig. 42–7. We can construct such a packet by superposing several sinusoidal waves with various wavelengths. Each individual component wave has no beginning or end but extends indefinitely in both directions. We choose the wavelengths and the amplitude of each component wave so that constructive interference occurs only in a small region of width Δx, as shown in the figure, and the interference is destructive everywhere else. It turns out that a rather broad wave packet can be obtained by superposing waves with a relatively small range of wavelengths; but to make a narrow packet requires a wider range of wavelengths. Yet a wide range of wavelengths also means a correspondingly wide range of values of momentum, because of the de Broglie relation. So again we see that a small uncertainty in position must be accompanied by a large uncertainty in momentum, and conversely.

The energy of a state can also have uncertainty.

In addition, the *energy* of a system, as well as its position and momentum, always has uncertainty. The uncertainty ΔE is found to depend on the time interval Δt during which the system remains in the given state. The relation is

$$\Delta E \, \Delta t \geq \frac{h}{2\pi}. \tag{42-17}$$

Thus a system that remains in a certain state for a long time can have a very well defined energy, but if it remains in that state for only a short time, the uncertainty in energy must be correspondingly greater.

A simple example of energy uncertainty

EXAMPLE 42–3   A sodium atom in one of the "resonance levels" shown in Fig. 41–5 remains in that state for an average time of 1.6×10^{-8} s before making a transition to the ground state by emitting a photon of wavelength 589 nm and energy 2.109 eV. What is the uncertainty in energy of the resonance level?

SOLUTION   From Eq. (42–17),

$$\Delta E = \frac{h}{2\pi \, \Delta t} = \frac{(6.626 \times 10^{-34} \text{ J·s})}{(2\pi)(1.6 \times 10^{-8} \text{ s})}$$
$$= 6.59 \times 10^{-27} \text{ J} = 4.11 \times 10^{-8} \text{ eV}.$$

The atom remains an indefinitely long time in the ground state, so there is *no* uncertainty there; the uncertainty of the resonance level energy and of the corresponding photon energy amounts to about two parts in 10^8. This irreducible uncertainty is called the *natural line width* of this particular spectrum line. Ordinarily the natural line width is much smaller than line broadening from other causes, such as collisions among atoms.

Let us now take a brief look at a quantum interpretation of a *two-slit* interference pattern. We studied these patterns in detail for light in Sections 39–2 and 39–3. In terms of photons, the intensity-distribution pattern must correspond to the numbers of photons striking various regions of the screen where the pattern is formed. In fact, it is possible to detect individual photons with a device called a *photomultiplier*. Using the setup shown in Fig. 42–9, we can place the photomultiplier at various positions for equal time intervals, count photons at each position, and plot out the intensity distribution. We find that *on the average,* the distribution of photons agrees with our predictions from Section 39–3.

If we now reduce the light intensity to a point where only a few photons per second pass through the slits, there is no way to predict where an individual photon will go. Thus the interference pattern has to be viewed as a *statistical distribution* of photons. It tells us how many will go in each direction, or the *probability* for an individual photon to go in each of various directions, but *not* where any particular photon will go.

It is tempting to think that each individual photon must pass through one or the other of the slits. But if this were the case, we would be able to record the interference pattern on film by opening one slit for a time, then closing it and opening the other. As we know, this does not work; to form an interference pattern, the waves from the two slits have to be *coherent*. To resolve this apparent paradox, we are forced to conclude that *every* photon goes partly through *both* slits, and that *each photon interferes only with itself.* If this sounds like nonsense, remember that the photon is *not* a single point in space; it is a conceptual framework to describe the quantization of energy in electromagnetic waves. There is nothing conceptually wrong with saying that every photon passes through both slits, and indeed this is the *only* consistent viewpoint!

Particle and wave aspects of a two-slit interference experiment

Is it possible for a photon to pass through both slits?

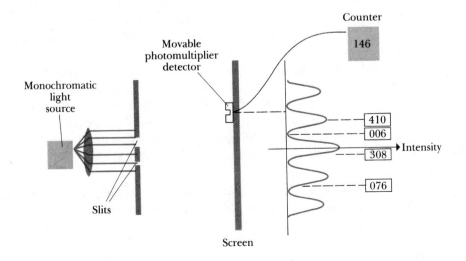

42–9 Two-slit interference pattern observed with photomultipliers. The curve shows the intensity distribution predicted by the wave picture, and the photon distribution is shown by the numbers of photons counted at various positions.

Finally, what happens when we do a two-slit interference experiment with electrons? Exactly the same thing as with photons! Again we can do a particle-counting experiment to trace out the interference pattern, as we did with photons, but we *cannot* predict where in the pattern an individual electron will land. Furthermore, again because of the coherence requirement, we have to assume that every electron goes through both slits! If the idea of a photon passing through both slits made you uncomfortable, you'll really hate this. But since the electron is not an entity described by a single geometric point, there is nothing conceptually wrong with having it pass through both slits. Like photons, each electron interferes only with itself.

42–5 WAVE FUNCTIONS

As we have seen, the dual wave-particle nature of electrons and other fundamental particles requires us to generalize the kinematic language we use to describe the position and motion of a particle. The classical notion of a particle as a point having at each instant a definite position in space (described by three coordinates) and a definite velocity (described by three components) must be replaced by more general language.

In making the needed generalizations, we are guided by the language of classical wave motion. When studying transverse waves on a string in Chapter 21, we described the motion of the string by specifying the position of each point in the string at each instant of time. This was done by means of a **wave function,** introduced in Section 21–3. If y represents the displacement from equilibrium at time t of a point on the string whose equilibrium position is a distance x from the origin, then the function $y = f(x, t)$ or, more briefly, $y(x, t)$ represents the displacement of point x at time t. If we know the wave function for a given motion, we know everything there is to know about the motion; from the function we can determine the shape of the string at any time, the slope at each point, the velocity and acceleration of each point, and any other needed information.

Similarly, in Chapter 23 we discussed a sound wave propagating in the x-direction. Letting p represent the variation in air pressure from its equilibrium value at any point, we write $p(x, t)$ as the pressure variation at any point x at any time t; again this is a *wave function*. If the wave is three-dimensional, we can describe p at a space point with coordinates (x, y, z) at any time t by means of a wave function $p(x, y, z, t)$, which contains all the space coordinates and time. The same pattern reappears in the description of *electromagnetic* waves in Section 35–5, where we use two wave functions to describe the electric and magnetic fields at any point in space, at any time.

Thus it is natural to use a wave function as the central element of our generalized language for describing particles. The symbol usually used for this wave function is Ψ, and it is, in general, a function of all the space coordinates and time. Just as the wave function $y(x, t)$ for mechanical waves on a string provides a complete description of the motion, the wave function $\Psi(x, y, z, t)$ for a particle contains all the information that can be known about the particle.

Two questions immediately arise. First, what is the *meaning* of the wave function Ψ for a particle? Second, how is Ψ determined for any given physical situation?

With reference to the first question, the wave function describes the distribution of the particle in space. It is related to the *probability* of finding the particle in each of various regions; the particle is most likely to be found in regions where Ψ is large, and so on. If the particle has charge, the wave function can be used to find the *charge density* at any point in space. In addition, from Ψ we can calculate the *average* position of the particle; its average velocity; and dynamic quantities such as momentum, energy, and angular momentum. The required techniques are far beyond the scope of this discussion, but they are well established and well supported by experimental results.

The answer to the second question is that the wave function must be one of a set of solutions of a certain differential equation called the **Schrödinger equation,** developed by Erwin Schrödinger in 1925. In principle we can set up a Schrödinger equation for any given physical situation, such as the electron in a hydrogen atom; the functions that are solutions of this equation represent various possible physical states of the system. Furthermore, it turns out for some systems that acceptable solutions exist only when some physical quantity, such as the energy of the system, has certain special values. Thus the solutions of the Schrödinger equation are also associated with *energy levels.* This discovery is of the utmost importance; before the development of the Schrödinger equation, there was no way to predict energy levels from any fundamental theory, except for the very limited success of the Bohr model for hydrogen.

Soon after it was developed, the Schrödinger equation was applied to the problem of the hydrogen atom. The predicted energy levels for the simplest model turned out to be identical to those from the Bohr model, Eq. (42–6), and thus agreed with experimental values from spectrum analysis. The energy levels are labeled with the quantum number n.

In addition, the solutions have *quantized* values of angular momentum; that is, only certain discrete values of the magnitude of angular momentum and its components are possible. Recall that quantization of angular momentum was put into the Bohr model as an *ad hoc* assumption with no fundamental justification; with the Schrödinger equation it appears automatically! Specifically, it is found that the magnitude L of the angular momentum of an electron in the hydrogen atom in a state with energy E_n and quantum number n must be given by

$$L = \sqrt{l(l + 1)}\left(\frac{h}{2\pi}\right), \qquad (42\text{--}18)$$

where l is zero or a positive integer no larger than $n - 1$. The *component* of L in a given direction, say the z-component L_z, can have only the set of values

$$L_z = m\frac{h}{2\pi}, \qquad (42\text{--}19)$$

where m can be zero or a positive or negative integer up to but no larger than l.

The quantity $h/2\pi$ appears so often in quantum mechanics that it is given a special symbol, $\hbar$. That is,

$$\hbar = \frac{h}{2\pi} = 1.054 \times 10^{-34} \text{ J·s}.$$

In terms of $\hbar$, the two preceding equations become

$$L = \sqrt{l(l + 1)}\,\hbar \qquad (l = 0, 1, 2, \ldots, n - 1) \qquad (42\text{--}20)$$

The wave functions must be solutions of a certain differential equation.

The energy levels come out of the Schrödinger equation automatically.

The quantization of angular momentum comes out automatically.

and

$$L_z = m\hbar \qquad (m = 0, \pm 1, \pm 2, \ldots, \pm l). \qquad (42\text{--}21)$$

Note that the component L_z can never be quite as large as L. For example, when $l = 4$ and $m = 4$, we find

$$L = \sqrt{4(4 + 1)}\,\hbar = 4.47\,\hbar,$$
$$L_z = 4\,\hbar.$$

This inequality arises from the uncertainty principle, which makes it impossible to predict the *direction* of the angular momentum vector with complete certainty. Thus the component of L in a given direction can never be quite as large as the magnitude L, except when $l = 0$ and both L and L_z are zero. Unlike the Bohr model, the Schrödinger equation gives values for the magnitude L of angular momentum that are *not* integer multiples of $\hbar$.

Another interesting feature of Eqs. (42–20) and (42–21) is that there are states for which the angular momentum is *zero*. This is a result that has no classical analog; in the Bohr model the electron always moved in an orbit and thus had nonzero angular momentum. But in the new mechanics we find states having zero angular momentum.

The possible wave functions for the hydrogen atom can be labeled according to the values of the three integers n, l, and m, called, respectively, the **principal quantum number,** the **angular momentum quantum number,** and the **magnetic quantum number.** For each energy level E_n, there are several distinct states having the same energy but different values of l and m, the only exception being the ground state $n = 1$, for which only $l = 0$, $m = 0$ is possible. Examination of the spatial extent of the wave functions shows that, in each case, the wave function is confined primarily to a region of space around the nucleus having a radius of the same order of magnitude as the corresponding Bohr radius. For increasing values of n, the electron is, on the average, farther away from the nucleus and hence has less negative potential energy and becomes less tightly bound.

The new mechanics described above, which we call quantum mechanics, is much more complex, both conceptually and mathematically, than Newtonian mechanics. However, quantum mechanics enables us to understand physical phenomena and to analyze physical problems for which classical mechanics is completely powerless. Added complexity is the price we pay for this greatly expanded understanding.

42–6 THE ZEEMAN EFFECT

In Section 41–1 we mentioned Faraday's suggestion that shifts of the spectrum wavelengths in a line spectrum might occur when the source is placed in a magnetic field. Faraday's spectroscopic techniques in 1862 were not refined enough to observe such shifts, but the Dutch physicist Pieter Zeeman, using improved instruments, was able in 1896 to detect small shifts, an effect now called the *Zeeman effect*.

We first analyze the interaction of atoms with a magnetic field using the Bohr model. This interaction can be represented by use of the concept of *magnetic moment*, which we studied in Section 30–8. To avoid confusion in notation, we denote magnetic moment by $\boldsymbol{\mu}$, the electron mass by m_e, and the magnetic quantum number by m. The interaction energy U of a magnetic moment $\boldsymbol{\mu}$ in a magnetic field $\boldsymbol{B}$ was derived in Section 30–8, Eq. (30–23). It is given by

$$U = -\boldsymbol{\mu} \cdot \boldsymbol{B}. \qquad (42\text{–}22)$$

Let us consider the electron in the first Bohr orbit ($n = 1$). The orbiting charge is equivalent to a current loop of radius r and area πr^2. The average charge per unit time passing a point of the orbit is the average current I, and this is given by e/τ, where τ is the time for one revolution: $\tau = 2\pi r/v$. Thus $I = ev/2\pi r$. The magnetic moment μ is given by

$$\mu = IA = \left(\frac{ev}{2\pi r}\right)\pi r^2 = \frac{evr}{2}. \qquad (42\text{–}23)$$

But according to the Bohr theory, the angular momentum $m_e vr$ in the $n = 1$ state is equal to $\hbar$, so

$$vr = \frac{\hbar}{m_e}$$

and

$$\mu = \frac{e}{2m_e}\hbar = \frac{1}{2}(1.758 \times 10^{11}\ \text{C·kg}^{-1})(1.054 \times 10^{-34}\ \text{J·s})$$
$$= 9.27 \times 10^{-24}\ \text{A·m}^2.$$

Zeeman effect in solar spectrum lines. The lines emitted by gaseous iron near a sunspot are broadened and split into doublets and triplets by the very strong magnetic field created by the sunspot. (Courtesy of Kitt Peak Observatory.)

EXAMPLE 42–4 Find the interaction potential energy when the hydrogen atom described above is placed in a magnetic field of 2 T.

SOLUTION According to Eq. (42–22), the interaction energy U when $\boldsymbol{B}$ and $\boldsymbol{\mu}$ are parallel is

$$U = -\mu B = -(9.27 \times 10^{-24}\ \text{A·m}^2)(2\ \text{T})$$
$$= -1.85 \times 10^{-23}\ \text{J}$$
$$= -1.16 \times 10^{-4}\ \text{eV}.$$

An example showing the order of magnitude of the interaction energy

When $\boldsymbol{\mu}$ and $\boldsymbol{B}$ are antiparallel, the energy is $+1.16 \times 10^{-4}$ eV. We note that these energies are *smaller* than the energy levels of the atom, by a factor of the order of 10^{-4}.

In the $n = 1$ state of the Bohr atom, the magnetic moment μ is equal to $e/2m_e$ times the angular momentum L. The ratio of magnetic moment to angular momentum is called the **gyromagnetic ratio,** and it can be shown that it has the value $e/2m_e$ for *every* Bohr orbit. It can also be shown that electrons described by the Schrödinger equation have this same ratio of μ to L.

We can generalize this discussion to states described by Schrödinger wave functions. Suppose the magnetic field $\boldsymbol{B}$ is directed along the $+z$-axis; then the interaction energy of the atom's magnetic moment with the field is given by

$$U = -\boldsymbol{\mu} \cdot \boldsymbol{B} = -\mu_z B. \qquad (42\text{–}24)$$

The ratio of magnetic moment to angular momentum is the same in the Schrödinger equation as for the Bohr model.

From the discussion above,

$$\mu_z = \frac{e}{2m_e}L_z. \qquad (42\text{–}25)$$

But $L_z = m\hbar$, with $m = 0, \pm 1, \pm 2, \ldots, \pm l$, so

$$\mu_z = \frac{e}{2m_e}L_z = m\frac{e\hbar}{2m_e}. \tag{42-26}$$

Finally, the interaction energy of Eq. (42-24) becomes

$$U = -\mu_z B = -m\frac{e\hbar B}{2m_e} \quad (m = 0, \pm 1, \pm 2, \ldots, \pm l). \tag{42-27}$$

The effect of the magnetic field is to shift the energy level by an amount U, given by Eq. (42-27), which depends on the orientation of the angular momentum, that is, on the value of the magnetic quantum number m. An energy level having a certain value of the quantum number l is thus split into $2l + 1$ sublevels, with adjacent sublevels differing in energy by $e\hbar B/2m_e$. Spectrum lines corresponding to transitions involving this energy level are correspondingly split and appear as a series of closely spaced spectrum lines replacing a single line. This effect, called the **Zeeman effect,** is a very direct experimental confirmation of the quantization of angular momentum.

EXAMPLE 42-5 An atom in a state having $l = 1$ emits a photon with wavelength 600 nm as it decays to its ground state (with $l = 0$). If the atom is placed in a magnetic field of magnitude $B = 1.0$ T, determine the shifts in the energy levels and in the wavelengths.

SOLUTION The energy of a 600-nm photon, as obtained in Section 41-4, is 3.31×10^{-19} J or 2.07 eV. The ground-state level has $l = 0$ and is not split by the field. The splitting of levels in the $l = 1$ state is given by Eq. (42-27):

$$U = -m\frac{e\hbar B}{2m_e}$$

$$= -m\frac{(1.60 \times 10^{-19}\,\text{C})(1.054 \times 10^{-34}\,\text{J·s})(1.00\,\text{T})}{2(9.11 \times 10^{-31}\,\text{kg})}$$

$$= -m(9.26 \times 10^{-24}\,\text{J}) = -m(5.79 \times 10^{-5}\,\text{eV}).$$

When $l = 1$, the possible values of m are -1, 0, and $+1$, and the three resulting levels are split by equal intervals of 5.79×10^{-5} eV. This is a small fraction of the photon energy: $(5.79 \times 10^{-5}\,\text{eV})/(2.07\,\text{eV}) = 2.80 \times 10^{-5}$. Thus we expect the corresponding *wavelength* shifts to be approximately $(2.80 \times 10^{-5})(600\,\text{nm}) = 0.017$ nm. The original 600-nm line is split into a triplet with wavelengths 599.983, 600.000, and 600.017 nm. This splitting is well within the limit of resolution of modern spectrometers.

42-7 ELECTRON SPIN

For the analysis of more complex atoms, and even for some details of the spectrum of hydrogen, we need one additional concept, **electron spin.** To illustrate this concept, we consider the motion of the earth around the sun. The earth travels in a nearly circular orbit and at the same time *rotates* on its axis. Each motion has its associated angular momentum, called the *orbital* and *spin* angular momentum, respectively, and the total angular momentum of the system is the sum of the two. If we were to model the earth as a single point, no spin angular momentum would be possible; but a more refined model, with an earth of finite size, includes the possibility of spin angular momentum.

This discussion can be translated to the language of the Bohr model. Suppose the electron is not a point charge moving in an orbit, but a small spinning sphere in orbit. Then the electron has not only orbital angular momentum but also additional angular momentum associated with the spin motion. Because the sphere carries an electric charge, the spinning motion leads to current loops and to a magnetic moment, as we discussed in Section 30–8. If this magnetic moment really exists, then it interacts with a magnetic field and there is an associated interaction energy. These effects are in addition to the interaction of the orbital magnetic moment with a magnetic field, which we discussed in Section 42–6. Thus there should be additional small shifts in the energy levels of the atom and in the wavelengths of the associated spectrum lines.

Such shifts *are* indeed observed in precise spectroscopic analysis; this and other experimental evidence have shown conclusively that the electron *does* have angular momentum and a magnetic moment that are not related to the orbital motion but are intrinsic to the particle itself. Like orbital angular momentum, spin angular momentum (usually denoted by S) is found to be *quantized*. Denoting the z-component of S by S_z, we find that the only possible values are

$$S_z = \pm \frac{1}{2}\hbar. \tag{42–28}$$

Spin angular momentum is quantized.

This relation is reminiscent of Eq. (42–21) for the z-component of orbital angular momentum, but the component is one-half of $\hbar$ instead of an *integral* multiple.

In quantum mechanics, where the Bohr orbits are superseded by wave functions, it is not really possible to *picture* electron spin. If the wave functions are visualized as clouds surrounding the nucleus, then we can imagine many tiny arrows distributed throughout the cloud, all pointing in the same direction, either all $+z$ or all $-z$. Of course, this picture should not be taken too seriously; there is no hope of actually seeing an atom's structure because it is thousands of times smaller than wavelengths of light and because interactions with light photons would seriously disturb the very structure we are trying to observe.

In any event, the concept of electron spin is well established by a variety of experimental evidence. To label completely the state of the electron in a hydrogen atom, we now need a fourth quantum number s to specify the electron spin orientation. If s can take the values $+1$ or -1, then the z-component of spin angular momentum is given by

$$S_z = \tfrac{1}{2}s\hbar \qquad (s = \pm 1). \tag{42–29}$$

The corresponding component of magnetic moment, which we again denote by μ_z, turns out to be related to S_z by

The gyromagnetic ratio for electron spin is twice as large as for orbital angular momentum.

$$\mu_z = \frac{e}{m_e}S_z, \tag{42–30}$$

where e and m_e are again the charge and mass of the electron, respectively. Note that for electron spin the gyromagnetic ratio is just *twice* the value for *orbital* angular momentum and magnetic moment. When the atom is placed in a magnetic field, the interaction of the electron spin magnetic moment with the field causes further splittings in energy levels and in the corresponding spectrum lines.

Somewhat more subtle is the level splitting caused by the electron spin magnetic moment even when there is *no* external field. In the Bohr model, an observer moving with the electron sees the positively charged nucleus moving around him; this moving charge causes a magnetic field at the location of the electron, and the resulting interaction energy with the spin magnetic moment causes a twofold splitting of this level, corresponding to the two possible orientations of electron spin.

Although the Bohr model is now known to be inadequate, a similar result can be derived from a more complete quantum-mechanical treatment based on the Schrödinger equation. The effect is called **spin-orbit coupling;** it is responsible for the small energy difference between the two closely spaced "resonance levels" of sodium shown in Fig. 41–5 and for the corresponding familiar doublet (589.0, 589.6 nm) in the spectrum of sodium.

The various line splittings resulting from these magnetic interactions are collectively called *fine structure*. There are additional, much smaller splittings resulting from the fact that the *nucleus* of the atom also has a magnetic moment and interacts with the orbital and spin angular momentum of the electrons. These effects are called *hyperfine structure*.

Magnetic interactions without an external field

Interactions with nuclear magnetic moments: additional, very small shifts in energy levels

SUMMARY

KEY TERMS
Bohr model
stable orbits
quantum number
ground state
excited state
de Broglie wavelength
Davisson-Germer experiment
electron diffraction
quantum mechanics
electron microscope
Heisenberg uncertainty principle
wave function
Schrödinger equation
principal quantum number
angular momentum quantum number
magnetic quantum number
gyromagnetic ratio
Zeeman effect
electron spin
spin-orbit coupling

In the Bohr model of the hydrogen atom, the electron travels in a circular orbit around the nucleus, with the centripetal force supplied by the $1/r^2$ electrical attraction. Bohr further assumed that the angular momentum must be an integral multiple of $h/2\pi$. The resulting orbit radii r_n are given by

$$r_n = \epsilon_0 \frac{n^2 h^2}{\pi m e^2}, \qquad (42\text{--}3)$$

and the energies are

$$E_n = K_n + U_n = -\frac{1}{\epsilon_0^2} \frac{m e^4}{8 n^2 h^2}. \qquad (42\text{--}6)$$

These equations agree with deductions from the Balmer formula and the quantum nature of light.

Electrons show wavelike properties in some situations; the first experimental observation of electron diffraction was the Davisson-Germer experiment. The wavelength λ of a particle with mass m and speed v is given by

$$\lambda = \frac{h}{mv}. \qquad (42\text{--}9)$$

The electron microscope uses an electron beam to form an enlarged image of an object. The wave nature of electrons is not an essential requirement for image formation, but the resolution is limited by diffraction effects. Electrons can have much shorter wavelengths than visible light, so the electron microscope has much greater resolution.

Analysis of a single-slit diffraction experiment with electrons shows that there are always uncertainties in position and momentum, related by the Heisenberg uncertainty principle:

$$\Delta x \, \Delta p_x \geq \frac{h}{2\pi}. \tag{42-16}$$

There is a corresponding uncertainty relation for energy: If a system exists in a certain energy state for a time Δt, the uncertainty ΔE in its energy is given by

$$\Delta E \, \Delta t \geq \frac{h}{2\pi}. \tag{42-17}$$

Analysis of interference and diffraction experiments for both photons and electrons leads us to a statistical interpretation of the result. The motion of an individual electron or photon cannot be predicted.

The new kinematic language needed to describe the position of a particle includes a wave function, which must be a solution of the Schrödinger equation. Solution of this equation for the hydrogen atom yields the same energy levels as the Bohr model and in addition shows that angular momentum and its component in a given direction are quantized, according to

$$L = \sqrt{l(l+1)}\hbar \qquad (l = 0, 1, 2, \ldots, n-1), \tag{42-20}$$

and

$$L_z = m\hbar \qquad (m = 0, \pm 1, \pm 2, \ldots, \pm l). \tag{42-21}$$

Three quantum numbers are used to identify the spatial states of the hydrogen atom; they are the principal, angular momentum, and magnetic quantum numbers, n, l, and m.

The Zeeman effect is the shifting of energy levels caused by interactions of an atom's magnetic moments with a magnetic field. The interaction energy associated with the orbital magnetic moment is

$$U = -\mu_z B = -m\frac{e\hbar B}{2m_e} \qquad (m = 0, \pm 1, \pm 2, \ldots, \pm l). \tag{42-27}$$

The electron also has an inherent or "spin" angular momentum S that can have a component $S_z = \pm\frac{1}{2}\hbar$ in a given direction; the associated magnetic moment is given by

$$\mu_z = \frac{e}{m_e}S_z. \tag{42-30}$$

The gyromagnetic ratio is the ratio of magnetic momentum to angular momentum. For orbital angular momentum it is $e/2m_e$; for spin angular momentum it is e/m_e.

QUESTIONS

42-1 In analyzing the absorption spectrum of hydrogen at room temperature, one finds absorption lines corresponding to wavelengths in the Lyman series but not to those in the Balmer series. Why not?

42-2 A singly ionized helium atom has one of its two electrons removed, and the energy levels of the remaining electron are closely related to those of the hydrogen atom. The nuclear charge for helium is $+2e$ instead of just $+e$;

exactly how are the energy levels related to those of hydrogen? How is the size of the ion in the ground state related to that of the hydrogen atom?

42-3 Consider the line spectrum emitted from a gas discharge tube such as a neon sign or a sodium-vapor or mercury-vapor lamp. It is found that when the pressure of the vapor is increased, the spectrum lines spread out, that is, are less sharp and less monochromatic. Why?

42–4 Suppose a two-slit interference experiment is carried out by using an electron beam. Would the same interference pattern result if one slit at a time is uncovered instead of both at once? If not, why not? Does not each electron go through one slit or the other? Or does every electron go through both slits? Does the latter possibility make sense?

42–5 Is the wave nature of electrons significant in the function of a television picture tube? For example, do diffraction effects limit the sharpness of the picture?

42–6 A proton and an electron have the same speed. Which has longer wavelength?

42–7 A proton and an electron have the same kinetic energy. Which has longer wavelength?

42–8 Does the uncertainty principle have anything to do with marksmanship? That is, is the accuracy with which a bullet can be aimed at a target limited by the uncertainty principle?

42–9 Is the Bohr model of the hydrogen atom consistent with the uncertainty principle?

42–10 If the energy of a system can have uncertainty, as stated by Eq. (42–17), does this mean that the principle of conservation of energy is no longer valid?

42–11 If quantum mechanics replaces the language of Newtonian mechanics, why do we not have to use wave functions to describe the motion of macroscopic objects such as baseballs and cars?

42–12 Why is analysis of the helium atom much more complex than that of the hydrogen atom, either in a Bohr type of model or using the Schrödinger equation?

42–13 Do gravitational forces play a significant role in atomic structure?

EXERCISES

$$e = 1.602 \times 10^{-19} \text{ C}$$
$$m = 9.110 \times 10^{-31} \text{ kg}$$
$$h = 6.626 \times 10^{-34} \text{ J·s}$$
$$N_A = 6.022 \times 10^{23} \text{ atoms·mol}^{-1}$$

Energy equivalent of 1 u = 931.5 MeV

$$\frac{e}{m} = 1.758 \times 10^{11} \text{ C·kg}^{-1}$$
$$k = 1.381 \times 10^{-23} \text{ J·K}^{-1}$$
$$1 \text{ eV} = 1.602 \times 10^{-19} \text{ J}$$
$$1 \text{ u} = 1.661 \times 10^{-27} \text{ kg}$$
$$\epsilon_0 = 8.854 \times 10^{-12} \text{ C}^2\text{·N}^{-1}\text{·m}^{-2}$$

Section 42–1 The Bohr Atom

42–1

a) Calculate the Bohr-model speed of the electron in a hydrogen atom in the $n = 1, 2$, and 3 states.

b) Calculate the orbital period in each of these states.

c) The average lifetime of the first excited state of a hydrogen atom is 10^{-8} s. How many orbits does an electron in an excited atom complete before returning to the ground state?

42–2 According to the Bohr model, the Rydberg constant R is equal to $me^4/8\epsilon_0^2h^3c$.

a) Calculate R in m^{-1} and compare with the experimental value.

b) Calculate the energy (in electronvolts) of a photon whose wavelength equals R^{-1}. (This quantity is known as the Rydberg energy.)

42–3 For a hydrogen atom in the ground state, determine in electronvolts:

a) the kinetic energy of the electron;

b) its potential energy;

c) its total energy;

d) the energy required to remove the electron completely.

e) What wavelength would a photon with the energy calculated in (d) have? In what region of the electromagnetic spectrum does it lie?

42–4 A hydrogen atom initially in the ground state absorbs a photon, which excites it to the $n = 4$ state. Determine the wavelength and frequency of the photon.

42–5 A singly ionized helium ion (a helium atom with one electron removed) behaves very much like a hydrogen atom, except that the nuclear charge is twice as great.

a) How do the energy levels differ in magnitude from those of the hydrogen atom?

b) Which spectral series for He^+ have lines in the visible spectrum? (Refer to Exercise 41–12.)

c) For a given value of n how does the radius of an orbit in He^+ relate to that for H?

Section 42–2 Wave Nature of Particles

Section 42–3 The Electron Microscope

42–6

a) An electron moves with a speed of 3×10^6 m·s^{-1}. What is its de Broglie wavelength?

b) A proton moves with the same speed. Determine its de Broglie wavelength.

42–7 Approximately what range of photon energies (in electronvolts) corresponds to the visible spectrum? Approximately what range of wavelengths would electrons in this energy range have?

42–8 For crystal diffraction experiments, wavelengths of the order of 0.1 nm are often appropriate. Find the energy, in electronvolts, for a particle with this wavelength if the particle is

a) a photon;   b) an electron.

42–9

a) What is the de Broglie wavelength of an electron accelerated through 500 V?

b) What is the de Broglie wavelength of a proton accelerated through the same potential difference?

42–10

a) What is the de Broglie wavelength of an electron that has been accelerated through a potential difference of 200 V?

b) Would this electron exhibit particlelike or wavelike characteristics on meeting an obstacle or opening 1 mm in diameter?

Section 42–4 Probability and Uncertainty

42–11

a) Suppose that the uncertainty in position of a particle is on the order of its de Broglie wavelength. Show that in this case the uncertainty in its momentum is on the order of its momentum.

b) Suppose the uncertainty in position of an electron is equal to the radius of the $n = 1$ Bohr orbit, about 0.5×10^{-10} m. Estimate the uncertainty in its momentum, and compare this with the magnitude of the momentum of the electron in the $n = 1$ Bohr orbit.

42–12 A certain atom has an energy level 2.0 eV above the ground state. When excited to this state, it remains on the average 2.0×10^{-6} s before emitting a photon and returning to the ground state.

a) What is the energy of the photon? Its wavelength?

b) What is the smallest possible uncertainty in energy of the photon?

c) Show that $|\Delta E/E| = |\Delta \lambda/\lambda|$ when $|\Delta \lambda/\lambda|$ is small. Use this to calculate the magnitude of the smallest possible uncertainty in the wavelength of the photon.

42–13 Suppose an unstable particle produced in a high-energy collision has a mass three times that of the proton and an uncertainty in mass that is 1% of the particle's mass. Assuming mass and energy are related by $E = mc^2$, estimate the lifetime of the particle.

Section 42–5 Wave Functions

42–14 Make a chart showing all the possible sets of quantum numbers l and m for the states of the electron in the hydrogen atom when $n = 3$. How many combinations are there?

42–15 Consider states with $l = 3$.

a) In units of $\hbar$, what is the largest possible value of L_z?

b) In units of $\hbar$, what is the value of L? Which is larger, L or the maximum possible L_z?

c) Assume a model where L is described as a classical vector. For each allowed value of L_z, what angle does the vector L make with the $+z$-axis?

Section 42–6 The Zeeman Effect

42–16 Consider an atom in an $l = 2$ state. In the absence of an external magnetic field, the states with different m have (approximately) the same energy. (One says that the states are *degenerate*.)

a) If the effect of electron spin can be ignored (which is not actually the case), calculate the splitting in eV of the m-levels when the atom is put in a 0.6-T magnetic field.

b) Which m-level will have the lowest energy?

c) Draw an energy-level diagram that shows the $l = 2$ levels with and without the external magnetic field.

Section 42–7 Electron Spin

42–17 If you treat an electron as a classical spherical particle with a radius of 1×10^{-17} m, what is the angular velocity necessary to produce a spin angular momentum of $\hbar$?

42–18 A hydrogen atom in the $n = 1$ state is placed in a magnetic field of magnitude 1.2 T. Find the interaction energy (in electronvolts) of the atom with the field due to the electron spin.

PROBLEMS

42–19 The negative μ-meson (or muon) has a charge equal to that of an electron but a mass about 207 times as great. Consider a hydrogenlike atom consisting of a proton and a muon. (For simplicity, assume that the muon orbits around the proton, which is stationary. In reality both revolve about the center of mass of the system.)

a) What is the ground-state energy (in electronvolts)?

b) What is the radius of the $n = 1$ Bohr orbit?

c) What is the wavelength of the radiation emitted in the transition from the $n = 2$ state to the $n = 1$ state?

42–20 Consider a hydrogenlike atom of nuclear charge Z. (Refer to Exercise 42–5.)

a) For what value of Z (rounded to the nearest integer value) is the Bohr speed of the electron in the ground state equal to 5% of the speed of light?

b) For what value of Z is the ionization energy of the ground state equal to 1% of the rest mass energy of the electron?

42–21 Refer to Example 42–4. Suppose a hydrogen atom makes a transition from the $n = 3$ state to the $n = 2$ state (the Balmer H_α line at 656.3 nm) while in a magnetic field of magnitude 2 T. If the magnetic moment of the atom is parallel to the field in both the initial and final states,

a) by how much is each energy level shifted from the zero-field value?

b) by how much is the wavelength of the spectrum line shifted?

42-22 A 10-kg satellite circles the earth once every 2 hr in an orbit having a radius of 8060 km.

a) Assuming that Bohr's angular-momentum postulate applies to satellites just as it does to an electron in the hydrogen atom, find the quantum number of the orbit of the satellite.

b) Show from Bohr's first postulate and Newton's law of gravitation that the radius of an earth-satellite orbit is directly proportional to the square of the quantum number, $r = kn^2$, where k is the constant of proportionality.

c) Using the result from part (b), find the distance between the orbit of the satellite in this problem and its next "allowed" orbit. (Calculate a numerical value.)

d) Comment on the possibility of observing the separation of the two adjacent orbits.

e) Do quantized and classical orbits correspond for this satellite? Which is the "correct" method for calculating the orbits?

42-23 The radii of atomic nuclei are of the order of 10^{-15} m.

a) Estimate the minimum uncertainty in the momentum of an electron if it is confined within a nucleus.

b) Take this uncertainty in momentum to be an estimate of the magnitude of the momentum. (Refer to Exercise 42-11.) Use the relativistic expression of Eq. (40-34) to obtain an estimate of the energy of an electron confined within a nucleus.

c) Compare the energy calculated in (b) to the Coulomb potential energy of a proton and an electron separated by 1×10^{-15} m. On the basis of your result, could there be electrons within the nucleus?

42-24 The average kinetic energy of a thermal neutron is $\frac{3}{2}kT$. What is the de Broglie wavelength associated with the neutrons in thermal equilibrium with matter at 300 K? (The mass of a neutron is approximately 1 u.)

42-25 In a certain television picture tube, the accelerating voltage is 20,000 V and the electron beam passes through an aperture 0.5 mm in diameter to a screen 0.3 m away.

a) What is the uncertainty in position of the point where the electrons strike the screen?

b) Does this uncertainty affect the clarity of the picture significantly? (Use nonrelativistic expressions for the motion of the electrons. This is fairly accurate and is certainly adequate for obtaining an estimate of uncertainty-principle effects.)

42-26 The π^0 meson is an unstable particle produced in high-energy particle collisions. Its mass is about 264 times that of the electron, and it exists for an average lifetime of 0.8×10^{-16} s before decaying into two gamma-ray photons. Assuming that the mass and energy of the particle are related by the Einstein relation $E = mc^2$, find the uncertainty in the mass of the particle and express it as a fraction of the mass.

42-27 Show that the total number of hydrogen-atom states (including different spin states) for a given value of the principal quantum number n is $2n^2$. (*Hint:* The sum of the first N numbers, $1 + 2 + 3 + \cdots + N$ is given by $N[N + 1]/2$.)

42-28 When a photon is emitted by an atom, the atom must recoil to conserve momentum. This means that the photon and the recoiling atom share the transition energy. For a hydrogen atom, calculate the correction due to recoil to the wavelength of the photon emitted when an electron in the $n = 5$ state returns to the ground state.

CHALLENGE PROBLEMS

42-29 You are entered in a contest to drop a marble of mass 40 g from the roof of a building onto a small target 30 m below. From uncertainty considerations, what is the typical distance by which you will miss the target, given that you aim with the highest possible precision? (*Hint:* The uncertainty Δx_f in the x-coordinate of the marble when it reaches the ground comes in part from the uncertainty Δx_i in the x-coordinate initially and in part from the initial uncertainty in v_x. The latter gives rise to an uncertainty in the horizontal motion of the marble as it falls. Δx_i and Δv_x are related by the uncertainty principle. A small Δx_i gives rise to a large Δv_x, and vice versa. Find the Δx_i that gives the smallest total uncertainty in x at the ground.)

42-30 The wave nature of particles results in the quantum mechanical situation that a particle confined in a box can assume only wavelengths that result in standing waves in the box, with nodes at the box walls.

a) Show that an electron confined in a one-dimensional box of length L will have energy levels given by

$$E_n = \frac{n^2 h^2}{8mL^2}.$$

(*Hint:* Recall that the relation between the de Broglie wavelength and the speed of a particle is $mv = h/\lambda$. The energy of the particle is $\frac{1}{2}mv^2$.)

b) If a hydrogen atom were modeled as a one-dimensional box with length equal to the Bohr radius, what would be the energy (in eV) of the ground state of the electron?

42–31 Consider a particle of mass m moving in a potential $U = \frac{1}{2}kx^2$, as in a mass-spring system. The total energy of the particle is $E = p^2/2m + \frac{1}{2}kx^2$. Assume that p and x are approximately related by the Heisenberg uncertainty principle, $px \sim \hbar$.

a) Calculate the minimum possible value of the energy E, and the value of x that gives this minimum E. This lowest possible energy, which is not zero, is called the *zero-point* energy.

b) For the x calculated in (a), what is the ratio of the kinetic to the potential energy of the particle?

42–32

a) Show that the frequency of revolution of an electron in its circular orbit in the Bohr model of the hydrogen atom is $f = me^4/4\epsilon_0{}^2 n^3 h^3$.

b) Show that when n is very large, the frequency of revolution equals the radiated frequency calculated from Eq. (42–7) for a transition from

$$n_1 = n + 1 \quad \text{to} \quad n_2 = n.$$

(This problem illustrates Bohr's *correspondence principle*, which is often used as a check on quantum calculations. When n is small, quantum physics gives results that are very different from those of classical physics. When n is large, the differences are not significant, and the two methods then "correspond." In fact, when Bohr first tackled the hydrogen atom problem, he sought to determine f as a function of n such that it would correspond to classical physics for large n.)

43

ATOMS, MOLECULES, AND SOLIDS

THE PRINCIPLES OF QUANTUM MECHANICS THAT WE STUDIED IN CHAPTER 42 provide the basis for understanding a wide variety of phenomena associated with the structure of atoms, molecules, and solid materials. One additional principle is needed, the *exclusion principle,* which states that two electrons may not occupy the same quantum-mechanical state. With this principle, the most important features of the structure and chemical behavior of multielectron atoms can be derived, including the periodic table of the elements.

The nature of *molecular bonds,* by which two or more atoms combine in a stable structure, can be understood on the basis of the electron configurations of the atoms. We will see that transitions among vibrational and rotational energy states of molecules give rise to *molecular spectra.* Interatomic forces are also responsible for the large-scale binding of atoms into solid structures. Several different types of interatomic interactions are possible, and many properties of solid materials can be understood at least qualitatively on the basis of the types of bonding that are present. *Semiconductors,* a particular class of solid materials, are discussed in some detail because of their inherent interest and their great practical importance in present-day technology.

43–1 THE EXCLUSION PRINCIPLE

How do we deal with the interactions among electrons in a multielectron atom?

The hydrogen atom is the simplest of all atoms, containing one electron and one proton. Analysis of atoms with more than one electron increases in complexity very rapidly; each electron interacts not only with the positively charged nucleus but also with all the other electrons. The motion of the electrons is governed by the Schrödinger equation, mentioned in Section 42–5, but the mathematical problem of finding appropriate solutions of this equation is so complex that it has not been accomplished exactly even for the helium atom, with two electrons.

Various approximation schemes can be used to apply the Schrödinger equation to many-electron atoms. The simplest and most drastic scheme is simply to ignore the interactions between electrons and to regard each electron as influenced only by the electric field of the nucleus, considered to be a point charge. A less drastic and more useful approximation is to think of all the electrons together as making up a charge cloud that is, on the average, *spherically symmetric*. We can then think of each individual electron as moving in the total electric field due to the nucleus and this averaged-out electron cloud. This model is called the **central-field approximation;** it provides a useful starting point for the understanding of atomic structure.

An additional principle is also needed, the *exclusion principle*. To understand the need for this principle, we consider the lowest-energy state or *ground state* of a many-electron atom. The central-field model suggests that each electron has a lowest-energy state (roughly corresponding to the $n = 1$ state for the hydrogen atom). We might expect that, in the ground state of a complex atom, all the electrons should be in this lowest state. If this is the case, then when we examine the behavior of atoms with increasing numbers of electrons, we should find gradual changes in physical and chemical properties of elements as the number of electrons in the atoms increases.

A variety of evidence shows conclusively that this is *not* what happens at all. For example, the elements fluorine, neon, and sodium have, respectively, 9, 10, and 11 electrons per atom. Fluorine is a *halogen* and tends strongly to form compounds in which each fluorine atom acquires an extra electron. Sodium, an *alkali metal*, forms compounds in which it *loses* an electron; and neon is an *inert gas*, forming no compounds at all. This and many other observations show that in the ground state of a complex atom the electrons *cannot* all be in the lowest-energy state.

The key to this puzzle, discovered by the Swiss physicist Wolfgang Pauli in 1925, is called the **Pauli exclusion principle,** or the *exclusion principle*. Briefly, it states that *no two electrons can occupy the same quantum-mechanical state*. Since different states correspond to different spatial distributions, including different distances from the nucleus, this means that in a complex atom there is not enough room for all the electrons in the states nearest the nucleus; some are forced into states farther away, having higher energies.

To apply the Pauli principle to atomic structure, we first review some results quoted in Sections 42–5 and 42–7. The quantum-mechanical state of the electron in the hydrogen atom is identified by the four quantum numbers n, l, m, and s, which determine the energy, angular momentum, and components of orbital and spin angular momentum in a particular direction. It turns out that we can still use this scheme when the electron moves not in the electric field of a point charge, as in the hydrogen atom, but in the electric field of *any spherically symmetric charge distribution*, as in the central-field approximation. One important difference is that the energy of a state is no longer given by Eq. (42–6); in general the energy corresponding to a given set of quantum numbers depends on *both* n and l, usually increasing with increasing l for a given value of n.

We can now make a list of all the possible sets of quantum numbers and thus of the possible states of electrons in an atom. Such a list is given in Table 43–1, which also indicates two alternative notations. It is customary to designate the value of l by a letter, according to this scheme:

A brief look at electron arrangement in multielectron atoms

A zoning ordinance for electrons in atomic systems: Let's not all try to crowd into the same space.

Describing one-electron states by use of four quantum numbers

$$l = 0: \quad s \text{ state}$$
$$l = 1: \quad p \text{ state}$$
$$l = 2: \quad d \text{ state}$$
$$l = 3: \quad f \text{ state}$$
$$l = 4: \quad g \text{ state}$$

TABLE 43–1

n	l	m	Spectroscopic Notation	Maximum Number of Electrons	Shell
1	0	0	$1s$	2	K
2	0	0	$2s$	2 } 8	L
2	1	-1			
2	1	0	$2p$	6	
2	1	1			
3	0	0	$3s$	2	
3	1	-1			
3	1	0	$3p$	6	
3	1	1			
3	2	-2			} 18 M
3	2	-1			
3	2	0	$3d$	10	
3	2	1			
3	2	2			
4	0	0	$4s$	2	
4	1	-1			
4	1	0	$4p$	6 } 32	N
4	1	1			
	etc.				

The electrons arrange themselves in concentric shells around the nucleus.

The origins of these letters are rooted in the early days of spectroscopy and need not concern us. A state for which $n = 2$ and $l = 1$ is called a $2p$ state, and so on, as shown in Table 43–1. This table also shows the relation between values of n and the x-ray levels ($K, L, M, \ldots$) described in Section 41–8. The $n = 1$ levels are designated as K; $n = 2$ as L; and so on. Because the average electron distance from the nucleus increases with n, each value of n corresponds roughly to a region of space around the nucleus in the form of a spherical **shell.** Hence we speak of the L shell as the region occupied by the electrons in the $n = 2$ states, and so on. States with the same n but different l are said to form *subshells,* such as the $3p$ subshell.

The exclusion principle can be stated in terms of quantum numbers.

We are now ready for a more precise statement of the exclusion principle: *In any atom, only one electron can occupy any given quantum state.* That is, no two electrons in an atom can have the same values of all four quantum numbers. Since each quantum state corresponds to a certain distribution of the electron "cloud" in space, the principle says, in effect: "Not more than two electrons (with opposite values of the spin quantum number s) can occupy the same region of space." We should not take this statement too seriously, since the clouds and associated wave functions describing electron distributions do not have definite sharp boundaries; but the exclusion principle limits the degree of overlap of electron wave functions that is permitted. The maximum number of electrons in each shell and subshell is shown in Table 43–1.

The exclusion principle plays an essential role in understanding the structure of complex atoms. In the next section we will see how the periodic table of the elements can be understood on the basis of this principle.

43–2 ATOMIC STRUCTURE

The number of electrons in an atom in its normal state is called the **atomic number,** denoted by Z. The nucleus contains Z protons and some number of neutrons. The proton and electron charges have the same magnitude but opposite sign, so in the normal atom the net electrical charge is zero. Because the electrons are attracted to the nucleus, we expect the quantum states corresponding to regions near the nucleus to have the lowest energies. We may imagine starting with a bare nucleus with Z protons, and adding electrons one by one until the normal complement of Z electrons for a neutral atom is reached. We expect the lowest-energy states, ordinarily those with the smallest values of n and l, to fill first, and we use successively higher states until all electrons are accommodated.

The chemical properties of an atom are determined principally by interactions involving the outermost electrons, so it is of particular interest to find out how these electrons are arranged. For example, when an atom has one electron considerably farther from the nucleus (on the average) than the others, this electron is relatively loosely bound. The atom tends to lose this electron and form what chemists call an *electrovalent* or *ionic* bond, with valence $+1$. This behavior is characteristic of the alkali metals lithium, sodium, potassium, and so on.

We now proceed to describe ground-state electron configurations for the first few atoms (in order of increasing Z). For hydrogen, the ground state is $1s$; the single electron is in the state $n = 1$, $l = 0$, $m = 0$, and $s = \pm 1$. In the helium atom ($Z = 2$), *both* electrons are in the $1s$ states, with opposite spins; this state is denoted as $1s^2$. For helium, the K shell is completely filled and all others are empty.

Lithium ($Z = 3$) has three electrons; in the ground state two are in the $1s$ state, and one in a $2s$ state. We denote this state as $1s^2 2s$. On the average, the $2s$ electron is considerably farther from the nucleus than the $1s$ electrons, as shown schematically in Fig. 43–1. Hence, according to Gauss's law, the *net* charge influencing the $2s$ electron is $+e$, rather than $+3e$ as it would be without the $1s$ electrons present. Thus the $2s$ electron is loosely bound, as the chemical behavior of lithium suggests. An alkali metal, lithium forms ionic compounds in which each atom loses an electron. In such compounds, lithium has a valence of $+1$.

Next is beryllium ($Z = 4$); its ground-state configuration is $1s^2 2s^2$, with two electrons in the L shell. Beryllium is the first of the *alkaline-earth* elements, forming ionic compounds in which they have a valence of $+2$.

Table 43–2 shows the ground-state electron configurations of the first thirty elements. The L shell can hold a total of eight electrons; we invite you to verify this from the rules in Section 43–1. At $Z = 10$ the K and L shells are filled and there are no electrons in the M shell. We expect this to be a particularly stable configuration, with little tendency to gain or lose electrons, and in fact this element is neon, a "noble gas" with no known compounds. The next element after neon is sodium ($Z = 11$), with filled K and L shells and one electron in the M shell. Thus its "filled-shell-plus-one-electron" structure resembles that of lithium; both are alkali metals. The element *before* neon is fluorine, with $Z = 9$. It has a vacancy in the L shell and might be expected to have an affinity for an electron, normally forming ionic compounds in which

Electron structure of complex atoms: filling up the lowest-energy states first

The chemical behavior of elements is determined mostly by the arrangement of outer electrons in their atoms.

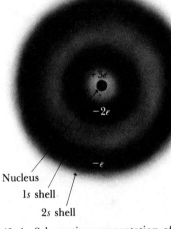

Nucleus
$1s$ shell
$2s$ shell

43–1 Schematic representation of charge distribution in a lithium atom. The nucleus has a charge $3e$; the two $1s$ electrons are closer to the nucleus than the $2s$ electron, which moves in a field approximately equal to that of a point charge of $3e - 2e$, or simply e.

When there are only one or two electrons in the outermost shell, they are only loosely bound.

All the noble gases have filled-shell electron configurations.

All the halogens have one vacancy in the outermost shell.

TABLE 43–2 Ground-state Electron Configurations

Element	Symbol	Atomic Number (Z)	Electron Configuration
Hydrogen	H	1	$1s$
Helium	He	2	$1s^2$
Lithium	Li	3	$1s^2 2s$
Beryllium	Be	4	$1s^2 2s^2$
Boron	B	5	$1s^2 2s^2 2p$
Carbon	C	6	$1s^2 2s^2 2p^2$
Nitrogen	N	7	$1s^2 2s^2 2p^3$
Oxygen	O	8	$1s^2 2s^2 2p^4$
Fluorine	F	9	$1s^2 2s^2 2p^5$
Neon	Ne	10	$1s^2 2s^2 2p^6$
Sodium	Na	11	$1s^2 2s^2 2p^6 3s$
Magnesium	Mg	12	$1s^2 2s^2 2p^6 3s^2$
Aluminum	Al	13	$1s^2 2s^2 2p^6 3s^2 3p$
Silicon	Si	14	$1s^2 2s^2 2p^6 3s^2 3p^2$
Phosphorus	P	15	$1s^2 2s^2 2p^6 3s^2 3p^3$
Sulfur	S	16	$1s^2 2s^2 2p^6 3s^2 3p^4$
Chlorine	Cl	17	$1s^2 2s^2 2p^6 3s^2 3p^5$
Argon	Ar	18	$1s^2 2s^2 2p^6 3s^2 3p^6$
Potassium	K	19	$1s^2 2s^2 2p^6 3s^2 3p^6 4s$
Calcium	Ca	20	$1s^2 2s^2 2p^6 3s^2 3p^6 4s^2$
Scandium	Sc	21	$1s^2 2s^2 2p^6 3s^2 3p^6 3d 4s^2$
Titanium	Ti	22	$1s^2 2s^2 2p^6 3s^2 3p^6 3d^2 4s^2$
Vanadium	V	23	$1s^2 2s^2 2p^6 3s^2 3p^6 3d^3 4s^2$
Chromium	Cr	24	$1s^2 2s^2 2p^6 3s^2 3p^6 3d^5 4s$
Manganese	Mn	25	$1s^2 2s^2 2p^6 3s^2 3p^6 3d^5 4s^2$
Iron	Fe	26	$1s^2 2s^2 2p^6 3s^2 3p^6 3d^6 4s^2$
Cobalt	Co	27	$1s^2 2s^2 2p^6 3s^2 3p^6 3d^7 4s^2$
Nickel	Ni	28	$1s^2 2s^2 2p^6 3s^2 3p^6 3d^8 4s^2$
Copper	Cu	29	$1s^2 2s^2 2p^6 3s^2 3p^6 3d^{10} 4s$
Zinc	Zn	30	$1s^2 2s^2 2p^6 3s^2 3p^6 3d^{10} 4s^2$

it has a valence of -1. This behavior is characteristic of the *halogens* (fluorine, chlorine, bromine, iodine, astatine), all of which have "filled-shell-minus-one" configurations.

The periodic table of the elements is a result of the numbers of electron states and the exclusion principle.

By similar analysis, we can understand all the regularities in chemical behavior exhibited by the **periodic table of the elements,** on the basis of electron configurations. A slight complication occurs with the M and N shells because the $3d$ and $4s$ subshells ($n = 3$, $l = 2$, and $n = 4$, $l = 0$, respectively) overlap in energy. Thus argon ($Z = 18$) has all the $1s$, $2s$, $2p$, $3s$, and $3p$ states filled, but in potassium ($Z = 19$) the additional electron goes into a $4s$ level rather than a $3d$ level. The next several elements have one or two electrons in the $4s$ states and increasing numbers in the $3d$ states. These elements are all metals with rather similar chemical and physical properties; they form the first *transition series,* starting with scandium ($Z = 21$) and ending with zinc ($Z = 30$), for which the $3d$ and $4s$ levels are filled.

Each vertical column in the periodic table has a characteristic outer-electron configuration.

Thus the similarity of elements in each *group* (vertical column) of the periodic table reflects corresponding similarity in outer-electron configuration. All the noble gases (helium, neon, argon, krypton, xenon, and radon) have filled-shell configurations. All the alkali metals (lithium, sodium, potassium, rubidium, cesium, and francium) have "filled-shell-plus-one" configurations. All the alkaline-earth metals (beryllium, magnesium, calcium, strontium, barium, and radium) have "filled-shell-plus-two" configurations, and all the halogens (fluorine, chlorine, bromine, iodine, and astatine) have "filled-shell-minus-one" structures.

This whole theory can be refined to account for differences between elements in a group and to account for various aspects of chemical behavior. Indeed, some physicists like to claim that all of chemistry is contained in the Schrödinger equation! This statement is a bit extreme, but we can see that even this qualitative discussion of atomic structure takes us a considerable distance toward understanding the atomic basis of many chemical phenomena. Next we look in more detail at the nature of the chemical bond.

PROBLEM-SOLVING STRATEGY: *Atomic structure*

1. Some numerology has to be mastered in counting the energy levels for electrons in the central-field approximation. There are four quantum numbers: n, l, m, and s, all integers; n is always positive, l can be zero or positive, and the other two can be positive or negative. Be sure you know how to count the number of levels in each shell and subshell; study Tables 43–1 and 43–2 carefully.

2. As in Chapter 42, familiarizing yourself with some numerical magnitudes is useful. Here are two examples to work out: The electrical potential energy of a proton and an electron 0.1 nm apart (typical of atomic dimensions) is 14.4 eV or 2.31×10^{-18} J. The moment of inertia of two protons 0.1 nm apart (about an axis through the center of mass, perpendicular to the line joining them) is 0.836×10^{-47} kg·m^2. Think of other examples like this and work them out, to help you know what kinds of magnitudes to expect in atomic physics.

3. As in Chapter 42, you will need to use both electronvolts and joules. The conversion 1 eV $= 1.602 \times 10^{-19}$ J and Planck's constant in eV, $h = 4.136 \times 10^{-15}$ eV·s, are handy. Nanometers are convenient for atomic and molecular dimensions, but don't forget to convert to meters in calculations.

43–3 DIATOMIC MOLECULES

As we learned in Section 43–2, the study of electron configurations in atoms provides valuable insight into the nature of the chemical bond; that is, the interactions that hold atoms together to form stable structures such as molecules and crystalline solids. There are several types of chemical bonds; the simplest to understand is the **ionic bond,** also called the *electrovalent* or *heteropolar* bond. The most familiar example is sodium chloride (NaCl), in which the sodium atom gives its one 3s electron to the chlorine atom, filling the vacancy in the 3p subshell of chlorine. Energy is required to make this transfer if the atoms are far apart, but the result is two ions, one positively charged and one negatively charged, that attract each other. As they come together, their potential energy decreases, so that the final bound state of Na$^+$Cl$^-$ has lower total energy than the state in which the two atoms are separated and neutral.

Removing the 3s electron from the sodium atom requires 5.1 eV of energy; this is called the **ionization energy** or *ionization potential* of sodium. Chlorine has an **electron affinity** of 3.8 eV. That is, the neutral chlorine atom can attract an extra electron; once this electron takes its place in the 3p level, 3.8 eV of energy are required to remove it. Thus, creating the separate Na$^+$ and Cl$^-$ ions requires a net expenditure of only $5.1 - 3.8$ eV $= 1.3$ eV. The negative potential energy associated with the mutual attraction of the ions is determined by the closeness to which they can approach each other; this in turn is determined by the electrical interactions and by the exclusion principle,

What holds atoms together in a molecule: the types of chemical bond

How much energy is required to gain or lose an electron?

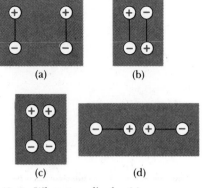

43–2 When two dipoles (a) are brought together, the interaction may be (b) attractive, or (c and d) repulsive.

A bond formed by sharing a pair of electrons between two atoms

The electrons in a covalent bond must have opposite spins.

An atom can form several covalent bonds with other atoms.

which forbids extensive overlap of the electron clouds of the two atoms. The minimum potential energy is -5.5 eV at a separation of 0.24 nm. This is more than enough to repay the initial investment in creating the ions. At lesser distances, the interaction becomes repulsive. Thus the net energy involved in creating the ions and letting them come together to the equilibrium separation of 0.24 nm is $(-5.5 + 1.3)$ or -4.2 eV, which is the binding energy of the molecule. To put it another way, 4.2 eV of energy are needed to dissociate the molecule into separate atoms.

Ionic bonds can involve more than one electron per atom. The alkaline-earth elements form ionic compounds in which each atom loses *two* electrons; an example is $Mg^{2+}Cl_2^-$. Loss of more than two electrons is relatively rare; instead a *different kind of bond* comes into operation.

The **covalent** or *homopolar* **bond** is characterized by a more nearly symmetric participation of the two atoms, as contrasted with the complete asymmetry involved in the electron-transfer process of the ionic bond. The simplest example of a covalent bond is the hydrogen molecule, a structure containing two protons and two electrons. As a preliminary to understanding this bond, consider first the interaction of two electric dipoles, as in Fig. 43–2. In (a) the dipoles are far apart; in (b) the like charges are farther apart than the unlike charges, and there is a net *attractive* force. In (c) and (d) the interaction is repulsive.

In a molecule the charges are not at rest, but Fig. 43–2b at least makes it seem plausible that if the electrons are localized primarily in the region between the protons, the electrons exert an attractive force on each proton that may more than counteract the repulsive interactions of the protons on each other and the electrons on each other. That is, in the covalent bond in the hydrogen molecule, the attractive interaction is supplied by a pair of electrons, one contributed by each atom, whose charge clouds are concentrated primarily in the region between the two atoms. Hence this bond may also be thought of as a shared-electron or electron-pair bond.

According to the exclusion principle, two electrons can occupy the same region of space only when they have opposite spin orientations. When the spins are parallel, the state that would be most favorable from energy considerations is forbidden by the exclusion principle, and the lowest-energy state permitted is one in which the electron clouds are concentrated *outside* the central region between atoms. The nuclei then repel each other, and the interaction is repulsive rather than attractive. Thus opposite spins are an essential requirement for an electron-pair bond, and no more than two electrons can participate in such a bond.

This is not to say, however, that an atom cannot have several electron-pair bonds. On the contrary, an atom having several electrons in its outermost shell can form covalent bonds with several other atoms. The bonding of carbon and hydrogen atoms, of central importance in organic chemistry, is an example. In the *methane* molecule (CH_4) the carbon atom is at the center of a regular tetrahedron, with a hydrogen atom at each corner. The carbon atom has four electrons in its *L* shell, and one of these electrons forms a covalent bond with each of the four hydrogen atoms, as shown in Fig. 43–3. Similar patterns occur in more complex organic molecules.

Ionic and covalent bonds represent two extremes in the nature of molecular bonds, but there is no sharp division between the two types. In many situations there is a *partial* transfer of one or more electrons (corresponding to a greater or smaller distortion of the electron wave functions) from one atom to another. As a result, many molecules having dissimilar atoms have electric dipole moments, that is, a preponderance of positive charge at one end and of negative charge at the other. Such molecules are said to be *polar*. Water molecules have large dipole moments that are responsible for the exceptionally large dielectric constant of liquid water.

The bonds discussed so far typically have energies in the range of 1 eV to 5 eV. They are called *strong bonds* to distinguish them from several types of much *weaker* bonds having energies typically of the order of 0.1 eV or less. One of these, the **van der Waals bond,** is an interaction between the electric dipole moments of two atoms or molecules. The existence of dipole moments in molecules was mentioned above. Even when an atom or molecule has no permanent dipole moment, fluctuating charge densities in the interior of the structure can lead to fluctuating dipole moments; these in turn can induce dipole moments in neighboring structures. The resulting dipole–dipole interaction can be attractive and can lead to weak bonding of atoms or molecules. For example, the low-temperature liquefaction and solidification of such molecules as H_2, O_2, and N_2 and of the noble gases is due to interaction of the induced-dipole type. Not much energy of thermal agitation is needed to break these weak bonds, so these substances exist in the liquid and solid states only at very low temperatures.

A second type of important weak bond is the **hydrogen bond,** which is analogous to the covalent bond. In the latter an electron pair serves to bind two positively charged structures together. In the hydrogen bond a hydrogen atom acts as the glue to bond two negatively charged structures. Again the bond energy is only about 0.1 eV, but nevertheless the hydrogen bond plays an essential role in many organic molecules, including, for example, cross-link bonding between the two strands of the famous double-helix DNA molecule.

All these bond types play roles in the structure of *solids* as well as of molecules. Indeed, a solid is in many respects a gigantic molecule. Still another type of bonding, the *metallic bond*, comes into play in the structure of metallic solids. We return to this subject in Section 43–5.

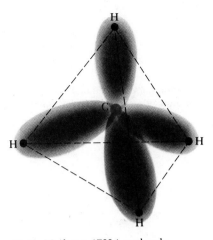

43–3 Methane (CH_4) molecule, showing four covalent bonds. The electron cloud between the central carbon atom and each of the four hydrogen nuclei represents the two electrons of a covalent bond. The hydrogen nuclei are at the corners of a regular tetrahedron.

The hydrogen bond: another type of shared-particle bond

43–4 MOLECULAR SPECTRA

The energy levels of atoms are associated with the kinetic energy of electron motion relative to the nucleus and with the potential energy of interaction of the electrons with the nucleus and with each other. Energy levels of *molecules* have additional features that result from motion of the *nuclei* of the atoms relative to each other. Associated with these energy levels are characteristic **molecular spectra.**

To keep things as simple as possible, we consider only *diatomic* molecules. Viewed as a rigid dumbbell, a diatomic molecule can *rotate* about an axis through its center of mass. According to classical mechanics, the kinetic energy K of a rigid body with moment of inertia I rotating with angular velocity ω can be expressed as $K = \frac{1}{2}I\omega^2$. This is Eq. (9–13); you may want to review its derivation in Section 9–5. The kinetic energy can also be expressed in terms of the magnitude L of angular momentum, which is given by $L = I\omega$. Combining this with the preceding expression, we can express the kinetic energy alternatively as $L^2/2I$. If the molecule rotates as a rigid body, there is no potential

The rotational kinetic energy of a diatomic molecule has energy levels.

energy; the kinetic energy is then equal to the total energy E:

$$E = \frac{L^2}{2I}. \qquad (43\text{-}1)$$

In a quantum-mechanical discussion of this molecular rotation, it is reasonable to assume that the angular momentum is quantized in the same way as for electrons in an atom, as given by Eq. (42–20):

$$L^2 = l(l + 1)\hbar^2 \qquad (l = 0, 1, 2, 3, \ldots). \qquad (43\text{-}2)$$

Combining Eqs. (43–1) and (43–2), we obtain the **rotational energy levels:**

$$E = l(l + 1)\left(\frac{\hbar^2}{2I}\right) \qquad (l = 0, 1, 2, 3, \ldots). \qquad (43\text{-}3)$$

A more detailed analysis, using the Schrödinger equation, confirms that the angular momentum is given by Eq. (43–2) and the energy levels by Eq. (43–3).

Considering the magnitudes involved, we note that for an oxygen molecule, $I = 5 \times 10^{-46}$ kg·m^2. Thus the constant $\hbar^2/2I$ in Eq. (43–3) is approximately equal to

$$\frac{\hbar^2}{2I} = \frac{(1.05 \times 10^{-34}\text{ J·s})^2}{2(5 \times 10^{-46}\text{ kg·m}^2)} = 1.10 \times 10^{-23}\text{ J} = 0.69 \times 10^{-4}\text{ eV}.$$

This energy is much *smaller* than typical atomic energy levels (of the order of a few eV) associated with optical spectra. The energies of photons emitted or absorbed in transitions among rotational levels are correspondingly small, and they fall in the far infrared region of the spectrum.

The rigid-dumbbell model of a diatomic molecule suggests that the distance between the two nuclei is constant. In fact, a more realistic model would represent the connection as a *spring* rather than a rigid rod. The atoms can undergo *vibrational* motion relative to the center of mass, and there are additional kinetic and potential energies associated with this motion. Application of the Schrödinger equation shows that the corresponding **vibrational energy levels** are given by

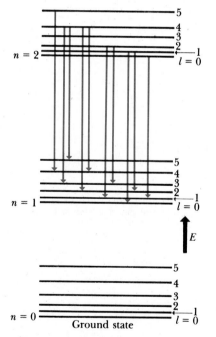

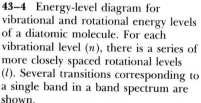

$$E = \left(n + \frac{1}{2}\right)hf \qquad (n = 0, 1, 2, 3, \ldots), \qquad (43\text{-}4)$$

where f is the frequency of vibration. For typical diatomic molecules this turns out to be of the order of 10^{13} Hz; thus the constant hf in Eq. (43–4) is of the order of

$$hf = (6.6 \times 10^{-34}\text{ J·s})(10^{13}\text{ s}^{-1}) = 6.6 \times 10^{-21}\text{ J}$$
$$= 0.041\text{ eV}.$$

Hence the vibrational energies, while still much smaller than those of atomic spectra, are typically considerably *larger* than the rotational energies.

43–4 Energy-level diagram for vibrational and rotational energy levels of a diatomic molecule. For each vibrational level (n), there is a series of more closely spaced rotational levels (l). Several transitions corresponding to a single band in a band spectrum are shown.

An energy-level diagram for a diatomic molecule has the general appearance of Fig. 43–4. For each value of n, there are many values of l, leading to a series of closely spaced levels. Transitions between different pairs of n values give different series of spectrum lines, and the resulting spectrum has the appearance of a series of *bands*. Each band corresponds to a particular vibrational transition, and each individual line in a band, to a particular rotational transition. A typical **band spectrum** is shown in Fig. 43–5.

43–5 Typical band spectrum (Courtesy of R. C. Herman.)

The same considerations can be applied to more complex molecules. A molecule with three or more atoms has several different kinds or *modes* of vibratory motion, each with its own set of energy levels, related to the frequency by Eq. (43–4). The resulting energy-level scheme and associated spectra can be quite complex, but the general considerations discussed above still apply. In nearly all cases the associated radiation lies in the infrared region of the electromagnetic spectrum. Analysis of molecular spectra has proved to be an extremely valuable analytical tool, providing a great deal of information about the strength and rigidity of molecular bonds and the structure of complex molecules.

Molecular spectroscopy: an important tool for investigating molecular structure

43–5 STRUCTURE OF SOLIDS

At ordinary temperatures and pressures, most materials are in the *solid state.* This is a condensed state of matter in which the interactions among the atoms or molecules are strong enough to give the material a definite volume and shape that change relatively little with stress. The distances between adjacent atoms in a solid are of the same order of magnitude as the diameters of the electron clouds of the atoms.

A solid may be **amorphous** or **crystalline.** We have mentioned crystalline solids briefly in Section 20–7, and we suggest you review that discussion. A crystalline solid is characterized by *long-range order,* which is a recurring pattern in the arrangement of atoms. This pattern is called the *crystal structure* or the *lattice structure.* Amorphous solids have more in common with liquids than with crystalline solids; liquids also have short-range order but not long-range order.

Some solid materials have a definite crystal structure; some don't.

The forces responsible for the regular arrangement of atoms in a crystal are, in some cases, the same as those involved in molecular bonds. Corresponding to the classes of strong molecular bonds are ionic and covalent crystals. The alkali halides, of which ordinary salt (NaCl) is the most common variety, are the most familiar **ionic crystals.** The positive sodium ions and the negative chlorine ions occupy alternate positions in a cubic crystal lattice, as in Fig. 43–6. The forces are the familiar Coulomb's-law forces between charged particles; these forces are not directional, and the particular arrangement in which the material crystallizes is determined by the relative size of the two ions.

Crystals can be classified in terms of the kinds of bonds that hold them together.

The simplest example of a **covalent crystal** is the *diamond structure,* a structure found in the diamond form of carbon and also in silicon, germanium, and tin. All these elements are in Group IV of the periodic table, with four electrons in the outermost shell. Each atom in this structure is situated at the center of a regular tetrahedron, with four nearest-neighbor atoms at the corners, and it forms a covalent bond with each of these atoms. These bonds are strongly directional because of the asymmetric electron distribution, leading to the tetrahedral structure.

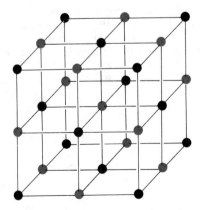

43–6 Symbolic representation of a sodium chloride crystal, with exaggerated distances between ions.

Metallic crystals: electron sharing on a vast scale

A third crystal type, which is less directly related to the chemical bond than are ionic or covalent crystals, is the **metallic crystal.** In this structure the outermost electrons are not localized at individual atomic lattice sites but are detached from their parent atoms and free to move through the crystal. The corresponding charge clouds (and their associated wave functions) extend over many atoms. Thus we can picture a metallic crystal roughly as an array of positive ions (atoms from which one or more electrons have been removed) immersed in a sea of electrons whose attraction for the positive ions holds the crystal together. This "sea" has many of the properties of a gas, and indeed we speak of the *electron-gas model* of metallic solids.

Close packing: How many atoms can you stuff into a given volume?

In a metallic crystal the situation is as though the atoms would like to form shared-electron bonds but do not have enough valence electrons. Instead, electrons are shared among *many* atoms. This bonding is not strongly directional in nature, and the shape of the crystal lattice is determined primarily by considerations of **close packing,** that is, the maximum number of atoms that can fit into a given volume. The two most common metallic crystal lattices, the face-centered cubic and the hexagonal close-packed, are shown in Fig. 43–7. In each of these lattices, each atom has 12 nearest neighbors.

As we mentioned in Section 43–3, van der Waals interactions and hydrogen bonding also play a role in the structure of some solids. In polyethylene and similar polymers, covalent bonding of atoms forms long-chain molecules, and hydrogen bonding forms cross links between adjacent chains. In solid water, both van der Waals forces and hydrogen bonds are significant, and together they determine the crystal structure of ice. Many other examples might be cited.

Crystals can have several kinds of imperfections in the lattice.

The discussion in this section has centered around *perfect crystals*, crystals in which the crystal lattice extends uninterrupted throughout the entire material. Real crystals show a variety of departures from this idealized structure. Materials are often *polycrystalline*, composed of many small single crystals bonded together at *grain boundaries*. Within a single crystal, *interstitial* atoms may occur in places where they do not belong, and there may be *vacancies*, lattice sites that should be occupied by an atom but are not. An imperfection of particular interest in semiconductors, to be discussed in Section 43–7, is the

Models illustrating the six basic crystallographic types: orthorhombic, hexagonal, monoclinic, tetragonal, triclinic, and isometric. (Photo by Chip Clark.)

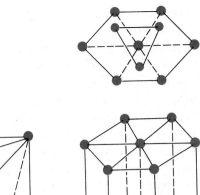

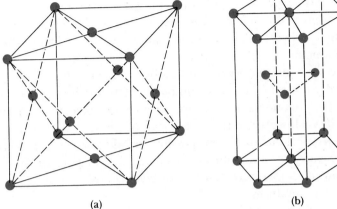

43–7 Close-packed crystal lattice structures. (a) Face-centered cubic; (b) hexagonal close-packed.

(a)

(b)

impurity atom, a foreign atom (e.g., arsenic in a silicon crystal) occupying a regular lattice site.

A more complex kind of imperfection is the *dislocation,* shown schematically in Fig. 43–8, in which one plane of atoms slips relative to another. The mechanical properties of metallic crystals are influenced strongly by the presence of dislocations. The ductility and malleability of some metals depend on the presence of dislocations that move through the lattice during plastic deformations.

43–6 PROPERTIES OF SOLIDS

We can understand many *macroscopic* properties of solids, including mechanical, thermal, electrical, magnetic, and optical properties, by considering their relation to the *microscopic* structure of the material. Various aspects of the relation of structure to properties are part of the vigorous program of research in the physics of solids being carried on throughout the world. Although we cannot discuss these topics in detail in a book such as this, a few examples will indicate the kinds of insights to be gained through study of the microscopic structure of solids.

We have already discussed, in Section 20–8, the subject of specific heat capacities of crystals. We applied the same principle of equipartition of energy as in the kinetic theory of gases. A simple analysis permits understanding of the empirical rule of Dulong and Petit on the basis of a microscopic model. This analysis has its limitations, to be sure; it does not include the energy of electron motion, which in metals makes a small additional contribution to specific heat; nor does it predict the temperature dependence of specific heats resulting from the *quantization* of the lattice-vibration energy discussed in Section 20–8. But these additional refinements *can* be included in the model to permit more detailed comparison of observed macroscopic properties with theoretical predictions.

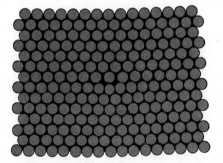

43–8 A dislocation. The concept is seen most easily by viewing the figure from various directions at a grazing angle with the page.

Understanding thermal and electrical properties on the basis of the microscopic structure of materials

The electrical resistivity of a material depends on the mobility or lack of mobility of electrons in the material. In a *metallic* crystal, the valence electrons are not bound to individual lattice sites but are free to move through the crystal. Thus we expect metals to be good conductors, and they usually are. In a *covalent* crystal, the valence electrons are tied up in the bonds responsible for the crystal structure and are therefore *not* free to move. Because no mobile charges are available for conduction, we expect such materials to be insulators. Similarly, an ionic crystal such as NaCl has no charges that are free to move, and solid NaCl is an insulator. When salt is melted, however, the ions are no longer locked to their individual lattice sites but are free to move, and *molten* NaCl is a good conductor.

There are, of course, no perfect conductors or insulators (except for superconductors at very low temperatures), although the resistivity of good insulators is greater than that of good conductors by an enormous factor, of the order of at least 10^{15}. This great difference is one of the factors that make extremely precise electrical measurements possible. In addition, the resistivity of all materials depends on temperature; in general, the large resistivity of insulators *decreases* with temperature, but that of good conductors usually *increases* at increased temperatures.

Two competing effects are responsible for this difference. In metals the *number* of electrons available for conduction is nearly independent of temperature, and the resistivity is determined by the frequency of collisions between electrons and the lattice. Roughly speaking, lattice vibrations increase with increased temperature, and the ion cores present a larger target area for collisions with electrons. Hence resistivities of metals usually increase with temperature. In insulators, the small amount of conduction that does take place is due to electrons that have gained enough energy from thermal motion of the lattice to break away from their "home" atoms and wander through the lattice. The number of electrons able to acquire the needed energy is very strongly temperature dependent; a twofold increase for a 10 C° temperature rise is typical. There is also increased scattering at higher temperatures, as with metals, but the increased number of carriers is a far larger effect. Thus insulators invariably become better conductors at higher temperatures.

A similar analysis can be made for *thermal* conductivity, which involves transport of microscopic mechanical energy rather than electric charge. The wave motion associated with lattice vibrations is one mechanism for energy transfer, and in metals the mobile electrons also carry kinetic energy from one region to another. This effect is much larger than that of the lattice vibrations, and so metals are usually much better thermal conductors than are electrical insulators, which have at most very few free electrons available to transport energy.

Optical properties are also related directly to microscopic structure. Good electrical conductors *cannot* be transparent to electromagnetic waves, for the electric fields of the waves induce currents in the material. These induced currents dissipate the wave energy into heat as the electrons collide with the atoms in the lattice. All transparent solid materials are very good insulators. Metals are good *reflectors* of radiation, however, and again the reason is the presence of free electrons at the surface of the material, which can move in response to the incident wave and generate a reflected wave. Reflection from the polished surface of an insulator is a somewhat more subtle phenomenon,

Why do some materials conduct better at higher temperatures, and others not so well?

Why are good electrical conductors also good thermal conductors?

Why are some materials transparent and others opaque?

dependent on polarization of the material; but again, it is possible to relate macroscopic properties to microscopic structure.

43–7 BAND THEORY OF SOLIDS

The concept of **energy bands** in solids offers us additional insight into some of the properties of solids discussed in Section 43–6. It also allows more detailed theoretical analysis of these properties on the basis of electron energy levels and wave functions.

In solid materials, electron energy levels form bands of allowed energies, separated by forbidden bands.

To introduce the idea, suppose we have a large number N of identical atoms, far enough apart so their interactions are negligible. Then every atom has the same energy-level diagram, showing the possible quantum states for the electrons and their associated energy levels. We can draw an energy-level diagram for the entire system; it looks just like the diagram for a single atom, except that the exclusion principle permits each state to be occupied by N electrons instead of just one.

Now we begin to push the atoms closer together. Because of the electrical interactions and the exclusion principle, the wave functions begin to distort, especially those of the outer, or *valence*, electrons. The energy levels also shift somewhat; some move upward, some downward, depending on the environment of each individual atom. Each valence-electron state for the system, formerly a sharp energy level that could accommodate N electrons, becomes spread out into a *band* containing N closely spaced levels, as shown in Fig. 43–9. Since N is ordinarily very large, of the order of Avogadro's number (10^{23}), we can think of the levels as forming a continuous distribution of energies within a band. Between adjacent energy bands are gaps or forbidden regions where there are *no* possible energy levels.

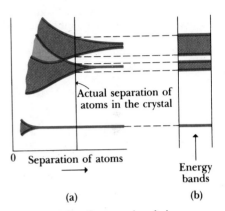

43–9 Origin of energy bands in a solid. (a) As the atoms are pushed together, the energy levels spread into bands. The vertical line shows the actual atomic spacing in the crystal lattice. (b) Symbolic representation of energy bands.

The inner electrons in an atom are affected much less by nearby atoms than the valence electrons, and their energy levels remain relatively sharp. Ordinarily the behavior of an atom in a solid structure can be modeled as that of a rigid, unchanging *ion core* consisting of the positively charged nucleus and the inner electrons, surrounded by the valence electrons, whose states and energies *are* altered significantly by interactions with neighboring atoms.

Now let us see what this has to do with electrical conductivity. In insulators and semiconductors the valence electrons completely fill the highest occupied band, called the **valence band;** the next higher band, called the **conduction band,** is completely empty. The gap separating the two may be of the order of 1 to 5 eV. This situation is shown in Fig. 43–10b. The electrons in the valence band are not free to move in response to an applied electric field; to move, an electron would have to go to a different quantum-mechanical state with slightly different energy, but all the neighboring states are already occupied. The only way an electron can move is to jump into the conduction band, which would require an additional energy of a few electronvolts. Ordinarily that much energy is not available; for electric fields of reasonable magnitude the potential difference between two points spaced one atom apart is very much *less* than 1 eV. There is no way such a field can do enough work on an electron to raise it across the gap into the conduction band. The situation is like a completely filled parking lot; none of the cars can move because there is no place to go. If a car could jump over the others, it could move!

A completely filled band is like a grid-locked traffic jam; nobody can move.

43–10 Three types of energy-band structure. (a) A conductor; there is a partially filled band, and electrons in this band are free to move when an electric field is applied. (b) An insulator; a completely full band is separated by a gap of several electronvolts from a completely empty band, and electrons in the full band cannot move. At finite temperatures a few electrons can reach the upper "conduction band." (c) A semiconductor; a completely filled band is separated by a small gap of 1 eV or less from an empty band; at finite temperatures substantial numbers of electrons can reach the upper "conduction band," where they are free to move.

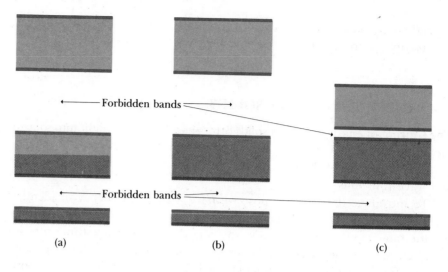

(a)   (b)   (c)

Some electrons can jump into the conduction band, where they are free to move.

Materials with partially filled bands are always conductors.

Dielectric breakdown: when an insulator becomes a conductor

At any finite temperature, however, the crystal lattice has some vibrational motion, and there is some probability that an electron can gain enough energy from *thermal* motion to jump to the conduction band. Once in the conduction band, an electron is free to move in response to an applied electric field because plenty of nearby empty states are available. At any finite temperature a few electrons are present in the conduction band, so no material is a perfect insulator. Furthermore, as the temperature increases, the population in the conduction band increases very rapidly. An increase of a factor of two for a temperature rise of 10 C° is typical.

With metals, the situation is different because the valence band is only partially filled. The metal sodium is a simple example. The energy-level diagram for sodium in Fig. 41–5 shows that for an isolated atom, the 3p "resonance-level" states for the valence electron are about 2.1 eV above the 3s ground state. (The wavelength of the familiar yellow sodium doublet corresponds to this energy difference.) But in the crystal lattice of solid sodium, the 3s and 3p *bands* spread out enough so that they actually overlap, forming a single band that is only $\frac{1}{4}$ filled. The situation is similar to the one shown in Fig. 43–10a. Electrons in states near the top of the filled portion of the band have many adjacent unoccupied states available, and they can easily gain or lose small amounts of energy in response to an applied electric field. Thus these electrons are mobile and can contribute to electrical and thermal conductivity. Partly filled bands are always characteristic of the metallic bond found in most metallic crystals. In the ionic NaCl crystal, however, there is no overlapping of bands; the valence band is completely filled, and solid sodium chloride is an insulator.

The band picture also adds insight to the phenomenon of *dielectric breakdown*, in which insulators subjected to a large enough electric field become conductors. If the electric field in a material is so large that there is a potential difference of a few volts over a distance comparable to atomic sizes (i.e., a field of the order of 10^{10} V·m^{-1}), then the field can do enough work on a valence electron to boost it over the forbidden region and into the conduction band. Real insulators usually have dielectric strengths much less than this because of structural imperfections that provide some energy states in the forbidden region.

The concept of energy bands is very useful in understanding the properties of semiconductors, which we will study in the next section.

43-8 SEMICONDUCTORS

As the name implies, a **semiconductor** has an electrical resistivity intermediate between those of good conductors and of good insulators. This is only one aspect of the behavior of this important class of materials, so vital to present-day electronics. To keep things as simple as possible, we will discuss mostly the elements germanium and silicon. These are the simplest semiconductors, but they illustrate the most important concepts.

Both silicon and germanium have four electrons in the outermost electron subshell, and both crystallize in the diamond structure described in Section 43-5. Each atom lies at the center of a regular tetrahedron, with four nearest neighbors at the corners and a covalent bond with each. Thus all the valence electrons are involved in the bonding, and the materials should be insulators. An unusually small amount of energy, however, is needed to break one of the bonds and set an electron free to roam around the lattice: 1.1 eV for silicon and only 0.7 eV for germanium. This corresponds to the energy gap between the valence and conduction bands, shown in Fig. 43-10c. So even at room temperature, a substantial number of electrons are dissociated from their parent atoms, and this number increases rapidly with temperature.

Furthermore, when an electron is removed from a covalent bond, it leaves a vacancy where there would ordinarily be an electron. An electron in a neighboring atom can drop into this vacancy, leaving the neighbor with the vacancy. In this way the vacancy, usually called a **hole,** can travel through the lattice and serve as an additional current carrier. A hole behaves like a positively charged particle. In a pure semiconductor, holes and electrons are always present in equal numbers; the resulting conductivity is called *intrinsic conductivity* to distinguish it from conductivity due to impurities, to be discussed later.

The parking-lot analogy mentioned in Section 43-7 is useful in clarifying the mechanism of conduction in a semiconductor. A crystal with no bonds broken is like a completely filled floor of a parking garage. No cars (electrons) can move because there is nowhere for them to go. But if one car is removed to the empty floor above, it can move freely, and the vacancy it leaves also permits cars to move on the nearly filled floor. This motion is most easily described in terms of *motion of the vacant space* from which the car has been removed. This corresponds to a vacancy or hole in the normally filled valence band.

Now suppose we mix into melted germanium a small amount of arsenic, a Group V element having *five* valence electrons. When one of these electrons is removed, the remaining electron structure is essentially that of germanium; the only difference is that it is scaled down in size by the insignificant factor $\frac{32}{33}$ because the arsenic nucleus contains a charge of $+33e$ rather than $+32e$. Thus an arsenic atom can take the place of a germanium atom in the lattice. Four of its five valence electrons form the necessary covalent bonds with the nearest neighbors; the fifth is very loosely bound, with a binding energy of only about 0.01 eV. This corresponds in the band picture to an energy level 0.01 eV below the bottom of the conduction band. Even at ordinary temperature this electron can very easily escape and wander about the lattice. The corresponding positive charge is associated with the nuclear charge and is *not* free to move, in contrast to the situation with electrons and holes in pure germanium.

Because at ordinary temperatures only a very small fraction of the valence electrons in germanium are able to escape their sites and participate in conduction, a concentration of arsenic atoms as small as one part in 10^{10} can

A crystal puller used to produce exceptionally pure single crystals of germanium. A seed crystal is pulled out of the melt in a quartz crucible heated by radiofrequency induction in an atmosphere of hydrogen. Slices of the crystal are used in making highly sensitive particle detectors. (Courtesy of Lawrence Berkeley Laboratory, University of California.)

Vacancies (holes) can act as current carriers.

The conductivity of a semiconductor can be changed drastically by trace impurities.

Impurity atoms can contribute extra electrons or steal electrons from the atoms of the host material.

increase the conductivity so drastically that conduction due to impurities becomes by far the dominant mechanism. In such a case the conductivity is due almost entirely to *negative* charge motion; the material is called an **n-type semiconductor** and is said to have *n*-type impurities.

Adding atoms of an element in Group III, with only *three* valence electrons, has an analogous effect. An example is gallium; placed in the germanium lattice, the gallium atom would like to form four covalent bonds, but it has only three outer electrons. It can, however, steal an electron from a neighboring germanium atom to complete the bonding. This leaves the neighboring atom with a *hole*, or missing electron, and this hole can then move through the lattice just as with intrinsic conductivity. In this case the corresponding negative charge is associated with the deficiency of positive charge of the gallium nucleus ($+31e$ instead of $+32e$), so it is *not* free to move. This situation is characteristic of **p-type semiconductors**, materials with *p*-type impurities. The two types of impurities, *n* and *p*, are also called *donors* and *acceptors*, respectively, and the deliberate addition of these impurity elements is called *doping*.

The assertion that in *n* and *p* semiconductors the current *is* actually carried by electrons and holes, respectively, can be verified by using the Hall effect (Section 30–10). The direction of the Hall emf is opposite in the two cases, and measurements of the Hall effect in various semiconductor materials confirm our analysis of the conduction mechanisms.

43–9 SEMICONDUCTOR DEVICES

Semiconductor devices are the heart and soul of modern electronics.

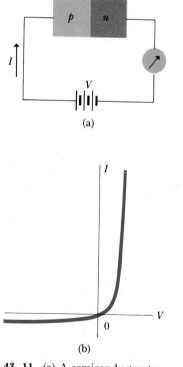

43–11 (a) A semiconductor *p-n* junction in a circuit; (b) graph showing the asymmetric voltage–current relationship given by Eq. (43–5).

Semiconductor devices are of tremendous importance in contemporary electronics. Although radio and television transmitting and receiving equipment originally relied on vacuum tubes, these have been almost completely replaced in the last two decades by transistors, diodes, integrated circuits, and other semiconductor devices. The only surviving vacuum tubes in radio and TV equipment are the picture tube in a TV receiver and the imaging tubes in TV cameras. Equally significant are large-scale integrated circuits that incorporate the equivalent of many thousands of transistors, capacitors, resistors, and diodes on a silicon chip less than 1 cm square. Such chips are at the heart of every pocket calculator, personal computer, and large mainframe computer.

The usefulness of semiconductor materials in these devices results directly from the fact that the conductivity of the material can be controlled within wide limits and changed from one region of a device to another by changing impurity concentrations, deposition of metallic and oxide layers, and etching of patterns to form current paths. The simplest example of controlling conductivity is the **p-n junction,** which is a crystal of germanium or silicon with *p*-type impurities in one region and *n*-type impurities in another. The two regions are separated by a boundary region called the *junction*. Originally such junctions were produced by growing a crystal, usually by pulling a seed crystal very slowly away from the surface of a melted semiconductor. If the concentration of impurities in the melt is changed as the crystal is grown, the result is a crystal with two or more regions of varying conductivity and conductivity type. There are now much better ways to fabricate *p-n* junctions, but we do not need to go into the details.

When a *p-n* junction is connected in an external circuit as shown in Fig. 43–11a, and the potential *V* across the device is varied, the behavior of the current is as shown in Fig. 43–11b. The device conducts much more readily in the direction $p \rightarrow n$ than in the reverse, in striking contrast to the behavior of materials that obey Ohm's law. In the language of electronics, such a one-way device is called a **diode.**

We can understand the behavior of a p-n junction diode at least roughly on the basis of the conductivity mechanisms in the two regions. When the p region is at higher potential than the n, holes in the p region flow into the n region and electrons in the n region move into the p region, so both contribute substantially to current. When the polarity is reversed, the resulting electric fields tend to push electrons from p to n and holes from n to p. But there are very few mobile electrons in the p region, only those associated with intrinsic conductivity and some that diffuse over from the n region. Similarly, there are very few holes in the n region. As a result, the current is much smaller than with the opposite polarity.

A more detailed analysis of this process, taking into account the effects of drift under the applied field and the diffusion that takes place even in the absence of a field, shows that the voltage–current relationship in a p-n junction diode is given by

$$I = I_0[\exp(eV/kT) - 1], \tag{43–5}$$

where I_0 is a constant characteristic of the device, e is the electron charge, k is Boltzmann's constant, and T is absolute temperature.

A **transistor** includes two p-n junctions in a "sandwich" configuration, which may be either p-n-p or n-p-n. A p-n-p transistor is shown in Fig. 43–12. The three regions are usually called the emitter, base, and collector, as shown. In the absence of current in the left loop of the circuit, there is only a very small current through the resistor R because the voltage across the base–collector junction is in the "reverse" direction, that is, the direction of small current flow, as in a simple p-n junction. But when a voltage is applied between emitter and base, as shown, the holes traveling from emitter to base can travel *through* the base to the second junction, where they come under the influence of the collector-to-base potential difference and thus flow on through the resistor.

Hence the current in the collector circuit is *controlled* by the current in the emitter circuit. Furthermore, since V_c may be considerably larger than V_e, the power dissipated in R may be much larger than that supplied to the emitter circuit by the battery V_e. Thus the device functions as a power amplifier. If the potential drop across R is greater than V_e, it may also be a voltage amplifier.

In this configuration the *base* is the common element between the "input" and "output" sides of the circuit. Another widely used arrangement is the *common-emitter* circuit, shown in Fig. 43–13. In this circuit the current in the collector side of the circuit is much larger than in the base side, and the result is current amplification.

Transistors have taken over most of the functions for which vacuum tubes were formerly used following the invention of the triode by de Forest in 1907. They offer many advantages over vacuum tubes, including mechanical ruggedness (since no vacuum container or fragile electrodes are involved), small size, long life (because they operate at relatively low temperatures and deteriorate very little with age), and efficiency (because no power is wasted in heating a cathode for thermionic emission). Invented in 1948, semiconductor devices have completely revolutionized the electronics industry, including applica-

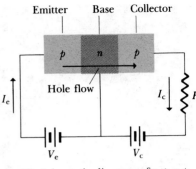

43–12 Schematic diagram of a p-n-p transistor and circuit. When $V_e = 0$, the current in the collector circuit is very small. When a potential V_e is applied between emitter and base, holes travel from emitter to base, as shown; when V_c is sufficiently large, most of them continue into the collector. The collector current I_c is controlled by the emitter current I_e.

Two p-n junctions form a transistor.

A transistor can be a voltage amplifier or a current amplifier.

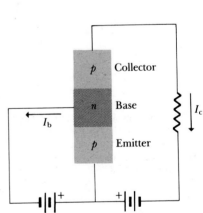

43–13 A common-emitter circuit. When $V_b = 0$, I_c is very small and most of the voltage V_c appears across the base–collector junction. As V_b increases, the base–collector potential decreases and more holes can diffuse into the collector; thus I_c increases. Ordinarily I_c is much larger than I_b.

A silicon wafer containing many identical individual chips. These will be separated and encapsulated in various kinds of envelopes as shown, depending on number of connections, heat-dissipation requirements, and other considerations. (Courtesy of Philips Electronics Co.)

Photocells and light sources using semiconductor devices

tions in communications, computer systems, control systems, and many other areas.

A further refinement in semiconductor technology is the **integrated circuit.** By successively depositing layers of material and etching patterns to define current paths, we can combine the functions of several transistors, capacitors, and resistors on a single square of semiconductor material that may be only a few millimeters on a side. A further elaboration of this idea leads to the *large-scale integrated circuit* and *very-large-scale integration* (VLSI). Starting on a base consisting of a silicon chip, various layers are built up, including evaporated-metal layers for conducting paths and silicon-dioxide layers for insulators and for dielectric layers in capacitors. Appropriate patterns are etched into each layer by use of photosensitive etch-resistant materials and optically reduced patterns. This makes it possible to build up a circuit containing the functional equivalent of many thousands of transistors, diodes, resistors, and capacitors on a single chip. These so-called MOS (metal-oxide-semiconductor) chips are the heart of pocket calculators and microprocessors, as well as of many large-scale computers. An example is shown in Fig. 43–14.

Semiconductors have many other practical applications. A thin slab of pure silicon or germanium can serve as a *photocell*. When the material is irradiated with light whose photons have at least as much energy as the gap between the valence and conduction bands, an electron in the valence band can absorb a photon and jump to the conduction band, where it contributes to the conductivity. Thus the conductivity increases when the material is exposed to light, and this change can be used to control the current in a circuit containing the device. Detectors for charged particles operate on the same principle. A high-energy charged particle passing through the semiconductor material interacts with the electrons, and as it loses energy some of the electrons are excited from the valence to the conduction band. The conductivity increases momentarily, causing a pulse of current in the external circuit. Solid-state detectors are widely used in nuclear and high-energy physics research.

The inverse of the semiconductor photocell is the light-emitting diode (LED), which acts as a source of light. When a diode such as that shown in Fig. 43–11a is given a large *forward* voltage, many holes are injected across the junction into the n region and electrons into the p region. When these minor-

43–14 A large-scale integrated circuit. This dime-sized chip is a 32-bit microprocessor chip containing the equivalent of 150,000 transistors. (Courtesy of AT&T Bell Laboratories.)

ity carriers recombine with majority carriers in the respective regions, they lose energy, which in some cases is radiated as photons of visible light. Light-emitting diodes are widely used for digital displays in clocks and electronic equipment.

43–10 SUPERCONDUCTIVITY

At very low temperatures (below about 20 K), some materials show a complete disappearance of all electrical resistance. Such materials are called **superconductors.** Superconductivity was discovered in 1911 by the Dutch physicist H. Kamerlingh-Onnes during experiments on the temperature dependence of resistivity of materials. This area of research served at the time as a testing ground for competing theories of the behavior of electrons in solids. Three years earlier, Kamerlingh-Onnes had succeeded in liquefying helium for the first time, thereby attaining lower temperatures than those available by any other means. Helium boils at 4.2 K at a pressure of 1 atm, and at lower temperatures under reduced pressure. Using liquid helium cooling, he found that when very pure solid mercury was cooled to 4.16 K, its resistivity suddenly dropped to zero.

> Strange things happen at very low temperatures.

Subsequently many other superconducting metals and alloys have been found. Each one has a characteristic superconducting transition temperature called its *critical temperature.* The highest critical temperature found so far is 23 K for a niobium-germanium alloy, and several niobium alloys have critical temperatures in the range 15 to 20 K.

> A phase transition to a state with zero resistance

Below the superconducting transition temperature, the resistivity of a material is believed to be *exactly* zero. Experimental upper limits are of the order of 10^{-25} $\Omega \cdot$m, compared with typical values of 10^{-8} $\Omega \cdot$m for good conductors such as silver and copper at ordinary temperatures. When a current is magnetically induced in a superconducting ring, the current continues without measurable decrease for many months!

A magnetic field can never exist inside a superconducting material. Any attempt to establish such a field results in induced eddy currents that exactly cancel the applied field everywhere inside the material. Related to this behavior is the fact that an applied magnetic field lowers the critical temperature, as shown in Fig. 43–15; a sufficiently strong field eliminates the superconducting transition completely. This transition can be regarded as a *phase transition* (although there is no associated heat of transition), and Fig. 43–15 is a *phase diagram.*

Although discovered in 1911, superconductivity was not well understood on a theoretical basis until 1957, when Bardeen, Cooper, and Schrieffer published the theory (now called the BCS theory) that was to earn them the Nobel Prize in 1972. It is difficult to describe the BCS theory in terms of elementary concepts. It is based on the existence of an *energy gap* in what would ordinarily be a continuum of electron energy states in a partly filled band. This gap is created through the collective motions of pairs of electrons with opposite spins and momenta. Once excited to energy states above the gap, electrons cannot decay back to their "normal" states below it by usual means, such as loss of energy by collisions with the ion cores in the crystal lattice. Thus these electrons become free to move through the lattice without any scattering by the ion cores.

A superconducting magnet used with a 12-foot-diameter hydrogen bubble chamber at Argonne National Laboratory. Its special niobium-titanium coils become superconducting at a temperature of about 4.5 K, and their electrical resistance becomes zero. The magnet weighs about 110 tons and produces a magnetic field of 1.8 T (18,000 gauss). A conventional electromagnet with similar characteristics would consume about 10 megawatts of electrical power. (Courtesy of Argonne National Laboratory.)

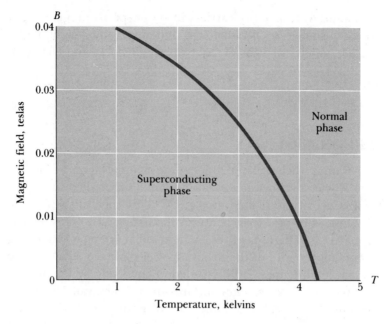

43-15 Phase diagram for pure mercury, showing the superconducting transition critical temperature and its dependence on magnetic field. Other superconducting materials have similar curves but with different scales.

Practical applications of superconductivity: powerful magnets, efficient power transmission, levitating trains

Many important and exciting applications of superconductors are under development. Superconducting electromagnets have been used in research laboratories for several years. Once a current is established in the coil of such a magnet, no additional power input is required because there is no resistive energy loss. The coils can also be made more compact because there is no need to provide channels for the circulation of cooling fluids. Hence superconducting magnets can attain very large fields much more easily and economically than conventional magnets; fields of the order of 10 T are fairly routine. These considerations also make superconductors attractive for long-distance electric-power transmission, an active area of development.

One of the most glamorous applications of superconductors is in the field of magnetic levitation. Imagine a superconducting ring mounted on a railroad car that runs on a magnetized rail. The current induced in the ring leads to a repulsive interaction with the rail, and levitation is possible. Magnetically levitated trains have been running on an experimental basis in Japan for several years, and similar development projects are also under way in Germany.

SUMMARY

KEY TERMS

central-field approximation
Pauli exclusion principle
shell
atomic number
periodic table of the elements
ionic bond
ionization energy
electron affinity
covalent bond
van der Waals bond

In the central-field approximation, each electron in an atom moves under the action of a spherically symmetric electric field caused by the nucleus and all the other electrons. In this approximation, the quantum state of each electron is identified by the four quantum numbers n, l, m, and s. The Pauli exclusion principle states that no two electrons can occupy the same quantum-mechanical state, that is, have the same four quantum numbers. The general nature of the structure of complex atoms, including the main features of the periodic table of the elements, can be understood on the basis of filling up successive energy levels as the atomic number Z increases.

In an ionic bond, one atom gives one or more electrons completely to another; the resulting positive and negative ions attract each other. In a covalent bond, one or more pairs of electrons occupy the space between atoms and cause an attractive interaction. To satisfy the exclusion principle, the two electrons in a pair must have opposite spins. The van der Waals bond is a much weaker dipole–dipole interaction, and the hydrogen bond uses the positive charge of the hydrogen nucleus to form an attractive interaction.

Diatomic molecules have sets of energy levels associated with vibrational and rotational motion; the rotational levels are usually much more closely spaced than the vibrational ones. Transitions among these levels result in molecular spectra. The energies are typically in the range of 0.0001 to 0.1 eV, and the corresponding photons are in the far infrared region of the spectrum.

Solids may be amorphous (having no long-range pattern) or crystalline (having a definite pattern extending over many atoms or molecules). The ionic, covalent, van der Waals, and hydrogen bonds play the same role in binding atoms or molecules into a solid structure as in molecules. An additional type of solid is the metallic solid, in which one or more electrons are detached from each atom and can wander freely through the lattice. Electrical and thermal conductivity of the various types of solids can be understood on the basis of the presence or absence of mobile electrons. The electron states can be further understood on the basis of energy bands, which are ranges of allowed electron energies, separated by ranges of forbidden energies.

Semiconductors have electrical conductivity intermediate between those of good conductors and good insulators. Their conductivity can be changed drastically by addition of minute quantities of impurities, which can lead to conduction predominantly by electrons (n-type) or by holes (p-type). Diodes and transistors contain p-n junctions, regions of transition between p-type and n-type material. A large number of junctions, capacitors, resistors, and conductors can be built up on a silicon chip by successive deposition of layers and etching.

Superconductivity is the complete disappearance of a material's electrical resistance at very low temperature. The transition to the superconducting state is a phase transition, and the temperature at which it occurs depends on the magnetic field. There can be no magnetic field inside a superconducting material.

hydrogen bond
molecular spectra
rotational energy levels
vibrational energy levels
band spectrum
amorphous solid
crystalline solid
ionic crystal
covalent crystal
metallic crystal
close packing
energy bands
valence band
conduction band
semiconductor
hole
n-type semiconductor
p-type semiconductor
p-n junction
diode
transistor
integrated circuit
superconductors

QUESTIONS

43–1 In the ground state of the helium atom, the electrons must have opposite spins. Why?

43–2 The central-field approximation is more accurate for alkali metals than for transition metals (Group IV of the periodic table). Why?

43–3 The outermost electron in the potassium atom is in a 4s state. What does this tell you about the relative positions of the 3d and 4s states for this atom?

43–4 A student asserted that any filled shell (i.e., all the levels for a given n occupied by electrons) must have zero total angular momentum and hence must be spherically symmetric. Do you believe this? What about a filled subshell (all values of m for given values of n and l)?

43–5 What factors determine whether a material is a conductor of electricity or an insulator?

43–6 The nucleus of a gold atom contains 79 protons. How would you expect the energy required to remove a 1s electron completely from a gold atom to compare with the energy required to remove the electron from a hydrogen atom? In what region of the electromagnetic spectrum would a photon of the appropriate energy lie?

43–7 Elements can be identified by their visible spectra. Could analogous techniques be used to identify compounds from their molecular spectra? In what region of the electromagnetic spectrum would the appropriate radiation lie?

43–8 The ionization energies of the alkali metals (i.e., the energy required to remove an outer electron) are in the range from 4 eV to 6 eV, while those of the inert gases are in the range from 15 eV to 25 eV. Why the difference?

43–9 The energy required to remove the 3s electron from a sodium atom in its ground state is about 5 eV. Would you expect the energy required to remove an additional electron to be about the same, or more or less? Why?

43–10 Electrical conductivities of most metals decrease gradually with increasing temperature, but the intrinsic conductivity of semiconductors always *increases* rapidly with increasing temperature. Why the difference?

43–11 What are some advantages of transistors compared to vacuum tubes, for electronic devices such as amplifiers? What are some disadvantages? Are there any situations in which vacuum tubes *cannot* be replaced by solid-state devices?

EXERCISES

Section 43–1 The Exclusion Principle

Section 43–2 Atomic Structure

43–1

a) List the different possible combinations of quantum numbers l and m for the $n = 5$ shell.

b) How many electrons can be placed in the $n = 5$ shell?

43–2 For germanium ($Z = 32$) make a list of the number of electrons in each state ($1s$, $2s$, $2p$, etc.).

43–3 Work the two examples described in Problem-Solving Strategy step 2 in Section 43–2. That is,

a) show that the magnitude of the electrical potential energy of a proton and an electron 0.1 nm apart is 14.4 eV, or 2.31×10^{-18} J;

b) show that the moment of inertia of two protons 0.1 nm apart about an axis through the center of mass and perpendicular to the line joining them is 0.836×10^{-47} kg·m^2.

Section 43–3 Diatomic Molecules

Section 43–4 Molecular Spectra

43–4

a) Calculate the electrical potential energy for a K$^+$ and a Br$^-$ ion separated by a distance of 0.29 nm, the equilibrium separation in the KBr molecule.

b) The ionization energy of the potassium atom is 4.3 eV. Atomic bromine has an electron affinity of 3.5 eV. Use these data and the results of (a) to *estimate* the binding energy of the KBr molecule. Do you expect the actual binding energy to be larger or smaller than your estimate?

43–5 Show that the frequencies in a pure rotation spectrum (no change in vibrational level) are all integer multiples of the quantity $\hbar/2\pi I$. (Assume that the rotational quantum number l changes by ± 1 in the rotational transitions.)

43–6 If the distance between atoms in a diatomic oxygen molecule is 0.20 nm, calculate the moment of inertia about an axis through the center of mass perpendicular to the line joining the atoms.

43–7 The distance between atoms in a hydrogen molecule is 0.074 nm.

a) Calculate the moment of inertia of a hydrogen molecule about an axis perpendicular to the line joining the nuclei, at its center.

b) Find the energies (in electronvolts) of the $l = 0$, $l = 1$, and $l = 2$ rotational states.

c) Find the wavelength and frequency of the photon emitted in the transition from $l = 2$ to $l = 0$.

43–8 The vibrational frequency of the hydrogen molecule is 1.29×10^{14} Hz.

a) What is the spacing of adjacent vibrational energy levels, in electronvolts?

b) What is the wavelength of radiation emitted in the transition from the $n = 2$ to $n = 1$ vibrational state of the hydrogen molecule?

c) From what initial values of n do transitions to the ground state of vibrational motion yield radiation in the visible spectrum (400 to 700 nm)?

Section 43–5 Structure of Solids

Section 43–6 Properties of Solids

43–9 The spacing of adjacent atoms in a sodium-chloride crystal is 0.282 nm. Calculate the density of sodium chloride.

43–10 Potassium bromide, KBr, has a density of 2.75×10^3 kg·m^{-3}, and the same crystal structure as NaCl.

a) Calculate the average spacing between adjacent atoms in a KBr crystal.

b) How does the value calculated in (a) compare with the spacing in NaCl? Is the relation between the two values qualitatively what you would expect?

Section 43–8 Semiconductors

Section 43–9 Semiconductor Devices

43–11

a) Suppose a piece of very pure germanium is to be used as a light detector by observing the increase in conductivity resulting from generation of electron-hole pairs by absorption of photons. If each pair requires 0.7 eV of energy, what is the maximum wavelength that can be detected? In what portion of the spectrum does it lie?

b) What are the answers to (a) if the material is silicon, with an energy requirement of 1.1 eV per pair, corresponding to the gap between valence and conduction bands in silicon?

PROBLEMS

43–12 For magnesium, the first ionization potential is 7.6 eV; the second (additional energy required to remove a second electron) is almost twice this, 15 eV; and the third ionization potential is much larger, about 80 eV. How can these numbers be understood?

43–13 The vibration frequency for the molecule HCl is 8.6×10^{13} Hz.

a) If this molecule can be regarded as a simple harmonic oscillator with the Cl atom stationary (because it is much more massive than the H atom), what is the effective "spring constant" k corresponding to the interatomic force?

b) What is the spacing between adjacent vibrational energy levels, in joules? In electronvolts?

c) What is the wavelength of a photon emitted in a transition between two adjacent vibrational levels? In what region of the spectrum does it lie?

43—14 In Problem 43–13, suppose the hydrogen atom is replaced by an atom of deuterium, an isotope of hydrogen with nuclear mass 2 u. The effective spring constant k is determined by the electron configuration, so it is the same as for the normal HCl molecule.

a) What is the vibrational frequency of this molecule?

b) What is the wavelength of light emitted in the transition $n = 2$ to $n = 1$? In what region of the spectrum does it lie?

43–15

a) For the sodium chloride molecule (NaCl) discussed at the beginning of Section 43–3, what is the maximum separation of the ions for stability if they may be regarded as point charges? That is, what is the largest separation for which the energy of $Na^+ + Cl^-$, calculated in this model, is lower than the energy of the two separate atoms Na and Cl?

b) Calculate this distance for the potassium bromide (KBr) molecule (Exercise 43–4).

43–16 The rotational spectrum of HCl contains the following wavelengths:

$$60.4 \ \mu m$$
$$69.0 \ \mu m$$
$$80.4 \ \mu m$$
$$96.4 \ \mu m$$
$$120.4 \ \mu m$$

Find the moment of inertia of the HCl molecule.

43–17 The dissociation energy of the hydrogen molecule (i.e., the energy required to separate the two atoms) is 4.72 eV. At what temperature is the average kinetic energy of a molecule equal to this energy?

CHALLENGE PROBLEMS

43–18

a) If we consider the hydrogen molecule (H_2) to be a simple harmonic oscillator with an equilibrium spacing of $r_0 = 0.074$ nm, estimate the vibrational energy-level spacing for H_2. (*Hint:* Estimate the force constant k by equating the net Coulomb repulsion of the protons, if the atoms move slightly closer together than r_0, with the "spring" force. That is, assume that the chemical binding force remains approximately constant as r is decreased slightly from r_0.)

b) Use the results of part (a) to calculate the vibrational energy-level spacing for the deuterium molecule, D_2. As in Exercises 43–13 and 43–14, assume that the spring constant is the same for these two molecules.

(In each case above, the spring constant is related to the angular frequency ω by $\omega = \sqrt{k/\mu}$, where

$\mu = m_A m_B / [m_A + m_B]$ is the *reduced mass* for the diatomic molecule whose atoms have masses m_A and m_B.)

43–19

a) Use the result of Problem 43–16 to calculate the equilibrium separation of the atoms in HCl. Assume a mass of the chlorine atom of 35 u.

b) Given that l changes by ± 1 in rotational transitions, what is the value of l for the upper level of the transition that gives rise to each of the wavelengths listed in Problem 43–16?

c) What will be the longest wavelength line in the rotational spectrum of HCl?

d) Calculate the wavelengths of the emitted light for the corresponding transitions in the DCl molecule. Assume that the equilibrium separation between the atoms is the same as for HCl.

NUCLEAR AND HIGH-ENERGY PHYSICS

EVERY ATOM CONTAINS AT ITS CENTER AN EXTREMELY DENSE, POSITIVELY charged *nucleus*, much smaller than the overall size of the atom but nevertheless containing most of its total mass. In this chapter we will review the experimental evidence for the existence of nuclei and then describe their most important properties in terms of the protons and neutrons of which they are composed. The stability or instability of a particular nuclear structure is determined by the competition between the attractive nuclear force among the protons and neutrons and the repulsive electrical interactions among the protons. Unstable nuclei *decay*, or transform themselves spontaneously into other structures, by a variety of decay processes. Structure-altering nuclear reactions can also be induced by bombardment of a nucleus with particles or other nuclei. Two classes of reactions of special interest are *fission* and *fusion*.

Protons and neutrons are not among the most fundamental of particles; they are believed to be composed of more basic entities called *quarks*, which also play a central role in understanding the relationships among the four types of interactions: strong, electromagnetic, weak, and gravitational. Research into the nature and interactions of fundamental particles has required the construction of very large experimental facilities such as particle accelerators, as scientists continue to probe more and more deeply into this most basic aspect of the nature of our physical universe.

44–1 THE NUCLEAR ATOM

In 1900 nobody knew what the inside of an atom was like.

The first experimental evidence for the existence of nuclei in atoms was provided by the **Rutherford scattering** experiments, carried out in 1910–1911 by Sir Ernest Rutherford and two of his students, Hans Geiger and Ernest Marsden, at Cambridge, England. The electron had been discovered 13 years earlier, in 1897, by Sir J. J. Thomson, and Millikan had measured its charge in 1910. It was also known by this time that all atoms except hydrogen contain more than one electron. Thomson had proposed a model of the atom that included a relatively large (of the order of 10^{-10} m) sphere of positive charge, with the electrons embedded in it like chocolate chips in a cookie.

Rutherford's experiment consisted of projecting other charged particles at the atoms under study. Observations of the ways the projected particles were deflected, or *scattered,* provided information about the internal structure and charge distribution of the atoms. The particle accelerators now in common use in high-energy physics laboratories had not yet been invented, and Rutherford's projectiles were alpha particles emitted from naturally radioactive elements. We now know that alpha particles are identical with the nuclei of helium atoms, two protons and two neutrons bound together. They are ejected from unstable nuclei with speeds of the order of 10^7 m·s^{-1}, and they can travel several centimeters through air, or 0.1 mm or so through solid matter, before they are brought to rest by collisions.

Rutherford's experimental setup is shown schematically in Fig. 44–1. A radioactive material at the left emits alpha particles. Thick lead screens stop all particles except those in a narrow *beam* defined by small holes. The beam then passes through a thin gold, silver, or copper foil and strikes a plate coated with zinc sulfide. A momentary flash, or *scintillation,* can be seen on the screen whenever it is struck by an alpha particle. (This is the same phenomenon that makes your TV screen glow when the electron beam strikes it.) The numbers of particles deflected through various angles can therefore be determined.

According to the Thomson model, the atoms of a solid are packed together like marbles in a box. An alpha particle can pass through a thin sheet of metal foil; thus if the Thomson model is correct, the alpha particle must actually penetrate the spheres of positive charge. Assuming for the moment that this happens, we can compute the deflection it would undergo. The Thomson atom is electrically neutral, so outside the atom no force would be exerted on the alpha particle. Within the atom, the electrical force would be due in part to the electrons and in part to the sphere of positive charge. The mass of an alpha particle, however, is about 7400 times that of an electron; momentum considerations show that the alpha particle can be scattered only a negligible amount by its interaction with the electrons. It is like driving a car through a hailstorm; the hailstones do not deflect the car much. Only interactions with the *positive* charge, which makes up most of the mass of the atom, can deflect the alpha particle appreciably.

Assuming the positive charge is distributed through the whole atom, as in the Thomson model, we can calculate the maximum deflection of an alpha

Exploring atomic structure by shooting atomic bullets into atoms

Primitive but crucial experiments in nuclear structure

44–1 The scattering of alpha particles by a thin metal foil.

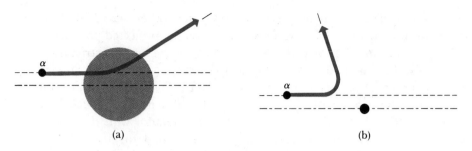

44–2 (a) Alpha particle scattered through a small angle by the Thomson atom. (b) Alpha particle scattered through a large angle by the Rutherford nuclear atom.

The nucleus is extremely small and extremely dense.

particle. It turns out that the interaction potential energy is much *smaller* than the kinetic energy of the alpha particles, and that the maximum deflection to be expected is only a few degrees, as Fig. 44–2a suggests.

The results were very different from this and were totally unexpected. Some alpha particles were scattered by nearly 180°, that is, almost straight backward. Rutherford wrote later:

> It was quite the most incredible event that ever happened to me in my life. It was almost as incredible as if you had fired a 15-inch shell at a piece of tissue paper and it came back and hit you.

Back to the drawing board! Suppose the positive charge, instead of being distributed through a sphere of atomic dimensions (of the order of 10^{-10} m), is all concentrated in a much *smaller* volume. Rutherford called this concentration of charge the **nucleus.** Then it would act like a point charge down to much smaller distances. The maximum repulsive force on the alpha particle would be much larger, and large-angle scattering would be possible, as in Fig. 44–2b. Rutherford again computed the numbers of particles expected to be scattered through various angles. Within the precision of his experiments, the computed and measured results agreed, down to distances of the order of 10^{-14} m. Thus his experiments confirmed the concept of the nucleus of the atom as a very small, very dense structure containing all the positive charge and most of the mass of the atom. They also showed that the size of the nucleus cannot be larger than 10^{-14} m.

44–2 PROPERTIES OF NUCLEI

As we have just seen, the most obvious feature of the atomic nucleus is its size, 20,000 to 200,000 times smaller than that of the atom itself. Since Rutherford's initial experiments, many additional scattering experiments have been performed, using high-energy protons, electrons, and neutrons as well as alpha particles. Although the "surface" of a nucleus is not a sharp boundary, these experiments have determined an approximate radius for each nucleus. The radius is found to depend on the mass, which in turn depends on the total number A of neutrons and protons in the nucleus, usually called the **mass number.** The radii of most nuclei are represented fairly well by the empirical equation

$$r = r_0 A^{1/3}, \tag{44–1}$$

where r_0 is an empirical constant equal to 1.2×10^{-15} m and is the same for all nuclei.

All nuclei have more or less the same density.

Since the volume of a sphere is proportional to r^3, Eq. (44–1) shows that the *volume* of a nucleus is proportional to A (i.e., to the total mass), and therefore that the mass per unit volume (proportional to A/r^3) is the same for all nuclei. That is, *all nuclei have approximately the same density.* This fact is of crucial importance in understanding nuclear structure.

Two additional important properties of nuclei are angular momentum and magnetic moment. The particles in the nucleus are in motion, just as the electrons in an atom are in motion. Associated with this motion is angular momentum, and, because circulating charge constitutes a current, there is also magnetic moment. Experimental evidence for the existence of nuclear angular momentum (often called **nuclear spin**) and nuclear magnetic moment came originally from spectroscopy. Some spectrum lines are found to be split into series of very closely spaced lines, called *hyperfine structure,* which can be understood on the basis of interactions between electrons and the nuclear magnetic moment. Detailed analysis indicates that the nuclear angular momentum is *quantized,* just as it is for electrons and for molecular rotation. The component of angular momentum in a specified axis direction is a multiple of $\hbar$, but some nuclei have *integral* multiples (as with orbital angular momentum of electrons) and some *half-integral* multiples of $\hbar$ (as with electron spin). Nuclear spin can be understood in detail by analysis of the motions of the particles making up the nucleus.

> Nuclei have spin angular momentum and magnetic moment.

Although it was once believed that nuclei were made of protons and electrons, the discovery of the neutron (discussed in Section 44–9) and many other experiments have established that the basic building blocks of the nucleus are the **proton** and the **neutron.** The total number of protons, equal in a neutral atom to the number of electrons, is the **atomic number** Z. The total number of **nucleons** (protons and neutrons) is called the mass number A. The number of neutrons, denoted by N, is called the **neutron number.** For any nucleus, these three quantities are related by

> Nuclei are made of protons and neutrons.

$$A = Z + N. \tag{44–2}$$

Table 44–1 lists values of A, Z, and N for several nuclei. As the table shows, some nuclei have the same Z but different N. Since the electron structure of an atom, which determines its chemical properties, depends on the charge of the nucleus, these are nuclei of the same element, but they have different masses and can be distinguished in precise experiments. Nuclei of a given element having different mass numbers are called **isotopes** of the element, and a single nuclear species (unique values of both Z and N) is called a **nuclide.** We studied the experimental investigation of *isotopes* through mass spectroscopy in Section 30–6, and we suggest you review that section.

Table 44–1 also shows the usual notation for individual nuclides; we write the symbol of the element, with a presubscript equal to Z and a presuperscript equal to the mass number A. This is redundant, since the element is determined by the atomic number, but the notation is a useful aid to memory.

> Notation for describing the constituents of nuclei

The total mass of a nucleus is always *less* than the total mass of its constituent parts because of the mass equivalent ($E = mc^2$) of the (negative) potential energy associated with the attractive forces that hold the nucleus together. This mass difference is sometimes called the **mass defect.** In fact, the best way to determine the total potential energy, or **binding energy,** of a nucleus is to compare its mass with the total mass of its constituents. The proton and neutron masses are

$$m_{\mathrm{p}} = 1.6726 \times 10^{-27} \ \mathrm{kg},$$

$$m_{\mathrm{n}} = 1.6750 \times 10^{-27} \ \mathrm{kg}.$$

TABLE 44–1   Compositions of Some Common Nuclei

Nucleus	Mass Number (Total Number of Nuclear Particles), A	Atomic Number (Number of Protons), Z	Neutron Number, $N = A - Z$
$^1_1$H	1	1	0
$^2_1$D	2	1	1
$^4_2$He	4	2	2
$^6_3$Li	6	3	3
$^7_3$Li	7	3	4
$^9_4$Be	9	4	5
$^{10}_5$B	10	5	5
$^{11}_5$B	11	5	6
$^{12}_6$C	12	6	6
$^{13}_6$C	13	6	7
$^{14}_7$N	14	7	7
$^{16}_8$O	16	8	8
$^{23}_{11}$Na	23	11	12
$^{65}_{29}$Cu	65	29	36
$^{200}_{80}$Hg	200	80	120
$^{235}_{92}$U	235	92	143
$^{238}_{92}$U	238	92	146

Since these values are nearly equal, it is not surprising that many nuclear masses are approximately integer multiples of the proton or neutron mass.

This observation suggests that we define a new mass unit equal to the proton or neutron mass. Instead, for reasons of precision of measurement, it

The atomic mass unit: a convenient unit for nuclear masses

has been found more convenient to define a new unit, called the **atomic mass unit** (u), as $\frac{1}{12}$ the mass of the neutral carbon atom having mass number $A = 12$. It is found that

$$1 \text{ u} = 1.660566 \times 10^{-27} \text{ kg.}$$

In atomic units, the masses of the proton, neutron, and electron are found to be

$$m_p = 1.007276 \text{ u,}$$

$$m_n = 1.008665 \text{ u,}$$

$$m_e = 0.000549 \text{ u.}$$

The masses of some common atoms, including their electrons, are shown in Table 44–2. The masses of the bare nuclei are obtained by subtracting Z

The energy equivalent of the atomic mass unit

times the electron mass. The energy equivalent of 1 u, which we need for calculations of mass defect and binding energy, is found from the relation $E = mc^2$:

$$E = (1.660566 \times 10^{-27} \text{ kg})(2.998 \times 10^8 \text{ m·s}^{-1})^2$$

$$= 1.492 \times 10^{-10} \text{ J}$$

$$= 931.5 \text{ MeV.}$$

TABLE 44–2 Atomic Masses of Light Elements

Element	Atomic Number, Z	Neutron Number, N	Atomic Mass, u	Mass Number, A
Hydrogen H	1	0	1.00783	1
Deuterium H	1	1	2.01410	2
Helium He	2	1	3.01603	3
Helium He	2	2	4.00260	4
Lithium Li	3	3	6.01512	6
Lithium Li	3	4	7.01600	7
Beryllium Be	4	5	9.01218	9
Boron B	5	5	10.01294	10
Boron B	5	6	11.00931	11
Carbon C	6	6	12.00000	12
Carbon C	6	7	13.00336	13
Nitrogen N	7	7	14.00307	14
Nitrogen N	7	8	15.00011	15
Oxygen O	8	8	15.99491	16
Oxygen O	8	9	16.99913	17
Oxygen O	8	10	17.99916	18

Source: Encyclopedia of Physics, Lerner and Trigg, eds., (Reading, Mass.: Addison-Wesley, 1981).

EXAMPLE 44–1 Find the mass defect, the total binding energy, and the binding energy per nucleon for the common isotope of carbon, ^{12}C.

SOLUTION The mass of the neutral carbon atom, including the nucleus and the six electrons, is, according to Table 44–2, 12.00000 u. The mass of the bare nucleus is obtained by subtracting the mass of the six electrons:

$$m = 12.00000 \text{ u} - (6)(0.000549 \text{ u})$$

$$= 11.996706 \text{ u}.$$

The total mass of the six protons and six neutrons in the nucleus is

$$(6)(1.007276 \text{ u}) + (6)(1.008665 \text{ u}) = 12.095646 \text{ u}.$$

The mass defect is therefore

$$12.095646 \text{ u} - 11.996706 \text{ u} = 0.09894 \text{ u}.$$

The energy equivalent of this mass is

$$(0.09894 \text{ u})(931.5 \text{ MeV·u}^{-1}) = 92.16 \text{ MeV}.$$

Thus the total binding energy for the 12 nucleons is 92.16 MeV. To pull the carbon nucleus completely apart into 12 separate nucleons would require a minimum of 92.16 MeV. The binding energy *per nucleon* is 1/12 of this amount, or 7.68 MeV per nucleon. Nearly all stable nuclei, from the lightest to the most massive, have binding energies in the range of 6 to 9 MeV per nucleon. Figure 44–3 is a graph of binding energy per nucleon, as a function of the mass number *A*.

The mass of a nucleus is always less than the total mass of its constituents.

The mass defect: mass equivalent of the binding energy

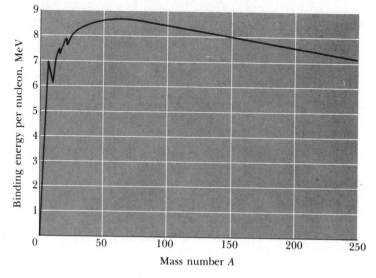

44–3 Binding energy per nucleon as a function of mass number A (the total number of nucleons). The curve reaches a peak of about 8.7 MeV at about $A = 60$, corresponding to the element iron. The spike at $A = 4$ shows the unusual stability of the α-particle structure.

Nuclear energy levels are a million times as large as atomic energy levels.

Characteristics of the nuclear force: What holds nucleons together?

Because the nucleus is a composite structure, it can have internal motion, with corresponding energy levels. Thus each nucleus has a set of allowed energy levels, corresponding to a *ground state* (state of lowest energy) and several *excited states*. Because of the strength of nuclear interactions, excitation energies of nuclei are typically of the order of 1 MeV, compared with a few eV for atomic energy levels. In ordinary physical and chemical transformations, the nucleus always remains in its ground state. When a nucleus is placed in an excited state, either by bombardment by high-energy particles or by a radioactive transformation, it can decay to the ground state by emission of one or more photons, called in this case **gamma rays,** or *gamma-ray photons.*

The forces that hold protons and neutrons together in the nucleus, despite the electrical repulsion of the protons, are not of any familiar sort but are unique to the nucleus. Some aspects of the **nuclear force** are still incompletely understood, but we can describe several qualitative features. First, it does not depend on charge; neutrons as well as protons must be bound, and the binding is found to be the same for both. Second, it must be of short range; otherwise the nucleus would pull in additional protons and neutrons. But within its range, the nuclear force must be stronger than electrical forces; otherwise the nucleus could never be stable. Third, the nearly constant density of nuclear matter and the nearly constant binding energy per nucleon indicate that a given nucleon cannot interact simultaneously with *all* the other nucleons in a nucleus, but only with those few in its immediate vicinity. This feature is again in contrast to the behavior of electrical forces, in which *every* proton in the nucleus repels every other one. This limitation on the maximum number of nucleons with which a nucleon can interact is called *saturation;* it is analogous in some respects to covalent bonding in solids. Finally, the nuclear force favors binding of *pairs* of protons or neutrons with opposite spins, and of *pairs of pairs,* a pair of protons and a pair of neutrons, with each pair having total spin zero. For example, the alpha particle is an exceptionally stable nuclear structure.

These qualitative features of the nuclear force are helpful in understanding the various kinds of nuclear instability, discussed in the following sections.

PROBLEM-SOLVING STRATEGY: Nuclear structure

1. As in Chapters 42 and 43, some familiarity with numerical magnitudes is helpful. The scale of things in nuclear structures is very different from that in atomic structures. The size of a nucleus is of the order of 10^{-15} m; the potential energy of interaction of two protons at this distance is 2.31×10^{-13} J or 1.44 MeV. Hence characteristic nuclear energies are of the order of a few MeV, rather than a few eV as with atoms. Protons and neutrons are about 1840 times as massive as electrons. The binding energy per nucleon is roughly 1% of the rest energy of a nucleon; compare this with the ionization energy of the hydrogen atom, which is only 0.003% of its rest energy. Angular momentum is of the same order of magnitude in both atoms and nuclei because it is determined by the value of Planck's constant. But magnetic moments of nuclei are typically much *smaller* than those of electrons in atoms because the nuclear gyromagnetic ratio (the ratio of magnetic moment to angular momentum) is $e/2m_p$ instead of $e/2m_e$, smaller than for orbital electron motion by a factor of the order of 2000. Check out all these numbers, and try to think of other magnitudes to check.

2. In energy calculations involving the mass defect, binding energies, and so on, the mass tables nearly always list the masses of neutral atoms, including their full complements of electrons. To get the mass of a bare nucleus, you have to subtract the masses of these electrons. In principle, you should also take into account the binding energies of the electrons; we won't worry about that in this book, but in very precise work it does have to be considered. Binding-energy calculations often involve subtracting two quantities that are nearly equal; to get enough precision in the difference you often have to carry five or six significant figures, if that many are available. If not, you may have to be content with an approximate result.

44–3 NUCLEAR STABILITY

Of about 1500 different nuclides now known, only about one-fifth are stable. The others are **radioactive;** this means they are unstable structures that decay to form other nuclides by emitting particles and electromagnetic radiation. The time scale of these decay processes ranges from a small fraction of a microsecond to billions of years. The stable nuclides are shown by dots on the graph in Fig. 44–4, where the neutron number N and charge number (or atomic number) Z are plotted. Such a chart is called a *Segrè chart,* after its inventor.

Why are some nuclear structures stable and others unstable?

The mass number A is equal to the sum $N + Z$, so a curve of constant A is a straight line perpendicular to the line $N = Z$. In general, lines of constant A pass through only one or two stable nuclides; that is, there are usually only one or two stable nuclides with a given mass number. The lines at $A = 20$, $A = 40$, $A = 60$, and $A = 80$ are examples. In four cases, these lines pass through *three* stable nuclides, namely, at $A = 96$, 124, 130, and 136. Only four stable nuclides have both odd Z and odd N:

$$^{2}_{1}\text{H}, \quad ^{6}_{3}\text{Li}, \quad ^{10}_{5}\text{B}, \quad ^{14}_{7}\text{N};$$

these are called *odd-odd nuclides*. Also, there is *no* stable nuclide with $A = 5$ or $A = 8$.

The points representing stable nuclides define a rather narrow stability region. For low mass numbers, $N/Z = 1$. This ratio increases and becomes about 1.6 at large mass numbers. Points to the right of the stability region represent nuclides that have too many protons, or not enough neutrons, to be stable. To the left of the stability region are the points representing nuclei with

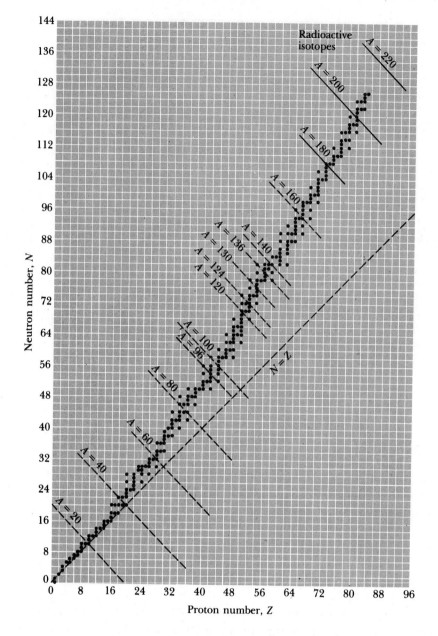

44–4 Segre chart, showing neutron number and proton number for stable nuclides.

A nucleus can be too big to be stable.

The neutron–proton balance has to be right for the nucleus to be stable.

too many neutrons or not enough protons. The graph also shows that there are maximum values of A and Z; no nuclide with A greater than 209 or Z greater than 83 is stable.

The stability of nuclei can be understood qualitatively on the basis of the nature of the nuclear force and the competition between the attractive nuclear force and the repulsive electrical force. As we mentioned in Section 44–2, the nuclear force favors *pairs* of nucleons and *pairs of pairs*. In the absence of electrical interactions, the most stable nuclei would be those having equal numbers of neutrons and protons, $N = Z$. The electrical repulsion shifts the balance to favor greater numbers of neutrons, but a nucleus with *too many* neutrons is unstable because not enough of them are paired with protons. A nucleus with too many *protons* has too much repulsive electrical interaction, compared with the attractive nuclear interaction, to be stable.

As the number of nucleons increases, the total energy of electrical interaction increases faster than that of the nuclear interaction. To understand this behavior, recall the discussion of electrostatic energy in Section 27–4. The energy of a capacitor with a charge Q is proportional to Q^2. It can be shown that to bring a total charge Q to form a uniform volume charge distribution inside a sphere of radius a requires a total energy $3Q^2/20\pi\epsilon_0 a$. Thus the (positive) electric potential energy in the nucleus increases approximately as Z^2, while the (negative) nuclear potential energy increases approximately as A, with corrections for pairing effects. At large A, the electric energy *per nucleon* grows faster than the nuclear energy per nucleon, until the point is reached where stability is impossible. Thus the competition of electric and nuclear forces accounts for the fact that the neutron–proton ratio in stable nuclei increases with Z, and also for the fact that there are maximum values of A and Z for stability. Unstable nuclei respond to these conditions in various ways; in the next several sections we discuss various types of decay of unstable nuclei.

44–4 RADIOACTIVE TRANSFORMATIONS

The study of naturally occurring radioactivity began in 1896, only one year after Röntgen discovered x-rays, when Becquerel found that uranium salts emit a radiation that seemed similar to x-rays. Intensive investigation in the following two decades by Marie and Pierre Curie, Rutherford, and many others revealed that the emissions consist of positively and negatively charged and neutral particles, which were christened **alpha, beta,** and **gamma particles.** It was later established that alpha particles are helium nuclei (two protons and two neutrons bound together), betas are high-energy electrons, and gammas are high-energy electromagnetic-wave photons. The Curies discovered the elements radium and polonium, and showed that the emission of radiation from radium, per unit mass, is of the order of a million times more intense than that from uranium.

Not all three particles are emitted at once by all radioactive elements. Some radioactive elements emit alphas, some emit betas, and gammas sometimes accompany one and sometimes the other. No change in physical or chemical environment, such as chemical reactions or heating or cooling, affects the rate of decay. As soon as the existence of the nucleus was established by Rutherford, it was suspected that radioactivity was a nuclear process. Emission of a charged particle from a nucleus leaves behind a nucleus with a different charge, which therefore must belong to a different chemical element. Thus radioactivity has to involve the transformation of one element into another, which is called *transmutation of elements.*

The first measurement of the charge of the alpha particle was made with a *Geiger counter,* a device that detects high-energy charged particles by the ionization they cause in a gas enclosed in a tube. The gas becomes temporarily conductive when a high-energy charged particle passes through. Rutherford and Geiger counted the number of alpha particles emitted from a radium source in a known time interval. They then allowed the alpha particles from the same source to fall upon a conducting plate and measured its rate of increase of charge. The charge on an alpha particle was found to be equal, within experimental error, to twice the charge on an electron, but opposite in sign. The *mass* of the alpha particle was determined by measuring the ratio of charge to mass by the electric and magnetic deflection method described in Section 30–5. When the result of this measurement was combined with the

Three kinds of radiation from unstable nuclei

A detector for charged particles emitted by nuclei

Identifying the alpha particle as a helium nucleus: an interesting detective mystery

charge of an alpha particle, the mass was found to be 6.62×10^{-27} kg, almost exactly four times the mass of a hydrogen atom.

Since a helium atom has a mass four times that of a hydrogen atom and, stripped of its two electrons (as a bare nucleus), has a charge equal in magnitude and opposite in sign to two electrons, it seemed certain that alpha particles are helium nuclei. To make the identification certain, however, Rutherford and Royds collected alpha particles in a glass discharge tube over a period of about six days and then established an electric discharge in the tube. In the spectrum of the emitted light they identified the characteristic helium spectrum and established without doubt that alpha particles *are* helium nuclei.

The speed of an alpha particle can be determined from the curvature of its path in a transverse magnetic field. The alpha particles emitted by radium, $^{226}_{88}$Ra, have speeds of about 1.5×10^7 m·s^{-1}. The corresponding kinetic energy is

$$K = \frac{1}{2}(6.62 \times 10^{-27} \text{ kg})(1.5 \times 10^7 \text{ m·s}^{-1})^2$$

$$= 7.4 \times 10^{-13} \text{ J} = 4.6 \times 10^6 \text{ eV}$$

$$= 4.6 \text{ MeV}.$$

This speed, although large, is only 5% of the speed of light, so the nonrelativistic kinetic-energy expression may be used. The energy is *larger* than typical energies of atomic electrons by a factor of the order of a million. Because of these large energies, alpha particles are capable of traveling several centimeters in air, or a few tenths or hundredths of a millimeter through solids, before they are brought to rest by collisions.

Beta particles are electrons moving very fast.

Beta particles are *negatively* charged and are therefore deflected in an electric or magnetic field. Deflection experiments similar to those described in Section 30–5 prove conclusively that beta particles have the same charge and mass as electrons. They are emitted with tremendous speeds, up to 0.9995 that of light. Thus relativistic relations must be used in analyzing their motion. Unlike alpha particles, which are emitted from a given nucleus with one speed or a few definite speeds, beta particles are emitted with various speeds within a range from zero up to a maximum that depends on the emitting nucleus.

The neutrino: a particle predicted from conservation principles long before it was actually observed

In order to satisfy conservation of energy and momentum in beta emission, it is necessary to assume that the emission of a beta particle is accompanied by the emission of another particle with no charge. This particle, called a **neutrino,** has zero rest mass and zero charge and therefore produces very little measurable effect, even in traversing the densest matter. Nevertheless, Reines and Cowan were able in 1953 to detect its existence in a series of extraordinary experiments. Subsequent investigation has shown that in fact there are at least three varieties of neutrinos, the one associated with beta decay and two others associated with the decay of unstable particles, the μ mesons and the τ particles.

Gamma rays are emitted during transitions in nuclear energy levels.

Gamma rays are not deflected by a magnetic field, so they cannot be charged particles. However, they are diffracted at the surface of a crystal in a manner similar to x-rays, but with extremely small angles of diffraction. Diffraction experiments of this sort led to the conclusion that gamma rays are actually electromagnetic waves of extremely short wavelength, of the order of $\frac{1}{100}$ that of x-rays, with correspondingly higher-energy photons.

The gamma-ray spectrum of any individual element is a *line spectrum;* this fact suggests that gamma emission is analogous to emission of spectrum lines from atoms. That is, a gamma-ray photon is emitted when a nucleus proceeds from a state of higher energy to one of lower energy. For example, alpha particles emitted from radium have a kinetic energy of either 4.879 MeV or 4.695 MeV. When a radium nucleus emits an alpha particle with the smaller energy, the resulting nucleus (which corresponds to the element *radon*) has *more* energy than if the higher-energy alpha particle had been emitted. Hence the radon nucleus is left in an excited state. It can then undergo a transition from this state to its ground state, emitting a gamma-ray photon of energy

$$(4.879 - 4.695) \text{ MeV} = 0.184 \text{ MeV}.$$

The *measured* energy of the gamma-ray photon is 0.189 MeV, in excellent agreement.

When a radioactive nucleus decays by alpha or beta emission, the resulting nucleus may also be unstable, and there may be a series of successive decays until a stable configuration is reached. The most abundant radioactive nucleus found on earth is that of uranium $^{238}_{92}\text{U}$, which undergoes a series of 14 decays, including eight alpha emissions and six beta emissions, terminating at the stable isotope of lead, $^{206}_{82}\text{Pb}$.

In alpha decay, the neutron number N and the charge number Z each decrease by two, and the mass number A decreases by four, corresponding to the values $N = 2$, $Z = 2$, $A = 4$ for the alpha particle. The situation in beta decay is less obvious; how can a nucleus composed of protons and neutrons emit an *electron*? The answer is that, in beta decay, a neutron in the nucleus is transformed into a proton, an electron, and a neutrino. We will study such transformations of fundamental particles in Section 44–10. The effect is to *increase* the charge number Z by one, decrease the neutron number N by one, and leave the mass number A unchanged. Finally, gamma emission leaves all three numbers unchanged. Both alpha and beta emissions are often accompanied by gamma emission.

The number of radioactive nuclei in any sample of radioactive material decreases continuously as some of the nuclei disintegrate. The *rate* at which the number decreases, however, varies widely for different kinds of nuclei. Let N or $N(t)$ represent the number of radioactive nuclei in a sample at time t, and let dN be the number that undergo transformations in a short time interval dt. Since every transformation results in a *decrease* in the number N, the corresponding change in N is $-dN$ and the rate of change of N is $-dN/dt$. The larger the number of nuclei in the sample, the larger the number that will undergo transformations, so the rate of change of N is proportional to N, or is equal to a constant λ multiplied by N. Therefore

$$\frac{dN}{dt} = -\lambda N. \tag{44–3}$$

The constant λ is called the **decay constant,** and it has different values for different nuclides. Clearly, a large value of λ corresponds to rapid decay, and conversely. If $N_0 = N(0)$ is the number of nuclei at time $t = 0$, then the solution of this differential equation is an exponential function:

$$N(t) = N_0 e^{-\lambda t}. \tag{44–4}$$

A graph of this function is shown in Fig. 44–5.

The **half-life** $t_{1/2}$ of a radioactive sample is defined as the time at which the number of radioactive nuclei has decreased to one-half the number at

Sometimes several successive decays occur before a stable configuration is reached.

The number of radioactive nuclei in a specimen decreases exponentially with time.

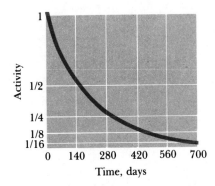

44–5 Decay curve for the radioactive element polonium. Polonium has a half-life of 140 days.

How long does it take for half of the unstable nuclei to decay?

$t = 0$. At this time,

$$e^{-\lambda t_{1/2}} = \frac{N}{N_0} = \frac{1}{2}.$$

Taking natural logarithms of both sides and solving for $t_{1/2}$, we find

$$\lambda t_{1/2} = \ln 2,$$

$$t_{1/2} = \frac{\ln 2}{\lambda} = \frac{0.693}{\lambda}. \tag{44-5}$$

Half of the original nuclei in a radioactive sample decay in a time interval $t_{1/2}$, half of those remaining at this time decay in a second interval $t_{1/2}$, and so on; thus the number remaining after successive intervals of $t_{1/2}$ is $N_0/2$, $N_0/4$, $N_0/8$, and so on.

The **mean lifetime** (or average lifetime) of a nucleus or of an unstable particle is related to the half-life $t_{1/2}$ as follows:

$$t_{\text{mean}} = \frac{1}{\lambda} = \frac{t_{1/2}}{\ln 2} = \frac{t_{1/2}}{0.693}. \tag{44-6}$$

In particle physics, the life of an unstable particle is usually described in terms of the mean lifetime rather than the half-life.

The *activity* of a sample is defined to be the number of disintegrations per unit time. A commonly used unit is the *curie*, abbreviated Ci, defined to be 3.70×10^{10} decays per second. This is approximately equal to the activity of one gram of radium. Since the number of disintegrations is proportional to the number of radioactive nuclei in the sample, the activity decreases exponentially with time in the same way as the number N. Figure 44–5 is a graph of the activity of polonium, $^{210}_{84}$Po, which has a half-life of 140 days.

How to characterize radioactive decay processes

The SI unit of activity is the *becquerel*, abbreviated Bq. One becquerel is one disintegration per second. Thus $1 \text{ Bq} = 1 \text{ s}^{-1}$, and $1 \text{ Ci} = 3.70 \times 10^{10} \text{ Bq}$.

In studying sequences of radioactive decays, the following questions are relevant:

1. What is the parent nucleus?
2. What particle is emitted from this nucleus?
3. What is the resulting nucleus (often called the *daughter nucleus*)?
4. What is the half-life of the parent nucleus?
5. Is the daughter nucleus radioactive, and if so, what are the answers to questions 2, 3, and 4 for this nucleus?

Extensive investigations of radioactive decays have been carried out in the last 75 years, and these questions have been answered for many nuclides. The results for any particular parent nuclide are most conveniently presented on a Segre chart such as that shown in Fig. 44–6. The neutron number N is plotted along the vertical axis and the atomic number (or charge number) Z on the horizontal axis. Unit increase of Z with unit decrease of N indicates beta emission; decrease of two in both N and Z indicates alpha emission. The half-lives are given either in years (y), days (d), hours (h), minutes (m), or seconds (s).

Figure 44–6 shows the uranium decay series, which begins with the common uranium isotope ^{238}U and ends with an isotope of lead, ^{206}Pb. Each arrow represents a decay in which an alpha or beta particle is emitted. The

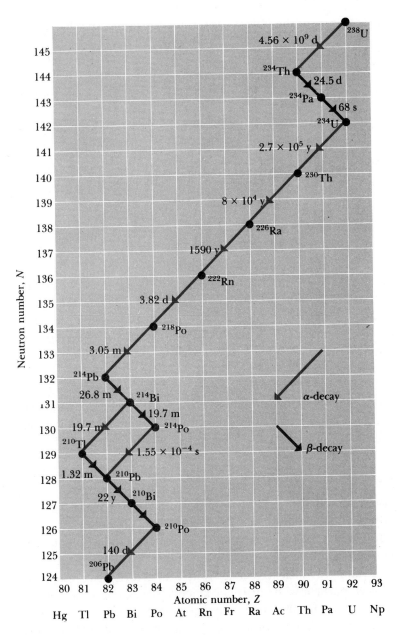

44–6 Segre chart showing the uranium ^{238}U decay series, terminating with the stable nuclide $^{206}_{82}$Pb.

decays can also be represented in equation form; the first two decays in the series are written as

$$^{238}\text{U} \longrightarrow {}^{234}\text{Th} + \alpha,$$

$$^{234}\text{Th} \longrightarrow {}^{234}\text{Pa} + \beta,$$

or, even more briefly, as

$$^{238}\text{U} \xrightarrow{\alpha} {}^{234}\text{Th},$$

$$^{234}\text{Th} \xrightarrow{\beta} {}^{234}\text{Pa}.$$

In some cases there are alternate modes of decay of a particular nucleus.

In the second decay a gamma emission follows the beta emission. The reason is that the beta decay leaves the daughter nucleus in an excited state, from which it decays to the ground state by emitting a photon.

An interesting feature of the ^{238}U decay series is the branching that occurs at ^{214}Bi. This nuclide decays to ^{210}Pb by emission of an alpha and a beta, which can occur in either order. We also note that the series includes unstable isotopes of several elements that also have stable isotopes, including Tl, Pb, and Bi. The unstable isotopes of these elements that occur in the ^{238}U series all have too many neutrons to be stable, as discussed in Section 44–3.

Three other decay series are known; two of these occur in nature, one starting with the uncommon isotope ^{235}U, the other with thorium (^{232}Th). The last series starts with neptunium (^{237}Np), an element not found in nature but produced in nuclear reactors. In each case the series continues until a stable nucleus is reached; for these series the final members are ^{207}Pb, ^{208}Pb, and ^{209}Bi, respectively.

An interesting application of radioactivity is the dating of archaeological and geological specimens by measuring the concentration of radioactive isotopes. The most familiar example is carbon dating. The unstable isotope ^{14}C is produced by nuclear reactions in the atmosphere, caused by cosmic-ray bombardment, and there is a small proportion of ^{14}C in the CO_2 in the atmosphere. Plants, which obtain their carbon from this source, contain the same proportion of ^{14}C as the atmosphere. When a plant dies, it stops taking in carbon, and the ^{14}C it has already taken in decays, with a half-life of 5568 years. Thus by measuring the proportion of ^{14}C in the remains, we can determine how long ago the organism died. Similar techniques are used with other isotopes for dating geologic specimens. A difficulty with carbon dating is that the ^{14}C concentration in the atmosphere changes with time, over long intervals.

Using radioisotopes for archaeological and geological dating

44–5 NUCLEAR REACTIONS

In Section 44–4 we studied the decay of unstable nuclei. The processes we described there were natural, spontaneous emission of an alpha or beta particle, sometimes followed by gamma emission. Nothing was done to initiate this emission, and nothing could be done to control it. The first step in artificially starting and controlling a nuclear reaction was Rutherford's suggestion in 1919 that a massive particle with sufficient kinetic energy might be able to penetrate a nucleus. The result would be either a new nucleus with greater atomic number and mass number or a disintegration of the original nucleus. Rutherford bombarded nitrogen with alpha particles and obtained an oxygen nucleus and a proton, according to the reaction

$$\,^{4}_{2}\text{He} + \,^{14}_{7}\text{N} \longrightarrow \,^{17}_{8}\text{O} + \,^{1}_{1}\text{H}. \qquad (44\text{–}7)$$

Rearranging nuclear structures by bombarding a nucleus with high-energy particles

Note that the sum of the initial atomic numbers is equal to the sum of the final atomic numbers, a condition imposed by conservation of charge. The sum of the initial mass numbers is also equal to the sum of the final mass numbers, but the initial rest mass is *not* equal to the final rest mass. Such a process is called a **nuclear reaction,** a rearrangement of nuclear components that results from bombardment by a particle rather than a spontaneous natural process. This experiment was the first nuclear reaction produced under laboratory conditions.

The difference between the rest masses before and after the reaction corresponds to the **reaction energy,** according to the mass–energy relation $E = mc^2$. If the sum of the final rest masses *exceeds* the sum of the initial rest masses, energy is *absorbed* in the reaction. If the final sum is *less than* the initial sum, energy is released in the form of kinetic energy of the final particles. (1 u is equivalent to 931.5 MeV.)

For example, in the nuclear reaction represented by Eq. (44–7), the rest masses of the various particles, in u, are found from Table 44–2 to be

$$
\begin{array}{ll}
{}^4_2\text{He} = 4.00260 \text{ u} & {}^{17}_8\text{O} = 16.99913 \text{ u} \\
\underline{{}^{14}_7\text{N} = 14.00307 \text{ u}} & \underline{{}^1_1\text{H} = 1.00783 \text{ u}} \\
\phantom{{}^{14}_7\text{N} = }18.00567 \text{ u} & \phantom{{}^1_1\text{H} = }18.00696 \text{ u}
\end{array}
$$

Some nuclear reactions absorb energy; some liberate energy.

(These values include nine electron masses in each case.) The total rest mass of the final products exceeds that of the initial particles by 0.00129 u, which is equivalent to 1.20 MeV. This amount of energy is *absorbed* in the reaction. If the initial particles do not have at least this much kinetic energy, the reaction cannot take place.

Here is another example. When lithium is bombarded by a proton, two alpha particles are produced:

$$
{}^1_1\text{H} + {}^7_3\text{Li} \longrightarrow {}^4_2\text{He} + {}^4_2\text{He}. \tag{44–8}
$$

The sum of the final masses is *smaller* than the sum of the initial values, as shown by the following data:

$$
\begin{array}{ll}
{}^1_1\text{H} = 1.00783 \text{ u} & {}^4_2\text{He} = 4.00260 \text{ u} \\
\underline{{}^7_3\text{Li} = 7.01600 \text{ u}} & \underline{{}^4_2\text{He} = 4.00260 \text{ u}} \\
\phantom{{}^7_3\text{Li} = }8.02383 \text{ u} & \phantom{{}^4_2\text{He} = }8.00520 \text{ u}
\end{array}
$$

(Four electron masses are included on each side.) Since the decrease in mass is 0.01863 u, 17.4 MeV of energy are liberated and appear as kinetic energy of the two separating alpha particles. We can verify this computation by observing the distance the alpha particles travel in air at atmospheric pressure before being brought to rest by collisions with molecules. We can then compare this value with the range of alpha particles of known energy. The range is found to be 8.31 cm. This corresponds to an alpha particle kinetic energy of 8.7 MeV. The energy of the two alphas together is therefore $2 \times 8.7 = 17.4$ MeV, the same value obtained from the mass decrease.

For positively charged particles such as the proton and the alpha particle to penetrate nuclei of the other atoms, they must have enough initial kinetic energy to overcome the potential-energy barrier caused by the repulsive electrostatic forces. For example, in the reaction of Eq. (44–8), if the lithium nucleus has a radius of the order of 2.3×10^{-15} m, as suggested by Eq. (44–1), the repulsive potential energy of the proton (charge $+e$) and the lithium nucleus (charge $+3e$) at this distance is

When the bombarding particles have positive charge, there is a threshold energy because of electrostatic repulsion.

$$
U = \frac{1}{4\pi\epsilon_0} \frac{(e)(3e)}{r} = \frac{3(9.0 \times 10^9 \text{ N·m}^2\text{·C}^{-2})(1.6 \times 10^{-19} \text{ C})^2}{2.3 \times 10^{-15} \text{ m}}
$$

$$
= 3.01 \times 10^{-13} \text{ J} = 1.88 \times 10^6 \text{ eV} = 1.88 \text{ MeV}.
$$

Even though energy is liberated in this reaction, the proton must have a minimum, or **threshold,** energy of about 1.9 MeV for the reaction to occur.

Absorption of *neutrons* by nuclei forms an important class of nuclear reactions. Heavy nuclei bombarded by neutrons in a nuclear reactor can undergo a series of neutron absorptions alternating with beta decays, in which the mass

Several new elements have been produced in nuclear physics laboratories; they are all unstable.

number A increases by as much as 25. The *transuranic elements*, elements having Z larger than 92, are produced in this way. These elements do not occur in nature. Seventeen transuranic elements, having Z up to 109 and A up to about 265, have been identified.

The analytical technique of *neutron activation analysis* uses similar reactions. When stable nuclei are bombarded by neutrons, some absorb neutrons and then undergo beta decay. The energies of the beta and gamma emissions depend on the parent nucleus and hence provide a means of identifying it. The presence of elements in quantities far too small for conventional chemical analysis can be detected in this way.

44–6 NUCLEAR FISSION

A nucleus can sometimes split into two massive fragments.

Up to this point, all the nuclear reactions we have considered involve the ejection of relatively light particles, such as alpha particles, beta particles, protons, or neutrons. This is not always the case, as Hahn and Strassman discovered in Germany in 1939. These scientists bombarded uranium ($Z = 92$) with neutrons, and after a careful chemical analysis they discovered barium ($Z = 56$) and krypton ($Z = 36$) among the products. Cloud chamber photographs showed the two heavy particles traveling in opposite directions with tremendous speed. In this process, the uranium nucleus is said to undergo **fission,** and the two pieces resulting from this split are called **fission fragments.** Measurement showed that an enormous amount of energy, 200 MeV, is released when the uranium nucleus splits up in this way. The rest mass of a uranium atom is greater than the sum of the rest masses of the fission products. The energy released during fission emerges as kinetic energy of the fission fragments. Uranium fission is usually accompanied by the release of a few free neutrons. Both of the two most abundant isotopes of uranium, ^{238}U and ^{235}U, may be split by neutron bombardment.

Energy is liberated when fission occurs.

When uranium undergoes fission, barium and krypton are not the only products. Over 100 different isotopes of more than 20 different elements have been detected among the fission products. All of these atoms are in the middle of the periodic table, however, with atomic numbers ranging from 34 to 58. Because the neutron–proton ratio needed for stability in this range is much *smaller* than that of the original uranium nucleus, the fission fragments always have too many neutrons for stability. A few free neutrons are liberated during fission, and the fission fragments undergo a series of beta decays (each of which increases Z by one and decreases N by one) until a stable nucleus is reached. During decay of the fission fragments, an average of 15 MeV of additional energy is liberated.

Fission can be triggered by neutron absorption.

Discovery of the facts that 200 MeV of energy are released when a uranium nucleus undergoes fission triggered by neutron bombardment, and that other neutrons are liberated from the uranium nucleus during fission, suggested the possibility of a **chain reaction;** that is, a self-sustaining series of events that, once started, continues until much of the uranium in a given sample is used up (provided the sample stays together). In the case of a uranium chain reaction, a neutron causes one uranium atom to undergo fission, during which a large amount of energy and several neutrons are emitted. These neutrons then cause fission in neighboring uranium nuclei, which also give out energy and more neutrons. The chain reaction may be made to pro-

Chain reactions: Neutrons emitted during fission trigger additional fissions, ad infinitum.

ceed slowly and in a controlled manner; the device for accomplishing this effect is called a **nuclear reactor.** If the chain reaction is fast and uncontrolled, the device is a bomb.

The probability of neutron absorption by a nucleus is much larger for low-energy (less than 1 eV) neutrons than for the higher-energy neutrons liberated during fission. In a nuclear reactor, the fissionable isotope is contained in *fuel elements;* neutrons emitted during fission are slowed down by collisions with nuclei in the surrounding material, so that they can cause further fissions. The material where the neutrons are slowed down is called the *moderator;* in reactors in nuclear-power plants it is often water. On the average each fission produces about 2.5 free neutrons, so 40% of the neutrons are needed to sustain a chain reaction. The *rate* of the reaction is controlled by inserting or withdrawing *control rods* made of elements (often cadmium) whose nuclei *absorb* neutrons without undergoing any additional reaction.

The most familiar application of nuclear reactors is for the generation of electric power. To illustrate some of the numbers involved, consider a hypothetical nuclear-power plant with a generating capacity of 1000 MW; this figure is typical of large plants currently being built. As noted above, the fission energy appears as kinetic energy of the fission fragments, and its immediate result is to heat the fuel elements and the surrounding water. This heat generates steam to drive turbines, which in turn drive the electrical generators. The turbines, being heat engines, are subject to the efficiency limitations imposed by the second law of thermodynamics, as we discussed in Chapter 19. In modern nuclear plants the overall efficiency is about one-third, so 3000 MW of thermal power from the fission reaction are needed to generate 1000 MW of electrical power.

It is easy to calculate how much uranium must undergo fission per unit time to provide 3000 MW of thermal power. Each second, we need 3000 MJ

What does the moderator in a nuclear reactor do?

How much uranium does a nuclear power plant use?

Control rods in a nuclear-fission reactor. The rods are made of a material that absorbs neutrons without undergoing any further nuclear reaction. Inserting or withdrawing the rods from the interior of the reactor controls the rate of the nuclear-fission reaction. (Courtesy of Southern California Edison.)

or 3000×10^6 J. Each fission provides 200 MeV, which is

$$200 \text{ MeV} = (200 \text{ MeV})(1.6 \times 10^{-13} \text{ J} \cdot \text{MeV}^{-1}) = 3.2 \times 10^{-11} \text{ J}.$$

Thus the number of fissions needed per second is

$$\frac{3000 \times 10^6 \text{ J}}{3.2 \times 10^{-11} \text{ J}} = 0.94 \times 10^{20}.$$

Each uranium atom has a mass of about $(235)(1.67 \times 10^{-27} \text{ kg}) = 3.9 \times 10^{-25}$ kg, so the mass of uranium needed per second is

$$(0.94 \times 10^{20})(3.9 \times 10^{-25} \text{ kg}) = 3.7 \times 10^{-5} \text{ kg} = 37 \text{ mg}.$$

In one day (86,400 seconds), the consumption of uranium is

$$(3.7 \times 10^{-5} \text{ kg} \cdot \text{s}^{-1})(86,400 \text{ s} \cdot \text{d}^{-1}) = 3.2 \text{ kg} \cdot \text{d}^{-1}.$$

For comparison, note that the 1000-MW coal-fired power plant described in Section 19–10 burns 10,600 tons (about 10^7 kg) of coal per day! Fission of one uranium nucleus liberates 200 MeV of energy, while combustion of one carbon atom yields about 2 eV.

Nuclear fission reactors have several other practical uses. Among these are the production of artificial radioactive isotopes for medical and other research; production of high-intensity neutron beams for research in nuclear structure; and production of fissionable transuranic elements such as plutonium from the common isotope ^{238}U. The last is the function of *breeder reactors*.

Additional energy is evolved by beta decay of the radioactive fission fragments.

As noted above, about 15 MeV of the energy liberated as a result of fission of a ^{235}U nucleus comes from the subsequent beta decay of the fission fragments rather than from the kinetic energy of the fragments themselves. This fact poses a serious problem with respect to control and safety of reactors. Even after the chain reaction has been completely stopped by insertion of control rods into the core, heat continues to be evolved by the beta decays, which cannot be stopped. For a 1000-MW reactor, this heat power amounts to about 200 MW, which, in the event of total loss of cooling water, is more than enough to cause a catastrophic "meltdown" of the reactor core and possible penetration of the containment vessel. The difficulty in achieving a "cold shutdown" following an accident at the Three Mile Island nuclear power plant in Pennsylvania in March 1979 was a result of the continued evolution of heat due to beta decays.

How big can a nucleus be?

Fission appears to set an upper limit on the production of transuranic nuclei, discussed in Section 44–5. When a nucleus with $Z = 109$ is bombarded with neutrons, fission occurs essentially instantaneously; no $Z = 110$ nucleus is formed even for a short time. There are theoretical reasons to expect that nuclei in the vicinity of $Z = 114$, $N = 184$, might be stable with respect to spontaneous fission. These numbers correspond to *filled shells* in the nuclear energy-level structure, analogous to the filled shells of electrons in the noble gases, as discussed in Section 43–2. Such nuclei, called *superheavy nuclei*, would still be unstable with respect to alpha emission, but they might live long enough to be identified. Attempts to produce superheavy nuclei in the laboratory have not been successful; whether they exist in nature is still an open question.

44-7 NUCLEAR FUSION

In any nuclear reaction where the total rest mass of the products is less than the original rest mass, energy is liberated. The fission of uranium, which we have just described, is an example of one type of energy-liberating reaction. Another type involves the *combination* of two light nuclei to form a nucleus that is more complex but whose rest mass is less than the sum of the rest masses of the original nuclei. These are called **fusion reactions.** Here are three examples of energy-liberating fusion reactions:

When two light nuclei combine, energy is evolved.

$$_1^1\text{H} + {}_1^1\text{H} \longrightarrow {}_1^2\text{H} + \beta^+ + \nu,$$
$$_1^2\text{H} + {}_1^1\text{H} \longrightarrow {}_2^3\text{He} + \gamma,$$
$$_2^3\text{He} + {}_2^3\text{He} \longrightarrow {}_2^4\text{He} + {}_1^1\text{H} + {}_1^1\text{H}.$$

In the first reaction, two protons combine to form a deuteron and a β^+ or *positron* (a positively charged electron, to be discussed in Section 44–9). In the second, a proton and a deuteron unite to form the light isotope of helium. For the third reaction to occur, the first two reactions must occur twice, in which case two nuclei of light helium unite to form ordinary helium. These fusion reactions, known as the *proton-proton* chain, are believed to take place in the interior of the sun and other stars.

The positrons produced during the first step of the proton-proton chain collide with electrons; mutual annihilation takes place, and their energy is converted into gamma radiation. The net effect of the chain, therefore, is the combination of four hydrogen nuclei into a helium nucleus and gamma radiation. The net amount of energy released may be calculated from the mass balance as follows:

Mass of four hydrogen atoms (including electrons)	= 4.03132 u
Mass of one helium plus two additional electrons	= 4.00370 u
Difference in mass	= 0.02762 u
	= 25.7 MeV

In the case of the sun, 1 g of its mass contains about 2×10^{23} protons. Hence, if all these protons were fused into helium, the energy released would be about 57,000 kWh. If the sun were to continue to radiate at its present rate, it would take about 30 billion years to exhaust its supply of protons.

For fusion of two nuclei to occur, they must come together to within the range of the nuclear force, typically of the order of 2×10^{-15} m. To do this, they must overcome the electrical repulsion of their positive charges. For two protons at this distance, the corresponding potential energy is of the order of 1.1×10^{-13} J or 0.7 MeV; this represents the initial *kinetic* energy the fusion nuclei must have.

Such energies are available at extremely high temperatures. According to Section 20–4, the average translational kinetic energy of a gas molecule at temperature T is $\frac{3}{2}kT$, where k is Boltzmann's constant. For this value to be equal to 1.1×10^{-13} J, the temperature must be of the order of 5×10^9 K. Not all the nuclei have to have this energy, but this calculation shows that the temperature must be of the order of millions of kelvins if any appreciable fraction of the nuclei are to have enough kinetic energy to surmount the electrical repulsion and achieve fusion.

Fusion reactions in the sun are the source of its energy and ours.

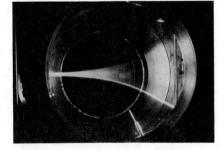

Photograph showing fusion fuel at Lawrence Livermore Laboratories, given a fan-shaped appearance by magnetic mirrors. The goal of the magnetic fusion energy program is the development of a nuclear-fusion reactor to generate electricity.· (Courtesy of Lawrence Livermore Laboratory.)

44–7 The Novette laser system at Lawrence Livermore National Laboratory, used for fusion research. This system went into operation in January 1983; it can deliver a power of 50×10^{12} W for a period of the order of 1 ns. (Courtesy Lawrence Livermore National Laboratory.)

Such temperatures occur in stars as a result of gravitational contraction and its associated liberation of gravitational potential energy. When the temperature gets high enough, fusion reactions occur, more energy is liberated, and the pressure of the resulting radiation prevents further contraction. Only after most of the hydrogen has been converted into helium do further contraction and an accompanying increase of temperature result. Conditions are then suitable for the formation of heavier elements.

Intensive efforts are under way in many laboratories to achieve controlled fusion reactions, which potentially represent an enormous energy resource. In one kind of experiment, a plasma is heated to extremely high temperature by an electrical discharge, while being contained by appropriately shaped magnetic fields. In another experiment, pellets of the material to be fused are heated by a high-intensity laser beam. One current laser experiment setup is shown in Fig. 44–7. Some of the reactions being studied are the following:

$$\begin{aligned}
{}_1^2\text{H} + {}_1^2\text{H} &\longrightarrow {}_1^3\text{H} + {}_1^1\text{H} + 4 \text{ MeV}, &(1) \\
{}_1^3\text{H} + {}_1^2\text{H} &\longrightarrow {}_2^4\text{He} + {}_0^1\text{n} + 17.6 \text{ MeV}, &(2) \\
{}_1^2\text{H} + {}_1^2\text{H} &\longrightarrow {}_2^3\text{He} + {}_0^1\text{n} + 3.3 \text{ MeV}, &(3) \\
{}_2^3\text{He} + {}_1^2\text{H} &\longrightarrow {}_2^4\text{He} + {}_1^1\text{H} + 18.3 \text{ MeV}. &(4)
\end{aligned}$$

In the first reaction, two deuterons combine to form tritium and a proton. In the second, the tritium nucleus combines with another deuteron to form helium and a neutron. The result of both these reactions together is the conversion of three deuterons into a helium-4 nucleus, a proton, and a neutron, with the liberation of 21.6 MeV of energy. Reactions (3) and (4) together achieve the same conversion. In a plasma containing deuterium, the two pairs of reactions occur with roughly equal probability. As yet no one has succeeded in producing these reactions under controlled conditions in such a way as to yield a net surplus of usable energy, but the practical problems do not appear to be insurmountable.

44–8 PARTICLE ACCELERATORS

Many important experiments in nuclear and high-energy physics during the last 60 years or so have made use of beams of charged particles, such as protons or electrons, that have been accelerated to high speeds. Any device that uses electric and magnetic fields to guide and accelerate a beam of charged particles is called a **particle accelerator.** In a sense, the cathode-ray tubes of Thomson and his contemporaries were the first accelerators. In more recent times accelerators have grown enormously in size, complexity, and energy range.

Particle accelerators: more and more powerful tools for research in fundamental-particle interactions

The **cyclotron,** developed in 1931 by Lawrence and Livingston at the University of California, is important historically because it was the first accelerator to use a magnetic field to guide particles in a nearly circular path so that they could be accelerated repeatedly by an electric field in a cyclic process.

In the cyclotron, shown schematically in Fig. 44–8, particles with mass m and charge q move inside a vacuum chamber in a uniform magnetic field B perpendicular to the plane of their trajectories. We learned in Section 30–4 that in such a situation, a particle with speed v moves in a circular path with a radius of curvature r given by

$$r = \frac{mv}{qB}. \tag{44–9}$$

The angular velocity ω of the particles is

$$\omega = \frac{v}{r} = \frac{qB}{m}. \tag{44–10}$$

Note that ω is independent of r.

Now we apply an alternating potential difference between the two hollow electrodes D_1 and D_2, which are called *dees*. If this potential difference has the same frequency as the particles' circular motions, it gives them a push twice each revolution, as they pass the gaps between the dees, boosting them into paths with larger radius and greater kinetic energy. The maximum radius is determined by the radius R of the electromagnet poles. We can find the corresponding maximum energy by solving Eq. (44–9) for v and substituting the

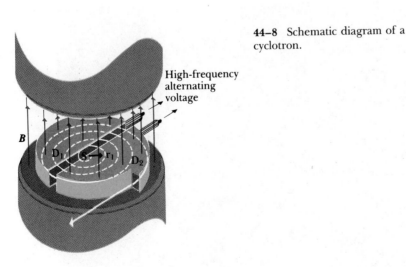

44–8 Schematic diagram of a cyclotron.

result into the relation $K = \frac{1}{2}mv^2$. The result is

$$K_{\text{max}} = \frac{1}{2}mv^2 = \frac{q^2B^2R^2}{2m}. \tag{44-11}$$

If the particles are protons,

$$\frac{q}{m} = \frac{1.60 \times 10^{-19}\text{ C}}{1.67 \times 10^{-27}\text{ kg}} = 9.58 \times 10^7\text{ C·kg}^{-1}.$$

Limitations on the maximum energy of particles in a cyclotron

Taking values typical of early cyclotrons, we assume $B = 1.5$ T and $R = 0.5$ m. Then from Eq. (44–11), the maximum kinetic energy is

$$K = \frac{1}{2}(1.67 \times 10^{-27}\text{ kg})(9.58 \times 10^7\text{ C·kg}^{-1})^2(1.5\text{ T})^2(0.5\text{ m})^2$$

$$= 4.31 \times 10^{-12}\text{ J} = 2.69 \times 10^7\text{ eV} = 26.7\text{ MeV}.$$

This energy, considerably larger than the average binding energy per nucleon, is enough to cause a variety of interesting nuclear reactions.

The energy attainable with the cyclotron is limited by relativistic effects. For Eq. (44–10) to be relativistically correct, m should be replaced by $m/(1 - v^2/c^2)^{1/2}$. As the particles speed up, their angular velocity *decreases;* if the decrease is appreciable, the particle motion is no longer in the correct phase relative to the alternating dee voltage. In a variation called the **synchrocyclotron,** the particles are accelerated in bursts, and for each burst the frequency of the alternating voltage is decreased at just the right rate to maintain the correct phase relation with the particles' motion. A practical limitation of the cyclotron is the expense of building very large electromagnets. The largest synchrocyclotron ever built has a vacuum chamber about 8 m in diameter and accelerates protons to energies of about 600 MeV. The synchrocyclotron, incidentally, provides a very direct confirmation of relativistic mechanics.

The synchrotron: a giant doughnut for accelerating particles to extremely high energies

To attain higher energies, another type of machine, called the **synchrotron,** is more practical. In a synchrotron, the vacuum chamber in which the particles move is in the form of a thin doughnut, called the *accelerating ring.* The particles are forced to move within this chamber by a series of magnets

Interior view of part of the Stanford Linear Accelerator (SLAC). (Courtesy of Stanford Linear Accelerator, Stanford University.)

44–9 An aerial view of the 800-GeV accelerator at the Fermi National Accelerator Laboratory, Batavia, Illinois. (Courtesy of Fermi National Accelerator Laboratory.)

placed around it. As the particles speed up, the magnetic field is increased so that the particles retrace the same trajectory over and over. The synchrotron located at the Fermi National Accelerator Laboratory (Fermilab) in Batavia, Illinois, can accelerate protons to an energy of 800 GeV (800×10^9 eV), and modifications are under way that will permit a maximum energy of 1000 GeV. The accelerating ring is 2 km in diameter, and the accelerator and associated facilities cost about $400 million to build. In each machine cycle, of a few seconds' duration, it accelerates approximately 10^{13} protons. An aerial view of the Fermilab accelerator is shown in Fig. 44–9.

As higher and higher energies are sought in the attempt to investigate new phenomena in particle interactions, a new problem emerges. In an experiment where a beam of high-energy particles collides with a stationary target, not all the kinetic energy of the incident particles is available to form new particle states. Because momentum must be conserved, the particles emerging from the collision must have some motion and thus some kinetic energy. The energy E_a available for creating new particles or particle configurations is the *difference* between initial and final kinetic energies.

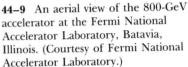

Available energy: a law of diminishing returns in accelerator energies

In the extreme-relativistic range, where the kinetic energies of the particles are large compared to their rest energies, this is a very severe limitation. When beam and target particles have equal mass, as with protons bombarding a hydrogen target, it can be shown from relativistic mechanics that the available energy E_a is related to the total energy E of the bombarding particle and to its mass m by

$$E_a = \sqrt{2mc^2E}. \tag{44–12}$$

For example, for the proton, $mc^2 = 931$ MeV $= 0.931$ GeV. If $E = 800$ GeV, as for the Fermilab accelerator, then

$$E_a = \sqrt{2(0.931 \text{ GeV})(800 \text{ GeV})} = 38.6 \text{ GeV},$$

and if $E = 1000$ GeV, $E_a = 43.1$ GeV. Thus, increasing the beam energy by 200 GeV increases the available energy by only 4.5 GeV.

This limitation may be circumvented in part by *colliding-beam* experiments, in which there is no stationary target but in which beams of particles and their antiparticles (such as electrons and positrons, or protons and antiprotons)

Colliding-beam experiments: making all the beam energy available

circulate in opposite directions in arrangements called *storage rings*. In regions where the rings intersect, the beams are focused sharply onto one another, and collisions can occur. Because the total momentum in such two-particle collisions is zero, the available energy E_a is the total kinetic energy of the two particles.

An example is the storage-ring facility at the Stanford Linear Accelerator Center (SLAC), where electron and positron beams collide with total available energy of up to 36 GeV. Other storage-ring facilities are located at DESY (German Electron Synchrotron) in Hamburg, West Germany (E_a up to 38 GeV total in electron-positron collisions), and at the Cornell Electron Storage Ring Facility (CESR), with 16 GeV maximum total available energy. At the CERN (European Council for Nuclear Research) laboratory in Geneva, Switzerland, construction has begun for a large electron-positron storage ring that will transport beams of particles with energies of 50 GeV or more, for a total available energy of at least 100 GeV. This facility is expected to have usable beams in 1989.

44–9 FUNDAMENTAL PARTICLES

The physics of fundamental particles has been a recognized field of research for only the past 50 years. The electron and proton were known by the turn of the century, but the existence of the neutron was not established definitely until 1930; its discovery is an interesting story and a useful illustration of nuclear reactions.

Discovery of the neutron: another interesting detective mystery

In 1930, two German physicists, Bothe and Becker, observed that when beryllium, boron, or lithium was bombarded by fast alpha particles, the bombarded material emitted something, either particles or electromagnetic waves, of much greater penetrating power than the original alpha particles. Further experiments in 1932 by I. Curie and Joliot in Paris confirmed these results, but all attempts to explain them in terms of gamma rays were unsuccessful. Chadwick in England repeated the experiments and found that they could be satisfactorily interpreted on the assumption that *uncharged* particles of mass approximately equal to that of the proton were emitted from the nuclei of the bombarded material. He called the particles *neutrons*. The emission of a neutron from a beryllium nucleus takes place according to the reaction

$$\,_2^4\text{He} + \,_4^9\text{Be} \longrightarrow \,_6^{12}\text{C} + \,_0^1\text{n}, \qquad (44\text{–}13)$$

where $\,_0^1\text{n}$ is the symbol for a neutron.

Because neutrons have no charge, they produce no ionization in their passage through gases. They are not deflected by the electric field around a nucleus and can be stopped only by colliding with a nucleus in a direct hit, in which case they may either undergo an elastic impact or penetrate the nucleus. We showed in Section 8–5 that if a particle collides elastically with a stationary particle of the same mass, the first particle stops and the second moves off with the same speed as the first. Since the proton and neutron masses are almost the same, fast neutrons can be stopped during collisions with hydrogen atoms in hydrogenous materials such as water or paraffin. A common laboratory method of obtaining slow neutrons is to surround the fast-neutron source with water or blocks of paraffin.

Detecting neutrons by the reactions they cause

Once the neutrons are moving slowly, they may be detected by means of the alpha particles they eject from the nucleus of a boron atom, according to

the reaction

$${}_{0}^{1}\text{n} + {}_{5}^{10}\text{B} \longrightarrow {}_{3}^{7}\text{Li} + {}_{2}^{4}\text{He}. \qquad (44\text{--}14)$$

The ejected alpha particle then produces ionization that may be detected in a Geiger counter or other particle detector.

The discovery of the neutron gave the first real clue to the structure of the nucleus. Before 1930, it had been thought that the total mass of a nucleus was due to protons only. We now know that a nucleus consists of both protons and neutrons (except hydrogen, whose nucleus consists of a lone proton) and that (1) the mass number A equals the total number of nuclear particles and (2) the atomic number Z equals the number of protons.

In the early days of particle physics, the *cloud chamber* and the *bubble chamber* were used to visualize paths of charged particles. In the cloud chamber, supercooled vapor condenses around a line of ions created by passage of a charged particle; the result is a visible track. In the bubble chamber, superheated liquid boils locally around a similar line of ions, creating a visible track of tiny bubbles. By placing either instrument in a magnetic field and measuring the radius of curvature of a particle trajectory, we can determine the momentum of the particle.

The positive electron, or **positron,** was first observed during an investigation of cosmic rays by Carl D. Anderson in 1932, in a historic cloud-chamber photograph reproduced in Fig. 44–10. The photograph was made with the cloud chamber in a magnetic field perpendicular to the plane of the paper. A lead plate crosses the chamber, and evidently the particle has passed through it. Since the curvature of the track is greater above the plate than below it, the velocity is less above than below; the inference is that the particle was moving upward, since it could not have *gained* energy going through the lead.

The *density* of droplets along the path is the same as would be expected if the particle were an electron. But the direction of the magnetic field and the direction of motion are consistent only with a particle of *positive* charge. Hence Anderson concluded that the track had been made by a positive electron, or *positron.* The mass of the positron is equal to that of an ordinary (negative) electron, and its charge is equal in magnitude but of opposite sign to that of the electron. Pairs of particles related to each other in this way are said to be *antiparticles* of each other.

Positrons have only a transitory existence; they do not form a part of ordinary matter. They are produced in high-energy collisions of charged particles or gamma rays with matter in a process called *pair production,* in which an ordinary electron and a positron are produced simultaneously. Electric charge is conserved in this process, but enough energy must be available to account for the energy equivalent of the rest masses of the two particles, about 0.5 MeV each. The inverse process, $e^{+}e^{-}$ *annihilation,* occurs when a positron and an electron collide. Both particles disappear, and two or three gamma-ray photons appear, with total energy $2mc^2$, where m is the electron rest mass. Decay into a *single* gamma is impossible because such a process cannot possibly conserve both energy and momentum.

Positrons also occur in the decay of some unstable nuclei. Recall that nuclei having too many neutrons for stability often emit a beta particle (electron), decreasing N by one and increasing Z by one. Similarly, a nucleus having *too few neutrons* for stability may respond by converting a proton to a neutron, emitting a positron, increasing N by one and decreasing Z by one. Such nu-

Historical instruments for visualizing paths of high-energy particles

Discovery of the positron: still another detective thriller

Electrons and positrons can be created or destroyed only in pairs.

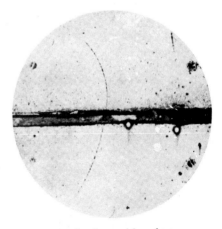

44–10 Track of a positive electron traversing a lead plate 6 mm thick. (Photograph by C. D. Anderson.)

clides do not occur in nature, but they can be produced artificially by neutron bombardment of stable nuclides in nuclear reactors. An example is the unstable nuclide $^{22}_{11}$Na, which has one less neutron that the stable $^{23}_{11}$Na. It emits a positron, leaving the stable nuclide $^{22}_{10}$Ne.

In 1935 the Japanese physicist Hideki Yukawa inferred, from theoretical considerations, the existence of a particle having a mass intermediate between that of the electron and the proton. A particle of intermediate mass, but *not* identical with that predicted by Yukawa, was discovered one year later by Anderson and Neddermeyer as a component of cosmic radiation. This particle is now known as a μ *meson* (or *muon*). The μ^- has charge equal to that of the electron, and its antiparticle the μ^+ has a positive charge of equal magnitude. The two particles have equal mass, about 207 times the electron mass. The muons are unstable; each decays into an electron of the same sign, plus two neutrinos, with a lifetime of about 2.2×10^{-6} s.

Yukawa first proposed the mesons as a basis for transmitting nuclear forces. He suggested that nucleons could interact by emitting and absorbing unstable particles, just as two basketball players interact by tossing the ball back and forth, or by snatching it away from each other. It was established soon after the discovery of the muons that they could not be Yukawa's particles because their interactions with nuclei were far too weak. But in 1947 *another* family of mesons was discovered, called π *mesons* or *pions*. There are three types, positive, negative, and neutral. The charged pions have masses of about 273 times the electron mass and decay into muons with the same sign, plus a neutrino, with a lifetime of about 2.6×10^{-8} s. The neutral pion has a smaller mass, about 264 electron masses, and decays, with a very short lifetime of about 0.8×10^{-16} s, into two gamma-ray photons. The pions interact strongly with nuclei, and they *are* the particles predicted by Yukawa.

In the years since 1947, *high-energy physics* has emerged as a distinct branch of physics. These years have witnessed the attainment of higher and higher energies in particle accelerators, the discovery of many new particles, and intensive efforts to understand the properties of these new particles and their interactions.

Along with higher energies and the creation of new particles has come the need for new and more sophisticated detectors. Since energy and momentum must be conserved in any reaction or decay process, most large present-day detectors are designed to handle the problems of mass, charge, and momentum identification. Early detectors such as scintillation counters, proportional tubes, and cloud and bubble chambers have been replaced by vertex detectors, electromagnetic calorimeters, wire proportional chambers, large solenoidal magnets, and various kinds of Cerenkov counters. Modern electronics, in the form of on-line computers and microprocessors, has followed, along with increased intensity of accelerator beams. Present-day electronic detectors can comfortably distinguish individual interactions (events) that occur one microsecond apart.

44–10 HIGH-ENERGY PHYSICS

It was recognized, even in the early years of high-energy physics, that fundamental particles are not *permanent* entities but can be created or destroyed in interactions with other particles. The earliest such interactions to be observed

More mysteries: prediction of a new particle and discovering a different one

Pions: Yukawa's predicted nuclear glue

Particle detectors have become very sophisticated.

All fundamental particles can be created and destroyed in appropriate circumstances.

were creation and destruction of electron-positron pairs. Such pairs are *created* in collisions of high-energy cosmic-ray particles with stationary targets; when an electron and a positron collide, both *disappear* and two or three gamma-ray photons are created to carry away the energy. This transitory nature of the fundamental particles may seem disturbing, but in one sense it is a welcome development. We have seen that photons and electrons (and indeed all particles) share the dual wave-particle nature discussed in Section 42–2, and photons are known to be created and destroyed (or emitted and absorbed) in atomic transitions. Thus it seems natural that other particles can also be created and destroyed.

For example, it was speculated as early as 1932 that there might be an *antiproton*, bearing the same relation to the ordinary proton as the positron does to the electron; that is, a particle with the same mass as the proton but negatively charged. Finally, in 1955 proton-antiproton pairs were created by impact on a stationary target of a beam of protons with kinetic energy 6 GeV (6×10^9 eV) from the Bevatron at the University of California at Berkeley.

Discovery of the antiproton: This time they knew what they were looking for.

In the years after 1960, as higher-energy accelerators and more sophisticated detectors were developed, a veritable blizzard of new unstable particles was identified. To describe them, we have to create a small blizzard of new terms. Initially the particles were classified according to *mass*. The particles having the smallest masses (electrons, muons, and their associated neutrinos) are called **leptons.** (The recently discovered τ particles are also classified as leptons, even though their masses are greater than those of nucleons; we will return to this point later.) Particles with masses between those of muons and nucleons are called **mesons.** Particles that resemble nucleons but are more massive are called **hyperons,** and nucleons and hyperons collectively are called **baryons.** Particles are further classified according to electric charge, spin, and two additional quantum numbers, *isospin* (the number that determines the number of different charges a particular type of particle can have) and *strangeness* (a number needed to account for the production and decay modes of certain particles). A partial list of some known particles is shown in Table 44–3. All particles having zero or integer spin (including photons and π and K mesons) are called **bosons,** and all particles having half-integer spin (including all leptons and baryons) are called **fermions.** These terms, derived from the names Bose and Fermi, refer to the different statistical energy-distribution functions describing the behaviors of the two classes of particles.

Several new terms used to classify particles

During this period it became clear that particles could also be classified in terms of the types of *interactions* in which they participate and in terms of the conservation laws associated with these interactions. In Section 5–1 we spoke briefly about kinds of interactions. There appear to be four classes of interactions; in order of decreasing strength, they are:

Classifying particles in terms of their interactions

1. Strong interactions,
2. Electromagnetic interactions,
3. Weak interactions,
4. Gravitational interactions.

Particles that experience strong interactions are called **hadrons;** these include all the mesons and baryons in Table 44–3. The strong interactions are responsible for the nuclear force and also for the creation of pions, heavy mesons, and hyperons in high-energy collisions. Electrons, muons, and neutrinos have *no* strong interactions.

TABLE 44–3 Some Known Particles and Their Properties

Particle	Mass (MeV/c^2)	Charge	Spin	Isopin	Strange-ness	Mean Lifetime (s)	Typical Decay Modes	Quark Content
e^-	0.511	-1	$\frac{1}{2}$	—	0	stable	—	—
ν_e	0 ($<5 \times 10^{-5}$)	0	$\frac{1}{2}$	—	0	stable	—	—
μ^-	105.7	-1	$\frac{1}{2}$	—	0	2.2×10^{-6}	$e^- \bar{\nu}_e \nu_\mu$	—
ν_μ	0 (<0.52)	0	$\frac{1}{2}$	—	0	stable	—	—
τ^-	1784	-1	$\frac{1}{2}$	—	0	5×10^{-13}	$\mu^- \bar{\nu}_\mu \nu_\tau$	—
ν_τ	0 (<250)	0	$\frac{1}{2}$	—	0	stable	—	—
π^0	135.0	0	0	1	0	0.83×10^{-16}	$\gamma\gamma$	$u\bar{u}, d\bar{d}$
π^+	139.6	$+1$	0	1	0	2.6×10^{-8}	$\mu^+ \nu_\mu$	$u\bar{d}$
π^-	139.6	-1	0	1	0	2.6×10^{-8}	$\mu^- \bar{\nu}_\mu$	$\bar{u}d$
K^+	493.7	$+1$	0	$\frac{1}{2}$	$+1$	1.24×10^{-8}	$\mu^+ \nu_\mu$	$u\bar{s}$
K^-	492.67	-1	0	$\frac{1}{2}$	-1	1.24×10^{-8}	$\mu^- \bar{\nu}_\mu$	$\bar{u}s$
η^0	548.8	0	0	0	0	$\sim 10^{-18}$	$\gamma\gamma$	$u\bar{u}, d\bar{d}, s\bar{s}$
p	938.3	$+1$	$\frac{1}{2}$	$\frac{1}{2}$	0	stable	—	uud
n	939.6	0	$\frac{1}{2}$	$\frac{1}{2}$	0	917	$pe^- \bar{\nu}_e$	udd
Λ	1115	0	$\frac{1}{2}$	0	-1	2.63×10^{-10}	$p\pi^-$ or $n\pi^0$	uds
Σ^+	1189	$+1$	$\frac{1}{2}$	1	-1	0.80×10^{-10}	$p\pi^0$ or $n\pi^+$	uus
Δ^{++}	1232	$+2$	$\frac{3}{2}$	$\frac{3}{2}$	0	$\sim 10^{-23}$	$p\pi^+$	uuu
Ξ^-	1321	-1	$\frac{1}{2}$	$\frac{1}{2}$	-2	1.64×10^{-10}	$\Lambda\pi^-$	dss
Ω^-	1672	-1	$\frac{3}{2}$	0	-3	0.82×10^{-10}	ΛK^-	sss
Λ_c^+	2273	1	$\frac{1}{2}$	0	0	$\sim 7 \times 10^{-13}$	$\Lambda\pi\pi$	udc

The *electromagnetic* interactions are those associated directly with electric charge. As noted previously, the electromagnetic interaction between two protons is weaker at distances of the order of nuclear dimensions than the strong interaction, but the electromagnetic interaction has longer range. Neutral particles other than photons have no electromagnetic interactions, with the exception of effects due to the magnetic moments of neutral baryons. These magnetic moments are believed to be associated with the emission and absorption of charged pions and heavy mesons.

The *weak* interaction is responsible for beta decay, such as the conversion of a neutron into a proton, an electron, and a neutrino. It is also responsible for the decay of many unstable particles (pions into muons, muons into electrons, Λ particles into protons, and so on). The *gravitational* interaction, although of central importance for the large-scale structure of celestial bodies, is not believed to be of significance in the analysis of fundamental-particle interactions. For example, the gravitational attraction of two electrons is smaller than their electrical repulsion by a factor of about 2.4×10^{-43}.

Classifying interactions in terms of the conservation principles that are obeyed or not obeyed

Several conservation laws are believed to be obeyed by *all* of the interactions mentioned above. These include the laws growing out of classical physics: energy, momentum, angular momentum, and electrical charge. In addition, several new quantities having no classical analog have been introduced to help characterize the properties of particles. These include *baryon number* (the number of baryons involved in an interaction, minus the number of antibaryons), *isospin* (used also to describe the charge independence of nuclear forces), *parity* (the comparative behavior of two systems that are mirror images of each other), *strangeness* (a quantum number used to classify particle production and decay reactions), and *lepton number* (the number of leptons involved in an interaction, minus the number of antileptons). Baryon number is conserved in *all* interactions; isospin is conserved in strong interactions but not in electromagnetic or weak interactions. Parity and strangeness are conserved in strong and electromagnetic interactions but not in weak interactions. Lepton number is thought to be conserved in all interactions. Thus the new conservation laws are not absolute but instead serve as a means for *classifying* interactions.

The large number of supposedly fundamental particles discovered since 1960 (well over a hundred) suggests strongly that these particles *do not* represent the most fundamental level of the structure of matter, but that there is at least one additional level of structure. There is now fairly general agreement among physicists concerning the nature of this level; the theory is based on a proposal made initially in 1964 by Gell-Mann and his collaborators. We cannot discuss this theory in detail, but the following is a very brief sketch of some of its features.

Leptons are indeed fundamental particles. In addition to the electrons and muons and their associated neutrinos, a third massive lepton called the *tau* (τ), having spin $\frac{1}{2}$ and mass 1784 MeV, was discovered in 1974. The τ neutrino has thus far escaped detection, but experiments designed to establish its existence are under way. In Table 44–3 upper limits on the three neutrino masses are given. Although zero-mass neutrinos are postulated in most theories, a small mass could be accommodated. If neutrinos do have mass, oscillations in which one type of neutrino changes into another type are possible, and these oscillations allow for experimental detection of finite neutrino mass. Experiments designed to detect neutrino oscillations are under way, but no positive results have yet been obtained. Meanwhile, neutrino research continues; an example is shown in Fig. 44–11.

It now appears that hadrons are *not* fundamental particles but are composite structures whose constituents are spin-$\frac{1}{2}$ fermions called **quarks.** In fact, all known hadrons can be constructed as follows: baryons are composed of three quarks (qqq), and mesons are composed of quark-antiquark pairs ($q\bar{q}$). No other combinations seem to be necessary. This scheme requires that quarks have properties not previously allowed for fundamental particles. For example, quarks have electric charge of magnitude $\frac{1}{3}$ and $\frac{2}{3}$ of the electron charge, which was previously thought to be the fundamental unit of electric charge. Quarks have a strong affinity for each other through a new kind of charge known as "color" charge. Thus color charge is responsible for strong interactions, and the force is known as the color force. The color force is mediated (transmitted) by exchange of color *gluons*, massless spin-one bosons that play the role in strong interactions that the pions played in the Yukawa theory of nuclear force and the photon plays in a quantum theory of electromagnetic interactions.

The weak and gravitational forces are also mediated by exchange of particles. In these cases the particles exchanged are the weak bosons ($W^{\pm}$ and Z^0) and the graviton, respectively. The theory of strong forces is known as **quantum chromodynamics** (QCD). Most QCD theories require that phenomena associated with the creation of quark-antiquark pairs make it impossible to observe a free, isolated quark. The binding energy between quarks is thought to be so strong that any stray quark or antiquark in matter will always be reabsorbed, and only baryons or mesons can emerge.

Quarks: A more fundamental level in the structure of matter.

44–11 Brookhaven National Laboratory's solar neutrino experiment, located 4900 feet underground in a gold mine in South Dakota in order to shield out cosmic rays and all other particles except neutrinos. The tank contains 100,000 gallons of perchloroethylene. Neutrinos from the interior of the sun are captured by ^{37}Cl nuclei, which then beta-decay into ^{37}A. The argon is then trapped and measured.

A unified theory of the interactions of all the strongly interacting particles (hadrons)

TABLE 44–4   Properties of Quarks

Symbol	Q/e	Spin	Baryon Number	Strange-ness	Charm	Bottom-ness	Top-ness
u	$\frac{2}{3}$	$\frac{1}{2}$	$\frac{1}{3}$	0	0	0	0
d	$-\frac{1}{3}$	$\frac{1}{2}$	$\frac{1}{3}$	0	0	0	0
s	$-\frac{1}{3}$	$\frac{1}{2}$	$\frac{1}{3}$	-1	0	0	0
c	$\frac{2}{3}$	$\frac{1}{2}$	$\frac{1}{3}$	0	$+1$	0	0
b	$-\frac{1}{3}$	$\frac{1}{2}$	$\frac{1}{3}$	0	0	$+1$	0
t	$\frac{2}{3}$	$\frac{1}{2}$	$\frac{1}{3}$	0	0	0	$+1$

These strange new quark properties have challenged many experimentalists. To date, isolated free quarks have not been observed; the observation of fractional electric charge has been claimed by a few experimenters and disputed by most others. Many indirect observations, however, lead us to believe that the quark structure of hadrons is correct and that quantum chromodynamics may aid in the understanding of the strong force.

Early quark theory suggested the existence of three types (flavors) of quarks; these were labeled **u** (up), **d** (down), and **s** (strange). (See Table 44–4.) Protons, neutrons, π and K mesons, and so on, can all be constructed from these three quarks. For convenience, we describe the charge Q of a particle as a multiple of the magnitude e of the electron charge. For example, a proton has $Q/e = +1$, baryon number $(B) = +1$, strangeness $(S) = 0$. The proton quark content is **uud** if the **u** quark has $Q/e = \frac{2}{3}$ and $B = \frac{1}{3}$ and the **d** quark has $Q/e = -\frac{1}{3}$ and $B = \frac{1}{3}$. The neutron would then have quark content **udd**, the π^+ meson **ud̄**, and the K^+ meson **us̄**. Antiparticles can easily be accommodated: $\bar{p} = \overline{\textbf{uud}}$, $\pi^- = \overline{\textbf{u}}\textbf{d}$, and so on. Particles can then be arranged according to quark content, and families of particles can be classified according to intrinsic orbital angular momentum, spin, and parity. Thus a state **qq̄**, for example, could represent particles in different families depending on the spin configuration of the quarks and the orbital angular momentum.

Because of the Pauli exclusion principle (Section 42–1), quarks are required to have a property that distinguishes one quark from another of the same flavor. This new property was labeled *color*; each quark flavor has three colors. This "color charge" is then responsible for the strong force between quarks. Figure 44–12 shows how we can picture the decay process $K^0 \rightarrow \pi^- + e^+ + \nu_e$, according to QCD theory.

For symmetry and other compelling reasons, theorists later predicted the existence of a fourth quark flavor. This quark was labeled **c** (charm); it was required to have $Q/e = \frac{2}{3}$, $B = \frac{1}{3}$, $S = 0$, and a new quantum number $C = +1$. Charm was discovered in 1974 at both the SLAC and Brookhaven accelerator laboratories by the observation of a meson of mass 3100 MeV. This meson, named ψ at SLAC and J at Brookhaven, was found to have several decay modes, decaying into e^+e^-, $\mu^+\mu^-$, or into hadrons. The mean lifetime was found to be $\sim 10^{-20}$ s. This is consistent with J/ψ being the ground state of a bound $c\bar{c}$ system, much like the way in which the hydrogen atom is a bound p-e system. Immediately after this finding, excited $c\bar{c}$ states or energy levels were observed; and finally (a few years later) isolated mesons having the charm quantum number were also observed. These mesons, D^0 ($c\bar{u}$) and D^+ ($c\bar{d}$), and their excited states are now firmly established, and a charmed baryon, Λ_c^+, has been observed.

More quarks: Six flavors and three colors

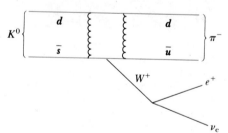

44–12 Diagram showing decay of the K^0 meson. The time sequence is shown from left to right. The initial K^0 meson consists of **d** and **s̄** quarks, bound together by gluon exchange (wiggly line). The quark **s̄** emits the weak boson W^+ and becomes quark **ū**. This is bound by gluon exchange to the **d** quark, forming the final π^- meson, while the W^+ decays into an electron and a neutrino.

In 1977 a meson of mass 9460 MeV, called upsilon (Y) was discovered at Brookhaven. Because it had properties similar to J/ψ, it was conjectured that the meson was really the bound system of a new quark b and its antiquark $\bar{b}$. Excited states of Y were soon observed, and the B^+ ($b\bar{u}$) and B^0 ($b\bar{d}$) mesons are now well established also.

Thus five flavors of quarks (u, d, s, c, b) are thought to exist along with six flavors of leptons (e, μ, τ, ν_e, ν_μ, and ν_τ). If we assume that quarks and leptons are the basic particles of matter, we can explain many of the strong and weak interactions of hadrons and mesons and the weak interactions of leptons. But it is easy to conjecture that nature is symmetric in its building blocks, and so a sixth quark is thought to exist. This quark, labeled t (top), should have $Q/e = \frac{2}{3}$, $B = \frac{1}{3}$, and a new quantum number, $T = 1$. No compelling experimental evidence requiring the existence of the t quark has yet been found. Table 44–4 lists some properties of the six quarks.

Particle theorists have long tried to combine all four forces of nature into a single unified theory, but with little success. Recently, the weak and electromagnetic forces were successfully unified by Weinberg, Salam, and Glashow. Thus there is a fundamental force in nature, the electro-weak force. The electro-weak theory was successfully verified in 1983 with the discovery of the weak force propagators, the Z^0 and $W^\pm$ bosons, by two experimental groups working at the $\bar{p}p$ collider at CERN, in Geneva. Not only were these particles observed, but their masses are as predicted in the electro-weak theory. It is possible that quantum electrodynamics, the theory of strong interaction, and the electro-weak theory, when correctly unified, will give a valid theory of strong, weak, and electromagnetic forces. At present the photon, lepton, quarks, Z^0, and $W^\pm$ particles have been established and provide the basis for attempts to construct this "grand unified" theory. Although such schemes are still speculative in nature, the entire area is a very active field of present-day theoretical and experimental research.

Quarks and leptons are the fundamental building blocks of matter. At least we think so.

Attempts to combine all four kinds of interactions into one comprehensive theory: the ultimate dream of the particle theorist.

SUMMARY

In Rutherford scattering, alpha particles collide with atoms; their scattering pattern reveals the charge distribution in the atom. Most of the mass and all the positive charge are concentrated in the nucleus, which is much smaller than the overall size of the atom.

All nuclei have nearly the same density, and the radius r of a nucleus is given approximately by

$$r = r_0 A^{1/3}. \qquad (44\text{–}1)$$

where $r_0 = 1.2 \times 10^{-15}$ m and A is the mass number, equal to the total number of protons and neutrons. The number of protons Z is the atomic number, and the number of neutrons is N. Thus

$$A = Z + N. \qquad (44\text{–}2)$$

Nuclei of a given element all have the same Z; those with different values of N are called isotopes, and a single species (single values of Z and N) is called a nuclide.

The mass of a nucleus is less than the total mass of its constituents because of the mass equivalent of the binding energy. Binding energy per nucleon is typically about 8 MeV. Nuclear masses are conveniently measured in terms of

KEY TERMS
Rutherford scattering
nucleus
mass number
nuclear spin
proton
neutron
atomic number
nucleons
neutron number
isotopes
nuclide
mass defect
binding energy
atomic mass unit
gamma rays
nuclear force
radioactive
alpha particle

beta particle

gamma particle

neutrino

decay constant

half-life

mean lifetime

nuclear reaction

reaction energy

threshold

fission

fission fragments

chain reaction

nuclear reactor

fusion reactions

particle accelerator

cyclotron

synchrocyclotron

synchrotron

positron

leptons

mesons

hyperons

baryons

bosons

fermions

hadrons

quarks

quantum chromodynamics

the atomic mass unit (u), equal to $\frac{1}{12}$ the mass of the neutral carbon atom with $A = 12$.

Nuclei have energy levels; each has a ground state and excited states. Gamma-ray photons are emitted and absorbed during transitions between states; excitation energies are typically of the order of 1 MeV. Nuclei are bound together by the nuclear force, which is short range, saturating, and favoring pairs of particles. Within its range, it is stronger than the electrical repulsion of the protons, and the stability or lack of stability of a nucleus is determined by the competition between the attractive nuclear forces and the repulsive electrical forces.

Unstable nuclei undergo spontaneous radioactive decay, emitting alpha or beta particles, sometimes followed by one or more gamma photons. The alpha particle is identical to the helium nucleus: two protons and two neutrons bound together. The beta particle is a high-energy electron. In alpha decay, N and Z both decrease by two; in beta decay, N decreases by one and Z increases by one.

If N_0 is the number of radioactive nuclei present at time $t = 0$, the number $N(t)$ at a later time t is given by

$$N(t) = N_0 e^{-\lambda t}, \qquad (44\text{--}4)$$

where λ is the decay constant, different for different nuclides. The half-life $t_{1/2}$ is the time for the number of nuclei to decrease to half the original number. These quantities and the mean lifetime t_{mean} are related by

$$t_{1/2} = \frac{\ln 2}{\lambda} \qquad (44\text{--}5)$$

and

$$t_{\text{mean}} = \frac{1}{\lambda}. \qquad (44\text{--}6)$$

When a naturally radioactive nucleus decays, a series of successive decays occurs until a stable configuration is reached. No nucleus with Z greater than 83 or A greater than 209 is stable.

A nuclear reaction is a rearrangement of the constituents of a nucleus, caused by bombardment by a particle or a photon. When the bombarding particle is positively charged, there is always a threshold energy because of electrostatic repulsion. The final kinetic energy of all parts of the system may be either greater or less than the initial energy, depending on the total mass change.

In nuclear fission, a heavy nucleus such as uranium or thorium splits into two fission fragments of nearly equal size. Fission can occur spontaneously or can be triggered by absorption of a neutron. The fission fragments have substantial kinetic energy, 200 MeV for uranium. A few free neutrons are released during fission, making a self-sustaining fission chain reaction possible.

In nuclear fusion, two light nuclei fuse into a single nucleus, with the liberation of energy.

Particle accelerators produce beams of high-energy particles for research in nuclear and high-energy physics. In cyclotrons and synchrotrons, particles are accelerated in circles under the action of electric and magnetic fields. In recent installations, storage rings permit the collision of two beams (particles and their antiparticles) traveling in opposite directions, which makes the total energy of the particles available to cause reactions.

Fundamental particles are classified as leptons, mesons, and baryons. Particles having strong interactions are called hadrons. Several new quantities, including isospin and strangeness, are used to characterize particle interactions. Interactions are classified as strong, electromagnetic, weak, or gravitational, although attempts are under way to unify these classes. Hadrons are not fundamental particles but are composed of six kinds of quarks, held together with an interaction mediated by gluons. The leptons (electrons, muons, and tau particles) are themselves fundamental particles.

QUESTIONS

44–1 How can you be sure that nuclei are not made of protons and electrons, rather than of protons and neutrons?

44–2 In calculations of nuclear binding energies such as those in the examples of Sections 44–2 and 44–5, should the binding energies of the *electrons* in the atoms be included?

44–3 If different isotopes of the same element have the same chemical behavior, how can they be separated?

44–4 In beta decay a neutron becomes a proton, an electron, and a neutrino. This decay also occurs with free neutrons, with a half-life of about 15 min. Could a free *proton* undergo a similar decay?

44–5 Since lead is a stable element, why doesn't the ^{238}U decay series shown in Fig. 44–6 stop at lead, ^{214}Pb?

44–6 In the ^{238}U decay chain shown in Fig. 44–6, some nuclides in the chain are found much more abundantly in nature than others, despite the fact that every ^{238}U nucleus goes through every step in the chain before finally becoming ^{206}Pb. Why are the abundances of the intermediate nuclides not all the same?

44–7 Radium has a half-life of about 1600 years. If the universe was formed five billion or more years ago, why is there any radium left now?

44–8 Why is the decay of an unstable nucleus unaffected by the *chemical* situation of the atom, such as the nature of the molecule in which it is bound, and so on?

44–9 Fusion reactions, which liberate energy, occur only with light nuclei, and fission reactions only with heavy

nuclei. A student asserted that this shows that the binding energy per nucleon increases with A at small A but decreases at large A and hence must have a maximum somewhere in between. Do you agree?

44–10 Nuclear power plants use nuclear fission reactions to generate steam to run steam-turbine generators. How does the nuclear reaction produce heat?

44–11 There are cases where a nucleus having too few neutrons for stability can capture one of the electrons in the K shell of the atom. What is the effect of this process on N, A, and Z? Is this the same effect as that of β^+ emission? Might there be situations where K capture is energetically possible while β^+ emission is not? Explain.

44–12 Is it possible that some parts of the universe contain antimatter whose atoms have nuclei made of antiprotons and antineutrons, surrounded by positrons? How could we detect this condition without actually going there? What problems might arise if we actually *did* go there?

44–13 When x-rays are used to diagnose stomach disorders such as ulcers, the patient first drinks a thick mixture of (insoluble) barium sulfate and water. What does this do? What is the significance of the choice of barium for this purpose?

44–14 Why are so many health hazards associated with fission fragments that are produced during fission of heavy nuclei?

EXERCISES

Section 44–1 The Nuclear Atom

44–1 A beam of alpha particles is incident on gold nuclei. A particular alpha particle comes in "head-on" and stops 1×10^{-14} m away from the center of a gold nucleus. Assume that the gold nucleus remains at rest.

a) Calculate the electrostatic potential energy of the alpha particle when it has stopped. Express your result in joules and in MeV.

b) What initial kinetic energy did the alpha particle have? Express in joules and in MeV.

c) What was the initial velocity of the alpha particle?

44–2 A 4.7-MeV alpha particle from a radium ^{226}Ra decay makes a head-on collision with a gold nucleus.

a) What is the distance of closest approach of the alpha particle to the center of the nucleus? Assume that the gold nucleus remains at rest.

b) What is the force on the alpha particle at the instant when it is at the distance of closest approach?

Section 44–2 Properties of Nuclei

44–3 How many protons and neutrons are there in a nucleus of

a) hydrogen, ^{1}H?

b) iron, ^{56}Fe?

c) gold, ^{197}Au?

44–4 Consider the three nuclei of Exercise 44–3.

a) Estimate the radius of each nucleus.

b) Estimate the surface area of each.

c) Estimate the volume of each.

d) Determine the mass density (in $kg \cdot m^{-3}$) for each.

e) Determine the particle density (in $particles \cdot m^{-3}$) for each.

44–5 Using the data in Table 44–2, calculate the binding energy of the deuterium nucleus. (Express your result in MeV.)

44–6 Calculate

a) the mass defect,

b) the binding energy,

c) the binding energy per nucleon, for the common isotope of oxygen, ^{16}O.

Section 44–3 Nuclear Stability
Section 44–4 Radioactive Transformations

44–7 Tritium is an unstable isotope of hydrogen, $^3_1$H; its mass, including one electron, is 3.01647 u.

a) Show that it must be unstable with respect to beta decay because $^3_2$He plus an emitted electron has less total mass.

b) Determine the total kinetic energy of the decay products, taking care to account for the electron masses correctly.

44–8 Radium (^{226}Ra) undergoes alpha emission, leading to radon (^{222}Rn). The masses, including all electrons in each atom, are 226.0254 u and 222.0163 u, respectively. Find the maximum kinetic energy that the emitted alpha particle can have.

44–9 The common isotope of uranium, ^{238}U, has a half-life of 4.50×10^9 years, decaying by alpha emission.

a) What is the decay constant?

b) What mass of uranium would be required for an activity of one curie?

c) How many alpha particles are emitted per second by 1 g of uranium?

44–10 An unstable isotope of cobalt, ^{60}Co, has one more neutron in its nucleus than the stable ^{59}Co and is a beta emitter with a half-life of 5.3 years. This isotope is widely used in medicine. A certain radiation source in a hospital contains 0.01 g of ^{60}Co.

a) What is the decay constant for this isotope?

b) How many atoms are in the source?

c) How many decays occur per second?

d) What is the activity of the source, in curies? How does this compare with the activity of an equal mass of radium?

44–11 The unstable isotope ^{40}K is used for dating rock samples. Its half-life is 2.4×10^8 years.

a) How many decays occur per second in a sample containing 2×10^{-6} g of ^{40}K?

b) What is the activity of the sample, in curies?

Section 44–5 Nuclear Reactions
Section 44–6 Nuclear Fission
Section 44–7 Nuclear Fusion

44–12 In the fission of one ^{238}U nucleus, 200 MeV of energy are released. Express this energy in joules per mole and compare with typical heats of combustion, which are on the order of $1.0 \times 10^5 \ J \cdot mol^{-1}$.

44–13 Consider the nuclear reaction

$$^2_1H + {}^9_4Be \longrightarrow {}^7_3Li + {}^4_2He.$$

a) How much energy is liberated?

b) Estimate the threshold energy for this reaction.

44–14 Consider the fusion reaction

$$^2H + {}^2H \longrightarrow {}^4He + energy.$$

a) Compute the energy liberated in this reaction, in MeV and in joules.

b) Compute the energy *per mole* of deuterium, remembering that the gas is diatomic, and compare with the heat of combustion of hydrogen, about $2.9 \times 10^5 \ J \cdot mol^{-1}$.

44–15. Consider the nuclear reaction

$$^2_1H + {}^{14}_7N \longrightarrow {}^6_3Li + {}^{10}_5B.$$

Is energy absorbed or liberated, and how much?

Section 44–8 Particle Accelerators

44–16 The magnetic field in a cyclotron that is accelerating protons is 0.8 T.

a) How many times per second should the potential across the dees reverse?

b) The maximum radius of the cyclotron is 0.25 m. What is the maximum velocity of the proton?

c) Through what potential difference would the proton have to be accelerated to give it the maximum cyclotron velocity?

d) What is the energy of the protons when they emerge? Express your result in joules and in MeV.

44–17 Deuterons in a cyclotron describe a circle of radius 32.0 cm just before emerging from the dees. The frequency of the applied alternating voltage is 10 MHz. Find

a) the magnetic field,

b) the energy and speed of the deuterons upon emergence.

Section 44-9 Fundamental Particles
Section 44-10 High-Energy Physics

44-18 If two gamma-ray photons are produced in e^+e^- annihilation, find the energy, frequency, and wavelength of each photon.

44-19 A neutral pion at rest decays into two gamma-ray photons. Find the energy, frequency, and wavelength of each photon.

44-20 Beams of π^- mesons are being used experimentally in the treatment of cancer. What is the minimum total energy a pion can release in a tumor?

44-21 Determine the electric charge, baryon number, strangeness, and charm quantum numbers for the following quark combinations:

a) uus, b) $c\bar{s}$, c) $\overline{ddu}$, d) $c\bar{b}$.

PROBLEMS

44-22 Compute the approximate density of nuclear matter, and compare your result with typical densities of ordinary matter.

44-23 The starship *Enterprise* is powered by the controlled combination of matter and antimatter. If the entire 100 kg antimatter fuel supply of the *Enterprise* were to combine with matter, how much energy would be released?

44-24 The results of activity measurements on a radioactive sample are given below.

a) Find the half-life.

b) How many radioactive nuclei were present in the sample at $t = 0$?

c) How many were present after 7 hr?

Time (hr)	Counts·s^{-1}
0.	20,000
0.5	14,800
1.0	11,000
1.5	8,130
2.0	6,020
2.5	4,460
3.0	3,300
4.0	1,810
5.0	1,000
6.0	550
7.0	300

44-25 A carbon specimen found in a cave believed to have been inhabited by cavemen contained ⅛ as much ^{14}C as an equal amount of carbon in living matter. Find the approximate age of the specimen.

44-26

a) What is the binding energy of the least strongly bound proton in $^{12}_{6}$C?

b) The least strongly bound neutron in $^{13}_{6}$C?

44-27 A free neutron at rest decays into a proton, an electron, and a neutrino, with a half-life of about 15 min. Calculate the total kinetic energy of the decay products.

44-28 A K^+ meson at rest decays into two π mesons.

a) What are the allowed combinations of π^0, π^+, and π^- as decay products?

b) Find the total kinetic energy of the π mesons.

44-29 A Λ hyperon at rest decays into a proton and a π^-.

a) Find the total kinetic energy of the decay products.

b) What fraction of the energy is carried off by each particle? (For simplicity, use nonrelativistic momentum and kinetic-energy expressions.)

44-30 The measured energy width of the ϕ meson is 4 MeV and its mass is 1020 MeV/c^2. Using the uncertainty principle, Eq. (42-17), estimate the lifetime of the ϕ meson.

44-31 A ϕ meson (Problem 44-30) at rest decays via $\phi \rightarrow K^+K^-$.

a) Find the kinetic energy of the K^+ meson.

b) Suggest a reason why the decay $\phi \rightarrow K^+K^-\pi^0$ has not been observed.

c) Suggest reasons why the decays $\phi \rightarrow K^+\pi^-$ and $\phi \rightarrow K^+\mu^-$ have not been observed.

CHALLENGE PROBLEMS

44-32 The results of activity measurements on a radioactive sample that is a mixture of radioactive elements is shown in the following table.

a) How many different nuclides are present in the mixture?

b) What are their half-lives?

c) How many nuclei of each type are initially present in the sample?

d) How many of each type are present at $t = 6$ hr?

Time (hr)	Counts·s^{-1}	Time (hr)	Counts·s^{-1}
0.	7500	5.0	414
0.5	4120	6.0	288
1.0	2570	7.0	201
1.5	1790	8.0	140
2.0	1350	9.0	98
2.5	1070	10.0	68
3.0	872	12.0	33
4.0	596		

44-33 ^{128}I is created by the irradiation of ^{127}I with neutrons. The half-life of ^{128}I is 25 min. A neutron beam is used that creates 1×10^6 ^{128}I nuclei per second.

a) Sketch a graph of the number of ^{128}I nuclei present as a function of time.

b) What is the activity of the sample 1, 10, 25, 50, 75, and 180 min after irradiation is begun?

c) What is the maximum number of ^{128}I atoms that can be created in the sample, after being irradiated for a long time? (This steady-state situation is called *saturation*.)

d) What therefore is the maximum activity that can be produced?

44-34 Radioisotopes are used in a variety of manufacturing and testing techniques. Wear measurements can be made by the following method. An automobile engine is produced using piston rings with a total mass of 100 g, which includes 10μ Ci of ^{59}Fe whose half-life is 45 days. The engine is test run for 1000 hr, after which the oil is drained and its activity measured. If the activity of the engine oil is 50 decays·s^{-1}, how much mass was worn from the piston rings per hour of operation?

APPENDIX A

THE INTERNATIONAL SYSTEM OF UNITS

The Système International d'Unités, abbreviated SI, is the system developed by the General Conference on Weights and Measures and adopted by nearly all the industrial nations of the world. It is based on the mksa (meter-kilogram-second-ampere) system. The following material is adapted from NBS Special Publication 330 (1981 edition) of the National Bureau of Standards.

Quantity	Name of unit	Symbol	Equivalent Units
SI Base Units			
length	meter	m	
mass	kilogram	kg	
time	second	s	
electric current	ampere	A	
thermodynamic temperature	kelvin	K	
luminous intensity	candela	cd	
amount of substance	mole	mol	
SI Derived Units			
area	square meter	m^2	
volume	cubic meter	m^3	
frequency	hertz	Hz	s^{-1}
mass density (density)	kilogram per cubic meter	$kg \cdot m^{-3}$	
speed, velocity	meter per second	$m \cdot s^{-1}$	
angular velocity	radian per second	$rad \cdot s^{-1}$	
acceleration	meter per second squared	$m \cdot s^{-2}$	
angular acceleration	radian per second squared	$rad \cdot s^{-2}$	
force	newton	N	$kg \cdot m \cdot s^{-2}$
pressure (mechanical stress)	pascal	Pa	$N \cdot m^{-2}$
kinematic viscosity	square meter per second	$m^2 \cdot s^{-1}$	
dynamic viscosity	newton-second per square meter	$N \cdot s \cdot m^{-2}$	
work, energy, quantity of heat	joule	J	$N \cdot m$
power	watt	W	$J \cdot s^{-1}$
quantity of electricity	coulomb	C	$A \cdot s$
potential difference, electromotive force	volt	V	$W \cdot A^{-1}, J \cdot C^{-1}$
electric field strength	volt per meter	$V \cdot m^{-1}$	$N \cdot C^{-1}$
electric resistance	ohm	Ω	$V \cdot A^{-1}$
capacitance	farad	F	$A \cdot s \cdot V^{-1}$

Quantity	Name of unit	Symbol	Equivalent Units
magnetic flux	weber	Wb	$V \cdot s$
inductance	henry	H	$V \cdot s \cdot A^{-1}$
magnetic flux density	tesla	T	$Wb \cdot m^{-2}$
magnetic field strength	ampere per meter	$A \cdot m^{-1}$	
magnetomotive force	ampere	A	
luminous flux	lumen	lm	$cd \cdot sr$
luminance	candela per square meter	$cd \cdot m^{-2}$	
illuminance	lux	lx	$lm \cdot m^{-2}$
wave number	1 per meter	m^{-1}	
entropy	joule per kelvin	$J \cdot K^{-1}$	
specific heat capacity	joule per kilogram kelvin	$J \cdot kg^{-1} \cdot K^{-1}$	
thermal conductivity	watt per meter kelvin	$W \cdot m^{-1} \cdot K^{-1}$	
radiant intensity	watt per steradian	$W \cdot sr^{-1}$	
activity (of a radioactive source)	becquerel	Bq	s^{-1}
radiation dose	gray	Gy	$J \cdot kg^{-1}$
radiation dose equivalent	sievert	Sv	$J \cdot kg^{-1}$
SI Supplementary Units			
plane angle	radian	rad	
solid angle	steradian	sr	

DEFINITIONS OF SI UNITS

meter (m)   The *meter* is the length equal to the distance traveled by light, in vacuum, in a time of 1/299,792,458 second.

kilogram (kg)   The *kilogram* is the unit of mass; it is equal to the mass of the international prototype of the kilogram. (The international prototype of the kilogram is a particular cylinder of platinum-iridium alloy that is preserved in a vault at Sèvres, France, by the International Bureau of Weights and Measures.)

second (s)   The *second* is the duration of 9,192,631,770 periods of the radiation corresponding to the transition between the two hyperfine levels of the ground state of the cesium-133 atom.

ampere (A)   The *ampere* is that constant current that, if maintained in two straight parallel conductors of infinite length, of negligible circular cross section, and placed 1 meter apart in vacuum, would produce between these conductors a force equal to 2×10^{-7} newton per meter of length.

kelvin (K)   The *kelvin*, unit of thermodynamic temperature, is the fraction 1/273.16 of the thermodynamic temperature of the triple point of water.

ohm (Ω)   The *ohm* is the electric resistance between two points of a conductor when a constant difference of potential of 1 volt, applied between these two points, produces in this conductor a current of 1 ampere, this conductor not being the source of any electromotive force.

coulomb (C)   The *coulomb* is the quantity of electricity transported in 1 second by a current of 1 ampere.

candela (cd)   The *candela* is the luminous intensity, in a given direction, of a source that emits monochromatic radiation of frequency 540×10^{12} hertz and that has a radiant intensity in that direction of 1/683 watt per steradian.

mole (mol)   The *mole* is the amount of substance of a system that contains as many elementary entities as there are carbon atoms in 0.012 kg of carbon 12. The elementary entities must be specified and may be atoms, molecules, ions, electrons, other particles, or specified groups of such particles.

newton (N)   The *newton* is that force that gives to a mass of 1 kilogram an acceleration of 1 meter per second per second.

joule (J)   The *joule* is the work done when the point of application of 1 newton is displaced a distance of 1 meter in the direction of the force.

watt (W)   The *watt* is the power that gives rise to the production of energy at the rate of 1 joule per second.

volt (V)   The *volt* is the difference of electric potential between two points of a conducting wire carrying a constant current of 1 ampere, when the power dissipated between these points is equal to 1 watt.

weber (Wb)   The *weber* is the magnetic flux that, linking a circuit of one turn, produces in it an electromotive force of 1 volt as it is reduced to zero at a uniform rate in 1 second.

lumen (lm)   The *lumen* is the luminous flux emitted in a solid angle of 1 steradian by a uniform point source having an intensity of 1 candela.

farad (F) The *farad* is the capacitance of a capacitor between the plates of which there appears a difference of potential of 1 volt when it is charged by a quantity of electricity equal to 1 coulomb.

henry (H) The *henry* is the inductance of a closed circuit in which an electromotive force of 1 volt is produced when the electric current in the circuit varies uniformly at a rate of 1 ampere per second.

radian (rad) The *radian* is the plane angle between two radii of a circle that cut off on the circumference an arc equal in length to the radius.

steradian (sr) The *steradian* is the solid angle that, having its vertex in the center of a sphere, cuts off an area of the surface of the sphere equal to that of a square with sides of length to the radius of the sphere.

SI Prefixes The names of multiples and submultiples of SI units may be formed by application of the prefixes listed in Table 1–1, page 6.

APPENDIX B
USEFUL MATHEMATICAL RELATIONS

ALGEBRA

$$a^{-x} = \frac{1}{a^x} \qquad\qquad a^{(x+y)} = a^x a^y \qquad\qquad a^{(x-y)} = \frac{a^x}{a^y}$$

Logarithms: If $\log a = x$, then $a = 10^x$. $\log a + \log b = \log (ab)$ $\log a - \log b = \log (a/b)$ $\log (a^n) = n \log a$

If $\ln a = x$, then $a = e^x$. $\ln a + \ln b = \ln (ab)$ $\ln a - \ln b = \ln (a/b)$ $\ln (a^n) = n \ln a$

Quadratic formula: If $ax^2 + bx + c = 0$, $x = \dfrac{-b \pm \sqrt{b^2 - 4ac}}{2a}$.

BINOMIAL THEOREM

$$(a + b)^n = a^n + na^{n-1}b + \frac{n(n-1)a^{n-2}b^2}{2!} + \frac{n(n-1)(n-2)a^{n-3}b^3}{3!} + \cdots$$

TRIGONOMETRY

In the right triangle ABC, $x^2 + y^2 = r^2$.

Definitions of the trigonometric functions: $\sin a = y/r$ $\cos a = x/r$ $\tan a = y/x$

Identities: $\sin^2 a + \cos^2 a = 1$ $\tan a = \dfrac{\sin a}{\cos a}$

$\sin 2a = 2 \sin a \cos a$ $\cos 2a = \cos^2 a - \sin^2 a = 2 \cos^2 a - 1$

$\sin \frac{1}{2}a = \sqrt{\dfrac{1 - \cos a}{2}}$ $\cos \frac{1}{2}a = \sqrt{\dfrac{1 + \cos a}{2}}$

$\sin (-a) = -\sin a$ $\sin (a \pm b) = \sin a \cos b \pm \cos a \sin b$

$\cos (-a) = \cos a$ $\cos (a \pm b) = \cos a \cos b \mp \sin a \sin b$

$\sin (a \pm \pi/2) = \pm\cos a$ $\sin a + \sin b = 2 \sin \frac{1}{2}(a + b) \cos \frac{1}{2}(a - b)$

$\cos (a \pm \pi/2) = \mp\sin a$ $\cos a + \cos b = 2 \cos \frac{1}{2}(a + b) \cos \frac{1}{2}(a - b)$

GEOMETRY

Circumference of circle of radius r: $C = 2\pi r$

Area of circle of radius r: $A = \pi r^2$

Volume of sphere of radius r: $V = 4\pi r^3/3$

Surface area of sphere of radius r: $A = 4\pi r^2$

Volume of cylinder of radius r and height h: $V = \pi r^2 h$

CALCULUS

Derivatives:

$$\frac{d}{dx}x^n = nx^{n-1}$$

$$\frac{d}{dx}\sin ax = a\cos ax$$

$$\frac{d}{dx}\cos ax = -a\sin ax$$

$$\frac{d}{dx}e^{ax} = ae^{ax}$$

$$\frac{d}{dx}\ln ax = \frac{1}{x}$$

Integrals:

$$\int x^n\,dx = \frac{x^{n+1}}{n+1}$$

$$\int \frac{dx}{x} = \ln x$$

$$\int \sin ax\,dx = -\frac{1}{a}\cos ax$$

$$\int \cos ax\,dx = \frac{1}{a}\sin ax$$

$$\int e^{ax}\,dx = \frac{1}{a}e^{ax}$$

$$\int \frac{dx}{\sqrt{a^2-x^2}} = \arcsin\frac{x}{a}$$

$$\int \frac{dx}{\sqrt{x^2+a^2}} = \ln(x+\sqrt{x^2+a^2})$$

$$\int \frac{dx}{x^2+a^2} = \frac{1}{a}\arctan\frac{x}{a}$$

$$\int \frac{dx}{(x^2+a^2)^{3/2}} = \frac{1}{a^2}\frac{x}{\sqrt{x^2+a^2}}$$

$$\int \frac{x\,dx}{(x^2+a^2)^{3/2}} = -\frac{1}{\sqrt{x^2+a^2}}$$

Power series (convergent for range of x shown):

$$\sin x = x - \frac{x^3}{3!} + \frac{x^5}{5!} - \frac{x^7}{7!} + \cdots \qquad (\text{all } x)$$

$$\cos x = 1 - \frac{x^2}{2!} + \frac{x^4}{4!} - \frac{x^6}{6!} + \cdots \qquad (\text{all } x)$$

$$\tan x = x + \frac{x^3}{3} + \frac{2x^5}{15} + \frac{17x^7}{315} + \cdots \qquad (|x| < \pi/2)$$

$$e^x = 1 + x + \frac{x^2}{2!} + \frac{x^3}{3!} + \cdots \qquad (\text{all } x)$$

$$\ln(1+x) = x - \frac{x^2}{2} + \frac{x^3}{3} - \frac{x^4}{4} + \cdots \qquad (|x| < 1)$$

APPENDIX C

THE GREEK ALPHABET

Name	Capital	Lowercase	Name	Capital	Lowercase
Alpha	A	α	Nu	N	ν
Beta	B	β	Xi	Ξ	ξ
Gamma	Γ	γ	Omicron	O	o
Delta	Δ	δ	Pi	Π	π
Epsilon	E	ϵ	Rho	P	ρ
Zeta	Z	ζ	Sigma	Σ	σ
Eta	H	η	Tau	T	τ
Theta	Θ	θ	Upsilon	Υ	υ
Iota	I	ι	Phi	Φ	ϕ
Kappa	K	κ	Chi	X	χ
Lambda	Λ	λ	Psi	Ψ	ψ
Mu	M	μ	Omega	Ω	ω

APPENDIX D

PERIODIC TABLE OF THE ELEMENTS

Period	IA	IIA	IIIB	IVB	VB	VIB	VIIB	VIIIB	VIIIB	VIIIB	IB	IIB	IIIA	IVA	VA	VIA	VIIA	Noble gases
1	1 H 1.008																	2 He 4.003
2	3 Li 6.941	4 Be 9.012											5 B 10.811	6 C 12.011	7 N 14.007	8 O 15.999	9 F 18.998	10 Ne 20.179
3	11 Na 22.990	12 Mg 24.305											13 Al 26.982	14 Si 28.086	15 P 30.974	16 S 32.064	17 Cl 35.453	18 Ar 39.948
4	19 K 39.098	20 Ca 40.08	21 Sc 44.956	22 Ti 47.90	23 V 50.942	24 Cr 51.996	25 Mn 54.938	26 Fe 55.847	27 Co 58.933	28 Ni 58.70	29 Cu 63.546	30 Zn 65.38	31 Ga 69.72	32 Ge 72.59	33 As 74.922	34 Se 78.96	35 Br 79.904	36 Kr 83.80
5	37 Rb 85.468	38 Sr 87.62	39 Y 88.906	40 Zr 91.22	41 Nb 92.906	42 Mo 95.94	43 Tc (99)	44 Ru 101.07	45 Rh 102.905	46 Pd 106.4	47 Ag 107.868	48 Cd 112.41	49 In 114.82	50 Sn 118.69	51 Sb 121.75	52 Te 127.60	53 I 126.905	54 Xe 131.30
6	55 Cs 132.905	56 Ba 137.33	57 La 138.905	72 Hf 178.49	73 Ta 180.948	74 W 183.85	75 Re 186.2	76 Os 190.2	77 Ir 192.22	78 Pt 195.09	79 Au 196.966	80 Hg 200.59	81 Tl 204.37	82 Pb 207.19	83 Bi 208.2	84 Po (210)	85 At (210)	86 Rn (222)
7	87 Fr (223)	88 Ra (226)	89 Ac (227)	104 Rf(?) (261)	105 Ha(?) (262)	106 (257)	107 (260)											

58 Ce 140.12	59 Pr 140.907	60 Nd 144.24	61 Pm (145)	62 Sm 150.35	63 Eu 151.96	64 Gd 157.25	65 Tb 158.925	66 Dy 162.50	67 Ho 164.930	68 Er 167.26	69 Tm 168.934	70 Yb 173.04	71 Lu 174.96
90 Th (232)	91 Pa (231)	92 U (238)	93 Np (239)	94 Pu (239)	95 Am (240)	96 Cm (242)	97 Bk (245)	98 Cf (246)	99 Es (247)	100 Fm (249)	101 Md (256)	102 No (254)	103 Lr (257)

For each element the average atomic mass of the mixture of isotopes occurring in nature is shown. For elements having no stable isotope, the approximate atomic mass of the most common isotope is shown in parentheses.

APPENDIX E
UNIT CONVERSION FACTORS

LENGTH

1 m = 100 cm = 1000 mm = 10^6 μm = 10^9 nm
1 km = 1000 m = 0.6214 mi
1 m = 3.281 ft = 39.37 in.
1 cm = 0.3937 in.
1 in. = 2.540 cm
1 ft = 30.48 cm
1 yd = 91.44 cm
1 mi = 5280 ft = 1.609 km
1 Å = 10^{-10} m = 10^{-8} cm = 10^{-1} nm
1 nautical mile = 6080 ft
1 light year = 9.461×10^{15} m

AREA

1 cm^2 = 0.155 in^2
1 m^2 = 10^4 cm^2 = 10.76 ft^2
1 in^2 = 6.452 cm^2
1 ft^2 = 144 in^2 = 0.0929 m^2

VOLUME

1 liter = 1000 cm^3 = $10^{-3}m^3$ = 0.03531 ft^3 = 61.02 in^3
1 ft^3 = 0.02832 m^3 = 28.32 liters = 7.477 gallons
1 gallon = 3.788 liters

TIME

1 min = 60 s
1 hr = 3600 s
1 da = 86,400 s
1 yr = 365.24 da = 3.156×10^7 s

ANGLE

1 rad = 57.30° = 180°/π
1° = 0.01745 rad = π/180 rad
1 revolution = 360° = 2π rad
1 rev·min^{-1} (rpm) = 0.1047 rad·s^{-1}

SPEED

1 m·s^{-1} = 3.281 ft·s^{-1}
1 ft·s^{-1} = 0.3048 m·s^{-1}
1 mi·min^{-1} = 60 mi·hr^{-1} = 88 ft·s^{-1}
1 km·hr^{-1} = 0.2778 m·s^{-1} = 0.6214 mi·hr^{-1}
1 mi·hr^{-1} = 1.466 ft·s^{-1} = 0.4470 m·s^{-1} = 1.609 km·hr^{-1}
1 furlong·$fortnight^{-1}$ = 1.662×10^{-4} m·s^{-1}

ACCELERATION

1 m·s^{-2} = 100 cm·s^{-2} = 3.281 ft·s^{-2}
1 cm·s^{-2} = 0.01 m·s^{-2} = 0.03281 ft·s^{-2}
1 ft·s^{-2} = 0.3048 m·s^{-2} = 30.48 cm·s^{-2}
1 mi·hr^{-1}·s^{-1} = 1.467 ft·s^{-2}

MASS

1 kg = 10^3 g = 0.0685 slug
1 g = 6.85×10^{-5} slug
1 slug = 14.59 kg
1 u = 1.661×10^{-27} kg
1 kg has a weight of 2.205 lb when g = 9.80 m·s^{-2}

FORCE

1 N = 10^5 dyn = 0.2248 lb
1 lb = 4.448 N = 4.448×10^5 dyn

PRESSURE

1 Pa = 1 N·m^{-2} = 1.451×10^{-4} lb·in^{-2} = 0.209 lb·ft^{-2}
1 bar = 10^5 Pa
1 lb·in^{-2} = 6891 Pa
1 lb·ft^{-2} = 47.85 Pa
1 atm = 1.013×10^5 Pa = 1.013 bar
   = 14.7 lb·in^{-2} = 2117 lb·ft^{-2}
1 mm Hg = 1 torr = 133.3 Pa

ENERGY

1 J = 10^7 ergs = 0.239 cal
1 cal = 4.186 J (based on 15° calorie)
1 ft·lb = 1.356 J
1 Btu = 1055 J = 252 cal = 778 ft·lb
1 eV = 1.602×10^{-19} J
1 kWh = 3.600×10^6 J

MASS–ENERGY EQUIVALENCE

1 kg $\leftrightarrow$ 8.988×10^{16} J
1 u $\leftrightarrow$ 931.5 MeV
1 eV $\leftrightarrow$ 1.073×10^{-9} u

POWER

1 W = 1 J·s^{-1}
1 hp = 746 W = 550 ft·lb·s^{-1}
1 Btu·hr^{-1} = 0.293 W

APPENDIX F
NUMERICAL CONSTANTS

FUNDAMENTAL PHYSICAL CONSTANTS

Name	Symbol	Value
Speed of light	c	2.9979×10^8 m·s^{-1}
Charge of electron	e	1.602×10^{-19} C
Gravitational constant	G	6.673×10^{-11} N·m^2·kg^{-2}
Planck's constant	h	6.626×10^{-34} J·s
Boltzmann's constant	k	1.381×10^{-23} J·K^{-1}
Avogadro's number	N_0	6.022×10^{23} molecules·mol^{-1}
Gas constant	R	8.314 J·mol^{-1}·K^{-1}
Mass of electron	m_e	9.110×10^{-31} kg
Mass of neutron	m_n	1.675×10^{-27} kg
Mass of proton	m_p	1.673×10^{-27} kg
Permittivity of free space	ϵ_0	8.854×10^{-12} C^2·N^{-1}·m^{-2}
	$1/4\pi\epsilon_0$	8.987×10^9 N·m^2·C^{-2}
Permeability of free space	μ_0	$4\pi \times 10^{-7}$ Wb·A^{-1}·m^{-1}

OTHER USEFUL CONSTANTS

Name	Symbol	Value
Mechanical equivalent of heat		4.186 J·cal^{-1} (15° calorie)
Standard atmospheric pressure	1 atm	1.013×10^5 Pa
Absolute zero	0 K	-273.15°C
Electronvolt	1 eV	1.602×10^{-19} J
Atomic mass unit	1 u	1.661×10^{-27} kg
Electron rest energy	mc^2	0.511 MeV
Energy equivalent of 1 u	Mc^2	931.5 MeV
Volume of ideal gas (0°C and 1 atm)	V	22.4 liter·mol^{-1}
Acceleration due to gravity (sea level, at equator)	g	9.78049 m·s^{-2}

ASTRONOMICAL DATA

Body	Mass, kg	Radius, m	Orbit radius, m	Orbit period
Sun	1.99×10^{30}	6.95×10^8	—	—
Moon	7.36×10^{22}	1.74×10^6	0.38×10^9	27.3 d
Mercury	3.28×10^{23}	2.57×10^6	5.8×10^{10}	88.0 d
Venus	4.82×10^{24}	6.31×10^6	1.08×10^{11}	224.7 d
Earth	5.98×10^{24}	6.38×10^6	1.49×10^{11}	365.3 d
Mars	6.34×10^{23}	3.43×10^6	2.28×10^{11}	687.0 d
Jupiter	1.88×10^{27}	7.18×10^7	7.78×10^{11}	11.86 y
Saturn	5.63×10^{26}	6.03×10^7	1.43×10^{12}	29.46 y
Uranus	8.61×10^{25}	2.67×10^7	2.87×10^{12}	84.02 y
Neptune	9.99×10^{25}	2.48×10^7	4.49×10^{12}	164.8 y
Pluto	5×10^{23}	4×10^5	5.90×10^{12}	247.7 y

ANSWERS TO ODD-NUMBERED PROBLEMS

CHAPTER 24

24–1 9.65×10^4 C

24–3 -1.78×10^{-6} C

24–5 a) 1.67×10^{-7} C, 1.67×10^{-7} C
b) 1.18×10^{-7} C, 2.36×10^{-7} C

24–7 0.119 m

24–9 a) 2.75×10^{26} b) 6.58×10^{15}
c) $2.39 \times 10^{-9}\%$

24–11 2.19×10^6 m·s^{-1}

24–13 a) 0 b) $(1/4\pi\epsilon_0)2xq^2/(a^2+x^2)^{3/2}$,
+x-direction d) $(1/4\pi\epsilon_0)2q^2/x^2$

24–15 a) $F = aq^2/2\pi\epsilon_0 x^3$ b) $F = aq^2/\pi\epsilon_0 y^3$

24–17 $F_x = -Q/(4\pi\epsilon_0 x\sqrt{a^2+x^2})$;
$F_y = (Q/4\pi\epsilon_0 a)[1/x - 1/\sqrt{a^2+x^2}]$

24–19 b) 1.36×10^{-6} C c) 31.7°

24–21 a) 1.06×10^{-10} m b) 1.09×10^6 m·s^{-1}

24–23 a) 9.00×10^{-7} N, away from the vacant corner
b) 8.61×10^{-7} N, toward the opposite corner

24–25 a) $(Q/2\pi\epsilon_0 a)[1/y - 1/\sqrt{a^2+y^2}]$ in the $-x$-direction; when $y \to \infty$, $F \to Qa/4\pi\epsilon_0 y^3$
b) $(Q/4\pi\epsilon_0 a)[1/(x-a) + 1/(x+a) - 2/x]$ in +x-direction; when $x \to \infty$, $F \to Qa/2\pi\epsilon_0 x^3$

24–27 b) $q_1 < 0$, $q_2 > 0$ c) 1.69×10^{-6} C d) 28.1 N

CHAPTER 25

25–1 7.20×10^4 N·C^{-1}, upward

25–3 4.74 m

25–5 4.0 N·C^{-1}, upward

25–7 5.57×10^{-11} N·C^{-1}

25–9 1.42×10^4 N·C^{-1}

25–11 a) 50.0 N·C^{-1},
+x-direction b) 10.8 N·C^{-1},+x-direction

25–13 a) 1.80×10^4 N·C^{-1}, $-x$-direction
b) 8.00×10^3 N·C^{-1}, +x-direction
c) 3.34×10^3 N·C^{-1}, 69.9° above $-x$-axis
d) 6.36×10^3 N·C^{-1}, $-x$-direction

25–15 3.54×10^{-11} C·m^{-2}

25–17 2.26×10^9 electrons

25–21 8.85×10^{-10} C

25–23 b) $q/4\pi\epsilon_0 r^2$

25–25 a) 1.42 cm b) 9.85 cm

25–27 b) $q_1 < 0$, $q_2 < 0$ c) 1.95×10^7 N·C^{-1}

25–29 $E_x = E_y = Q/2\pi^2\epsilon_0 a^2$

25–31 -2.66×10^{-10} C

25–33 a) $\lambda/2\pi\epsilon_0 r$ b) $\lambda/2\pi\epsilon_0 r$ d) $-\lambda$ inner, $+\lambda$ outer

25–35 c) $(4 - 3r/R)Qr/4\pi\epsilon_0 R^3$ d) $Q/4\pi\epsilon_0 R^2$, both expressions

25–37 a) $8Q/5\pi R^3$; 6.37×10^{23} C·m^{-3}
b) $r \le R/2$: $8Qr/15\pi\epsilon_0 R^3$;
$R/2 \le r \le R$: $(4 - 3r/R)4Qr/15\pi\epsilon_0 R^3$
$-Q/60\pi\epsilon_0 r^2$; $r \ge R$: $Q/4\pi\epsilon_0 r^2$ c) 4/15
e) 9.68×10^{-23} s

CHAPTER 26

26–1 a) -0.0225 J b) 12.2 m·s^{-1}

26–3 a) point a b) 1500 V·m^{-1} c) 1.20×10^{-4} J

26–5 a) 1.80×10^3 V b) 0 c) 4.50×10^{-5} J

26–7 b) $q/2\pi\epsilon_0 a$ e) $\pm\sqrt{3}a$

26–9 a) 5000 V·m^{-1} b) 1.0×10^{-6} J
c) 1.0×10^{-6} J

26–11 2.5 mm

26–13 a) 180 V b) 180 V

26–15 a) 0 b) -1.00×10^{-3} J c) 2.30×10^{-3} J

26–17 a) Inside: $q(3 - r^2/R^2)/8\pi\epsilon_0 R = (\rho/2\epsilon_0)(R^2 - r^2/3)$
outside: $q/4\pi\epsilon_0 r = \rho R^3/3\epsilon_0 r$

26–21 a) 4.33×10^{-4} m·s^{-1} (downward)
b) -2.75×10^{-4} m·s^{-1} (upward)

26–23 1.44 MeV

26–25 a) 5.11×10^3 V b) $0.141c$
c) 9.39×10^6 V; $0.141c$

26–27 a) 0.703 cm   b) 19.4°   c) 4.92 cm
26–29 a) -4.5×10^{-5} J   b) 3.0×10^5 V·m^{-1}
c) -1.5×10^4 V
26–31 a) Remains at rest   b) Oscillates about the origin, along the y-axis   c) Accelerates away from the origin along the x-axis
26–33 a) 2.18×10^{-5} m   b) Electron's is 42.8 times the proton's   c) Equal
26–35 a) -90 V; -315 V   b) $+9.0 \times 10^{-7}$ J
26–37 a) 1.00×10^5 V·m$^{-4/3}$
b) $-4/3\ Cx^{1/3}$
c) 3.39×10^{-15} N
26–39 a) Inside: $\lambda(R^2 - r^2)/4\pi\epsilon_0 R^2$
outside: $-\lambda \ln (r/R)/2\pi\epsilon_0$
26–41 a) $(Q/4\pi\epsilon_0 a) \ln [(x + a)/x]$
b) $(Q/4\pi\epsilon_0 a) \ln [a + \sqrt{a^2 + y^2}]/y]$
c) Point P: $Q/4\pi\epsilon_0 x$;   point Q: $Q/4\pi\epsilon_0 y$
26–43 2
26–45 b) $Q(2R^2 - 2r^2 + r^3/R)/4\pi\epsilon_0 R^3$
26–47 a) 867 m·s^{-1} c) -7.50 J   d) Will not escape because $E_r < 0$ e) 12.0 mm
26–49 a) $r \le a$: $E_r = (\rho_0 r/3\epsilon_0)(1 - r/a)$, $E_\theta = 0$, $E_\phi = 0$
$r \ge a$: $E = 0$
b) $r \le a$: $\rho(r) = (\rho_0/3)(3 - 4r/a)$
$r \ge a$: $\rho(r) = 0$

CHAPTER 27

27–1 a) 400 V   b) 113 cm^2 c) 2.00×10^6 V·m^{-1}
d) 1.77×10^{-5} C·m^{-2}
27–3 4.25×10^{-4} C
27–5 a) $Q_1 = 1.44 \times 10^{-4}$ C, $Q_2 = 2.16 \times 10^{-4}$ C
b) $V_1 = V_2 = 36.0$ V
27–7 a) $Q_1 = 1.92 \times 10^{-5}$ C, $Q_2 = 1.92 \times 10^{-5}$ C,
$Q_3 = 3.84 \times 10^{-5}$ C, $Q_4 = 5.76 \times 10^{-5}$ C
b) $V_1 = 9.6$ V, $V_2 = 9.6$ V, $V_3 = 19.2$ V, $V_4 = 28.8$ V
c) 19.2 V
27–9 a) 500 V   b) 1.25×10^{-4} J   c) 0.113 m^2
d) 800 V
27–11 0.0692 V·m^{-3}
27–13 a) $xq^2/2\epsilon_0 A$ b) $(x + dx)q^2/2\epsilon_0 A$ d) E is due to superposition of fields of each plate; must calculate force of one plate due to electric field from other plate alone
27–15 1.69 m^2
27–17 a) 0.71×10^{-6} C·m^{-2} b) 1.67
27–19 a) 3.42   b) 7.08×10^{-8} C
27–21 a) 1.77×10^{-11} F   b) 8.85×10^{-10} C
c) 2500 V·m^{-1} d) 2.21×10^{-8} J
27–23 a) 1 μF   b) $Q_2 = 6.0 \times 10^{-4}$ C, $Q_3 = 9.0 \times 10^{-4}$ C   c) 100 V
27–25 a) 1 μF: $q = 1.2 \times 10^{-3}$ C, $V = 1200$ V
2 μF: $q = 2.4 \times 10^{-3}$ C, $V = 1200$ V
b) 1 μF: $q = 4.0 \times 10^{-4}$ C, $V = 400$ V
2 μF: $q = 8.0 \times 10^{-4}$ C, $V = 400$ V
27–27 a) 2.4×10^{-5} b) 1.44×10^{-4} J   c) 3.6 V
d) 1.30×10^{-4} J
27–29 a) 5.53×10^{-11} F   b) 1.66×10^{-8} C
c) 2.49×10^{-6} J   d) 0.498 J·m^{-3}
27–31 a) $(Q/4\pi\epsilon_0)(1/r_a - 1/r_b)$
27–35 b) 1.77×10^{-9} F

27–37 b) 14 μF   c) 72 μF: 504 μC, 7 V;   27 μF: 270 μC, 10 V;
18 μF: 234 μC, 13 V;   6 μF: 18 μC, 3 V;
28 μF: 252 μC, 9 V;   21 μF: 252 μC, 12 V

CHAPTER 28

28–1 a) 0.020 A   b) 2.74×10^{-6} m·s^{-1}
28–3 a) 603 C   b) 121 A
28–5 0.261 Ω
28–7 a) 1.32×10^{-4} Ω   b) 1.35 cm
28–9 a) 99.5 Ω   b) 0.0148 Ω
28–11 28.5°C
28–13 $\mathscr{E} = 1.52$ V, $r = 0.10$ Ω
28–15 a) 0.050 Ω   b) 0.15 Ω   c) 0.0060 Ω
28–17 b) Yes   c) 4.36 Ω
28–19 a) EJ b) $J^2\rho$ c) E^2/ρ
28–21 a) 21.8 Ω   b) 5.50 A   c) 555 W
28–23 a) 2.16×10^6 J   b) 0.05221   c) 2.0 hr
28–25 a) 1.82 A   b) 3.31×10^4 W   c) 20×10^6 Ω
28–27 9.6×10^{-4} A
28–29 a) 2.40×10^{-8} Ω·m   b) 20.0 A
c) 3.90×10^{-4} m·s^{-1}
28–33 $R/9$
28–35 a) 0.50 Ω   b) 10.0 V
28–37 a) 15.0 V   b) 3.24×10^6 J   c) 6.48×10^5 J
d) 0.60 Ω   e) 1.94×10^6 J   f) 6.48×10^5 J
g) Energy dissipated in internal resistance of battery
28–39 a) 0.40 A   b) 1.6 W   c) In 12-V battery; 4.8 W   d) In 8-V battery; 3.2 W
28–41 b) $a = 8.04 \times 10^{-5}$ Ω·m·K$^{0.146}$; $n = 0.146$
c) Freezing point: 3.54×10^{-5} Ω·m; boiling point: 3.38×10^{-5} Ω·m

CHAPTER 29

29–1 a) 36.0 Ω   b) 3.33 A   c) 2.00 A through 60 Ω, 1.33 A through 90 Ω
29–3 a) 0.545 Ω   b) 1 Ω: 12 A; 2 Ω: 6 A; 3 Ω: 4 A
c) 22 A   d) 12 V for each   e) 1 Ω: 144 W; 2 Ω: 72 W; 3 Ω: 48 W
29–5 a) 141 V   b) 4.5 W   c) Two in series, connected in parallel with two others in series
29–7 a) 2.0 A   b) 5.0 Ω   c) 42.0 V   d) 3.5 A
29–9 a) 0.222 V   b) 0.464 A
29–11 9.96 Ω
29–13 $R_1 = 2.99 \times 10^3$ Ω, $R_2 = 1.20 \times 10^4$ Ω, $R_3 = 1.35 \times 10^5$ Ω
3 V: $R = 3.0 \times 10^3$ Ω;
15 V: $R = 1.50 \times 10^4$ Ω;
150 V: $R = 1.50 \times 10^5$ Ω
29–15 15,000 Ω: 10.9 V; 150,000 Ω: 109 V
29–17 a) 0, 3.93×10^{-4} C, 6.32×10^{-4} C, 8.65×10^{-4} C, 1.00×10^{-3} C (or, more precisely, 0.99995×10^{-3} C)   b) 1.00×10^{-4} A, 6.07×10^{-5} A, 3.68×10^{-5} A; 1.35×10^{-5} A, 4.54×10^{-9} A   c) 10.0 s   d) 6.93 s
29–19 a) 0.5 s   b) 1.0 s
29–21 a) 1.00×10^{-8} C; 2.82×10^6 V·m^{-1}; 8.47×10^3 V
b) 5.65×10^{11} V·m^{-1}·s^{-1} no

c) $J_D = 5.00$ A·m^{-2}; $I_D = 2.00 \times 10^{-3}$ A; equal

29–23 a) 3.15×10^9 J = 876 kW h   b) $87.60

29–25 a) Toaster: 12.5 A; frypan: 10.0 A; lamp: 0.833 A
b) Yes ($I = 23.3$ A)

29–29 a) Two in series in parallel with two in series
b) 0.5 W

29–31 27 W

29–33 $I_1 = 0.848$ A, $I_2 = 2.14$ A, $I_3 = 0.171$ A

29–35 a) -12 V   b) 3 A   c) -12 V   d) 12/7 A,
from b to a e) 4.5 Ω   f) 4.2 Ω

29–37 a) -6 V   b) b c) $+6$ V   d) 54 μ C, from b
to a

29–39 a) 1200 Ω   b) 30 V

29–41 4.2×10^5 Ω

29–43 55 Ω

29–45 b) No current through galvanometer   c) 9.52 V
d) No

29–47 a) CV^2 b) $\frac{1}{2}CV^2$ c) $\frac{1}{2}CV^2$ d) 50%,
independent of R

29–49 a) 5.81×10^6 A·m^{-2} b) 3.34×10^{-10} A·m^{-2}

29–55 a) $I_1 = 1$ A, $I_2 = 3$ A, $I_3 = 2$ A   b) 6 V
c) 12 Ω   d) 78 V

29–57 a) $I_1 = 0.300$ A, $I_2 = 0.500$ A, $I_3 = 0.200$ A
b) $I_1 = 0.541$ A, $I_2 = 0.0773$ A, $I_3 = -0.464$ A
c) $I_1 = -0.287$ A, $I_2 = 0.192$ A, $I_3 = 0.479$ A
d) $I_1 = 0.0462$ A, $I_2 = 0.231$ A, $I_3 = 0.185$ A
f) $I_1 = 0.0833$ A, $I_2 = 0.644$ A, $I_3 = 0.561$ A

CHAPTER 30

30–1 Negative

30–3 a) $F = +(2.5 \times 10^{-3}$ N$)k$
b) $F = -(2.5 \times 10^{-3}$ N$)i - (1.5 \times 10^{-3}$ N$)j$

30–5 a: $-qv\,Bk$; b: $qv\,Bj$; c: 0; d: $-(qv\,B/\sqrt{2})j$;
e: $-(qv\,B/\sqrt{2})(j+k)$; f: $-(qv\,B/\sqrt{3})(j+k)$

30–7 a) 1.14×10^{-3} T, into page   b) 1.57×10^{-8} s

30–9 a) 2.89×10^7 m·s^{-1} b) 4.34×10^{-8} s
c) 8.68×10^6 V

30–11 3.97×10^{-3} T

30–13 a) 1.70×10^7 m·s^{-1} c) 4.83×10^{-3} m

30–15 21

30–17 24 A

30–19 a) -0.030 Nk b) -0.025 Nj c) 0
d) 0.015 Nj e) 0.020 Nj + 0.045 Nk

30–21 a) 0.181 N·m   b) Angle of 30° between B and
normal to the coil

30–23 a) $\Gamma = NIAB$, $-x$-direction $U = 0$
b) $\Gamma = 0$; $U = -NIAB$
c) $\Gamma = NIAB$, $+x$-direction; $U = 0$
d) $\Gamma = 0$; $U = +NIAB$

30–25 a) 0.80 A   b) 3.7 A   c) 113 V   d) 417 W

30–27 a) 8.44×10^{-4} m·s^{-1} b) 1.27×10^{-3} V·m^{-1},
$+z$-direction   c) 2.53×10^{-5} V

30–29 a) 1.2×10^3 V·m^{-1}, $+z$-direction
b) 1.2×10^3 V·m^{-1}, $+z$-direction

30–31 0.5 T, $-y$-direction

30–33 Number for alpha particle = 3.64×10^3 times
number for electron

30–35 1.25 mm

30–37 a) 4.0×10^{-6} C   b) $(1.2i + 1.6j) \times 10^{15}$ m·s^{-2}
c) 1.25 cm   d) 6.37×10^7 Hz
e) (R, 0, 37.7 cm)

30–39 a) ab: -1.2 Nk; bc: -1.2 Nj; cd: $+1.2$ N$(j+k)$;
de: -1.2 Nj; ef: 0   b) -1.2 Nj

30–41 0.0132 T, in $+y$-direction

30–43 a) 4.32 m·s^{-1} b) 7.69 A   c) 0.195 Ω

30–45 a) 1.0×10^{-2} J·T^{-1}, $+z$-direction   b) $B_x = 0.8$ T,
$B_y = 0.6$ T, $B_z = 2.4$ T

30–49 a) 1.0 m   b) 1.26×10^{-6} s   c) 0.0318 m
d) 0.161 m

30–51 a) $v = lB\Delta q/m$ c) 7.34×10^{-5} s   d) 0.0168 C
e) 2.11×10^3 m·s^{-1}

CHAPTER 31

31–1 a) 0   b) -1.60×10^{-5} Tk c) -1.60×10^{-5} Tj
d) -5.66×10^{-6} $T(j+k)$

31–3 2.97×10^{-6} T, into the page

31–5 $\mu_0 I/4R$, into page (no)

31–7 a) 1.6×10^{-5} T, west   b) Yes

31–9 a) 6.0×10^{-6} T   b) 2.0×10^{-6} T

31–11 a) $\mu_0 I/\pi a$ b) $\mu_0 I/3\pi a$ c) 0   d) $2\mu_0 I/3\pi a$

31–13 a) 7.5 A   b) Opposite

31–15 1.02 cm

31–17 a) 5.03×10^{-4} T   b) 4.50×10^{-5} T

31–19 6.0×10^{-4} T

31–21 a) $\mu_0 I/2\pi r$ b) 0

31–23 χ varies inversely with Kelvin temperature

31–25 a) 0.0426 A   b) 0.0115 A

31–27 2.40×10^{-20} N, away from the wire

31–29 b) $\mu_0 Ix/\pi(x^2 + a^2)$ d) $x = \pm a$

31–31 a) 2 A, out of page   b) 2.13×10^{-6} T, to the
right   c) 2.06×10^{-6} T, 39.0° below horizontal
and to the left

31–33 7.2×10^{-4} N, toward the wire

31–35 $(2\mu_0 I/\pi ab)\sqrt{a^2 + b^2}$, into page

31–37 a) $1/2\, N\mu_0 Ia^2(1/[a^2 + (x + a/2)^2]^{3/2}$
$+ 1/[a^2 + (x - a/2)^2]^{3/2})$
b) $8\,N\mu_0 I/\sqrt{125}a$
c) 1.50×10^{-3} T

31–39 a) 398   b) 397

31–41 a) $3I/2\pi R^3$ b) (i) $\mu_0 Ir^2/2\pi R^3$; (ii) $\mu_0 I/2\pi r$

31–43 b) $\mu_0 I_0/2\pi r$ c) $I_0 r^2(2 - r^2/a^2)/a^2$
d) $\mu_0 I_0 r(2 - r^2/a^2)/2\pi a^2$

31–45 $Qn\mu_0/a$

31–47 $\mu_0 I/4\pi a$, out of page

31–49 b) 0.286 m·s^{-1} c) 4.16×10^{-3} m

CHAPTER 32

32–1 a) 5.0 m·s^{-1} b) 2.0 A   c) 0.96 N, to the left

32–3 a) 0.30 V   b) 0.30 V   c) b

32–7 8.79 V

32–9 a) 0.737 rev·s^{-1} b) 4.32×10^3 N·m

32–11 5.0×10^{-2} T

32–13 7.54×10^{-6} V

32–15 a) Clockwise concentric circles
b) 5.0×10^{-3} V·m^{-1}, tangent to the ring and
clockwise; 3.14×10^{-3} V   c) 1.57×10^{-3} A
d) 0

32–17 a) Right to left   b) Right to left   c) Left to right

32–19 b) RF/B^2l^2

32–21 a) 3.14 V   b) 3.14 V

32–23 a) 2.00×10^{-2} W·b   b) 0.200 V
c) 2.00×10^{-3} N·m

32–25 a) 0.0471 V   b) a to b

32–27 Point a: $(qr/2)\, dB/dt$, to the left in the plane of the figure
Point b: $(qr/2)\, dB/dt$, upward in the plane of the figure
Point c: 0

32–29 20

32–31 a) $(\mu_0 Iv/2\pi)\ln(1 + l/d)$ b) a c) 0

32–33 a) $B_0\pi a^2 e^{-t/\tau}\cos\omega t$
b) $B_0\pi a^2 e^{-t/\tau}[(1/\tau)\cos\omega t + \omega\sin\omega t]$ c) 62.8 A
d) 3.09×10^{-3} s   e) 1.47×10^{-3} s; 11.7 V

32–35 a) $B\,v/\lambda\theta$ b) $B^2v^2l/\lambda\theta$ d) 294°
e) a: $(Bv/\lambda)(2\pi/\theta[2\pi - \theta])$; b: $B^2v^2l2\pi/\lambda\theta(2\pi - \theta)$;
d: 188°

32–37 b) 0   c) 4.0×10^{-3} V   d) 2.0×10^{-3} A
e) 1.0×10^{-3} V; a

CHAPTER 33

33–1 7.90×10^{-4} H

33–3 a) 5.0×10^{-4} V; yes   b) 5.0×10^{-4} V

33–5 0.10 V, in the direction of the current

33–7 a) 1.00×10^{-3} H   b) 2.25×10^{-3} H if both coils are wound in same sense, 2.5×10^{-4} H if one coil is wound in opposite sense to other

33–9 387

33–13 a) 4.00 A·s^{-1} b) 2.00 A·s^{-1} c) 0.659 A
d) 2.00 A

33–17 0.158 pF

33–21 a) 141 rad·s^{-1} b) 61.6 Ω

33–23 a) 5.0 H   b) 250 Wb   c) 0.020

33–25 b) $\mathcal{E}_L$ is proportional to di/dt

33–27 1.33×10^{-6} T

33–31 a) 2.40 J   b) 4.41 J   c) 2.02 J
d) (a) + (c) = (b)

33–33 a) $4L$ b) $2L$ c) $\omega/2$

33–35 Rate of decrease of energy in capacitor equals rate of storage in inductor plus rate of dissipation in resistor.

33–37 a) $i_R = (0.10 \text{ A})e^{-30t}$; $i_{R_0} = 0.40$ A;
$i_{s_2} = 0.40 \text{ A} - (0.10 \text{ A})e^{-30t}$
b) 0.326 A, to the right

33–41 a) $i_1(t) = (\mathcal{E}/R_1)(1 - e^{-R_1t/L})$; $i_2(t) = (\mathcal{E}/R_2)e^{-t/R_2C}$;
$q_2(t) = \mathcal{E}C(1 - e^{-t/R_2C})$ b) $i_1(0) = 0$; $i_2(0) =$
0.010 A   c) $i_1(\infty) = 2.0$ A; $i_2(\infty) = 0$; several time
constants   d) 1.97×10^{-3} s   e) 9.81 m A
f) 0.277 s

33–43 a) $D(L - L_0)/(L_f - L_0)$
b) One-quarter: 1.2505 H; one-half: 1.2510 H;
three-quarters: 1.2514 H
c) One-quarter: 1.2500 H; one-half: 1.2500 H;
three-quarters: 1.2500 H
d) Completely ineffective for mercury, marginal for oxygen

CHAPTER 34

34–1 a) 377 Ω   b) 2.65×10^{-3} H   c) 2.65×10^3 Ω
d) 2.65×10^{-3} F

34–3 a) 5.0×10^{-3} A   b) 5.0×10^{-2} A   c) 0.50 A

34–5 a) 5.0×10^{-2} A   b) 5.0×10^{-3} A
c) 5.0×10^{-4} A

34–7 $V_C = Q/C$, whereas $\mathcal{E}_L = L\, di/dt$

34–9 a) 583 Ω   b) 0.0857 A   c) 25.7 V, 42.9 V
d) −59.1°; lags

34–11 a) 514 Ω; −38.9°; lags   b) 506 Ω; +37.8°; leads

34–13 a) 25.0 W   b) 25.0 W   c) 0   d) 0
e) (a) = (b) + (c) + (d)

34–15 a) 745 rad·s^{-1} b) 1   c) $V_1 = 35.4$ V,
$V_2 = 79.1$ V, $V_3 = 79.1$ V, $V_4 = 0$, $V_5 = 35.4$ V
d) 745 rad·s^{-1} e) 0.354 A

34–17 a) 3160 rad·s^{-1} c) 0.60 A   d) 0.60 A

34–19 a) 10   b) 6.0 A   c) 72.0 W   d) 200 Ω

34–21 a) 31.6   b) 3.16 V

34–23 0.0183 H

34–25 47.2 Ω

34–27 a) Inductor   b) 0.106 H

34–29 a) 1.59×10^3 Hz; 1.00×10^4 rad·s^{-1} b) 1.00 A
c) 1.00 A   d) 0.10 A   e) 0.10 A
f) 5.00×10^{-4} J; 5.00×10^{-4} J

34–31 0; $I_0/\sqrt{3}$

34–33 a) $[LC]^{-1/2}$ b) $[LC - R^2C^2/2]^{-1/2}$
c) $[1/LC - R^2/2L^2]^{1/2}$

34–35 a) 400 Ω + (300 Ω)i b) 300 Ω − (400 Ω)i
c) 350 Ω − (50 Ω)i d) 0.280 A + (0.040 A)i
e) 0.283 A; 8.13°   f) 0.160 A − (0.120 A)i
g) 0.200 A; −36.9°   h) 0.120 A + (0.160 A)i
i) 0.200 A; 53.1

CHAPTER 35

35–1 300 m

35–3 a) 294 m   b) 4.80×10^{-3} V·m^{-1}

35–7 a) 6.67×10^{-11} T   b) 1.67×10^4 W
c) 100 km

35–9 a) 1.56×10^{-14} kg·m^{-2}·s^{-1} b) 4.67×10^{-6} Pa

35–11 a) 3.0×10^6 m·s^{-1} b) 0.030 m

35–13 a) 1.50×10^{-9} m = 1.50×10^{-3} microns =
1.50 nm = 15.0 Å   b) 5.35×10^{-7} m =
0.535 microns = 535 nm = 5350 Å

35–17 a) $(1/2)r\mu_0 n\, di/dt$, in tangential direction
b) $(1/2)\mu_0 n^2 ri\, di/dt$, radially inward

35–19 3.19 V

35–21 6.14×10^4 V·m^{-1}, 2.05×10^{-4} T

35–23 a) Reflective; gives twice the radiation pressure
b) 24.1 mi^2

35–25 2.90×10^{11} eV·s^{-1}; doesn't apply

CHAPTER 36

36–1 a) 4.48×10^9 m   b) 15.0 s

36–3 34.8°

36–5 a) 54.7°   b) 82.8°

36–7 1.87

36–9 a) 2.00×10^8 m·s^{-1} b) 333 nm

36–13 a) 14.5°   b) Air

36–15   a) $I_0/4$ b) Linearly polarized, parallel to axis of second polarizer

36–17   a) Transmitted intensities are $I_0/2$, $I_0/4$, $I_0/8$. In each case the light is linearly polarized, along the axis of the polarizer.
        b) $I_0/2$ for first filter, 0 for the next.

36–19   54.7°

36–21   a) 1.60   b) 32.0°

36–23   Elliptical

36–27   1.69

36–29   30°

36–31   1.30

36–33   a) $\frac{1}{8}I_0(\sin 2\theta)^2$ b) 45°

36–35   a) 35.0°   b) 10.1 W·m^{-2}, 19.9 W·m^{-2}

36–37 l-leucine: $\theta = -0.110\,c$, where c is the concentration in grams/100 ml
        l-glutamic: $\theta = 0.124\,c$

36–39   b) 856 nm

36–43   b) 37.2°   c) 1.75°

CHAPTER 37

37–1   80 cm to right of mirror; 6 cm

37–3   3.61 cm

37–5   b) 60 cm in front of mirror, 10 cm, inverted, real

37–9   b) 4.44 cm behind mirror, 0.833 cm, erect, virtual

37–11   6.0 cm

37–15   1.35

37–17   10 cm to left of vertex, +1/3

37–19   Half the observer's height

37–21   a) 5.0 cm   b) 9.88 cm

37–23   a) $-5\,\text{cm} < s < 0$ b) Erect

37–25   2.00

37–27   a) 0.667 cm   b) Independent of the radius

37–29   a) As object point moves closer to mirror, image moves away   b) (i) 20.00 cm, 19.89 cm; (ii) $-1/3$, $-1/9$; (iii) transverse dimension is 0.33 cm, longitudinal dimension is 0.11 cm

37–31   a) -0.0331 m·s^{-1} b) -0.444 m·s^{-1}

37–33   b) 11.4°

CHAPTER 38

38–1 $s = 30$ cm: a) 15 cm   b) $-1/2$ c) real   d) inverted
        $s = 20$ cm: a) 20 cm   b) -1 c) real   d) inverted
        $s = 15$ cm: a) 30 cm   b) -2 c) real   d) inverted
        $s = 5$ cm: a) -10 cm   b) $+2$ c) virtual   d) erect

38–3 $s = 30$ cm: a) -7.50 cm   b) $+1/4$ c) virtual   d) erect
        $s = 20$ cm: a) -6.67 cm   b) $+1/3$ c) virtual   d) erect
        $s = 15$ cm: a) -6.00 cm   b) $+2/5$ c) virtual   d) erect
        $s = 5$ cm: a) -3.33 cm   b) $+2/3$ c) virtual   d) erect

38–5   a) -5.0 cm; diverging   b) 0.40 cm; erect

38–7   7.21 cm in front of lens

38–9   4.34 cm

38–11   a) Image formed by first surface   b) +30 cm   c) Real   d) At ∞

38–13   4.5 cm from center of sphere

38–15   60 cm to right of third lens

38–17   2 cm to left of first lens and 2 cm to right of second lens

38–19   a) 50 cm   b) 200 cm

38–21   0.667 cm

38–23   a) 7.14 cm   b) 3.5 mm

38–25   a) 35 mm   b) 200 mm

38–27   a) 2.86 cm   b) $1/50\,s$

38–29   a) 10.2 cm   b) No

38–31   a) 1.74 cm   b) -11.3 c) -113

38–33   a) -5.0 b) 4.0 cm

38–35   15.7 cm

38–37   Convex, 4.66 m

38–39   4.0 cm

38–41 -15 cm

38–43   a) 0.254 m   b) 0.254 m

38–45   a) 7.55 m   b) 5.05 m

38–47 $4R$ from center of sphere

38–49   0.67 cm above page

38–51   2.0 cm

38–53   2.97 cm; elongated

38–55   a) 361   b) 23.3 cm

38–57   b) -20 cm   c) Galilean telescope is 80 cm long, telescope in Exercise 38-33 is 120 cm long.

38–59   a) $4f$

38–61   a) 103 cm from lens   b) Virtual   c) Inverted   d) 13.2 mm

CHAPTER 39

39–1   1.00 mm

39–3   545 nm

39–5 $\lambda/2d$; independent of m

39–7   0.833 mm

39–9 2.94×10^{-4} rad = 0.0168°

39–11   a) 7.14×10^{-8} m   b) Red

39–15   533 nm

39–17   17.5°, 36.9°, 64.2°

39–19   12.5°

39–21   2000 m

39–23   Smaller by a factor of 5/6

39–25   No difference

39–27   0.750 mm

39–29   480 nm

39–31   2.79 mm

39–33   a) Shifted downward on screen
        b) $I(\theta) = I_0 \cos^2[(\pi/\lambda)\{d \sin \theta + L(1 - 1/n)\}]$
        c) $\sin \theta = [m\lambda - L(1 - 1/n)n)]/d$

39–35   0.10 mm

39–37   500 nm

39–39   119 km

39–41   a) 25.6°   b) 10.2°   c) 5.08°

39–43   3.80 fringes·min^{-1}

CHAPTER 40

40–1   Bolt at A

40–3   a) $0.990c$ b) 4.75×10^3 m

40–5 333 m

40–11 2.60×10^8 m·s^{-1}

40–15 a) $0.866c$ b) $0.996c$

40–17 a) Nonrelativistic: 4.56×10^{-15} J; relativistic: 4.97×10^{-15} J; ratio = 0.916
b) Nonrelativistic: 3.57×10^{-14} J; relativistic: 1.47×10^{-13} J; ratio = 0.243

40–19 111 kg

40–25 7.38×10^7 m·s^{-1}

40–27 a) 3.38×10^{-6} b) 5.37×10^4 MeV

40–29 a) 7.14×10^3 m·s^{-1} b) No

40–31 a) $0.995c$ b) 1.0%

40–33 a) 1.80×10^5 eV   b) 6.91×10^5 eV
c) 2.02×10^8 m·s^{-1} d) 2.52×10^8 m·s^{-1}

40–35 168 keV

40–37 2.55×10^8 m·s^{-1}; receding

40–43 a) 600 MeV   b) 2.15 times larger

40–45 c) 5.78×10^{-9} s

CHAPTER 41

41–1 a) 2.42×10^{20} Hz   b) 1.24×10^{-12} m
c) 1000 times as large

41–3 5.00×10^{14} Hz; 1.10×10^{-27} kg·m·s^{-1}; 3.31×10^{-19} J = 2.07 eV

41–5 a) 6.63×10^{-26} J = 4.14×10^{-7} eV
b) 7.55×10^{29} photons·s^{-1}

41–7 1.72×10^6 m·s^{-1}

41–9 289 nm

41–11 a) 6.17×10^{14} Hz   b) 486 nm

41–13 387 nm

41–15 1.96 eV = 3.14×10^{-19} J;   633 nm

41–17 1.59×10^{15} photons·s^{-1}

41–23 37.4°

41–25 a) 1.04 eV   b) 1197 nm   c) 2.51×10^{14} Hz
d) 4.14×10^{-7} eV   f) No

41–27 0.344 V greater

41–29 a) 4.59×10^{14} Hz   b) 652 nm   c) 1.89 eV
d) 6.59×10^{-34} J·s

41–31 a) 12.8 eV   b) 6; $n = 4 \rightarrow 3$, 1875 nm; $n = 4 \rightarrow 2$, 486 nm; $n = 4 \rightarrow 1$, 97.2 nm; $n = 3 \rightarrow 2$, 656 nm; $n = 3 \rightarrow 1$, 103 nm; $n = 2 \rightarrow 1$, 122 nm

41–33 7730 K

41–35 a) 4.85×10^{-3} nm   b) 2.56×10^5 eV

41–37 a) 1.13×10^3 eV   b) 11 Å

CHAPTER 42

42–1 a) 2.19×10^6 m·s^{-1}, 1.09×10^6 m·s^{-1}, 7.29×10^5 m·s^{-1}
b) 1.52×10^{-16} s, 1.22×10^{-15} s, 4.10×10^{-15} s
c) 8.22×10^6

42–3 a) 13.6 eV   b) −27.2 eV   c) −13.6 eV
d) 13.6 eV   e) 91.2 nm; ultraviolet

42–5 a) 4 times larger   b) Paschen, Brackett, Pfund
c) Smaller by a factor of 2

42–7 1.77 eV to 3.10 eV; 0.696 nm to 0.922 nm

42–9 a) 5.48×10^{-2} nm   b) 1.28×10^{-3} nm

42–11 b) 2×10^{-24} kg·m·s^{-1}; 2×10^{-24} kg·m·s^{-1}

42–13 2.34×10^{-23} s

42–15 a) $3\hbar$ b) $3.46\hbar$; L is larger   c) $m = 3$, 30.0°; $m = 2$, 54.7°; $m = 1$, 73.2°; $m = 0$, 90.0°; $m = -1$, 106.8°; $m = -2$, 125.3°; $m = -3$, 150.0°

42–17 2.89×10^{30} rad·s^{-1}

42–19 a) −2820 eV   b) 2.56×10^{-13} m   c) 0.586 nm

42–21 a) $n = 3$, -3.48×10^{-4} eV; $n = 2$, -2.32×10^{-4} eV
b) Increased by 0.0403 nm

42–23 a) 1.05×10^{-19} kg·m·s^{-1} b) 197 MeV
c) −1.44 MeV; no

42–25 a) 1.0×10^{-8} m   b) No

42–29 1.61×10^{-16} m

42–31 a) $\hbar\omega$ b) 1

CHAPTER 43

43–1 a) $l = 0$, $m = 0$; $l = 1$, $m = 0, \pm 1$; $l = 2$, $m = 0, \pm 1, \pm 2$; $l = 3$, $m = 0, \pm 1, \pm 2, \pm 3$; $l = 4$, $m = 0, \pm 1, \pm 2, \pm 3, \pm 4$ b) 50

43–7 a) 4.58×10^{-48} kg·m^2 b) 0; 0.0151 eV; 0.0454 eV   c) 27.3 µm, 1.10×10^{13} Hz

43–9 2.16×10^3 kg·m^{-3}

43–11 a) 1770 nm;   infrared   b) 1126 nm; infrared

43–13 489 N·m^{-1} b) 5.70×10^{-20} J; 0.356 eV
c) 3490 nm; infrared

43–15 a) 1.11 nm   b) 1.80 nm

43–17 3.65×10^4 K

43–19 a) 0.129 nm   b) 8, 7, 6, 5, 4   c) 484 µm
d) 118 µm, 134 µm, 157 µm, 188 µm, 234 µm

CHAPTER 44

44–1 a) 3.64×10^{-12} J, 22.8 MeV   b) 3.64×10^{-12} J, 22.8 MeV   c) 3.31×10^7 m·s^{-1}

44–3 a) 1 proton, 1 neutron   b) 28 protons, 28 neutrons   c) 79 protons, 118 neutrons

44–5 2.23 MeV

44–7 b) 0.410 MeV

44–9 a) 4.88×10^{-18} s^{-1} b) 3.00×10^3 kg
c) 1.24×10^4 alpha-particles

44–11 a) 2.76 decays·s^{-1} b) 7.45×10^{-11} Ci

44–13 a) 7.15 MeV   b) 2.31 MeV

44–15 10.1 MeV is absorbed

44–17 a) 1.30 T   b) 4.19 MeV, 2.01×10^7 m·s^{-1}

44–19 67.5 MeV, 1.63×10^{22} Hz, 1.84×10^{-5} nm

44–21 a) $Q = +e$, $B = 1$, $S = -1$, $C = 0$
b) $Q = +e$, $B = 0$, $S = 1$, $C = 1$
c) $Q = 0$, $B = -1$, $S = 0$, $C = 0$
d) $Q = +e$, $B = 0$, $S = 0$, $C = 1$

44–23 1.80×10^{19} J

44–25 1.67×10^4 yr

44–27 0.782 MeV

44–29 a) 37 MeV   b) Proton: 13%, π^-: 87%

44–31 a) 16.3 MeV   b) Not energetically allowed
c) Violate conservation of strangeness

44–33 a) $N = (\alpha/\lambda)(1 - e^{-\lambda t})$ b) 2.73×10^4 counts·s^{-1}, 2.42×10^5 counts·s^{-1}, 5.00×10^5 counts·s^{-1}, 7.50×10^5 counts·s^{-1}, 8.75×10^5 counts·s^{-1}, 9.93×10^5 counts·s^{-1} c) 2.16×10^9 atoms
d) 1.0×10^6 counts·s^{-1}

INDEX

Aberration, 895–897
  chromatic, 896
  monochromatic, 896
  spherical, 896
Absorption, of light, 980–981
Absorption spectrum, 991
ac circuits, 788–805
Accelerator, particle, 1081–1085
Accommodation, 897
Activity, 1072
Airy, Sir George, 940
Airy disk, 940
Alkali metal, 1039–1040
Alkaline-earth element, 1039–1040
Alpha particle, 531, 1062, 1069
Alphabet, Greek, A-6
Alternating-current generator, 747, 788
Alternator, 747, 788
Ammeter, 660
Amorphous solid, 1045
Ampere, Andre, 626
Ampere, definition, 626, 721–722
Ampere's law, 724–725, 814
  applications, 726–728
Amplitude, current, 789
  simple harmonic motion, 264
  voltage, 788
amu, 694, 1064
Analog computer, 778
Analyzer, 851
Anderson, Carl D., 1085
Angular frequency, 788
Angular magnification, of microscope, 902
  of telescope, 903
Angular momentum, orbital, 1028
  quantization, 1025
  spin, 1029
Angular momentum quantum
  number, 1026

Angular size, 899
Anode, 982
Answers to odd-numbered problems, A-10–A-15
Antiparticle, 1085
Aqueous humor, 897
Arc lamp, 838
Astigmatism, 896, 898
Aston, Francis, 693
Astronomical data, table, A-9
Astronomical telescope, 903
Atomic mass, 1064–1065
  table, 1065
Atomic mass unit, 694, 1064
Atomic number, 531, 1063
Atomic spectrum, 989–991
Atomic structure, 1039–1041
Automotive wiring, 667–670
Average value, in ac circuit, 796–798

Back emf, in motor, 702
Bainbridge mass spectrometer, 694
Balmer, Johann, 986
Balmer series, 986–987, 1010
Band spectrum, 1044
Band theory of solids, 1049–1050
Baryon, 1087
Baryon number, 1088
Battery, 632–635
Beam splitter, 926
Becquerel, 1072
Beta particle, 1069
Binding energy, 1063, 1065
Binocular, 904
Biot and Savart, law of, 717–718
Birefringence, 853
Bohr model, of hydrogen atom, 530, 538, 1007–1012
Bohr radius, 1009
Boltzmann constant, 995–996
Boson, 1087

Bottom quark, 1091
Bound charge, 616
Brackett series, 986–987, 1010
Bragg reflection, 939
Brake, eddy-current, 756
Branch point, 656
Brewster's law, 852–853
Bubble chamber, 691, 1085

Camera, 900–901
Cancellation, 915
Capacitance, 605
  equivalent, 607–609
Capacitive reactance, 790
Capacitor, 604–607
  in ac circuit, 789–790
  electrolytic, 611
  energy in, 609–611
  parallel-plate, 605–607
  in series and parallel, 607–609
Carbon dating, 1074
CAT scanner, 1001
Catadioptric lens, 904
Cathode, 982
Cathode-ray tube, 594–597
Center of curvature, 870
Central-field approximation, 1037
Chain reaction, 1076–1077
Charge, electric, 529
  on conductor, 566–568
  conservation of, 532, 627
  of electron, 591–593
Charge density, linear, 553
  surface, 554
Charm, 1090
Chromatic aberration, 896
Circle of least confusion, 896
Circuit, complete, 632
  electric, energy and power in, 640–644
Circuit breaker, 669

Circular polarization, 824, 855
Closely packed crystal, 1046
Cloud chamber, 691, 1085
Coherence, 913–916, 992
Colliding-beam accelerator, 1083–1084
Color, and wavelength, 827
Color code, wiring, 670
Coma, 896
Common-emitter circuit, 1053
Commutator, 747
Complete circuit, 627, 632–633
Compound microscope, 902
Compound motor, 702
Compression ratio, 429
Compton scattering, 999–1001
Computer, analog, 778
Computerized axial tomography, 1001
Condenser. See Capacitor
Condenser lens, 901
Conduction, electrical, microscopic
        view, 644–646
Conduction band, 1049
Conductivity, intrinsic, 1051
Conductor, electrical, 532
Conservation, of electric charge, 532,
        627
    of mass and energy, 970
Conservation laws, 1088
Conservative force, 576
Constructive interference, 915
Continuous spectrum, 981, 994–996
Conversion factors, unit, A-8
Coolidge x-ray tube, 997
Correspondence principle, 973
Coulomb's law, 534–540
    and Gauss's law, 562
Coulomb, 535
Coulomb, Augustin de, 534
Covalent bond, 1042
Covalent crystal, 1045
Critical angle, for total internal
        reflection, 845
Critical damping, 778
Critical temperature, for
        superconductivity, 1055
Crystal lattice, 938–939
Crystal types, 1044–1046
Crystalline solid, 1045
Curie, 1072
Current, displacement, 666–667
    electric, 625
Current amplitude, 789
Current balance, 722
Current density, 627
Current loop, force and torque on,
        697–701
Curvature of field, 896
Cyclotron, 1081–1082
Cylindrical lens, 899

Damped oscillation, electrical,
        778–780
Damping, 778–780
D'Arsonval galvanometer, 659–662,
        701
Davisson-Germer experiment, 1013

De Broglie wavelength, 1013
Decay constant, in R–L circuit, 774
    nuclear, 1071
Decay series, 1072–1074
Defects of vision, 898–899
Deflecting plate, 595
Deflection, full-scale, 659
Del, 590
Density, of nucleus, 1062–1063
Derivative, table, A-5
Destructive interference, 915
Diamagnetism, 729
Diamond crystal structure, 1045
Diaphragm, 900
Diatomic molecule, 1041–1043
Dichroism, 853
Dielectric, 611–615
    Gauss's law in, 617–618
    molecular model, 615–617
Dielectric breakdown, 611, 1050
Dielectric constant, 612
    table, 612
Dielectric strength, 584, 615
Diffraction, 928–943
    from circular aperture, 939–942
    from circular obstacle, 929
    from edge, 929
    of electrons, 1013, 1020–1024
    Fraunhofer, 930
    Fresnel, 930
    single-slit, 930–934
    x-ray, 938–939
Diffraction grating, 934–937
Diffuse reflection, 844
Dilation, of time, 959
Diode, 639
    light-emitting, 1054
    semiconductor, 1053
Diopter, 899
Dipole, electric, 549
    oscillating, 828
Dipole moment, electric, 550
Direct-current generator, 747
Dislocation, 1047
Dispersion, 847
Displacement, electric, 618
Displacement current, 666–667
    and electromagnetic waves, 814
    and magnetic field, 730–732
Distortion, 896
Diverging lens, 889
Doppler effect, relativistic, 971–972

Eddy current, 756–757
Eddy-current brake, 756
Electric charge, 529
    on conductor, 566–568
Electric circuit, energy and power in,
        640–644
Electric dipole, 549
Electric dipole moment, 550
Electric displacement, 618
Electric field, 546–555
    of conducting cylinder, 563
    between conducting planes, 565
    of conducting sphere, 562

    in conductor, 547
    induced, 752–754
    of infinite-plane charge, 564
    of line charge, 563
Electric field intensity, 547
Electric field line, 556
Electric flux, 560, 732
    and displacement current, 667
Electric intensity, 547
Electric potential energy, 575–579
Electrolytic capacitor, 611
Electromagnet, superconducting, 1056
Electromagnetic induction, 742–757
Electromagnetic interactions, 1087
Electromagnetic spectrum, 826–828
Electromagnetic wave, 758, 803–828
    in matter, 820–821
    speed of, 815–817
Electromotive force, 632–635
    motional, 743–748
    of rotating disk, 747–748
    in rotating loop, 746–747
    self-induced, 769–771
    sinusoidal, 788
Electron, 530
    e/m, 692–693
    charge of, 591–593
Electron affinity, 1041
Electron configuration, 1039–1041
    table, 1040
Electron diffraction, 1013, 1020–1024
Electron gas, 645
Electron gun, 595
Electron microscope, 943, 1016–1019
    scanning, 1018
Electron spin, 1028–1030
Electronvolt, 593–594, 982–983
Electrostatic lens, 1016
Electrostatic shielding, 567
Electrostatics, 529–540
Electrovalent bond, 1039–1042
Elementary particles, 1084–1091
    table, 1088
Elliptical polarization, 855
emf. See Electromotive force
Emission, of light, 980–981
    spontaneous, 992
    stimulated, 992
Endoscope, 846
Energy, in capacitor, 609–611
    in electric circuit, 640–644
    in electromagnetic wave, 817–820
    in inductor, 771–772
    relativistic, 967–970
    potential, elastic, electrical, 575–579
    of magnetic dipole, 699
    rest, 969–970
Energy density, in electric field, 610
    in electromagnetic wave, 817
    in magnetic field, 772–773
Energy gap, 1049, 1055
Energy level, 988–989
Equipotential surface, 587–589
esu, 536
Ether, 927–928, 956
Excited state, 990, 1011

of nucleus, 1066
Exclusion principle, 1036–1038
Eye, 897
Eyepiece lens, 902

Far point, 897
Farad, 605
Faraday, Michael, 556, 742
Faraday disk dynamo, 747–748
Faraday's law, 748–754, 814
Fermilab, 1083
Fermion, 1087
Ferrite, 757
Ferromagnetism, 729
Fiber optics, 847
Field, electric, 546–555
    vector, 547
Field line, electric, 556
    magnetic, 688–690, 715
Field point, 550, 714
Filled shell, 998
Film, interference in, 922–925
Fine structure, 1030
Fission, nuclear, 1076–1078
Fission fragment, 1076
Fizeau, 839
Fluorescent lamp, 838
Fluorescent screen, 595
Flux, electric, 560, 732
    and displacement current, 667
Flux, magnetic, 688–690, 748
Flux density, 690
Focal length, spherical mirror,
    874–875
    thin lens, 887–888, 893–894
Focal point, spherical mirror, 874–875
    thin lens, 887–888, 893–894
Focus, spherical mirror, 874–875
    thin lens, 887–888, 893–894
    virtual, 874
Force, magnetic, between moving
    charges, 716
    on conductor, 695–701
    on current loop, 697–701
Fovea centralis, 897
Franck–Hertz experiment, 989
Fraunhofer, Joseph, 930
Fraunhofer diffraction, 930–934
Fraunhofer line, 991
Free charge, 616
Frequency, angular, 788
    of L–C circuit, 777
Fresnel, Augustin, 930
Fresnel diffraction, 930–931
Fuel cell, 632
Fuel element, 1077
Full-scale deflection, 659
Full-wave rectifier, 797
Fundamental constants, table, A-9
Fundamental particles, 1084–1091
    table, 1088
Fuse, 668
Fusion, nuclear, 1079–1080

Galilean coordinate transformation,
    957

Galilean telescope, 911
Galvanometer, d'Arsonval, 659–662,
    701
Gamma ray, 827, 1066, 1069
Gauss, 686
Gauss, Karl F., 557
Gauss's law, 557–566, 814
    in dielectric, 617–618
Geiger counter, 1069
General theory of relativity, 973–974
Generator, 747
    electric, 632
    homopolar, 747–748
Geometrical optics, 841, 913
Gradient, 589–590
Grain boundary, 1046
Grand unified theory, 1091
Gravitational interaction, 1087
Greek alphabet, A-6
Grimaldi, 837
Ground side, of power line, 667
Ground state, 990, 1011
    of nucleus, 1066
Ground-fault interrupter, 669
Grounding conductor, 669
Gyromagnetic ratio, 1027, 1029

Hadron, 1087
Half-life, of nucleus, 1071
Hall effect, 703–804
    in semiconductors, 1052
Halogen, 1040
Harmonic motion, damped, 778–780
Heisenberg uncertainty principle,
    1021–1023
Henry, definition, 768
Henry, Joseph, 742, 768
Hertz, Heinrich, 816
Heteropolar bond, 1039–1042
High-energy physics, 1086–1091
Hole, 1051
Hole conduction, 704
Holography, 943–945
Homopolar bond, 1042
Homopolar generator, 747–748
Hot side, of power line, 667
Household wiring, 667–670
Huygens' principle, 857–859
Hydrogen atom, Bohr model,
    1007–1012
Hydrogen bond, 1042
Hydrogen molecule, 1042
Hyperfine structure, 1030
Hyperon, 1087
Hyperopia, 898
Hysteresis, 729

Image, 868
    as object, 893
    inverted, 888
    virtual, thin lens, 891
Impedance, 794
Impedance matching, 805
Impurity atom, 1046
Incandescent lamp, 838
Index of refraction, 820, 842–844

and Huygens' principle, 859
    table, 843
    and wavelength, 844
Induced charge, 533
Induced current, 743
Induced electric field, 752–754
Induced emf, 743
    in motor, 702
Inductance, 769–771
    in ac circuit, 791–792
    mutual, 767–768
Induction, electromagnetic, 742–757
    electrostatic, 533
Inductive reactance, 792
Inductor, 769
    energy in, 771–772
Inertial frame of reference, 955
Interference, 913–928
Infrared, 827
Insulator, electrical, 532
Integral, table, A-5
Integrated circuit, 1053
Intensity, of electromagnetic wave,
    818, 822–823
    in interference pattern, 919–922
    light, 919
    strong, 1087
    weak, 1087
Interactions, kinds of, 1087
Interference, 913–928
    constructive, 915
    destructive, 915
    quantum interpretation, 1023
    thin films, 922–925
    two-source, 917–922
Interferometer, Michelson, 926–928
Internal reflection, total, 845–848
Internal resistance, 634
International System of Units,
    A-1–A-3
Interstitial atom, 1046
Intrinsic conductivity, 1051
Invariance, of physical laws, 955–956,
    970–971
Inverse photoelectric effect, 999
Inverted image, 888
Ion, 530
Ionic bond, 1039–1042
Ionic crystal, 1045
Ionization energy, 1041
Ionization potential, 1041
Isospin, 1087–1088
Isotope, 693, 1063

Jupiter, satellites of, 839

Kamerlingh-Onnes, H., 1055
Kerr effect, 856
Kilohm, 630
Kinetic energy, relativistic, 967–970
Kirchhoff, Gustav, 656
Kirchhoff's loop rule, 637, 656
Kirchhoff's point rule, 656

Lamps, 838
Land, Edwin, 853

Laser, 838–839, 992–994
Lateral magnification, microscope, 902
  plane mirror, 869
  plane refracting surface, 878–879
  spherical mirror, 871–872
  spherical refracting surface, 880
L–C circuit, 775–778
Length, relativity of, 961–963
Lens, 887–889
  condenser, 901
  cylindrical, 899
  diverging, 889
  electrostatic, 1016
  magnetic, 1016
  telephoto, 901
  wide-angle, 901
Lens aberrations, 895–897
Lens aperture, 900
Lensmaker's equation, 893–894
Lenz's law, 754–756
Lepton, 1087, 1089
Lepton number, 1089
Levitation, magnetic, 696, 1056
Light, emission and absorption, 980–981
  monochromatic, 828
  nature of, 837–838
  polarization, 848–853
  speed of, 536, 716–717, 839–840, 956
  wavelengths, 826–828
Light pipe, 846
Light-emitting diode, 1054
Limit of resolution, 942
Line integral, 580
Line spectrum, 981, 985–987
Linear charge density, 553
Linear polarization, 824
Linear superposition, of light, 914
Lines of force, electric, 556
  magnetic, 688
Loop, in circuit, 656
Lorentz transformation, 963–965
L–R circuit, 772–775
L–R–C series circuit, 778–786, 793–794
Lyman series, 986–987, 1010

Magnetic declination, 684
Magnetic dipole, 699
Magnetic field, 684–688
  of circular loop, 722–724
  of current element, 717–718
  and displacement current, 730–732
  of earth, 684
  energy density in, 772, 821
  measurement, 751–752
  of moving charge, 714–717
  of solenoid, 726–727
  of straight conductor, 719–720, 726
  of toroidal solenoid, 728
  units, 715–716
Magnetic field lines, 715
Magnetic flux, 688–690, 748–750
Magnetic force, on conductor, 695–701

  on moving charge, 684–688
  between moving charges, 716
  between parallel conductors, 721–722
Magnetic lens, 1016
Magnetic levitation, 696, 1056
Magnetic lines of force, 688
Magnetic materials, 728–730
Magnetic moment, 698–699, 1026–1029
Magnetic monopole, 683
Magnetic pole, 683, 720
Magnetic quantum number, 1026
Magnetic susceptibility, 729
  table, 729
Magnetic-field line, 688–690
Magnetism, 683–684
Magnetization, 728–730
Magnetizing current, 804
Magnification, lateral. *See* Lateral magnification
Magnifier, 899–900
Magnifying glass, 899–900
Malus, Etienne, 851
Malus's law, 851
Marconi, Guglielmo, 815
Mass defect, 1063
Mass number, of isotope, 694
  of nucleus, 1062
Mass spectroscopy, 693–694
Massless particle, 970
Mathematical relations, A-4
Maxwell, 690
Maxwell, James C., 666
Maxwell's equations, 757–758, 814
Mean free path, 645
Mean lifetime, of nucleus, 1072
Megohm, 630
Meson, 1087
Metallic crystal, 1046
Metastable state, 993
Meter, definition, 927
Meters, for ac circuit measurements, 797
Methane, 1042
Michelson, Albert A., 840
Michelson interferometer, 926–928
Michelson–Morley experiment, 927–928, 956
Microampere, 626
Microfarad, 606
Microscope, compound, 902
  electron, 1016–1019
  scanning, 1018
Milliampere, 626
Millikan, Robert A., 591
Millikan oil-drop experiment, 591–593
Moderator, 1077
Molecular model, of dielectric, 615–617
Molecular rotation and vibration, 1043–1045
Molecular spectrum, 1043–1045
Molecule, diatomic, 1041–1043
  polar, 616
Moment, electric-dipole, 550

  magnetic, 698–699
Momentum, in electromagnetic wave, 819
  of photon, 984–985
  relativistic, 966–967
Monochromatic aberration, 896
Monochromatic light, 828, 913–916
Motional electromotive force, 743–748
Motor, direct-current, 702
Muon, 1086, 1088
Mutual inductance, 767–768
Myopia, 898

Near point, 897
Negative ion, 530
Negative resistance, 780
Neutrino, 1070, 1089
Neutron, 530, 1063, 1084–1085
  thermal, 1015
Neutron activation analysis, 1076
Neutron number, 1063
Newton's rings, 924
Newton's third law, in electrostatics, 534
Nodal plane, 825
Non-electrostatic field, 753
Nonlinear optics, 914
Nonreflective coatings, 924
Normal mode, of electromagnetic wave, 826
n-type semiconductor, 1051
Nuclear fission, 1076–1078
Nuclear force, 1066
Nuclear fusion, 1079–1080
Nuclear reaction, 1074–1076
Nuclear reactor, 1077
Nuclear spin, 1063
Nucleus, 530, 1060–1065
  stability of, 1067–1074
Nuclide, 1063
Numerical constants, table, A-9

Object, virtual, 893
  thin lens, 891
Objective lens, 902
Odd-odd nucleus, 1067
Ohm, definition, 630
Ohm, Georg, 629
Ohmmeter, 662
Ohm's law, 629–630, 639
Oil-drop experiment, Millikan, 591–593
Open circuit, 633, 669
Optic axis, spherical mirror, 870
  thin lens, 888
Optic nerve, 897
Optical activity, 856
Orbital angular momentum, 1028
Oscillation, electrical, 775–780
Oscillator, 788
Oscilloscope, 594–597
Overdamping, 778

p–n junction, 1052–1053
p-type semiconductor, 1052

Pair production, 1085
Pairing, in nuclear force, 1066, 1068
Parallel, capacitors, 607–609
  resistors, 653–655
Parallel resonance, 803
Parallel-plate capacitor, 605–607
Paramagnetism, 729
Paraxial approximation, 871, 878, 880, 895
Paraxial ray, 871, 878, 880, 895
Parity, 1088
Particle, fundamental, 1084–1091
  table, 1088
Particle accelerator, 1081–1085
Paschen series, 986–987, 1010
Pauli, Wolfgang, 1037
Pauli exclusion principle, 1036–1038
Periodic table of elements, 1040, A-7
Permeability, 729
Permittivity, 613
Pfund series, 986–987, 1010
Phase, in light interference, 916
Phase angle, in interference, 916, 920, 925, 932
  in $L-R-C$ circuit, 790, 795
Phase difference, 790
Phase shift, in reflection, 923
Phase transition, superconducting, 1055
Phasor, 788–789
Phasor diagram, for $L-R-C$ circuit, 793–794
  single-slit diffraction, 932
  two-source interference, 920
Photocell, semiconductor, 1054
Photoelasticity, 856
Photoelectric effect, 981–985
  inverse, 999
Photomultiplier, 1023
Photon, 838, 983
Physical constants, table, A-9
Physical optics, 913
Picofarad, 606
Pion, 1086, 1088
Planck, Max, 995
Planck radiation law, 995–996
Planck's constant, 983, 1008
Plane wave, electromagnetic, 821–824
Polar molecule, 616
Polarization, circular and elliptical, 855
  of dielectric, 613
  of electromagnetic wave, 824
  linear, of light, 848–853
  waves on a string, 848
Polarizer, 848–853
Polarizing angle, 852
Polarizing axis, 849
Polarizing filter, 848–853
Polaroid, 853
Polycrystalline substance, 1046
Population inversion, 992
Porro prism, 846
Positive ion, 530
Positron, 691, 1085
Positron annihilation, 1085
Potential, electric, 580–583

Potential difference, 580, 659
Potential energy, electrical, 575–579
  of magnetic dipole, 699
Potential gradient, 589–590
Potentiometer, 635
Power, in ac circuits, 798–800
  in electric circuit, 640–644
  in electromagnetic wave, 817–820
  in transformer, 805
Power distribution systems, electric, 667–670
Power factor, 799
Poynting vector, 818
Presbyopia, 897
  table, 898
Primary winding, 804
Principal quantum number, 1026
Principal ray, thin lens, 890
  spherical mirror, 875
Principle of relativity, 955
Principle of superposition, optics, 914
Printed circuit, 656
Probability and uncertainty, 1019–1024
Projector, 901–902
Proton, 530, 1063
Pupil, 897

Quantization, of angular momentum, 1025
  of spin angular momentum, 1029
Quantum, 838, 983
Quantum chromodynamics, 1089
Quantum condition, Bohr, 1009, 1014
Quantum mechanics, 980, 1007–1030
Quantum number, 1011, 1026
Quark, 1089–1091
  table, 1090
Quarter-wave plate, 856

Radiation, from antenna, 828
  thermal, 838
Radiation law, Planck, 995–996
Radiation pressure, 819
Radio receiver, 801–802
Radio wave, 827
Radioactive decay series, 1072–1074
Radioactivity, 1067–1074
Radiocarbon dating, 1074
Rayleigh's criterion, 941–942
Rays, and waves, 840–841
$R-C$ series circuit, 662–666
Reactance, 794
  capacitive, 790
  inductive, 792
Reaction, nuclear, 1074–1077
Reaction energy, 1074
Real image, 868
Rectifier circuit, 797
Reflecting telescope, 904
Reflection, diffuse, 844
  of electromagnetic wave, 824, 828
  and Huygens' principle, 858–859
  of light, 841–844
  at plane surface, 868–870, 877–879

specular, 844
  at spherical surface, 870–874, 879–882
Refractive index. See Index of refraction
Reinforcement, 915
Relative velocity, in relativity, 964–965
Relativity, 955–974
  general theory, 973–974
  of length, 961–963
  and Newtonian mechanics, 973
  special theory, 955–974
  of time, 958–961
Relaxation time, of $L-R$ circuit, 772–775
  of $R-C$ circuit, 664
Resistance, 630–632
  in ac circuit, 789
  internal, 634
  measurement of, 661–662
  negative, 780
Resistivity, 628–629, 1048
  table, 628
  temperature coefficient, 629
    table, 629
  temperature dependence, 1048
Resistor, 630–632
  in series and parallel, 653–655
Resolution, of optical instruments, 941–943
  of electron microscope, 1016
Resolving power, 941–942
Resonance, parallel, 803
  series, 800–803
Resonance curve, 802
Resonance radiation, 991
Response curve, 802
Rest energy, 969–970
Reversed image, 870
Rheostat, 635
Right-hand rule, for magnetic force, 685
$R-L$ circuit, 772–775
rms value, in ac circuit, 796–798
Roemer, Olaf, 839
Röntgen, Wilhelm, 996
Root-mean-square value, in ac circuit, 796–798
Rotation, molecular, 1043–1045
Rutherford nuclear atom, 1062
Rutherford scattering, 1008, 1060–1062
Rutherford, Sir Ernest, 1060
Rydberg constant, 986

Satellite, of Jupiter, 839
Saturation, of nuclear force, 1066
Saturation magnetization, 729
Scanning electron microscope, 1018
Scattering, light, 854–855
  x-ray, 996–1001
Schrödinger equation, 1025
Scintillation, 1061
Search coil, 751–752
Secondary winding, 804

Segre chart, 1067–1068
Self-inductance, 769–771
  in ac circuit, 791–792
Semiconductor, 1051–1055
Semiconductor devices, 1052–1055
Series, capacitors in, 607–609
Series, resistors in, 653–655
Series motor, 702
Series resonance, 800–803
Shell, electrons in atom, 998–1038
Shielding, electrostatic, 567
Short circuit, 634, 669
Shunt motor, 702
Shunt resistor, 660
Shutter, 900
SI electrical units, 535
SI units, A-1–A-3
Sign rules, lenses and mirrors, 883
  Kirchhoff's laws, 637, 656–657
  for reflection, 869
Simultaneity, relative nature of,
    957–958
Single-slit diffraction, 930–934
Sinusoidal emf, 788
Sinusoidal wave, electromagnetic,
    821–824
Sky, light from, 854
Slip ring, 747
Snell, Willebrord, 842
Snell's law, 842
Solar cell, 632
Solenoid, 700, 726–727
  toroidal, 728
Solid, band theory, 1049–1050
  properties of, 1047–1049
  structure of, 1045–1047
Source point, 550, 714
Spectrometer, 936
Spectrum, absorption, 991
  atomic, 989–991
  band, 1044
  continuous, 981, 994–996
  electromagnetic, 826–828
  line, 981, 985–987
  molecular, 1043–1045
  visible, 827
  x-ray, 999
Spectrum line, 914
Specular reflection, 844
Speed, of electromagnetic wave,
    815–817
  of light, 716–717, 839–840, 956
Spherical aberration, 871, 896
Spin, electron, 1028–1030
  nuclear, 1063
Spin angular momentum, 1028
Spin-orbit coupling, 1029
Spontaneous emission, 992
Stability, of nuclei, 1067–1074
Stable orbit, 1008
Standing wave, electromagnetic,
    824–826
Statcoulomb, 536
Stefan–Boltzmann constant, 994

Stefan–Boltzmann law, 994–996
Stimulated emission, 992
Stokes's law, 592
Stopping potential, 983
Storage battery, 642–643
Storage ring, 1084
Strangeness, 1087–1088
Strong interaction, 1087
Subshell, electrons in atom, 1038
Superconducting electromagnet, 1056
Superconductivity, 1055–1056
Superheavy nucleus, 1078
Superposition principle, electrical
    forces, 535
  electric fields, 551
  light, 914
  magnetic fields, 717
Surface charge density, 554
Susceptibility, magnetic, 729
  table, 729
Synchrocyclotron, 1082
Synchrotron, 1082–1084

Telephoto lens, 901
Telescope, 903–905
  Galilean, 911
Television receiver, 801–802
Temperature coefficient of resistivity,
    629
  table, 629
Tesla, definition, 686
Tesla, Nikolai, 686
Test charge, 547
Thermal conductivity, 1048
Thermal neutron, 1015
Thermal radiation, 838
Thermocouple, 632
Thermonuclear reaction, 1079–1080
Thin film, interference in, 922–925
Thin lens, 887–889
Thomson, J. J., 692
Thomson model of atom, 1061
Three Mile Island, 1078
Threshold, for nuclear reaction, 1075
Threshold frequency, 982
Time, relativity of, 958–961
Time constant, of R–C circuit, 664
  of R–L circuit, 774
Time dilation, 959
Top quark, 1091
Toroidal solenoid, 728
Torque, on current loop, 697–701
Total internal reflection, 845–848
Transformer, 804–805
Transient current, 625
Transistor, 1053
Transition series, 1040
Transuranic element, 1076
Transverse wave, electromagnetic, 816
Tuning, of string instruments, 501
Turns ratio, of transformer, 805
Twin paradox, 961

Ultraviolet, 827
Uncertainty principle, 1021–1023
Underdamping, 780
Unit conversions, table, A-8
Units, International System, A-1–A-3
  definitions, A-2–A-3

Vacancy, 1046
Valence band, 1049
Van de Graaff generator, 633
Van der Waals bond, 1043
Vector field, 547, 685
Velocity, relative, in relativity,
    964–965
Velocity selector, 694
Vertex, of spherical mirror, 870
Vibration, molecular, 1044–1045
Virtual focus, 874
Virtual image, 868
  thin lens, 891
Virtual object, 893
  thin lens, 891
Visible spectrum, 827
Vitreous humor, 897
Voltage amplitude, 788
Voltmeter, 581, 661

Wave, electromagnetic, 758, 803–828
  speed of, 815–817
Wave function, for electromagnetic
    wave, 822
  in quantum mechanics, 1024–1026
Wave nature of particles, 1012–1014
Wave number, 920
Wave packet, 1019
Wave pulse, 1019
Wave speed, electromagnetic waves,
    816, 820–821
Weak interaction, 1087
Weber, definition, 690
Weber, Wilhelm, 690
Wide-angle lens, 901
Wien displacement law, 995
Wimshurst generator, 633
Wire, current capacity, 668
Work function, 983

X-rays, 981
  applications, 1001
  biological effects, 1001
  production and scattering,
    996–1001
X-ray diffraction, 938–939
X-ray spectrum, 999

Young, Thomas, 917
Young's experiment, 917–922
Yukawa, Hideki, 1086

Zeeman, Pieter, 1026
Zeeman effect, 1026–1029

APPENDIX F

NUMERICAL CONSTANTS

FUNDAMENTAL PHYSICAL CONSTANTS

Name	Symbol	Value
Speed of light	c	$2.9979 \times 10^8 \text{ m} \cdot \text{s}^{-1}$
Charge of electron	e	$1.602 \times 10^{-19} \text{ C}$
Gravitational constant	G	$6.673 \times 10^{-11} \text{ N} \cdot \text{m}^2 \cdot \text{kg}^{-2}$
Planck's constant	h	$6.626 \times 10^{-34} \text{ J} \cdot \text{s}$
Boltzmann's constant	k	$1.381 \times 10^{-23} \text{ J} \cdot \text{K}^{-1}$
Avogadro's number	N_0	$6.022 \times 10^{23} \text{ molecules} \cdot \text{mol}^{-1}$
Gas constant	R	$8.314 \text{ J} \cdot \text{mol}^{-1} \cdot \text{K}^{-1}$
Mass of electron	m_e	$9.110 \times 10^{-31} \text{ kg}$
Mass of neutron	m_n	$1.675 \times 10^{-27} \text{ kg}$
Mass of proton	m_p	$1.673 \times 10^{-27} \text{ kg}$
Permittivity of free space	ϵ_0	$8.854 \times 10^{-12} \text{ C}^2 \cdot \text{N}^{-1} \cdot \text{m}^{-2}$
	$1/4\pi\epsilon_0$	$8.987 \times 10^9 \text{ N} \cdot \text{m}^2 \cdot \text{C}^{-2}$
Permeability of free space	μ_0	$4\pi \times 10^{-7} \text{ Wb} \cdot \text{A}^{-1} \cdot \text{m}^{-1}$

OTHER USEFUL CONSTANTS

Name	Symbol	Value
Mechanical equivalent of heat		$4.186 \text{ J} \cdot \text{cal}^{-1}$ (15° calorie)
Standard atmospheric pressure	1 atm	$1.013 \times 10^5 \text{ Pa}$
Absolute zero	0 K	$-273.15°\text{C}$
Electronvolt	1 eV	$1.602 \times 10^{-19} \text{ J}$
Atomic mass unit	1 u	$1.661 \times 10^{-27} \text{ kg}$
Electron rest energy	mc^2	0.511 MeV
Energy equivalent of 1 u	Mc^2	931.5 MeV
Volume of ideal gas (0°C and 1 atm)	V	$22.4 \text{ liter} \cdot \text{mol}^{-1}$
Acceleration due to gravity (sea level, at equator)	g	$9.78049 \text{ m} \cdot \text{s}^{-2}$

ASTRONOMICAL DATA

Body	Mass, kg	Radius, m	Orbit radius, m	Orbit period
Sun	1.99×10^{30}	6.95×10^8	—	—
Moon	7.36×10^{22}	1.74×10^6	0.38×10^9	27.3 d
Mercury	3.28×10^{23}	2.57×10^6	5.8×10^{10}	88.0 d
Venus	4.82×10^{24}	6.31×10^6	1.08×10^{11}	224.7 d
Earth	5.98×10^{24}	6.38×10^6	1.49×10^{11}	365.3 d
Mars	6.34×10^{23}	3.43×10^6	2.28×10^{11}	687.0 d
Jupiter	1.88×10^{27}	7.18×10^7	7.78×10^{11}	11.86 y
Saturn	5.63×10^{26}	6.03×10^7	1.43×10^{12}	29.46 y
Uranus	8.61×10^{25}	2.67×10^7	2.87×10^{12}	84.02 y
Neptune	9.99×10^{25}	2.48×10^7	4.49×10^{12}	164.8 y
Pluto	5×10^{23}	4×10^5	5.90×10^{12}	247.7 y